Finite Element Structural Analysis

Prentice-Hall International Series
in Civil Engineering and Engineering Mechanics

N. M. Newmark and William J. Hall, Editors

Finite Element Structural Analysis

T. Y. YANG

Professor of Aeronautics and Astronautics
Dean of Engineering
Purdue University

Prentice-Hall, Inc. Englewood Cliffs, N.J. 07632

Library of Congress Cataloging-in-Publication Data

Yang, T. Y.
Finite element structural analysis.
(Prentice-Hall international series in civil engineering and engineering mechanics)
Includes bibliographies and index.
1. Structures, Theory of. 2. Finite element method. I. Title. II. Series.
TA645.Y36 1986 624.1'7 85-12278
ISBN 0-13-317116-7

Editorial/production supervision and
interior design: *Nancy G. Follender*
Cover design: *Joseph Curcio*
Manufacturing buyer: *Rhett Conklin*

Printed in the United States of America

10 9 8 7 6 5 4 3 2 1

ISBN 0-13-317116-7 01

Prentice-Hall International (UK) Limited, *London*
Prentice-Hall of Australia Pty. Limited, *Sydney*
Prentice-Hall Canada Inc., *Toronto*
Prentice-Hall Hispanoamericana, S.A. *Mexico*
Prentice-Hall of India Private Limited, *New Delhi*
Prentice-Hall of Japan, Inc., *Tokyo*
Prentice-Hall of Southeast Asia Pte. Ltd., *Singapore*
Editora Prentice-Hall do Brasil, Ltda., *Rio de Janeiro*
Whitehall Books Limited, *Wellington, New Zealand*

To my wife
Dehleen

Contents

Preface

This book is based on the classroom materials developed by the author during a period of fifteen years. The materials are for two three-credit semester courses (or three one-quarter courses), one at the junior or senior level and the other at the senior or graduate level. This book is designed to include the following features:

1. It is simple and self-contained. To understand the materials, the reader needs only to have the background of a sophomore-level strength of materials course. Basic references have been well interpreted and digested in the text. Minimal instructional aid or discussion is needed.
2. The level of explanation for each subject goes deeper than that found in most texts. This is a book that undergraduates will feel comfortable reading. Ample illustrative examples, figures, and problems are given. Although it is perfectly suitable as a research reference, it can comfortably be used for a sequence of two undergraduate courses.
3. Throughout the text, examples with figures and numbers are given generously. Physical interpretation and practical application are emphasized. Abstract mathematics and sudden interruptions of explanations are avoided.
4. This book can easily be understood by a practicing engineer who graduated from college a long time ago and has become gradually less familiar with mathematics. The book is also suited for those practicing engineers who have not been exposed to finite elements either in industry or in college.

After the general introduction in Chapter 1, an elementary review of matrix mathematics is given in Chapter 2. Chapter 3 reviews the basic structural theorems at a junior level. Chapter 4 introduces truss analysis and Chapter 5 introduces beam and frame analyses. Chapter 6 introduces tapered and curved beam elements, and Chapter 7 extends the formulations for bars, beams, and frames for free-vibration analysis. Chapter 8 extends the formulations for bars, beams, and frames for buckling and large deflection analyses. Chapters 1 through 7 are suited for a required course at the junior level in aeronautical, civil, and mechanical engineering disciplines. Chapter 8 can be taught in either the first or second course.

Chapter 9 introduces all types of plane stress and plane strain elements, together with solution procedures, a computer program, and practical sample solutions. Chapter 10 introduces axisymmetric solid elements and three-dimensional elements of various shapes. Chapter 11 discusses in detail the methods of numerical integration and introduces systematically a variety of curved isoparametric elements. Chapter 12 introduces various kinds of flat plate elements in bending. Chapter 13 gives Fortran programs, user's manuals, and sample input and output data for static and vibration analyses of plane trusses and frames, and static analysis using the constant strain triangle and the 16-degree-of-freedom rectangle in bending. Chapters 8 through 13 or 9 through 13 are suited for a second dual-level course (senior and graduate), required for those with a structural major and optional for those with a structural minor.

The author devotes this book to his wife, Dehleen, for her infinite patience and encouragement, and also to his daughters, Maria, 14, and Martha, 11, to ease Daddy's guilt from spending too little play and homework time with them.

The author is indebted to Richard H. Gallagher, his major professor during his Ph.D. studies at Cornell. It was Professor Gallagher's technical guidance, inspiration, professional example, and continuous encouragement that led the author, with confidence, to pursue a career in engineering education.

The author is indebted to many of his undergraduate and graduate students, including Rakesh K. Kapania and Sunil Saigal. Special thanks go to Kapania for his assistance in searching the literature, generating numerical data, and for providing valuable in-depth discussions and comments.

The technical typing and grammatical corrections by Nancy A. Stivers are gratefully acknowledged.

The author most graciously acknowledges the patience, care, and quality of editing and production of this book by Nancy G. Follender of Prentice-Hall.

T. Y. Yang

CHAPTER 1

Introduction

The rapid development of computers has completely revolutionalized research and practice in every scientific and engineering field. Conventional computers have rapidly branched out into supercomputers, minicomputers, and graphics computers. The dream that every engineering office and home would have a computer terminal and/or a microcomputer has become a reality. Personal computers in the 1980s are as popular as pocket calculators were in the 1970s and slide rules in the 1960s. Microcomputers are rapidly evolving to have unprecedented memory, speed and graphics capabilities. Following this trend, analysis and design methods that provide computerized solutions to scientific and engineering problems have rapidly been developing for increasingly routine daily use. In this book we focus on one such significantly developed method, the *finite element method.* Although this method is applicable to many scientific and engineering fields, we deal only with the field of structural analysis and design.

The finite element method has long been a fertile research field. It has also been increasingly used as a research tool for numerical experiment. Most importantly, the finite element method has now become a predominant structural analysis and design tool which is used routinely by structural engineers. Because of the need for structural engineers to be familiar with the finite element method, it is necessary that structural engineering students take an introductory finite element course in their junior or senior year. This introductory course is gradually replacing the required traditional courses on structural analysis in aeronautical, civil, mechanical, and other engineering disciplines. Furthermore, it has become a trend that a second finite element course be taught at the dual level (senior and graduate), required for those with a structural major and optional for those with a structural minor. This book provides more than enough material for these two courses. Furthermore, this easy-to-read self-contained book is suitable for those instructors who have been more used to conventional structural analysis methods. This book is ideal for practicing structural engineers who have limited or no background in finite elements and wish to learn the subject through self-study.

The finite element method in structural analysis is a technique that first discretizes a structure into a set or different sets of structural components, each set with a similar geometric pattern and physical assumption. Each pattern of such components is called a specific kind of *finite element*, which appears to have been first so termed by Clough [1.1]. Each kind of finite element has a specific type of structural shape and is interconnected with the adjacent elements by *nodal points.*

Acting at each nodal point are *nodal forces* and the node is subjected to displacements (degrees of freedom). In a general sense, these nodal physical quantities are not limited to being forces and displacements, such as in the cases of thermal, fluid, electrical, and other problems. Thus for each element a standard set of simultaneous equations can be formulated to relate these physical quantities. Physically assembling these elements to form the whole structure is equivalent to superimposing these element equations mathematically. The result is a large set of simultaneous equations which are suited for solution by computer. Upon implementing the loading and boundary conditions (for structural problems), the assembled set of equations can be solved and the unknown parameters found. Substituting these values back to each element formulation provides the distributions of stress and displacement everywhere within each element.

1.1 BRIEF REVIEW OF FINITE ELEMENT HISTORY

Using the methods of discretization and numerical approximation to solve scientific and engineering problems is a fact of life. The concept of finite elements stems from the idea of discretization and numerical approximation. If we were to identify the evidence of the earliest finite element concept, we probably could loosely trace it back to the geometric approximations of pyramids by Egyptians some 5000 years ago. If we consider, for example, the numerical approximation of π as a starting point of finite elements [1.2], we could find some interesting historical records in China, Egypt, and Greece.

The records show that the Chinese started approximating π in the first century A.D. A value of π equal to 3.1547 was evidenced in the design of a cylindrical volume measurer. In the second century A.D. astronomer Chang Heng of the Eastern Han Dynasty approximated π as 3.1466 (730/232) and $\sqrt{10}$. In approximately A.D. 230, Wang Fan of the Country of Eastern Wu used

π as 3.1556 (142/45). In the dynasty of Western Jihn (A.D. 265–317), Liu Hui in his comment on *Mathematics—Nine Chapters* used a regular polygon inscribed in a circle to approximate the circumference and he found π to be 3.1416 (3927/1250) using a polygon of 3072 equal sides (finite elements). Thus some Chinese have since referred to π as Hui's ratio. In the dynasties of Former Sung and Southern Chi, mathematician Tzu Tzong Tze (A.D. 429–500) determined that $3.1415926 < \pi < 3.1415927$. To have obtained such an accuracy was equivalent to finding the side length of a polygon of 12,288 equal sides and the area of a polygon of 24,576 equal sides, both inscribed in a circle.

The Ahmes paper shows that by 1500 B.C., the Egyptians were using $\sqrt{10}$ for π. A still earlier papyrus, now in Moscow, indicates that the Egyptians used the correct formula for the volume of a pyramid and the area of a circle by about 1800 B.C. [1.2]. Archimedes (287?–212 B.C.), one of the greatest of the early mathematicians and inventors, used finite elements to determine the volume of solids [1.2].

More rigorously, if we consider some approximate solutions to elasticity and structural problems as the starting point of the finite element method, we can refer to a historical account of the developments by Timoshenko [1.3]. If we consider the emergence of the concept of frame analysis as the starting point of the finite element method, we can trace back to the works by Maxwell [1.4], Castigliano [1.5], and Mohr [1.6], among others, during the period 1850–1875. An account of the development starting from this period can be found in, for example, the text by Gallagher [1.7].

In 1915, Maney [1.8] in the United States presented the method of slope deflection, expressing the moments in terms of deflections and slopes at the rigid joints of frame structures. Such a formulation is in precisely the same form as the stiffness equations given in Chapter 5. A similar development was put forth by Ostenfeld [1.9] in Denmark. In 1929, Cross [1.10] made public his method of moment distribution for frame analysis. The method relaxes the joint moments very quickly and simply for an approximate solution or, with a little more labor, extends to any degree of exactness desired. The method dominated the practice in frame analysis and design for the following 35 years.

Parallel to the earlier works on the analysis of frame structures, the concept of using lattice analogy to solve continuum mechanics problems also began to take form (see, for example, Refs. 1.11 to 1.15). In the early 1940s, Courant [1.16] proposed using piecewise polynomial interpolation functions to formulate triangular subregions as a special Rayleigh–Ritz variational method to achieve approximate solutions.

Today's practical finite element method is essentially a by-product of computers. The rapid development of the finite element method has kept close pace with the swift development of computers ever since their inception in the early 1950s. Two early classic publications in the mid-1950s by Argyris and Kelsey [1.17] and Turner, Clough, Martin, and Topp [1.18] merged the initial concepts of discretized frame analysis and continuum analysis and kicked off the explosive development in finite element method. Many reviews accounting for the growth of the finite element method are available (see, for example, Ref. 1.19). A thorough chronological review of the fundamental developments of the finite element method is given in this text.

Following the linear static formulation for each finite element, extensions for practical applications have continued to include the various physical effects and fields of vibration and dynamic response, buckling and postbuckling, geometrical and material nonlinearities, thermal effect, fluid–structural interaction, aeroelasticity, structure–acoustics interaction, fracture, laminated composites, wave propagation, structural dynamics and control interaction of aircraft and space structures, random dynamic response, and others. Following these developments, both research and commercially oriented computer programs have become available. The use of special-purpose and general-purpose finite element programs has become routine practice in structural engineering offices.

A current major research thrust is directed to computer-integrated graphics, analysis, design, manufacturing, and automation. The finite element method occupies a predominant role and will have a significant impact on this research direction. Many fertile research areas within the domain of finite elements lie in this broad cross-disciplinary field.

With the growing availability of microcomputers, it becomes apparent that a trend toward tailoring finite element methods and programs to fit microcomputers is in progress.

1.2 SOME SAMPLE APPLICATIONS

In this book we introduce in great detail the various kinds of elements: truss bars; beam and frame members; tapered and curved beams; plane stress and plane strain triangles and rectangles; tetrahedra and hexahedra; and flat rectangles, triangles, and quadrilaterals in bending. The curved-thin-shell type of element is, however, beyond the scope of this text. The effects of vibration, buckling, and large deflection are included in appropriate sections for general illustration.

To illustrate the applicability of these elements in practical applications, a few example figures are shown. Figure 1.1 shows a structure of a fuselage and a wing modeled using beam, plate, and shell elements. A detailed model showing a special component with a cutout illustrates the versatility of finite elements. This model can be used for the analysis of static stress, free vibration, landing impact response, panel and wing flutter, and optimization for minimum weight and full strength.

Figure 1.2 shows an example of a multi-story hotel building modeled using beam, column, and plate elements. This model can be used for static, free vibration, earthquake response, static equivalent, and random wind response analyses [1.20].

Figure 1.3 shows the outside configuration of a 1200-MW fossil fuel steam generator hung and tied to a supporting steel frame structure. The whole system is modeled using beam, column, and plate elements with 1860 degrees of freedom. The figure shows that the structure is vibrating in the third mode with a natural frequency of 1.1 Hz. The natural frequencies and mode shapes are used for earthquake response analysis [1.21].

Figure 1.4 shows the modeling of a column-supported cooling tower with a dimple type of local imperfection. The columns are modeled using column elements. The hyperboloidal shell is modeled using quadrilateral curved shell elements. To model the local imperfect area, smaller quadrilaterals are needed. Quadrilaterals of large and small sizes are connected by curved triangular "filler" elements. This model can be used for earthquake and random wind response analyses [1.22, 1.23]. If initial imperfection is taken into account, the behavior can become geometrically nonlinear.

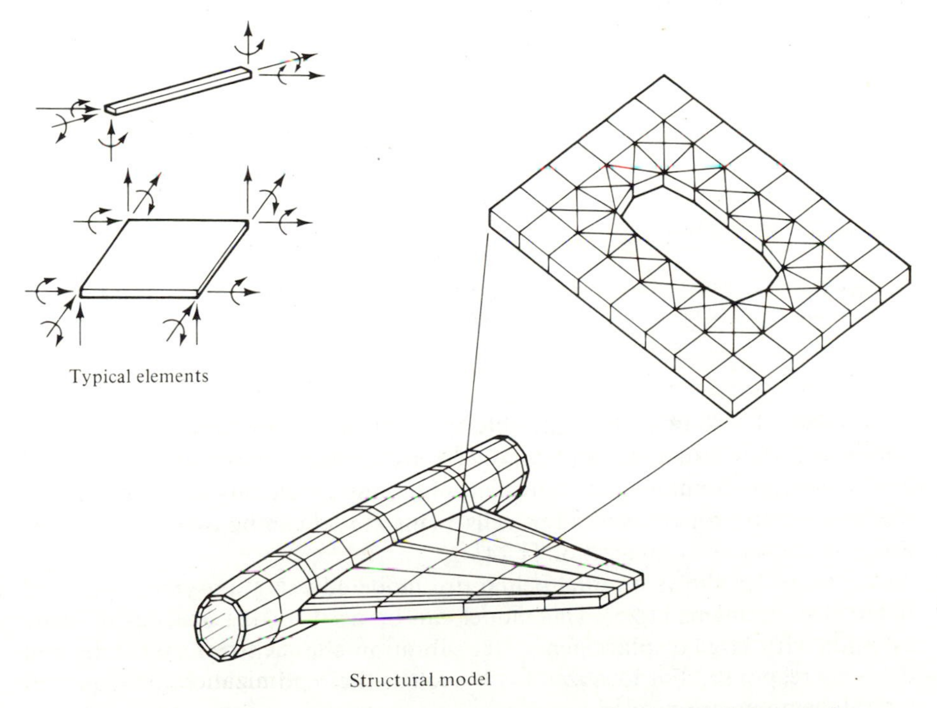

Figure 1.1 Fuselage and a wing modeled using beam, plate, and shell elements for static, dynamic, flutter, and optimization analyses. (Courtesy of Harvey G. McComb, Jr., NASA Langley Research Center.)

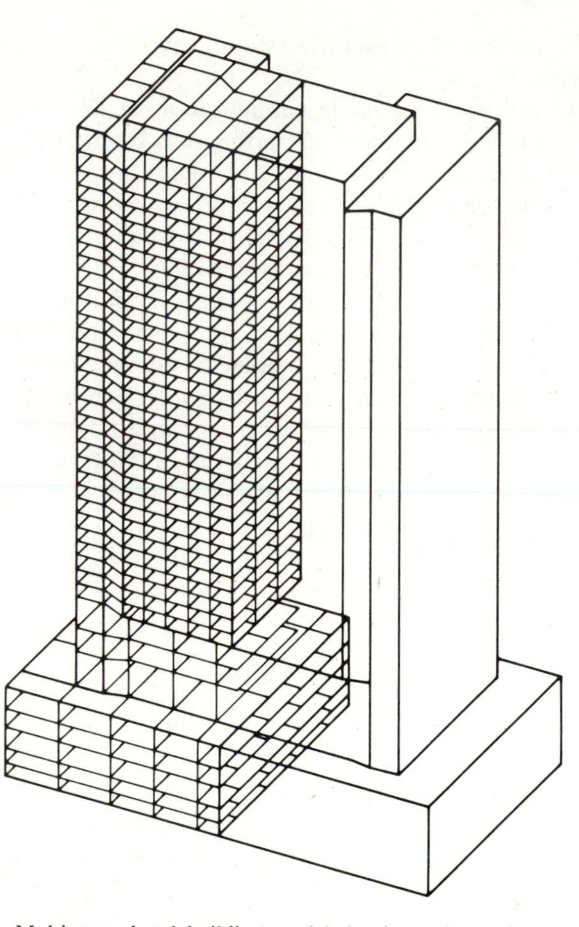

Figure 1.2 Multi-story hotel building modeled using column, beam, and plate elements for static, earthquake, and wind response analyses.

Figure 1.5 shows a large flexible space structure and a single repetitive lattice cell. The structure can be modeled by a huge number of truss bars or a relatively small number of equivalent plate finite elements, each possessing the continuum properties of a few cells. This model can be used for dynamic analysis and control integration [1.24].

Figure 1.6 shows a radial belted tire modeled using axisymmetric laminated shell elements [1.25]. This model can be used for the analysis of static inflation with large displacements, free vibration about the inflated state, and dynamic response. For localized loads such as the road contact force, quadrilateral elements are needed.

To illustrate the effectiveness of graphical display that can help us perceive and model structures of complex geometry, Fig. 1.7 shows some

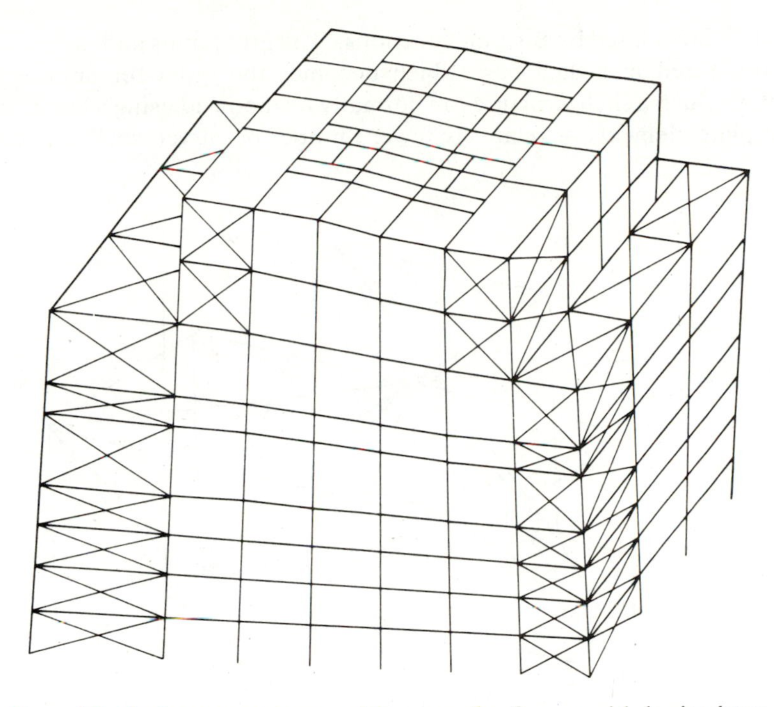

Figure 1.3 Coal steam-generator and its supporting frame modeled using beam, column, and plate elements with 1860 d.o.f.'s vibrating in the third mode with a natural frequency of 1.1 Hz.

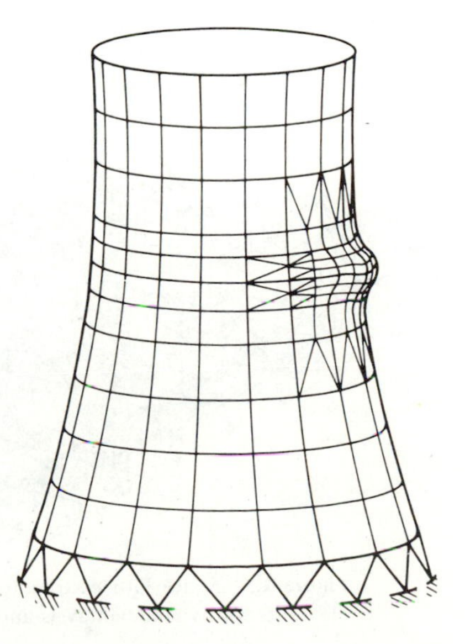

Figure 1.4 Imperfect cooling tower modeled using column, quadrilateral, and triangular curved shell elements for earthquake and random wind response analyses.

curved surfaces fitted by B-spline functions. The grid points and their coordinates are stored in a data base which becomes the input for finite element modeling. Such a shell structure can normally be modeled using a large number of flat plate elements or relatively fewer number of curved shell elements.

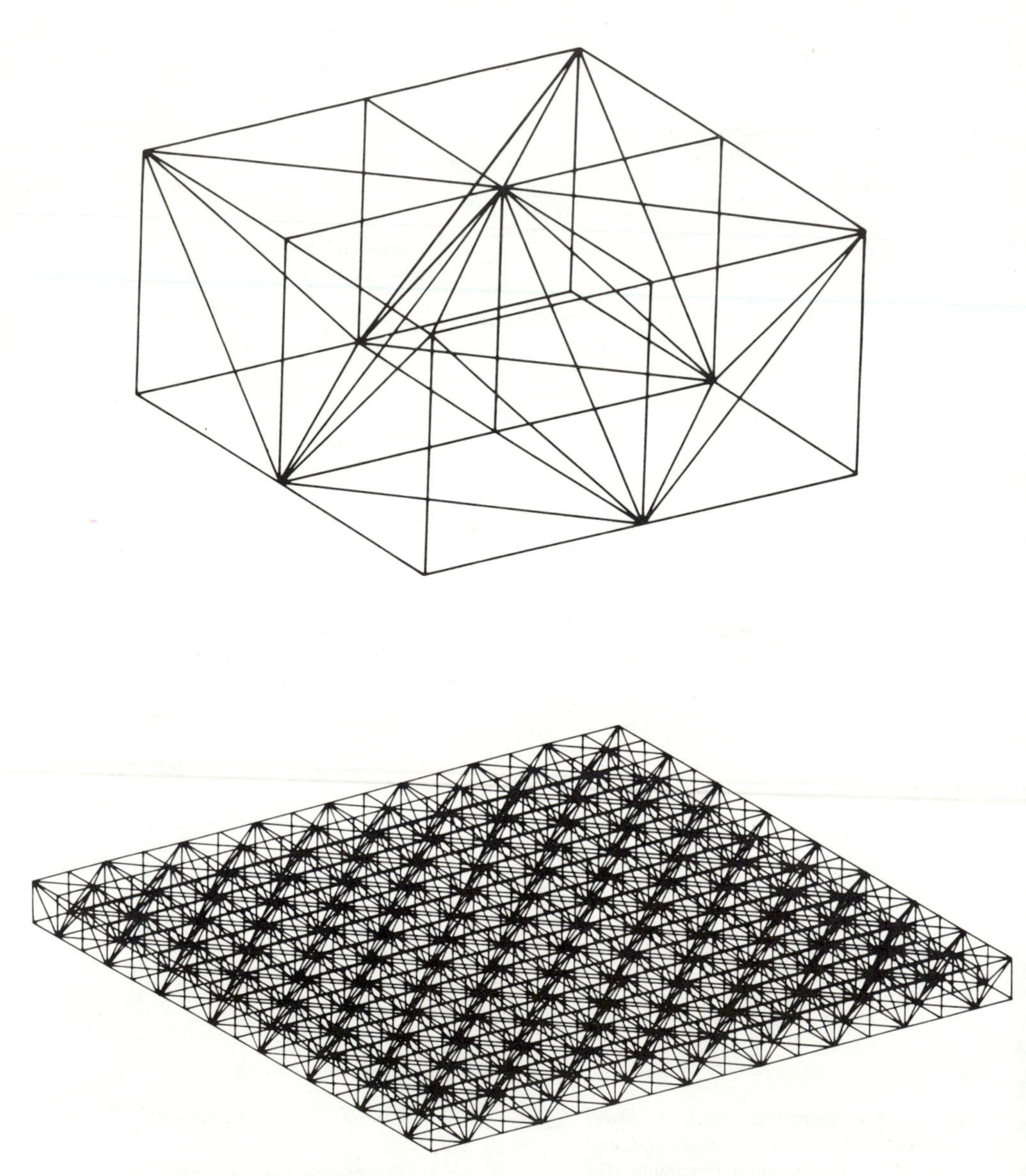

Figure 1.5 Space lattice structure modeled by truss bars or equivalent plate elements for dynamic analysis and control integration.

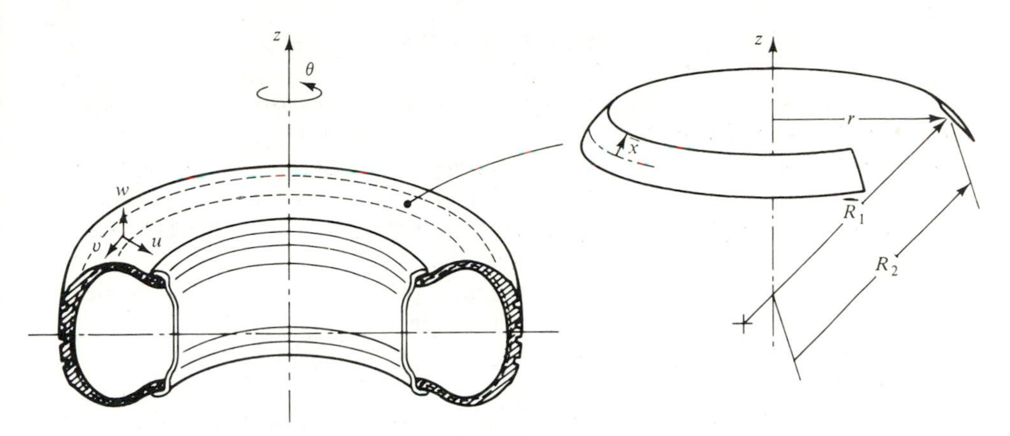

Figure 1.6 Radial belted tire modeled using axisymmetric laminated shell elements for large deflection and vibration analyses.

Figure 1.7 Computer graphical display of curved surfaces with gridpoints for shell finite element modeling. (Courtesy of Computer-Aided Design and Graphics Laboratory, Purdue University.)

REFERENCES

1.1. Clough, R. W., "The Finite Element in Plane Stress Analysis," *Proceedings*, 2nd ASCE Conference on Electronic Computation, Pittsburgh, Pa., Sept. 1960.

1.2. Martin, H. C., and Carey, G. F., *Introduction to Finite Element Analysis*, McGraw-Hill Book Company, New York, 1973, p. 2.

1.3. Timoshenko, S. P., *History of Strength of Materials*, McGraw-Hill Book Company, New York, 1953.

1.4. Maxwell, J. C., "On the Calculations of the Equilibrium and Stiffness of Frames," *Philosophical Transactions*, Vol. 27, No. 4, 1864, p. 294.

1.5. Castigliano, A., *Théorie de l'équilibre des systèmes élastiques*, Turin, 1879 (English translation by Dover Publications).

1.6. Mohr, O., "Beitrag zur Theorie der Holz- und Eisen Konstruktionen," *Zeitschrift des Architekten und Ingenieur Verienes zu Hannover*, 1868.

1.7. Gallagher, R. H., *Finite Element Analysis—Fundamentals*, Prentice-Hall, Inc., Englewood Cliffs, N.J., 1975.

1.8. Maney, G. B., *Studies in Engineering*, No. 1, University of Minnesota, Minneapolis, 1915.

1.9. Ostenfeld, A., *Die Deformationsmethode*, Springer-Verlag, Berlin, 1926.

1.10. Cross, H., "Continuity as a Factor in Reinforced Concrete Design," *Proceedings*, ACI, 1929, and "Analysis of Continuous Frames by Distributing Fixed-End Moments," *Transactions*, ASCE, Vol. 96, 1932, pp. 1-10.

1.11. Wieghardt, K., "Über einen Grenzübergang der Elastizitätslehre und seine Anwendung auf die Statik hochgradig statisch unbestimmter Fachwerke," *Verhandlungen des Vereins z. Beförderung des Gewerbef leisses, Abhandlungen*, Vol. 85, 1906, pp. 139-176.

1.12. Riedel, W., "Beiträge zur Lösung des ebenen Problems eines elastischen Körpers mittels der Airyschen Spannungsfunktion," *Zeitschrift für Angewandte Mathematik und Mechanik*, Vol. 7, No. 3, 1927, pp. 169-188.

1.13. Hrenikoff, A., "Solution of Problems in Elasticity by the Framework Method," *Journal of Applied Mechanics*, Vol. 8, 1941, pp. 169-175.

1.14. McHenry, D., "A Lattice Analogy for the Solution of Plane Stress Problems," *Journal of the Institute of Civil Engineers*, Vol. 21, 1943, pp. 59-82.

1.15. Newmark, N. M., "Numerical Methods of Analysis in Bars, Plates and Elastic Bodies," in *Numerical Methods in Analysis in Engineering*, ed. L. E. Grinter, Macmillan Publishing Company, New York, 1949.

1.16. Courant, R., "Variational Methods for the Solution of Problems of Equilibrium and Vibrations," *Bulletin of the American Mathematical Society*, Vol. 49, 1943, pp. 1-23.

1.17. Argyris, J. H., and Kelsey, S., *Energy Theorems and Structural Analysis*, Butterworths Scientific Publications, London, 1960 (collection of papers published in *Aircraft Engineering* in 1954 and 1955).

1.18. Turner, M. J., Clough, R. W., Martin, H. C., and Topp, L. J., "Stiffness and Deflection Analysis of Complex Structures," *Journal of Aeronautical Sciences*, Vol. 23, No. 9, 1956, pp. 805-823.

1.19. Zienkiewicz, O. C., "The Finite Element Method: From Intuition to Generality," *Applied Mechanics Review*, Vol. 23, No. 23, 1970, pp. 249–256.

1.20. Hua, L. C., "Structural Analysis for a 40-Story Building," NASTRAN: User's Experiences, NASA TM X-2637, 1972, pp. 421–427.

1.21. Yang, T. Y., and Baig, M. I., "Seismic Analysis of Fossil-Fuel Boiler Structures," *Journal of the Structural Division*, ASCE, Vol. 105, No. ST12, 1979, pp. 2511–2528.

1.22. Yang, T. Y., and Kapania, R. K., "Random Response of Finite Element Cooling Towers," *Journal of Engineering Mechanics*, ASCE, Vol. 110, No. 4, 1984, pp. 589–609.

1.23. Kapania, R. K., and Yang, T. Y., "Time Domain Random Wind Response of Cooling Towers," *Journal of Engineering Mechanics*, ASCE, Vol. 110, No. 10, 1984, pp. 1524–1543.

1.24. Lamberson, S. E., and Yang, T. Y., "Continuum Plate Finite Elements for Vibration Analysis and Feedback Control of Space Lattice Structures," Symposium on Advances and Trends in Structures and Dynamics, Washington, D.C., Oct. 1984.

1.25. Hunckler, C. J., Yang, T. Y., and Soedel, W., "A Geometrically Nonlinear Shell Finite Element for Tire Vibration Analysis," *Computers and Structures*, Vol. 17, No. 2, 1983, pp. 217–226.

CHAPTER 2

Matrix Algebra and Linear Equations

2.1 MATRICES

A rectangular matrix [**A**] of order $m \times n$ (m rows by n columns) is defined as

$$[\mathbf{A}] = [\mathbf{a}_{ij}] = \begin{bmatrix} a_{11} & a_{12} & \cdots & a_{1n} \\ a_{21} & a_{22} & \cdots & a_{2n} \\ \vdots & \vdots & & \vdots \\ a_{m1} & a_{m2} & \cdots & a_{mn} \end{bmatrix} \tag{2.1}$$

where a_{ij} is the element at the ith row and jth column.

If $m = n$, it is a square matrix.
If $m = 1$, it is a row matrix.
If $n = 1$, it is a column matrix.

2.2 ROW AND COLUMN MATRICES

A matrix consisting of a single row is called a *row matrix.* It will be represented by semibrackets, as follows:

$$\lfloor \mathbf{A} \rfloor = \lfloor a_{11} \quad a_{12} \quad a_{13} \quad \cdots \quad a_{1n} \rfloor \tag{2.2}$$

A matrix consisting of a single column is called a *column matrix*. It will be represented by braces, as follows:

$$\{\mathbf{A}\} = \begin{Bmatrix} a_{11} \\ a_{21} \\ a_{31} \\ \vdots \\ a_{m1} \end{Bmatrix} \tag{2.3}$$

2.3 ADDITION AND SUBTRACTION OF MATRICES

Additions and subtractions can be performed only on matrices of the same order. Such operations are performed by adding or subtracting corresponding elements. For example,

$$\begin{bmatrix} a_{11} & a_{12} \\ a_{21} & a_{22} \end{bmatrix} \pm \begin{bmatrix} b_{11} & b_{12} \\ b_{21} & b_{22} \end{bmatrix} = \begin{bmatrix} a_{11} \pm b_{11} & a_{12} \pm b_{12} \\ a_{21} \pm b_{21} & a_{22} \pm b_{22} \end{bmatrix} \tag{2.4}$$

Matrix addition and subtraction are *commutative*. For example,

$$[\mathbf{A}] - [\mathbf{B}] = -[\mathbf{B}] + [\mathbf{A}] \tag{2.5}$$

They are also *associative*. For example,

$$([\mathbf{A}] + [\mathbf{B}]) - [\mathbf{C}] = [\mathbf{A}] + ([\mathbf{B}] - [\mathbf{C}]) \tag{2.6}$$

2.4 SCALAR MULTIPLIERS

A matrix of order 1×1 is a *scalar*. A scalar is a quantity. Let k be a scalar; then

$$k[\mathbf{A}] = \begin{bmatrix} ka_{11} & ka_{12} & \cdots & ka_{1n} \\ ka_{21} & ka_{22} & \cdots & ka_{2n} \\ ka_{31} & \cdot & \cdots & \cdot \\ \vdots & \vdots & \vdots & \vdots \\ ka_{m1} & ka_{m2} & \cdots & ka_{mn} \end{bmatrix} \tag{2.7}$$

For example, if

$$[\mathbf{A}] = \begin{bmatrix} 1 & 5 \\ 3 & 7 \end{bmatrix} \qquad [\mathbf{B}] = \begin{bmatrix} 2 & -4 \\ 6 & 8 \end{bmatrix} \qquad [\mathbf{C}] = \begin{bmatrix} 1 & -1 \\ 0 & 2 \end{bmatrix}$$

then

$$2[\mathbf{A}] + 3[\mathbf{B}] - 4[\mathbf{C}] = \begin{bmatrix} 4 & 2 \\ 24 & 30 \end{bmatrix} = 2\begin{bmatrix} 2 & 1 \\ 12 & 15 \end{bmatrix}$$

2.5 MATRIX MULTIPLICATION

If matrix [**A**] is of order $m \times p$, matrix [**B**] is of order $p \times n$, and matrix [**C**] is the product of matrices [**A**] and [**B**], matrix [**C**] is then of order $m \times n$. The element at the ith row and jth column of matrix [**C**] is obtained by the following procedure:

$$c_{ij} = \sum_{k=1}^{p} a_{ik} b_{kj}$$

$$= a_{i1}b_{1j} + a_{i2}b_{2j} + a_{i3}b_{3j} + \cdots + a_{ip}b_{pj} \tag{2.8}$$

For example,

$$\underset{2 \times 2}{[\mathbf{C}]} = \underset{2 \times 3}{[\mathbf{A}]} \underset{3 \times 2}{[\mathbf{B}]} \tag{2.9a}$$

can be written in detail as

$$\begin{bmatrix} c_{11} & c_{12} \\ c_{21} & c_{22} \end{bmatrix} = \begin{bmatrix} a_{11} & a_{12} & a_{13} \\ a_{21} & a_{22} & a_{23} \end{bmatrix} \begin{bmatrix} b_{11} & b_{12} \\ b_{21} & b_{22} \\ b_{31} & b_{32} \end{bmatrix}$$

$$= \begin{bmatrix} a_{11}b_{11} + a_{12}b_{21} + a_{13}b_{31} & a_{11}b_{12} + a_{12}b_{22} + a_{13}b_{32} \\ a_{21}b_{11} + a_{22}b_{21} + a_{23}b_{31} & a_{21}b_{12} + a_{22}b_{22} + a_{23}b_{32} \end{bmatrix} \tag{2.9b}$$

The matrix multiplication procedure can also be explained by using the following example in diagram form. If

$$[\mathbf{A}] = \begin{bmatrix} 1 & 2 & -1 & 0 & 1 & 2 \\ 0 & 0 & -1 & 1 & 2 & 2 \\ 1 & 1 & -2 & 1 & 0 & 0 \end{bmatrix} \quad [\mathbf{B}] = \begin{bmatrix} 1 & 0 & 1 & 2 & 3 \\ 2 & 2 & 0 & 1 & 0 \\ 3 & 2 & 1 & 1 & 2 \\ 3 & 2 & 2 & 1 & 1 \\ 2 & 1 & 1 & 2 & 3 \\ 1 & 0 & 1 & 1 & 2 \end{bmatrix} \tag{2.10a}$$

and

$$\underset{3 \times 5}{[\mathbf{C}]} = \underset{3 \times 6}{[\mathbf{A}]} \underset{6 \times 5}{[\mathbf{B}]} \tag{2.10b}$$

then

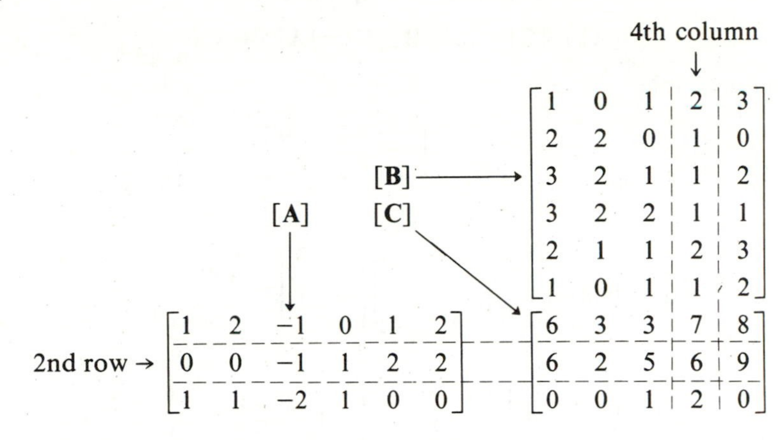

The schematic layout shows how any element in the product is related to the row and column from which it is formed. For example,

$$c_{24} = \sum_{k=1}^{6} a_{2k} b_{k4} = 6 \tag{2.10c}$$

Some notes for matrix multiplication

1. Matrices are in general not commutative in multiplication.

$$[\mathbf{A}][\mathbf{B}] \neq [\mathbf{B}][\mathbf{A}] \tag{2.11}$$

For example,

$$[\mathbf{A}][\mathbf{B}] = \begin{bmatrix} 3 & 2 \\ 1 & 4 \end{bmatrix} \begin{bmatrix} 2 & 1 \\ 1 & 2 \end{bmatrix} = \begin{bmatrix} 8 & 7 \\ 6 & 9 \end{bmatrix}$$

$$[\mathbf{B}][\mathbf{A}] = \begin{bmatrix} 2 & 1 \\ 1 & 2 \end{bmatrix} \begin{bmatrix} 3 & 2 \\ 1 & 4 \end{bmatrix} = \begin{bmatrix} 7 & 8 \\ 5 & 10 \end{bmatrix}$$

2. Two matrices [**A**] and [**B**] can be multiplied together only if they are *conformable*; that is, the number of columns in matrix [**A**] must be equal to the number of rows in matrix [**B**].

$$\underset{M \times P}{[\mathbf{A}]} \; \underset{L \times N}{[\mathbf{B}]} = \underset{M \times N}{[\mathbf{C}]} \quad \text{if and only if } P = L \tag{2.12}$$

$$\underset{M \times P}{[\mathbf{A}]} \; \underset{L \times Q}{[\mathbf{B}]} \; \underset{R \times N}{[\mathbf{C}]} = \underset{M \times N}{[\mathbf{D}]} \quad \text{if and only of } Q = R \text{ and } P = L \tag{2.13}$$

where [**C**] is premultiplied by [**B**] and [**B**] is premultiplied by [**A**]. [**A**] is postmultiplied by [**B**].

3. The matrices are associative in multiplication.

$$[\mathbf{A}][\mathbf{B}][\mathbf{C}] = ([\mathbf{A}][\mathbf{B}])[\mathbf{C}] = [\mathbf{A}]([\mathbf{B}][\mathbf{C}]) \tag{2.14}$$

For example,

$$[\mathbf{A}] = \begin{bmatrix} 1 & -1 & 1 \\ 2 & 0 & 1 \end{bmatrix} \qquad [\mathbf{B}] = \begin{bmatrix} 1 & -1 & 0 \\ 0 & 1 & -1 \\ 1 & 1 & 1 \end{bmatrix} \qquad [\mathbf{C}] = \begin{bmatrix} 1 & 0 \\ 0 & 1 \\ 1 & 1 \end{bmatrix}$$

$$([\mathbf{A}][\mathbf{B}])[\mathbf{C}] = \begin{bmatrix} 2 & -1 & 2 \\ 3 & -1 & 1 \end{bmatrix} \begin{bmatrix} 1 & 0 \\ 0 & 1 \\ 1 & 1 \end{bmatrix} = \begin{bmatrix} 4 & 1 \\ 4 & 0 \end{bmatrix}$$

$$[\mathbf{A}]([\mathbf{B}][\mathbf{C}]) = \begin{bmatrix} 1 & -1 & 1 \\ 2 & 0 & 1 \end{bmatrix} \begin{bmatrix} 1 & -1 \\ -1 & 0 \\ 2 & 2 \end{bmatrix} = \begin{bmatrix} 4 & 1 \\ 4 & 0 \end{bmatrix}$$

2.6. FORTRAN STATEMENTS FOR MULTIPLYING AND PRINTING MATRICES

The Fortran statements describing the multiplication of

$$\underset{M \times N}{[\mathbf{C}]} = \underset{M \times L}{[\mathbf{A}]} \; \underset{L \times N}{[\mathbf{B}]} \tag{2.15}$$

are given as follows:

```
DO 10 I = 1,M
DO 10 J = 1,N
C(I,J) = 0.0
DO 10 K = 1,L
C(I,J) = C(I,J) + A(I,K) * B(K,J)
```

It is important to note that the matrices must be sufficiently dimensioned by DIMENSION statements.

To print a matrix [**A**] of order $M \times N$, the following Fortran statements may be used:

```
WRITE(6,1)
FORMAT(55X, 10H**********)
WRITE(6,2)
FORMAT(55X, 10H*MATRIX A*)
WRITE(6,1)
DO 10 I = 1,M
WRITE(6,3)   (A(I,J),J=1,N)
FORMAT(11E12.4)
```

The number "6" is the output tape number. The statements are based on a computer that allows us to print 132 characters on each line. The READ statements for inputting a matrix are similar to the WRITE statements.

2.7 TRANSPOSE OF MATRIX

The *transpose* of a matrix is obtained by interchanging the rows and columns of the matrix. If matrix [**A**] is of order $m \times n$, its transposed matrix $[\mathbf{A}]^T$ is of order $n \times m$. For example, if

$$\underset{2\times 3}{[\mathbf{A}]} = \begin{bmatrix} a_{11} & a_{12} & a_{13} \\ a_{21} & a_{22} & a_{23} \end{bmatrix}$$

then

$$\underset{3\times 2}{[\mathbf{A}]^T} = \begin{bmatrix} a_{11} & a_{21} \\ a_{12} & a_{22} \\ a_{13} & a_{23} \end{bmatrix} \tag{2.16}$$

From the definition of the transpose of matrix, it is obvious that

$$([\mathbf{A}]^T)^T = [\mathbf{A}] \tag{2.17}$$

$$[\mathbf{A}]^T + [\mathbf{B}]^T = ([\mathbf{A}] + [\mathbf{B}])^T \tag{2.18}$$

If matrix [**A**] is of order $m \times n$ and matrix [**B**] is of order $n \times l$, the two matrices are conformable and we can form product [**A**][**B**]. For the transposed matrices $[\mathbf{A}]^T$ and $[\mathbf{B}]^T$, the orders are $n \times m$ and $l \times n$, respectively. We can no longer form the product for $[\mathbf{A}]^T[\mathbf{B}]^T$ but we can form the product for $[\mathbf{B}]^T[\mathbf{A}]^T$. So we know that

$$([\mathbf{A}][\mathbf{B}])^T \neq [\mathbf{A}]^T[\mathbf{B}]^T \tag{2.19}$$

Let us first consider the product [**C**] = [**A**][**B**] in diagram form, as shown in Fig. 2.1. We then consider another product, $[\mathbf{D}] = [\mathbf{B}]^T[\mathbf{A}]^T$, in a similar diagram form, as shown in Fig. 2.2. It is seen in the two figures that the

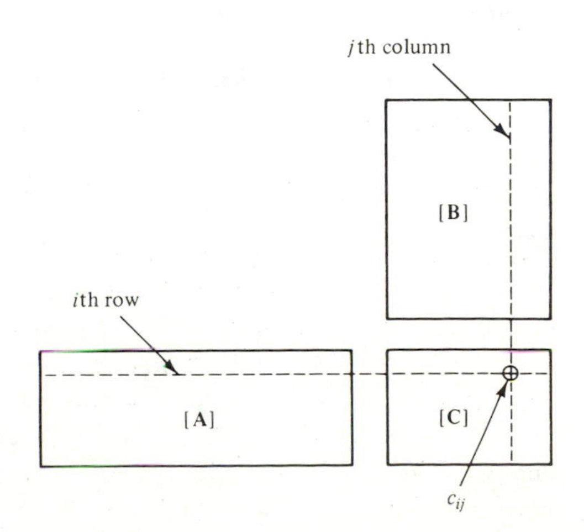

Figure 2.1 Diagrammatic explanation of the procedure for obtaining [**C**] = [**A**][**B**].

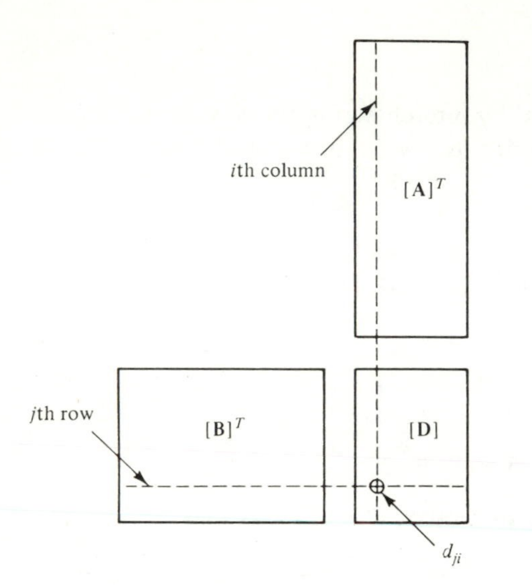

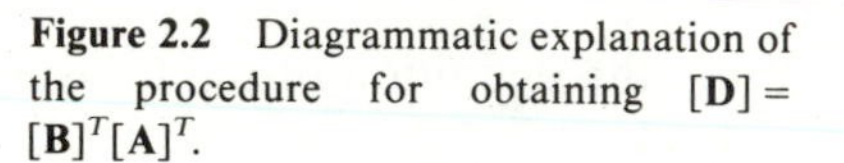
Figure 2.2 Diagrammatic explanation of the procedure for obtaining $[\mathbf{D}] = [\mathbf{B}]^T[\mathbf{A}]^T$.

elements c_{ij} and d_{ji} are both obtained as the sum of products of a pair of identical row and column (or column and row). Therefore, these two elements are equal. By definition, we found that

$$[\mathbf{C}]^T = [\mathbf{D}]$$

We then conclude that

$$([\mathbf{A}][\mathbf{B}])^T = [\mathbf{B}]^T[\mathbf{A}]^T \tag{2.20}$$

This formula can be generalized to transpose the product of more than two, say four, matrices by using the associative law,

$$\begin{aligned}([\mathbf{A}][\mathbf{B}][\mathbf{C}][\mathbf{D}])^T &= [\mathbf{D}]^T([\mathbf{A}][\mathbf{B}][\mathbf{C}])^T \\ &= [\mathbf{D}]^T[\mathbf{C}]^T([\mathbf{A}][\mathbf{B}])^T \\ &= [\mathbf{D}]^T[\mathbf{C}]^T[\mathbf{B}]^T[\mathbf{A}]^T \end{aligned} \tag{2.21}$$

Example 2.1

If

$$[\mathbf{A}] = \begin{bmatrix} 2 & 1 & 0 \\ 1 & 2 & 1 \end{bmatrix} \qquad [\mathbf{B}] = \begin{bmatrix} 1 & 0 & 1 & 1 \\ 1 & 1 & 0 & -1 \\ 0 & 1 & -1 & 0 \end{bmatrix} \qquad [\mathbf{C}] = [\mathbf{A}][\mathbf{B}]$$

find $[\mathbf{C}]^T$.

Solution 1

$$[\mathbf{C}] = [\mathbf{A}][\mathbf{B}] = \begin{bmatrix} 2 & 1 & 0 \\ 1 & 2 & 1 \end{bmatrix} \begin{bmatrix} 1 & 0 & 1 & 1 \\ 1 & 1 & 0 & -1 \\ 0 & 1 & -1 & 0 \end{bmatrix}$$

$$= \begin{bmatrix} 3 & 1 & 2 & 1 \\ 3 & 3 & 0 & -1 \end{bmatrix}$$

$$[\mathbf{C}]^T = \begin{bmatrix} 3 & 3 \\ 1 & 3 \\ 2 & 0 \\ 1 & -1 \end{bmatrix}$$

Solution 2

$$[\mathbf{C}]^T = [\mathbf{B}]^T[\mathbf{A}]^T = \begin{bmatrix} 1 & 1 & 0 \\ 0 & 1 & 1 \\ 1 & 0 & -1 \\ 1 & -1 & 0 \end{bmatrix} \begin{bmatrix} 2 & 1 \\ 1 & 2 \\ 0 & 1 \end{bmatrix} = \begin{bmatrix} 3 & 3 \\ 1 & 3 \\ 2 & 0 \\ 1 & -1 \end{bmatrix}$$

2.8 SPECIAL MATRICES

Square matrix. If $m = n$ for matrix $\underset{m \times n}{[\mathbf{A}]}$, then $[\mathbf{A}]$ is a *square matrix.* For a *symmetrical matrix,*

$$a_{ij} = a_{ji} \qquad \text{and} \qquad [\mathbf{A}] = [\mathbf{A}]^T \tag{2.22}$$

It will be shown in Chapter 3 through the use of Maxwell reciprocal theorem that all the stiffness and flexibility matrices for finite elements are symmetrical.

For an *antisymmetrical matrix,*

$$a_{ij} = -a_{ji} \qquad \text{when } i \neq j \tag{2.23}$$

Zero or Null matrix. If all the elements in matrix $[\mathbf{A}]$ are zero, it is a *zero matrix.*

Diagonal matrix (restricted to square matrix). If $a_{ij} = 0$ for $i \neq j$ and $a_{ij} \neq 0$ for $i = j$, matrix $[\mathbf{A}]$ is a *diagonal matrix.* If a diagonal matrix $[\mathbf{A}]$ is of order, say, 5×5, we have

$$\underset{5 \times 5}{[\mathbf{A}]} = \begin{bmatrix} a_{11} & 0 & 0 & 0 & 0 \\ 0 & a_{22} & 0 & 0 & 0 \\ 0 & 0 & a_{33} & 0 & 0 \\ 0 & 0 & 0 & a_{44} & 0 \\ 0 & 0 & 0 & 0 & a_{55} \end{bmatrix} \tag{2.24}$$

If [**A**] is a diagonal matrix and

$$[\mathbf{A}]\{\mathbf{x}\} = \{\mathbf{c}\}$$

the solution for the unknown vector becomes

$$x_i = \frac{c_i}{a_{ii}} \qquad i = 1, 2, 3, \cdots$$

Identity matrix (unit matrix). The *identity matrix* is a special form of diagonal matrix where all the diagonal elements are equal to unity. It is denoted by [**I**]. If [**I**] is of order, say, 5 × 5, we have

$$\underset{5\times 5}{[\mathbf{I}]} = \begin{bmatrix} 1 & 0 & 0 & 0 & 0 \\ 0 & 1 & 0 & 0 & 0 \\ 0 & 0 & 1 & 0 & 0 \\ 0 & 0 & 0 & 1 & 0 \\ 0 & 0 & 0 & 0 & 1 \end{bmatrix} \tag{2.25}$$

If matrices [**A**] and [**I**] are both of the same order,

$$[\mathbf{I}][\mathbf{A}] = [\mathbf{A}][\mathbf{I}] = [\mathbf{A}] \tag{2.26}$$

Scalar matrix. A *scalar matrix* is a special form of diagonal matrix where all the diagonal elements are equal to a scalar. If a scalar matrix [**A**] is of order, say, 5 × 5, and the scalar is, say, 4, we have

$$\underset{5\times 5}{[\mathbf{A}]} = \begin{bmatrix} 4 & 0 & 0 & 0 & 0 \\ 0 & 4 & 0 & 0 & 0 \\ 0 & 0 & 4 & 0 & 0 \\ 0 & 0 & 0 & 4 & 0 \\ 0 & 0 & 0 & 0 & 4 \end{bmatrix} = 4[\mathbf{I}] \tag{2.27}$$

Triangular matrix (restricted to square matrix). If in matrix [**A**] all the elements above the main diagonal are zero (i.e., $a_{ij} = 0$ for $i < j$), the matrix is called a *lower triangular matrix*. For example,

$$\begin{bmatrix} 1 & 0 & 0 & 0 & 0 \\ 2 & 2 & 0 & 0 & 0 \\ 3 & 1 & 3 & 0 & 0 \\ 0 & 2 & 1 & 4 & 0 \\ 2 & 4 & 2 & 1 & 2 \end{bmatrix}$$

If in matrix [A] all the elements below the main diagonal are zero (i.e., $a_{ij} = 0$ for $i > j$), the matrix is called an *upper triangular matrix.* For example,

$$\begin{bmatrix} 2 & 1 & 3 & 2 & 3 \\ 0 & 2 & 7 & 1 & 0 \\ 0 & 0 & 4 & 5 & 4 \\ 0 & 0 & 0 & 2 & 2 \\ 0 & 0 & 0 & 0 & 1 \end{bmatrix}$$

If [A] is a lower triangular matrix and we have a set of matrix equations,

$$[\mathbf{A}]\{\mathbf{x}\} = \{\mathbf{c}\} \tag{2.28}$$

we can first find x_1 as c_1/a_{11}. We can then find $x_2, x_3, \cdots$ sequentially by the method of substitution.

2.9 MATRIX PARTITION (Submatrices)

The array of elements in a matrix may be divided into smaller arrays by horizontal and vertical dash lines. Such a matrix is then referred to as a *partitioned matrix*, and the smaller arrays are called *submatrices.* For example,

$$[\mathbf{A}] = \left[\begin{array}{cc:c} a_{11} & a_{12} & a_{13} \\ a_{21} & a_{22} & a_{23} \\ \hdashline a_{31} & a_{32} & a_{33} \end{array}\right] = \left[\begin{array}{c:c} A_{11} & A_{12} \\ \hdashline A_{21} & A_{22} \end{array}\right] \tag{2.29}$$

where

$$[\mathbf{A}_{11}] = \begin{bmatrix} a_{11} & a_{12} \\ a_{21} & a_{22} \end{bmatrix} \qquad [\mathbf{A}_{12}] = \begin{Bmatrix} a_{13} \\ a_{23} \end{Bmatrix}$$

$$[\mathbf{A}_{21}] = [a_{31} \quad a_{32}] \qquad [\mathbf{A}_{22}] = a_{33}$$

2.10 ORTHOGONAL MATRIX (Restricted to Square Matrix)

If the transpose of a square matrix is equal to its inverse, the matrix is an *orthogonal matrix.* The obvious advantage of an orthogonal matrix is that its inverse can easily be obtained simply by transposing the matrix.

Let us consider a square matrix of order, say, 3×3,

$$[\mathbf{A}] = \begin{bmatrix} a_{11} & a_{12} & a_{13} \\ a_{21} & a_{22} & a_{23} \\ a_{31} & a_{32} & a_{33} \end{bmatrix} \tag{2.30}$$

The property of orthogonality implies that

$$[\mathbf{A}][\mathbf{A}]^T = [\mathbf{A}][\mathbf{A}]^{-1} = [\mathbf{I}] \tag{2.31a}$$

or

$$\begin{bmatrix} a_{11} & a_{12} & a_{13} \\ a_{21} & a_{22} & a_{23} \\ a_{31} & a_{32} & a_{33} \end{bmatrix} \begin{bmatrix} a_{11} & a_{21} & a_{31} \\ a_{12} & a_{22} & a_{32} \\ a_{13} & a_{23} & a_{33} \end{bmatrix} = \begin{bmatrix} 1 & 0 & 0 \\ 0 & 1 & 0 \\ 0 & 0 & 1 \end{bmatrix} \tag{2.31b}$$

By evaluating the three diagonal elements in the identity matrix, we obtain

$$\begin{cases} a_{11}^2 + a_{12}^2 + a_{13}^2 = 1 \\ a_{21}^2 + a_{22}^2 + a_{23}^2 = 1 \\ a_{31}^2 + a_{32}^2 + a_{33}^2 = 1 \end{cases} \tag{2.32a}$$

By evaluating the three elements either above the diagonal or below the diagonal, we obtain

$$\begin{cases} a_{11}a_{21} + a_{12}a_{22} + a_{13}a_{23} = 0 \\ a_{21}a_{31} + a_{22}a_{32} + a_{23}a_{33} = 0 \\ a_{31}a_{11} + a_{32}a_{12} + a_{33}a_{13} = 0 \end{cases} \tag{2.32b}$$

Thus we reach the following conditions of orthogonality for a matrix:

1. The sum of the squares of each element in each row (or column) must be equal to unity.
2. The sum of the products of corresponding elements between any two rows (or columns) must be equal to zero.

2.11 COORDINATE TRANSFORMATION MATRIX

In the finite element formulation, it is often required to transform the formulation from a set of local coordinates to a more general set of global coordinates. The *matrices for coordinate transformation* have the special property of being orthogonal, and such a property simplifies the transformation procedure.

Three examples of coordinate transformation matrices are given below. These three coordinate transformation matrices are important because they will be used in the following chapters.

Example 2.2 Two-dimensional coordinate transformation

A set of old coordinates (x, y) and a set of new coordinates (x', y') are shown in Fig. 2.3. The new coordinates are resulted from a counterclockwise rotation of the old coordinates by an angle θ.

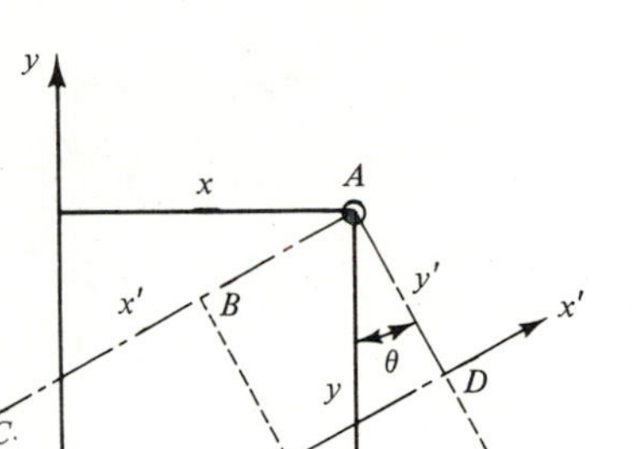

Figure 2.3 Rotation of a set of two-dimensional Cartesian coordinates by an angle θ.

For an arbitrary point A, the new coordinates (x', y') can be related to the old ones as follows:

$$\begin{aligned} x' &= \overline{CB} + \overline{BA} = x \cos\theta + y \sin\theta \\ y' &= \overline{AE} - \overline{DE} = y \cos\theta - x \sin\theta \end{aligned} \tag{2.33a}$$

or in matrix form

$$\begin{Bmatrix} x' \\ y' \end{Bmatrix} = [T] \begin{Bmatrix} x \\ y \end{Bmatrix} \tag{2.33b}$$

where

$$[\mathbf{T}] = \begin{bmatrix} \cos\theta & \sin\theta \\ -\sin\theta & \cos\theta \end{bmatrix} \tag{2.33c}$$

Matrix [T] is a coordinate transformation matrix. This matrix satisfies the conditions of orthogonality that

$$\begin{aligned} \cos^2\theta + \sin^2\theta &= 1 \\ (-\sin\theta)^2 + \cos^2\theta &= 1 \\ (-\sin\theta)(\cos\theta) + (\cos\theta)(\sin\theta) &= 0 \end{aligned} \tag{2.33d}$$

Thus we know that this coordinate transformation matrix [T] is an orthogonal matrix.

Example 2.3 Two-dimensional coordinate transformation matrix for axial-flexural beam element

An axial-flexural beam element, oriented along the *local* x'-coordinate axis, is shown in Fig. 2.4. The x' axis is oriented at an angle θ from the *global* coordinate x axis. At joint 2 of the element, there exists an axial force X'_2, a transverse shear force Y'_2, and a bending moment M'_2. These forces and moment can be transformed to those corresponding to the global coordinates X_2, Y_2, and M_2, respectively, as shown in

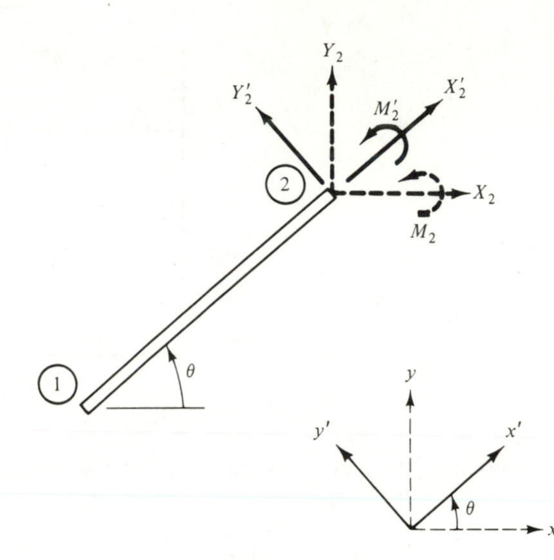

Figure 2.4 Transformation of joint force vectors from local coordinates (x', y') to global coordinates (x, y).

Fig. 2.4. The relations are as follows:

$$\begin{Bmatrix} X_2' \\ Y_2' \\ M_2' \end{Bmatrix} = \begin{bmatrix} \cos\theta & \sin\theta & 0 \\ -\sin\theta & \cos\theta & 0 \\ 0 & 0 & 1 \end{bmatrix} \begin{Bmatrix} X_2 \\ Y_2 \\ M_2 \end{Bmatrix}$$

$$= [\mathbf{T}] \begin{Bmatrix} X_2 \\ Y_2 \\ M_2 \end{Bmatrix} \tag{2.34}$$

It is again seen that the transformation matrix [T] is an orthogonal matrix.

Example 2.4 Three-dimensional coordinate transformation matrix

A set of old coordinates (x, y, z) and a set of new coordinates (x', y', z') are shown in Fig. 2.5. Let **i**, **j**, **k** and **i**′, **j**′, **k**′ be unit vectors in the directions of the respective axes, and let A be a general point in space having coordinates (x, y, z) and (x', y', z') in the respective systems. We can write for vector OA

$$\mathbf{R} = OA = x\mathbf{i} + y\mathbf{j} + z\mathbf{k} \tag{2.35a}$$

The components x', y', z' are obtained from the dot products of vector **R** with the unit vectors **i**′, **j**′, **k**′.

$$\begin{Bmatrix} x' \\ y' \\ z' \end{Bmatrix} = \begin{Bmatrix} \mathbf{R}\cdot\mathbf{i}' \\ \mathbf{R}\cdot\mathbf{j}' \\ \mathbf{R}\cdot\mathbf{k}' \end{Bmatrix} = \begin{bmatrix} \mathbf{i}\cdot\mathbf{i}' & \mathbf{j}\cdot\mathbf{i}' & \mathbf{k}\cdot\mathbf{i}' \\ \mathbf{i}\cdot\mathbf{j}' & \mathbf{j}\cdot\mathbf{j}' & \mathbf{k}\cdot\mathbf{j}' \\ \mathbf{i}\cdot\mathbf{k}' & \mathbf{j}\cdot\mathbf{k}' & \mathbf{k}\cdot\mathbf{k}' \end{bmatrix} \begin{Bmatrix} x \\ y \\ z \end{Bmatrix}$$

$$= \begin{bmatrix} l_1 & m_1 & n_1 \\ l_2 & m_2 & n_2 \\ l_3 & m_3 & n_3 \end{bmatrix} \begin{Bmatrix} x \\ y \\ z \end{Bmatrix} = [\mathbf{T}] \begin{Bmatrix} x \\ y \\ z \end{Bmatrix} \tag{2.35b}$$

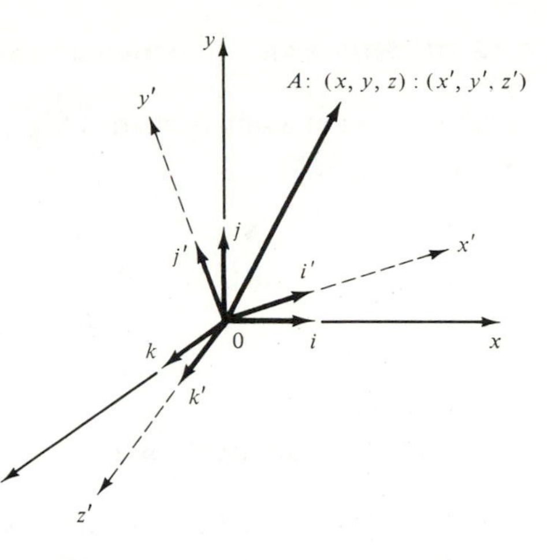

Figure 2.5 Unit vectors in two sets of three-dimensional Cartesian coordinates.

where l_1, m_1, n_1; l_2, m_2, n_2; l_3, m_3, n_3 are the direction cosines of the x'; y'; z' axes relative to the x, y, z axes, respectively. Matrix [T] is the transformation matrix.

It is necessary to show that the transformation matrix [T] is orthogonal. We first write the new unit vectors **i**′, **j**′, **k**′ as components of the old unit vectors **i**, **j**, **k**;

$$\begin{aligned} \mathbf{i}' &= l_1\mathbf{i} + m_1\mathbf{j} + n_1\mathbf{k} \\ \mathbf{j}' &= l_2\mathbf{i} + m_2\mathbf{j} + n_2\mathbf{k} \\ \mathbf{k}' &= l_3\mathbf{i} + m_3\mathbf{j} + n_3\mathbf{k} \end{aligned} \tag{2.35c}$$

Because the two coordinate systems are rectangular systems, we have

$$\begin{aligned} \mathbf{i}' \cdot \mathbf{i}' &= \mathbf{j}' \cdot \mathbf{j}' = \mathbf{k}' \cdot \mathbf{k}' \\ &= l_1^2 + m_1^2 + n_1^2 \\ &= l_2^2 + m_2^2 + n_2^2 \\ &= l_3^2 + m_3^2 + n_3^2 \\ &= 1 \end{aligned} \tag{2.35d}$$

Also,

$$\begin{aligned} \mathbf{i}' \cdot \mathbf{j}' &= \mathbf{j}' \cdot \mathbf{k}' = \mathbf{k}' \cdot \mathbf{i}' \\ &= l_1 l_2 + m_1 m_2 + n_1 n_2 \\ &= l_2 l_3 + m_2 m_3 + n_2 n_3 \\ &= l_3 l_1 + m_3 m_1 + n_3 n_1 \\ &= 0 \end{aligned} \tag{2.35e}$$

The conditions of orthogonality are thus obtained for the transformation matrix [T].

2.12 DETERMINANT (Restricted to Square Matrix)

A determinant of a square matrix $[\mathbf{A}]_{m\times m}$ is a scalar value function denoted by

$$\det[\mathbf{A}] = |A| = \begin{vmatrix} a_{11} & a_{12} & \cdots & a_{1m} \\ a_{21} & a_{22} & \cdots & a_{2m} \\ \vdots & \vdots & & \vdots \\ a_{m1} & a_{m2} & \cdots & a_{mm} \end{vmatrix} \tag{2.36}$$

2.12.1 Minor and Cofactor

The first *minor* of a determinant $\det[\mathbf{A}]$, corresponding to the element a_{ij}, is defined as the determinant obtained by omission of the ith row and jth column of $\det[\mathbf{A}]$. Let us denote this minor by $\bar{M}_{ij}$. For example, if

$$|A| = \begin{vmatrix} 1 & 3 & 5 & 7 & 9 \\ 2 & 4 & 6 & 8 & 2 \\ 1 & 2 & 3 & 3 & 2 \\ 2 & 1 & 2 & 1 & 2 \\ 3 & 4 & 3 & 4 & 3 \end{vmatrix}$$

the first minor for a_{32} is defined as

$$\bar{M}_{32} = \begin{vmatrix} 1 & 5 & 7 & 9 \\ 2 & 6 & 8 & 2 \\ 2 & 2 & 1 & 2 \\ 3 & 3 & 4 & 3 \end{vmatrix}$$

If the first minor $\bar{M}_{ij}$ is multiplied by $(-1)^{i+j}$, it becomes the *cofactor* of the term a_{ij}. Thus

$$\bar{A}_{ij} = (-1)^{i+j}\bar{M}_{ij}$$

For example,

$$\bar{A}_{32} = (-1)^{3+2}\bar{M}_{32} = -\begin{vmatrix} 1 & 5 & 7 & 9 \\ 2 & 6 & 8 & 2 \\ 2 & 2 & 1 & 2 \\ 3 & 3 & 4 & 3 \end{vmatrix}$$

2.12.2 Expansion of Determinant (Laplace Expansion Formula)

The determinant of a matrix [**A**] can be found by the repetitive use of the Laplace expansion formula:

$$\underset{m \times m}{\det\ [\mathbf{A}]} = \sum_{k=1}^{m} a_{ik}\bar{A}_{ik} \qquad (i \text{ can be any row and } i \text{ is not to be summed}) \tag{2.37a}$$

or

$$\underset{m \times m}{\det\ [\mathbf{A}]} = \sum_{k=1}^{m} a_{kj}\bar{A}_{kj} \qquad (j \text{ can be any column and } j \text{ is not to be summed}) \tag{2.37b}$$

Det [**A**] is written in terms of the sum of the products of the elements in the ith row (or jth column) and their corresponding cofactors. For example,

$$\begin{aligned}
\det [\mathbf{A}] &= \begin{vmatrix} a_{11} & a_{12} & a_{13} \\ a_{21} & a_{22} & a_{23} \\ a_{31} & a_{32} & a_{33} \end{vmatrix} \\
&= a_{11}\bar{A}_{11} + a_{12}\bar{A}_{12} + a_{13}\bar{A}_{13} \\
&= a_{11}(-1)^{1+1}\begin{vmatrix} a_{22} & a_{23} \\ a_{32} & a_{33} \end{vmatrix} + a_{12}(-1)^{1+2}\begin{vmatrix} a_{21} & a_{23} \\ a_{31} & a_{33} \end{vmatrix} + a_{13}(-1)^{1+3}\begin{vmatrix} a_{21} & a_{22} \\ a_{31} & a_{32} \end{vmatrix} \\
&= a_{11}a_{22}(-1)^{1+1}a_{33} + a_{11}a_{23}(-1)^{1+2}a_{32} - a_{12}a_{21}(-1)^{1+1}a_{33} \\
&\quad - a_{12}a_{23}(-1)^{1+2}a_{31} + a_{13}a_{21}(-1)^{1+1}a_{32} + a_{13}a_{22}(-1)^{1+2}a_{31} \\
&= a_{11}a_{22}a_{33} + a_{12}a_{23}a_{31} + a_{13}a_{21}a_{32} - a_{13}a_{22}a_{31} - a_{11}a_{23}a_{32} - a_{12}a_{21}a_{33}
\end{aligned}$$

In other words, the foregoing operation is in the following "familiar" diagram form:

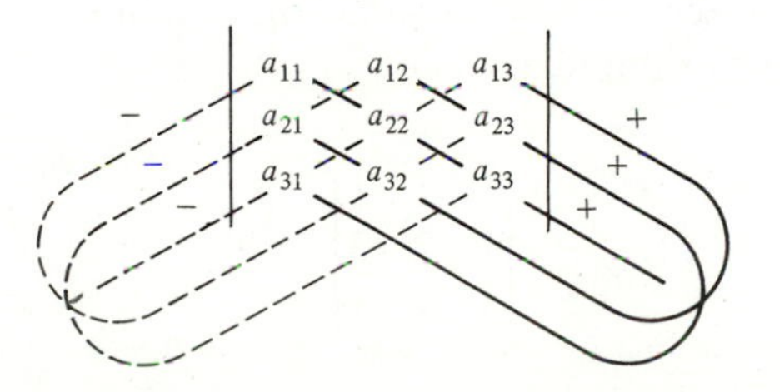

Example 2.5

Find

$$\begin{vmatrix} 1 & 2 & 3 \\ 3 & 1 & 0 \\ 0 & 1 & 1 \end{vmatrix}$$

Solution. By the diagram method,

$$\begin{vmatrix} 1 & 2 & 3 \\ 3 & 1 & 0 \\ 0 & 1 & 1 \end{vmatrix} = 1 + 0 + 9 - 0 - 0 - 6 = 4$$

or by the Laplace expansion formula,

$$\begin{vmatrix} 1 & 2 & 3 \\ 3 & 1 & 0 \\ 0 & 1 & 1 \end{vmatrix} = 1(-1)^2\begin{vmatrix} 1 & 0 \\ 1 & 1 \end{vmatrix} + 2(-1)^3\begin{vmatrix} 3 & 0 \\ 0 & 1 \end{vmatrix} + 3(-1)^4\begin{vmatrix} 3 & 1 \\ 0 & 1 \end{vmatrix}$$

$$= \begin{vmatrix} 1 & 0 \\ 1 & 1 \end{vmatrix} - 2\begin{vmatrix} 3 & 0 \\ 0 & 1 \end{vmatrix} + 3\begin{vmatrix} 3 & 1 \\ 0 & 1 \end{vmatrix}$$

$$= 1(-1)^2 \cdot 1 + 0 - 2 \cdot 3(-1)^2 \cdot 1 + 3 \cdot 3(-1)^2 \cdot 1 + 3 \cdot 1(-1)^3 \cdot 0$$

$$= 1 - 6 + 9$$

$$= 4$$

2.12.3 Properties of the Determinant

1. The determinant of a matrix and the determinant of the transpose of that matrix are equal, $|\mathbf{A}| = |\mathbf{A}|^T$.
2. $|([\mathbf{A}][\mathbf{B}])| = |\mathbf{A}|\,|\mathbf{B}|$.
3. Interchanging any two rows or columns changes the sign of the determinant.
4. If all the elements in a row, or a column, in a matrix are zeros, the determinant is zero.
5. If two rows or two columns in a matrix are identical, the value of the determinant is zero. This is a sufficient condition, but not a necessary condition. For example,

$$\begin{vmatrix} 1 & 1 & 1 \\ 1 & 2 & 3 \\ 3 & 2 & 1 \end{vmatrix} = 0$$

6. If matrix [**B**] is obtained from matrix [**A**] by adding a multiple of one row of [**A**] to another (or a multiple of one column to another), then $|\mathbf{A}| = |\mathbf{B}|$. For example,

$$|\mathbf{A}| = \begin{vmatrix} 1 & 1 & 2 \\ 2 & 1 & 1 \\ 1 & 2 & 1 \end{vmatrix} = 4$$

If the second row is replaced by the sum of the second row and twice of the first row, we have

$$|\mathbf{B}| = \begin{vmatrix} 1 & 1 & 2 \\ 4 & 3 & 5 \\ 1 & 2 & 1 \end{vmatrix} = 4 = |\mathbf{A}|$$

2.13 MATRIX INVERSION (Restricted to Square Matrix)

There are many methods available for matrix inversion. Three popular methods are illustrated here. Their properties are also discussed.

2.13.1 The Adjoint Method

Definition of adjoint matrix: The *adjoint* of a matrix [**A**], written as adj [**A**], is of the form

$$\text{adj}\,[\mathbf{A}] = \begin{bmatrix} \bar{A}_{11} & \bar{A}_{21} & \cdots & \bar{A}_{m1} \\ \bar{A}_{12} & \bar{A}_{22} & \cdots & \bar{A}_{m2} \\ \vdots & \vdots & & \vdots \\ \bar{A}_{1m} & \bar{A}_{2m} & \cdots & \bar{A}_{mm} \end{bmatrix} = [\bar{A}_{ij}]^T \tag{2.38}$$

which is defined as the transpose of the matrix of cofactors.

Example 2.6

Find adj [**A**] for

$$[\mathbf{A}] = \begin{bmatrix} 3 & 2 & 1 \\ 2 & 4 & 2 \\ 3 & 1 & 2 \end{bmatrix}$$

Solution. First we form the cofactors:

$$\bar{A}_{11} = 6 \qquad \bar{A}_{12} = 2 \qquad \bar{A}_{13} = -10$$

$$\bar{A}_{21} = -3 \qquad \bar{A}_{22} = 3 \qquad \bar{A}_{23} = 3$$

$$\bar{A}_{31} = 0 \qquad \bar{A}_{32} = -4 \qquad \bar{A}_{33} = 8$$

Then

$$\text{adj}\,[\mathbf{A}] = \begin{bmatrix} 6 & -3 & 0 \\ 2 & 3 & -4 \\ -10 & 3 & 8 \end{bmatrix}$$

2.13.2 Derivation of the Equation for Matrix Inversion

The *inverse* of a matrix [**A**], expressed by $[\mathbf{A}]^{-1}$, is defined such that

$$[\mathbf{A}][\mathbf{A}]^{-1} = [\mathbf{I}]$$

First let a matrix [**P**] of order $m \times m$ be defined as

$$[\mathbf{P}] = [\mathbf{A}]\,\text{adj}\,[\mathbf{A}]$$

The term at the ith row and jth column of the matrix [**P**] is

$$p_{ij} = \sum_{k=1}^{m} a_{ik}\bar{A}_{jk}$$

If $i = j$,

$$p_{ij} = p_{ii} = \sum_{k=1}^{m} a_{ik}\bar{A}_{ik} = \det\,[\mathbf{A}]$$

which is precisely the Laplace expansion equation for finding the determinant. Thus

$$p_{11} = p_{22} = p_{33} \cdots = p_{mm} = |A|$$

If $i \neq j$,

$$p_{ij} = \sum_{k=1}^{m} a_{ik}\bar{A}_{jk} = 0$$

The proof of $p_{ij} = 0$ will be given subsequently. From the explanation above, it is readily seen that

$$[\mathbf{P}] = \begin{bmatrix} |A| & & & \text{all zeros} \\ & |A| & & \\ & & |A| \,. & \\ \text{all zeros} & & & \cdot \; |A| \end{bmatrix}$$

$$[\mathbf{P}] = |A|[\mathbf{I}]$$

$$[\mathbf{A}]\frac{\text{adj}\,[\mathbf{A}]}{|\mathbf{A}|} = [\mathbf{I}]$$

We premultiply both sides by $[\mathbf{A}]^{-1}$:

$$[\mathbf{I}]\frac{\text{adj}\,[\mathbf{A}]}{|\mathbf{A}|} = [\mathbf{A}]^{-1}[\mathbf{I}]$$

Finally, we arrived at the equation for matrix inversion:

$$[\mathbf{A}]^{-1} = \frac{\text{adj}\,[\mathbf{A}]}{\det\,[\mathbf{A}]} \tag{2.39}$$

Example 2.7

Invert the matrix [A] as defined in Example 2.6 in Section 2.13.1.

$$[\mathbf{A}]^{-1} = \frac{\text{adj}\,[\mathbf{A}]}{\det\,[\mathbf{A}]} = \begin{bmatrix} 6 & -3 & 0 \\ 2 & 3 & -4 \\ -10 & 3 & 8 \end{bmatrix} \div 12$$

$$= \begin{bmatrix} \frac{1}{2} & -\frac{1}{4} & 0 \\ \frac{1}{6} & \frac{1}{4} & -\frac{1}{3} \\ -\frac{5}{6} & \frac{1}{4} & \frac{2}{3} \end{bmatrix}$$

Proof that $p_{ij} = 0$ when $i \neq j$. Let [A] be a 3×3 matrix. We can evaluate p_{ij} for both $i = j$ and $i \neq j$.

Case 1: For $i = j = 1$,

$$p_{ij} = p_{11} = \sum_{k=1}^{3} a_{1k}\bar{A}_{1k} = a_{11}\bar{A}_{11} + a_{12}\bar{A}_{12} + a_{13}\bar{A}_{13}$$

$$= a_{11}\begin{vmatrix} a_{22} & a_{23} \\ a_{32} & a_{33} \end{vmatrix} - a_{12}\begin{vmatrix} a_{21} & a_{23} \\ a_{31} & a_{33} \end{vmatrix} + a_{13}\begin{vmatrix} a_{21} & a_{22} \\ a_{31} & a_{32} \end{vmatrix}$$

$$= \det \begin{bmatrix} a_{11} & a_{12} & a_{13} \\ a_{21} & a_{22} & a_{23} \\ a_{31} & a_{32} & a_{33} \end{bmatrix} \tag{2.40a}$$

Case 2: For $i = 2 \neq j = 1$,

$$p_{ij} = p_{21} = \sum_{k=1}^{3} a_{2k}\bar{A}_{1k} = a_{21}\bar{A}_{11} + a_{22}\bar{A}_{12} + a_{23}\bar{A}_{13}$$

$$= a_{21}\begin{vmatrix} a_{22} & a_{23} \\ a_{32} & a_{33} \end{vmatrix} - a_{22}\begin{vmatrix} a_{21} & a_{23} \\ a_{31} & a_{33} \end{vmatrix} + a_{23}\begin{vmatrix} a_{21} & a_{22} \\ a_{31} & a_{32} \end{vmatrix}$$

$$= \det \begin{vmatrix} a_{21} & a_{22} & a_{23} \\ a_{21} & a_{22} & a_{23} \\ a_{31} & a_{32} & a_{33} \end{vmatrix} = \text{zero} \tag{2.40b}$$

Since the first row and the second row are identical, we conclude that $p_{ij} = 0$ for $i \neq j$.

In general, to evaluate p_{ij} with $i \neq j$ is equivalent to finding the determinant of a matrix with two identical rows of elements a_{ik} with $k = 1$ to m.

2.13.3 Gauss–Jordan Elimination Method

This method can be explained by performing an example. Let it be desired to find the inverse of the matrix

$$[\mathbf{A}] = \begin{bmatrix} 3 & 2 & 1 \\ 2 & 4 & 2 \\ 3 & 1 & 2 \end{bmatrix} \tag{2.41}$$

TABLE 2.1 Matrix Inversion by Gauss–Jordan Elimination

Row	[A]			[I]			Explanation
[1]	3	2	1	1	0	0	
[2]	2	4	2	0	1	0	
[3]	3	1	2	0	0	1	
[4]	1	$\frac{2}{3}$	$\frac{1}{3}$	$\frac{1}{3}$	0	0	$[1] \div a_{11} = [1] \div 3$
[5]	0	$\frac{8}{3}$	$\frac{4}{3}$	$-\frac{2}{3}$	1	0	$[2] - a_{21}[4] = [2] - 2[4]$
[6]	0	-1	1	-1	0	1	$[3] - a_{31}[4] = [3] - 3[4]$
[7]	0	1	$\frac{1}{2}$	$-\frac{1}{4}$	$\frac{3}{8}$	0	$[5] \div a_{52} = [5] \div \frac{8}{3}$
[8]	0	0	$\frac{3}{2}$	$-\frac{5}{4}$	$\frac{3}{8}$	1	$[6] - (-1)[7]$
[9]	0	0	1	$-\frac{5}{6}$	$\frac{1}{4}$	$\frac{2}{3}$	$[8] \div a_{83} = [8] \div \frac{3}{2}$
[10]	0	1	0	$\frac{1}{6}$	$\frac{1}{4}$	$-\frac{1}{3}$	$[7] - \frac{1}{2}[9]$
[11]	1	0	0	$\frac{1}{2}$	$-\frac{1}{4}$	0	$[4] - \frac{2}{3}[10] - \frac{1}{3}[9]$

In Table 2.1 the identity matrix [**I**] is subjected to the same process as matrix [**A**]. Row [1] is first divided by a_{11} to obtain row [4]. Row [4] is then multiplied by a_{i1} $(i = 2, 3)$ successively and subtracted from the remaining equations. Thus all elements but a_{11} in the first column are eliminated. Such a process (called *pivot condensation*) is then performed to eliminate all elements but a_{12} and a_{22} in the second column. The process is continued until matrix [**A**] is reduced to an upper triangular matrix with 1's along the diagonal (see rows [4], [7], and [9]). We then start from row [9] backward to eliminate the last element from row [7] and last two elements from row [4]. We finally obtain an identity matrix as given in rows [11], [10], [9],

$$\begin{bmatrix} 1 & 0 & 0 \\ 0 & 1 & 0 \\ 0 & 0 & 1 \end{bmatrix} \begin{bmatrix} \frac{1}{2} & -\frac{1}{4} & 0 \\ \frac{1}{6} & \frac{1}{4} & -\frac{1}{3} \\ -\frac{5}{6} & \frac{1}{4} & \frac{2}{3} \end{bmatrix}$$

The foregoing process is equivalent to premultiplying the left-hand-side matrix [**A**] by $[\mathbf{A}]^{-1}$ to obtain an identity matrix [**I**]. The same process applied to the right-hand-side identity matrix is equivalent to performing

$$[\mathbf{I}][\mathbf{A}]^{-1} = [\mathbf{A}]^{-1}$$

Thus the inverse of the matrix is obtained as

$$[\mathbf{A}]^{-1} = \begin{bmatrix} \frac{1}{2} & -\frac{1}{4} & 0 \\ \frac{1}{6} & \frac{1}{4} & -\frac{1}{3} \\ -\frac{5}{6} & \frac{1}{4} & \frac{2}{3} \end{bmatrix} \tag{2.42}$$

2.13.4 Cholesky's Method

The Cholesky method for inverting a matrix is explained step by step using the following example:

$$[\mathbf{A}] = \begin{bmatrix} 3 & 2 & 1 \\ 2 & 4 & 2 \\ 3 & 1 & 2 \end{bmatrix} \tag{2.43}$$

Step 1: First, let us define

$[\mathbf{L}]$ = lower triangular matrix

$[\mathbf{T}]$ = upper triangular matrix with 1's along the diagonal

We then find matrices $[\mathbf{L}]$ and $[\mathbf{T}]$ such that

$$[\mathbf{A}] = [\mathbf{L}][\mathbf{T}] \tag{2.44a}$$

or

$$\begin{bmatrix} 3 & 2 & 1 \\ 2 & 4 & 2 \\ 3 & 1 & 2 \end{bmatrix} = \begin{bmatrix} l_{11} & 0 & 0 \\ l_{21} & l_{22} & 0 \\ l_{31} & l_{32} & l_{33} \end{bmatrix} \begin{bmatrix} 1 & t_{12} & t_{13} \\ 0 & 1 & t_{23} \\ 0 & 0 & 1 \end{bmatrix} \tag{2.44b}$$

The nine unknown elements in matrices $[\mathbf{L}]$ and $[\mathbf{T}]$ are obtained by evaluating the nine known values in matrix $[\mathbf{A}]$. We evaluate these values in a columnwise direction starting from the first column. Thus for the first column, we obtain

$$3 = l_{11} \qquad 2 = l_{21} \qquad 3 = l_{31}$$

For the second column, we obtain

$$\begin{cases} 2 = l_{11}t_{12} = 3t_{12} & t_{12} = \frac{2}{3} \\ 4 = l_{21}t_{12} + l_{22} = (2)(\frac{2}{3}) + l_{22} & l_{22} = \frac{8}{3} \\ 1 = l_{31}t_{12} + l_{32} = (3)(\frac{2}{3}) + l_{32} & l_{32} = -1 \end{cases}$$

For the third column,

$$\begin{cases} 1 = l_{11}t_{13} = 3t_{13} & t_{13} = \frac{1}{3} \\ 2 = l_{21}t_{13} + l_{22}t_{23} = (2)(\frac{1}{3}) + \frac{8}{3}t_{23} & t_{23} = \frac{1}{2} \\ 2 = l_{31}t_{13} + l_{32}t_{23} + l_{33} = (3)(\frac{1}{3}) + (-1)(\frac{1}{2}) + l_{33} & l_{33} = \frac{3}{2} \end{cases}$$

We find

$$[\mathbf{L}] = \begin{bmatrix} 3 & 0 & 0 \\ 2 & \frac{8}{3} & 0 \\ 3 & -1 & \frac{3}{2} \end{bmatrix} \qquad [\mathbf{T}] = \begin{bmatrix} 1 & \frac{2}{3} & \frac{1}{3} \\ 0 & 1 & \frac{1}{2} \\ 0 & 0 & 1 \end{bmatrix} \tag{2.45}$$

Step 2: Procedure for obtaining $[\mathbf{A}]^{-1}$:

$$[\mathbf{A}] = [\mathbf{L}][\mathbf{T}]$$

$$[\mathbf{A}][\mathbf{T}]^{-1} = [\mathbf{L}][\mathbf{T}][\mathbf{T}]^{-1} = [\mathbf{L}]$$

$$[\mathbf{A}]^{-1}[\mathbf{A}][\mathbf{T}]^{-1} = [\mathbf{A}]^{-1}[\mathbf{L}]$$

$$[\mathbf{T}]^{-1} = [\mathbf{A}]^{-1}[\mathbf{L}] \tag{2.46}$$

In Eq. (2.46) we already have matrix $[\mathbf{L}]$. If we know $[\mathbf{T}]^{-1}$, we can find $[\mathbf{A}]^{-1}$ by following the similar procedure in step 1. Because of the triangular nature, the inverse of matrix $[\mathbf{T}]$ can easily be found by simple backward substitution. The inverse $[\mathbf{T}]^{-1}$ is also an upper triangular matrix with 1's along the diagonal.

Step 3: Find $[\mathbf{T}]^{-1}$ from the equation

$$[\mathbf{I}] = [\mathbf{T}][\mathbf{T}]^{-1}$$

or

$$\begin{bmatrix} 1 & 0 & 0 \\ 0 & 1 & 0 \\ 0 & 0 & 1 \end{bmatrix} = \begin{bmatrix} 1 & \frac{2}{3} & \frac{1}{3} \\ 0 & 1 & \frac{1}{2} \\ 0 & 0 & 1 \end{bmatrix} \begin{bmatrix} u_{11} & u_{12} & u_{13} \\ u_{21} & u_{22} & u_{23} \\ u_{31} & u_{32} & u_{33} \end{bmatrix} \tag{2.47}$$

The nine elements in matrix $[\mathbf{T}]^{-1}$ are obtained by evaluating the nine known values in the identity matrix $[\mathbf{I}]$. We evaluate these values in row-wise direction and start from the last row. Thus for the last row, we have

$$u_{31} = u_{32} = 0 \qquad u_{33} = 1$$

For the second row,

$$0 = u_{21} + \tfrac{1}{2}u_{31} \qquad u_{21} = 0$$

$$1 = u_{22} + \tfrac{1}{2}u_{32} \qquad u_{22} = 1$$

$$0 = u_{23} + \tfrac{1}{2}u_{33} \qquad u_{23} = -\tfrac{1}{2}$$

For the first row,

$$1 = u_{11} + \tfrac{2}{3}u_{21} + \tfrac{1}{3}u_{31} \qquad u_{11} = 1$$

$$0 = u_{12} + \tfrac{2}{3}u_{22} + \tfrac{1}{3}u_{32} \qquad u_{12} = -\tfrac{2}{3}$$

$$0 = u_{13} + \tfrac{2}{3}u_{23} + \tfrac{1}{3}u_{33} \qquad u_{13} = 0$$

we have

$$[\mathbf{T}]^{-1} = \begin{bmatrix} 1 & -\frac{2}{3} & 0 \\ 0 & 1 & -\frac{1}{2} \\ 0 & 0 & 1 \end{bmatrix} \tag{2.48}$$

Step 4: Find $[\mathbf{A}]^{-1}$ from the equation

$$[\mathbf{T}]^{-1} = [\mathbf{A}]^{-1}[\mathbf{L}] \tag{2.49a}$$

or

$$\begin{bmatrix} 1 & -\frac{2}{3} & 0 \\ 0 & 1 & -\frac{1}{2} \\ 0 & 0 & 1 \end{bmatrix} = \begin{bmatrix} b_{11} & b_{12} & b_{13} \\ b_{21} & b_{22} & b_{23} \\ b_{31} & b_{32} & b_{33} \end{bmatrix} \begin{bmatrix} 3 & 0 & 0 \\ 2 & \frac{8}{3} & 0 \\ 3 & -1 & \frac{3}{2} \end{bmatrix} \tag{2.49b}$$

The nine elements in matrix $[\mathbf{A}]^{-1}$ are obtained by evaluating the nine elements in matrix $[\mathbf{T}]^{-1}$. We evaluate these values in a column-wise direction starting from the last column. Thus for the third column,

$$b_{13} = 0 \qquad b_{23} = -\tfrac{1}{2} \div \tfrac{3}{2} = -\tfrac{1}{3} \qquad b_{33} = \tfrac{2}{3}$$

For the second column,

$$-\tfrac{2}{3} = \tfrac{8}{3}b_{12} - b_{13} \qquad b_{12} = -\tfrac{1}{4}$$

$$1 = \tfrac{8}{3}b_{22} - b_{23} \qquad b_{22} = \tfrac{1}{4}$$

$$0 = \tfrac{8}{3}b_{32} - b_{33} \qquad b_{32} = \tfrac{1}{4}$$

For the first column,

$$1 = 3b_{11} + 2b_{12} + 3b_{13} \qquad b_{11} = \tfrac{1}{2}$$

$$0 = 3b_{21} + 2b_{22} + 3b_{23} \qquad b_{21} = \tfrac{1}{6}$$

$$0 = 3b_{31} + 2b_{32} + 3b_{33} \qquad b_{31} = -\tfrac{5}{6}$$

Finally, we obtain

$$[\mathbf{A}]^{-1} = \begin{bmatrix} \frac{1}{2} & -\frac{1}{4} & 0 \\ \frac{1}{6} & \frac{1}{4} & -\frac{1}{3} \\ -\frac{5}{6} & \frac{1}{4} & \frac{2}{3} \end{bmatrix} \tag{2.50}$$

2.13.5 Properties of the Three Methods for Matrix Inverse

1. The adjoint method is the easiest one to use for a matrix of order less than 5. Throughout this text, the matrices for all the noncomputer examples are smaller than 5×5. This method is indeed the one to use. This method is convenient when the matrix elements contain variables instead of pure numbers, as is the case for most examples.

2. The Gauss–Jordan method is a well-known method of elimination for the solution of simultaneous equations. It has been adopted by IBM for the matrix inverse subroutine MINV in the System/360 Scientific Subroutine Package [2.1]. Such a subroutine is usually implemented in the IBM 360 series computers.
3. Cholesky's method is a perfectly general systematic procedure for matrix inversion. The entire numerical process is simple substitution. It is used, for example, by Argonne National Laboratory in the subroutine LINEQ1 in EISPACK. The EISPACK package is well implemented in many CDC computers.

2.14 LINEAR EQUATIONS

A set of linear simultaneous equations can usually be solved by one of the following two popular methods.

2.14.1 Matrix Inverse

Consider, for example, the following three linear equations with three unknowns:

$$\begin{aligned} 3x_1 + 2x_2 + x_3 &= 10 \\ 2x_1 + 4x_2 + 2x_3 &= 16 \\ 3x_1 + x_2 + 2x_3 &= 11 \end{aligned} \tag{2.51a}$$

In matrix form, we have

$$\begin{bmatrix} 3 & 2 & 1 \\ 2 & 4 & 2 \\ 3 & 1 & 2 \end{bmatrix} \begin{Bmatrix} x_1 \\ x_2 \\ x_3 \end{Bmatrix} = \begin{Bmatrix} 10 \\ 16 \\ 11 \end{Bmatrix} \tag{2.51b}$$

or in symbolic form,

$$[\mathbf{A}]\{\mathbf{x}\} = \{\mathbf{c}\} \tag{2.51c}$$

The solution is obtained by the use of the matrix inverse:

$$\begin{aligned} \begin{Bmatrix} x_1 \\ x_2 \\ x_3 \end{Bmatrix} &= [\mathbf{A}]^{-1}\{\mathbf{c}\} \\ &= \begin{bmatrix} \frac{1}{2} & -\frac{1}{4} & 0 \\ \frac{1}{6} & \frac{1}{4} & -\frac{1}{3} \\ -\frac{5}{6} & \frac{1}{4} & \frac{2}{3} \end{bmatrix} \begin{Bmatrix} 10 \\ 16 \\ 11 \end{Bmatrix} = \begin{Bmatrix} 1 \\ 2 \\ 3 \end{Bmatrix} \end{aligned} \tag{2.52}$$

2.14.2 Triangular Matrix and Back Substituting

The example above can also be solved by first obtaining an upper triangular matrix with 1's along the diagonal. This is done by the methods of elimination described in Table 2.1. From rows [4], [7], and [9], we have

$$\begin{bmatrix} 1 & \frac{2}{3} & \frac{1}{3} \\ 0 & 1 & \frac{1}{2} \\ 0 & 0 & 1 \end{bmatrix} \begin{Bmatrix} x_1 \\ x_2 \\ x_3 \end{Bmatrix} = \begin{Bmatrix} c_1 \\ c_2 \\ c_3 \end{Bmatrix} = \begin{Bmatrix} \frac{10}{3} \\ \frac{7}{2} \\ 3 \end{Bmatrix} \tag{2.53}$$

From row [4], $c_1 = \frac{10}{3}$.
From rows [5, 7], $c_2 = [16 - (2)(\frac{10}{3})]/\frac{8}{3} = \frac{7}{2}$.
From rows [6–9], $c_3 = [11 - (3)(\frac{10}{3}) - (-1)(\frac{7}{2})]/\frac{3}{2} = 3$.

We first find from the third equation,

$$x_3 = 3 \tag{2.54a}$$

Substituting x_3 into the second equation gives

$$x_2 = c_2 - \tfrac{1}{2}x_3 = 2 \tag{2.54b}$$

Substituting x_3 and x_2 into the first equation gives

$$x_1 = c_1 - \tfrac{2}{3}x_2 - \tfrac{1}{3}x_3 = 1 \tag{2.54c}$$

2.15 USE OF FORTRAN SUBROUTINES FOR MATRIX INVERSION

Nowadays, there is a library of subroutines available in almost every computer system. Matrix inversion subroutines are among the most popular. To perform matrix inversion, we can simply call such subroutines.

Alternatively, we may choose to use our own subroutines for matrix inversion together with our main program. For such purposes two example programs are given below. Both subroutines are based on the popular Gauss–Jordan method.

Example 2.8 Subroutine MINV

MINV is a subroutine available in the IBM Scientific Subroutine Package (SSP) for the 360 series systems [2.1]. The subroutine contains 90 noncommented statements. To use this subroutine, we simply have to use the following statement:

```
CALL MINV (A, N, D, L, M)
```

where A = before the call statement, [**A**] is the matrix to be inverted; after the call statement, [**A**] is the inverse matrix
N = order of the square matrix [**A**]
D = determinant fed back from the subroutine
L = column matrix of order N for use in the subroutine
M = column matrix of order N for use in the subroutine

It is important to note that because a storage-compression technique is used in this SSP, matrices [**A**], {**L**}, and {**M**} must be dimensioned exactly, that is, neither greater nor smaller than N.

```
DIMENSION A(N,N), L(N), M(N)
```

Example 2.9 Subroutine MATINV

Subroutine MATINV can be found in Ref. 2.2. It contains 72 noncommented statements. The usage is as follows:

```
CALL MATINV (A, N, B, M, D, ID)
```

where A = before the call statement, [**A**] is the matrix to be inverted; after the call statement, [**A**] is the inverse matrix
N = order of matrix [**A**]
B = before the call statement, {**B**} is the vector of constants of the simultaneous equations; after the call statement, {**B**} is the solution vector of unknowns
M = number of sets of constant vector
D = determinant fed back from the subroutine; when ID = 1, [**A**] is a nonsingular matrix, and when ID = 2, [**A**] is a singular matrix

In the main program, the matrices must be dimensioned as

```
DIMENSION A(L,L), B(L,1)
```

It is important to note that L must be equal to or greater than that which appeared in the dimension statements for A and B in the subroutine.

2.16 EIGENVALUES AND EIGENVECTORS

In the analysis of free-vibration and buckling problems, the finite element formulations are in the form of a set of simultaneous homogeneous equations; that is, the constant vector contains only zeros. It is of the form

$$\underset{m \times m}{[\mathbf{A}]}\ \underset{m \times 1}{\{\mathbf{x}\}} - \lambda\ \underset{m \times m}{[\mathbf{B}]}\ \underset{m \times 1}{\{\mathbf{x}\}} = \underset{m \times 1}{\{\mathbf{0}\}} \tag{2.55}$$

where the m values of λ are the *eigenvalues*, and corresponding to each eigenvalue λ, we have a solution for {**x**} which is the *eigenvector*. In the free-vibration problem, eigenvalues are the squares of natural frequencies and corresponding to each eigenvalue, the eigenvector gives the nondimensional mode shape. In the buckling problem, the eigenvalues are the buckling loads and the eigenvectors give the corresponding mode shapes. Matrix [**A**] is the *stiffness matrix* and matrix [**B**] is the *mass matrix* for a vibration problem and the *incremental stiffness matrix* for a buckling problem.

Equation (2.55) can also be written in such a form that matrix [**B**] is replaced by an identity matrix [**I**]. This is done simply by premultiplying by the matrix $[\mathbf{B}]^{-1}$ on both sides:

$$[\mathbf{C}]\{\mathbf{x}\} - \lambda[\mathbf{I}]\{\mathbf{x}\} = \{\mathbf{0}\} \tag{2.56}$$

It is important to note that matrix [**C**] is the product of two symmetrical matrices [**A**] and $[\mathbf{B}]^{-1}$. It is commonly a nonsymmetrical matrix.

Two methods for the solution of eigenvalues and eigenvectors for the equations above are given next.

2.16.1 Direct Solution

Let it be desired to solve for the following equations:

$$\begin{bmatrix} 3 & 2 & 1 \\ 2 & 2 & 1 \\ 0 & 1 & 1 \end{bmatrix} \begin{Bmatrix} x_1 \\ x_2 \\ x_3 \end{Bmatrix} - \lambda \begin{bmatrix} 1 & 0 & 0 \\ 0 & 1 & 0 \\ 0 & 0 & 1 \end{bmatrix} \begin{Bmatrix} x_1 \\ x_2 \\ x_3 \end{Bmatrix} = \begin{Bmatrix} 0 \\ 0 \\ 0 \end{Bmatrix} \tag{2.57a}$$

We first add the two matrices:

$$\begin{bmatrix} 3-\lambda & 2 & 1 \\ 2 & 2-\lambda & 1 \\ 0 & 1 & 1-\lambda \end{bmatrix} \begin{Bmatrix} x_1 \\ x_2 \\ x_3 \end{Bmatrix} = \begin{Bmatrix} 0 \\ 0 \\ 0 \end{Bmatrix} \tag{2.57b}$$

or in symbolic form,

$$[\mathbf{D}]\{\mathbf{x}\} = \{\mathbf{0}\} \tag{2.57c}$$

The value of x_1 (or x_2, or x_3) may be solved by using Cramer's rule,

$$x_1 = \frac{\begin{vmatrix} 0 & 2 & 1 \\ 0 & 2-\lambda & 1 \\ 0 & 1 & 1-\lambda \end{vmatrix}}{|D|} = \frac{0}{|D|} \tag{2.58}$$

In order to have a solution for x_1, determinant $|D|$ must be equal to zero. This is a necessary and sufficient condition. We thus set $|D| = 0$, which yields the "nontrivial solution"

$$\begin{vmatrix} 3-\lambda & 2 & 1 \\ 2 & 2-\lambda & 1 \\ 0 & 1 & 1-\lambda \end{vmatrix} = 0$$

$$\lambda^3 - 6\lambda^2 + 6\lambda - 1 = 0 \tag{2.59}$$

From this cubic equation we find three roots for the three eigenvalues:

$$\begin{aligned} \lambda_1 &= 4.7913 \\ \lambda_2 &= 1.0 \\ \lambda_3 &= 0.2087 \end{aligned} \tag{2.60}$$

Corresponding to each eigenvalue, there is an eigenvector (x_1, x_2, x_3). If we assign a value to any of the three unknowns, say $x_3 = 1$, we can find the relative nondimensional magnitudes of the other two variables x_1 and x_2. Thus for $\lambda_3 = 0.2087$, $x_3 = 1$, the original equations become

$$\begin{aligned} (3 - 0.2087)x_1 + 2x_2 + 1 &= 0 \\ 2x_1 + (2 - 0.2087)x_2 + 1 &= 0 \\ 0 + x_2 + (1 - 0.2087)(1) &= 0 \end{aligned} \tag{2.61}$$

Solving any two of the three equations gives the eigenvector

$$\begin{Bmatrix} x_1 \\ x_2 \\ x_3 \end{Bmatrix} = \begin{Bmatrix} 0.2087 \\ -0.7913 \\ 1.000 \end{Bmatrix} \tag{2.62a}$$

We can also find for $\lambda_2 = 1.0$ and $x_3 = 1$,

$$\begin{Bmatrix} x_1 \\ x_2 \\ x_3 \end{Bmatrix} = \begin{Bmatrix} -0.5 \\ 0.0 \\ 1.0 \end{Bmatrix} \tag{2.62b}$$

For $\lambda_1 = 4.7913$ and $x_3 = 1$,

$$\begin{Bmatrix} x_1 \\ x_2 \\ x_3 \end{Bmatrix} = \begin{Bmatrix} 4.7913 \\ 3.7913 \\ 1 \end{Bmatrix} \quad \text{or} \quad \begin{Bmatrix} 1.000 \\ 0.7913 \\ 0.2087 \end{Bmatrix} \tag{2.62c}$$

2.16.2 Iteration method (Power Method)

For buckling problem, usually only the lowest eigenvalue (the critical buckling load) is needed. For vibration problem, the lowest natural frequency is usually of significant interest. There is a simple iteration method that gives the largest eigenvalue and the relative eigenvector. If we divide the eigenvalue equation by λ, we obtain

$$[\mathbf{B}]\{\mathbf{x}\} - \mu[\mathbf{A}]\{\mathbf{x}\} = \{\mathbf{0}\} \tag{2.63}$$

where $\mu = 1/\lambda$. This means that the largest eigenvalue for μ is the lowest value for λ, that is, the lowest buckling load or frequency. Using this method let us now solve for the same equations in Section 2.16.1. We first assume the initial values for the eigenvector to be

$$\lfloor \mathbf{x} \rfloor = \lfloor 1 \quad 0 \quad 0 \rfloor \tag{2.64a}$$

Then we perform the simple multiplication

$$[\mathbf{C}]\{\mathbf{x}\} = \begin{bmatrix} 3 & 2 & 1 \\ 2 & 2 & 1 \\ 0 & 1 & 1 \end{bmatrix} \begin{Bmatrix} 1 \\ 0 \\ 0 \end{Bmatrix} = \begin{Bmatrix} 3 \\ 2 \\ 0 \end{Bmatrix} = 3 \begin{Bmatrix} 1.0 \\ 0.6667 \\ 0 \end{Bmatrix} \tag{2.64b}$$

We perform multiplication again using the normalized eigenvector.

$$\begin{bmatrix} 3 & 2 & 1 \\ 2 & 2 & 1 \\ 0 & 1 & 1 \end{bmatrix} \begin{Bmatrix} 1.0 \\ 0.6667 \\ 0 \end{Bmatrix} = \begin{Bmatrix} 4.3334 \\ 3.3334 \\ 0.6667 \end{Bmatrix} = 4.3334 \begin{Bmatrix} 1.0 \\ 0.7692 \\ 0.1539 \end{Bmatrix}$$

A similar process yields

$$4.6923 \begin{Bmatrix} 1.0 \\ 0.7868 \\ 0.1967 \end{Bmatrix} \quad 4.7705 \begin{Bmatrix} 1.0 \\ 0.7904 \\ 0.2062 \end{Bmatrix} \quad 4.7870 \begin{Bmatrix} 1.0 \\ 0.7911 \\ 0.2082 \end{Bmatrix} \quad 4.7904 \begin{Bmatrix} 1.0 \\ 0.7912 \\ 0.2086 \end{Bmatrix}$$

$$4.7910 \begin{Bmatrix} 1.0 \\ 0.7914 \\ 0.2087 \end{Bmatrix} \quad 4.7915 \begin{Bmatrix} 1.0 \\ 0.7913 \\ 0.2087 \end{Bmatrix} \quad 4.7913 \begin{Bmatrix} 1.0 \\ 0.7913 \\ 0.2087 \end{Bmatrix} \tag{2.64c}$$

Finally, we obtain the converged largest eigenvalue 4.7913 and the corresponding eigenvector: 1.0; 0.7913; 0.2087.

This method converges rapidly when the two largest eigenvalues are not too close. This method fails when the two largest eigenvalues have the same modulus. It also fails when the initial eigenvector is orthogonal to the expected eigenvector.

To write a computer program for this straightforward iteration method is very simple. We have only to multiply the matrix by the eigenvector using the Fortran statements described in Section 2.6 and compare the eigenvectors between cycles of iteration for convergence.

This iteration method, often called the *power method*, can be used to extract the remaining eigenvalues and eigenvectors as well, assuming that there are no repeated eigenvalues. The second highest eigenvalue λ_2 and the corresponding eigenvector can be obtained based on the matrix $[\mathbf{D}] = [\mathbf{C}] - \lambda_1[\mathbf{I}]$.

Similarly, once λ_1 and λ_2 are known, the third highest eigenvalue λ_3 and the corresponding eigenvector can be obtained based on $[\mathbf{D}] = ([\mathbf{C}] - \lambda_1[\mathbf{I}])([\mathbf{C}] - \lambda_2[\mathbf{I}])$, and so on. The choice of the starting eigenvector is important. A detailed description of the method and a Fortran program are given in Ref. 2.3.

We can now find the remaining eigenvalues and eigenvectors for the example above. Based on $[\mathbf{D}] = [\mathbf{C}] - \lambda_1[\mathbf{I}]$ with $\lambda_1 = 4.7913$, we can find, by iteration, the following solution:

$$-3.7913\begin{Bmatrix} -0.5 \\ 0.0 \\ 1.0 \end{Bmatrix}$$

where $\lambda_2 - \lambda_1 = -3.7913$ or $\lambda_2 = 1.0$. Based on $[\mathbf{D}] = ([\mathbf{C}] - \lambda_1[\mathbf{I}])([\mathbf{C}] - \lambda_2[\mathbf{I}])$, we can find, by iteration, the following solution:

$$3.6262\begin{Bmatrix} 0.2087 \\ -0.7913 \\ 1.0 \end{Bmatrix}$$

where $(\lambda_3 - \lambda_1)(\lambda_3 - \lambda_2) = 3.6262$ or $\lambda_3 = 0.2087$. These solutions are the same as those obtained by the direct method.

2.17 USE OF FORTRAN SUBROUTINES FOR SOLVING EIGENVALUE EQUATIONS

The subroutines for solving eigenvalue equations are usually implemented in the computer system libraries. The purpose and use of such subroutines are demonstrated by using subroutines NROOT and EIGEN (available in Ref. 2.1) as an example. NROOT has 69 and EIGEN has 105 noncommented statements. We have only to call NROOT, which in turn calls EIGEN. The two subroutines have to be used simultaneously.

Purpose: Subroutine NROOT calculates the eigenvalues λ_i and the matrix of eigenvectors of a real square nonsymmetric matrix of the special form $[\mathbf{B}]^{-1}[\mathbf{A}]$, where both $[\mathbf{B}]$ and $[\mathbf{A}]$ are real symmetrical matrices and $[\mathbf{B}]$ is positive definite. The eigenvalue problem is in the form

$$\Big[[\mathbf{A}] - \lambda[\mathbf{B}]\Big]\{\mathbf{x}\} = \{\mathbf{0}\} \tag{2.65}$$

Subroutine EIGEN computes eigenvalues and eigenvectors of a real symmetrical matrix.

Use:

```
DIMENSION A(M,M), B(M,M), VALUE(M), VECTOR(M,M)
      .
      .
      .
CALL NROOT (M, A, B, VALUE, VECTOR)
```

where M = order of the square matrices [**A**] and [**B**] and column matrix {**x**}
A = input matrix ($M \times M$), real and symmetric
B = input matrix ($M \times M$), real, symmetric, positive definite
VALUE = output column matrix containing M eigenvalues
VECTOR = output matrix ($M \times M$) containing M sets of eigenvectors stored in M columns, respectively

It is important to note that because of the storage compression technique used in IBM SSP, matrices [**A**] and [**B**] must be dimensioned to be exactly (M, M). These two subroutines are usually implemented in IBM 360 systems.

2.18 IMSL LIBRARY SUBROUTINES

IMSL subroutine library is a package available for lease from International Mathematical and Statistical Libraries, Inc. It is a popular library available in common computer systems.

IMSL Library 3 now has 32 subroutines for the analysis of linear algebraic equations. Each subroutine has its special feature. The features include matrix inversion, linear equations solution, band storage, symmetric or general matrices, space economizer, real or complex matrices, and high accuracy.

Among the 29 subroutines for real and complex eigenvalue analysis, IMSL Library 3 has EIGRF, which can calculate eigenvalues and eigenvectors of a real general matrix, and EIGZF, which can calculate eigenvalues and eigenvectors of the system $([\mathbf{A}] - \lambda[\mathbf{B}])\{\mathbf{x}\} = 0$.

REFERENCES

2.1. *System/360 Scientific Subroutine Package (360-CM-03X) Version II.* Programmer's Manual, IBM Application Program, H20-0205-2, White Plains, N.Y., 1967, p. 55 and pp. 31 and 56.

2.2. McCormick, J. M., and Salvadori, M. G., *Numerical Methods in FORTRAN*, Prentice-Hall, Inc., Englewood Cliffs, N.J., 1965, p. 306.

2.3. Carnaham, B., Luther, H. A., and Wilkes, J. O., *Applied Numerical Methods*, John Wiley & Sons, Inc., New York, 1969, pp. 226–235.

PROBLEMS

2.1. Three matrices are defined as follows:

In matrix $[\mathbf{A}]_{6\times7}$, $a_{ij} = 5.6 * (i + j)$.
In matrix $[\mathbf{B}]_{7\times9}$, $b_{ij} = 2.5 * i + 0.8 * j$.
In matrix $[\mathbf{C}]_{8\times9}$, $c_{ij} = 7.1 * i - 3.5 * j$.

Write a Fortran program to determine **(a)** [**D**] = [**A**][**B**], and **(b)** [**E**] = [**C**][**B**]^T[**A**]^T. Write matrices [**A**], [**B**], [**C**], [**D**], and [**E**], respectively, by a neat FORMAT statement. *Note*:

1. Do not forget the DIMENSION statement.
2. Do not mix the floating mode with the integer mode.

2.2. Solve the following set of seven linear simultaneous equations by using either one of the two subroutines suggested in Section 2.15 or by using the subroutine available in the computer system library.

$$\begin{bmatrix} 3.0 & 1.5 & 1.0 & 0.0 & 0.0 & 0.0 & 0.0 \\ 1.5 & 2.0 & 0.5 & 1.0 & 0.0 & 0.1 & 0.0 \\ 1.0 & 0.5 & 3.0 & 0.0 & 3.0 & 0.0 & 0.0 \\ 0.0 & 1.0 & 0.0 & 1.0 & 0.5 & 2.0 & 0.0 \\ 0.0 & 0.0 & 3.0 & 0.5 & 2.0 & 1.0 & 0.1 \\ 0.0 & 0.1 & 0.0 & 2.0 & 1.0 & 5.0 & 0.0 \\ 1.0 & 0.0 & 0.0 & 0.0 & 0.1 & 0.0 & 1.0 \end{bmatrix} \begin{Bmatrix} x_1 \\ x_2 \\ x_3 \\ x_4 \\ x_5 \\ x_6 \\ x_7 \end{Bmatrix} = \begin{Bmatrix} 3.5 \\ 6.5 \\ 18.5 \\ 16.0 \\ 21.1 \\ 35.1 \\ 6.4 \end{Bmatrix}$$

or symbolically,

$$[\mathbf{A}]\{\mathbf{x}\} = \{\mathbf{B}\}$$

The output should include the following:

1. WRITE the matrix [**A**].
2. WRITE the constant vector {**B**}.
3. WRITE the determinant of matrix [**A**].
4. WRITE the inverse of matrix [**A**].
5. WRITE the solution vector {**x**}.
6. WRITE the matrix [**I**], which is found as the product of the matrices [**A**] and [**A**]$^{-1}$.
7. WRITE appropriate titles and leave reasonable spaces for each of the items above. The F-format is suggested for output.

2.3. Invert the following matrices by all of the following three methods: the adjoint method; the Gauss–Jordan method; and Cholesky's method. The three methods should give the same results. The work must be done by hand or calculator, not by computer.

$$\begin{bmatrix} 1 & 3 & 1 \\ 3 & 2 & 3 \\ 1 & 3 & 2 \end{bmatrix} \quad \begin{bmatrix} 2 & 2 & 1 \\ 1 & 6 & 3 \\ 0 & 2 & 5 \end{bmatrix}$$

$$\begin{bmatrix} 24 & 0 & -12 & -6 \\ 0 & 8 & 6 & 2 \\ -12 & 6 & 24 & 0 \\ -6 & 2 & 0 & 8 \end{bmatrix} \quad \begin{bmatrix} 6 & 2 & 1 & 0 \\ 5 & 4 & 0 & 1 \\ 2 & 2 & 3 & 2 \\ 1 & 1 & 1 & 3 \end{bmatrix}$$

2.4. For a beam with both ends fixed as shown in Fig. P2.4, the equations of motion for a three-element model are obtained as follows (these will be explained in Chapter 7):

$$\left[\frac{EI}{L}\begin{bmatrix} \frac{24}{L^2} & 0 & -\frac{12}{L^2} & \frac{6}{L} \\ 0 & 8 & -\frac{6}{L} & 2 \\ -\frac{12}{L^2} & -\frac{6}{L} & \frac{24}{L^2} & 0 \\ \frac{6}{L} & 2 & 0 & 8 \end{bmatrix} - \omega^2\frac{\rho AL}{420}\begin{bmatrix} 312 & 0 & 54 & -13L \\ 0 & 8L^2 & 13L & -3L^2 \\ 54 & 13L & 312 & 0 \\ -13L & -3L^2 & 0 & 8L^2 \end{bmatrix}\right]\begin{Bmatrix} v_1 \\ \theta_1 \\ v_2 \\ \theta_2 \end{Bmatrix} = \begin{Bmatrix} 0 \\ 0 \\ 0 \\ 0 \end{Bmatrix}$$

where v_1 and v_2 are the deflections and θ_1 and θ_2 are the slopes at points 1 and 2, respectively. Their sign conventions are shown in Fig. P2.4.

The beam is an 8I23 steel I beam with the following properties:

$A = 6.71$ in.^{2} $\quad I = 64.2$ in.^{4} $\quad L = 120$ in. $\quad E = 30 \times 10^6$ psi
$\rho = 0.000733$ lb-sec^2/in.^{4}

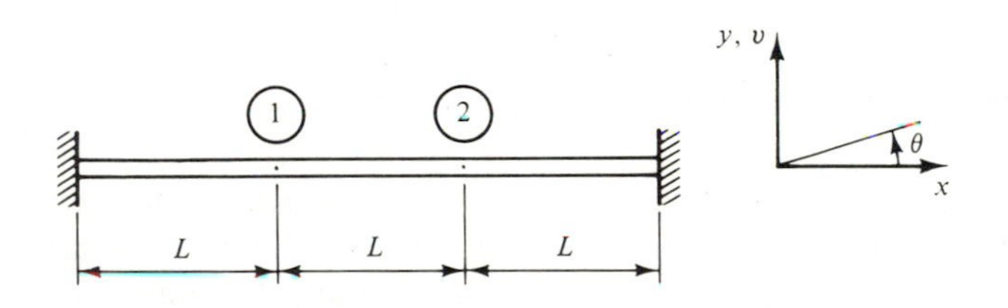

Figure P2.4 Clamped-clamped beam modeled by three beam elements.

Find the four natural frequencies (ω_1, ω_2, ω_3, and ω_4) and plot the shapes of the two modes corresponding to ω_1 and ω_2. If the subroutine for solving for eigenvalues and eigenvectors of a real nonsymmetric matrix is not available in the system library, use NROOT and EIGEN as described in Section 2.17.

2.5. Determine the eigenvalues and the corresponding eigenvectors of the following systems of equations by hand calculation.

$$\left[\begin{bmatrix} 10 & 2 & 1 \\ 2 & 10 & 1 \\ 2 & 1 & 10 \end{bmatrix} - \lambda\begin{bmatrix} 1 & 0 & 0 \\ 0 & 1 & 0 \\ 0 & 0 & 1 \end{bmatrix}\right]\begin{Bmatrix} x_1 \\ x_2 \\ x_3 \end{Bmatrix} = \begin{Bmatrix} 0 \\ 0 \\ 0 \end{Bmatrix}$$

$$\left[\begin{bmatrix} 11 & 6 & -2 \\ -2 & 18 & 1 \\ -12 & 24 & 13 \end{bmatrix} - \lambda\begin{bmatrix} 1 & 0 & 0 \\ 0 & 1 & 0 \\ 0 & 0 & 1 \end{bmatrix}\right]\begin{Bmatrix} x_1 \\ x_2 \\ x_3 \end{Bmatrix} = \begin{Bmatrix} 0 \\ 0 \\ 0 \end{Bmatrix}$$

2.6. Determine the largest eigenvalue and the corresponding eigenvector of the two systems of equations in Problem 2.5 using the iteration method.

CHAPTER 3

Basic Structural Theorems

3.1 PRINCIPLE OF SUPERPOSITION

The *principle of superposition* states that when a linear structure is subjected to a number of loads, the combined effect of these loads is equal to the sum of the effects of each load applied separately in any sequence. This principle can be described by the following example.

A linear structure is shown in Fig. 3.1. The structure is subjected to, say, three concentrated loads P_1, P_2, and P_3. Corresponding to the same locations and same directions as the three loads, there are three displacement components with magnitudes q_1, q_2, and q_3, respectively.

According to the principle of superposition, the displacements can be written as

$$\begin{aligned} q_1 &= f_{11}P_1 + f_{12}P_2 + f_{13}P_3 \\ q_2 &= f_{21}P_1 + f_{22}P_2 + f_{23}P_3 \\ q_3 &= f_{31}P_1 + f_{32}P_2 + f_{33}P_3 \end{aligned} \tag{3.1a}$$

where f_{ij} is the flexibility coefficient or the influence coefficient that defines the displacement at i due to a unit load at j. Equation (3.1a) can be written in matrix form as

$$\begin{Bmatrix} q_1 \\ q_2 \\ q_3 \end{Bmatrix} = \begin{bmatrix} f_{11} & f_{12} & f_{13} \\ f_{21} & f_{22} & f_{23} \\ f_{31} & f_{32} & f_{33} \end{bmatrix} \begin{Bmatrix} P_1 \\ P_2 \\ P_3 \end{Bmatrix} \tag{3.1b}$$

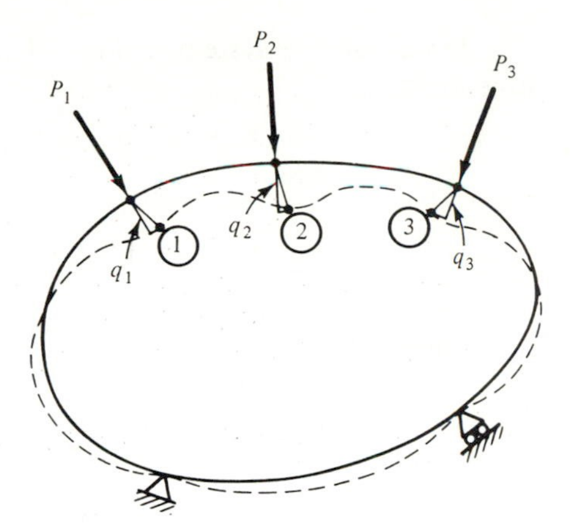

Figure 3.1 Linear elastic body under three concentrated loads.

or in symbolic form as

$$\{\mathbf{q}\} = [\mathbf{f}]\{\mathbf{P}\} \tag{3.1c}$$

where [**f**] is called the *flexibility matrix* or the *influence coefficient matrix.*

Equation (3.1c) can also be written in inverse form,

$$\{\mathbf{P}\} = [\mathbf{k}]\{\mathbf{q}\} \tag{3.2a}$$

with

$$[\mathbf{k}] = [\mathbf{f}]^{-1} \tag{3.2b}$$

where [**k**] is called the *stiffness matrix.* "Stiffness" and "flexibility" are words with opposite meanings. As is evident from Eq. (3.2b), the stiffness matrix is the inverse of the flexibility matrix.

3.2 WORK DONE BY A LOAD SYSTEM

In the derivation of stiffness matrix equations of equilibrium for a finite element or a system of finite elements, the most common approach is to minimize the total potential energy. This approach requires the formulation of the work done by the external loads and the strain energy stored in the element or the system.

In deriving the expression for the work done by a system of loads, it is assumed that each load is applied gradually from zero to its final value and that each load-displacement relationship is linear. Thus the work done by each load is equal to the triangular area covered by the load-displacement line.

If we use the system in Fig. 3.1 as an example, the work done by the three loads is

$$W = \tfrac{1}{2}P_1q_1 + \tfrac{1}{2}P_2q_2 + \tfrac{1}{2}P_3q_3$$

$$= \tfrac{1}{2}\lfloor P_1 \quad P_2 \quad P_3 \rfloor \begin{Bmatrix} q_1 \\ q_2 \\ q_3 \end{Bmatrix} \tag{3.3a}$$

or symbolically,

$$W = \tfrac{1}{2}\lfloor P \rfloor \{q\} \tag{3.3b}$$

3.3 MAXWELL–BETTI RECIPROCAL THEOREM

The reciprocal theorem was first introduced by J. C. Maxwell in 1864 for a particular case and a general proof of the theorem was given by E. Betti in 1872 (see Ref. 3.1).

Figure 3.2a shows a cantilever beam subjected to two concentrated loads P_1 and P_2 applied at two arbitrary points 1 and 2, respectively. Let it be defined that

f_{11} = deflection at point 1 due to a unit load at point 1

f_{12} = deflection at point 1 due to a unit load at point 2

f_{21} = deflection at point 2 due to a unit load at point 1

f_{22} = deflection at point 2 due to a unit load at point 2

According to the reciprocal theorem,

$$f_{12} = f_{21} \tag{3.4}$$

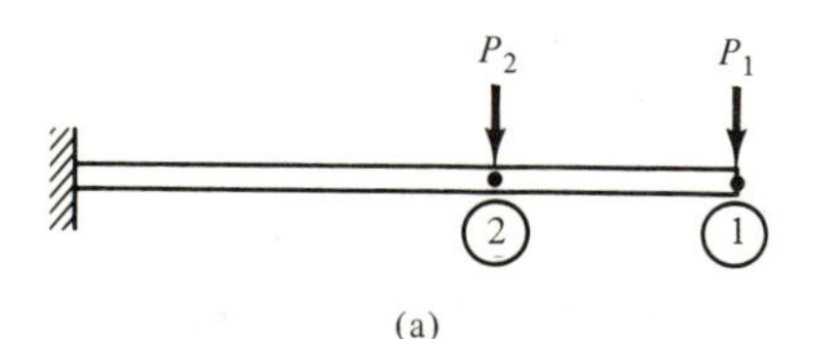

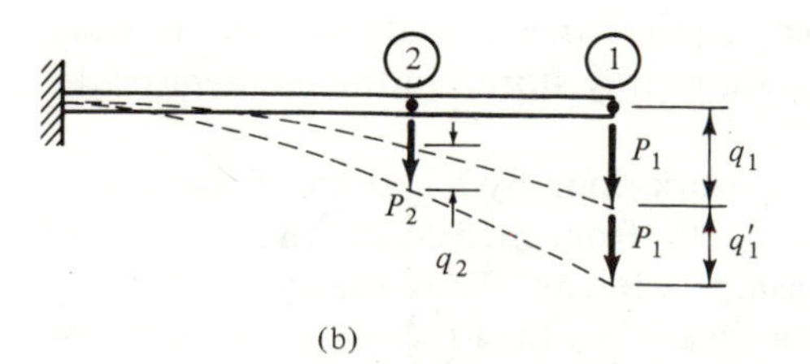

Figure 3.2 (a) Cantilever beam under two concentrated loads; (b) P_1 applied before P_2.

Equation (3.4) can be proven by applying the two loads in two different sequences.

Loading sequence 1: The load P_1 is first applied. As shown in Fig. 3.2b, the work done is

$$W_1 = \tfrac{1}{2}P_1 q_1 = \tfrac{1}{2}P_1 f_{11} P_1 \tag{3.5}$$

The load P_2 is then applied. As also shown in Fig. 3.2b, the additional work done is

$$\begin{aligned} W_2 &= \tfrac{1}{2}P_2 q_2 + P_1 q_1' \\ &= \tfrac{1}{2}P_2(f_{22}P_2) + P_1(f_{12}P_2) \end{aligned} \tag{3.6}$$

Work W_1 and work W_2 are equal to the areas covered by the two load-displacement lines as shown in Fig. 3.3. It is seen in this figure that after the final value of P_1 is reached, P_1 is maintained constant throughout the additional deflection q_1'. Thus the additional area covered by the load-displacement line is $P_1 q_1'$ instead of $\tfrac{1}{2}P_1 q_1'$.

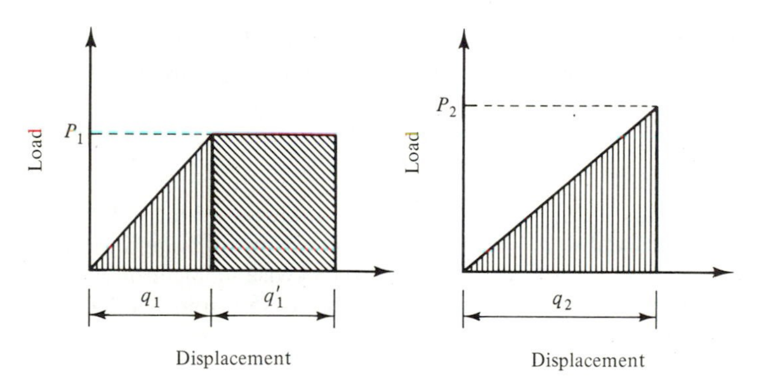

Figure 3.3 Load-displacement lines with P_2 applied after the final value of P_1 is reached.

The total work is obtained by summing up W_1 and W_2:

$$W = \tfrac{1}{2}(P_1^2 f_{11} + P_2^2 f_{22}) + P_1 P_2 f_{12} \tag{3.7}$$

Loading sequence 2: If P_2 is applied first, the work done is

$$W_1 = \tfrac{1}{2}P_2 q_2 = \tfrac{1}{2}P_2 f_{22} P_2 \tag{3.8}$$

The load P_1 is then applied; the additional work done is

$$\begin{aligned} W_2 &= \tfrac{1}{2}P_1 q_1 + P_2 q_2' \\ &= \tfrac{1}{2}P_1 f_{11} P_1 + P_2 f_{21} P_1 \end{aligned} \tag{3.9}$$

The total work done is

$$W = \tfrac{1}{2}(P_1^2 f_{11} + P_2^2 f_{22}) + P_1 P_2 f_{21} \tag{3.10}$$

When the structural behavior is linear, the total work done is independent of the sequence of loading. By equating the two work expressions obtained in Eqs. (3.7) and (3.10), it is shown that

$$f_{12} = f_{21} \tag{3.4}$$

Physically, Eq. (3.4) states that the deflection at point 1 due to a unit load at point 2 is equal to the deflection at point 2 due to a unit load at point 1.

From the example of the cantilever beam, we can generalize the reciprocal theorem as follows:

1. The concentrated loads can be in terms of forces (pounds) and/or moments (inch-pounds).
2. The corresponding displacements can be in terms of direct displacements (inches) and/or rotational displacements (radians).
3. As a result of the generalized loads and displacements, the flexibility coefficients can have the dimensions in./lb, in./in.-lb, rad/lb, rad/in.-lb, and so on.
4. Equation (3.4) can be generalized as $f_{ij} = f_{ji}$. Thus the reciprocal theorem shows that the structural flexibility matrix is symmetric, and so is the stiffness matrix.

Example 3.1

An arbitrary beam is shown in Fig. 3.4a. When a downward load of 100 lb is applied at B, the deflections and slopes at A, B, C, and D are measured and are given in Table 3.1. Find the deflection at B due to a counterclockwise moment of 300 in.-lb applied at B and a downward force of 60 lb applied at C as shown in Fig. 3.4b.

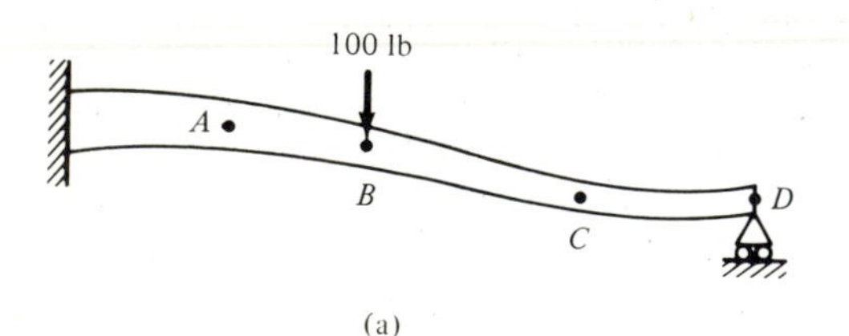

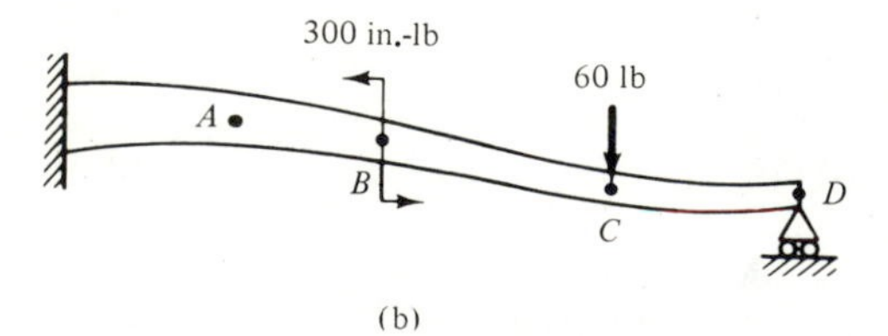

Figure 3.4 (a) Arbitrary beam loaded at B; (b) the same beam loaded at B and C.

TABLE 3.1 Measured Deflections and Slopes Due to a Downward Load of 100 Pounds at B

Location	A	B	C	D
Clockwise slope, θ (rad)	0.006	0.021	−0.006	−0.032
Downward deflection, q (in.)	0.060	0.240	0.560	0.0

Solution. From Table 3.1 we first find the needed flexibility coefficients.

$$\text{slope } \theta_B = f_{BB}P_B \quad \text{or} \quad 0.021 \text{ rad} = (f_{BB})(100 \text{ lb})$$

$$f_{BB} = 0.00021 \text{ rad/lb}$$

$$\text{deflection } q_C = f_{CB}P_B \quad \text{or} \quad 0.56 \text{ in.} = (f_{CB})(100 \text{ lb})$$

$$f_{CB} = 0.0056 \text{ in./lb}$$

From the reciprocal theorem

$$f_{BC} = f_{CB}$$

Thus we can find the deflection at B due to the loads as shown in Fig. 3.4b.

$$\begin{aligned} q_B &= f_{BB}M_B + f_{BC}P_C \\ &= (0.00021 \text{ rad/lb})(-300 \text{ in.-lb}) + (0.0056 \text{ in./lb})(60 \text{ lb}) \\ &= 0.273 \text{ in.} \end{aligned}$$

The negative sign in front of the bending moment 300 in.-lb indicates that it is applied in a direction opposite to the positive direction (clockwise) defined for θ_B in Table 3.1.

3.4 ENERGY THEOREMS

3.4.1 Principle of Virtual Work (Due to Virtual Displacement)

One particle. In variational mechanics, we imagine displacements to occur when in reality no such displacements exist. These fictitious displacements are called *virtual displacements* and the work done by these imaginary displacements is referred to as the *virtual work.*

The virtual displacements are assumed as small so that there will be no significant changes in geometry. Thus the forces may also be assumed to remain unchanged during the virtual displacements.

Let us consider a particle undergoing a virtual displacement $\delta \mathbf{q}$ as shown in Fig. 3.5a. The particle is subjected to a system of, say, three forces P_1, P_2, and P_3, respectively. The resulting virtual work can be expressed as

$$\delta W = P_1\, \delta q_1 + P_2\, \delta q_2 + P_3\, \delta q_3 \tag{3.11}$$

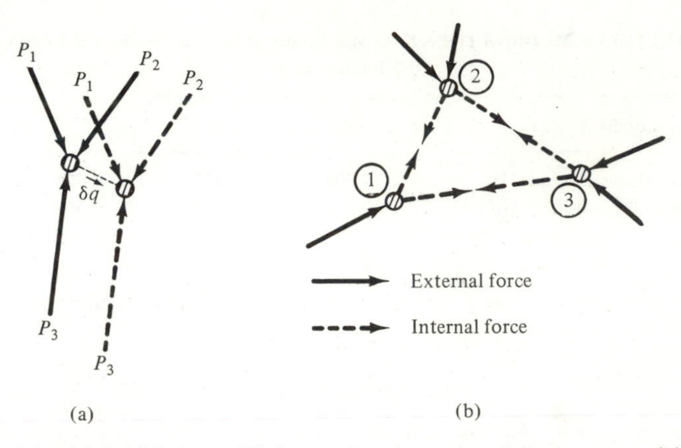

Figure 3.5 (a) Particle in equilibrium undergoing a virtual displacement; (b) three particles in equilibrium.

where δ is a variational operator denoting a virtual quantity and δq_1, δq_2, and δq_3 are the three components of the virtual displacement $\delta \mathbf{q}$ in the P_1, P_2, and P_3 directions, respectively.

Because the three forces P_1, P_2, and P_3 are in equilibrium, their vectorial sum must vanish and so does the work done for a virtual displacement,

$$\delta W = 0 \tag{3.12}$$

Alternatively, if it is known that $\delta W = 0$, it follows that the vectorial sum of the forces is zero and the forces are in equilibrium.

Theorem. A necessary and sufficient condition for the equilibrium of a particle is that the virtual work done by the forces acting on the particle vanishes for any virtual displacement.

A number of particles. The internal forces of an elastic continuum system represent the actions of the particles on each other and therefore occur as pairs of equal and opposite forces. This phenomenon is shown in Fig. 3.5b for a three-particle system.

If the system is in equilibrium, Eq. (3.12) can be generalized to that for three particles,

$$\sum_{j=1}^{3} (\delta W_j^e + \delta W_j^i) = 0 \tag{3.13}$$

where δW_j^e = virtual work of external forces for particle j

δW_j^i = virtual work of internal forces for particle j

For an elastic continuum system, there are infinite number of particles. Equation (3.13) may be further generalized as

$$\delta W^e + \delta W^i = 0 \tag{3.14}$$

where W is, for simplicity, defined as the summation of the work for the system of infinite number of particles.

Continuum (elastic structure). A *continuum* is defined as a domain in which matter exists at every point. We may think of a continuum as consisting of an infinite number of particles.

Theorem. A necessary and sufficient condition for the equilibrium of a system of particles or a continuum system is that the virtual work done by the external forces plus the virtual work done by the internal forces vanish for any virtual displacement.

For an elastic continuum system or an elastic structure, the work done by the internal forces W^i is equal and opposite to the strain energy U stored in the system during the deformation:

$$W^i = -U \tag{3.15}$$

Substituting Eq. (3.15) in Eq. (3.14) and simply using W to represent the external work W^e results in

$$\delta W = \delta U \tag{3.16}$$

which states that the external virtual work done due to a virtual displacement is equal to the internal virtual strain energy caused by the same virtual displacement. This statement is the *principle of virtual work* (due to virtual displacement).

3.4.2 Castigliano's First Theorem

Let us consider a structure subjected to a set of n external forces $P_1, P_2, \ldots, P_i, \ldots, P_n$ and apply a virtual displacement δq. This virtual displacement is allowed to take place in the structure in such a manner that δq is continuous everywhere but vanishes at all points of loading except under P_i. Thus only the δq_i absorb work from P_i; the other external forces do not do any work.

Due to δq_i, the virtual work done is $P_i\,\delta q_i$. According to the principle of virtual work as described in Eq. (3.16), this virtual work equals the virtual strain energy:

$$\delta U = P_i\,\delta q_i$$

or, in the limit,

$$P_i = \frac{\partial U}{\partial q_i} \tag{3.17a}$$

If instead of virtual displacement δq_i, we introduce a virtual rotation $\delta\theta_i$ at the point where a moment M_i is acting, Eq. (3.17a) can be written as

$$M_i = \frac{\partial U}{\partial \theta_i} \tag{3.17b}$$

Equations (3.17a) and (3.17b) are known as *Castigliano's first theorem,* after Alberto Castigliano, who published the theorem in his thesis for the engineer's degree in 1879 (Ref. 3.2). It is noted that this theorem is the foundation for the derivation of the finite element stiffness equations, especially for sophisticated elements for which the stress-strain equilibrium approach becomes difficult. It is also important to note that this theorem also applies when the structural behavior is nonlinear.

3.4.3 Principle of Complementary Virtual Work (Due to Virtual Force)

Figure 3.6 shows a load-displacement relation curve. We define W^* and U^* as the complementary work and complementary strain energy, respectively. Again δ is variational operator denoting a virtual quantity. From this figure we see that

$$\delta W^* = q\,\delta P \tag{3.18}$$

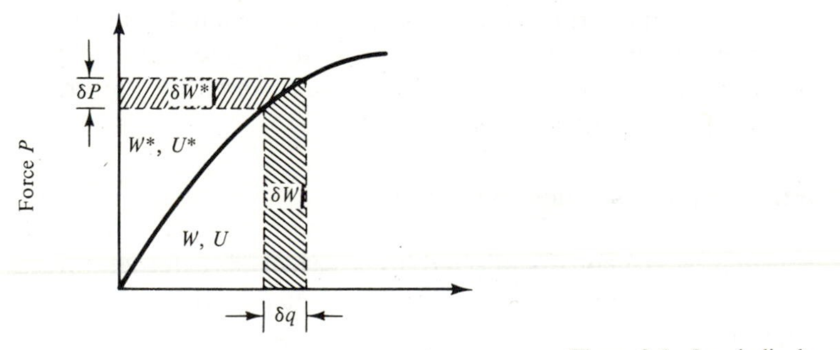

Figure 3.6 Load-displacement curve and the associated works and energies.

The principle of virtual work also holds for the case of complementary work and strain energy:

$$\delta W^* = \delta U^* \tag{3.19}$$

In this case the virtual work is that due to a virtual force.

3.4.4 Castigliano's Second Theorem

Let us again consider a structure subjected to a set of n external forces, $P_1, P_2, \ldots, P_i, \ldots, P_n$. If a virtual increment δP_i is given to the external force

P_i, the complementary work will be increased by δW^*. The only force that does any complementary work is P_i, because the other forces are not changed. Thus the increase in complementary energy can be obtained, following Eq. (3.18), as

$$\delta W^* = q_i \, \delta P_i \tag{3.20}$$

Based on the principle of complementary virtual work as given in Eq. (3.19),

$$\delta U^* = q_i \, \delta P_i \tag{3.21}$$

Assuming the structural behavior to be *linear*, the load-displacement curve in Fig. 3.6 becomes a straight line. Then

$$\delta U = \delta U^* = q_i \, \delta P_i \tag{3.22a}$$

In the limit, Eq. (3.22a) becomes

$$q_i = \frac{\partial U}{\partial P_i} \tag{3.22b}$$

which is known as *Castigliano's second theorem.*

For linear structural behavior, the principle of superposition can be used. Equation (3.22b) can, alternatively, be proven by applying δP_i first and afterward all the external forces. The load δP_i first produces an infinitesimal displacement so that the corresponding work is a small quantity of second order and can be neglected. When all the forces P_1, P_2, P_3, . . . are then applied, the work (or strain energy) in addition to that produced by all the forces is

$$\delta U = q_i \, \delta P_i \tag{3.22a}$$

which arrives at the same conclusion as given in Eqs. (3.22a) and (3.22b).

If we consider a moment M_i and a rotation θ_i instead of P_i and q_i, respectively, Eq. (3.22b) can be generalized to

$$\theta_i = \frac{\partial U}{\partial M_i} \tag{3.22c}$$

In applying Castigliano's theorems, we must be careful that:

1. The displacement q_i (or rotation θ_i) must be referred to the same location at which the concentrated force P_i (or moment M_i) is acting.
2. q_i (or θ_i) must be referred to the same direction as P_i (or M_i).
3. The strain energy U must be formulated for the entire structure based on all the external loads.

In some literature, Castigliano's first theorem is often referred to as the theorem of virtual work and Castigliano's second theorem is often simply referred to as Castigliano's theorem, and on occasion, as Castigliano's first theorem. In that case, the principle of least work is referred to as Castigliano's second theorem. The principle of least work will be described in connection with Example 3.5.

3.4.5 Strain Energy Expressions

Castigliano's first theorem is the foundation for deriving finite element stiffness matrices. Its application will be demonstrated in subsequent chapters.

Castigliano's second theorem is very useful in finding displacements and analyzing statically indeterminate structures. Its application is demonstrated in this chapter by a variety of examples. Before demonstrating examples, we must introduce the strain energy expressions for several basic structural elements. It is assumed that the reader is familiar with the derivations of these expressions.

For a beam in bending:

$$U = \int_0^l \frac{M^2\,dx}{2EI} \tag{3.23a}$$

For a beam in shear:

$$U = \int_0^l \frac{V^2\,dx}{2GA} \tag{3.23b}$$

For a bar in tension:

$$U = \int_0^l \frac{S^2\,dx}{2EA} \tag{3.23c}$$

For a bar in torsion:

$$U = \int_0^l \frac{T^2\,dx}{2GJ} \tag{3.23d}$$

For a panel in shear:

$$U = \iint_{\text{Area}} \frac{q^2\,dx\,dy}{2Gt} \tag{3.23e}$$

where M = bending moment (in.-lb)
E = modulus of elasticity (psi)
I = moment of inertia (in.^{4})
V = shearing force (lb)
$G = E/2(1+\nu)$, shear modulus (psi)
A = cross-sectional area (in.^{2})
S = axial force (lb)
T = twisting moment (in.-lb)

J = torsional moment of inertia dependent on the geometry of the cross section (in.[4])
q = shear flow (lb/in.)
t = thickness of the panel or skin (in.)
l = length of the bar or beam (in.)

Example 3.2

Find (a) the vertical displacement v and (b) the horizontal displacement u at joint B for the two-bar truss shown in Fig. 3.7a.

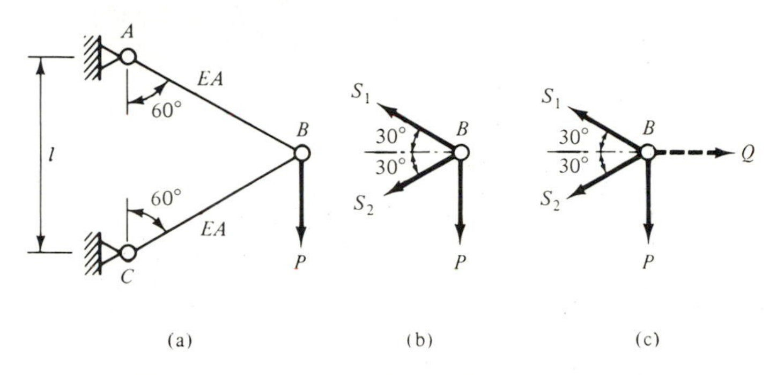

Figure 3.7 (a) Two-bar truss; (b) joint B as a free body; (c) fictitious horizontal force Q applied at joint B.

Solution. (a) For the vertical displacement v at B, let us consider joint B as a free body, as shown in Fig. 3.7b.

$$\sum F_{\text{horizontal}} = \sum F_h = 0$$

$$S_1 \cos 30° + S_2 \cos 30° = 0 \qquad S_1 = -S_2$$

$$\sum F_{\text{vertical}} = \sum F_v = 0$$

$$S_1 \sin 30° - S_2 \sin 30° = P$$

Hence

$$S_1 = -S_2 = P$$

and the strain energy for the truss is

$$U = \frac{S_1^2 l}{2EA} + \frac{S_2^2 l}{2EA} = \frac{P^2 l}{EA}$$

From Castigliano's theorem

$$v = \frac{\partial U}{\partial P} = \frac{2Pl}{EA}$$

(b) In order to apply Castigliano's theorem to find a horizontal displacement u at joint B, a horizontal force must be acting at B. Let us apply a fictitious force Q at

B horizontally. Considering joint B as a free body as shown in Fig. 3.7c, we have

$$\sum F_h = 0 \qquad S_1 \cos 30^\circ + S_2 \cos 30^\circ = Q$$

$$\sum F_v = 0 \qquad S_1 \sin 30^\circ - S_2 \sin 30^\circ = P$$

which gives

$$S_1 = P + \frac{Q}{\sqrt{3}} \qquad \text{and} \qquad S_2 = -P + \frac{Q}{\sqrt{3}}$$

The strain energy for the truss is

$$U = \frac{S_1^2 L}{2EA} + \frac{S_2^2 L}{2EA} = \frac{L}{2EA}\left[\left(P + \frac{Q}{\sqrt{3}}\right)^2 + \left(-P + \frac{Q}{\sqrt{3}}\right)^2\right]$$

$$u = \frac{\partial U}{\partial Q} = \frac{2L}{2EA}\left[\left(P + \frac{Q}{\sqrt{3}}\right)\frac{1}{\sqrt{3}} + \left(-P + \frac{Q}{\sqrt{3}}\right)\frac{1}{\sqrt{3}}\right]$$

$$= \frac{2QL}{3EA}$$

After the execution of Castigliano's theorem, we can set the fictitious load Q equal to zero. We thus find

$$u = 0$$

Example 3.3

Find the deflection v at the midspan of a simply supported beam subjected to a uniformly distributed load w as shown in Fig. 3.8.

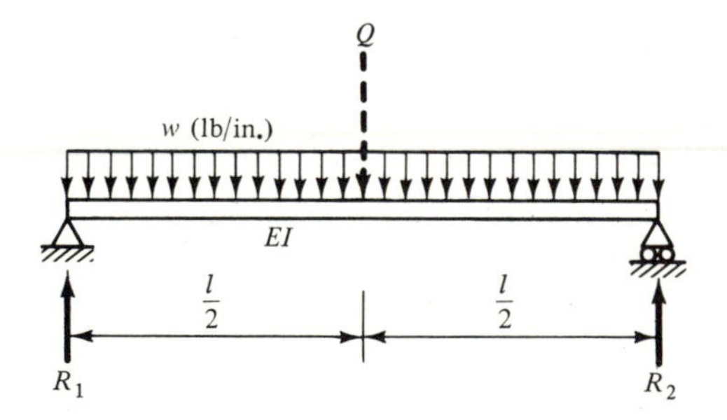

Figure 3.8 Simply supported beam under a uniformly distributed load w.

Solution. We first apply a downward fictitious force Q at midspan. The bending moment at x from the left end is

$$M = R_1 x - \frac{wx^2}{2} = \left(\frac{Q}{2} + \frac{wl}{2}\right)x - \frac{wx^2}{2}$$

$$\text{strain energy } U = 2\int_0^{l/2} \frac{M^2\,dx}{2EI}$$

Following Castigliano's theorem,

$$v = \frac{\partial U}{\partial Q} = \frac{4}{2EI}\int_0^{l/2} M\frac{\partial M}{\partial Q}\,dx$$

$$= \frac{2}{EI}\int_0^{l/2}\left[\left(\frac{Q}{2}+\frac{wl}{2}\right)x - \frac{wx^2}{2}\right]\frac{x}{2}\,dx$$

$$= \frac{1}{EI}\left(\frac{Ql^3}{48}+\frac{5wl^4}{384}\right)$$

Dropping out the fictitious load Q, we finally obtain

$$v = \frac{5wl^4}{384EI}$$

If the beam is subjected to a concentrated load P at midspan but no distributed load, the deflection at midspan can be obtained by dropping out w and replacing Q by P:

$$v = \frac{Pl^3}{48EI}$$

Example 3.4

As shown in Fig. 3.9, a shaft with one end free and the other end rigidly fixed is subjected to a variable twisting moment indicated by the ordinates to the curve. Find the angle of twist of the section at a distance $l/2$ from the free end.

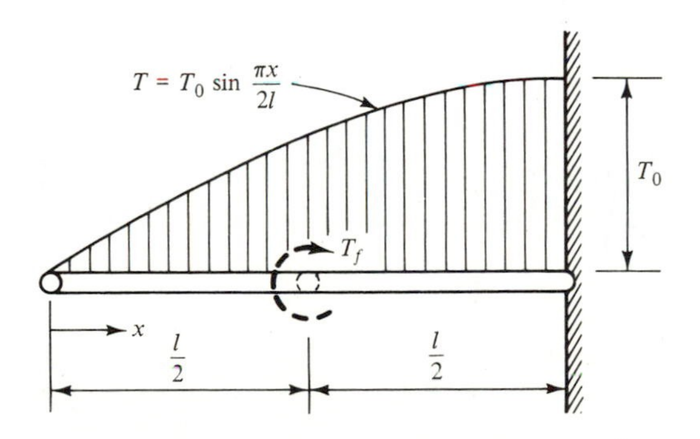

Figure 3.9 Cantilever shaft subjected to a distributed torque.

Solution. We first apply a fictitious twisting moment T_f at a distance $l/2$ from the free end. The strain energy is

$$U = \int_0^{l/2}\frac{T^2\,dx}{2GJ} + \int_{l/2}^{l}\frac{(T+T_f)^2\,dx}{2GJ}$$

The angle of twist is obtained as

$$\theta = \frac{\partial U}{\partial T_f} = 0 + \frac{2}{2GJ}\int_{l/2}^{l} (T + T_f)(1)\, dx$$

$$= \frac{1}{GJ}\int_{l/2}^{l} T_0 \sin\frac{\pi x}{2l}\, dx = \frac{\sqrt{2}T_0 l}{\pi GJ}$$

Example 3.5

Figure 3.10a shows a cantilever beam supported by a truss bar at the hinged joint 2. The bar can be considered as a spring with an elastic constant of $k = EA/L$ (lb/in.). Find the compressive axial force R in the truss bar.

Solution. We can separate the truss bar and the cantilever beam as shown in Fig. 3.10b. Thus the beam is subjected to an upward force R at joint 2, whereas the truss bar is subjected to a downward force R at joint 2.

For the cantilever beam, the moment equation is

$$M = Rx - \frac{wx^2}{2} \qquad \text{and} \qquad \frac{\partial M}{\partial R} = x$$

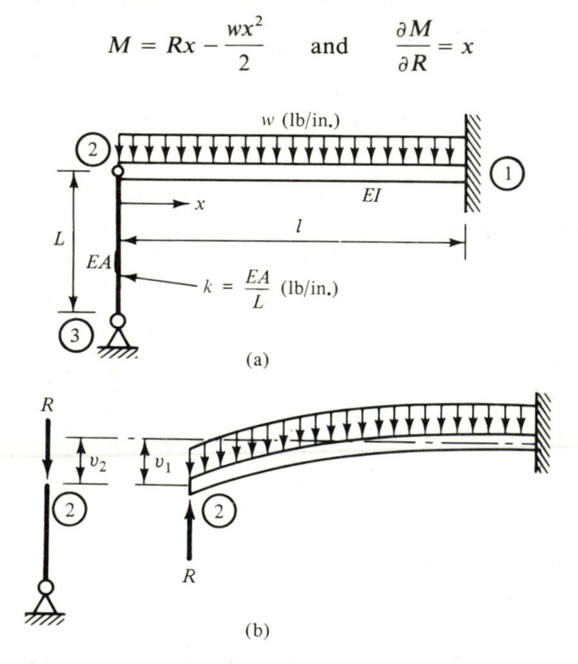

Figure 3.10 (a) Uniformly loaded beam-truss structure; (b) deflections of the beam and the truss bar at their common joint (2), respectively.

Thus the displacement at joint 2 is

$$v_1 = \frac{\partial U_1}{\partial R} = \frac{2}{2EI}\int_0^l M\frac{\partial M}{\partial R}\, dx$$

$$= \frac{1}{EI}\int_0^l \left(Rx - \frac{wx^2}{2}\right)(x)\, dx = \frac{Rl^3}{3EI} - \frac{wl^4}{8EI}$$

For the truss bar with elastic spring constant of k or EA/L and axial force of R, the strain energy is, from Eq. (3.23c),

$$U_2 = \frac{R^2 L}{2EA}$$

The displacement at joint 2 is

$$v_2 = \frac{\partial U_2}{\partial R} = \frac{RL}{EA} \quad \text{or} \quad \frac{R}{k}$$

Because the two forces R, one acting to the beam and the other to the bar, are opposite in direction, so do the two corresponding displacements v_1 and v_2. Compatibility between the beam and the bar at joint 2 requires that

$$v_1 + v_2 = 0 \tag{3.24}$$

or

$$\frac{Rl^3}{3EI} - \frac{wl^4}{8EI} + \frac{RL}{EA} = 0$$

$$R = \frac{3Awl^4}{8(l^3 A + 3LI)} \quad \text{lb}$$

If we let U represent the strain energy of the total structure ($U_1 + U_2$), Eq. (3.24) can be thought of as arriving at by setting

$$\frac{\partial U}{\partial R} = 0 \tag{3.25}$$

Equation (3.25) can be generalized as that the derivative of the total strain energy U with respect to a force (R_i) or a moment (M_i) in any redundant member of a statically indeterminate structure is zero. In other words, R_i or M_i has such a value that U is a minimum. Equation (3.25) is known as the *principle of least work*, also referred to as Castigliano's second theorem in some literature.

Example 3.6

Figure 3.11a shows a panel-stiffener or skin-rib structure. For the three vertical stiffening bars, $EA = 2.5 \times 10^6$ lb. For the four horizontal stiffening bars, $EA = 1.6 \times 10^6$ lb. For the two panels, $G = 4 \times 10^6$ psi and $t = 0.032$ in. Assuming that the panel does not buckle, find the vertical deflection v at point A due to a load of 10,000 lb.

Solution. Based on Castligliano's theorem, we can find the contributions of deflection due to the strain energy of each structural member and then sum these contributions to obtain the total displacement at point A. To formulate the strain energy for each member, we must discretize the structure and consider every member as a free body. The results of the equilibrium forces acting on each member are shown in Fig. 3.11b. Each stiffening bar is acting as an axial force member. The value of the shear flow q is the same everywhere in the two panels.

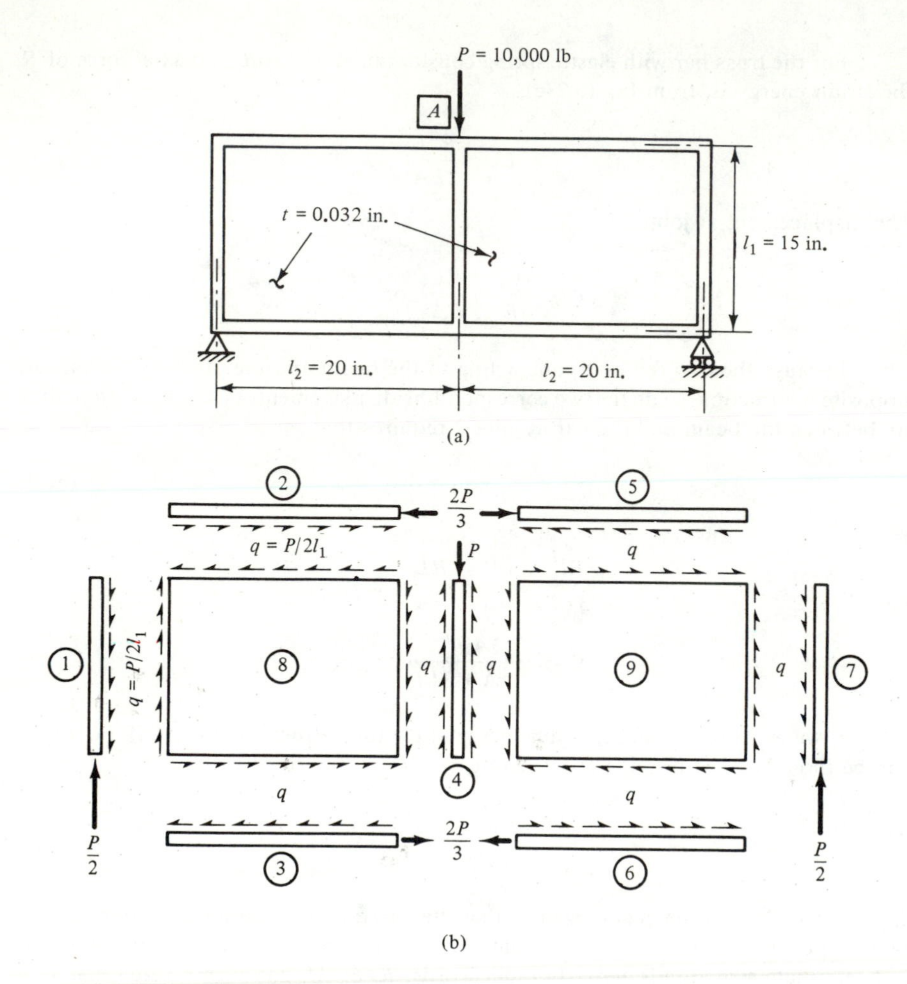

Figure 3.11 (a) Panel-stiffener structure; (b) discretized free bodies (shear flow q is the same everywhere).

Member 1: At a distance x from the top end of the stiffening bar (1), the axial force is

$$S = -\frac{Px}{2l_1} \quad \text{and} \quad \frac{\partial S}{\partial P} = \frac{-x}{2l_1}$$

$$v_1 = \frac{\partial U}{\partial P} = \frac{\partial}{\partial P} \int_0^{l_1} \frac{S^2\,dx}{2(EA)_1} = \frac{2}{2(EA)_1} \int_0^{l_1} S\left(\frac{\partial S}{\partial P}\right) dx$$

$$= \frac{1}{(EA)_1} \int_0^{l_1} \left(-\frac{Px}{2l_1}\right)\left(-\frac{x}{2l_1}\right) dx = \frac{Pl_1}{12(EA)_1} = 0.005 \text{ in.}$$

Member 4: At a distance x from the bottom end of the stiffening bar (4), the axial force is

$$S = -\frac{Px}{l_1} \qquad \text{and} \qquad \frac{\partial S}{\partial P} = -\frac{x}{l_1}$$

$$v_4 = \frac{\partial U}{\partial P} = \frac{1}{(EA)_4} \int_0^{l_1} \left(-\frac{Px}{l_1}\right)\left(-\frac{x}{l_1}\right) dx$$

$$= \frac{Pl_1}{3(EA)_4} = 0.020 \text{ in.}$$

Member 2: At a distance x from the left end of the stiffening bar (2), the axial force is

$$S = -\frac{Px}{2l_1} \qquad \text{and} \qquad \frac{\partial S}{\partial P} = -\frac{x}{2l_1}$$

$$v_2 = \frac{\partial U}{\partial P} = \frac{1}{(EA)_2} \int_0^{l_2} \left(-\frac{Px}{2l_1}\right)\left(-\frac{x}{2l_1}\right) dx$$

$$= \frac{Pl_2^3}{12(EA)_2 l_1^2} = 0.0185 \text{ in.}$$

Member 8: The rectangular panel is under constant shear flow everywhere.

$$q = \frac{P}{2l_1} \qquad \text{and} \qquad \frac{\partial q}{\partial P} = \frac{1}{2l_1}$$

$$v_8 = \frac{\partial U}{\partial P} = \frac{\partial}{\partial P} \int_0^{l_1} \int_0^{l_2} \frac{q^2\, dx\, dy}{2Gt}$$

$$= \frac{2}{2Gt} \int_0^{l_1} \int_0^{l_2} \frac{P}{2l_1}\left(\frac{1}{2l_1}\right) dx\, dy$$

$$= \frac{Pl_2}{4Gtl_1} = 0.0260 \text{ in.}$$

Because this structure is symmetrical, we do not have to repeat the calculations for the members that have the same strain energies. The total deflection at point A is

$$v = v_1 + v_2 + v_3 + \cdots + v_9$$

$$= 2v_1 + v_4 + 4v_2 + 2v_8 = 0.156 \text{ in.}$$

REFERENCES

3.1. Timoshenko, S. P., *Strength of Materials*, Part I: *Elementary Theory and Problems*, D. Van Nostrand Company, Inc., New York, 1958, pp. 336 and 352.

3.2. Timoshenko, S. P., *History of Strength of Materials*, McGraw-Hill Book Company, New York, 1953, p. 289.

PROBLEMS

3.1. A frame structure is shown in Fig. P3.1a. For an upward load of 100 lb, the deflections w and the two rotations $\partial w/\partial x$ and $\partial w/\partial y$ for points A, B, C, and D are given as follows. The positive sign conventions are also shown in the figure. Find the vertical deflections at A due to a system of loadings as given in Fig. P3.1b.

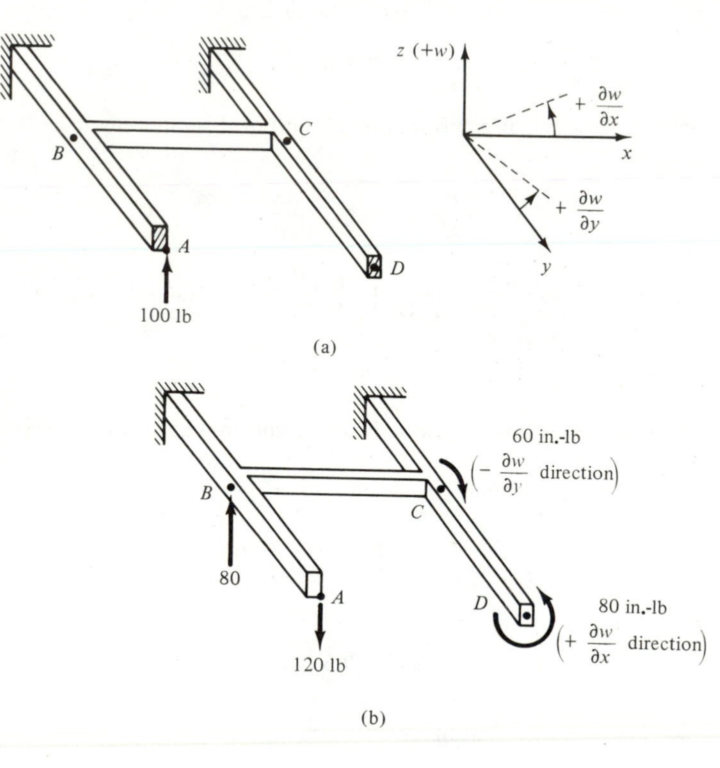

Figure P3.1

Location	A	B	C	D
Deflection, w (in.)	1.200	0.380	0.120	0.340
Rotation, $\partial w/\partial x$ (rad)	0.056	0.021	−0.008	−0.008
Rotation, $\partial w/\partial y$ (rad)	0.066	0.028	0.010	0.010

3.2. A delta-wing panel has been loaded with 160 lb at point 1, as shown in Fig. P3.2a. The data for the vertical deflection w, the rotation about the x axis θ_x, and the rotation about the y axis θ_y have been measured for every grid point. Some of the selected data are shown below. Positive sign conventions for w, θ_x, and θ_y are also shown in Fig. P3.2a. Find the deflection at point 1 due to the loads as shown in Fig. P3.2b.

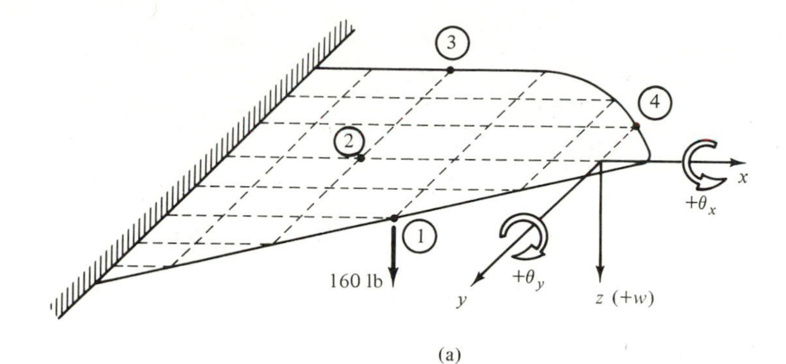

(a)

(b)

Figure P3.2

Location	1	2	3	4
Deflection, w (in.)	1.080	0.420	0.380	1.92
Rotation, θ_x (rad)	0.038	0.018	0.012	0.024
Rotation, θ_y (rad)	0.080	0.046	0.042	0.096

3.3. Find the vertical displacement at joint B of the steel truss as shown in Fig. P3.3. The modulus of elasticity for steel is $E = 30 \times 10^6$ psi. The cross-sectional areas for the bars are $A_1 = 0.75$ in.^{2}, $A_2 = A_5 = 0.6$ in.^{2}, and $A_3 = A_4 = 0.5$ in.^{2}.

3.4. In the analysis of an aircraft skin-stringer structure as shown in Fig. P3.4a, we can simplify the structure to be a one-dimensional bar and spring system with both ends fixed, as shown in Fig. P3.4b. The bending stiffness of the stringers is accounted for by the spring with elastic constant of k (lb/in.) and the in-plane stiffness of the skin is accounted for by the bar with axial spring constant of EA/l (lb/in.). Let α and ΔT be the thermal expansion coefficient and the rise in temperature, respectively, find the reaction force at the end of the bar or the compression force in the spring.

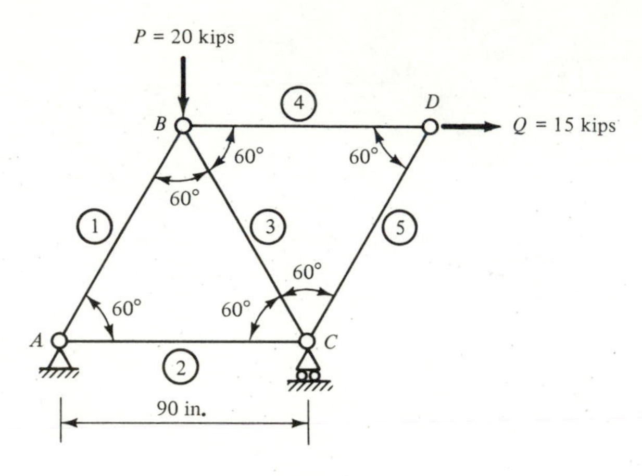

Figure P3.3

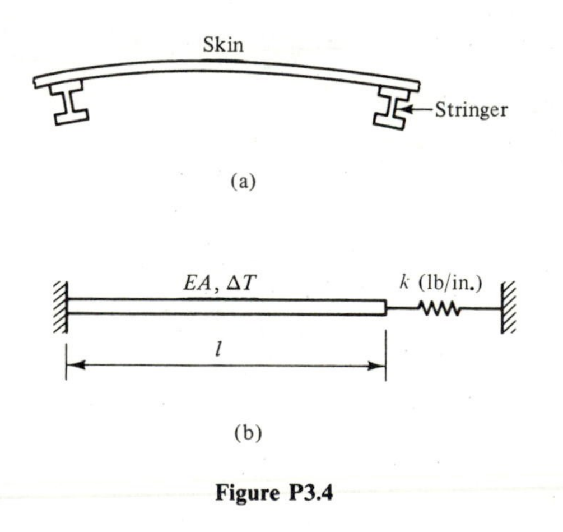

Figure P3.4

3.5. For a simply supported beam with length l and under a uniformly distributed load w (lb/in.), find the deflection and slope at a section at $l/4$ from the left end.

3.6. Find the vertical and horizontal displacements at point A for the structural system of an L-shape frame and an elastic spring as shown in Fig. P3.6. The frame has a constant bending rigidity EI. The strain energy due to axial force in member BC need not be considered.

3.7. Find the horizontal deflection and the angle of rotation at the free end of a quarter ring beam with one end fixed as shown in Fig. P3.7. The beam has constant bending rigidity EI and $EI \ll EA$. *Note:* $\cos^2 \theta = (1 + \cos 2\theta)/2$.

3.8. Find the radial deflections at sections 1 and 2 for the diametrically loaded circular beam as shown in Fig. P3.8. *Hints:* (1) Problem 3.7 is helpful; (2) because of symmetry, we have only to consider half of the ring beam and the rotation at section 2 vanishes ($\theta_2 = 0$).

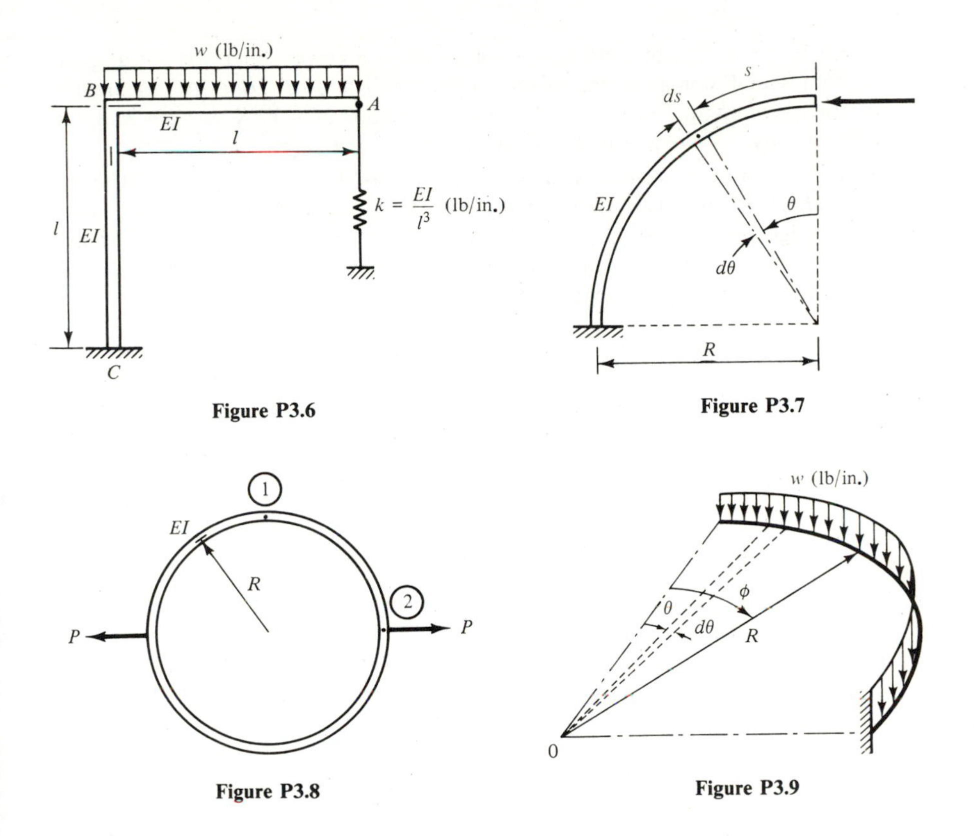

Figure P3.6

Figure P3.7

Figure P3.8

Figure P3.9

3.9. Figure P3.9 shows a horizontal circular beam with one end fixed and the other end free. The beam is of constant bending and torsional rigidities EI and GJ, respectively, and is under a uniformly distributed load w (lb/in.). Find the vertical deflection at the free end. One application of the curved beams can be

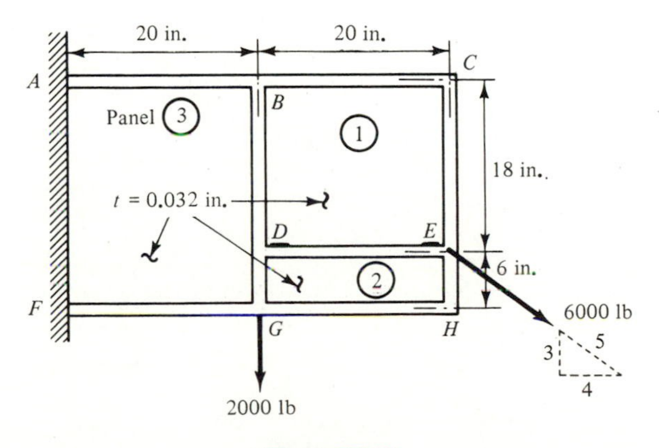

Figure P3.10

found in the design of a bridge to connect two highways that are not aligned. Balconies and stairways are among other applications. *Hint:* At an angular distance ϕ from the free end, the bending moment is $M = \int_0^\phi R \sin(\phi - \theta) wR \, d\theta$ and the twisting moment is $T = \int_0^\phi R[1 - \cos(\phi - \theta)] wR \, d\theta$.

3.10. An aluminum panel-stringer structure is loaded as shown in Fig. P3.10. The two stringers $\overline{ABC}$ and $\overline{FGH}$ are fixed at A and F, respectively. Panel 3 is also fixed at the edge AF. For all the stringers and the stiffener bars, $EA = 2 \times 10^6$ lb and for all panels or skins, $G = 4 \times 10^6$ psi and $t = 0.032$ in. Find the horizontal and vertical deflection components at E.

CHAPTER 4

Truss Bar Elements

4.1 TRUSSES

A *truss* is a structure composed of straight bars connected at their points of intersection by means of momentless joints called *pins* or *hinges*. All loadings are assumed to be applied only at these points of intersection. Thus each straight bar is subjected only to axial force, not to bending nor twisting moments. The stress corresponding to the axial force in the bar is called the *primary stress* in the bar.

The truss bars are often connected by riveted or welded joints. The rigidity of the joints indeed induces bending or twisting moments to the joints and consequently to the bars. Such moments superimpose additional stresses to the bars, which are called *secondary stresses*. However, if the bars are carefully arranged so that their center lines meet in one point at each joint, we shall find that the presence of secondary stresses due to the rigidity of the joints usually does not greatly affect the magnitudes of the primary stresses. Thus it becomes a common practice to assume that the joints are pinned or hinged (i.e., momentless).

The plane truss is one of the most important of all structural forms. Examples of application can be found in bridge trusses, roof trusses, truss beams and girders, and truss walls. The simplest plane truss is the one that consists of three bars forming a triangle. Such a triangular truss is stable, whereas a four-bar rectangular truss is unstable. If at each of any two joints of a triangular truss an additional bar is connected lying in the plane of the original truss and the two bars are joined at their outer ends, we have a five-bar, four-joint truss with two triangles. If this process is continued, we will have a stable truss with triangular net. Thus for a stable truss, it can be shown that

$$n = 2j - 3 \tag{4.1}$$

where n is the number of bars and j is the number of joints. A plane truss with fewer bars will be unstable. One with more bars will be statically indeterminate. The bars in excess of the number required for a stable truss are called *redundant bars.*

Three-dimensional or space trusses are also an important structural form. Examples of application can be found in roof trusses for arenas or coliseums, truss structures for supporting huge steam generators, truss towers for supporting power line cables, and truss structures for large, flexible space satellites and the like.

A truss bar with hinged ends can be considered as the simplest possible form of structural finite element. The concepts of stiffness matrix equations, assemblage, and solution procedure can best be illustrated in the case of trusses.

4.2. STIFFNESS EQUATIONS FOR A TRUSS BAR ELEMENT IN LOCAL COORDINATES

A truss bar element with length L, modulus of elasticity E, and cross-sectional area A lying along a local x axis is shown in Fig. 4.1. The two ends or joints are referred to as *nodal points* and are numbered 1 and 2, respectively. There are two forces X_1 and X_2 acting in the x direction at nodal points 1 and 2, respectively. Corresponding to the two forces, there are two respective displacements u_1 and u_2. These displacements are often called *degrees of freedom.* There are a total of two degrees of freedom for this truss bar element.

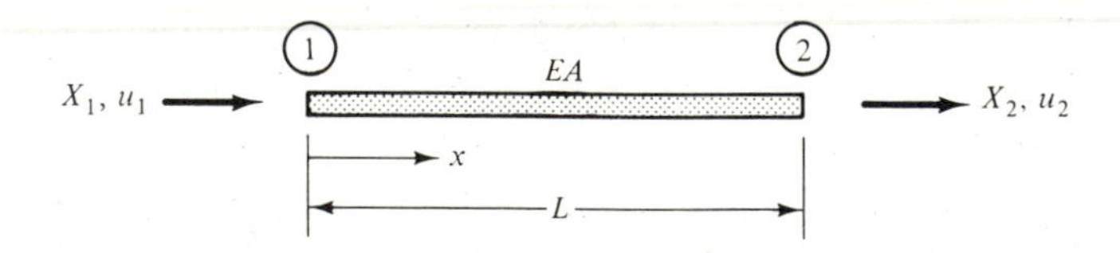

Figure 4.1 Truss bar element lying along the local x axis.

It is necessary to derive a set of two equations in matrix form to relate the two forces X_1 and X_2 to the two displacements u_1 and u_2. The derivation can be done either by an energy approach or a stress–strain equilibrium approach. The energy approach is more general and more powerful, especially for sophisticated types of finite elements. The stress–strain equilibrium approach is simple and physically clear. But it can be applied only to simple finite elements. To use the energy approach, it is necessary first to define a displacement function for the element.

4.2.1 Displacement Function and Shape Functions

For a bar element with constant axial stress or constant axial strain, the axial displacement $u(x)$ at a distance x from nodal point 1 may be assumed as varying linearly with x, or

$$u(x) = a_1 + a_2 x \tag{4.2}$$

where a_1 and a_2 are two constants to be determined by the two nodal point conditions,

At $x = 0$,

$$u(x) = u(0) = u_1 = a_1$$

At $x = L$,

$$u(x) = u(L) = u_2 = a_1 + a_2 L$$

$$a_2 = \frac{u_2 - u_1}{L}$$

Substituting the results for a_1 and a_2 into Eq. (4.2) and rearranging the equation gives the final form of displacement function:

$$u(x) = f_1(x)u_1 + f_2(x)u_2 \tag{4.3a}$$

with

$$f_1(x) = 1 - \frac{x}{L} \quad \text{and} \quad f_2(x) = \frac{x}{L} \tag{4.3b}$$

where $f_1(x)$ and $f_2(x)$ describe the distribution or shape of the displacement associated with the degrees of freedom u_1 and u_2, respectively. They are referred to as *shape functions.* Physical insight into the displacement function (4.3a) and the shape functions (4.3b) can be explained by the following examples.

First, let us check Eqs. (4.3a) and (4.3b) by setting $x = 0$ and L, respectively. For $x = 0$, we find $f_1(x) = 1$ and $f_2(x) = 0$, which yield $u(0) = u_1$. For $x = L$, we find $f_1(x) = 0$ and $f_2(x) = 1$, which yield $u(L) = u_2$.

Next, let it be required to find the displacement u at $x = L/4$ and $x = L/2$ for a bar element with $u_1 = 0.04$ in. and $u_2 = 0.08$ in. For this case, Eq. (4.3a) becomes

$$u(x) = \left(1 - \frac{x}{L}\right)(0.04) + \frac{x}{L}(0.08)$$

With the values of $x = L/4$ and $L/2$, we have

$$u\left(\frac{L}{4}\right) = 0.05 \text{ in.}$$

$$u\left(\frac{L}{2}\right) = 0.06 \text{ in.}$$

4.2.2 Stiffness Equations by Energy Method

For the case of a uniaxial stress or strain, the strain is defined as

$$\epsilon = \lim_{\Delta x \to 0} \frac{u(x + \Delta x) - u(x)}{\Delta x} = \frac{\partial u}{\partial x} \tag{4.4}$$

where Δx denotes an infinitesimal length. For the present axial force bar element, the axial strain can be obtained by substituting Eq. (4.2) or (4.3a) into Eq. (4.4):

$$\epsilon = a_1 \tag{4.5a}$$

or

$$\epsilon = f_1'(x)u_1 + f_2'(x)u_2 \tag{4.5b}$$

Equation (4.5a) states that the strain is a constant. The primes in Eq. (4.5b) indicate derivative with respect to x.

The axial force is

$$S = \sigma A = E \in A = EA\frac{\partial u}{\partial x}$$

$$= EA[f_1'(x)u_1 + f_2'(x)u_2] \tag{4.6}$$

The strain energy expression is obtained by substituting Eq. (4.6) into Eq. (3.23c),

$$U = \int_0^L \frac{S^2\,dx}{2EA} = \frac{EA}{2}\int_0^L (f_1'u_1 + f_2'u_2)^2\,dx \tag{4.7}$$

Applying Castigliano's theorem gives

$$X_1 = \frac{\partial U}{\partial u_1} = \frac{2EA}{2}\int_0^L (f_1'u_1 + f_2'u_2)(f_1')\,dx$$

$$= \left(EA\int_0^L f_1'f_1'\,dx\right)u_1 + \left(EA\int_0^L f_1'f_2'\,dx\right)u_2$$

and

$$X_2 = \frac{\partial U}{\partial u_2} = \left(EA\int_0^L f_2'f_1'\,dx\right)u_1 + \left(EA\int_0^L f_2'f_2'\,dx\right)u_2$$

or in a matrix form,

$$\begin{Bmatrix} X_1 \\ X_2 \end{Bmatrix} = \begin{bmatrix} k_{11} & k_{12} \\ k_{21} & k_{22} \end{bmatrix} \begin{Bmatrix} u_1 \\ u_2 \end{Bmatrix} \tag{4.8a}$$

or symbolically,

$$\{\mathbf{X}\} = [\mathbf{k}]\{\mathbf{u}\} \tag{4.8b}$$

where $[\mathbf{k}]$ is called the *stiffness matrix*, in which the stiffness coefficients are defined as

$$k_{ij} = EA \int_0^L f_i'(x) f_j'(x)\, dx \tag{4.9}$$

with $i = 1$ to 2 and $j = 1$ to 2. Mathematically, Eq. (4.9) shows that $k_{ij} = k_{ji}$, so that the stiffness matrix is symmetric. Physically, we have already shown that the stiffness matrix is symmetric due to the reciprocal theorem. As will be seen in subsequent chapters, Eq. (4.9) is in a form quite typical for obtaining stiffness coefficients for a finite element with given shape functions.

Substituting the shape functions (4.3b) into Eq. (4.9), the stiffness equations are finally found:

$$\begin{Bmatrix} X_1 \\ X_2 \end{Bmatrix} = \frac{EA}{L} \begin{bmatrix} 1 & -1 \\ -1 & 1 \end{bmatrix} \begin{Bmatrix} u_1 \\ u_2 \end{Bmatrix} \tag{4.10}$$

The term EA/L is the axial stiffness of the bar. The bar behaves like a spring with spring constant equal to EA/L in the units of pounds per inch.

4.2.3 Stiffness Equations by Stress–Strain Equilibrium Method

Let us first assume that the bar element is fixed at nodal point 1 but free at nodal point 2, as shown in Fig. 4.2a. This means that $u_1 = 0$.

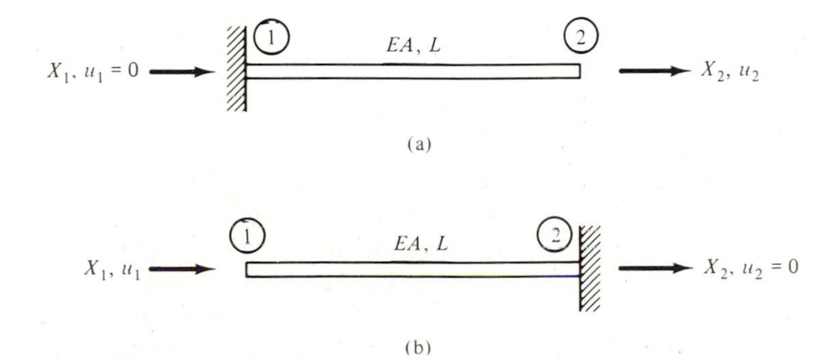

Figure 4.2 (a) Bar element fixed at nodal point 1; (b) bar element fixed at nodal point 2.

Because EA/L is the spring constant in pounds per inch and $u_1 = 0$,

$$X_2 = \frac{EA}{L} u_2 \tag{4.11a}$$

Equilibrium in the x direction requires that

$$X_1 = -X_2 = -\frac{EA}{L} u_2 \tag{4.11b}$$

Let us then assume that the bar element is fixed at nodal point 2 but free at nodal point 1, as shown in Fig. 4.2b. This means that $u_2 = 0$. Then

$$X_1 = \frac{EA}{L} u_1 \tag{4.11c}$$

Equilibrium in the x direction requires that

$$X_2 = -X_1 = -\frac{EA}{L} u_1 \tag{4.11d}$$

Superimpose the two cases: combining Eqs. (4.11a) to (4.11d), stiffness equations identical to that given in Eq. (4.10) are obtained.

4.3 STIFFNESS EQUATIONS FOR A TRUSS BAR ELEMENT ORIENTED ARBITRARILY IN A TWO-DIMENSIONAL PLANE

Having derived the stiffness equations for a horizontal truss bar element lying along a local coordinate axis, we now consider a general case of a bar element oriented arbitrarily in a two-dimensional plane. We can derive the stiffness equations by using an energy approach combined with a coordinate transformation approach, a pure coordinate transformation approach, or a stress–strain equilibrium approach.

4.3.1 Energy Method Combined with Coordinate Transformation

A truss bar element lying along an $\bar{x}$ axis is shown in Fig. 4.3. The $\bar{x}$ axis is a local coordinate axis and the axes (x, y) are the global or reference coordinate axes. The $\bar{x}$ axis is oriented at an arbitrary angle ϕ measured counterclockwise from the x axis. To distinguish between the two coordinate systems, bars are added to all the symbols referring to the local $\bar{x}$ axis. In the global coordinate system, each nodal point has a horizontal force X, a vertical force Y, a horizontal displacement u, and a vertical displacement v. Thus each element has a total of four degrees of freedom: u_1, v_1, u_2, and v_2.

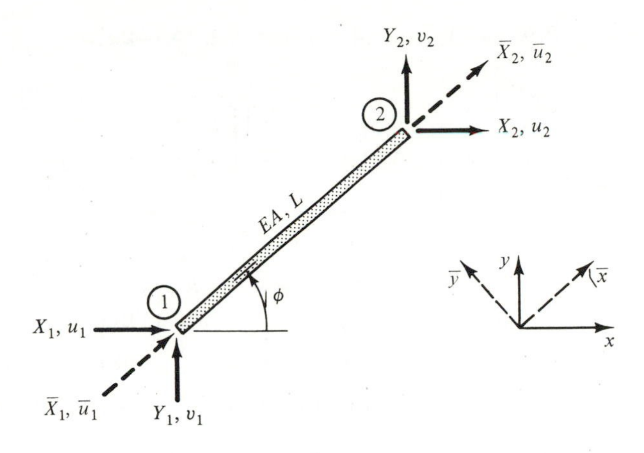

Figure 4.3 Local $(\bar{x}, \bar{y})$ and reference (x, y) coordinate systems for a truss bar element.

From Fig. 4.3 we see at nodal point 1,

$$\bar{u}_1 = u_1 \cos \phi + v_1 \sin \phi \tag{4.12a}$$

and at nodal point 2,

$$\bar{u}_2 = u_2 \cos \phi + v_2 \sin \phi \tag{4.12b}$$

If we use the symbols

$$\lambda = \cos \phi \qquad \text{and} \qquad \mu = \sin \phi$$

Equations (4.12a) and (4.12b) can be written as

$$\begin{Bmatrix} \bar{u}_1 \\ \bar{u}_2 \end{Bmatrix} = \begin{bmatrix} \lambda & \mu & 0 & 0 \\ 0 & 0 & \lambda & \mu \end{bmatrix} \begin{Bmatrix} u_1 \\ v_1 \\ u_2 \\ v_2 \end{Bmatrix} \tag{4.13}$$

As explained in Chapter 3 in connection with Eqs. (3.3a) and (3.3b), the strain energy or the work done is equal to one-half of the nodal forces multiplied by the corresponding displacements,

$$\begin{aligned} U &= \tfrac{1}{2}(\bar{X}_1 \bar{u}_1 + \bar{X}_2 \bar{u}_2) \\ &= \tfrac{1}{2} \begin{Bmatrix} \bar{X}_1 \\ \bar{X}_2 \end{Bmatrix}^T \begin{Bmatrix} \bar{u}_1 \\ \bar{u}_2 \end{Bmatrix} \end{aligned} \tag{4.14}$$

Equation (4.10) can be rewritten for an element in a local $\bar{x}$ axis,

$$\begin{Bmatrix} \bar{X}_1 \\ \bar{X}_2 \end{Bmatrix} = \frac{EA}{L} \begin{bmatrix} 1 & -1 \\ -1 & 1 \end{bmatrix} \begin{Bmatrix} \bar{u}_1 \\ \bar{u}_2 \end{Bmatrix} \tag{4.15}$$

Substituting Eqs. (4.13) and (4.15) into Eq. (4.14) gives

$$U = \frac{EA}{2L}\begin{Bmatrix} u_1 \\ v_1 \\ u_2 \\ v_2 \end{Bmatrix}^T \begin{bmatrix} \lambda & 0 \\ \mu & 0 \\ 0 & \lambda \\ 0 & \mu \end{bmatrix} \begin{bmatrix} 1 & -1 \\ -1 & 1 \end{bmatrix} \begin{bmatrix} \lambda & \mu & 0 & 0 \\ 0 & 0 & \lambda & \mu \end{bmatrix} \begin{Bmatrix} u_1 \\ v_1 \\ u_2 \\ v_2 \end{Bmatrix}$$

$$= \frac{EA}{2L}\begin{Bmatrix} u_1 \\ v_1 \\ u_2 \\ v_2 \end{Bmatrix}^T \begin{bmatrix} \lambda^2 & \lambda\mu & -\lambda^2 & -\lambda\mu \\ \lambda\mu & \mu^2 & -\lambda\mu & -\mu^2 \\ -\lambda^2 & -\lambda\mu & \lambda^2 & \lambda\mu \\ -\lambda\mu & -\mu^2 & \lambda\mu & \mu^2 \end{bmatrix} \begin{Bmatrix} u_1 \\ v_1 \\ u_2 \\ v_2 \end{Bmatrix} \tag{4.16}$$

The stiffness equations can be obtained by using Castigliano's theorem:

$$\begin{Bmatrix} X_1 \\ Y_1 \\ X_2 \\ Y_2 \end{Bmatrix} = \begin{Bmatrix} \dfrac{\partial U}{\partial u_1} \\ \dfrac{\partial U}{\partial v_1} \\ \dfrac{\partial U}{\partial u_2} \\ \dfrac{\partial U}{\partial v_2} \end{Bmatrix} = \frac{EA}{L} \begin{bmatrix} \lambda^2 & \lambda\mu & -\lambda^2 & -\lambda\mu \\ \lambda\mu & \mu^2 & -\lambda\mu & -\mu^2 \\ -\lambda^2 & -\lambda\mu & \lambda^2 & \lambda\mu \\ -\lambda\mu & -\mu^2 & \lambda\mu & \mu^2 \end{bmatrix} \begin{Bmatrix} u_1 \\ v_1 \\ u_2 \\ v_2 \end{Bmatrix} \tag{4.17}$$

This stiffness matrix has an easy-to-memorize pattern. We can simply remember the stiffness matrix as

$$[\boldsymbol{k}] = \frac{EA}{L}\begin{bmatrix} k_0 & -k_0 \\ -k_0 & k_0 \end{bmatrix} \tag{4.18a}$$

where

$$[\mathbf{k}_0] = \begin{bmatrix} \lambda^2 & \lambda\mu \\ \lambda\mu & \mu^2 \end{bmatrix} \tag{4.18b}$$

4.3.2 Coordinate Transformation Method

As shown in Fig. 4.3, the two-dimensional bar element has a total of four degrees of freedom. If we create an additional nonexistent force $\bar{Y}$ and a displacement $\bar{v}$ in the $\bar{y}$ direction at each nodal point, we can relate the four degrees of freedom in local coordinate directions to those in global coordinate directions as

$$\begin{Bmatrix} \bar{u}_1 \\ \bar{v}_1 \\ \bar{u}_2 \\ \bar{v}_2 \end{Bmatrix} = \begin{bmatrix} \lambda & \mu & 0 & 0 \\ -\mu & \lambda & 0 & 0 \\ 0 & 0 & \lambda & \mu \\ 0 & 0 & -\mu & \lambda \end{bmatrix} \begin{Bmatrix} u_1 \\ v_1 \\ u_2 \\ v_2 \end{Bmatrix} \tag{4.19a}$$

or symbolically,

$$\{\bar{\mathbf{q}}\} = [\mathbf{T}]\{\mathbf{q}\} \tag{4.19b}$$

where $[\mathbf{T}]$ is the coordinate transformation matrix. This matrix was also derived in Example 2.3.

We can relate the four nodal forces in local coordinate directions to those in global coordinate directions by the same coordinate transformation matrix as that derived for displacements:

$$\{\bar{\mathbf{F}}\} = [\mathbf{T}]\{\mathbf{F}\} \tag{4.19c}$$

Because the transformation matrix $[\mathbf{T}]$ is of order 4×4, it is necessary to augment the stiffness matrix for local coordinates as given in Eq. (4.10) from order 2×2 to 4×4. This can be done simply by putting zeros in the additional rows and columns corresponding to the nonexistent forces $\bar{Y}_1$, $\bar{Y}_2$ and displacements $\bar{v}_1$, $\bar{v}_2$.

$$\begin{Bmatrix} \bar{X}_1 \\ \bar{Y}_1 \\ \bar{X}_2 \\ \bar{Y}_2 \end{Bmatrix} = \frac{EA}{L} \begin{bmatrix} 1 & 0 & -1 & 0 \\ 0 & 0 & 0 & 0 \\ -1 & 0 & 1 & 0 \\ 0 & 0 & 0 & 0 \end{bmatrix} \begin{Bmatrix} \bar{u}_1 \\ \bar{v}_1 \\ \bar{u}_2 \\ \bar{v}_2 \end{Bmatrix} \tag{4.20a}$$

or symbolically,

$$\{\bar{\mathbf{F}}\} = [\bar{\mathbf{k}}]\{\bar{\mathbf{q}}\} \tag{4.20b}$$

Substituting Eqs. (4.19b) and (4.19c) into Eq. (4.20b) gives

$$[\mathbf{T}]\{\mathbf{F}\} = [\bar{\mathbf{k}}][\mathbf{T}]\{\mathbf{q}\}$$

or

$$\{\mathbf{F}\} = [\mathbf{T}]^{-1}[\bar{\mathbf{k}}][\mathbf{T}]\{\mathbf{q}\}$$

Because the transformation matrix $[\mathbf{T}]$ is an orthogonal matrix (i.e., its inverse is the same as its transpose), we have

$$\{\mathbf{F}\} = [\mathbf{k}]\{\mathbf{q}\} \tag{4.21a}$$

with

$$[\mathbf{k}] = [\mathbf{T}]^T[\bar{\mathbf{k}}][\mathbf{T}] \tag{4.21b}$$

This stiffness matrix is in a congruent transformation form. It is in precisely the same explicit form as that given in Eq. (4.17).

4.3.3 Stress–Strain Equilibrium Method

Let us first assume that the bar element is fixed by a hinge at nodal point 1 and supported by a roller at nodal point 2, as shown in Fig. 4.4. The roller

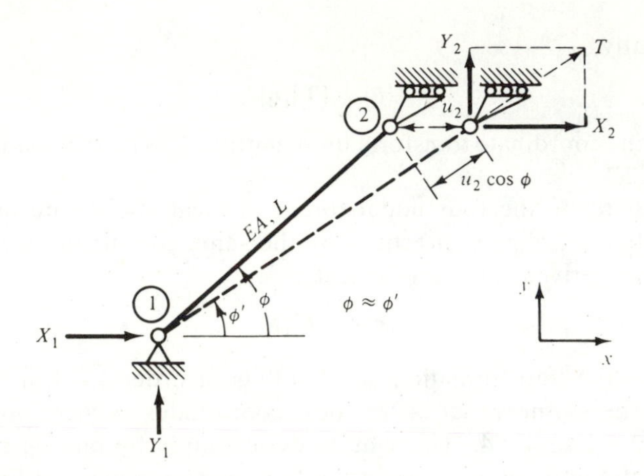

Figure 4.4 Two-dimensional truss bar element with $u_1 = v_1 = v_2 = 0$ but $u_2 \neq 0$.

is restrained from vertical movement but allowed to move horizontally. Such conditions mean that

$$u_1 = v_1 = v_2 = 0 \qquad \text{and} \qquad u_2 \neq 0$$

Let us now move nodal point 2 by an amount u_2, as shown in Fig. 4.4. It is assumed that u_2 is so small that the angle ϕ remains unchanged. The bar is stretched with

$$\text{elongation} = u_2 \cos \phi = \lambda u_2$$

The axial force equivalent to such elongation is

$$T = (\text{spring constant})(\text{elongation}) = \frac{EA}{L} \lambda u_2$$

The axial tensile force at nodal point 2 has two components:

$$\begin{aligned} X_2 &= T \cos \phi = \frac{EA}{L} \lambda^2 u_2 \\ Y_2 &= T \sin \phi = \frac{EA}{L} \lambda \mu u_2 \end{aligned} \tag{4.22a}$$

Considering the bar as a free body and setting up the equilibrium equations gives

$$\begin{aligned} X_1 &= -X_2 = -\frac{EA}{L} \lambda^2 u_2 \\ Y_1 &= -Y_2 = -\frac{EA}{L} \lambda \mu u_2 \end{aligned} \tag{4.22b}$$

Equations (4.22a) and (4.22b) can be written in the following form:

$$\begin{Bmatrix} X_1 \\ Y_1 \\ X_2 \\ Y_2 \end{Bmatrix} = \frac{EA}{L} \begin{Bmatrix} -\lambda^2 \\ -\lambda\mu \\ \lambda^2 \\ \lambda\mu \end{Bmatrix} \{\mathbf{u}_2\} \tag{4.23}$$

It is readily seen that Eq. (4.23) yields precisely the third column of the stiffness matrix given in Eq. (4.17).

By the same procedure, we can derive

$$\begin{Bmatrix} \text{1st column} \\ \text{2nd column} \\ \text{4th column} \end{Bmatrix} \text{ of } [\mathbf{k}] \text{ by setting } \begin{Bmatrix} u_1 = u_1,\ v_1 = u_2 = v_2 = 0 \\ v_1 = v_1,\ u_1 = u_2 = v_2 = 0 \\ v_2 = v_2,\ u_1 = v_1 = u_2 = 0 \end{Bmatrix}$$

This derivation and the subsequent procedure for assemblage and solution are due to Turner et al. [4.1]. This pioneering work was done during the emerging stage of electronic digital computers. The reader is referred also to Ref. 4.2.

4.3.4 Properties of the Two-Dimensional Bar Element Stiffness Equations

Equilibrium. Let us consider the two-dimensional truss bar element shown in Fig. 4.3 as a free body. The stiffness equations for this element are given in Eq. (4.17). We see in the stiffness matrix that the coefficients in the first row are the same but opposite in sign as the coefficients in the third row. The same relation holds between the second and fourth rows. If we multiply out the matrix equations, we find that

$$X_1 = -X_2 \qquad \text{and} \qquad Y_1 = -Y_2$$

So the equilibrium in both the x and y directions for the free body is satisfied.

We then take the moment about nodal point 1,

$$\begin{aligned} \sum M_1 &= X_2 L\mu - Y_2 L\lambda \\ &= EA[(-\lambda^2\mu + \lambda^2\mu)u_1 + (-\lambda\mu^2 + \lambda\mu^2)v_1 \\ &\quad + (\lambda^2\mu - \lambda^2\mu)u_2 + (\lambda\mu^2 - \lambda\mu^2)v_2] \\ &= 0 \end{aligned}$$

which satisfies the moment equilibrium conditions.

Singularity. In the stiffness matrix $[\mathbf{k}]$, the first and third rows are the same but opposite in sign, as are the second and fourth rows. Thus $[\mathbf{k}]$ is a singular matrix. It cannot be inverted and a solution to Eq. (4.17) is not possible.

Physically, singular stiffness matrix means that the element, without any support, is an unstable free body. The element can become stable and the stiffness matrix can become nonsingular if the element is properly supported. One of the ways to stabilize the element is to support it as shown in Fig. 4.4. In that case, $u_1 = v_1 = v_2 = 0$. If we consider only the third equation in the stiffness equations (4.17), we find that

$$\{\mathbf{X}_2\} = \frac{\lambda^2 EA}{L}\{\mathbf{u}_2\}$$

The reduced stiffness matrix is of order 1×1 and is no longer singular.

4.3.5 Equation for Axial Force

If we know the nodal displacements for a bar element, either given or solved, the axial force can be obtained directly from Eq. (4.17). Let us multiply out the third and fourth equations in Eq. (4.17),

$$X_2 = \frac{EA}{L}[\lambda^2(u_2 - u_1) + \lambda\mu(v_2 - v_1)]$$

$$Y_2 = \frac{EA}{L}[\lambda\mu(u_2 - u_1) + \mu^2(v_2 - v_1)]$$

If we use S to designate tensile axial force, then

$$\begin{aligned} S &= X_2\lambda + Y_2\mu \\ &= \frac{EA}{L}(\lambda^2 + \mu^2)[\lambda(u_2 - u_1) + \mu(v_2 - v_1)] \\ &= \frac{EA}{L}[\lambda(u_2 - u_1) + \mu(v_2 - v_1)] \\ &= \text{(spring constant)(total axial elongation)} \end{aligned}$$

where $(u_2 - u_1)$ and $(v_2 - v_1)$ are, respectively, the horizontal and vertical components of the axial elongation.

In matrix form, the tensile axial force is given by

$$S = \frac{EA}{L}\lfloor \lambda \quad \mu \rfloor \begin{Bmatrix} u_2 - u_1 \\ v_2 - v_1 \end{Bmatrix} \tag{4.24}$$

If the value obtained for S is negative, the axial force is in compression.

4.4 STIFFNESS EQUATIONS FOR A TRUSS BAR ELEMENT ORIENTED ARBITRARILY IN THREE-DIMENSIONAL SPACE

Let us now consider a bar element oriented arbitrarily in three-dimensional space. There are two sets of Cartesian coordinates: the local coordinates $(\bar{x}, \bar{y}, \bar{z})$ and the global or reference coordinates (x, y, z). The local $\bar{x}$ axis coincides with the axis of the bar. Let l_1, m_1, n_1; l_2, m_2, n_2; l_3, m_3, n_3 be the direction cosines of the $\bar{x}$; $\bar{y}$; $\bar{z}$ axes relative to the x; y; z axes, respectively. It has been shown in Example 2.4 that

$$\begin{Bmatrix} \bar{x} \\ \bar{y} \\ \bar{z} \end{Bmatrix} = \begin{bmatrix} l_1 & m_1 & n_1 \\ l_2 & m_2 & n_2 \\ l_3 & m_3 & n_3 \end{bmatrix} \begin{Bmatrix} x \\ y \\ z \end{Bmatrix} \tag{4.25}$$

The bar element in space has three forces and three displacement degrees of freedom (X, Y, Z and u, v, w) at each nodal point in the global x, y, z directions, respectively. The element in local coordinates has only an axial force $\bar{X}$ and an axial degree of freedom $\bar{u}$ at each nodal point. Let us first augment the 2×2 stiffness matrix equations in local coordinates to 6×6 stiffness matrix equations by creating two more nonexistent forces $\bar{Y}, \bar{Z}$ and displacement degrees of freedom $\bar{v}, \bar{w}$ at each nodal point in the local $\bar{y}, \bar{z}$ directions, respectively. Thus Eq. (4.10) becomes

$$\begin{Bmatrix} \bar{X}_1 \\ \bar{Y}_1 \\ \bar{Z}_1 \\ \bar{X}_2 \\ \bar{Y}_2 \\ \bar{Z}_2 \end{Bmatrix} = \frac{EA}{L} \begin{bmatrix} 1 & 0 & 0 & -1 & 0 & 0 \\ 0 & 0 & 0 & 0 & 0 & 0 \\ 0 & 0 & 0 & 0 & 0 & 0 \\ -1 & 0 & 0 & 1 & 0 & 0 \\ 0 & 0 & 0 & 0 & 0 & 0 \\ 0 & 0 & 0 & 0 & 0 & 0 \end{bmatrix} \begin{Bmatrix} \bar{u}_1 \\ \bar{v}_1 \\ \bar{w}_1 \\ \bar{u}_2 \\ \bar{v}_2 \\ \bar{w}_2 \end{Bmatrix} \tag{4.26a}$$

or symbolically,

$$\{\bar{\mathbf{F}}\} = [\bar{\mathbf{k}}]\{\bar{\mathbf{q}}\} \tag{4.26b}$$

Based on Eq. (4.25), the six nodal degrees of freedom in local coordinates can be related to that in global coordinates as

$$\begin{Bmatrix} \bar{u}_1 \\ \bar{v}_1 \\ \bar{w}_1 \\ \bar{u}_2 \\ \bar{v}_2 \\ \bar{w}_2 \end{Bmatrix} = \frac{EA}{L} \begin{bmatrix} l_1 & m_1 & n_1 & 0 & 0 & 0 \\ l_2 & m_2 & n_2 & 0 & 0 & 0 \\ l_3 & m_3 & n_3 & 0 & 0 & 0 \\ 0 & 0 & 0 & l_1 & m_1 & n_1 \\ 0 & 0 & 0 & l_2 & m_2 & n_2 \\ 0 & 0 & 0 & l_3 & m_3 & n_3 \end{bmatrix} \begin{Bmatrix} u_1 \\ v_1 \\ w_1 \\ u_2 \\ v_2 \\ w_2 \end{Bmatrix} \tag{4.27a}$$

or symbolically,

$$\{\bar{\mathbf{q}}\} = [\mathbf{T}]\{\mathbf{q}\} \tag{4.27b}$$

where [T] is the transformation matrix. It is an orthogonal matrix (i.e., its transpose is its inverse).

The six nodal forces in local coordinates can be related to those in global coordinates by the same matrix,

$$\{\bar{\mathbf{F}}\} = [\mathbf{T}]\{\mathbf{F}\} \tag{4.28}$$

By using the same procedure as that used in deriving Eq. (4.21a) for the two-dimensional case [i.e., by substituting Eqs. (4.27b) and (4.28) into Eq. (4.26b)], we obtain for the three-dimensional case

$$\{\mathbf{F}\} = [\mathbf{k}]\{\mathbf{q}\} \tag{4.29a}$$

with

$$[\mathbf{k}] = [\mathbf{T}]^T[\bar{\mathbf{k}}][\mathbf{T}] \tag{4.29b}$$

In an explicit form, Eq. (4.29a) is obtained as

$$\begin{Bmatrix} X_1 \\ Y_1 \\ Z_1 \\ X_2 \\ Y_2 \\ Z_2 \end{Bmatrix} = \frac{EA}{L}\begin{bmatrix} k_0 & -k_0 \\ -k_0 & k_0 \end{bmatrix}\begin{Bmatrix} u_1 \\ v_1 \\ w_1 \\ u_2 \\ v_2 \\ w_2 \end{Bmatrix} \tag{4.30a}$$

with

$$[\mathbf{k}_0] = \begin{bmatrix} l_1^2 & l_1 m_1 & n_1 l_1 \\ l_1 m_1 & m_1^2 & m_1 n_1 \\ n_1 l_1 & m_1 n_1 & n_1^2 \end{bmatrix} \tag{4.30b}$$

The subscript is 1 throughout matrix $[\mathbf{k}_0]$. For simplicity we can drop out the subscript

$$[\mathbf{k}_0] = \begin{bmatrix} l^2 & lm & nl \\ lm & m^2 & mn \\ nl & mn & n^2 \end{bmatrix} \tag{4.30c}$$

where l, m, n are the direction cosines between the $\bar{x}$ axis and the x, y, z axes, respectively.

It is observed that this 6×6 stiffness matrix is in a pattern similar to the 4×4 stiffness matrix in Eq. (4.17). This matrix is also a singular matrix.

If we know the nodal displacements for this element, either given or solved, the three force components X_2, Y_2, Z_2 at nodal point 2 can readily be obtained from Eq. (4.30a). The tensile axial force S is the resultant of these three forces. Letting l, m, n be the direction cosines of the element axis relative to the x, y, z axes, respectively, and making use of the fact that $l^2 + m^2 + n^2 = 1$ yields

$$S = X_2 l + Y_2 m + Z_2 n$$

$$= \frac{EA}{L} \lfloor l|m|n \rfloor \begin{Bmatrix} u_2 - u_1 \\ v_2 - v_1 \\ w_2 - w_1 \end{Bmatrix} \tag{4.31}$$

which is in a form similar to Eq. (4.24).

4.5 METHODS OF ASSEMBLAGE AND SOLUTION DEMONSTRATED BY A THREE-BAR TRUSS

4.5.1 Assemblage

To demonstrate the method of assemblage, an example of a three-bar truss as shown in Fig. 4.5a is chosen. All bars have the same length l and axial rigidity EA. The truss has three nodal points and six degrees of freedom. To formulate the total system, we must formulate the individual element first. Table 4.1 lists the direction cosines and sines for the three elements.

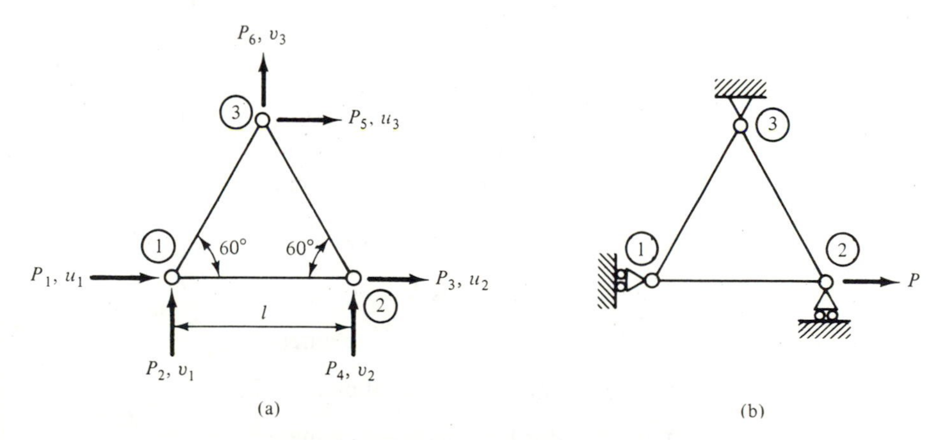

Figure 4.5 (a) Three-node, six-degree-of-freedom, three-bar truss; (b) boundary and loading conditions for the truss.

For element 1-2, we have, from Eq. (4.17) and Table 4.1,

$$\begin{Bmatrix} X_1 \\ Y_1 \\ X_2 \\ Y_2 \end{Bmatrix} = \frac{EA}{l} \begin{bmatrix} 1 & 0 & -1 & 0 \\ 0 & 0 & 0 & 0 \\ -1 & 0 & 1 & 0 \\ 0 & 0 & 0 & 0 \end{bmatrix} \begin{Bmatrix} u_1 \\ v_1 \\ u_2 \\ v_2 \end{Bmatrix} \tag{4.32a}$$

TABLE 4.1. Direction Cosines and Sines for Each Element

Element	ϕ	λ	μ	λ^2	$\lambda\mu$	μ^2
1-2	0°	1	0	1	0	0
2-3	120°	$-\frac{1}{2}$	$\frac{\sqrt{3}}{2}$	$\frac{1}{4}$	$-\frac{\sqrt{3}}{4}$	$\frac{3}{4}$
1-3	60°	$\frac{1}{2}$	$\frac{\sqrt{3}}{2}$	$\frac{1}{4}$	$\frac{\sqrt{3}}{4}$	$\frac{3}{4}$

For element 2-3,

$$\begin{Bmatrix} X_2 \\ Y_2 \\ X_3 \\ Y_3 \end{Bmatrix} = \frac{EA}{4l} \begin{bmatrix} 1 & -\sqrt{3} & -1 & \sqrt{3} \\ -\sqrt{3} & 3 & \sqrt{3} & -3 \\ -1 & \sqrt{3} & 1 & -\sqrt{3} \\ \sqrt{3} & -3 & -\sqrt{3} & 3 \end{bmatrix} \begin{Bmatrix} u_2 \\ v_2 \\ u_3 \\ v_3 \end{Bmatrix} \tag{4.32b}$$

For element 1-3,

$$\begin{Bmatrix} X_1 \\ Y_1 \\ X_3 \\ Y_3 \end{Bmatrix} = \frac{EA}{4l} \begin{bmatrix} 1 & \sqrt{3} & -1 & -\sqrt{3} \\ \sqrt{3} & 3 & -\sqrt{3} & -3 \\ -1 & -\sqrt{3} & 1 & \sqrt{3} \\ -\sqrt{3} & -3 & \sqrt{3} & 3 \end{bmatrix} \begin{Bmatrix} u_1 \\ v_1 \\ u_3 \\ v_3 \end{Bmatrix} \tag{4.32c}$$

We must bear in mind that the X's and Y's are the internal nodal forces. When the three sets of element stiffness equations are assembled, the sum of such internal forces in either the x or the y direction at each nodal point is equal to the external load P applied in the same direction. Thus

$$\begin{aligned} P_1 &= X_1 \text{ of element 1-2 plus } X_1 \text{ of element 1-3} \\ P_2 &= Y_1 \text{ of element 1-2 plus } Y_1 \text{ of element 1-3} \\ P_3 &= X_2 \text{ of element 1-2 plus } X_2 \text{ of element 2-3} \\ P_4 &= Y_2 \text{ of element 1-2 plus } Y_2 \text{ of element 2-3} \\ P_5 &= X_3 \text{ of element 1-3 plus } X_3 \text{ of element 2-3} \\ P_6 &= Y_3 \text{ of element 1-3 plus } Y_3 \text{ of element 2-3} \end{aligned} \tag{4.33}$$

We must also bear in mind that at each nodal point the displacements u and v at the element level are the same as those after assemblage.

With the understanding of the two rules above, the method of assemblage is simply first to multiply out Eqs. (4.32a) to (4.32c) to obtain the internal forces X and Y for each element and then to sum them up as indicated in

Eq. (4.33):

$$\begin{Bmatrix} P_1 \\ P_2 \\ P_3 \\ P_4 \\ P_5 \\ P_6 \end{Bmatrix} = \frac{EA}{4l} \begin{bmatrix} 4+1 & \sqrt{3} & -4 & 0 & -1 & -\sqrt{3} \\ \sqrt{3} & 3 & 0 & 0 & -\sqrt{3} & -3 \\ -4 & 0 & 4+1 & -\sqrt{3} & -1 & \sqrt{3} \\ 0 & 0 & -\sqrt{3} & 3 & \sqrt{3} & -3 \\ -1 & -\sqrt{3} & -1 & \sqrt{3} & 1+1 & \sqrt{3}-\sqrt{3} \\ -\sqrt{3} & -3 & \sqrt{3} & -3 & \sqrt{3}-\sqrt{3} & 3+3 \end{bmatrix} \begin{Bmatrix} u_1 \\ v_1 \\ u_2 \\ v_2 \\ u_3 \\ v_3 \end{Bmatrix} \tag{4.34}$$

The foregoing assemblage procedure is equivalent to move three 4×4 element stiffness matrices into a 6×6 total stiffness matrix as illustrated in Fig. 4.6. For elements 1-2 and 2-3, each 4×4 stiffness matrix remains as a block during assemblage. For element 1-3, the 4×4 stiffness matrix is separated into four blocks (2×2 submatrices) during assemblage. The positions where each block moves into depend on the related degree of freedom numbers. In portions where two blocks overlap, the element stiffness coefficients are superimposed.

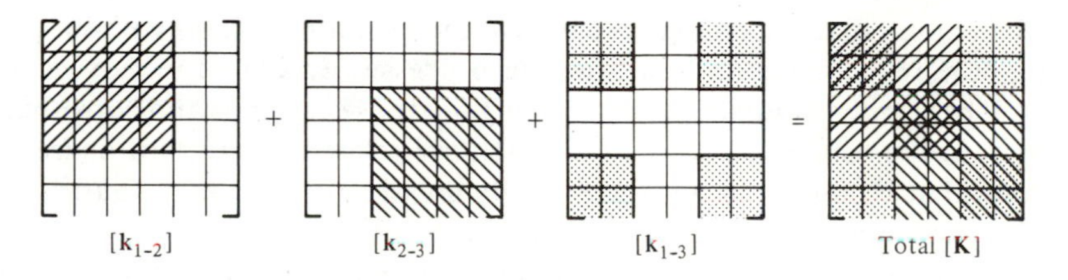

Figure 4.6 Assemblage of the three element stiffness matrices.

4.5.2 Solution for the Truss Problem

Let it be assumed that the truss is supported and loaded as that shown in Fig. 4.5b. We are to find the values of the six displacement degrees of freedom, the four reaction forces, and the three bar axial forces.

The supporting conditions are $u_1 = v_2 = u_3 = v_3 = 0$. Only v_1 and u_2 remain to be solved. Let us rearrange Eq. (4.34) so that v_1 and u_2 and the corresponding P_2 and P_3 appear first in their respective columns.

First, let us rearrange the positions of rows as

$$\begin{Bmatrix} P_2 \\ P_3 \\ \hdashline P_1 \\ P_4 \\ P_5 \\ P_6 \end{Bmatrix} = \frac{EA}{4l} \begin{bmatrix} \sqrt{3} & 3 & 0 & 0 & -\sqrt{3} & -3 \\ -4 & 0 & 5 & -\sqrt{3} & -1 & \sqrt{3} \\ \hdashline 5 & \sqrt{3} & -4 & 0 & -1 & -\sqrt{3} \\ 0 & 0 & -\sqrt{3} & 3 & \sqrt{3} & -3 \\ -1 & -\sqrt{3} & -1 & \sqrt{3} & 2 & 0 \\ -\sqrt{3} & -3 & \sqrt{3} & -3 & 0 & 6 \end{bmatrix} \begin{Bmatrix} u_1 \\ v_1 \\ u_2 \\ v_2 \\ u_3 \\ v_3 \end{Bmatrix} \tag{4.35a}$$

Then let us rearrange the positions of the columns as

$$\begin{Bmatrix} P_2 = 0 \\ P_3 = P \\ \hline P_1 = ? \\ P_4 = ? \\ P_5 = ? \\ P_6 = ? \end{Bmatrix} = \frac{EA}{4l} \left[\begin{array}{cc:cccc} 3 & 0 & \sqrt{3} & 0 & -\sqrt{3} & -3 \\ 0 & 5 & -4 & -\sqrt{3} & -1 & \sqrt{3} \\ \hdashline \sqrt{3} & -4 & 5 & 0 & -1 & -\sqrt{3} \\ 0 & -\sqrt{3} & 0 & 3 & \sqrt{3} & -3 \\ -\sqrt{3} & -1 & -1 & \sqrt{3} & 2 & 0 \\ -3 & \sqrt{3} & -\sqrt{3} & -3 & 0 & 6 \end{array}\right] \begin{Bmatrix} v_1 = ? \\ u_2 = ? \\ \hline u_1 = 0 \\ v_2 = 0 \\ u_3 = 0 \\ v_3 = 0 \end{Bmatrix} \quad (4.35b)$$

Multiplying out the first two rows gives

$$\begin{Bmatrix} 0 \\ P \end{Bmatrix} = \frac{EA}{4l} \begin{bmatrix} 3 & 0 \\ 0 & 5 \end{bmatrix} \begin{Bmatrix} v_1 \\ u_2 \end{Bmatrix} \quad (4.36a)$$

which gives

$$\begin{Bmatrix} v_1 \\ u_2 \end{Bmatrix} = \frac{4l}{15EA} \begin{bmatrix} 5 & 0 \\ 0 & 3 \end{bmatrix} \begin{Bmatrix} 0 \\ P \end{Bmatrix} = \frac{4Pl}{5EA} \begin{Bmatrix} 0 \\ 1 \end{Bmatrix} \quad (4.36b)$$

It is noted that Eq. (4.36a) can be obtained from Eq. (4.34) without going through the rearranging and partitioning processes described in Eqs. (4.35a) and (4.35b). This is done simply by crossing out those columns and corresponding rows in the total stiffness matrix which are related to the zero degrees of freedom; that is, Eq. (4.36a) is obtained by crossing out the first, fourth, fifth, and sixth columns and rows in the total 6×6 stiffness matrix.

Having found v_1 and u_2, we can find the reaction forces by multiplying out the third to sixth rows in Eq. (4.35b):

$$\begin{Bmatrix} P_1 \\ P_4 \\ P_5 \\ P_6 \end{Bmatrix} = \frac{EA}{4l} \begin{bmatrix} \sqrt{3} & -4 \\ 0 & -\sqrt{3} \\ -\sqrt{3} & -1 \\ -3 & \sqrt{3} \end{bmatrix} \begin{Bmatrix} v_1 \\ u_2 \end{Bmatrix} = \frac{P}{5} \begin{Bmatrix} -4 \\ -\sqrt{3} \\ -1 \\ \sqrt{3} \end{Bmatrix} \quad (4.37)$$

Now we have all six P's. They satisfy equilibrium if we consider the whole truss as a free body.

The axial forces $S_{1\text{-}2}$, $S_{2\text{-}3}$, and $S_{3\text{-}1}$ for the three bar elements are found by using Eq. (4.24):

$$S_{1\text{-}2} = \frac{EA}{l} \lfloor \lambda_{1\text{-}2} \quad \mu_{1\text{-}2} \rfloor \begin{Bmatrix} u_2 - u_1 \\ v_2 - v_1 \end{Bmatrix}$$

$$= \frac{EA}{l} \lfloor 1 \quad 0 \rfloor \begin{Bmatrix} \dfrac{4Pl}{5EA} \\ 0 \end{Bmatrix} = \frac{4P}{5} \quad \text{(tension)}$$

$$S_{2\text{-}3} = \frac{EA}{l} \lfloor \lambda_{2\text{-}3} \quad \mu_{2\text{-}3} \rfloor \begin{Bmatrix} u_3 - u_2 \\ v_3 - v_2 \end{Bmatrix}$$

$$= \frac{EA}{l} \begin{bmatrix} -\frac{1}{2} & \frac{\sqrt{3}}{2} \end{bmatrix} \begin{Bmatrix} \dfrac{-4Pl}{5EA} \\ 0 \end{Bmatrix} = \frac{2P}{5} \quad \text{(tension)}$$

$$S_{1\text{-}3} = \frac{EA}{l} \lfloor \lambda_{1\text{-}3} \quad \mu_{1\text{-}3} \rfloor \begin{Bmatrix} u_3 - u_1 \\ v_3 - v_1 \end{Bmatrix}$$

$$= \frac{EA}{l} \begin{bmatrix} \frac{1}{2} & \frac{\sqrt{3}}{2} \end{bmatrix} \begin{Bmatrix} 0 \\ 0 \end{Bmatrix} = 0$$

Having obtained the axial forces, the reaction forces may be found by considering equilibrium of each joint instead of using Eq. (4.37). Thus the formulation of the 6×6 total stiffness matrix equations (4.35b) may not be necessary. In fact, only the assemblage of the 2×2 stiffness matrix equations (4.36a) is necessary.

4.6 SOME BASIC TREATMENTS OF STIFFNESS EQUATIONS

After we have been familiar with the solution procedure of Section 4.5, it is easier to understand some basic procedures of treatment of stiffness equations as explained in the following symbolic form.

4.6.1 Treatment of Boundary Conditions

Let it be assumed that the stiffness matrix equations assembled for a finite element structural system are obtained as

$$\{\mathbf{P}\} = [\mathbf{K}]\{\mathbf{q}\} \tag{4.38}$$

Based on the prescribed zero displacement boundary conditions, Eq. (4.38) may be rearranged and partitioned as demonstrated in obtaining Eqs. (4.35b):

$$\begin{Bmatrix} \mathbf{P}_1 \\ \mathbf{P}_2 \end{Bmatrix} = \begin{bmatrix} \mathbf{K}_{11} & \mathbf{K}_{12} \\ \mathbf{K}_{21} & \mathbf{K}_{22} \end{bmatrix} \begin{Bmatrix} \mathbf{q}_1 = ? \\ \mathbf{q}_2 = 0 \end{Bmatrix} \tag{4.39}$$

where $\{\mathbf{q}_1\}$ contains the unknown unconstrained nodal degrees of freedom, $\{\mathbf{q}_2\}$ contains the constrained nodal degrees of freedom (zeros), $\{\mathbf{P}_1\}$ contains the external loads related to the unconstrained nodal degrees of freedom, and $\{\mathbf{P}_2\}$ contains the unknown reaction forces related to the constrained nodal degrees of freedom.

Multiplying out Eqs. (4.39) gives

$$\{\mathbf{P}_1\} = [\mathbf{K}_{11}]\{\mathbf{q}_1\} \tag{4.40a}$$

so that

$$\{\mathbf{q}_1\} = [\mathbf{K}_{11}]^{-1}\{\mathbf{P}_1\} \tag{4.40b}$$

and the unknown reaction forces are given by

$$\{\mathbf{P}_2\} = [\mathbf{K}_{21}]\{\mathbf{q}_1\} = [\mathbf{K}_{21}][\mathbf{K}_{11}]^{-1}\{\mathbf{P}_1\} \tag{4.40c}$$

It is a common and efficient practice to formulate only the smaller matrix $[\mathbf{K}_{11}]$ instead of the whole matrix $[\mathbf{K}]$ based on the zero-displacement conditions. Matrix $[\mathbf{K}_{11}]$ in Eq. (4.40a) is sufficient to yield all the unknown degrees of freedom so that the reaction forces can be obtained by using the necessary element stiffness equations instead of matrix $[\mathbf{K}_{21}]$. In computer programming, to formulate only $[\mathbf{K}_{11}]$ instead of $[\mathbf{K}]$ is essential in saving DIMENSION and computer storage.

4.6.2 Reduction Procedure

In an actual problem, not every unconstrained nodal degree of freedom is subjected to external load. The load vector $\{\mathbf{P}_1\}$ in Eq. (4.40a) may contain a certain amount of zeros. Based on these zeros, a reduction procedure may be developed.

For simplicity, let us rewrite Eq. (4.40a) by dropping out the subscripts,

$$\{\mathbf{P}\} = [\mathbf{K}]\{\mathbf{q}\} \tag{4.41}$$

where the vector $\{\mathbf{q}\}$ may no longer contain zero degrees of freedom.

With the knowledge of the zero loads, we may rearrange and partition Eq. (4.41) as

$$\begin{Bmatrix} \mathbf{P}_1 \\ \mathbf{P}_2 = 0 \end{Bmatrix} = \begin{bmatrix} \mathbf{K}_{11} & \mathbf{K}_{12} \\ \mathbf{K}_{21} & \mathbf{K}_{22} \end{bmatrix} \begin{Bmatrix} \mathbf{q}_1 \\ \mathbf{q}_2 \end{Bmatrix} \tag{4.42}$$

where $\{\mathbf{P}_1\}$ contains the actually applied loads and possibly some selected zeros, $\{\mathbf{P}_2\}$ is a zero load vector whose size can be chosen to be less than the actual number of zero loads, $\{\mathbf{q}_1\}$ contains the degrees of freedom corresponding to the loads $\{\mathbf{P}_1\}$, and $\{\mathbf{q}_2\}$ contains the degrees of freedom corresponding to the zero loads $\{\mathbf{P}_2\}$.

Multiplying out Eq. (4.42) gives

$$\{\mathbf{P}_1\} = [\mathbf{K}_{11}]\{\mathbf{q}_1\} + [\mathbf{K}_{12}]\{\mathbf{q}_2\} \tag{4.43a}$$

$$\{\mathbf{P}_2\} = \{\mathbf{0}\} = [\mathbf{K}_{21}]\{\mathbf{q}_1\} + [\mathbf{K}_{22}]\{\mathbf{q}_2\} \tag{4.43b}$$

From Eq. (4.43b), we can write

$$\{\mathbf{q}_2\} = -[\mathbf{K}_{22}]^{-1}[\mathbf{K}_{21}]\{\mathbf{q}_1\} \tag{4.43c}$$

It is noted that matrix $[\mathbf{K}_{22}]$ is a square-symmetric nonsingular submatrix. Unless $\{\mathbf{P}_1\}$ and $\{\mathbf{P}_2\}$ are of the same size, matrix $[\mathbf{K}_{21}]$ is a nonsquare matrix. If $[\mathbf{K}_{21}]$ is square, it is nonsymmetric and often singular.

Substituting Eq. (4.43c) into Eq. (4.43a) yields

$$\{\mathbf{P}_1\} = [\bar{\mathbf{K}}]\{\mathbf{q}_1\} \tag{4.44a}$$

with

$$\{\bar{\mathbf{K}}\} = [\mathbf{K}_{11}] - [\mathbf{K}_{12}][\mathbf{K}_{22}]^{-1}[\mathbf{K}_{21}] \tag{4.44b}$$

Hence

$$\{\mathbf{q}_1\} = [\bar{\mathbf{K}}]^{-1}\{\mathbf{P}_1\} \tag{4.44c}$$

After the displacement vector $\{\mathbf{q}_1\}$ is obtained, a back-substitution of $\{\mathbf{q}_1\}$ into Eq. (4.43c) will give $\{\mathbf{q}_2\}$.

By the use of this reduction procedure, we do not have to invert the whole matrix $[\mathbf{K}]$. Instead, we invert two smaller matrices, $[\mathbf{K}_{22}]$ and $[\bar{\mathbf{K}}]$. The sum of the sizes of $[\mathbf{K}_{22}]$ and $[\bar{\mathbf{K}}]$ is equal to that of $[\mathbf{K}]$. For a large set of stiffness equations, this reduction procedure can result in a considerable saving of computing time.

4.6.3 Joints with Prescribed Displacements Instead of Loads

In certain problems, a nodal point may be subjected to prescribed displacements instead of loads. Forced fit during structural construction is an example. Sometimes, certain nodal displacements may be known from measurement.

Let us start with Eqs. (4.41), for which the boundary conditions have already been accounted for. The equations are rearranged and partitioned according to the prescribed displacement conditions,

$$\begin{Bmatrix} \mathbf{P}_1 \\ \mathbf{P}_2 \end{Bmatrix} = \begin{bmatrix} \mathbf{K}_{11} & \mathbf{K}_{12} \\ \mathbf{K}_{21} & \mathbf{K}_{22} \end{bmatrix} \begin{Bmatrix} \mathbf{q}_1 \\ \mathbf{q}_2 = \bar{\mathbf{q}} \end{Bmatrix} \tag{4.45}$$

where $\{\mathbf{q}_1\}$ contains the unknown unconstrained nodal degrees of freedom, $\{\mathbf{q}_2\} = \{\bar{\mathbf{q}}\}$ contains the prescribed nodal degrees of freedom, $\{\mathbf{P}_1\}$ contains the known external loads corresponding to $\{\mathbf{q}_1\}$, and $\{\mathbf{P}_2\}$ contains the unknown forces required to produce the prescribed degrees of freedom $\{\mathbf{q}_2\}$.

Multiplying out Eqs. (4.45) gives

$$\{\mathbf{P}_1\} = [\mathbf{K}_{11}]\{\mathbf{q}_1\} + [\mathbf{K}_{12}]\{\bar{\mathbf{q}}\} \tag{4.46a}$$

$$\{\mathbf{P}_2\} = [\mathbf{K}_{21}]\{\mathbf{q}_1\} + [\mathbf{K}_{22}]\{\bar{\mathbf{q}}\} \tag{4.46b}$$

The unknown degrees of freedom may be found from Eq. (4.46a):

$$\{\mathbf{q}_1\} = [\mathbf{K}_{11}]^{-1}(\{\mathbf{P}_1\} - [\mathbf{K}_{12}]\{\bar{\mathbf{q}}\})$$

The unknown forces corresponding to the prescribed degrees of freedom may be obtained by back-substitution of $\{\mathbf{q}_1\}$ into Eq. (4.46b).

4.6.4 Initial and Thermal Forces in Truss Bars

During construction of a truss structure, a certain bar could be fabricated with improper length. Thus the bar is forced into its position between two joints by applying some initial extension or compression. Under such a condition, some axial forces are induced in the bars even in the absence of external loads. Such a condition may also arise when the structure is subjected to a temperature change.

The stress analysis for such a condition is divided into two steps and the results of the two steps are then superimposed.

Step 1: Let all the joints be assumed as restrained from displacements and the axial forces developed in all bars due to initial extensions (or compressions) or temperature changes are determined. The axial force for a typical bar element is given by

$$S = -EA\frac{\Delta L}{L} \tag{4.47a}$$

where the overlength ΔL is positive when the bar is longer than the distance L between its two joints. The axial force S is positive when in tension. The negative sign is introduced to indicate that the bar is in compression if its overlength is suppressed by restraining the joints in their original positions.

In the case of a temperature rise ΔT,

$$S = -EA\alpha\,\Delta T \tag{4.47b}$$

where α is the coefficient of thermal expansion (strain per degree) and ΔT is positive when the temperature rises. Again, the negative sign is introduced to indicate that the bar is in compression if its thermal expansion due to positive ΔT is suppressed by restraining the joints in their original positions.

For a typical element as shown in Fig. 4.3, the nodal forces required to restrain the nodal points or to maintain the initial axial forces S as obtained in Eq. (4.47a) or (4.47b) are determined as

$$\begin{Bmatrix} X_1 \\ Y_1 \\ X_2 \\ Y_2 \end{Bmatrix} = S\begin{Bmatrix} -\lambda \\ -\mu \\ \lambda \\ \mu \end{Bmatrix} = EA\left(\frac{\Delta L}{L}\right)\begin{Bmatrix} \lambda \\ \mu \\ -\lambda \\ -\mu \end{Bmatrix} \quad \text{or} \quad EA\alpha\,\Delta T\begin{Bmatrix} \lambda \\ \mu \\ -\lambda \\ -\mu \end{Bmatrix} \tag{4.48}$$

This equation shows that for a bar with overlength or temperature rise, the restraining forces X, Y are both positive at nodal point 1 and negative at nodal point 2.

Step 2: To eliminate the restraining forces obtained in Eq. (4.48) for all bars, the total truss system is analyzed for nodal point loads which are equal in magnitude but opposite in sign to these restraining forces. Thus the external nodal point loads, due to both step 1 and step 2, cancel each other so that the structure is without any external loads due to misfitting. The final results in displacements and axial forces are the sums of those obtained in both steps 1 and 2.

Because the nodal displacements in step 1 are all zeros, the displacement obtained in step 2 define the deformed configuration of the misfitted structure. The final axial force in each bar is the sum of the forces obtained in both steps.

4.7 EXAMPLES

Example 4.1 *Truss bar and spring combination*

A three-bar, one-spring truss is shown in Fig. 4.7. The axial rigidity EA and length l for each bar are the same. The spring has a stiffness of EA/l. Find the displacements at joint 1 and the axial force in each bar.

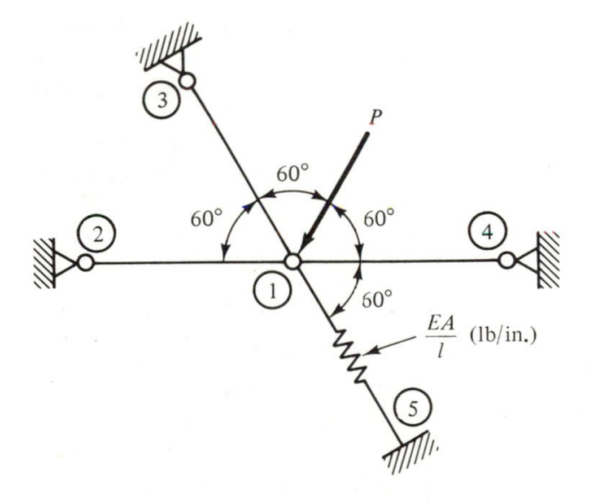

Figure 4.7 Three-bar, one-spring truss.

Solution. Actually, the spring is equivalent to a truss bar. If the truss is rotated counterclockwise by 30°, it becomes a symmetrical four-bar truss with load P acting along the vertical line of symmetry. There is no horizontal displacement u_1. The truss has only one degree of freedom, v_1, along the direction of P.

The total stiffness equation may be assembled as

$$\{-\mathbf{P}\} = [k_{1\text{-}2} + k_{1\text{-}3} + k_{1\text{-}4} + k_{1\text{-}5}(\text{spring})]\{v_1\}$$

where the negative P indicates that the load is downward.

In the element stiffness matrix for each bar as given in Eq. (4.17), only the term k_{22} or μ^2 contributes to the total stiffness matrix; thus

$$\{-\mathbf{P}\} = \frac{EA}{l}[\sin^2(-150°) + \sin^2(150°) + \sin^2(30°) + \sin^2(-30°)]\{v_1\}$$

$$= \frac{EA}{l}(4\sin^2 30°)\{v_1\} = \frac{EA}{l}\{v_1\}$$

$$v_1 = -\frac{Pl}{EA} \text{ (in the direction of load } P\text{)}$$

From Eq. (4.24), the axial force for element 1-4 is

$$S_{1\text{-}4} = \frac{EA}{l}\lfloor \cos 30° \quad \sin 30° \rfloor \begin{Bmatrix} 0 \\ \dfrac{Pl}{EA} \end{Bmatrix} = \frac{P}{2} \quad \text{(tension)}$$

From symmetry we know that

$$S_{1\text{-}3} = S_{1\text{-}4} = -S_{1\text{-}2} = -S_{1\text{-}5} = \frac{P}{2}$$

Example 4.2 Indeterminate truss solved by reduction method

A symmetrical indeterminate truss subjected to a pair of parting forces is shown in Fig. 4.8. All bars have the same axial rigidity EA. There is no connection between the two diagonal bars. Find the displacements at all joints and the axial force for each bar.

Solution. The truss is composed of six bar elements, four nodal points, and eight degrees of freedom, as shown in Fig. 4.8. The element stiffness matrices are obtained based on Eq. (4.17) and Table 4.2.

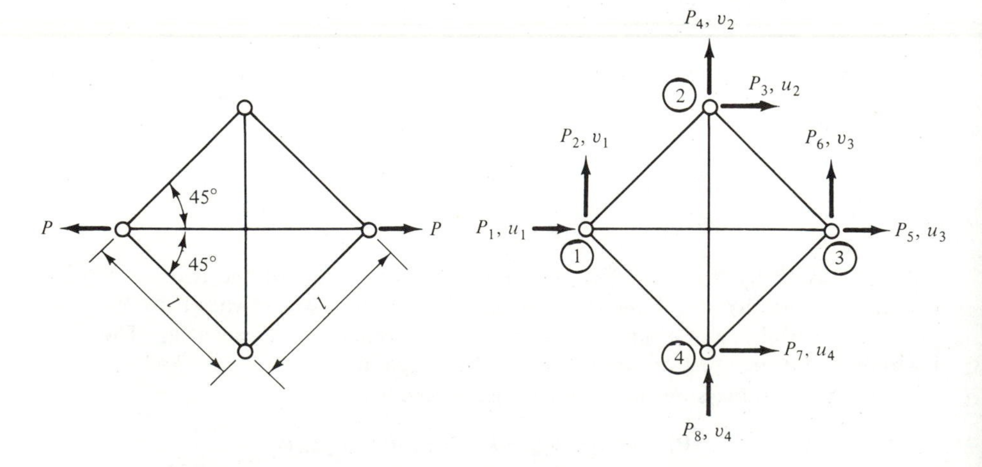

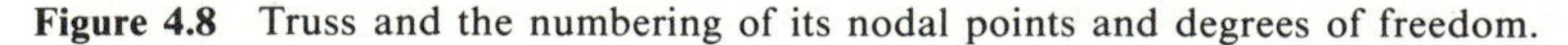

Figure 4.8 Truss and the numbering of its nodal points and degrees of freedom.

TABLE 4.2 Direction Cosines and Sines for Each Element

Element	ϕ	λ	μ	λ^2	μ^2	$\lambda\mu$	Length
1-2	45°	$\frac{1}{\sqrt{2}}$	$\frac{1}{\sqrt{2}}$	$\frac{1}{2}$	$\frac{1}{2}$	$\frac{1}{2}$	l
2-3, 1-4	−45°	$\frac{1}{\sqrt{2}}$	$-\frac{1}{\sqrt{2}}$	$\frac{1}{2}$	$\frac{1}{2}$	$-\frac{1}{2}$	l
3-4	−135°	$-\frac{1}{\sqrt{2}}$	$-\frac{1}{\sqrt{2}}$	$\frac{1}{2}$	$\frac{1}{2}$	$\frac{1}{2}$	l
1-3	0°	1	0	1	0	0	$\sqrt{2}l$
2-4	−90°	0	−1	0	1	0	$\sqrt{2}l$

For element 1-2,

$$\begin{Bmatrix} X_1 \\ Y_1 \\ X_2 \\ Y_2 \end{Bmatrix} = \frac{EA}{2l} \begin{bmatrix} 1 & 1 & -1 & -1 \\ 1 & 1 & -1 & -1 \\ -1 & -1 & 1 & 1 \\ -1 & -1 & 1 & 1 \end{bmatrix} \begin{Bmatrix} u_1 \\ v_1 \\ u_2 \\ v_2 \end{Bmatrix}$$

For element $i - j$ with $(i, j) = (2, 3)$ and $(1, 4)$,

$$\begin{Bmatrix} X_i \\ Y_i \\ X_j \\ Y_j \end{Bmatrix} = \frac{EA}{2l} \begin{bmatrix} 1 & -1 & -1 & 1 \\ -1 & 1 & 1 & -1 \\ -1 & 1 & 1 & -1 \\ 1 & -1 & -1 & 1 \end{bmatrix} \begin{Bmatrix} u_i \\ v_i \\ u_j \\ v_j \end{Bmatrix}$$

For element 3-4,

$$\begin{Bmatrix} X_3 \\ Y_3 \\ X_4 \\ Y_4 \end{Bmatrix} = \frac{EA}{2l} \begin{bmatrix} 1 & 1 & -1 & -1 \\ 1 & 1 & -1 & -1 \\ -1 & -1 & 1 & 1 \\ -1 & -1 & 1 & 1 \end{bmatrix} \begin{Bmatrix} u_3 \\ v_3 \\ u_4 \\ v_4 \end{Bmatrix}$$

For element 1-3,

$$\begin{Bmatrix} X_1 \\ Y_1 \\ X_3 \\ Y_3 \end{Bmatrix} = \frac{EA}{\sqrt{2}l} \begin{bmatrix} 1 & 0 & -1 & 0 \\ 0 & 0 & 0 & 0 \\ -1 & 0 & 1 & 0 \\ 0 & 0 & 0 & 0 \end{bmatrix} \begin{Bmatrix} u_1 \\ v_1 \\ u_3 \\ v_3 \end{Bmatrix}$$

For element 2-4,

$$\begin{Bmatrix} X_2 \\ Y_2 \\ X_4 \\ Y_4 \end{Bmatrix} = \frac{EA}{\sqrt{2}l} \begin{bmatrix} 0 & 0 & 0 & 0 \\ 0 & 1 & 0 & -1 \\ 0 & 0 & 0 & 0 \\ 0 & -1 & 0 & 1 \end{bmatrix} \begin{Bmatrix} u_2 \\ v_2 \\ u_4 \\ v_4 \end{Bmatrix}$$

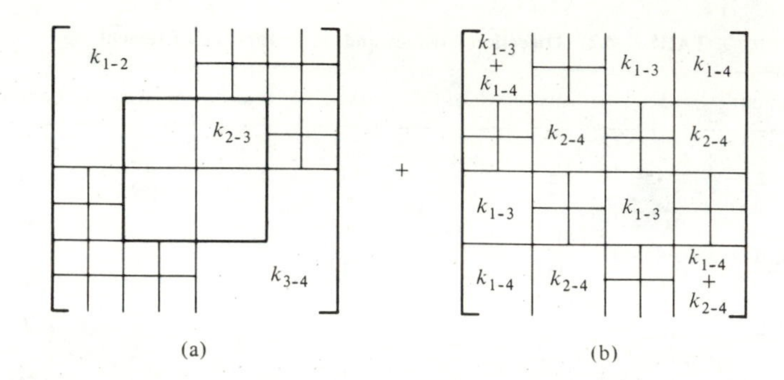

Figure 4.9 Positions the six element stiffness matrices move into during assemblage.

Figure 4.9 shows the positions where the six 4×4 element stiffness matrices move into during assemblage. Both 8×8 matrices shown in Fig. 4.9 are to be superimposed. Because the associated degrees of freedom in the total stiffness matrix are sequenced from 1 to 8, each angle ϕ in Table 4.2 is measured at i for element $i-j$ with $i<j$ so that during assemblage the sequence in each element stiffness matrix need not be reversed. After assemblage, we have

$$\begin{Bmatrix} P_1 \\ P_2 \\ P_3 \\ P_4 \\ P_5 \\ P_6 \\ P_7 \\ P_8 \end{Bmatrix} = \frac{EA}{2l} \begin{bmatrix} 2+\sqrt{2} & 1-1 & -1 & -1 & -\sqrt{2} & 0 & -1 & 1 \\ 1-1 & 1+1 & -1 & -1 & 0 & 0 & 1 & -1 \\ -1 & -1 & 1+1 & 1-1 & -1 & 1 & 0 & 0 \\ -1 & -1 & 1-1 & 2+\sqrt{2} & 1 & -1 & 0 & -\sqrt{2} \\ -\sqrt{2} & 0 & -1 & 1 & 2+\sqrt{2} & 1-1 & -1 & -1 \\ 0 & 0 & 1 & -1 & 1-1 & 1+1 & -1 & -1 \\ -1 & 1 & 0 & 0 & -1 & -1 & 1+1 & 1-1 \\ 1 & -1 & 0 & -\sqrt{2} & -1 & -1 & 1-1 & 2+\sqrt{2} \end{bmatrix} \begin{Bmatrix} u_1 \\ v_1 \\ u_2 \\ v_2 \\ u_3 \\ v_3 \\ u_4 \\ v_4 \end{Bmatrix}$$

Due to symmetry,

$$v_1 = u_2 = v_3 = u_4 = 0$$

The 8×8 matrix can be reduced by crossing out the second, third, sixth, and seventh rows and columns, respectively. We have

$$\begin{Bmatrix} P_1 = -P \\ P_4 = 0 \\ P_5 = P \\ P_8 = 0 \end{Bmatrix} = \frac{EA}{2l} \begin{bmatrix} 2+\sqrt{2} & -1 & -\sqrt{2} & 1 \\ -1 & 2+\sqrt{2} & 1 & -\sqrt{2} \\ -\sqrt{2} & 1 & 2+\sqrt{2} & -1 \\ 1 & -\sqrt{2} & -1 & 2+\sqrt{2} \end{bmatrix} \begin{Bmatrix} u_1 \\ v_2 \\ u_3 \\ v_4 \end{Bmatrix} \tag{4.49}$$

Equation (4.49) can be rearranged and partitioned as

$$\begin{Bmatrix} -P \\ P \\ \hline 0 \\ 0 \end{Bmatrix} = \frac{EA}{2l} \left[\begin{array}{cc|cc} 2+\sqrt{2} & -\sqrt{2} & -1 & 1 \\ -\sqrt{2} & 2+\sqrt{2} & 1 & -1 \\ \hline -1 & 1 & 2+\sqrt{2} & -\sqrt{2} \\ 1 & -1 & -\sqrt{2} & 2+\sqrt{2} \end{array}\right] \begin{Bmatrix} u_1 \\ u_3 \\ \hline v_2 \\ v_4 \end{Bmatrix} \tag{4.50}$$

Equation (4.50) may now be solved by using the reduction procedure described in Section 4.6.2.

$$\begin{Bmatrix} -P \\ P \end{Bmatrix} = \frac{EA}{2l}\begin{bmatrix} 2+\sqrt{2} & -\sqrt{2} \\ -\sqrt{2} & 2+\sqrt{2} \end{bmatrix}\begin{Bmatrix} u_1 \\ u_3 \end{Bmatrix} + \frac{EA}{2l}\begin{bmatrix} -1 & 1 \\ 1 & -1 \end{bmatrix}\begin{Bmatrix} v_2 \\ v_4 \end{Bmatrix} \tag{4.51a}$$

$$\begin{Bmatrix} 0 \\ 0 \end{Bmatrix} = \frac{EA}{2l}\begin{bmatrix} -1 & 1 \\ 1 & -1 \end{bmatrix}\begin{Bmatrix} u_1 \\ u_3 \end{Bmatrix} + \frac{EA}{2l}\begin{bmatrix} 2+\sqrt{2} & -\sqrt{2} \\ -\sqrt{2} & 2+\sqrt{2} \end{bmatrix}\begin{Bmatrix} v_2 \\ v_4 \end{Bmatrix} \tag{4.51b}$$

From Eq. (4.51b),

$$\begin{aligned}\begin{Bmatrix} v_2 \\ v_4 \end{Bmatrix} &= \frac{-1}{4+4\sqrt{2}}\begin{bmatrix} 2+\sqrt{2} & \sqrt{2} \\ \sqrt{2} & 2+\sqrt{2} \end{bmatrix}\begin{bmatrix} -1 & 1 \\ 1 & -1 \end{bmatrix}\begin{Bmatrix} u_1 \\ u_3 \end{Bmatrix} \\ &= \frac{1}{2+2\sqrt{2}}\begin{bmatrix} 1 & -1 \\ -1 & 1 \end{bmatrix}\begin{Bmatrix} u_1 \\ u_3 \end{Bmatrix} \end{aligned}\tag{4.51c}$$

Substituting Eq. (4.51c) into Eq. (4.51a) yields

$$\begin{aligned}\begin{Bmatrix} -P \\ P \end{Bmatrix} &= \frac{EA}{2l}\left\{\begin{bmatrix} 2+\sqrt{2} & -\sqrt{2} \\ -\sqrt{2} & 2+\sqrt{2} \end{bmatrix} + \frac{1}{1+\sqrt{2}}\begin{bmatrix} -1 & 1 \\ 1 & -1 \end{bmatrix}\right\}\begin{Bmatrix} u_1 \\ u_3 \end{Bmatrix} \\ &= \frac{EA}{2l}\begin{bmatrix} 3 & -1 \\ -1 & 3 \end{bmatrix}\begin{Bmatrix} u_1 \\ u_3 \end{Bmatrix}\end{aligned}$$

Hence

$$\begin{Bmatrix} u_1 \\ u_3 \end{Bmatrix} = \frac{l}{4EA}\begin{bmatrix} 3 & 1 \\ 1 & 3 \end{bmatrix}\begin{Bmatrix} -P \\ P \end{Bmatrix} = \frac{Pl}{2EA}\begin{Bmatrix} -1 \\ 1 \end{Bmatrix} \tag{4.52a}$$

and substituting Eq. (4.52a) into Eq. (4.51c) gives

$$\begin{Bmatrix} v_2 \\ v_4 \end{Bmatrix} = \frac{Pl}{2EA}\begin{Bmatrix} 1-\sqrt{2} \\ -1+\sqrt{2} \end{Bmatrix} \tag{4.52b}$$

which shows that $u_1 = -u_3$ and $v_2 = -v_4$.

The axial forces for the six bar elements are found by using Eq. (4.24):

$$S_{1\text{-}2} = \frac{EA}{l}\left\lfloor \frac{1}{\sqrt{2}} \quad \frac{1}{\sqrt{2}} \right\rfloor \begin{Bmatrix} \dfrac{Pl}{2EA} \\ \dfrac{(1-\sqrt{2})Pl}{2EA} \end{Bmatrix} = \frac{(\sqrt{2}-1)P}{2} \quad \text{(tension)}$$

$$S_{1\text{-}3} = \frac{EA}{\sqrt{2}l}\lfloor 1 \quad 0 \rfloor \begin{Bmatrix} \dfrac{Pl}{EA} \\ 0 \end{Bmatrix} = \frac{\sqrt{2}P}{2} \quad \text{(tension)}$$

$$S_{2\text{-}4} = \frac{EA}{\sqrt{2}l}\lfloor 0 \quad -1 \rfloor \begin{Bmatrix} 0 \\ \dfrac{(\sqrt{2}-1)Pl}{EA} \end{Bmatrix} = -\frac{(2-\sqrt{2})P}{2} \quad \text{(compression)}$$

Due to symmetry, we know that

$$S_{1\text{-}2} = S_{2\text{-}3} = S_{3\text{-}4} = S_{1\text{-}4} = \frac{(\sqrt{2}-1)P}{2}$$

We can easily find that equilibrium for each joint due to the external loads and the axial forces in the bars is satisfied.

Example 4.3 Alternative solution of Example 4.2 based on known relations among displacements

Because the geometry of the truss and the loadings in Example 4.2 are both symmetrical, the assembled total stiffness equations (4.49) may be solved based on the displacement relations as defined in the displacement column of the following equations:

$$\begin{Bmatrix} -P \\ 0 \\ P \\ 0 \end{Bmatrix} = \frac{EA}{2l} \begin{bmatrix} 2+\sqrt{2} & -1 & -\sqrt{2} & 1 \\ -1 & 2+\sqrt{2} & 1 & -\sqrt{2} \\ -\sqrt{2} & 1 & 2+\sqrt{2} & -1 \\ 1 & -\sqrt{2} & -1 & 2+\sqrt{2} \end{bmatrix} \begin{Bmatrix} u_1 \\ v_2 \\ u_3 = -u_1 \\ v_4 = -v_2 \end{Bmatrix} \tag{4.53}$$

Replacing u_3 by $-u_1$ and v_4 by $-v_2$ and multiplying out Eqs. (4.53) yields

$$\begin{Bmatrix} -P \\ 0 \\ P \\ 0 \end{Bmatrix} = \frac{EA}{2l} \begin{bmatrix} 2+2\sqrt{2} & -2 \\ -2 & 2+2\sqrt{2} \\ -2-2\sqrt{2} & 2 \\ 2 & -2-2\sqrt{2} \end{bmatrix} \begin{Bmatrix} u_1 \\ v_2 \end{Bmatrix} \tag{4.54}$$

The foregoing reduction is done by simply subtracting column 3 from column 1 to form column 1 and subtracting column 4 from column 2 to form column 2 in the stiffness matrix in Eqs. (4.53). In the resulting four equations (4.54), the first set of two equations the same as the second set of two equations. We can choose either set.

$$\begin{Bmatrix} -P \\ 0 \end{Bmatrix} = \frac{EA}{2l} \begin{bmatrix} 2+2\sqrt{2} & -2 \\ -2 & 2+2\sqrt{2} \end{bmatrix} \begin{Bmatrix} u_1 \\ v_2 \end{Bmatrix}$$

Then

$$\begin{Bmatrix} u_1 \\ v_2 \end{Bmatrix} = \frac{l}{4(1+\sqrt{2})EA} \begin{bmatrix} 2+2\sqrt{2} & 2 \\ 2 & 2+2\sqrt{2} \end{bmatrix} \begin{Bmatrix} -P \\ 0 \end{Bmatrix}$$

$$= \frac{Pl}{2EA} \begin{Bmatrix} -1 \\ 1-\sqrt{2} \end{Bmatrix} = \begin{Bmatrix} -u_3 \\ -v_4 \end{Bmatrix}$$

which are the same as those obtained in Example 4.2.

Example 4.4 Symmetrical seven-bar truss

A symmetrical truss under a concentrated load P at joint 3 is shown in Fig. 4.10. All bars have the same axial rigidity EA and length l. Find the displacements at all joints and the axial forces in all bars.

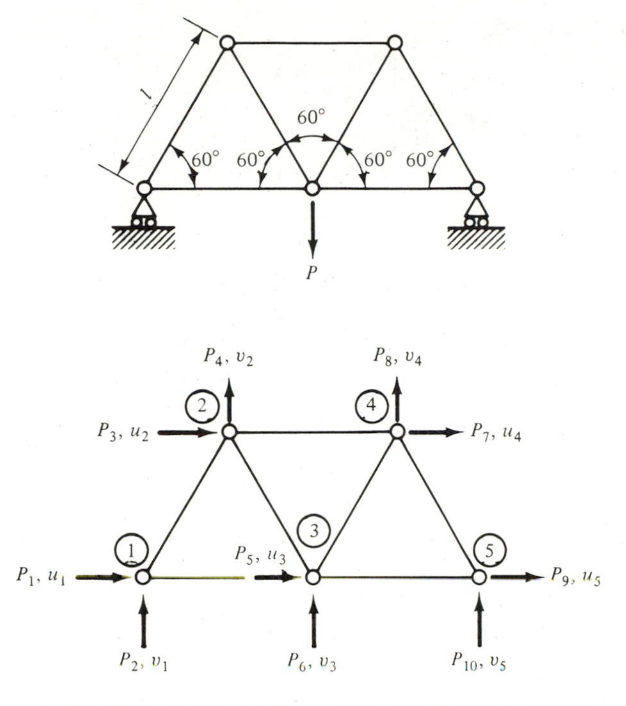

Figure 4.10 Truss and the numbering of its nodal points and degrees of freedom.

Solution. The seven-element truss has five nodal points and 10 degrees of freedom, as numbered in Fig. 4.10. The element stiffness matrices are obtained based on Eq. (4.17) and Table 4.3.

For elements 1-2 and 3-4,

$$[\mathbf{k}] = \frac{EA}{4l}\begin{bmatrix} 1 & \sqrt{3} & -1 & -\sqrt{3} \\ \sqrt{3} & 3 & -\sqrt{3} & -3 \\ -1 & -\sqrt{3} & 1 & \sqrt{3} \\ -\sqrt{3} & -3 & \sqrt{3} & 3 \end{bmatrix}$$

TABLE 4.3 Direction Cosines and Sines for Each Element

Element	ϕ	λ	μ	λ^2	μ^2	$\lambda\mu$	Length
1-2, 3-4	60°	$\frac{1}{2}$	$\frac{\sqrt{3}}{2}$	$\frac{1}{4}$	$\frac{3}{4}$	$\frac{\sqrt{3}}{4}$	l
2-3, 4-5	−60°	$\frac{1}{2}$	$\frac{-\sqrt{3}}{2}$	$\frac{1}{4}$	$\frac{3}{4}$	$\frac{-\sqrt{3}}{4}$	l
1-3, 2-4, 3-5	0°	1	0	1	0	0	l

For elements 2-3 and 4-5,

$$[\mathbf{k}] = \frac{EA}{4l}\begin{bmatrix} 1 & -\sqrt{3} & -1 & \sqrt{3} \\ -\sqrt{3} & 3 & \sqrt{3} & -3 \\ -1 & \sqrt{3} & 1 & -\sqrt{3} \\ \sqrt{3} & -3 & -\sqrt{3} & 3 \end{bmatrix}$$

For elements 1-3, 2-4, and 3-5,

$$[\mathbf{k}] = \frac{EA}{4l}\begin{bmatrix} 4 & 0 & -4 & 0 \\ 0 & 0 & 0 & 0 \\ -4 & 0 & 4 & 0 \\ 0 & 0 & 0 & 0 \end{bmatrix}$$

Because $v_1 = u_3 = v_5 = 0$, we have to assemble a set of seven stiffness equations:

$$\begin{Bmatrix} P_1 \\ P_3 \\ P_4 \\ P_6 \\ P_7 \\ P_8 \\ P_9 \end{Bmatrix} = \frac{EA}{4l}\begin{bmatrix} 1+4 & -1 & -\sqrt{3} & 0 & 0 & 0 & 0 \\ -1 & 1+1+4 & \sqrt{3}-\sqrt{3} & \sqrt{3} & -4 & 0 & 0 \\ -\sqrt{3} & \sqrt{3}-\sqrt{3} & 3+3 & -3 & 0 & 0 & 0 \\ 0 & \sqrt{3} & -3 & 3+3 & -\sqrt{3} & -3 & 0 \\ 0 & -4 & 0 & -\sqrt{3} & 1+1+4 & \sqrt{3}-\sqrt{3} & -1 \\ 0 & 0 & 0 & -3 & \sqrt{3}-\sqrt{3} & 3+3 & \sqrt{3} \\ 0 & 0 & 0 & 0 & -1 & \sqrt{3} & 1+4 \end{bmatrix}\begin{Bmatrix} u_1 \\ u_2 \\ v_2 \\ v_3 \\ u_4 \\ v_4 \\ u_5 \end{Bmatrix}$$

The total 7×7 stiffness matrix can be reduced to a 7×4 matrix by using the following antisymmetrical and symmetrical conditions:

$$u_1 = -u_5 \qquad u_2 = -u_4 \qquad v_2 = v_4$$

The reduction is done by subtracting column 7 from column 1 to form column 1, subtracting column 5 from column 2 to form column 2, adding column 6 to column 3 to form column 3, and preserving column 4.

$$\begin{Bmatrix} P_1 = 0 \\ P_3 = 0 \\ P_4 = 0 \\ P_6 = -P \\ P_7 = 0 \\ P_8 = 0 \\ P_9 = 0 \end{Bmatrix} = \frac{EA}{4l}\begin{bmatrix} 5 & -1 & -\sqrt{3} & 0 \\ -1 & 10 & 0 & \sqrt{3} \\ -\sqrt{3} & 0 & 6 & -3 \\ 0 & 2\sqrt{3} & -6 & 6 \\ 1 & -10 & 0 & -\sqrt{3} \\ -\sqrt{3} & 0 & 6 & -3 \\ -5 & 1 & \sqrt{3} & 0 \end{bmatrix}\begin{Bmatrix} u_1 \\ u_2 \\ v_2 \\ v_3 \end{Bmatrix}$$

We can choose either the first four equations or the last four equations to solve for the displacements. Let us choose the first four equations.

$$\begin{Bmatrix} P_1 = 0 \\ P_3 = 0 \\ \hline P_4 = 0 \\ P_6 = -P \end{Bmatrix} = \frac{EA}{4l}\left[\begin{array}{cc|cc} 5 & -1 & -\sqrt{3} & 0 \\ -1 & 10 & 0 & \sqrt{3} \\ \hline -\sqrt{3} & 0 & 6 & -3 \\ 0 & 2\sqrt{3} & -6 & 6 \end{array}\right]\begin{Bmatrix} u_1 \\ u_2 \\ \hline v_2 \\ v_3 \end{Bmatrix} \tag{4.55}$$

After the foregoing reduction, the 4×4 stiffness matrix is no longer symmetric. Multiplying out the first two equations in Eq. (4.55) gives

$$\begin{Bmatrix} u_1 \\ u_2 \end{Bmatrix} = \frac{\sqrt{3}}{49} \begin{bmatrix} 10 & -1 \\ 1 & -5 \end{bmatrix} \begin{Bmatrix} v_2 \\ v_3 \end{Bmatrix} \tag{4.56}$$

Multiplying out the last two equations and making use of the relation in Eq. (4.56) gives

$$\begin{Bmatrix} P_4 = 0 \\ P_6 = -P \end{Bmatrix} = \frac{6EA}{49l} \begin{bmatrix} 11 & -6 \\ -12 & 11 \end{bmatrix} \begin{Bmatrix} v_2 \\ v_3 \end{Bmatrix}$$

The solution is

$$\begin{Bmatrix} v_2 \\ v_3 \end{Bmatrix} = \frac{l}{6EA} \begin{bmatrix} 11 & 6 \\ 12 & 11 \end{bmatrix} \begin{Bmatrix} 0 \\ -P \end{Bmatrix} = \frac{Pl}{6EA} \begin{Bmatrix} -6 \\ -11 \end{Bmatrix}$$

and

$$\begin{Bmatrix} u_1 \\ u_2 \end{Bmatrix} = \frac{\sqrt{3}}{49} \begin{bmatrix} 10 & -1 \\ 1 & -5 \end{bmatrix} \left(\frac{Pl}{6EA}\right) \begin{Bmatrix} -6 \\ -11 \end{Bmatrix} = \frac{\sqrt{3}Pl}{6EA} \begin{Bmatrix} -1 \\ 1 \end{Bmatrix}$$

The axial forces in the bars are

$$S_{1\text{-}2} = \frac{EA}{l} \left\lfloor \frac{1}{2} \quad \frac{\sqrt{3}}{2} \right\rfloor \begin{Bmatrix} u_2 - u_1 \\ v_2 - v_1 \end{Bmatrix} = -\frac{\sqrt{3}P}{3} = S_{4\text{-}5} \quad \text{(compression)}$$

$$S_{2\text{-}3} = \frac{EA}{l} \left\lfloor \frac{1}{2} \quad \frac{-\sqrt{3}}{2} \right\rfloor \begin{Bmatrix} u_3 - u_2 \\ v_3 - v_2 \end{Bmatrix} = \frac{\sqrt{3}P}{3} = S_{3\text{-}4} \quad \text{(tension)}$$

$$S_{2\text{-}4} = \frac{EA}{l} \lfloor 1 \quad 0 \rfloor \begin{Bmatrix} u_4 - u_2 \\ v_4 - v_2 \end{Bmatrix} = -\frac{\sqrt{3}P}{3} \quad \text{(compression)}$$

$$S_{1\text{-}3} = \frac{EA}{l} \lfloor 1 \quad 0 \rfloor \begin{Bmatrix} u_3 - u_1 \\ v_3 - v_1 \end{Bmatrix} = \frac{\sqrt{3}P}{6} = S_{3\text{-}5} \quad \text{(tension)}$$

The reader should check all these axial forces to see whether they satisfy equilibrium at all joints.

Example 4.5 Statically indeterminate truss with a joint sliding along a slope

A four-bar, five-joint statically indeterminate truss with four joints fixed and one joint confined to slide along a 45° slope is shown in Fig. 4.11. All bars have the same axial rigidity EA. Find the reaction force normal to the slope, the displacements at joint 1, and the axial forces in all bars. Table 4.4 lists the direction cosines and sines for each element.

Solution. Because four out of five joints are fixed, the structure has only two degrees of freedom. The total stiffness equations can be assembled as

$$\begin{Bmatrix} P_1 \\ P_2 \end{Bmatrix} = \begin{bmatrix} K_{11} & K_{12} \\ K_{21} & K_{22} \end{bmatrix} \begin{Bmatrix} u_1 \\ v_1 \end{Bmatrix}$$

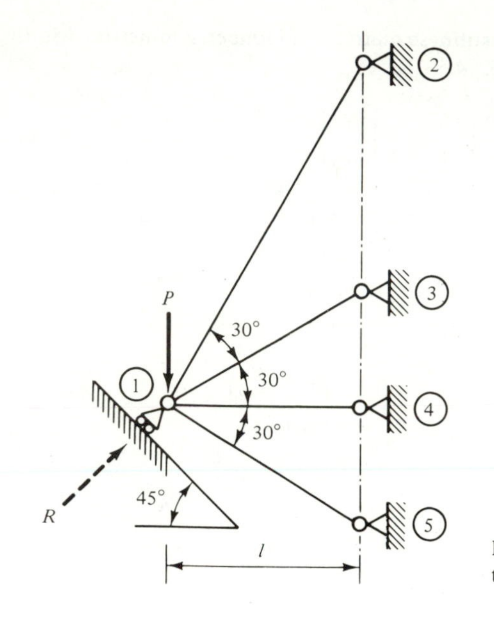

Figure 4.11 Truss with a joint confined to slide along a slope.

where

$$K_{11} = \sum \frac{EA}{L} \lambda^2 = \frac{9+6\sqrt{3}}{8} \frac{EA}{l}$$

$$K_{12} = K_{21} = \sum \frac{EA}{L} \lambda\mu = \frac{\sqrt{3}}{8} \frac{EA}{l}$$

$$K_{22} = \sum \frac{EA}{L} \mu^2 = \frac{3+2\sqrt{3}}{8} \frac{EA}{l}$$

Because joint 1 is confined to slide along a 45° slope as shown in Fig. 4.11, we have

$$v_1 = -u_1$$

TABLE 4.4 Direction Cosines and Sines for Each Element

Element	ϕ	λ	μ	λ^2	μ^2	$\lambda\mu$	Length
1-2	60°	$\frac{1}{2}$	$\frac{\sqrt{3}}{2}$	$\frac{1}{4}$	$\frac{3}{4}$	$\frac{\sqrt{3}}{4}$	$2l$
1-3	30°	$\frac{\sqrt{3}}{2}$	$\frac{1}{2}$	$\frac{3}{4}$	$\frac{1}{4}$	$\frac{\sqrt{3}}{4}$	$2l/\sqrt{3}$
1-4	0°	1	0	1	0	0	l
1-5	−30°	$\frac{\sqrt{3}}{2}$	$\frac{-1}{2}$	$\frac{3}{4}$	$\frac{1}{4}$	$\frac{-\sqrt{3}}{4}$	$2l/\sqrt{3}$

Because there is a reaction force R normal to the slope, the external loads are

$$P_1 = R\cos 45^\circ = \frac{R}{\sqrt{2}}$$

$$P_2 = R\sin 45^\circ - P = \frac{R}{\sqrt{2}} - P$$

The stiffness equations become

$$\begin{Bmatrix} \dfrac{R}{\sqrt{2}} \\ \dfrac{R}{\sqrt{2}} - P \end{Bmatrix} = \frac{EA}{8l}\begin{bmatrix} 9+6\sqrt{3} & \sqrt{3} \\ \sqrt{3} & 3+2\sqrt{3} \end{bmatrix}\begin{Bmatrix} u_1 \\ v_1 = -u_1 \end{Bmatrix}$$

or

$$\begin{Bmatrix} \dfrac{R}{\sqrt{2}} \\ \dfrac{R}{\sqrt{2}} - P \end{Bmatrix} = \frac{EA}{8l}\begin{Bmatrix} 9+5\sqrt{3} \\ -3-\sqrt{3} \end{Bmatrix}\{u_1\}$$

The two equations have two unknowns, R and u_1. Reaction force R can be eliminated by subtracting the second equation from the first:

$$P = (12+6\sqrt{3})\frac{EA}{8l}u_1$$

the displacements of joint 1 are

$$u_1 = -v_1 = \frac{4}{6+3\sqrt{3}}\frac{Pl}{EA} = \frac{4(2-\sqrt{3})}{3}\frac{Pl}{EA}$$

and the reaction force normal to the slope is obtained from either of the two stiffness equations:

$$R = \frac{\sqrt{2}(3+\sqrt{3})}{6}P$$

The axial forces in the bars are

$$S_{1\text{-}2} = \frac{EA}{2l}\begin{bmatrix} \frac{1}{2} & \frac{\sqrt{3}}{2} \end{bmatrix}\begin{Bmatrix} u_2 - u_1 \\ v_2 - v_1 \end{Bmatrix} = \frac{(3\sqrt{3}-5)P}{3} \quad \text{(tension)}$$

$$S_{1\text{-}3} = \frac{\sqrt{3}EA}{2l}\begin{bmatrix} \frac{\sqrt{3}}{2} & \frac{1}{2} \end{bmatrix}\begin{Bmatrix} u_3 - u_1 \\ v_3 - v_1 \end{Bmatrix} = -\frac{(9-5\sqrt{3})P}{3} \quad \text{(compression)}$$

$$S_{1\text{-}4} = \frac{EA}{l}\lfloor 1 \quad 0 \rfloor\begin{Bmatrix} u_4 - u_1 \\ v_4 - v_1 \end{Bmatrix} = -\frac{4(2-\sqrt{3})P}{3} \quad \text{(compression)}$$

$$S_{1\text{-}5} = \frac{\sqrt{3}EA}{2l}\begin{bmatrix} \frac{\sqrt{3}}{2} & -\frac{1}{2} \end{bmatrix}\begin{Bmatrix} u_5 - u_1 \\ v_5 - v_1 \end{Bmatrix} = -\frac{(3-\sqrt{3})P}{3} \quad \text{(compression)}$$

The reader should check these axial forces to see whether or not they satisfy equilibrium at joint 1.

Example 4.6 Truss bar with a temperature rise

A three-bar truss is shown in Fig. 4.12. All bars have the same axial stiffness *EA*. If the temperature of bar 1-2 is increased by an amount of ΔT above a certain specified uniform temperature of the truss, find the displacements of joint 1 and the axial forces in the three bars.

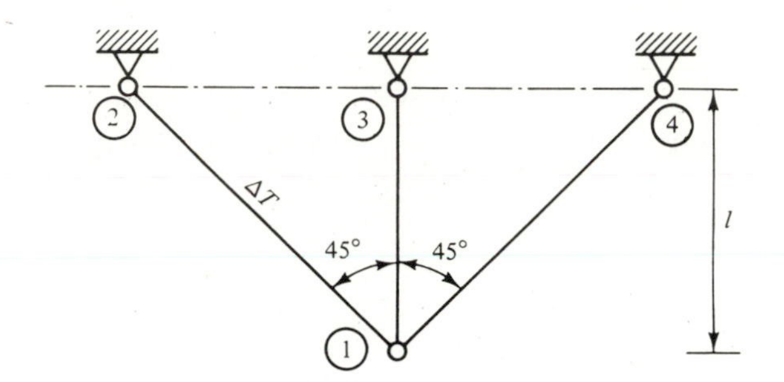

Figure 4.12 Three-bar truss with bar 1–2 subjected to a temperature rise ΔT.

Solution. This problem can be solved according to the two steps described in Section 4.6.4.

Step 1: Assuming that joint 1 is restrained, the axial forces developed in the three bars due to the temperature rise ΔT in bar 1-2 are, from Eq. 4.47b,

$$S'_{1\text{-}3} = S'_{1\text{-}4} = 0$$

$$S'_{1\text{-}2} = -EA\alpha\,\Delta T$$

where the primes are used to indicate that the forces are for step 1 and the negative sign indicates that bar 1-2 is in compression when joint 1 is restrained.

The nodal forces required to restrain nodal point 1 are, from Eq. (4.48),

$$X_1 = S'_{1\text{-}2}(-\cos 135°) = \frac{-EA\alpha\,\Delta T}{\sqrt{2}}$$
$$Y_1 = S'_{1\text{-}2}(-\sin 135°) = \frac{EA\alpha\,\Delta T}{\sqrt{2}} \tag{4.57}$$

If there is more than one bar with temperature rise, the restraining forces must be summed up at each common joint.

Step 2: Based on Table 4.5 and Eq. (4.17), the total stiffness equations are assembled as

$$\begin{Bmatrix} P_1 = \dfrac{EA\alpha\,\Delta T}{\sqrt{2}} \\ P_2 = \dfrac{-EA\alpha\,\Delta T}{\sqrt{2}} \end{Bmatrix} = \frac{EA}{\sqrt{2}l}\begin{bmatrix} 1 & 0 \\ 0 & 1+\sqrt{2} \end{bmatrix}\begin{Bmatrix} u_1 \\ v_1 \end{Bmatrix}$$

TABLE 4.5 Direction Cosines and Sines for Each Element

Element	ϕ	λ	μ	λ^2	$\lambda\mu$	μ^2	Length
1-2	135°	$\frac{-1}{\sqrt{2}}$	$\frac{1}{\sqrt{2}}$	$\frac{1}{2}$	$\frac{-1}{2}$	$\frac{1}{2}$	$\sqrt{2}l$
1-3	90°	0	1	0	0	1	l
1-4	45°	$\frac{1}{\sqrt{2}}$	$\frac{1}{\sqrt{2}}$	$\frac{1}{2}$	$\frac{1}{2}$	$\frac{1}{2}$	$\sqrt{2}l$

where the external loads are equal in magnitude but opposite in sign to the restraining forces as given in Eq. (4.57). The final solution for displacements at joint 1 is thus obtained:

$$\begin{Bmatrix} u_1 \\ v_1 \end{Bmatrix} = \frac{l}{EA}\begin{bmatrix} \sqrt{2} & 0 \\ 0 & 2-\sqrt{2} \end{bmatrix}\begin{Bmatrix} \dfrac{EA\alpha\,\Delta T}{\sqrt{2}} \\ -\dfrac{EA\alpha\,\Delta T}{\sqrt{2}} \end{Bmatrix} = \alpha\,\Delta Tl\begin{Bmatrix} 1 \\ 1-\sqrt{2} \end{Bmatrix}$$

The axial forces due to these displacements are

$$S''_{1\text{-}2} = \frac{EA}{\sqrt{2}l}\left\lfloor -\frac{1}{\sqrt{2}} \quad \frac{1}{\sqrt{2}} \right\rfloor \begin{Bmatrix} u_2-u_1 \\ v_2-v_1 \end{Bmatrix} = \frac{\sqrt{2}}{2}EA\alpha\,\Delta T$$

$$S''_{1\text{-}3} = \frac{EA}{l}\lfloor 0 \quad 1 \rfloor \begin{Bmatrix} u_3-u_1 \\ v_3-v_1 \end{Bmatrix} = (\sqrt{2}-1)EA\alpha\,\Delta T$$

$$S''_{1\text{-}4} = \frac{EA}{\sqrt{2}l}\left\lfloor \frac{1}{\sqrt{2}} \quad \frac{1}{\sqrt{2}} \right\rfloor \begin{Bmatrix} u_4-u_1 \\ v_4-v_1 \end{Bmatrix} = -\frac{2-\sqrt{2}}{2}EA\alpha\,\Delta T$$

where the double primes are used to indicate that the forces are for step 2.

The final axial forces in the three bars are obtained by superimposing the forces obtained in both steps,

$$S_{1\text{-}2} = S'_{1\text{-}2} + S''_{1\text{-}2} = -\frac{2-\sqrt{2}}{2}EA\alpha\,\Delta T \quad \text{(compression)}$$

$$S_{1\text{-}3} = S'_{1\text{-}3} + S''_{1\text{-}3} = (\sqrt{2}-1)EA\alpha\,\Delta T \quad \text{(tension)}$$

$$S_{1\text{-}4} = S'_{1\text{-}4} + S''_{1\text{-}4} = -\frac{2-\sqrt{2}}{2}EA\alpha\,\Delta T \quad \text{(compression)}$$

The three axial forces satisfy equilibrium at joint 1 while joint 1 is subjected to no external loads.

4.8 CONCLUDING REMARKS

In this chapter we have demonstrated the basic procedures for element formulation, assemblage, and solution of the finite element stiffness method by the use of truss bar elements and various kinds of examples. The sign conventions, numbering system, assemblage steps, and methods for solving simultaneous equations are introduced in a consistent and systematic fashion which is ideally suited for computer programming.

The advantages of the truss bar element formulations and the methods of solution introduced in this chapter are summarized as follows:

1. The orientation and length of each bar element can be arbitrary (i.e., the geometry of the truss structure can be arbitrary).
2. The boundary conditions can be arbitrary. For example, a joint can be allowed to displace along a locus such as that demonstrated in Example 4.5.
3. The loading conditions can be arbitrary. The given conditions can also be a mixture of loadings and displacements or merely displacements at any joint.
4. Indeterminacy or redundancy due to extra bars presents no difficulty in the assemblage. Thus the indeterminate structures can be solved just as straightforwardly as the determinate ones with little or no additional effort.
5. More fixed boundary conditions mean fewer degrees of freedom and consequently, less computational effort. This is contrary to the common traditional method, such as the method of least work, where more fixed boundary conditions mean higher degrees of indeterminacy and more complication in the solution of the problem.
6. The special class of problem of truss with misfit or temperature gradient can be solved as straightforwardly as the ordinary cases.

For the fundamental development of the methods described in this chapter, the reader is referred to Ref. 4.1. For the method of superposition that treats the thermal stress problems, the reader is referred to any good text on elasticity (for example, Ref. 4.3).

A Fortran program for the static analysis of plane trusses, a user's manual, and the input and output data for an example of a five-bar truss are given in Section 13.1.

REFERENCES

4.1. Turner, M. J., Clough, R. W., Martin, H. C., and Topp, L. J., "Stiffness and Deflection Analysis of Complex Structures," *Journal of Aeronautical Science*, Vol. 23, No. 9, Sept. 1956, pp. 805–823.

4.2. Martin, H. C., *Introduction to Matrix Methods of Structural Analysis*, McGraw-Hill Book Company, New York, 1966, Chap. 3.

4.3. Timoshenko, S. P., and Goodier, J. N., *Theory of Elasticity*, 2nd ed., McGraw-Hill Book Company, New York, 1951, Chap. 14.

PROBLEMS

4.1. A symmetrical truss is shown in Fig. P4.1. Each member has the same EA. Find the displacements of joint 1 and all the member axial forces. Also check the equilibrium of joint 1.

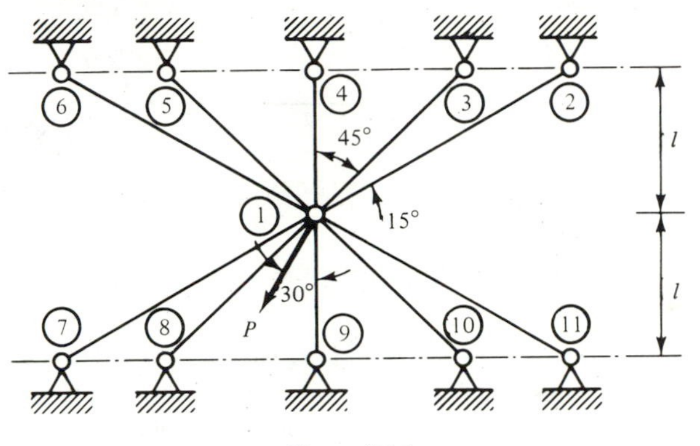

Figure P4.1

4.2. A symmetrical steel truss is shown in Fig. P4.2. The modulus of elasticity for steel is $E = 30 \times 10^6$ psi. The cross-sectional areas are $A_{1\text{-}2} = A_{2\text{-}4} = 2.0 \text{ in.}^2$, $A_{1\text{-}3} = A_{3\text{-}4} = 1.2 \text{ in.}^2$, and $A_{2\text{-}3} = 0.8 \text{ in.}^2$. Find the displacements at joints 2 and 3, all the member axial forces, and the reaction forces at joint 1 or 4. Also check the equilibrium of joints 2 and 3.

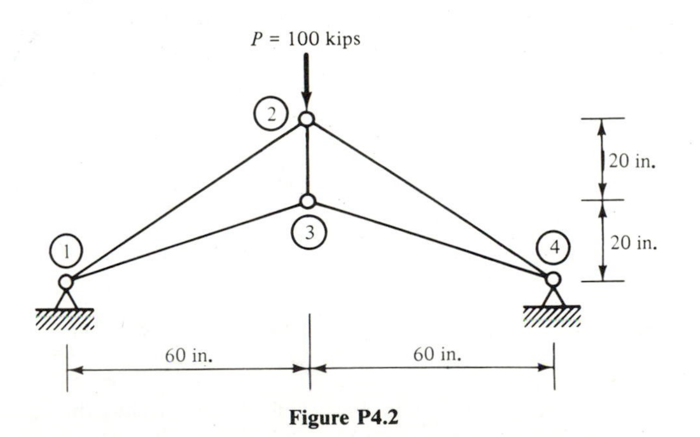

Figure P4.2

4.3. A three-bar aluminum truss is shown in Fig. P4.3. The modulus of elasticity for aluminum is $E = 10 \times 10^6$ psi. The cross-sectional areas are $A_{1\text{-}2} = 0.36 \text{ in.}^2$, $A_{2\text{-}3} = 0.4 \text{ in.}^2$, and $A_{1\text{-}3} = 0.6 \text{ in.}^2$. Find **(a)** the total assembled stiffness equations, **(b)** all the nodal displacements, **(c)** the axial stress (in psi) in each bar, **(d)** the reaction forces at joints 1 and 2, and **(e)** the equilibrium at joint 3. *Note*: A computer or at least a calculator is needed for inverting the 3×3 stiffness matrix.

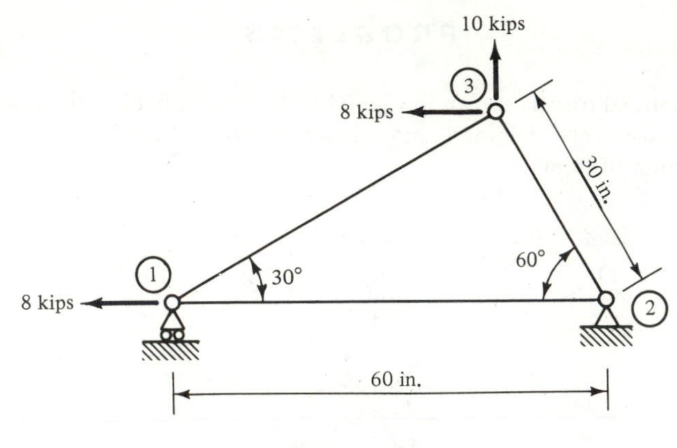

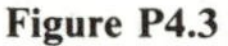
Figure P4.3

4.4. A five-bar statically indeterminate truss is shown in Fig. P4.4. Each bar has the same EA. Find nodal displacements, member forces, and reaction forces. Also check equilibrium at every joint. The reduction procedure is suggested.

4.5. The symmetrical truss as shown in Fig. P4.5 is loaded in an antisymmetrical way. Each bar has the same EA. The stiffness matrix can be reduced from 4×4 to 2×2. Find the nodal displacements, member forces, and the reaction forces. Also check the equilibrium of every joint.

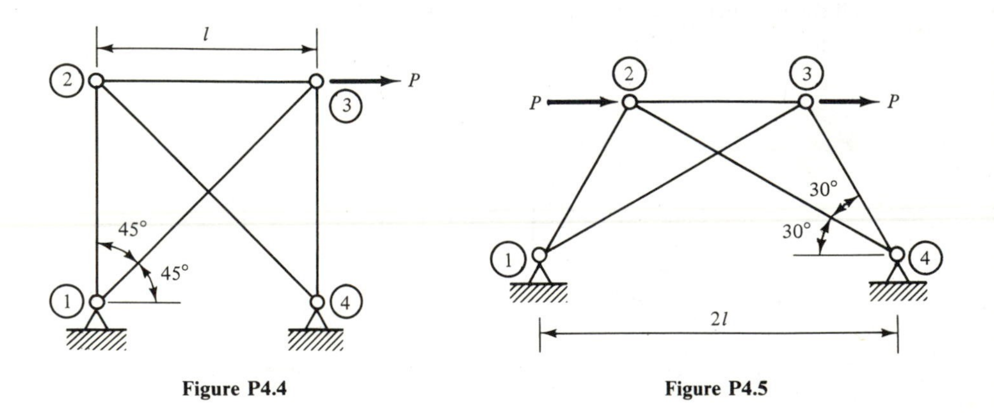

Figure P4.4 **Figure P4.5**

4.6. A symmetrical truss is shown in Fig. P4.6. Each bar has the same EA. The stiffness matrix can be reduced to 3×3 by making use of symmetry. Find the nodal displacements, member forces, and the reaction forces. Also check the equilibrium at every joint.

4.7. A symmetrical truss is shown in Fig. P4.7. Each bar has the same EA. This problem is similar to Problem 4.6 with the addition of four bars but no extra degrees of freedom. Find the nodal displacements, member forces, and reaction forces. Also check the equilibrium of every joint.

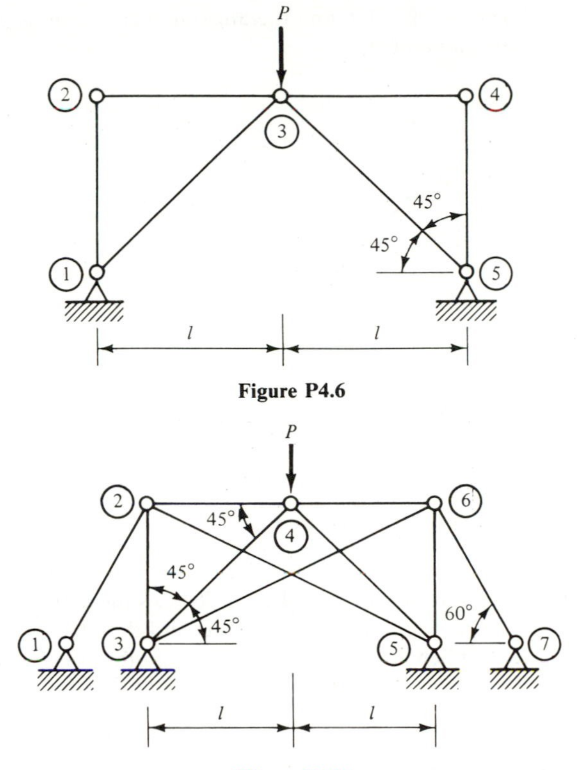

Figure P4.6

Figure P4.7

4.8. The truss as shown in Fig. P4.8 has eight degrees of freedom which can be reduced to four by using symmetrical conditions. The four equations can be solved by using reduction procedure. Each bar has the same EA. Find the nodal displacements, member forces, and reaction forces. Also check the equilibrium of every joint.

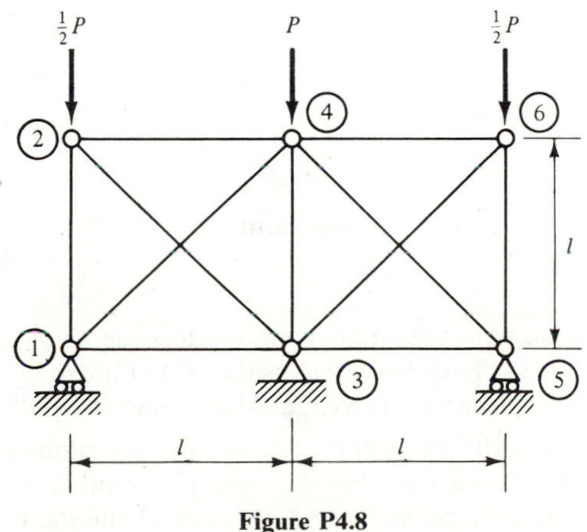

Figure P4.8

4.9. For the truss shown in Fig. P4.9, find the angle α such that u_3 is maximum. Also find the maximum value of u_3.

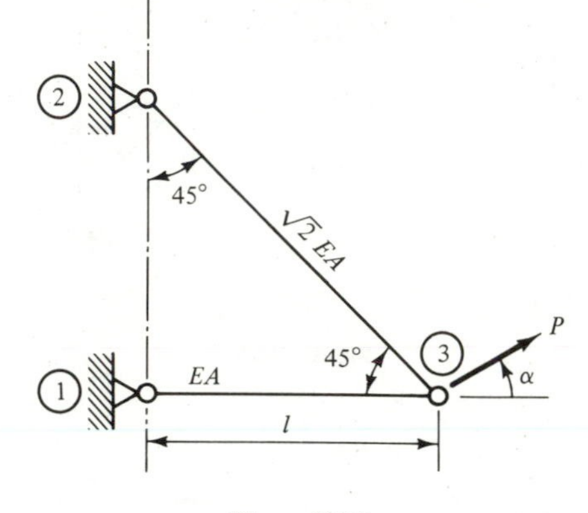

Figure P4.9

4.10. Figure P4.10 shows a truss with joint 4 confined to slide along a 60° slope. Each bar has the same EA. Find the displacements and the reaction force R at joint 4. Also find the member forces and check the equilibrium of joint 4.

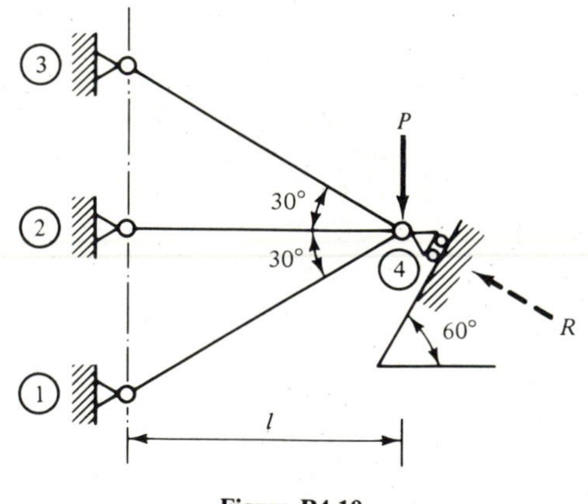

Figure P4.10

4.11. A symmetrical truss with two joints confined to slide along two separate slopes is shown in Fig. P4.11. Each bar has the same EA. Find the nodal displacements, member forces, and the reaction force R. Also check the equilibrium at every joint.

4.12. A symmetrical truss with two joints confined to slide along two separate slopes is shown in Fig. P4.12. Each bar has the same EA. Find the nodal displacements, member forces, and the reaction force R. Also check the equilibrium at every joint.

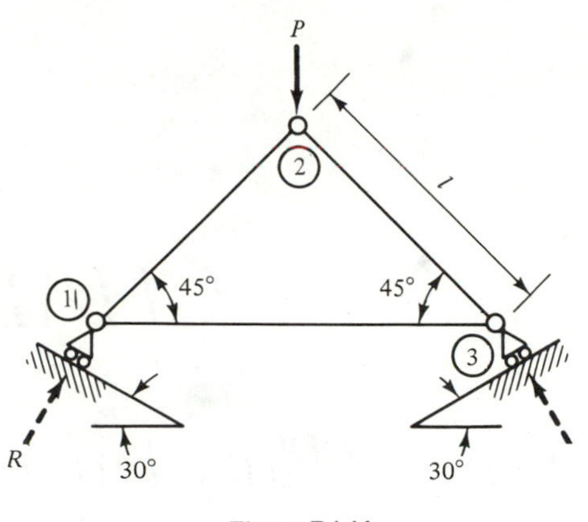

Figure P4.11

4.13. Assuming that the temperature in bars 1-2 and 1-11 of the truss shown in Fig. P4.1 is increased by ΔT degrees above a certain uniform temperature of the truss, that external load P is absent, and that each bar has the same EA and α, find the displacements at joint 1 and all the member forces. Also check the equilibrium of joint 1.

4.14. Assuming that the temperatures in bars 1-3 and 2-4 of the truss shown in Fig. P4.4 are increased and decreased by ΔT degrees, respectively, that there is no external load, and that each bar has the same EA and α, find the nodal displacements and member forces. Also check the equilibrium of every joint.

4.15. Assuming that the temperature in bar 1-4 of the truss shown in Fig. P4.10 is increased by ΔT degrees, that external load P is absent, and that each bar has the same EA and α, find the nodal displacements, member forces, and reaction force R. Also check the equilibrium of joint 4.

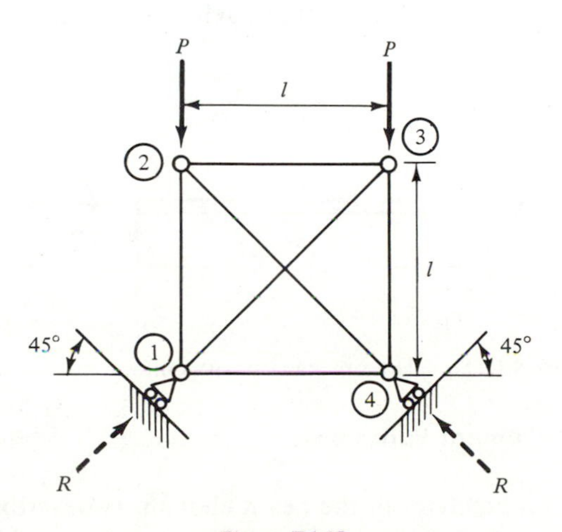

Figure P4.12

CHAPTER 5

Beam and Plane Frame Elements

5.1 UNIFORM STRAIGHT BEAM ELEMENT

A straight beam element of uniform cross section is shown in Fig. 5.1. The longitudinal axis of the element lies along the x axis. The element has constant moment of inertia I, modulus of elasticity E, and length L. The element is assumed to have two degrees of freedom at each end (nodal point): a transverse deflection v and an angle of rotation or slope θ (or $\partial v/\partial x$). Corresponding to the two degrees of freedom, v and θ, a transverse shear force Y and a bending moment M, respectively, act at each nodal point.

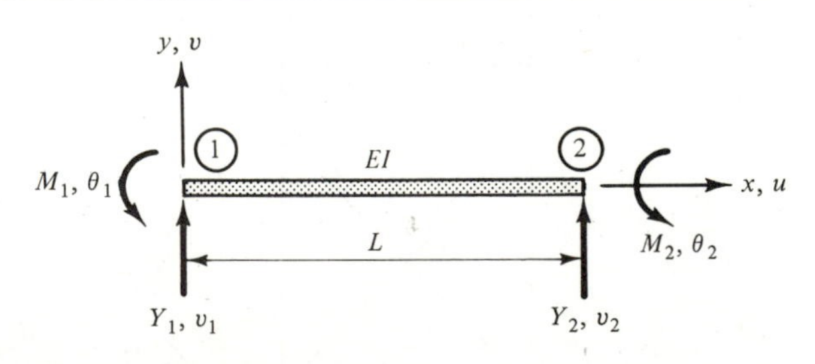

Figure 5.1 Straight beam element with uniform cross section.

5.1.1 Displacement Function

The deflection behavior of the beam element is described by a displacement function $v(x)$. It is desirable that this function satisfies the differential equation of equilibrium for a beam element in the unloaded region,

$$\frac{\partial^4 v}{\partial x^4} = 0 \tag{5.1}$$

The solution to Eq. (5.1) is a cubic polynomial function of x,

$$v(x) = a_1 + a_2 x + a_3 x^2 + a_4 x^3 \tag{5.2}$$

where the constants a_1, a_2, a_3 and a_4 are obtained by using the conditions at both nodal points,

$$\begin{aligned} v = v_1 \quad &\text{and} \quad \frac{\partial v}{\partial x} = \theta_1 \qquad \text{at } x = 0 \\ v = v_2 \quad &\text{and} \quad \frac{\partial v}{\partial x} = \theta_2 \qquad \text{at } x = L \end{aligned} \tag{5.3}$$

Application of boundary conditions (5.3) yields

$$\begin{Bmatrix} v_1 \\ \theta_1 \\ v_2 \\ \theta_2 \end{Bmatrix} = \begin{bmatrix} 1 & 0 & 0 & 0 \\ 0 & 1 & 0 & 0 \\ 1 & L & L^2 & L^3 \\ 0 & 1 & 2L & 3L^2 \end{bmatrix} \begin{Bmatrix} a_1 \\ a_2 \\ a_3 \\ a_4 \end{Bmatrix} \tag{5.4}$$

In inverse form, Eq. (5.4) becomes

$$\begin{Bmatrix} a_1 \\ a_2 \\ a_3 \\ a_4 \end{Bmatrix} = \frac{1}{L^3} \begin{bmatrix} L^3 & 0 & 0 & 0 \\ 0 & L^3 & 0 & 0 \\ -3L & -2L^2 & 3L & -L^2 \\ 2 & L & -2 & L \end{bmatrix} \begin{Bmatrix} v_1 \\ \theta_1 \\ v_2 \\ \theta_2 \end{Bmatrix} \tag{5.5}$$

or symbolically,

$$\{\mathbf{a}\} = [\mathbf{T}]\{\mathbf{q}\} \tag{5.5a}$$

Substituting the solution for a's in Eq. (5.5) into Eq. (5.2) gives

$$\begin{aligned} v(x) = v_1 + x\theta_1 - \frac{3x^2}{L^2} v_1 - \frac{2x^2}{L}\theta_1 + \frac{3x^2}{L^2} v_2 - \frac{x^2}{L}\theta_2 \\ + \frac{2x^3}{L^3} v_1 + \frac{x^3}{L^2}\theta_1 - \frac{2x^3}{L^3} v_2 + \frac{x^3}{L^2}\theta_2 \end{aligned}$$

After rearranging, we obtain the final form:

$$v(x) = f_1(x)v_1 + f_2(x)\theta_1 + f_3(x)v_2 + f_4(x)\theta_2 \tag{5.6}$$

where

$$
\begin{aligned}
f_1(x) &= 1 - 3\left(\frac{x}{L}\right)^2 + 2\left(\frac{x}{L}\right)^3 \\
f_2(x) &= x - 2\left(\frac{x^2}{L}\right) + \left(\frac{x^3}{L^2}\right) \\
f_3(x) &= 3\left(\frac{x}{L}\right)^2 - 2\left(\frac{x}{L}\right)^3 \\
f_4(x) &= -\left(\frac{x^2}{L}\right) + \left(\frac{x^3}{L^2}\right)
\end{aligned}
\tag{5.7}
$$

The functions obtained in Eq. (5.7) are called *shape functions.* If we vary the value of x from 0 to L, we obtain four curves as shown in Fig. 5.2 for the four shape functions, respectively.

Physically, each of the four shape functions represents the deflection curve for the beam element produced by setting the corresponding degree of freedom to be one and the rest degrees of freedom to be zero. In other

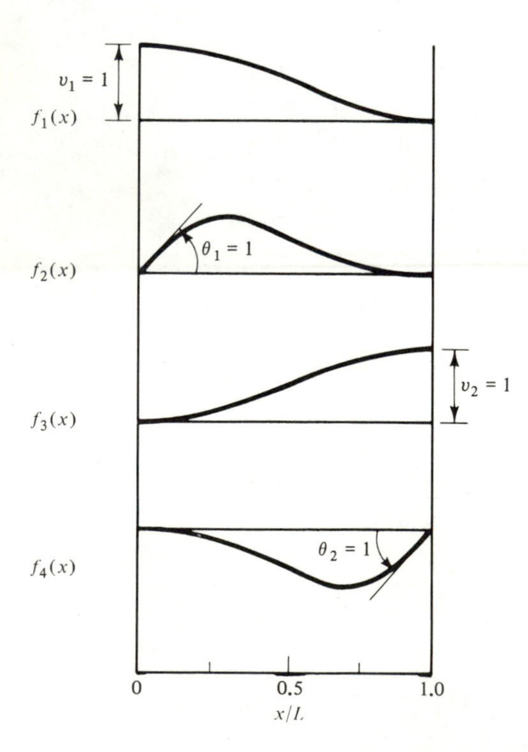

Figure 5.2 Plots of the four shape functions associated with v_1, θ_1, v_2, and θ_2 for a beam element.

words, by setting

$$v_1 = 1 \quad \text{and} \quad \theta_1 = v_2 = \theta_2 = 0$$

$$\theta_1 = 1 \quad \text{and} \quad v_1 = v_2 = \theta_2 = 0$$

$$v_2 = 1 \quad \text{and} \quad v_1 = \theta_1 = \theta_2 = 0$$

$$\theta_2 = 1 \quad \text{and} \quad v_1 = \theta_1 = v_2 = 0$$

we obtain $f_1(x)$, $f_2(x)$, $f_3(x)$, and $f_4(x)$, respectively.

It is seen that the deflection curve represented by the displacement function (5.6) is obtained as the linear superposition of the curves produced by the four degrees of freedom.

5.1.2 Stiffness Equations

The stiffness equations for this beam element can be obtained by using Castigliano's theorem, introduced in Chapter 3:

$$F_i = \frac{\partial U}{\partial q_i} \tag{5.8}$$

where F_i is the nodal force or moment, q_i the nodal displacement or rotation degree of freedom, subscript i the degree of freedom number, and U the strain energy.

The strain energy for a beam element with uniform cross section has been given in Chapter 3 as

$$U = \frac{EI}{2} \int_0^L \left(\frac{\partial^2 v}{\partial x^2} \right)^2 dx \tag{5.9}$$

Equation (5.9) indicates that we need the expression

$$\frac{\partial^2 v}{\partial x^2} = f_1''(x)v_1 + f_2''(x)\theta_1 + f_3''(x)v_2 + f_4''(x)\theta_2$$

where

$$\begin{aligned} f_1''(x) &= -\frac{6}{L^2} + 12\frac{x}{L^3} \\ f_2''(x) &= -\frac{4}{L} + 6\frac{x}{L^2} \\ f_3''(x) &= \frac{6}{L^2} - 12\frac{x}{L^3} \\ f_4''(x) &= -\frac{2}{L} + 6\frac{x}{L^2} \end{aligned} \tag{5.10}$$

Substituting the displacement function (5.6) into the strain energy expression (5.9) and then performing partial differentiation of the energy expression with respect to the degree of freedom v_1 as indicated in Castigliano's theorem (5.8) gives

$$
\begin{aligned}
Y_1 = \frac{\partial U}{\partial v_1} &= \frac{EI}{2}\int_0^L 2\left(\frac{\partial^2 v}{\partial x^2}\right)\frac{\partial}{\partial v_1}\left(\frac{\partial^2 v}{\partial x^2}\right) dx \\
&= EI\int_0^L [f_1''(x)v_1 + f_2''(x)\theta_1 + f_3''(x)v_2 + f_4''(x)\theta_2]f_1''(x)\,dx \\
&= k_{11}v_1 + k_{12}\theta_1 + k_{13}v_2 + k_{14}\theta_2
\end{aligned}
\tag{5.11}
$$

where

$$
\begin{aligned}
k_{11} &= EI\int_0^L f_1''(x)f_1''(x)\,dx \qquad k_{12} = EI\int_0^L f_1''(x)f_2''(x)\,dx \\
k_{13} &= EI\int_0^L f_1''(x)f_3''(x)\,dx \qquad k_{14} = EI\int_0^L f_1''(x)f_4''(x)\,dx
\end{aligned}
\tag{5.12}
$$

Equations (5.12) can be generalized to the standard form

$$
k_{ij} = EI\int_0^L f_i''(x)f_j''(x)\,dx \tag{5.13}
$$

which gives the stiffness matrix for the beam element. As an example, let $i = j = 1$ and we find that

$$
\begin{aligned}
k_{11} &= EI\int_0^L \left(-\frac{6}{L^2} + 12\frac{x}{L^3}\right)^2 dx \\
&= \frac{EI}{L^4}\left(36x - \frac{72x^2}{L} + \frac{48x^3}{L^2}\right)_0^L \\
&= 12\frac{EI}{L^3}
\end{aligned}
$$

By using the same procedure, we obtain the stiffness equations,

$$
\begin{Bmatrix} Y_1 \\ M_1 \\ Y_2 \\ M_2 \end{Bmatrix} = \frac{EI}{L}\begin{bmatrix} \frac{12}{L^2} & \frac{6}{L} & -\frac{12}{L^2} & \frac{6}{L} \\ \frac{6}{L} & 4 & -\frac{6}{L} & 2 \\ -\frac{12}{L^2} & -\frac{6}{L} & \frac{12}{L^2} & -\frac{6}{L} \\ \frac{6}{L} & 2 & -\frac{6}{L} & 4 \end{bmatrix}\begin{Bmatrix} v_1 \\ \theta_1 \\ v_2 \\ \theta_2 \end{Bmatrix} \tag{5.14}
$$

or symbolically,

$$\{\mathbf{F}\} = [\mathbf{k}]\{\mathbf{q}\} \tag{5.14a}$$

The method using shape functions to derive the stiffness matrix as described in Eq. (5.13) is general. It is particularly systematic and convenient for deriving the stiffness equations for the sophisticated finite elements such as plate and shell elements with known shape functions. It is also systematic and convenient for deriving the matrices that account for the effects of inertial and axial or middle surface forces.

For those finite elements with high numbers of degrees of freedom such as in the case of plates and shells, the solution for shape functions may become cumbersome. It is customary to use existing mathematical polynomial functions (e.g., Hermitian polynomials) as the shape functions. However, for those displacement functions whose shape functions are hard to find, we can simply operate on the constants (called the generalized coordinates) instead of the shape functions. Such a procedure is demonstrated by using the present beam element as an example.

We first rewrite the expression $\partial^2 v/\partial x^2$ based on the original form of the displacement function (5.2),

$$\frac{\partial^2 v}{\partial x^2} = 2a_3 + 6a_4 x \tag{5.15}$$

Substituting Eq. (5.15) into the strain energy expression (5.9) gives

$$\begin{aligned} U &= \frac{EI}{2}\int_0^L (2a_3 + 6a_4x)^2\,dx \\ &= 2EI(a_3^2L + 3a_3a_4L^2 + 3a_4^2L^3) \\ &= \tfrac{1}{2}\lfloor a_1 \quad a_2 \quad a_3 \quad a_4 \rfloor \begin{bmatrix} 0 & 0 & 0 & 0 \\ 0 & 0 & 0 & 0 \\ 0 & 0 & 4EIL & 6EIL^2 \\ 0 & 0 & 6EIL^2 & 12EIL^3 \end{bmatrix} \begin{Bmatrix} a_1 \\ a_2 \\ a_3 \\ a_4 \end{Bmatrix} \\ &= \tfrac{1}{2}\lfloor \mathbf{a} \rfloor [\bar{\mathbf{k}}]\{\mathbf{a}\} \end{aligned} \tag{5.16}$$

The elements in matrix $[\bar{\mathbf{k}}]$ are obtained by using the following partial differentiation procedure:

$$\bar{k}_{ij} = \frac{\partial^2 U}{\partial a_i\,\partial a_j} \qquad i, j = 1, 2, 3, 4 \tag{5.17}$$

Substituting Eq. (5.5a) for the constants into Eq. (5.16), we obtain

$$U = \tfrac{1}{2}\lfloor q \rfloor [\mathbf{T}]^T [\bar{\mathbf{k}}][\mathbf{T}]\{\mathbf{q}\} \tag{5.18}$$

Alternatively, we can write the strain energy expression as one-half of the sum of the products of the nodal forces and the corresponding nodal degrees of freedom:

$$U = \tfrac{1}{2}\lfloor v_1 \quad \theta_1 \quad v_2 \quad \theta_2 \rfloor \begin{Bmatrix} Y_1 \\ M_1 \\ Y_2 \\ M_2 \end{Bmatrix}$$

$$= \tfrac{1}{2}\lfloor \mathbf{q} \rfloor \{\mathbf{F}\} \tag{5.19}$$

Substituting the stiffness equations (5.14a) for the nodal forces, we have

$$U = \tfrac{1}{2}\lfloor q \rfloor [\mathbf{k}]\{\mathbf{q}\} \tag{5.20}$$

Although in separate forms, the two strain energy expressions in Eqs. (5.18) and (5.20) should be the same. Thus we conclude that

$$[\mathbf{k}] = [\mathbf{T}]^T[\bar{\mathbf{k}}][\mathbf{T}]$$

$$= \frac{EI}{L^6}\begin{bmatrix} L^3 & 0 & -3L & 2 \\ 0 & L^3 & -2L^2 & L \\ 0 & 0 & 3L & -2 \\ 0 & 0 & -L^2 & L \end{bmatrix} \begin{bmatrix} 0 & 0 & 0 & 0 \\ 0 & 0 & 0 & 0 \\ 0 & 0 & 4L & 6L^2 \\ 0 & 0 & 6L^2 & 12L^3 \end{bmatrix} \begin{bmatrix} L^3 & 0 & 0 & 0 \\ 0 & L^3 & 0 & 0 \\ -3L & -2L^2 & 3L & -L^2 \\ 2 & L & -2 & L \end{bmatrix}$$

$$= \frac{EI}{L}\begin{bmatrix} \frac{12}{L^2} & \frac{6}{L} & -\frac{12}{L^2} & \frac{6}{L} \\ \frac{6}{L} & 4 & -\frac{6}{L} & 2 \\ -\frac{12}{L^2} & -\frac{6}{L} & \frac{12}{L^2} & -\frac{6}{L} \\ \frac{6}{L} & 2 & -\frac{6}{L} & 4 \end{bmatrix} \tag{5.21}$$

which is the same as that obtained in Eq. (5.14) by using the shape functions.

5.1.3 Properties of the Beam Element Stiffness Equations

Equilibrium. If we consider the beam element as shown in Fig. 5.1 as a free body, it must be in equilibrium under the four nodal forces which are related to the four nodal degrees of freedom by the stiffness matrix in Eq. (5.14).

It is seen in the stiffness matrix [**k**] that the first row and the third row are exactly identical but with opposite signs. This means that forces Y_1 and Y_2 are equal in magnitude but opposite in direction. So the equilibrium in the transverse direction of the beam is satisfied.

We then take moment about nodal point 2,

$$\begin{aligned}\sum M_2 &= Y_1 L - M_1 - M_2 \\ &= (\text{first row})(L) - (\text{second row}) - (\text{fourth row}) \\ &= \frac{EI}{L}\left[\left(\frac{12}{L} - \frac{6}{L} - \frac{6}{L}\right)v_1 + (6 - 4 - 2)\theta_1 \right. \\ &\quad \left. + \left(-\frac{12}{L} + \frac{6}{L} + \frac{6}{L}\right)v_2 + (6 - 2 - 4)\theta_2\right] \\ &= 0\end{aligned}$$

which again satisfies moment equilibrium condition.

Singular matrix. Because the first and third rows in the stiffness matrix [**k**] are only different by a sign, [**k**] is a singular matrix. There is no inverse for matrix [**k**] and consequently there is no solution.

The fact that the element stiffness matrix is singular can be interpreted physically. The element, without support, is an unstable free body. It will be stable and the stiffness matrix will no longer be singular if it is properly supported. For example, if the element is clamped at nodal point (1) (i.e., $v_1 = \theta_1 = 0$), there are only two degrees of freedom, v_2 and θ_2, left. The stiffness equations then reduce to

$$\begin{Bmatrix} Y_2 \\ M_2 \end{Bmatrix} = \frac{EI}{L}\begin{bmatrix} \dfrac{12}{L^2} & -\dfrac{6}{L} \\ -\dfrac{6}{L} & 4 \end{bmatrix}\begin{Bmatrix} v_2 \\ \theta_2 \end{Bmatrix}$$

where the stiffness matrix is no longer singular for the supported stable element.

From the reasons explained here, we know that the stiffness matrix for a finite element is always singular.

5.2 BEAM ELEMENT ORIENTED ARBITRARILY IN TWO-DIMENSIONAL PLANE (PLANE FRAME ELEMENT)

In the case of plane frame structures, the beam elements are no longer horizontal. They can be arbitrarily oriented in a two-dimensional plane. Such elements are subjected to axial force, shear force, and bending moment. To cope with the plane frame problem, an element must possess three degrees of freedom at each nodal point: two displacement components u and v in the x and y directions, respectively, and an angle of rotation θ. Such an element is shown in Fig. 5.3. Corresponding to the degrees of freedom u, v, θ, there

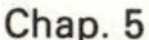

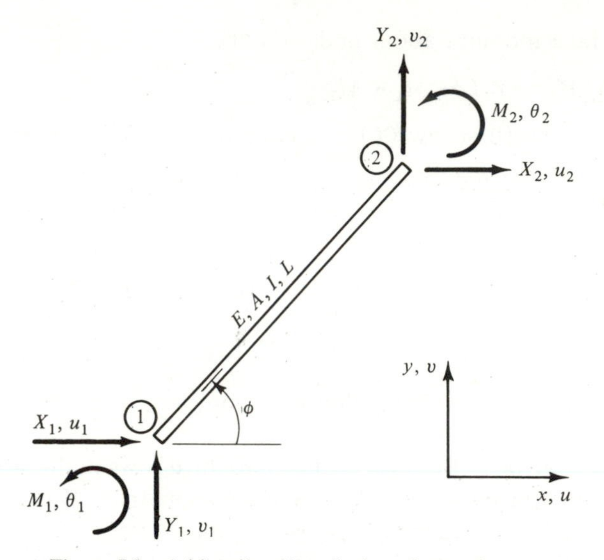

Figure 5.3 Arbitrarily oriented plane frame element.

are two forces X, Y, and a bending moment M, respectively, at each nodal point. This element is oriented arbitrarily at a counterclockwise angle ϕ with the global x axis. It has modulus of elasticity E, cross-sectional area A, moment of inertia I, and length L.

To derive the 6×6 stiffness matrix for this element, we first combine the 2×2 stiffness matrix for the truss bar element and the 4×4 stiffness matrix for the beam element to a 6×6 matrix. We then transform this 6×6 matrix from the local coordinates to the global coordinates (general reference coordinates).

5.2.1 Axial–Flexural Beam Element along the Local *X* Direction

When a truss bar element and a beam element are combined, we obtain the axial-flexural beam element shown in Fig. 5.4. The stiffness

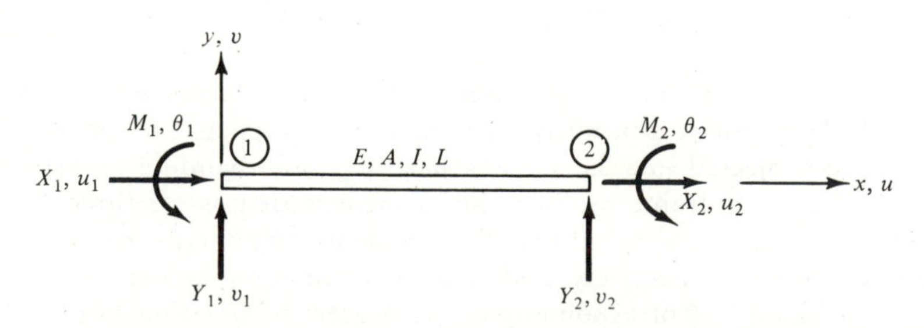

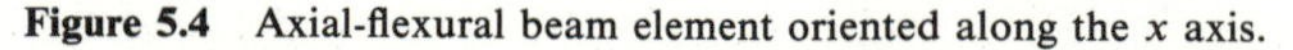

Figure 5.4 Axial-flexural beam element oriented along the x axis.

equations are

$$\begin{Bmatrix} X_1 \\ X_2 \\ \hline Y_1 \\ M_1 \\ Y_2 \\ M_2 \end{Bmatrix} = \left[\begin{array}{cc|cccc} \frac{EA}{L} & -\frac{EA}{L} & 0 & 0 & 0 & 0 \\ -\frac{EA}{L} & \frac{EA}{L} & 0 & 0 & 0 & 0 \\ \hline 0 & 0 & \frac{12EI}{L^3} & \frac{6EI}{L^2} & -\frac{12EI}{L^3} & \frac{6EI}{L^2} \\ 0 & 0 & \frac{6EI}{L^2} & \frac{4EI}{L} & -\frac{6EI}{L^2} & \frac{2EI}{L} \\ 0 & 0 & -\frac{12EI}{L^3} & -\frac{6EI}{L^2} & \frac{12EI}{L^3} & -\frac{6EI}{L^2} \\ 0 & 0 & \frac{6EI}{L^2} & \frac{2EI}{L} & -\frac{6EI}{L^2} & \frac{4EI}{L} \end{array}\right] \begin{Bmatrix} u_1 \\ u_2 \\ \hline v_1 \\ \theta_1 \\ v_2 \\ \theta_2 \end{Bmatrix} \tag{5.22}$$

This stiffness matrix is composed of four submatrices among which two are zero submatrices. The two nonzero submatrices are along the main diagonal. One is associated with the axial behavior and the other is associated with the flexural behavior. Such an arrangement of the submatrices indicates that the axial stiffness submatrix and the flexural stiffness submatrix are uncoupled. In other words, the solution for the axial displacements and the solution for the transverse deflections and rotations can be carried out separately and independently.

When this element is oriented in the two-dimensional plane with an angle ϕ with the x axis, the stiffness matrix in Eq. (5.22) must undergo a coordinate transformation procedure. In that case, the axial and flexural submatrices are no longer uncoupled.

For the convenience of assemblage, it is desirable to number all the degrees of freedom at each nodal point in a certain sequence. For this purpose, we rearrange Eq. (5.22) as

$$\begin{Bmatrix} X_1 \\ Y_1 \\ M_1 \\ X_2 \\ Y_2 \\ M_2 \end{Bmatrix} = \begin{bmatrix} \frac{EA}{L} & 0 & 0 & -\frac{EA}{L} & 0 & 0 \\ 0 & \frac{12EI}{L^3} & \frac{6EI}{L^2} & 0 & -\frac{12EI}{L^3} & \frac{6EI}{L^2} \\ 0 & \frac{6EI}{L^2} & \frac{4EI}{L} & 0 & -\frac{6EI}{L^2} & \frac{2EI}{L} \\ -\frac{EA}{L} & 0 & 0 & \frac{EA}{L} & 0 & 0 \\ 0 & -\frac{12EI}{L^3} & -\frac{6EI}{L^2} & 0 & \frac{12EI}{L^3} & -\frac{6EI}{L^2} \\ 0 & \frac{6EI}{L^2} & \frac{2EI}{L} & 0 & -\frac{6EI}{L^2} & \frac{4EI}{L} \end{bmatrix} \begin{Bmatrix} u_1 \\ v_1 \\ \theta_1 \\ u_2 \\ v_2 \\ \theta_2 \end{Bmatrix} \tag{5.23}$$

or symbolically,

$$\{\bar{\mathbf{F}}\} = [\bar{\mathbf{k}}]\{\bar{\mathbf{q}}\} \tag{5.23a}$$

where the bars indicate that the formulation is with reference to a set of local coordinates.

5.2.2 Coordinate Transformation

An axial–flexural beam element with its longitudinal axis lying along the $\bar{x}$ axis is shown in Fig. 5.5. The axes $(\bar{x}, \bar{y})$ are the local coordinate axes and (x, y) are the global or reference coordinate axes. The $\bar{x}$ axis is oriented at an angle ϕ measured counterclockwise from the x axis. To distinguish between the two coordinate systems, bars are added to all the symbols for the nodal forces and degrees of freedom that are with reference to the local coordinates.

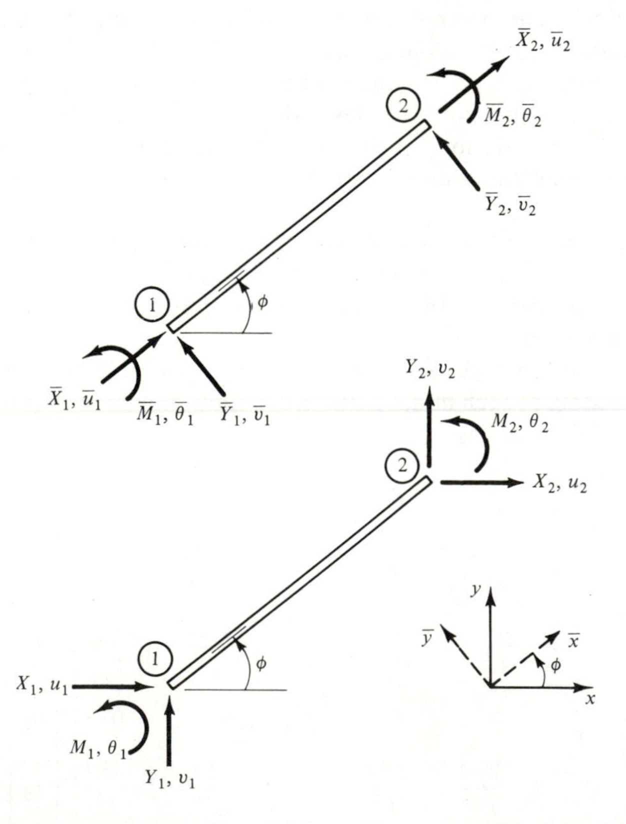

Figure 5.5 Transformation of an axial-flexural beam element from local $(\bar{x}, \bar{y})$ to global (x, y) coordinates.

The six nodal forces and bending moments in a local coordinate system are transformed into those in a global coordinate system:

$$\begin{Bmatrix} \bar{X}_1 \\ \bar{Y}_1 \\ \bar{M}_1 \\ \hline \bar{X}_2 \\ \bar{Y}_2 \\ \bar{M}_2 \end{Bmatrix} = \left[\begin{array}{ccc|ccc} \lambda & \mu & 0 & 0 & 0 & 0 \\ -\mu & \lambda & 0 & 0 & 0 & 0 \\ 0 & 0 & 1 & 0 & 0 & 0 \\ \hline 0 & 0 & 0 & \lambda & \mu & 0 \\ 0 & 0 & 0 & -\mu & \lambda & 0 \\ 0 & 0 & 0 & 0 & 0 & 1 \end{array}\right] \begin{Bmatrix} X_1 \\ Y_1 \\ M_1 \\ \hline X_2 \\ Y_2 \\ M_2 \end{Bmatrix} \tag{5.24}$$

where λ and μ are direction cosines and sines defined as

$$\begin{aligned} \lambda &= \cos \phi \\ \mu &= \sin \phi \end{aligned} \tag{5.25}$$

Equation (5.24) can be written in symbolic form as

$$\{\bar{\mathbf{F}}\} = [\mathbf{T}]\{\mathbf{F}\} \tag{5.26}$$

This transformation matrix was derived in Chapter 2. It has been proven to be an orthogonal matrix, that is,

$$[\mathbf{T}]^{-1} = [\mathbf{T}]^T \tag{5.27}$$

Equation (5.26) can be written in inverse form,

$$\{\mathbf{F}\} = [\mathbf{T}]^{-1}\{\bar{\mathbf{F}}\} = [\mathbf{T}]^T\{\bar{\mathbf{F}}\} \tag{5.28}$$

In the meantime, we can also transform the six nodal degrees of freedom in the local coordinate system into those in the global coordinate system:

$$\{\bar{\mathbf{q}}\} = [\mathbf{T}]\{\mathbf{q}\} \tag{5.29}$$

where the transformation matrix $[\mathbf{T}]$ is the same as that for the nodal forces and bending moments as obtained in Eq. (5.24).

Substituting Eq. (5.26) for the vector $\{\bar{\mathbf{F}}\}$ and Eq. (5.29) for the vector $\{\bar{\mathbf{q}}\}$ into the stiffness Eq. (5.23a) in a local coordinate system, we obtain

$$[\mathbf{T}]\{\mathbf{F}\} = [\bar{\mathbf{k}}][\mathbf{T}]\{\mathbf{q}\} \tag{5.30}$$

Making use of the orthogonal property of matrix $[\mathbf{T}]$, we obtain

$$\{\mathbf{F}\} = [\mathbf{T}]^T[\bar{\mathbf{k}}][\mathbf{T}]\{\mathbf{q}\} \tag{5.31}$$

The product $[\mathbf{T}]^T[\bar{\mathbf{k}}][\mathbf{T}]$ is called a *congruent transformation.* Substituting Eq. (5.23) for matrix $[\bar{\mathbf{k}}]$ and Eq. (5.24) for matrix $[\mathbf{T}]$ into Eq. (5.31), we finally obtain the stiffness matrix for a plane frame element in global coordinate systems.

5.2.3 Stiffness Equations

The stiffness equations for a plane frame element oriented at an angle ϕ measured counterclockwise from the x axis are of the following form:

$$\begin{Bmatrix} X_1 \\ Y_1 \\ M_1 \\ X_2 \\ Y_2 \\ M_2 \end{Bmatrix} = \frac{EI}{L} \begin{bmatrix} R\lambda^2 + \frac{12}{L^2}\mu^2 & & & & & \\ \left(R - \frac{12}{L^2}\right)\lambda\mu & R\mu^2 + \frac{12}{L^2}\lambda^2 & & & \text{symmetric} & \\ -\frac{6}{L}\mu & \frac{6}{L}\lambda & 4 & & & \\ -R\lambda^2 - \frac{12}{L^2}\mu^2 & \left(-R + \frac{12}{L^2}\right)\lambda\mu & \frac{6}{L}\mu & R\lambda^2 + \frac{12}{L^2}\mu^2 & & \\ \left(-R + \frac{12}{L^2}\right)\lambda\mu & -R\mu^2 - \frac{12}{L^2}\lambda^2 & -\frac{6}{L}\lambda & \left(R - \frac{12}{L^2}\right)\lambda\mu & R\mu^2 + \frac{12}{L^2}\lambda^2 & \\ -\frac{6}{L}\mu & \frac{6}{L}\lambda & 2 & \frac{6}{L}\mu & -\frac{6}{L}\lambda & 4 \end{bmatrix} \begin{Bmatrix} u_1 \\ v_1 \\ \theta_1 \\ u_2 \\ v_2 \\ \theta_2 \end{Bmatrix} \tag{5.32}$$

where R is defined as the ratio between the cross-sectional area and the moment of inertia (A/I).

5.3 APPLICATION OF BEAM ELEMENTS

The method of assemblage and procedure of solution for using beam elements are introduced with the aid of illustrative examples.

Beam Example 1 General procedure. Using two beam elements to model the beam structure as shown in Fig. 5.6a, find the deflection shape, the reaction forces and moments, and the shear force and bending moment diagrams.

We first idealize the beam by using two beam elements as shown in Fig. 5.6b. The total system has three nodal points and six degrees of freedom. To formulate for the total system, we must formulate the individual elements first.

For element 1-2, we have

$$\begin{Bmatrix} Y_1 \\ M_1 \\ Y_2 \\ M_2 \end{Bmatrix} = \frac{EI}{L} \begin{bmatrix} \frac{12}{L^2} & & \text{symmetric} & \\ \frac{6}{L} & 4 & & \\ -\frac{12}{L^2} & -\frac{6}{L} & \frac{12}{L^2} & \\ \frac{6}{L} & 2 & -\frac{6}{L} & 4 \end{bmatrix} \begin{Bmatrix} v_1 \\ \theta_1 \\ v_2 \\ \theta_2 \end{Bmatrix} \tag{5.33}$$

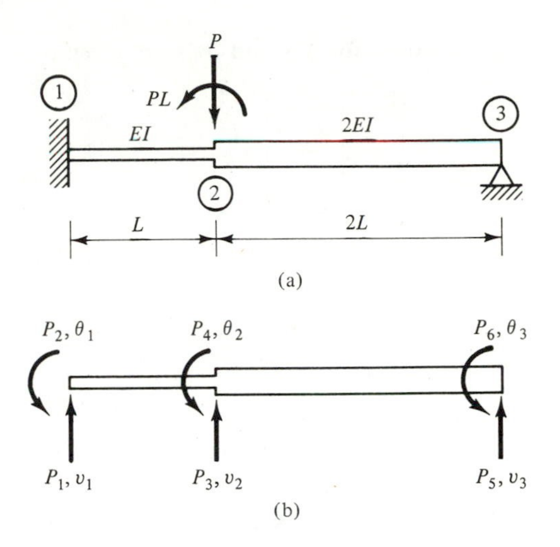

Figure 5.6 Beam and two-element modeling.

For element 2-3, we have

$$\begin{Bmatrix} Y_2 \\ M_2 \\ Y_3 \\ M_3 \end{Bmatrix} = \frac{EI}{L} \begin{bmatrix} \frac{3}{L^2} & \text{symmetric} & & \\ \frac{3}{L} & 4 & & \\ -\frac{3}{L^2} & -\frac{3}{L} & \frac{3}{L^2} & \\ \frac{3}{L} & 2 & -\frac{3}{L} & 4 \end{bmatrix} \begin{Bmatrix} v_2 \\ \theta_2 \\ v_3 \\ \theta_3 \end{Bmatrix} \tag{5.34}$$

It must be borne in mind that the Y's and M's are the internal forces and moments. When the two sets of element stiffness equations are assembled, the sum of such internal forces or moments at each nodal point are equal to the externally applied loads P's at the same nodal point as shown in Fig. 5.6b. Thus

$$\begin{aligned} P_1 &= Y_1 \\ P_2 &= M_1 \\ P_3 &= Y_2 \text{ of element 1-2} + Y_2 \text{ of element 2-3} \\ P_4 &= M_2 \text{ of element 1-2} + M_2 \text{ of element 2-3} \\ P_5 &= Y_3 \\ P_6 &= M_3 \end{aligned} \tag{5.35}$$

It must also be borne in mind that the deflections v and rotations θ at each nodal point are still the same as those for the assembled system.

With the understanding of these two ground rules, the method of assemblage is simply to sum up the Y's and M's as indicated in Eq. (5.35).

The equations for Y's and M's are given in Eqs. (5.33) and (5.34).

$$\begin{Bmatrix} P_1 \\ P_2 \\ P_3 \\ P_4 \\ P_5 \\ P_6 \end{Bmatrix} = \frac{EI}{L} \begin{bmatrix} \frac{12}{L^2} & & & & & \\ \frac{6}{L} & 4 & & \text{symmetric} & & \\ -\frac{12}{L^2} & -\frac{6}{L} & \frac{12}{L^2}+\frac{3}{L^2} & & & \\ \frac{6}{L} & 2 & -\frac{6}{L}+\frac{3}{L} & 4+4 & & \\ 0 & 0 & -\frac{3}{L^2} & -\frac{3}{L} & \frac{3}{L^2} & \\ 0 & 0 & \frac{3}{L} & 2 & -\frac{3}{L} & 4 \end{bmatrix} \begin{Bmatrix} v_1 \\ \theta_1 \\ v_2 \\ \theta_2 \\ v_3 \\ \theta_3 \end{Bmatrix} \tag{5.36}$$

The dashed lines bracket the two element stiffness matrices. The overlapped portion contains the stiffness terms that resulted from the superposition of the terms in the two individual element stiffness matrices.

We have three boundary support conditions:

$$v_1 = \theta_1 = v_3 = 0 \tag{5.37}$$

and the following external loading conditions:

$$\begin{aligned} P_1 &= \text{unknown reaction force at point 1} \\ P_2 &= \text{unknown reaction moment at point 1} \\ P_3 &= -P \text{ (opposite to the positive } y\text{-direction)} \\ P_4 &= PL \\ P_5 &= \text{unknown reaction force at point 3} \\ P_6 &= 0 \end{aligned} \tag{5.38}$$

Following the boundary conditions (5.37), we rearrange Eq. (5.36) such that v_1, θ_1, v_3 (or P_1, P_2, P_5) are together. We first rearrange the sequence of the rows:

$$\begin{Bmatrix} P_3 \\ P_4 \\ P_6 \\ P_1 \\ P_2 \\ P_5 \end{Bmatrix} = \frac{EI}{L} \begin{bmatrix} -\frac{12}{L^2} & -\frac{6}{L} & \frac{15}{L^2} & -\frac{3}{L} & -\frac{3}{L^2} & \frac{3}{L} \\ \frac{6}{L} & 2 & -\frac{3}{L} & 8 & -\frac{3}{L} & 2 \\ 0 & 0 & \frac{3}{L} & 2 & -\frac{3}{L} & 4 \\ \frac{12}{L^2} & \frac{6}{L} & -\frac{12}{L^2} & \frac{6}{L} & 0 & 0 \\ \frac{6}{L} & 4 & -\frac{6}{L} & 2 & 0 & 0 \\ 0 & 0 & -\frac{3}{L^2} & -\frac{3}{L} & \frac{3}{L^2} & -\frac{3}{L} \end{bmatrix} \begin{Bmatrix} v_1 \\ \theta_1 \\ v_2 \\ \theta_2 \\ v_3 \\ \theta_3 \end{Bmatrix} \tag{5.39}$$

We then rearrange the sequence of the columns:

$$\begin{Bmatrix} P_3 = -P \\ P_4 = PL \\ P_6 = 0 \\ \hline P_1 = ? \\ P_2 = ? \\ P_5 = ? \end{Bmatrix} = \frac{EI}{L} \left[\begin{array}{ccc|ccc} \frac{15}{L^2} & -\frac{3}{L} & \frac{3}{L} & -\frac{12}{L^2} & -\frac{6}{L} & -\frac{3}{L^2} \\ -\frac{3}{L} & 8 & 2 & \frac{6}{L} & 2 & -\frac{3}{L} \\ \frac{3}{L} & 2 & 4 & 0 & 0 & -\frac{3}{L} \\ \hline -\frac{12}{L^2} & \frac{6}{L} & 0 & \frac{12}{L^2} & \frac{6}{L} & 0 \\ -\frac{6}{L} & 2 & 0 & \frac{6}{L} & 4 & 0 \\ -\frac{3}{L^2} & -\frac{3}{L} & -\frac{3}{L} & 0 & 0 & \frac{3}{L^2} \end{array}\right] \begin{Bmatrix} v_2 = ? \\ \theta_2 = ? \\ \theta_3 = ? \\ \hline v_1 = 0 \\ \theta_1 = 0 \\ v_3 = 0 \end{Bmatrix} \tag{5.40}$$

Now we have six equations that can solve for six unknowns, three on each side of the equations. Multiplying out Eq. (5.40) gives

$$\begin{Bmatrix} -P \\ PL \\ 0 \end{Bmatrix} = \frac{EI}{L} \begin{bmatrix} \frac{15}{L^2} & -\frac{3}{L} & \frac{3}{L} \\ -\frac{3}{L} & 8 & 2 \\ \frac{3}{L} & 2 & 4 \end{bmatrix} \begin{Bmatrix} v_2 \\ \theta_2 \\ \theta_3 \end{Bmatrix} \tag{5.41}$$

and

$$\begin{Bmatrix} P_1 \\ P_2 \\ P_5 \end{Bmatrix} = \frac{EI}{L} \begin{bmatrix} -\frac{12}{L^2} & \frac{6}{L} & 0 \\ -\frac{6}{L} & 2 & 0 \\ -\frac{3}{L^2} & -\frac{3}{L} & -\frac{3}{L} \end{bmatrix} \begin{Bmatrix} v_2 \\ \theta_2 \\ \theta_3 \end{Bmatrix} \tag{5.42}$$

It is noted that Eq. (5.41) can be obtained by simply crossing out the first, second, and fifth rows as well as columns in the system stiffness matrix in Eq. (5.36). Thus we do not have to go through the tedious process of rearranging rows and columns to obtain such reduced equations.

In conventional structural theory, boundary constraints increase the number of degrees of redundancy and consequently complicate the problem by increasing the number of equations. In the finite element method, boundary constraints reduce the number of degrees of freedom and consequently simplify the problem by decreasing the number of equations.

Equation (5.41) may be solved by the method of matrix inverse. The adjoint method given in Sec. 2.13.1 is recommended.

$$\begin{Bmatrix} v_2 \\ \theta_2 \\ \theta_3 \end{Bmatrix} = \frac{L^3}{276EI}\begin{bmatrix} 28 & \dfrac{18}{L} & -\dfrac{30}{L} \\ \dfrac{18}{L} & \dfrac{51}{L^2} & -\dfrac{39}{L^2} \\ -\dfrac{30}{L} & -\dfrac{39}{L^2} & \dfrac{111}{L^2} \end{bmatrix}\begin{Bmatrix} -P \\ PL \\ 0 \end{Bmatrix}$$

$$= \frac{PL^3}{276EI}\begin{Bmatrix} -10 \\ \dfrac{33}{L} \\ -\dfrac{9}{L} \end{Bmatrix} \tag{5.43}$$

Equations (5.37) and (5.43) provide the solution for all six degrees of freedom. The corresponding deflection curve for the beam is shown in Fig. 5.7. At this point, if we want to know the deflection at a certain point, we can simply substitute the coordinate value of the point and the nodal degree of freedom values into the displacement function (5.6) of the element.

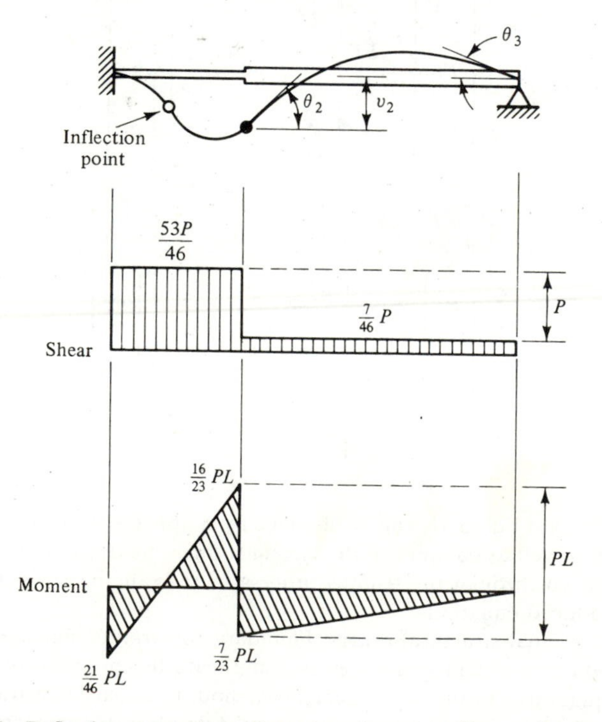

Figure 5.7 Deflection curve, shear diagram, and moment diagram for the beam in Fig. 5.6.

The internal shear forces and bending moments are obtained by substituting the known quantities of nodal point degrees of freedom into each individual element stiffness matrix.

For element 1-2, we have, from Eq. (5.33),

$$\begin{Bmatrix} Y_1 \\ M_1 \\ Y_2 \\ M_2 \end{Bmatrix} = \frac{EI}{L}\begin{bmatrix} & -\frac{12}{L^2} & \frac{6}{L} \\ & -\frac{6}{L} & 2 \\ \text{not needed} & \frac{12}{L^2} & -\frac{6}{L} \\ & -\frac{6}{L} & 4 \end{bmatrix}\begin{Bmatrix} 0 \\ 0 \\ -\frac{10PL^3}{276EI} \\ \frac{33PL^2}{276EI} \end{Bmatrix} = \begin{Bmatrix} \frac{53P}{46} \\ \frac{21PL}{46} \\ \frac{-53P}{46} \\ \frac{16PL}{23} \end{Bmatrix} \tag{5.44}$$

For element 2-3, we have, from Eq. (5.34),

$$\begin{Bmatrix} Y_2 \\ M_2 \\ Y_3 \\ M_3 \end{Bmatrix} = \frac{EI}{L}\begin{bmatrix} \frac{3}{L^2} & \frac{3}{L} & -\frac{3}{L^2} & \frac{3}{L} \\ \frac{3}{L} & 4 & -\frac{3}{L} & 2 \\ -\frac{3}{L^2} & -\frac{3}{L} & \frac{3}{L^2} & -\frac{3}{L} \\ \frac{3}{L} & 2 & -\frac{3}{L} & 4 \end{bmatrix}\begin{Bmatrix} \frac{-10PL^3}{276EI} \\ \frac{33PL^2}{276EI} \\ 0 \\ \frac{-9PL^2}{276EI} \end{Bmatrix} = \begin{Bmatrix} \frac{7P}{46} \\ \frac{7PL}{23} \\ -\frac{7P}{46} \\ 0 \end{Bmatrix} \tag{5.45}$$

The sign conventions of the Y's and M's are defined in Fig. 5.1. With the results obtained in Eqs. (5.44) and (5.45), we can plot the shear force and bending moment diagrams in Fig. 5.7. It is seen that the amount of drop in shear force at nodal point 2 is caused by the external load P and it is equal to P. The amount of drop in bending moment at point 2 is equal to the external moment PL. The point of zero moment is the point of inflection which is customarily termed as *hinge* by structural engineers.

The external reaction forces and bending moments at the supports can be obtained from Eq. (5.42):

$$\begin{Bmatrix} P_1 \\ P_2 \\ P_5 \end{Bmatrix} = \frac{EI}{L}\begin{bmatrix} -\frac{12}{L^2} & \frac{6}{L} & 0 \\ -\frac{6}{L} & 2 & 0 \\ -\frac{3}{L^2} & -\frac{3}{L} & -\frac{3}{L} \end{bmatrix}\begin{Bmatrix} -10 \\ \frac{33}{L} \\ -\frac{9}{L} \end{Bmatrix}\frac{PL^3}{276EI} = \begin{Bmatrix} \frac{53P}{46} \\ \frac{21PL}{46} \\ -\frac{7P}{46} \end{Bmatrix}$$

which, of course, agree with those obtained for Y_1, M_1, and Y_3, respectively, in Eqs. (5.44) and (5.45). The use of Eq. (5.42) is actually not necessary. In the solution of the stiffness equations for a system, it is a general practice simply to eliminate the rows and columns that correspond to the zero degrees of freedom instead of rearranging and partitioning the matrix as done in Eqs. (5.39) through (5.42).

Beam Example 2 Method of matrix partitioning and reduction. Using three elements to model the beam structure as shown in Fig. 5.8, find the deflection shape, the shear force and bending moment diagrams, and the reaction forces and moment.

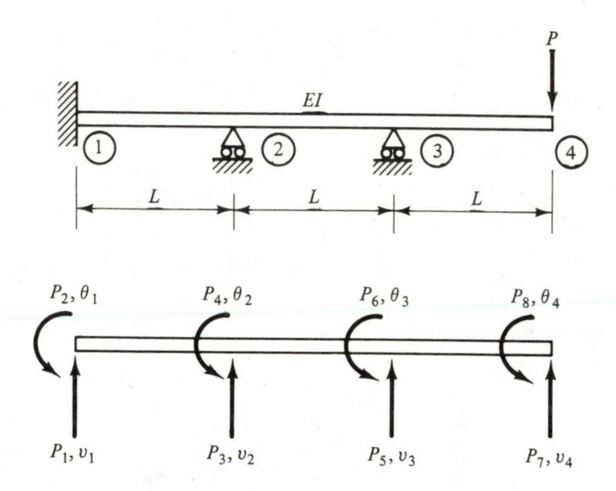

Figure 5.8 Beam and three-element modeling.

For this problem, the three-element stiffness matrices are the same. The total stiffness equations are assembled as

$$\begin{Bmatrix} P_1 \\ P_2 \\ P_3 \\ P_4 \\ P_5 \\ P_6 \\ P_7 \\ P_8 \end{Bmatrix} = \frac{EI}{L} \begin{bmatrix} \frac{12}{L^2} & & & & & \text{symmetric} & & \\ \frac{6}{L} & 4 & & & & & & \\ -\frac{12}{L^2} & -\frac{6}{L} & \frac{24}{L^2} & & & & & \\ \frac{6}{L} & 2 & 0 & 8 & & & & \\ 0 & 0 & -\frac{12}{L^2} & -\frac{6}{L} & \frac{24}{L^2} & & & \\ 0 & 0 & \frac{6}{L} & 2 & 0 & 8 & & \\ 0 & 0 & 0 & 0 & -\frac{12}{L^2} & -\frac{6}{L} & \frac{12}{L^2} & \\ 0 & 0 & 0 & 0 & \frac{6}{L} & 2 & -\frac{6}{L} & 4 \end{bmatrix} \begin{Bmatrix} v_1 \\ \theta_1 \\ v_2 \\ \theta_2 \\ v_3 \\ \theta_3 \\ v_4 \\ \theta_4 \end{Bmatrix} \qquad (5.46)$$

Boundary and loading conditions are

$$\begin{aligned} v_1 = \theta_1 = v_2 = v_3 = 0 \\ P_4 = P_6 = P_8 = 0 \\ P_7 = -P \end{aligned} \tag{5.47}$$

The total stiffness matrix may be reduced from 8×8 to 4×4 by crossing out the rows and columns corresponding to the zero degrees of freedom.

$$\begin{Bmatrix} P_4 = 0 \\ P_6 = 0 \\ \hline P_7 = -P \\ P_8 = 0 \end{Bmatrix} = \frac{EI}{L} \left[\begin{array}{cc|cc} 8 & 2 & 0 & 0 \\ 2 & 8 & -\frac{6}{L} & 2 \\ \hline 0 & -\frac{6}{L} & \frac{12}{L^2} & -\frac{6}{L} \\ 0 & 2 & -\frac{6}{L} & 4 \end{array} \right] \begin{Bmatrix} \theta_2 \\ \theta_3 \\ \hline v_4 \\ \theta_4 \end{Bmatrix} \tag{5.48}$$

It is recommended to assemble the element stiffness equations and take into account the zero degrees of freedom simultaneously [i.e., to form Eq. (5.48) directly without forming Eq. (5.46)]. Such a step is crucial in computer programming because the size of DIMENSION can be reduced and consequently core storage can be saved.

Equation (5.48) can be solved by inverting the 4×4 stiffness matrix. Instead, we can use a partition scheme by taking advantage of the zero-loading conditions so that we only have to invert two matrices of smaller sizes. Such a reduction procedure may result in easier hand computation or less computer time. To illustrate this procedure, we first write the partitioned matrix equations (5.48) in symbolic form:

$$\begin{Bmatrix} \bar{P}_1 = 0 \\ \hline \bar{P}_2 \end{Bmatrix} = \left[\begin{array}{c|c} K_{11} & K_{12} \\ \hline K_{21} & K_{22} \end{array} \right] \begin{Bmatrix} \bar{Q}_1 \\ \hline \bar{Q}_2 \end{Bmatrix} \tag{5.49}$$

Multiplying out Eq. (5.49) yields

$$\{\mathbf{0}\} = [\mathbf{K}_{11}]\{\bar{\mathbf{Q}}_1\} + [\mathbf{K}_{12}]\{\bar{\mathbf{Q}}_2\} \tag{5.50}$$

$$\{\bar{\mathbf{P}}_2\} = [\mathbf{K}_{21}]\{\bar{\mathbf{Q}}_1\} + [\mathbf{K}_{22}]\{\bar{\mathbf{Q}}_2\} \tag{5.51}$$

Equation (5.50) can be rewritten as

$$\{\bar{\mathbf{Q}}_1\} = -[\mathbf{K}_{11}]^{-1}[\mathbf{K}_{12}]\{\bar{\mathbf{Q}}_2\} \tag{5.52}$$

It is remembered that we should invert $[\mathbf{K}_{11}]$ instead of $[\mathbf{K}_{12}]$. Submatrix $[\mathbf{K}_{12}]$ is off-diagonal and it may well be a singular or non-square matrix. Substituting Eq. (5.52) into Eq. (5.51) gives

$$\begin{aligned} [\bar{\mathbf{P}}_2] &= [-[\mathbf{K}_{21}][\mathbf{K}_{11}]^{-1}[\mathbf{K}_{12}] + [\mathbf{K}_{22}]]\{\bar{\mathbf{Q}}_2\} \\ &= [\bar{\mathbf{K}}]\{\bar{\mathbf{Q}}_2\} \end{aligned} \tag{5.53}$$

where $[\bar{\mathbf{K}}]$ is the reduced matrix.

It is seen that in the reduction method, we have only to invert the submatrix $[\mathbf{K}_{11}]$ and reduced matrix $[\bar{\mathbf{K}}]$ instead of the full matrix $[\mathbf{K}]$. Substituting the submatrices defined in Eq. (5.48) into Eq. (5.53), we obtain

$$[\bar{\mathbf{K}}] = -\frac{EI}{L}\begin{bmatrix} 0 & -\frac{6}{L} \\ 0 & 2 \end{bmatrix}\begin{bmatrix} \frac{2}{15} & -\frac{1}{30} \\ -\frac{1}{30} & \frac{2}{15} \end{bmatrix}\begin{bmatrix} 0 & 0 \\ -\frac{6}{L} & 2 \end{bmatrix} + \frac{EI}{L}\begin{bmatrix} \frac{12}{L^2} & -\frac{6}{L} \\ -\frac{6}{L} & 4 \end{bmatrix}$$

$$= \frac{EI}{L}\begin{bmatrix} \frac{36}{5L^2} & -\frac{22}{5L} \\ -\frac{22}{5L} & \frac{52}{15} \end{bmatrix} \tag{5.54}$$

From Eqs. (5.53) and (5.54),

$$\{\bar{\mathbf{Q}}_2\} = \begin{Bmatrix} v_4 \\ \theta_4 \end{Bmatrix} = [\bar{\mathbf{K}}]^{-1}\{\bar{\mathbf{P}}_2\} = \frac{L^3}{EI}\begin{bmatrix} \frac{13}{21} & \frac{11}{14L} \\ \frac{11}{14L} & \frac{9}{7L^2} \end{bmatrix}\begin{Bmatrix} -P \\ 0 \end{Bmatrix}$$

$$\begin{Bmatrix} v_4 \\ \theta_4 \end{Bmatrix} = -\frac{PL^2}{EI}\begin{Bmatrix} \frac{13L}{21} \\ \frac{11}{14} \end{Bmatrix} \tag{5.55}$$

The rest of the degrees of freedom are obtained by substituting Eq. (5.55) into Eq. (5.52):

$$\begin{Bmatrix} \theta_2 \\ \theta_3 \end{Bmatrix} = \begin{bmatrix} -\frac{1}{5L} & \frac{1}{15} \\ \frac{4}{5L} & -\frac{4}{15} \end{bmatrix}\begin{Bmatrix} v_4 \\ \theta_4 \end{Bmatrix}$$

$$= \frac{PL^2}{EI}\begin{Bmatrix} \frac{1}{14} \\ -\frac{2}{7} \end{Bmatrix} \tag{5.56}$$

Thus we can construct the deflection curve shown in Fig. 5.9.

The internal shear forces and bending moments are found by substituting the foregoing deflections and rotations into the element stiffness equations. We find for elements 1-2, 2-3, and 3-4, respectively,

$$\begin{Bmatrix} Y_1 \\ M_1 \\ Y_2 \\ M_2 \end{Bmatrix} = \frac{P}{7}\begin{Bmatrix} 3 \\ L \\ -3 \\ 2L \end{Bmatrix} \qquad \begin{Bmatrix} Y_2 \\ M_2 \\ Y_3 \\ M_3 \end{Bmatrix} = \frac{P}{7}\begin{Bmatrix} -9 \\ -2L \\ 9 \\ -7L \end{Bmatrix} \qquad \begin{Bmatrix} Y_3 \\ M_3 \\ Y_4 \\ M_4 \end{Bmatrix} = \begin{Bmatrix} P \\ PL \\ -P \\ 0 \end{Bmatrix}$$

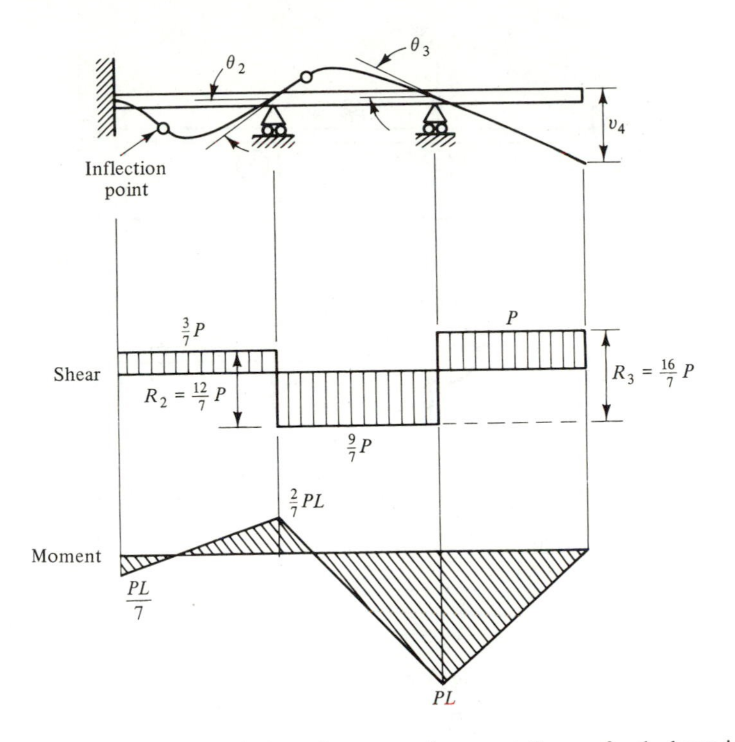

Figure 5.9 Deflection curve, shear diagram, and moment diagram for the beam in Fig. 5.8.

The resulting shear and moment diagrams are plotted in Fig. 5.9. These diagrams provided the reaction forces and moments:

$$P_1 = Y_1 = \frac{3P}{7} \quad \text{upward}$$

$$P_2 = M_1 = \frac{PL}{7} \quad \text{counterclockwise}$$

$$P_3 = \frac{12P}{7} \quad \text{downward}$$

$$P_5 = \frac{16P}{7} \quad \text{upward}$$

Beam Example 3 Treatment of symmetrical and antisymmetrical loading and supporting conditions. Using three elements to model the beam structure as shown in Fig. 5.10, find the deflection shape, the shear force and bending moment diagrams, and the reaction forces and moments.

For this problem, the total stiffness matrix equations are precisely the same as that given in Eq. (5.46). Because of the antisymmetrical type of loads, the boundary

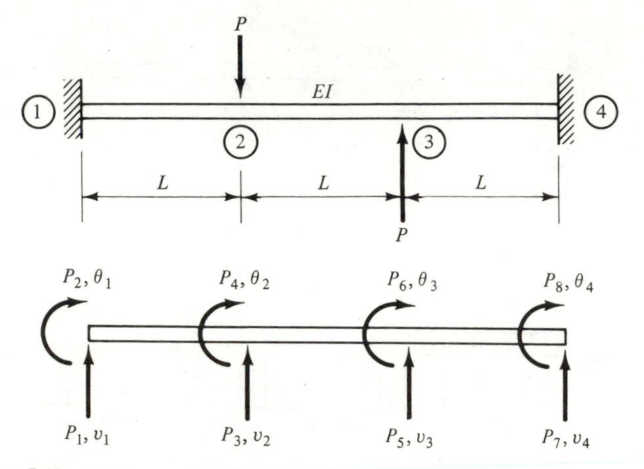

Figure 5.10 Antisymmetrically loaded beam and three-element modeling.

and loading conditions can be specified as

$$
\begin{aligned}
&v_1 = \theta_1 = v_4 = \theta_4 = 0 \\
&v_2 = -v_3 \\
&\theta_2 = \theta_3 \\
&P_4 = P_6 = 0 \\
&P_3 = -P_5 = -P
\end{aligned}
\tag{5.57}
$$

We first remove the four rows and columns that are related to the four zero degrees of freedom given in Eq. (5.57):

$$
\begin{Bmatrix} P_3 = -P \\ P_4 = 0 \\ P_5 = P \\ P_6 = 0 \end{Bmatrix} = \frac{EI}{L} \begin{bmatrix} \frac{24}{L^2} & 0 & -\frac{12}{L^2} & \frac{6}{L} \\ 0 & 8 & -\frac{6}{L} & 2 \\ -\frac{12}{L^2} & -\frac{6}{L} & \frac{24}{L^2} & 0 \\ \frac{6}{L} & 2 & 0 & 8 \end{bmatrix} \begin{Bmatrix} v_2 \\ \theta_2 \\ v_3 = -v_2 \\ \theta_3 = \theta_2 \end{Bmatrix}
\tag{5.58}
$$

Replacing v_3 by $-v_2$, θ_3 by θ_2, and multiplying out Eq. (5.58) yields

$$
\begin{Bmatrix} -P \\ 0 \\ P \\ 0 \end{Bmatrix} = \frac{EI}{L} \begin{bmatrix} \frac{36}{L^2} & \frac{6}{L} \\ \frac{6}{L} & 10 \\ -\frac{36}{L^2} & -\frac{6}{L} \\ \frac{6}{L} & 10 \end{bmatrix} \begin{Bmatrix} v_2 \\ \theta_2 \end{Bmatrix}
\tag{5.59}
$$

The process to treat the antisymmetrical conditions $(v_3 = -v_2)$ and $(\theta_3 = \theta_2)$ is simply to subtract the third column from the first and add the fourth column to the second in the matrix in Eq. (5.58). In the resulting four equations (5.59), we see that the first set of two equations are parallel to the second set of two equations. We can choose either set. For the first set,

$$\begin{Bmatrix} -P \\ 0 \end{Bmatrix} = \frac{EI}{L} \begin{bmatrix} \frac{36}{L^2} & \frac{6}{L} \\ \frac{6}{L} & 10 \end{bmatrix} \begin{Bmatrix} v_2 \\ \theta_2 \end{Bmatrix}$$

The solution is

$$\begin{Bmatrix} -v_3 \\ \theta_3 \end{Bmatrix} = \begin{Bmatrix} v_2 \\ \theta_2 \end{Bmatrix} = \frac{PL^2}{162EI} \begin{Bmatrix} -5L \\ 3 \end{Bmatrix} \tag{5.60}$$

From this example we can generalize the method of reduction of stiffness matrix whenever we can relate two degrees of freedom. For example, if we know

$$q_7 = (-6)q_2$$

The method of reduction is simply to

1. Eliminate the seventh row.
2. Add $(-6) \times$(seventh column) to the second column.
3. Eliminate the seventh column.

The internal shear forces and bending moments are found by substituting the foregoing deflections and rotations into the element stiffness equations. We find for elements 1-2, 2-3, and 3-4, respectively,

$$\begin{Bmatrix} Y_1 \\ M_1 \\ Y_2 \\ M_2 \end{Bmatrix} = \frac{P}{27} \begin{Bmatrix} 13 \\ 6L \\ -13 \\ 7L \end{Bmatrix} \qquad \begin{Bmatrix} Y_2 \\ M_2 \\ Y_3 \\ M_3 \end{Bmatrix} = \frac{P}{27} \begin{Bmatrix} -14 \\ -7L \\ 14 \\ -7L \end{Bmatrix} \qquad \begin{Bmatrix} Y_3 \\ M_3 \\ Y_4 \\ M_4 \end{Bmatrix} = \frac{P}{27} \begin{Bmatrix} 13 \\ 7L \\ -13 \\ 6L \end{Bmatrix}$$

The deflection curve, shear force, and bending moment diagrams are plotted in Fig. 5.11. The reactions, forces, and moments are given by Y_1, M_1, Y_4, and M_4.

Beam Example 4 Alternative solution for Beam Example 3. For symmetrical or antisymmetrical problems, it is a common practice to analyze only one portion of the structure cut out from the lines of symmetry or antisymmetry. For Beam Example 3, we can model only half of the beam by using two elements as shown in Fig. 5.12.

The boundary conditions for the present antisymmetrical case are

$$v_1 = \theta_1 = v_3 = 0$$

By imposing such boundary conditions, we can reduce the total stiffness equations from 6×6 to 3×3. Thus we have only to assemble a 3×3 stiffness matrix by collecting the pertinent terms in the two element stiffness matrices instead of forming the larger

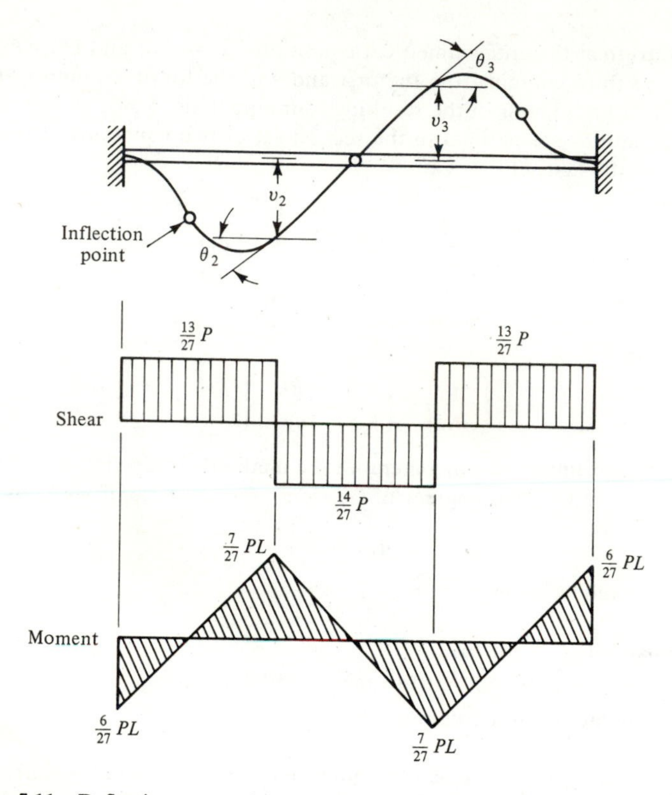

Figure 5.11 Deflection curve, shear diagram, and moment diagram for the beam in Fig. 5.10.

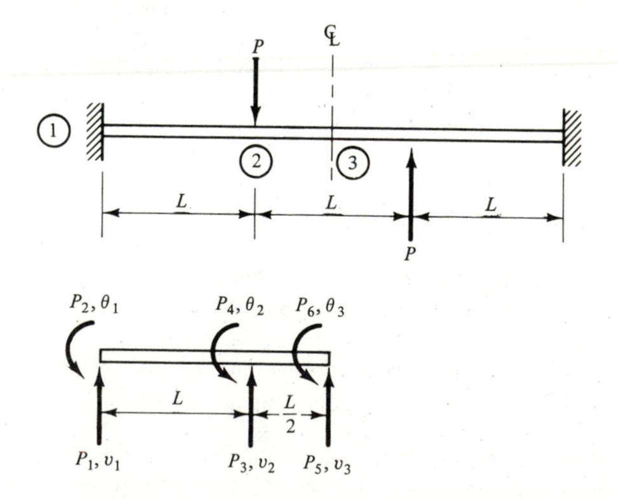

Figure 5.12 Two-element modeling for half of the beam shown in Fig. 5.10.

but unnecessary 6×6 stiffness matrix.

$$\begin{Bmatrix} P_3 \\ P_4 \\ P_6 \end{Bmatrix} = EI \begin{bmatrix} \frac{12}{L^3}+\frac{96}{L^3} & -\frac{6}{L^2}+\frac{24}{L^2} & \frac{24}{L^2} \\ -\frac{6}{L^2}+\frac{24}{L^2} & \frac{4}{L}+\frac{8}{L} & \frac{4}{L} \\ \frac{24}{L^2} & \frac{4}{L} & \frac{8}{L} \end{bmatrix} \begin{Bmatrix} v_2 \\ \theta_2 \\ \theta_3 \end{Bmatrix}$$

or

$$\begin{Bmatrix} -P \\ 0 \\ 0 \end{Bmatrix} = \frac{EI}{L} \begin{bmatrix} \frac{108}{L^2} & \frac{18}{L} & \frac{24}{L} \\ \frac{18}{L} & 12 & 4 \\ \frac{24}{L} & 4 & 8 \end{bmatrix} \begin{Bmatrix} v_2 \\ \theta_2 \\ \theta_3 \end{Bmatrix}$$

The solution is obtained by inverting the stiffness matrix,

$$\begin{Bmatrix} v_2 \\ \theta_2 \\ \theta_3 \end{Bmatrix} = \frac{L^3}{2592EI} \begin{bmatrix} 80 & & \\ -\frac{48}{L} & \text{not needed} & \\ -\frac{216}{L} & & \end{bmatrix} \begin{Bmatrix} -P \\ 0 \\ 0 \end{Bmatrix} = \frac{PL^2}{162EI} \begin{Bmatrix} -5L \\ 3 \\ \frac{27}{2} \end{Bmatrix}$$

where the results for v_2 and θ_2 agree with those obtained in Eq. (5.60). Substituting these results into the two-element stiffness matrices, respectively, we find

$$\begin{Bmatrix} Y_1 \\ M_1 \\ Y_2 \\ M_2 \end{Bmatrix} = \frac{P}{27} \begin{Bmatrix} 13 \\ 6L \\ -13 \\ 7L \end{Bmatrix} \quad \text{and} \quad \begin{Bmatrix} Y_2 \\ M_2 \\ Y_3 \\ M_3 \end{Bmatrix} = \frac{P}{27} \begin{Bmatrix} -14 \\ -7L \\ 14 \\ 0 \end{Bmatrix}$$

which provides us with precisely the same shear and moment diagrams as those shown in Fig. 5.11.

Beam Example 5 Treatment of elastic supports. Let us consider the same problem as Beam Example 1 with the addition of two elastic springs: an extensional spring with constant α (lb/in.) and a rotational spring with constant β (in.-lb/rad). The values for the elastic constants are assumed as

$$\alpha = \frac{8EI}{L^3} \qquad \beta = \frac{4EI}{L} \tag{5.61}$$

The beam and its two-element modeling are shown in Fig. 5.13.

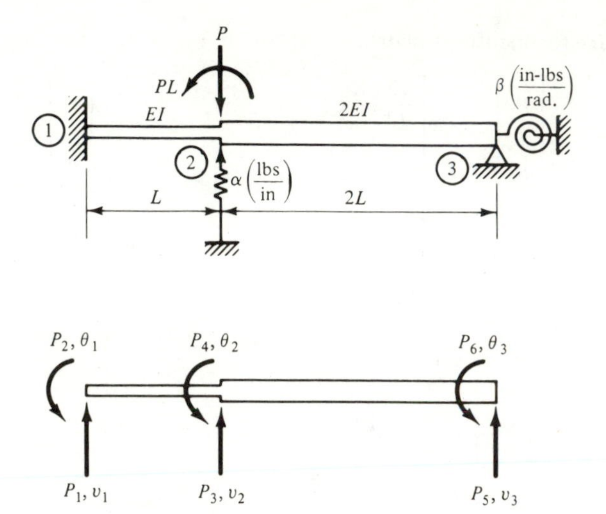

Figure 5.13 Beam with elastic supports and two-element modeling.

The effect of the two elastic springs is to produce a resistant force of $-\alpha v_2$ and a resistant moment $-\beta\theta_3$ in addition to the existing external loads. Such elastic force and moment always have a negative sign which means that they are acting in the directions opposite (resistant) to the positive directions of v and θ, respectively.

Without considering the springs, the stiffness matrix for the beam system is given in Eq. (5.41). Taking the spring effect into account, we add the spring force and moment to the existing loads in Eq. (5.41):

$$\begin{Bmatrix} -P-\alpha v_2 \\ PL \\ 0-\beta\theta_3 \end{Bmatrix} = \frac{EI}{L}\begin{bmatrix} \dfrac{15}{L^2} & -\dfrac{3}{L} & \dfrac{3}{L} \\ -\dfrac{3}{L} & 8 & 2 \\ \dfrac{3}{L} & 2 & 4 \end{bmatrix}\begin{Bmatrix} v_2 \\ \theta_2 \\ \theta_3 \end{Bmatrix} \tag{5.62}$$

Replacing the spring force and moment to the right-hand side of the equations (5.62) results in

$$\begin{Bmatrix} -P \\ PL \\ 0 \end{Bmatrix} = \begin{bmatrix} \dfrac{15EI}{L^3}+\alpha & -\dfrac{3EI}{L^2} & \dfrac{3EI}{L^2} \\ -\dfrac{3EI}{L^2} & \dfrac{8EI}{L} & \dfrac{2EI}{L} \\ \dfrac{3EI}{L^2} & \dfrac{2EI}{L} & \dfrac{4EI}{L}+\beta \end{bmatrix}\begin{Bmatrix} v_2 \\ \theta_2 \\ \theta_3 \end{Bmatrix} \tag{5.63}$$

It is seen that in order to include the effect of a spring associated with the degree of freedom i, we simply have to add the spring constant to the stiffness term k_{ii} along the main diagonal of the stiffness matrix. We see in Eq. (5.62) that the stiffness terms along the main diagonal have the same physical meaning as the elastic spring constants. They are always positive.

Substituting Eq. (5.61) into Eq. (5.63), we can solve for the unknowns:

$$\begin{Bmatrix} v_2 \\ \theta_2 \\ \theta_3 \end{Bmatrix} = \frac{L^3}{240EI} \begin{bmatrix} 12 & \dfrac{6}{L} & -\dfrac{6}{L} \\ \dfrac{6}{L} & \dfrac{35}{L^2} & -\dfrac{11}{L^2} \\ -\dfrac{6}{L} & -\dfrac{11}{L^2} & \dfrac{35}{L^2} \end{bmatrix} \begin{Bmatrix} -P \\ PL \\ 0 \end{Bmatrix}$$

$$= \frac{PL^2}{240EI} \begin{Bmatrix} -6L \\ 29 \\ -5 \end{Bmatrix}$$

The internal shear forces and bending moments are obtained by substituting the nodal displacements and rotations into the stiffness equations for elements 1-2 and 2-3, respectively,

$$\begin{Bmatrix} Y_1 \\ M_1 \\ Y_2 \\ M_2 \end{Bmatrix} = \frac{P}{120} \begin{Bmatrix} 123 \\ 47L \\ -123 \\ 76L \end{Bmatrix} \quad \text{and} \quad \begin{Bmatrix} Y_2 \\ M_2 \\ Y_3 \\ M_3 \end{Bmatrix} = \frac{P}{120} \begin{Bmatrix} 27 \\ 44L \\ -27 \\ 10L \end{Bmatrix}$$

The resulting deflection curve and shear and moment diagrams are shown in Fig. 5.14. The force at point 2 and the bending moment at point 3 due to the springs are

$$-\alpha v_2 = -\frac{8EI}{L^3}\left(-\frac{PL^3}{40EI}\right) = \frac{P}{5} \quad \text{upward}$$

$$-\beta\theta_3 = -\frac{4EI}{L}\left(-\frac{PL^2}{48EI}\right) = \frac{PL}{12} \quad \text{clockwise}$$

which agree with the amount of jump at point 2 in the shear diagram and the amount of jump at point 3 in the moment diagram.

Beam Example 6 Beam with prescribed displacements. In engineering application, there are cases where beams are given initial deflections and rotations due to circumstances such as thermal expansion or misfit. For such type of problems, we have to partition the total stiffness matrix into four submatrices based on two groups of quantities: the known degrees of freedom (or the corresponding unknown reaction forces) and the known forces and moments (or the corresponding unknown degrees of freedom).

An example is shown in Fig. 5.15. Let it be assumed that a small hole is punched through the middepth point of cross-section 2. This point is then pushed upward by a deflection of Δ and held by a momentless pin. It is desired to find (a) the slope at nodal point 2, (b) the force required to produce the deflection Δ at point 2, and (c) the shear and moment diagrams.

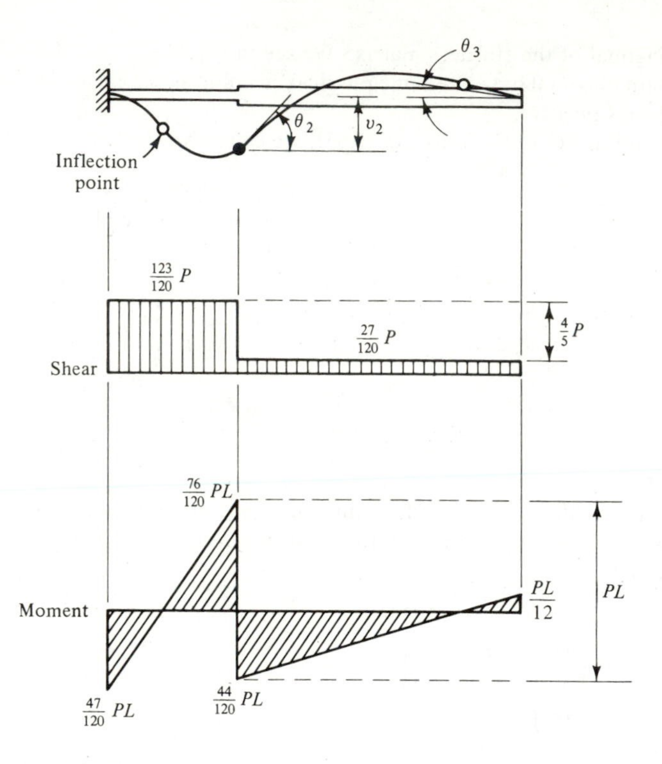

Fig. 5.14 Deflection curve, shear diagram, and moment diagram for the beam in Fig. 5.13.

Two elements are used to idealize the beam. The boundary conditions are

$$v_1 = \theta_1 = v_3 = \theta_3 = 0$$

plus the special condition that

$$v_2 = \Delta$$

By imposing the boundary conditions, we can reduce the total stiffness matrix from 6×6 to 2×2. Thus we only have to assemble a 2×2 stiffness matrix by collecting the pertinent terms in the two-element stiffness matrices instead of assembling the tedious

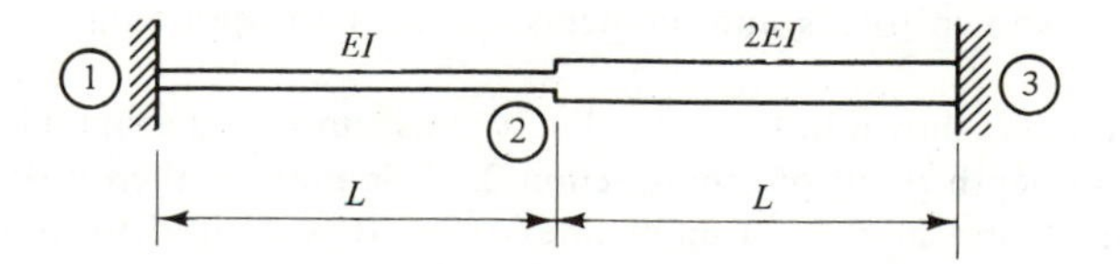

Figure 5.15 Beam with initial deflection at midspan.

and unnecessary 6×6 stiffness matrix.

$$\begin{Bmatrix} P_3 \\ P_4 \end{Bmatrix} = \frac{EI}{L} \begin{bmatrix} \frac{12}{L^2} + \frac{24}{L^2} & -\frac{6}{L} + \frac{12}{L} \\ -\frac{6}{L} + \frac{12}{L} & 4+8 \end{bmatrix} \begin{Bmatrix} v_2 \\ \theta_2 \end{Bmatrix}$$

or

$$\begin{Bmatrix} P_3 = ? \\ P_4 = 0 \end{Bmatrix} = \frac{EI}{L} \begin{bmatrix} \frac{36}{L^2} & \frac{6}{L} \\ \frac{6}{L} & 12 \end{bmatrix} \begin{Bmatrix} \Delta \\ \theta_2 \end{Bmatrix} \tag{5.64}$$

Multiplying out Eq. (5.64) yields

$$P_3 = \frac{36EI}{L^3}\Delta + \frac{6EI}{L^2}\theta_2 \tag{5.65}$$

$$0 = \frac{6EI}{L^2}\Delta + \frac{12EI}{L}\theta_2 \tag{5.66}$$

Equation (5.66) gives

$$\theta_2 = -\frac{\Delta}{2L}$$

which is the slope (clockwise) at point 2. Substituting the solution for θ_2 into Eq. (5.65) gives

$$P_3 = \frac{33EI\Delta}{L^3}$$

which is the force required to produce an upward deflection Δ at point 2.

The internal shear forces and bending moments are obtained by substituting the nodal displacements and rotations into the two element stiffness matrices, respectively.

$$\begin{Bmatrix} Y_1 \\ M_1 \\ Y_2 \\ M_2 \end{Bmatrix} = \frac{EI\Delta}{L^3} \begin{Bmatrix} -15 \\ -7L \\ 15 \\ -8L \end{Bmatrix} \quad \text{and} \quad \begin{Bmatrix} Y_2 \\ M_2 \\ Y_3 \\ M_3 \end{Bmatrix} = \frac{EI\Delta}{L^3} \begin{Bmatrix} 18 \\ 8L \\ -18 \\ 10L \end{Bmatrix}$$

The deflection curve and the shear and moment diagrams are shown in Fig. 5.16. The negative slope at point 2 shows that element 1-2 is bent with more curvature than element 2-3. This agrees with the physical reality that the former element has less bending rigidity than the latter. The jump at point 2 in the shear diagram has a value of $33EI\Delta/L^3$, which is the force required to hold point 2 at a distance Δ from its original position.

Beam Example 7 Beam connected with truss bars. An example of a beam connected to truss bars is shown in Fig. 5.17. The beam is assumed to be inextensible with a bending rigidity *EI*. The truss bars are assumed to have hinge connections at both ends. The axial rigidity *EA* of both truss bars is assumed to be equal to $24EI/L^2$.

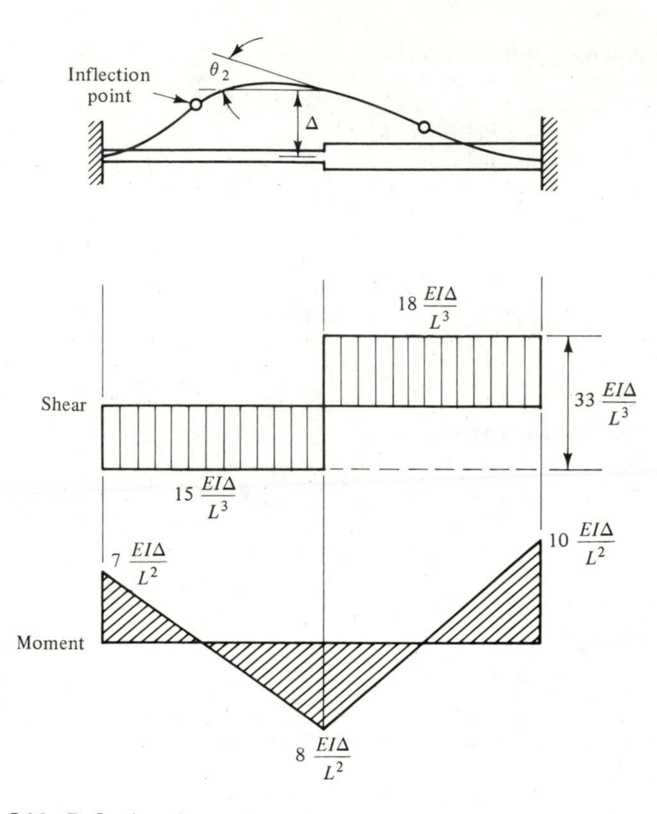

Figure 5.16 Deflection shape, shear diagram, and moment diagram for the beam in Fig. 5.15.

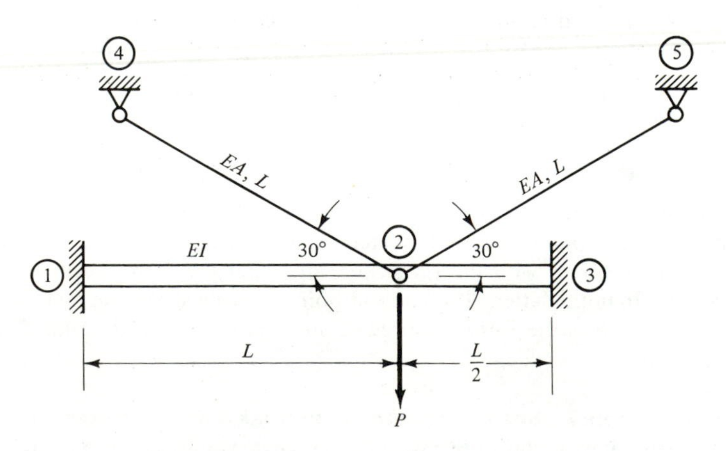

Figure 5.17 Beam connected to two truss bars.

To formulate the stiffness matrix for such a problem that has two types of elements, we simply superimpose the stiffness terms for both types of elements that are associated with the common pairs of degrees of freedom. To demonstrate the procedure, let us first formulate the total stiffness equations for the beam modeled by two elements.

Imposing the boundary conditions that

$$v_1 = \theta_1 = v_3 = \theta_3 = 0$$

we have

$$\begin{Bmatrix} P_3' \\ P_4' \end{Bmatrix} = \frac{EI}{L} \begin{bmatrix} \frac{12}{L^2} + \frac{96}{L^2} & -\frac{6}{L} + \frac{24}{L} \\ -\frac{6}{L} + \frac{24}{L} & 4+8 \end{bmatrix} \begin{Bmatrix} v_2 \\ \theta_2 \end{Bmatrix} \tag{5.67}$$

We then formulate the total stiffness equations for the truss bar elements. Imposing the boundary conditions and the inextensible beam condition that

$$u_4 = v_4 = u_5 = v_5 = u_2 = 0$$

we obtain, from Eq. (4.17),

$$\{\mathbf{P}_3''\} = \frac{EA}{L}[\sin^2(150°) + \sin^2(30°)]\{v_2\} \tag{5.68}$$

Superimposing Eqs. (5.67) and (5.68), we obtain the stiffness matrix for the total system,

$$\begin{Bmatrix} P_3 = P_3' + P_3'' \\ P_4 = P_4' \end{Bmatrix} = \begin{bmatrix} \frac{108EI}{L^3} + \frac{EA}{2L} & \frac{18EI}{L^2} \\ \frac{18EI}{L^2} & \frac{12EI}{L} \end{bmatrix} \begin{Bmatrix} v_2 \\ \theta_2 \end{Bmatrix} \tag{5.69}$$

where the sum of the unbalanced vertical force P_3' at beam joint 2 and the unbalanced vertical force P_3'' at truss joint 2 must be equal to the net external load P_3, or $-P$ in this case.

Equation (5.69) can be solved by setting $EA = 24EI/L^2$,

$$\begin{Bmatrix} v_2 \\ \theta_2 \end{Bmatrix} = \frac{L^3}{1116EI} \begin{bmatrix} 12 & -\frac{18}{L} \\ -\frac{18}{L} & \frac{120}{L^2} \end{bmatrix} \begin{Bmatrix} -P \\ 0 \end{Bmatrix} = \frac{PL^2}{186EI} \begin{Bmatrix} -2L \\ 3 \end{Bmatrix}$$

The axial forces in the two truss bars are obtained by using Eq. (4.24),

$$S_{24} = \frac{EA}{L} \lfloor \cos(150°) \quad \sin(150°) \rfloor \begin{Bmatrix} u_4 - u_2 \\ v_4 - v_2 \end{Bmatrix} = \frac{4P}{31}$$

$$S_{25} = S_{24} = \frac{4P}{31} \quad \text{(both in tension)}$$

The internal shear forces and bending moments for the two beam elements are

$$\begin{Bmatrix} Y_1 \\ M_1 \\ Y_2 \\ M_2 \end{Bmatrix} = \frac{P}{31} \begin{Bmatrix} 7 \\ 3L \\ -7 \\ 4L \end{Bmatrix} \quad \text{and} \quad \begin{Bmatrix} Y_2 \\ M_2 \\ Y_3 \\ M_3 \end{Bmatrix} = \frac{P}{31} \begin{Bmatrix} -20 \\ -4L \\ 20 \\ -6L \end{Bmatrix}$$

The resulting deflection curve, shear, and moment diagrams are shown in Fig. 5.18. The unbalanced force at point 2 due to the load P and the vertical components of the axial forces in two truss bars is equal to $27P/31$. This force is equal to the amount of jump in the shear force diagram for the beam.

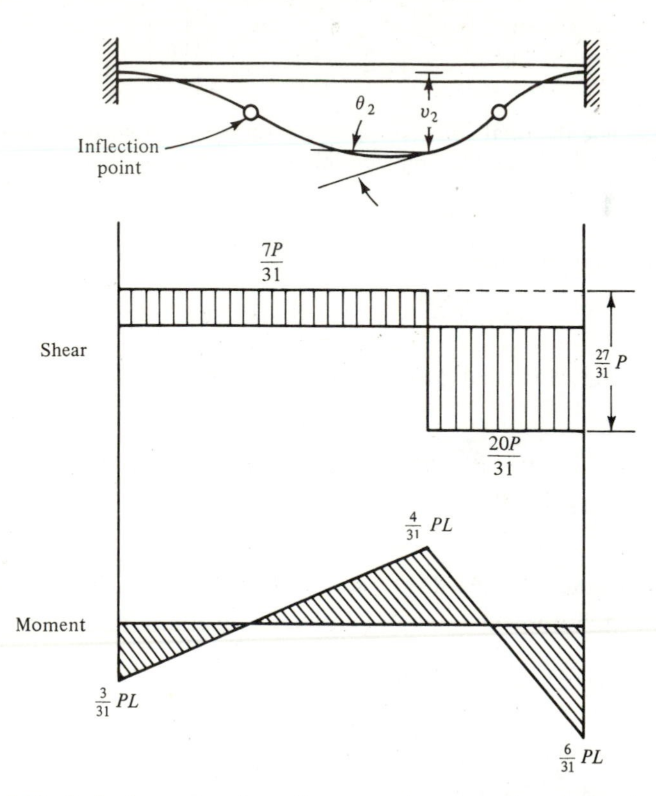

Figure 5.18 Deflection curve, shear diagram, and moment diagram for the beam in Fig. 5.17.

5.4 BEAMS UNDER DISTRIBUTED LOADS

In the examples illustrated in Sec. 5.3, the forces and bending moments applied are all concentrated at the nodal points. In practical engineering applications, the loads are not always concentrated. For such cases, we must transform the distributed loads to concentrated loads so that we can apply them at the nodal points. Two common methods are suggested as follows.

5.4.1 Work-Equivalent Loads

In the work-equivalent load method, we set the work produced by the unknown nodal concentrated loads to be equal to the work produced by the actual distributed load. This method is convenient when the distributed load can be described by a mathematical function.

Considering a beam element, the work done by the equivalent but unknown nodal loads is in the form

$$W = \tfrac{1}{2}\lfloor Y_1 \quad M_1 \quad Y_2 \quad M_2 \rfloor \begin{Bmatrix} v_1 \\ \theta_1 \\ v_2 \\ \theta_2 \end{Bmatrix} \tag{5.70}$$

On the other hand, the work done by the distributed loads due to the deflection of the beam element can be obtained as

$$W = \tfrac{1}{2} \int_0^L p(x) v(x)\, dx \tag{5.71}$$

where the deflection function is defined in Eq. (5.6) as

$$v(x) = \lfloor f_1(x) \quad f_2(x) \quad f_3(x) \quad f_4(x) \rfloor \begin{Bmatrix} v_1 \\ \theta_1 \\ v_2 \\ \theta_2 \end{Bmatrix} \tag{5.6}$$

The work given in Eq. (5.70) should be equal to the work given in Eq. (5.71). Thus we have

$$\begin{Bmatrix} Y_1 \\ M_1 \\ Y_2 \\ M_2 \end{Bmatrix} = \begin{Bmatrix} \int_0^L p(x) f_1(x)\, dx \\ \int_0^L p(x) f_2(x)\, dx \\ \int_0^L p(x) f_3(x)\, dx \\ \int_0^L p(x) f_4(x)\, dx \end{Bmatrix} \tag{5.72}$$

or in general form,

$$F_i = \int_0^L p(x) f_i(x)\, dx \tag{5.73}$$

where i denotes the degree-of-freedom number. We see in Eq. (5.73) that the work-equivalent load associated with a certain degree of freedom i is obtained by integrating the product of the distributed load function and the i-th shape function over the element length. Such a definition can be generalized to other types of finite elements, such as plate and shell finite elements.

Because such loads are found on the basis of functions that are consistent with the assumed shape functions, they are often called *consistent loads.*

Beam Example 8 Beam under uniformly distributed load. Figure 5.19 shows a beam with both ends simply supported and under uniformly distributed load p. Let it be desired to find the slope at the end (1) and deflection at the midspan (2).

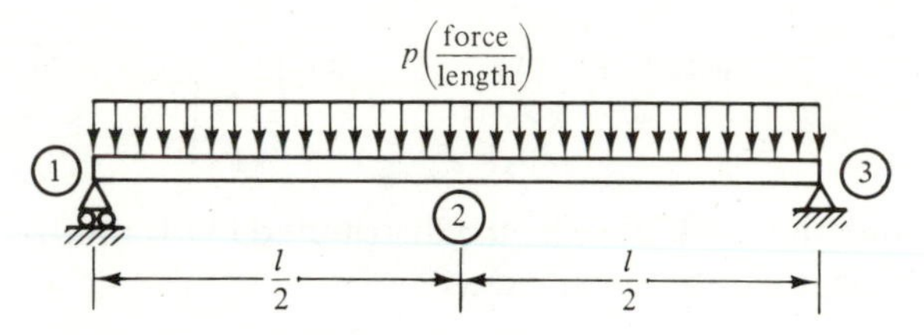

Figure 5.19 Beam under uniformly distributed load.

Because of symmetry, we can model only half of the beam and use only one beam element with length $l/2$. The boundary conditions for this beam element are

$$v_1 = \theta_2 = 0$$

The stiffness equation for this case is

$$\begin{Bmatrix} M_1 \\ Y_2 \end{Bmatrix} = EI \begin{bmatrix} \dfrac{8}{l} & -\dfrac{24}{l^2} \\ -\dfrac{24}{l^2} & \dfrac{96}{l^3} \end{bmatrix} \begin{Bmatrix} \theta_1 \\ v_2 \end{Bmatrix} \tag{5.74}$$

The work-equivalent loads for the downward uniform distributed load $-p$ can be obtained by using Eq. (5.72):

$$M_1 = \int_0^{l/2} -p\left[x - 4\left(\frac{x^2}{l}\right) + 4\frac{x^3}{l^2}\right] dx = -\frac{pl^2}{48}$$

$$Y_2 = \int_0^{l/2} -p\left[12\left(\frac{x}{l}\right)^2 - 16\left(\frac{x}{l}\right)^3\right] dx = -\frac{pl}{4}$$

Equation (5.74) can be solved as

$$\begin{Bmatrix} \theta_1 \\ v_2 \end{Bmatrix} = \frac{l^3}{24EI} \begin{bmatrix} \dfrac{12}{l^2} & \dfrac{3}{l} \\ \dfrac{3}{l} & 1 \end{bmatrix} \begin{Bmatrix} -\dfrac{pl^2}{48} \\ -\dfrac{pl}{4} \end{Bmatrix}$$

$$= -\frac{pl^3}{24EI} \begin{Bmatrix} 1 \\ \dfrac{5l}{16} \end{Bmatrix} \tag{5.75}$$

which agrees with the solution obtained using Castigliano's theorem, demonstrated in Example 3.3 of Chapter 3.

Beam Example 9 Cantilever beam under linearly varying load. Figure 5.20 shows a cantilever beam under a linearly varying distributed load

$$p(x) = -p_0\left(\frac{x}{L}\right) \tag{5.76}$$

Let it be desired to find the slope and deflection at the free end.

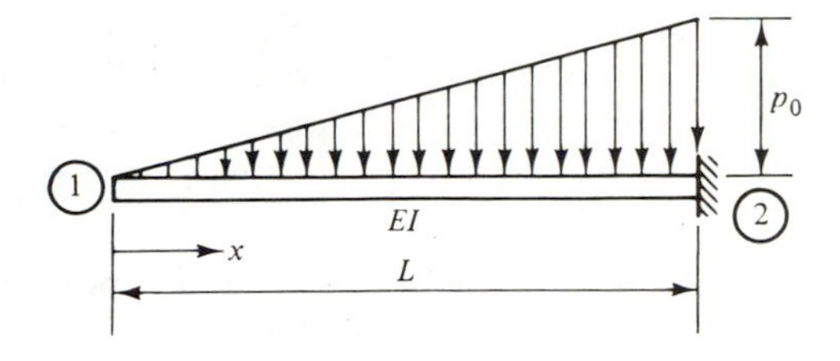

Figure 5.20 Cantilever beam under linearly distributed load.

We can use one element to model this beam. Taking into account the fixed-end conditions, we have the following simple stiffness equations:

$$\begin{Bmatrix} Y_1 \\ M_1 \end{Bmatrix} = \frac{EI}{L}\begin{bmatrix} \frac{12}{L^2} & \frac{6}{L} \\ \frac{6}{L} & 4 \end{bmatrix}\begin{Bmatrix} v_1 \\ \theta_1 \end{Bmatrix}$$

The two work-equivalent loads can be found by using Eq. (5.72):

$$Y_1 = \int_0^L -\frac{p_0 x}{L}\left[1 - 3\left(\frac{x}{L}\right)^2 + 2\left(\frac{x}{L}\right)^3\right] dx = -\frac{3p_0 L}{20}$$

$$M_1 = \int_0^L -\frac{p_0 x}{L}\left[x - 2\left(\frac{x^2}{L}\right) + \frac{x^3}{L^2}\right] dx = -\frac{p_0 L^2}{30}$$

Thus we have the solution

$$\begin{Bmatrix} v_1 \\ \theta_1 \end{Bmatrix} = \frac{L^3}{12EI}\begin{bmatrix} 4 & -\frac{6}{L} \\ -\frac{6}{L} & \frac{12}{L^2} \end{bmatrix}\begin{Bmatrix} -\frac{3p_0 L}{20} \\ -\frac{p_0 L^2}{30} \end{Bmatrix}$$

$$= \frac{p_0 L^3}{EI}\begin{Bmatrix} -\frac{L}{30} \\ \frac{1}{24} \end{Bmatrix} \tag{5.77}$$

It must be noted that the shape functions used to find the work equivalent loads for Beam Examples 8 and 9 are obtained as the exact solution to the differential equation of beam (5.1). This is why the answers we obtained by using only one element are exact.

5.4.2. Statically Equivalent Loads

In the statically equivalent load method, we have only to find the concentrated loads that are in statical equilibrium with the distributed loads. This is done by simply assigning the total distributed loads over one half of the element to one nodal point and those over the other half of the element to the other nodal point. Although this approximate method appears to be crude, it is convenient and versatile. It can cope with complex and arbitrarily distributed loading conditions in an approximate sense. This method provides results with good accuracy when sufficient number of elements is used.

Beam Example 10 Beam under uniformly distributed load. Let us consider the same simply supported, uniformly loaded beam shown in Fig. 5.19. Because of symmetry, we have to model only one half of the beam.

First, let us model the left half of the beam by one element 1-2 with length $L = l/2$. The total load on the beam element is pL. We apply $pL/2$ to each of the two nodal points. The stiffness equations are

$$\begin{Bmatrix} 0 \\ -\dfrac{pL}{2} \end{Bmatrix} = \frac{EI}{L} \begin{bmatrix} 4 & -\dfrac{6}{L} \\ -\dfrac{6}{L} & \dfrac{12}{L^2} \end{bmatrix} \begin{Bmatrix} \theta_1 \\ v_2 \end{Bmatrix}$$

Hence

$$\begin{Bmatrix} \theta_1 \\ v_2 \end{Bmatrix} = \frac{L^3}{12EI} \begin{bmatrix} \dfrac{12}{L^2} & \dfrac{6}{L} \\ \dfrac{6}{L} & 4 \end{bmatrix} \begin{Bmatrix} 0 \\ -\dfrac{pL}{2} \end{Bmatrix} \tag{5.78}$$

Substituting L by $l/2$ in Eq. (5.78) and multiplying out the two equations gives

$$\begin{Bmatrix} \theta_1 \\ v_2 \end{Bmatrix} = -\frac{pl^3}{EI} \begin{Bmatrix} \dfrac{1}{32} \\ \dfrac{l}{96} \end{Bmatrix} \tag{5.79}$$

Compared with the exact solutions given in Eq. (5.75), the one-element results for slope θ_1 and deflection v_2 are obviously smaller in absolute magnitude. The errors are 25% for θ_1 and 20% for v_2, respectively.

Let us then model the left half of the beam by using two beam elements, 1-2 and 2-3, each with length $L = l/4$. For each element we apply a concentrated load $pL/2$ at each end. The stiffness equations are

$$\begin{Bmatrix} 0 \\ -pL \\ 0 \\ -\dfrac{pL}{2} \end{Bmatrix} = \frac{EI}{L} \left[\begin{array}{c|c|c|c} 4 & -\dfrac{6}{L} & 2 & 0 \\ \hline -\dfrac{6}{L} & \dfrac{24}{L^2} & 0 & -\dfrac{12}{L^2} \\ \hline 2 & 0 & 8 & -\dfrac{6}{L} \\ \hline 0 & -\dfrac{12}{L^2} & -\dfrac{6}{L} & \dfrac{12}{L^2} \end{array}\right] \begin{Bmatrix} \theta_1 \\ v_2 \\ \theta_2 \\ v_3 \end{Bmatrix}$$

The equations can be rearranged and partitioned as

$$\begin{Bmatrix} 0 \\ 0 \\ \hline -pL \\ -\dfrac{pL}{2} \end{Bmatrix} = \frac{EI}{L} \left[\begin{array}{cc|cc} 4 & 2 & -\dfrac{6}{L} & 0 \\ 2 & 8 & 0 & -\dfrac{6}{L} \\ \hline -\dfrac{6}{L} & 0 & \dfrac{24}{L^2} & -\dfrac{12}{L^2} \\ 0 & -\dfrac{6}{L} & -\dfrac{12}{L^2} & \dfrac{12}{L^2} \end{array}\right] \begin{Bmatrix} \theta_1 \\ \theta_2 \\ \hline v_2 \\ v_3 \end{Bmatrix}$$

By the method of substitution described in Eqs. (5.49) through (5.53), we obtain

$$\begin{Bmatrix} \theta_1 \\ \theta_2 \end{Bmatrix} = \frac{3}{7L} \begin{bmatrix} 4 & -1 \\ -1 & 2 \end{bmatrix} \begin{Bmatrix} v_2 \\ v_3 \end{Bmatrix} \tag{5.80}$$

and

$$\begin{Bmatrix} -pL \\ -\dfrac{pL}{2} \end{Bmatrix} = \frac{6EI}{7L^3} \begin{bmatrix} 16 & -11 \\ -11 & 8 \end{bmatrix} \begin{Bmatrix} v_2 \\ v_3 \end{Bmatrix} \tag{5.81}$$

The solution is

$$\begin{Bmatrix} v_2 \\ v_3 \end{Bmatrix} = \frac{L^3}{6EI} \begin{bmatrix} 8 & 11 \\ 11 & 16 \end{bmatrix} \begin{Bmatrix} -pL \\ -\dfrac{pL}{2} \end{Bmatrix} = -\frac{pL^4}{12EI} \begin{Bmatrix} 27 \\ 38 \end{Bmatrix} \tag{5.82}$$

At this stage, we can substitute $l/4$ for L and obtain the final solution:

$$\begin{Bmatrix} v_2 \\ v_3 \\ \theta_1 \\ \theta_2 \end{Bmatrix} = -\frac{pl^3}{EI} \begin{Bmatrix} \dfrac{9l}{1024} \\ \dfrac{19l}{1536} \\ \dfrac{5}{128} \\ \dfrac{7}{256} \end{Bmatrix}$$

Compared to the exact solutions, we find that by using two elements the errors are reduced to $6\frac{1}{4}\%$ for θ_1 and 5.00% for v_3, respectively.

It is expected that the errors will be reduced even further as the number of elements used is increased. With the aid of a computer, results are obtained for cases with various numbers of elements and are compared with exact solution in Table 5.1. The negative signs in the table indicate that all the results are less than exact solutions in absolute magnitude. It is seen that for a three-element model (for half of the beam) the deflection reaches a satisfactory level of accuracy.

TABLE 5.1 Percentage Errors of Maximum Deflection and Slope for a Uniformly Loaded Simply Supported Beam

Number of Elements for Half the Beam	Error in Midspan Deflection (%)	Error in Slope at End (%)
1	−20.000	−25.000
2	−5.000	−6.250
3	−2.220	−2.776
4	−1.250	−1.563
5	−0.800	−1.000
6	−0.560	−0.688
7	−0.408	−0.510
8	−0.313	−0.391
9	−0.247	−0.309
10	−0.200	−0.250

5.5 APPLICATION OF PLANE FRAME ELEMENTS

The method of assemblage and procedure of solutions for using plane frame elements are introduced with the aid of illustrative examples.

Frame Example 1 Inextensible square frame. A square frame subjected to a pair of parting forces P is shown in Fig. 5.21. It is assumed that the frame members are inextensible and that the right angles at the joints are preserved. Let it be desired

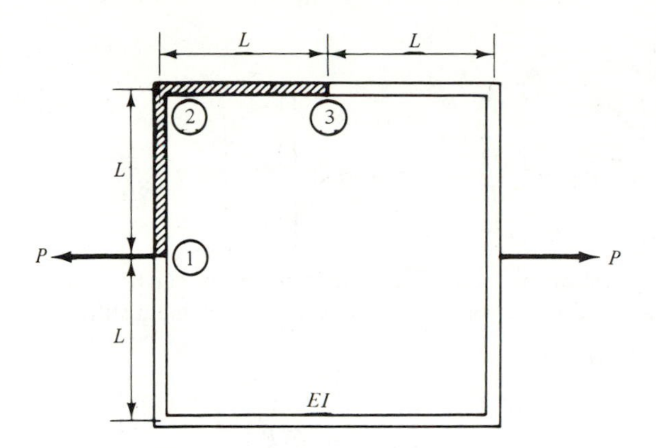

Figure 5.21 Inextensible square frame.

to find the deflection shape, the shearing force diagram, and the bending moment diagram.

Because the square frame is a structure of double symmetry, only a quadrant need be modeled and two elements, 1-2 and 2-3, are used. The reason that we assume the members to be inextensible is to eliminate the axial degrees of freedom so that we can obtain a simpler example for demonstrating the necessary procedure. The present formulation for the plane frame elements is by no means to be limited by such an assumption.

The boundary conditions for the two-element model shown in Fig. 5.21 are

$$v_1 = \theta_1 = v_2 = u_3 = \theta_3 = 0 \qquad \text{due to symmetry}$$

and

$$u_2 = 0 \qquad \text{due to inextensibility}$$

As a result, we have only to formulate a set of three stiffness equations which are assembled from the two element stiffness equations defined in Eq. (5.32). The direction sines and cosines are

$$\lambda = 0 \quad \text{and} \quad \mu = 1 \qquad \text{for element 1-2}$$

$$\lambda = 1 \quad \text{and} \quad \mu = 0 \qquad \text{for element 2-3}$$

Thus we have

$$\begin{Bmatrix} -\dfrac{P}{2} \\ 0 \\ 0 \end{Bmatrix} = \frac{EI}{L} \begin{bmatrix} \dfrac{12}{L^2} & -\dfrac{6}{L} & 0 \\ -\dfrac{6}{L} & 4+4 & -\dfrac{6}{L} \\ 0 & -\dfrac{6}{L} & \dfrac{12}{L^2} \end{bmatrix} \begin{Bmatrix} u_1 \\ \theta_2 \\ v_3 \end{Bmatrix}$$

The solution is

$$\begin{Bmatrix} u_1 \\ \theta_2 \\ v_3 \end{Bmatrix} = \frac{L^3}{288EI}\begin{bmatrix} 60 & & \\ \dfrac{72}{L} & \text{not needed} & \\ 36 & & \end{bmatrix}\begin{Bmatrix} -\dfrac{P}{2} \\ 0 \\ 0 \end{Bmatrix} = -\frac{PL^2}{48EI}\begin{Bmatrix} 5L \\ 6 \\ 3L \end{Bmatrix} \tag{5.83}$$

The internal shearing forces and bending moments can be obtained by substituting the resulting nodal displacements and slopes into the element stiffness equations (5.32). For element 1-2, we have

$$\begin{Bmatrix} X_1 \\ Y_1 \\ M_1 \\ X_2 \\ Y_2 \\ M_2 \end{Bmatrix} = \frac{EI}{L}\begin{bmatrix} \dfrac{12}{L^2} & & -\dfrac{6}{L} \\ 0 & & 0 \\ -\dfrac{6}{L} & & 2 \\ -\dfrac{12}{L^2} & \text{not needed} & \dfrac{6}{L} \\ 0 & & 0 \\ -\dfrac{6}{L} & & 4 \end{bmatrix}\left(-\frac{PL^2}{48EI}\right)\begin{Bmatrix} 5L \\ 0 \\ 0 \\ 0 \\ 0 \\ 6 \end{Bmatrix} = \begin{Bmatrix} -\dfrac{P}{2} \\ 0 \\ \dfrac{3PL}{8} \\ \dfrac{P}{2} \\ 0 \\ \dfrac{PL}{8} \end{Bmatrix}$$

For element 2-3, we have

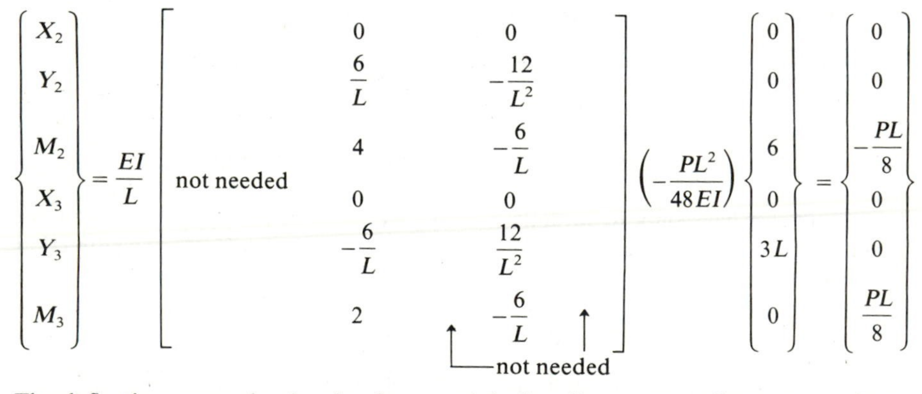

$$\begin{Bmatrix} X_2 \\ Y_2 \\ M_2 \\ X_3 \\ Y_3 \\ M_3 \end{Bmatrix} = \frac{EI}{L}\begin{bmatrix} & 0 & 0 & \\ & \dfrac{6}{L} & -\dfrac{12}{L^2} & \\ & 4 & -\dfrac{6}{L} & \\ \text{not needed} & 0 & 0 & \\ & -\dfrac{6}{L} & \dfrac{12}{L^2} & \\ & 2 & -\dfrac{6}{L} & \end{bmatrix}\left(-\frac{PL^2}{48EI}\right)\begin{Bmatrix} 0 \\ 0 \\ 6 \\ 0 \\ 3L \\ 0 \end{Bmatrix} = \begin{Bmatrix} 0 \\ 0 \\ -\dfrac{PL}{8} \\ 0 \\ 0 \\ \dfrac{PL}{8} \end{Bmatrix}$$

The deflection curve, the shearing force, and the bending moment diagrams are shown in Fig. 5.22.

It is important to note that the forces X_2 and X_3 in element 2-3 are obtained as zero, which obviously should be equal to $-P/2$ and $P/2$, respectively. This is caused by the imposition of inextensible assumption that $u_2 = 0$. Let us now disregard the condition of inextensibility.

Frame Example 2 Extensible square frame. If the members in the square frame shown in Fig. 5.21 are not assumed to be inextensible, we have one extra degree of freedom ($u_2 \neq 0$) in addition to what we had in Frame Example 1. Thus we have a

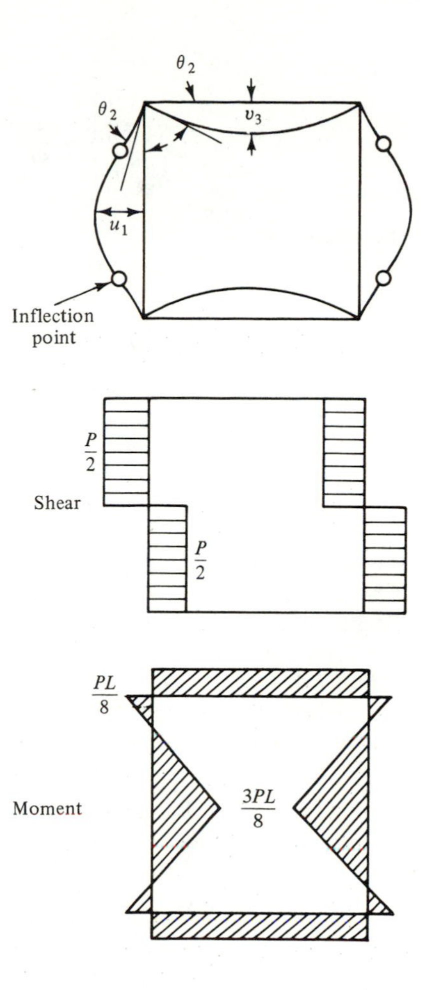

Figure 5.22 Deflection shape, shear diagram, and moment diagram for the frame in Fig. 5.21.

set of four stiffness equations.

$$\begin{Bmatrix} -\dfrac{P}{2} \\ 0 \\ \hdashline 0 \\ 0 \end{Bmatrix} = \frac{EI}{L} \left[\begin{array}{cc:cc} \dfrac{12}{L^2} & -\dfrac{12}{L^2} & -\dfrac{6}{L} & 0 \\ -\dfrac{12}{L^2} & \dfrac{12}{L^2} + \dfrac{A}{I} & \dfrac{6}{L} & 0 \\ \hdashline -\dfrac{6}{L} & \dfrac{6}{L} & 8 & -\dfrac{6}{L} \\ 0 & 0 & -\dfrac{6}{L} & -\dfrac{12}{L^2} \end{array} \right] \begin{Bmatrix} u_1 \\ u_2 \\ \hdashline \theta_2 \\ v_3 \end{Bmatrix}$$

By using the method of partitioning and substitution as described in Eqs. (5.49) to (5.53), we obtain

$$\begin{Bmatrix} \theta_2 \\ v_3 \end{Bmatrix} = \frac{3}{5L} \begin{bmatrix} 2 & -2 \\ L & -L \end{bmatrix} \begin{Bmatrix} u_1 \\ u_2 \end{Bmatrix}$$

and

$$\begin{Bmatrix} -\dfrac{P}{2} \\ 0 \end{Bmatrix} = \frac{24EI}{5L^3} \begin{bmatrix} 1 & -1 \\ -1 & 1 + \dfrac{5AL^2}{24I} \end{bmatrix} \begin{Bmatrix} u_1 \\ u_2 \end{Bmatrix}$$

The solution is

$$\begin{Bmatrix} u_1 \\ u_2 \end{Bmatrix} = \frac{L}{EA} \begin{bmatrix} 1 + \dfrac{5AL^2}{24I} & 1 \\ 1 & 1 \end{bmatrix} \begin{Bmatrix} -\dfrac{P}{2} \\ 0 \end{Bmatrix} = \begin{Bmatrix} -\dfrac{PL}{2EA} - \dfrac{5PL^3}{48EI} \\ -\dfrac{PL}{2EA} \end{Bmatrix}$$

and

$$\begin{Bmatrix} \theta_2 \\ v_3 \end{Bmatrix} = -\frac{PL^2}{48EI} \begin{Bmatrix} 6 \\ 3L \end{Bmatrix} \tag{5.84}$$

The solutions for θ_2 and v_3 are the same as those obtained in the inextensible case [Eq. (5.83)]. The solutions for u_1 and u_2 are, however, different. It is seen that when the effect of axial rigidity EA is included in the stiffness matrix of element 2-3 by setting $u_2 \neq 0$, the magnitudes of the displacements u_1 and u_2 are both increased by an amount of $PL/2EA$.

When the new values for the displacements and rotations are substituted into the two element stiffness equations, respectively, we find

$$\begin{Bmatrix} X_1 \\ Y_1 \\ M_1 \\ X_2 \\ Y_2 \\ M_2 \end{Bmatrix} = \begin{Bmatrix} -\dfrac{P}{2} \\ 0 \\ \dfrac{3PL}{8} \\ \dfrac{P}{2} \\ 0 \\ \dfrac{PL}{8} \end{Bmatrix} \quad \text{and} \quad \begin{Bmatrix} X_2 \\ Y_2 \\ M_2 \\ X_3 \\ Y_3 \\ M_3 \end{Bmatrix} = \begin{Bmatrix} -\dfrac{P}{2} \\ 0 \\ -\dfrac{PL}{8} \\ \dfrac{P}{2} \\ 0 \\ \dfrac{PL}{8} \end{Bmatrix}$$

Compared with those found in the inextensible case, X_2 and X_3 now possess values of $-P/2$ and $P/2$, respectively, instead of zero. The axial forces for this problem are thus properly included.

Frame Example 3 Antisymmetrical Z frame. Figure 5.23 shows an antisymmetrical Z frame under two concentrated loads P at points 2 and 3. All members have the same length L, axial rigidity EA, and bending rigidity EI. Let it be required to find the deflection shape and the shear force and bending moment diagrams by using three frame elements.

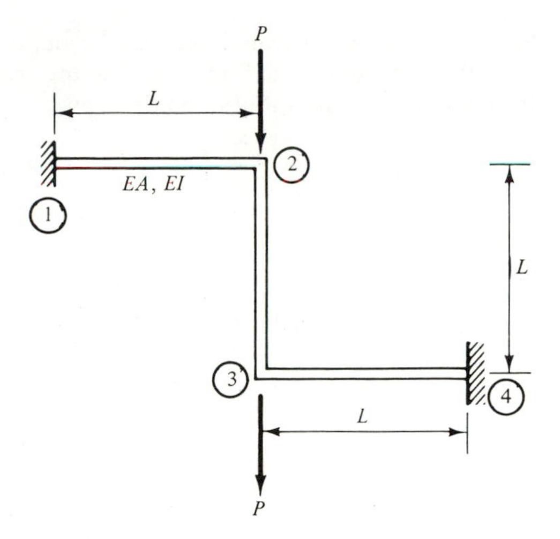

Figure 5.23 Antisymmetrical Z frame.

The fixed conditions at points 1 and 4 provided that

$$u_1 = v_1 = \theta_1 = u_4 = v_4 = \theta_4 = 0$$

We have a set of six stiffness equations:

$$\begin{Bmatrix} 0 \\ -P \\ 0 \\ 0 \\ -P \\ 0 \end{Bmatrix} = \frac{EI}{L} \begin{bmatrix} R+\frac{12}{L^2} & 0 & \frac{6}{L} & -\frac{12}{L^2} & 0 & \frac{6}{L} \\ 0 & R+\frac{12}{L^2} & -\frac{6}{L} & 0 & -R & 0 \\ \frac{6}{L} & -\frac{6}{L} & 8 & -\frac{6}{L} & 0 & 2 \\ -\frac{12}{L^2} & 0 & -\frac{6}{L} & R+\frac{12}{L^2} & 0 & -\frac{6}{L} \\ 0 & -R & 0 & 0 & R+\frac{12}{L^2} & \frac{6}{L} \\ \frac{6}{L} & 0 & 2 & -\frac{6}{L} & \frac{6}{L} & 8 \end{bmatrix} \begin{Bmatrix} u_2 \\ v_2 \\ \theta_2 \\ u_3 \\ v_3 \\ \theta_3 \end{Bmatrix}$$

This is a problem of antisymmetrical frame under a pair of loads acting along the axis of antisymmetry. The displacements and rotations have the following relations:

$$u_2 = -u_3$$

$$v_2 = v_3$$

$$\theta_2 = -\theta_3$$

Such antisymmetrical conditions can be imposed by subtracting the fourth column from the first, adding the fifth column to the second, and subtracting the sixth column from the third in the stiffness matrix above:

$$\begin{Bmatrix} 0 \\ -P \\ 0 \\ 0 \\ -P \\ 0 \end{Bmatrix} = \frac{EI}{L} \begin{bmatrix} R+\dfrac{24}{L^2} & 0 & 0 \\ 0 & \dfrac{12}{L^2} & -\dfrac{6}{L} \\ \dfrac{12}{L} & -\dfrac{6}{L} & 6 \\ -R-\dfrac{24}{L^2} & 0 & 0 \\ 0 & \dfrac{12}{L^2} & -\dfrac{6}{L} \\ \dfrac{12}{L} & \dfrac{6}{L} & -6 \end{bmatrix} \begin{Bmatrix} u_2 \\ v_2 \\ \theta_2 \end{Bmatrix}$$

The first or fourth equation gives

$$u_2 = 0$$

The second and third (or fifth and sixth) equations give the same results:

$$\begin{Bmatrix} v_2 \\ \theta_2 \end{Bmatrix} = \frac{PL^3}{6EI} \begin{Bmatrix} -1 \\ -\dfrac{1}{L} \end{Bmatrix}$$

It is seen that the axial rigidity EA (or R) does not appear in the solution. Clearly, they can be disregarded at the beginning of the formulation. This is due to the fact that all three members carry only bending moments and shear forces, no axial force.

The deflection shape is shown in Fig. 5.24. The forces and moments for element 1-2 are obtained by multiplying its stiffness matrix by the six nodal degrees of freedom.

$$\begin{Bmatrix} X_1 \\ Y_1 \\ M_1 \\ X_2 \\ Y_2 \\ M_2 \end{Bmatrix} = P \begin{Bmatrix} 0 \\ 1 \\ \dfrac{2L}{3} \\ 0 \\ -1 \\ \dfrac{L}{3} \end{Bmatrix}$$

Because of antisymmetry, the shear forces and bending moments for element 1-2 are sufficient for us to plot the shear and moment diagrams for the whole frame as shown in Fig. 5.24.

Frame Example 4 Inextensible square frame connected to truss bars. Figure 5.25 shows an inextensible square frame braced by four truss bars whose

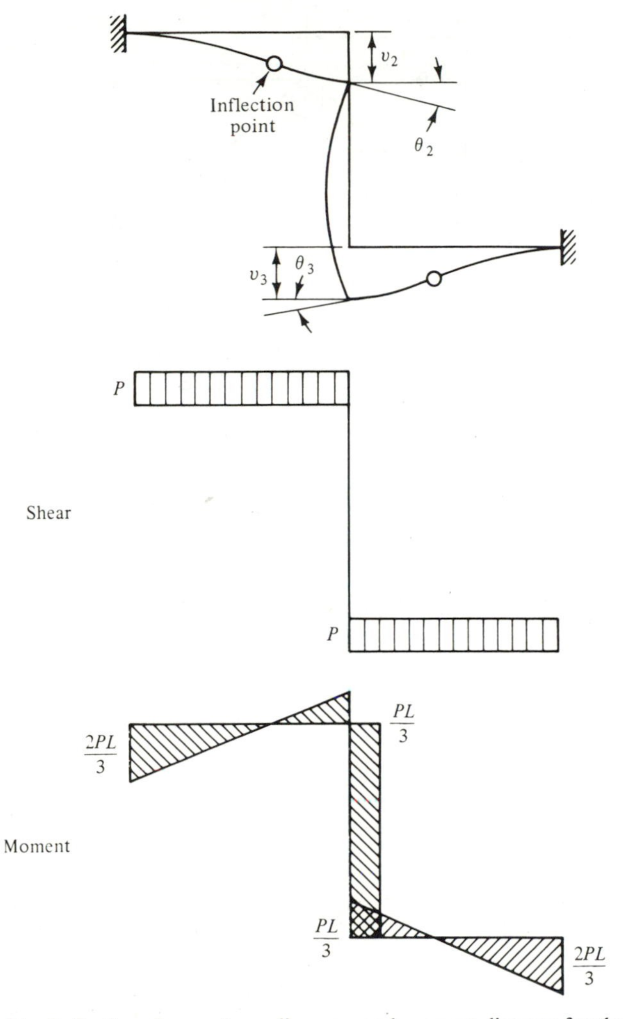

Figure 5.24 Deflection shape, shear diagram, and moment diagram for the frame in Fig. 5.23.

ends are connected to the frame by hinges. The square frame has constant bending rigidity EI. For simplicity of solution, all truss bars are assumed to have the same axial rigidity with $EA = 12\sqrt{2}EI/L^2$. The frame is loaded by a pair of parting forces P. Let it be required to find the deflection shape and the shear force and bending moment diagrams for the frame, and the axial forces in the truss bars.

Because of symmetry, only a quadrant need be analyzed. The quadrant contains two frame elements, 1-2 and 2-3, and a truss bar element, 1-3, as shown in Fig. 5.25. The boundary conditions for the quadrant are

$$v_1 = \theta_1 = u_3 = \theta_3 = 0 \qquad \text{due to symmetry}$$

$$u_2 = v_2 = 0 \qquad \text{due to inextensibility of the frame member}$$

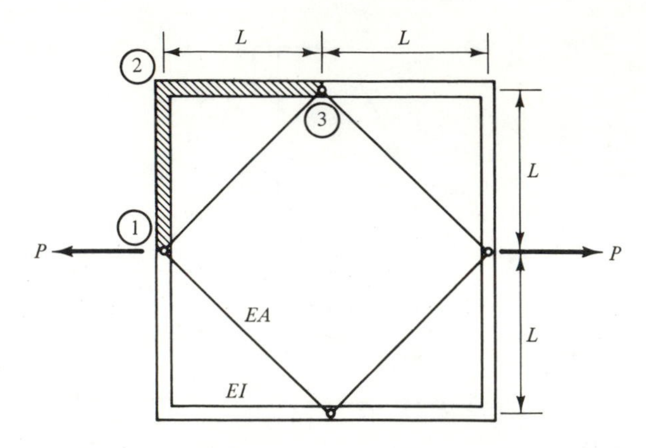

Figure 5.25 Inextensible square frame braced by truss bars.

As a result, we have only to assemble a set of three stiffness equations for the two frame elements and a set of two stiffness equations for the truss bar element.

For the two-frame-element system,

$$\begin{Bmatrix} X_1' \\ 0 \\ Y_3' \end{Bmatrix} = \frac{EI}{L}\begin{bmatrix} \frac{12}{L^2} & -\frac{6}{L} & 0 \\ -\frac{6}{L} & 8 & -\frac{6}{L} \\ 0 & -\frac{6}{L} & \frac{12}{L^2} \end{bmatrix}\begin{Bmatrix} u_1 \\ \theta_2 \\ v_3 \end{Bmatrix} \tag{5.85}$$

For the truss bar element with length $\sqrt{2}L$ and axial rigidity $12\sqrt{2}EI/L^2$, we have, from Eq. (4.17),

$$\begin{Bmatrix} X_1'' \\ Y_3'' \end{Bmatrix} = \frac{EA}{\sqrt{2}L}\begin{bmatrix} \frac{1}{2} & -\frac{1}{2} \\ -\frac{1}{2} & \frac{1}{2} \end{bmatrix}\begin{Bmatrix} u_1 \\ v_3 \end{Bmatrix} = \frac{6EI}{L^3}\begin{bmatrix} 1 & -1 \\ -1 & 1 \end{bmatrix}\begin{Bmatrix} u_1 \\ v_3 \end{Bmatrix} \tag{5.86}$$

Due to symmetry, the loading conditions for the quadrant are

$$X_1' + X_1'' = \text{external load} = -\frac{P}{2}$$

$$Y_3' + Y_3'' = 0$$

The total stiffness equations are obtained by the superposition of Eqs. (5.85) and (5.86).

$$\begin{Bmatrix} -\frac{P}{2} \\ 0 \\ 0 \end{Bmatrix} = \frac{EI}{L}\begin{bmatrix} \frac{18}{L^2} & -\frac{6}{L} & -\frac{6}{L^2} \\ -\frac{6}{L} & 8 & -\frac{6}{L} \\ -\frac{6}{L^2} & -\frac{6}{L} & \frac{18}{L^2} \end{bmatrix}\begin{Bmatrix} u_1 \\ \theta_2 \\ v_3 \end{Bmatrix}$$

The solution is

$$\begin{Bmatrix} u_1 \\ \theta_2 \\ v_3 \end{Bmatrix} = \frac{L^3}{576EI}\begin{bmatrix} 108 & & \\ \dfrac{144}{L} & \text{not needed} & \\ 84 & & \end{bmatrix}\begin{Bmatrix} -\dfrac{P}{2} \\ 0 \\ 0 \end{Bmatrix} = -\frac{PL^2}{96EI}\begin{Bmatrix} 9L \\ 12 \\ 7L \end{Bmatrix}$$

The deflection curve for the structure is shown in Fig. 5.26. The internal shear forces and bending moments are obtained by substituting the results for displacements and

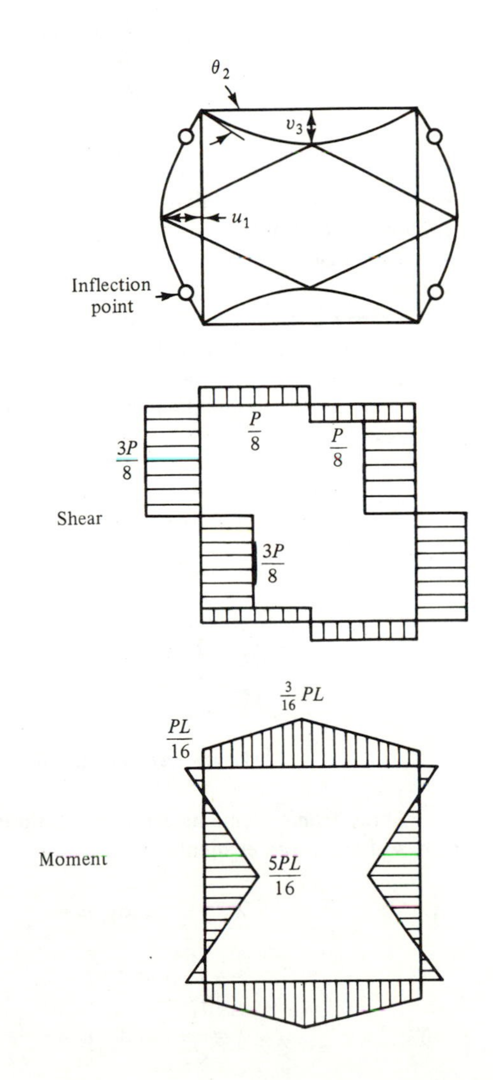

Figure 5.26 Deflection shape, shear diagram, and moment diagram for the frame in Fig. 5.25.

rotations into the two element stiffness equations, respectively,

$$\begin{Bmatrix} X_1 \\ Y_1 \\ M_1 \\ X_2 \\ Y_2 \\ M_2 \end{Bmatrix} = \frac{P}{16} \begin{Bmatrix} -6 \\ 0 \\ 5L \\ 6 \\ 0 \\ L \end{Bmatrix} \quad \text{and} \quad \begin{Bmatrix} X_2 \\ Y_2 \\ M_2 \\ X_3 \\ Y_3 \\ M_3 \end{Bmatrix} = \frac{P}{16} \begin{Bmatrix} 0 \\ 2 \\ -L \\ 0 \\ -2 \\ 3L \end{Bmatrix}$$

The shear force and bending moment diagrams are plotted in Fig. 5.26.

The axial force in the truss bar is found by using Eq. (4.24),

$$S_{1-3} = \frac{EA}{\sqrt{2}L} \lfloor \cos 45^\circ \quad \sin 45^\circ \rfloor \frac{PL^3}{96EI} \begin{Bmatrix} 9 \\ -7 \end{Bmatrix}$$

$$= \frac{\sqrt{2}P}{8} \quad \text{(tension)}$$

In the shear diagram we can see that there is a jump of $3P/4$ at nodal point 1. This force is equal to the unbalanced force between the external load P and the horizontal components $P/4$ of the tensile forces in the two truss bars. Such tensile forces in truss bars also cause a jump of shear force of $P/4$ at nodal point 3.

Frame Example 5 Portal frame with inclined columns. Figure 5.27 shows a portal frame with columns inclined at 45°. All three members have the same length L, axial rigidity EA, and bending rigidity EI. A horizontal load P is applied at joint 2. Let it be desired to find the deflection shape and the shear force and bending moment diagrams.

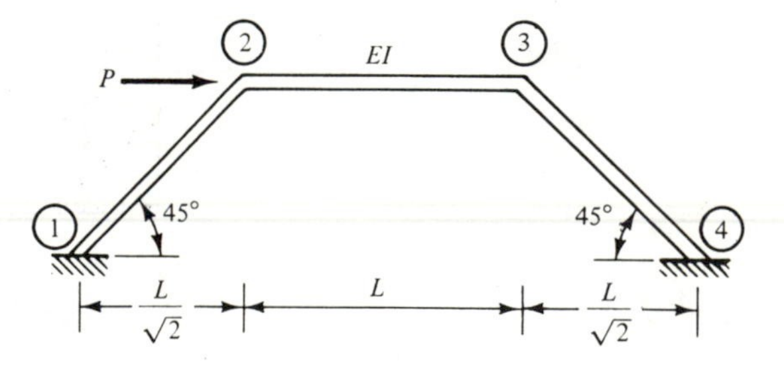

Figure 5.27 Portal frame with inclined columns.

Three frame elements are used to model the structure. The direction sines and cosines of the three elements are

$$\lambda = \frac{1}{\sqrt{2}} \quad \text{and} \quad \mu = \frac{1}{\sqrt{2}} \qquad \text{for element 1-2}$$

$$\lambda = 1 \quad \text{and} \quad \mu = 0 \qquad \text{for element 2-3}$$

$$\lambda = \frac{1}{\sqrt{2}} \quad \text{and} \quad \mu = -\frac{1}{\sqrt{2}} \qquad \text{for element 3-4}$$

The fixed conditions at two bases provide that

$$u_1 = v_1 = \theta_1 = u_4 = v_4 = \theta_4 = 0$$

We have a total of six stiffness equations for the three-element system:

$$\begin{Bmatrix} P_1 \\ 0 \\ 0 \\ P_2 \\ 0 \\ 0 \end{Bmatrix} = \frac{EI}{L} \begin{bmatrix} \frac{3R}{2}+\frac{6}{L^2} & & & & \text{symmetric} & \\ \frac{R}{2}-\frac{6}{L^2} & \frac{R}{2}+\frac{18}{L^2} & & & & \\ \frac{3\sqrt{2}}{L} & \frac{6-3\sqrt{2}}{L} & 8 & & & \\ -R & 0 & 0 & \frac{3R}{2}+\frac{6}{L^2} & & \\ 0 & -\frac{12}{L^2} & -\frac{6}{L} & -\frac{R}{2}+\frac{6}{L^2} & \frac{R}{2}+\frac{18}{L^2} & \\ 0 & \frac{6}{L} & 2 & \frac{3\sqrt{2}}{L} & \frac{3\sqrt{2}-6}{L} & 8 \end{bmatrix} \begin{Bmatrix} u_2 \\ v_2 \\ \theta_2 \\ u_3 \\ v_3 \\ \theta_3 \end{Bmatrix}$$

where $R = A/I$.

Because this is a symmetrical frame and the load is perpendicular to the axis of symmetry, it is expected that the deflection shape will be antisymmetrical, as shown in Fig. 5.28. The displacements and rotations at points 2 and 3 can be related as follows:

$$u_2 = u_3$$

$$v_2 = -v_3$$

$$\theta_2 = \theta_3$$

The three antisymmetrical conditions are imposed by adding the fourth column to the first, subtracting the fifth column from the second, and adding the sixth column to the third, respectively, in the foregoing stiffness matrix.

$$\begin{Bmatrix} P_1 \\ 0 \\ 0 \\ P_3 \\ 0 \\ 0 \end{Bmatrix} = \frac{EI}{L} \begin{bmatrix} \frac{R}{2}+\frac{6}{L^2} & \frac{R}{2}-\frac{6}{L^2} & \frac{3\sqrt{2}}{L} \\ \frac{R}{2}-\frac{6}{L^2} & \frac{R}{2}+\frac{30}{L^2} & \frac{12-3\sqrt{2}}{L} \\ \frac{3\sqrt{2}}{L} & \frac{12-3\sqrt{2}}{L} & 10 \\ \frac{R}{2}+\frac{6}{L^2} & \frac{R}{2}-\frac{6}{L^2} & \frac{3\sqrt{2}}{L} \\ -\frac{R}{2}+\frac{6}{L^2} & -\frac{R}{2}-\frac{30}{L^2} & \frac{3\sqrt{2}-12}{L} \\ \frac{3\sqrt{2}}{L} & \frac{12-3\sqrt{2}}{L} & 10 \end{bmatrix} \begin{Bmatrix} u_2 \\ v_2 \\ \theta_2 \end{Bmatrix} \tag{5.87}$$

It is seen in Eq. (5.87) that the first set of three equations and the second set of three equations are identical provided that the two horizontal loads at points 2 and 3 are

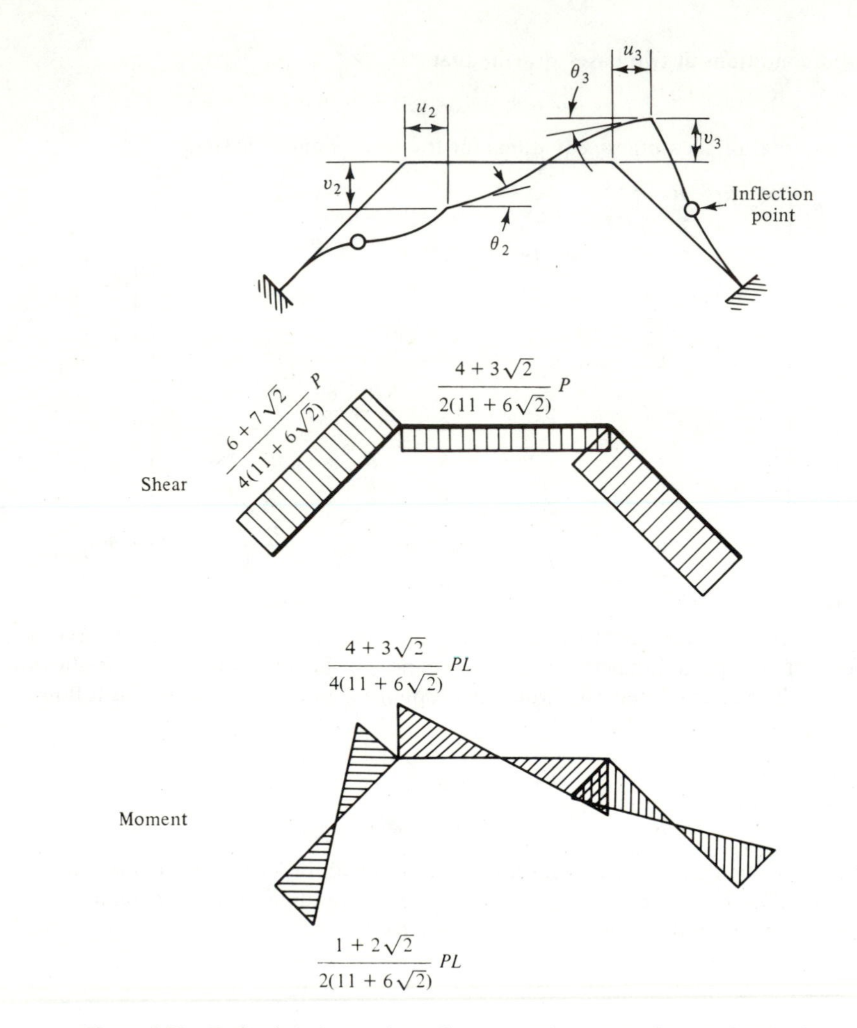

Figure 5.28 Deflection shape, shear diagram, and moment diagram for the frame in Fig. 5.27.

equal; that is,

$$P_1 = P_3$$

To satisfy this requirement, we split the load P and make

$$P_1 = P_3 = \frac{P}{2}$$

The split of the load makes the loading condition antisymmetrical, which does not, of course, change the problem.

Let us adopt the first set of three equations in Eqs. (5.87) to obtain our solution. The determinant of the 3×3 stiffness matrix, excluding EI/L, is

$$|K| = (132 + 72\sqrt{2})\frac{R}{L^2} + \frac{144}{L^4}$$

Using the adjoint method of inverse, we obtain the solution

$$\begin{Bmatrix} u_2 \\ v_2 \\ \theta_2 \end{Bmatrix} = \frac{L/EI}{|K|}\begin{bmatrix} 5R + \dfrac{138 + 72\sqrt{2}}{L^2} & & \\ -5R + \dfrac{42 + 36\sqrt{2}}{L^2} & \text{not needed} & \\ \dfrac{(6 - 3\sqrt{2})R}{L} - \dfrac{72(1 + \sqrt{2})}{L^3} & & \end{bmatrix}\begin{Bmatrix} \dfrac{P}{2} \\ 0 \\ 0 \end{Bmatrix}$$

and

$$\begin{Bmatrix} u_2 \\ v_2 \\ \theta_2 \end{Bmatrix} = \frac{PL^2/EI}{288 + 24(11 + 6\sqrt{2})RL^2}\begin{Bmatrix} 5RL^3 + 6(23 + 12\sqrt{2})L \\ -5RL^3 + 6(7 + 6\sqrt{2})L \\ 3(2 - \sqrt{2})RL^2 - 72(1 + \sqrt{2}) \end{Bmatrix} \tag{5.88}$$

Because this is an antisymmetrical problem, the results obtained in Eq. (5.88) are sufficient for us to plot the deflection curve as shown in Fig. 5.28.

Substituting the results obtained for nodal displacements and rotations into the stiffness equations for elements 1-2 and 2-3, respectively, we obtain the nodal forces and bending moments. For element 1-2, we have

$$\begin{Bmatrix} X_1 \\ Y_1 \\ M_1 \\ X_2 \\ Y_2 \\ M_2 \end{Bmatrix} = \frac{EI}{L}\begin{bmatrix} -\dfrac{R}{2} - \dfrac{6}{L^2} & -\dfrac{R}{2} + \dfrac{6}{L^2} & -\dfrac{3\sqrt{2}}{L} \\ -\dfrac{R}{2} + \dfrac{6}{L^2} & -\dfrac{R}{2} - \dfrac{6}{L^2} & \dfrac{3\sqrt{2}}{L} \\ \dfrac{3\sqrt{2}}{L} & -\dfrac{3\sqrt{2}}{L} & 2 \\ \dfrac{R}{2} + \dfrac{6}{L^2} & \dfrac{R}{2} - \dfrac{6}{L^2} & \dfrac{3\sqrt{2}}{L} \\ \dfrac{R}{2} - \dfrac{6}{L^2} & \dfrac{R}{2} + \dfrac{6}{L^2} & -\dfrac{3\sqrt{2}}{L} \\ \dfrac{3\sqrt{2}}{L} & \dfrac{-3\sqrt{2}}{L} & 4 \end{bmatrix}\begin{Bmatrix} u_2 \\ v_2 \\ \theta_2 \end{Bmatrix}$$

$$= \frac{P}{48 + 4(11 + 6\sqrt{2})RL^2}\begin{Bmatrix} -24 - 2(11 + 6\sqrt{2})RL^2 \\ 24 - 2(4 + 3\sqrt{2})RL^2 \\ 2(1 + 2\sqrt{2})(6 + RL^2)L \\ 24 + 2(11 + 6\sqrt{2})RL^2 \\ -24 + 2(4 + 3\sqrt{2})RL^2 \\ [-12 + (4 + 3\sqrt{2})RL^2]L \end{Bmatrix}$$

or

$$\begin{Bmatrix} X_1 \\ Y_1 \\ M_1 \\ X_2 \\ Y_2 \\ M_2 \end{Bmatrix} = \begin{Bmatrix} -\dfrac{P}{2} \\ \dfrac{[12-(4+3\sqrt{2})RL^2]P}{24+2(11+6\sqrt{2})RL^2} \\ \dfrac{(1+2\sqrt{2})(6+RL^2)PL}{24+2(11+6\sqrt{2})RL^2} \\ -X_1 \\ -Y_1 \\ \dfrac{[-12+(4+3\sqrt{2})RL^2]PL}{48+4(11+6\sqrt{2})RL^2} \end{Bmatrix} \tag{5.89}$$

For element 2-3, we have

$$\begin{Bmatrix} X_2 \\ Y_2 \\ M_2 \\ X_3 \\ Y_3 \\ M_3 \end{Bmatrix} = \frac{P}{48+4(11+6\sqrt{2})RL^2} \begin{Bmatrix} 0 \\ 24-2(4+3\sqrt{2})RL^2 \\ [12-(4+3\sqrt{2})RL^2]L \\ 0 \\ -24+2(4+3\sqrt{2})RL^2 \\ [12-(4+3\sqrt{2})RL^2]L \end{Bmatrix} \tag{5.90}$$

For element 3-4,

$$\begin{Bmatrix} X_3 \\ Y_3 \\ M_3 \\ X_4 \\ Y_4 \\ M_4 \end{Bmatrix} = \begin{Bmatrix} \dfrac{P}{2} \\ Y_1 \\ M_2 \\ -\dfrac{P}{2} \\ -Y_1 \\ M_1 \end{Bmatrix} \tag{5.91}$$

where the forces Y_1 and bending moments M_1 and M_2 are given in Eq. (5.89) for element 1-2.

It is desirable at this point to check equilibrium by considering the entire frame as a free body:

$$\sum F_x = P - \frac{P}{2} - \frac{P}{2} = 0$$

$$\sum F_y = Y_1 + Y_4 = Y_1 - Y_1 = 0$$

$$\sum M_{①} = P\left(\frac{L}{\sqrt{2}}\right) - Y_4(1+\sqrt{2})L - M_1 - M_4$$

$$= P\left(\frac{L}{\sqrt{2}}\right) + Y_1(1+\sqrt{2})L - 2M_1$$

Based on the expression for Y_1 and M_1 given in Eq. (5.89), we can easily show that

$$\sum M_{①} = 0$$

For a basic bending problem such as this, the effect of axial displacements in the members may contribute very little to the deflection. It is customary to assume that the members are inextensible,

$$RL^2 = \frac{AL^2}{I} = \infty$$

Based on inextensible assumption, we can obtain a modified set of results for displacements, rotations, forces, and bending moments as those obtained in Eqs. (5.88) to (5.91). This is done by simply dividing both the numerators and denominators of those resulting expressions by RL^2 and then setting RL^2 equal to infinity.

For displacements and rotations we have

$$\begin{Bmatrix} u_2 \\ v_2 \\ \theta_2 \end{Bmatrix} = \frac{PL^2/EI}{24(11+6\sqrt{2})} \begin{Bmatrix} 5L \\ -5L \\ 3(2-\sqrt{2}) \end{Bmatrix} \tag{5.92}$$

It is seen that u_2 and v_2 are equal in magnitude but opposite in sign. It means that nodal point 2 is deflected in a direction at 45° with the horizontal line or perpendicular to the axis of element 1-2. Within the limit of small deflection theory, such a conclusion agrees with the assumption that element 1-2 is inextensible.

For element 1-2, we have

$$\begin{Bmatrix} X_1 \\ Y_1 \\ M_1 \\ X_2 \\ Y_2 \\ M_2 \end{Bmatrix} = \begin{Bmatrix} -\dfrac{P}{2} \\ \dfrac{-(4+3\sqrt{2})P}{2(11+6\sqrt{2})} \\ \dfrac{(1+2\sqrt{2})PL}{2(11+6\sqrt{2})} \\ -X_1 \\ -Y_1 \\ \dfrac{(4+3\sqrt{2})PL}{4(11+6\sqrt{2})} \end{Bmatrix} \tag{5.93}$$

For element 2-3, we have

$$\begin{Bmatrix} X_2 \\ Y_2 \\ M_2 \\ X_3 \\ Y_3 \\ M_3 \end{Bmatrix} = \frac{P}{4(11+6\sqrt{2})} \begin{Bmatrix} 0 \\ -2(4+3\sqrt{2}) \\ -(4+3\sqrt{2})L \\ 0 \\ 2(4+3\sqrt{2}) \\ -(4+3\sqrt{2})L \end{Bmatrix} \tag{5.94}$$

Based on Eqs. (5.93), (5.94), and (5.91), we can plot the shear force and bending moment diagrams as shown in Fig. 5.28.

It is noted that in this problem the inclined columns are not perpendicular to the loads. They are subjected to transverse as well as axial loads. In order to transmit axial load, the axial rigidity EA (or R) must be included in the stiffness matrix such as we have done in this problem. If EA is not included, the solution will be incorrect. For example, the supposedly downward displacement v_2 in Eq. (5.88) will be positive if R is set to zero.

Frame Example 6 Two-story frame under uniformly distributed load. Figure 5.29 shows a two-story frame structure under uniformly distributed load p on the second floor. The lengths and bending rigidities are marked in the figure. Let it be desired to find the deflection shape and the shear force and bending moment diagrams by using six frame elements.

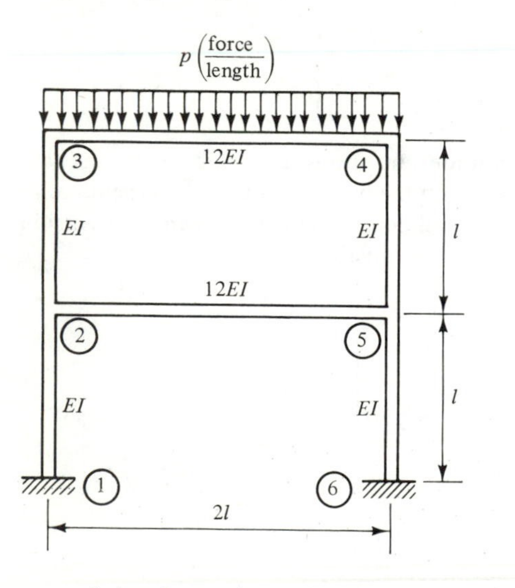

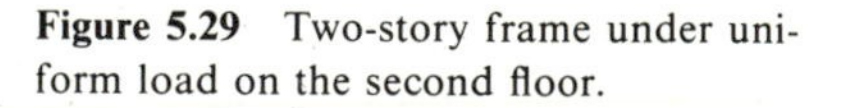
Figure 5.29 Two-story frame under uniform load on the second floor.

It has been shown in Sec. 5.4.1 that work-equivalent loads or consistent loads concentrated at nodal points can be obtained to replace the uniformly distributed load p. Using Eq. (5.7) for shape functions and Eq. (5.72) for work equivalent loads, we find for element 3-4,

$$\begin{Bmatrix} Y_3 \\ M_3 \\ Y_4 \\ M_4 \end{Bmatrix} = \frac{pL^2}{12} \begin{Bmatrix} -\dfrac{6}{L} \\ -1 \\ -\dfrac{6}{L} \\ 1 \end{Bmatrix} = pl^2 \begin{Bmatrix} -\dfrac{1}{l} \\ -\dfrac{1}{3} \\ -\dfrac{1}{l} \\ \dfrac{1}{3} \end{Bmatrix} \tag{5.95}$$

where the length for the element is $2l$.

For this symmetrical bending problem, the axial displacements in vertical columns do not contribute to bending and those in horizontal floor members contribute little to bending. The customary assumption of inextensibility may not alter the solution but does simplify the stiffness matrix. For the six-element model shown in Fig. 5.29, we have the following zero-displacement conditions:

$$u_1 = v_1 = \theta_1 = u_2 = v_2 = u_3 = v_3 = u_4$$
$$= v_4 = u_5 = v_5 = u_6 = v_6 = \theta_6 = 0$$

We have a set of four stiffness equations:

$$\begin{Bmatrix} 0 \\ -\dfrac{pl^2}{3} \\ \dfrac{pl^2}{3} \\ 0 \end{Bmatrix} = \frac{EI}{L} \begin{bmatrix} 4+4+24 & 2 & 0 & 12 \\ 2 & 4+24 & 12 & 0 \\ 0 & 12 & 24+4 & 2 \\ 12 & 0 & 2 & 4+4+24 \end{bmatrix} \begin{Bmatrix} \theta_2 \\ \theta_3 \\ \theta_4 \\ \theta_5 \end{Bmatrix} \tag{5.96}$$

It is noted that because of the assumption of inextensibility, the forces Y_3 and Y_4 found in Eq. (5.95) are not needed in the formulation (5.96).

To impose the symmetrical conditions that

$$\theta_2 = -\theta_5$$
$$\theta_3 = -\theta_4$$

we simply subtract the fourth column from the first and the third column from the second in the stiffness matrix in Eq. (5.96):

$$\begin{Bmatrix} 0 \\ -\dfrac{pl^2}{3} \\ \dfrac{pl^2}{3} \\ 0 \end{Bmatrix} = \frac{EI}{l} \begin{bmatrix} 20 & 2 \\ 2 & 16 \\ -2 & -16 \\ -20 & -2 \end{bmatrix} \begin{Bmatrix} \theta_2 \\ \theta_3 \end{Bmatrix}$$

The first (or the last) two equations give the solution:

$$\begin{Bmatrix} \theta_2 \\ \theta_3 \end{Bmatrix} = \frac{pl^3}{474EI} \begin{Bmatrix} 1 \\ -10 \end{Bmatrix} \tag{5.97}$$

The internal shear forces and bending moments are obtained by substituting the solution for rotations and the zero displacements into the individual stiffness equations for elements 1-2, 2-3, 2-5, and 3-4, respectively.

$$\begin{Bmatrix} X_1 \\ Y_1 \\ M_1 \\ X_2 \\ Y_2 \\ M_2 \end{Bmatrix} = \frac{pl}{237} \begin{Bmatrix} -3 \\ 0 \\ l \\ 3 \\ 0 \\ 2l \end{Bmatrix} \qquad \begin{Bmatrix} X_2 \\ Y_2 \\ M_2 \\ X_3 \\ Y_3 \\ M_3 \end{Bmatrix} = \frac{pl}{237} \begin{Bmatrix} 27 \\ 0 \\ -8l \\ -27 \\ 0 \\ -19l \end{Bmatrix}$$

$$\begin{Bmatrix} X_2 \\ Y_2 \\ M_2 \\ X_5 \\ Y_5 \\ M_5 \end{Bmatrix} = \frac{2pl^2}{79} \begin{Bmatrix} 0 \\ 0 \\ 1 \\ 0 \\ 0 \\ -1 \end{Bmatrix} \qquad \begin{Bmatrix} X_3 \\ Y_3 \\ M_3 \\ X_4 \\ Y_4 \\ M_4 \end{Bmatrix} = \frac{20pl^2}{79} \begin{Bmatrix} 0 \\ 0 \\ -1 \\ 0 \\ 0 \\ 1 \end{Bmatrix} \tag{5.98}$$

At this point it should be remembered that element 3-4 is subjected to a uniformly distributed load. The bending moments M_3 and M_4 obtained in Eq. (5.98) for element 3-4 should be superimposed by the fixed-end moments (work equivalent moments) caused by the uniformly distributed load. The real bending moment at point 3 acting on element 3-4 is

$$M_3 = \tfrac{1}{3}pl^2 - \tfrac{20}{79}pl^2 = \tfrac{19}{237}pl^2 \tag{5.99}$$

Such a superposition procedure is explained in Fig. 5.30. By the same token, we know the real shear force at point 3 acting on element 3-4 is

$$Y_3 = 0 + \frac{pL}{2} = pl \tag{5.100}$$

Because the computations are carried out by computers in practical applications, it is popular to use statically equivalent concentrated loads. In that case, such a superposition procedure is no longer needed. The simplicity is, however, gained at the expense of the need for more elements.

Because of symmetry, the shear forces and bending moments obtained in Eqs. (5.98) through (5.100) are sufficient for us to plot the shear and moment diagrams for the whole frame as shown in Fig. 5.31. Due to the assumption of inextensibility, the axial forces are not obtainable from Eq. (5.98). It is obvious that the columns 1-3 and 4-6 are under axial compressive forces of *pl*. The beam 2-5 is under an axial tensile force of $10pl/79$ and the beam 3-4 is under an axial compressive force of $9pl/79$. These two values are obtained from the jumps in shear force at points 2 and 3, respectively.

Frame Example 7 Arch with variable moment of inertia. Figure 5.32 shows a parabolic arch with both ends fixed and under uniformly distributed load. The

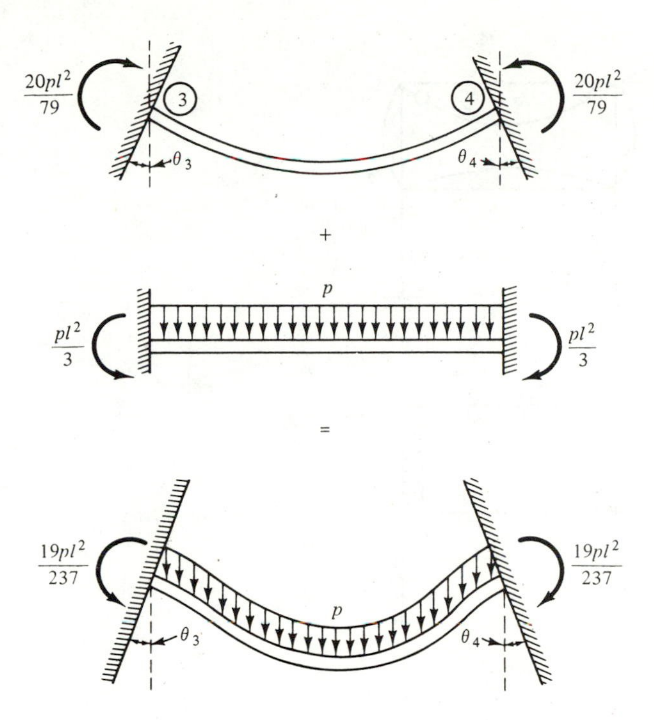

Figure 5.30 Superposition of bending moments due to end rotations and uniformly distributed load p.

centroidal axis of the arch is represented by the equation

$$y = h\left[1 - \left(\frac{x}{l}\right)^2\right]$$

where h is the rise and $2l$ is the width between two supports. The moment of inertia of the arch is defined as

$$\begin{aligned} I(x) &= I_c \sec \alpha \\ &= I_c\left[1 + \left(\frac{dy}{dx}\right)^2\right]^{1/2} \\ &= I_c\left(1 + \frac{4x^2h^2}{l^4}\right)^{1/2} \end{aligned}$$

where α is the slope of the tangent to the parabolic curve and I_c is the moment of inertia at the crown of the arch. Such design is widely used in the arch bridge and other types of constructions.

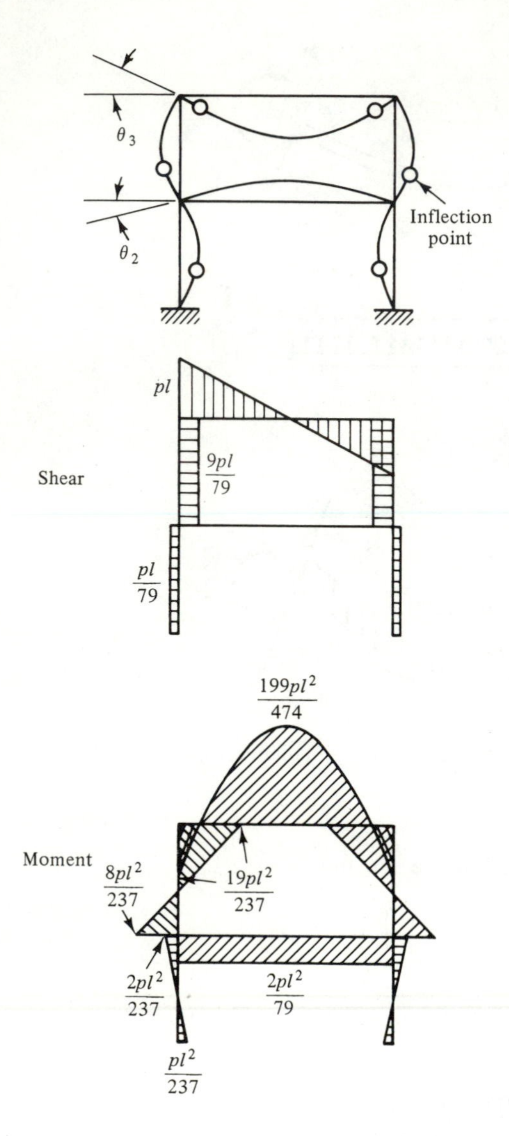

Figure 5.31 Deflection shape, shear diagram, and bending moment diagram for the frame in Fig. 5.29.

In this example, the parameters are assumed to have the following values for a steel box arch:

$$l = 83 \text{ ft} - 4 \text{ in.} = 1000 \text{ in.}$$

$$h = \frac{l}{2} = 500 \text{ in.}$$

$$I_c = 10^5 \text{ in.}^4$$

$$p = 6000 \text{ lb/ft} = 500 \text{ lb/in.}$$

$$E = 30 \times 10^6 \text{ psi}$$

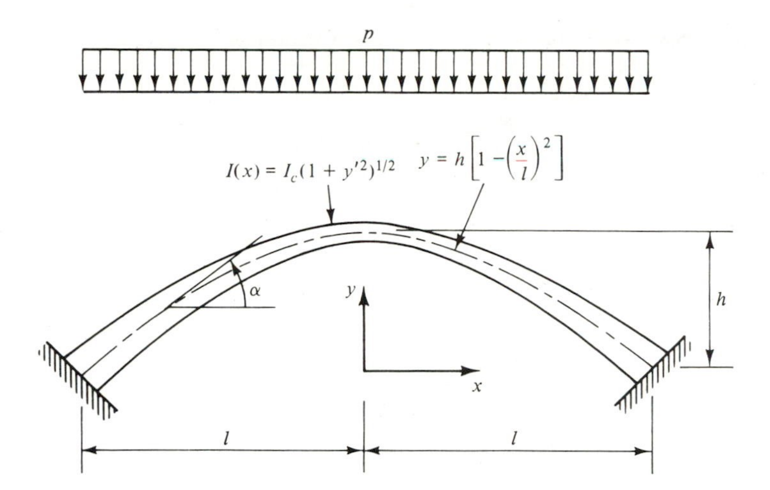

Figure 5.32 Parabolic arch with variable cross section.

For this symmetrical problem, we have to model only half of the arch. We use six different modelings with 2, 4, 6, 8, 12, and 16 elements, respectively. In each of the six models, the nodal points are equally spaced along the x direction. The constant moment of inertia for each element is approximated by the average of their values at both ends. Statically equivalent concentrated loads are applied.

For the case of inextensible arch, an exact solution is available in Ref. 5.1. The solution gives the reaction forces and moments at the ends and the crown which are listed in Table 5.2. By setting $A = 1000I$, the inextensible frame element solution is obtained and given in Table 5.2. It is seen that both the deflections and bending moments for this problem are so small that they can be considered as zero everywhere. This is actually a pure axial force problem without bending. It is seen in Table 5.2 that the results for reaction forces converge monotonically. In fact, these results are within 3% of accuracy at the eight-element level. A more realistic value of $I(x)/1000$ is then used for the cross-sectional area A. The results are presented in Table 5.3. It is seen that the deflections and bending moments are considerably larger in the extensible case.

TABLE 5.2 Results for the Arch with Variable Moment of Inertia (Inextensible Case)

Number of Elements for Half Arch	Horizontal Reaction at End (kips)	Vertical Reaction at End (kips)	Bending Moment at End (kip-in.)	Bending Moment at Crown (kip-in.)	Deflection at Crown (10^{-7} in.)
2	442	442	0.002	0.018	5.82
4	470	470	0.012	0.067	6.98
6	480	480	0.020	0.148	7.65
8	485	485	0.027	0.261	8.13
12	490	490	0.041	0.585	8.81
16	492	492	0.055	0.040	9.29
Ref. 5.1	500	500	0	0	

TABLE 5.3 Results for the Arch with Variable Moment of Inertia (Extensible Case)

Number of Elements for Half Arch	Horizontal Reaction at End (kips)	Vertical Reaction at End (kips)	Bending Moment at End (kip-in.)	Bending Moment at Crown (kip-in.)	Deflection at Crown (in.)
2	451	451	1,595	11,489	0.433
4	467	467	6,971	11,707	0.377
6	476	476	9,634	10,466	0.375
8	481	481	11,913	10,066	0.375
12	485	485	15,264	9,790	0.375
16	488	488	17,412	9,697	0.374

5.6 CONCLUDING REMARKS

The formulation and solution procedures illustrated in this chapter are based on the stiffness method. The method finds displacements and rotations first, which are then used to find the internal forces and bending moments at each nodal point.

Because the zero displacements and rotations result in reduction and simplification of the stiffness matrix, this method is particularly advantageous for structures with high number of constraints or high degrees of redundancy.

The formulation and solution procedures are systematically presented so that they can be straightforwardly adopted for computer programming. A Fortran program for the static analysis of plane frames, a user's manual, and the input and output data for an example of a stairway frame are given in Sec. 13.1.

By proper rearrangement, the stiffness formulations for beam element [Eq. (5.14)] and frame element [Eq. (5.32)] can be reduced to the same form as the well-known conventional slope-deflections equations (see, for example, Ref. 5.2). In this chapter, the formulations and the notations are the same as those presented by Martin [5.3].

REFERENCES

5.1. Parcel, R. B., and Moorman, R. B. B., *Analysis of Statically Indeterminate Structures*, John Wiley & Sons, Inc., New York, 1955, p. 430.

5.2. Maney, G. A., *Engineering Studies*, No. 1, University of Minnesota, 1915.

5.3. Martin, H. C., *Introduction to Matrix Methods of Structural Analysis*, McGraw-Hill Book Company, New York, 1966, Chaps. 4 and 8.

PROBLEMS

5.1. Derive all 16 terms in the stiffness matrix for a beam element using the equation

$$k_{ij} = EI \int_0^L f_i''(x) f_j''(x)\, dx$$

5.2. Using the beam element stiffness formulation given in Eq. (5.14), find **(a)** the displacements and rotations at all nodal points, **(b)** the shear force diagrams, and **(c)** the bending moment diagram for the problems given in Fig. P5.2. Use work-equivalent loads for distributed loads. Unless otherwise specified, all beam members are assumed as inextensible with bending rigidity EI.

5.3. Using the plane frame element stiffness formulation given in Eq. (5.32), find **(a)** the displacements and rotations at all nodal points, **(b)** the shear force diagram, and **(c)** the bending moment diagram for the problems given in Fig. P5.3. Use work-equivalent loads for distributed loads. Unless otherwise defined, all frame members are assumed as inextensible with bending rigidity EI. All right angles are assumed to be preserved after bending.

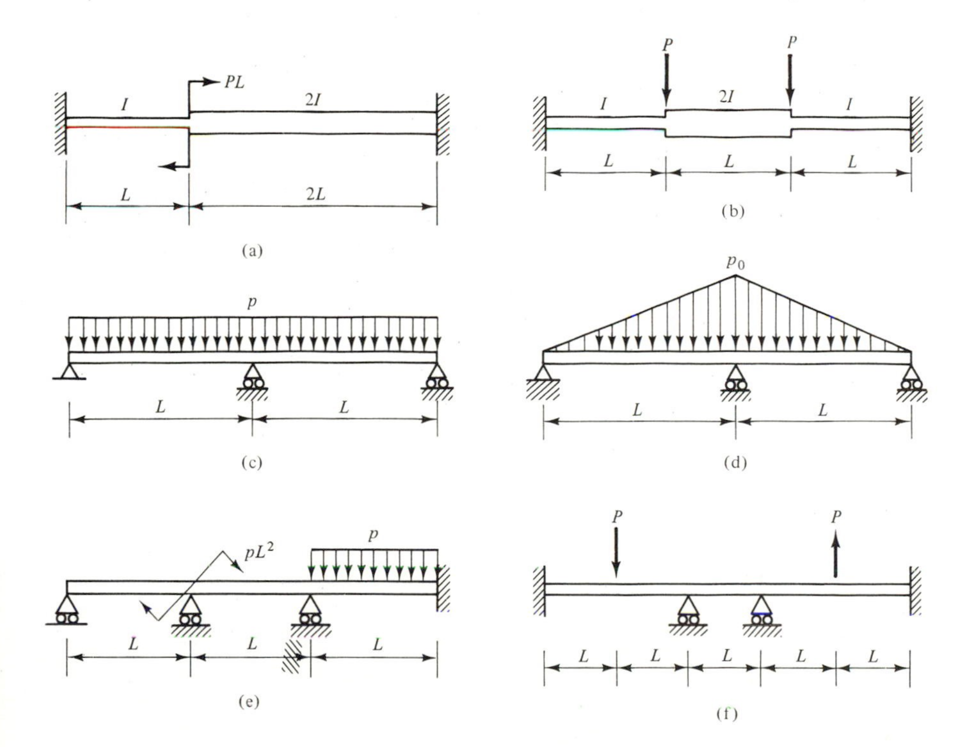

Figure P5.2

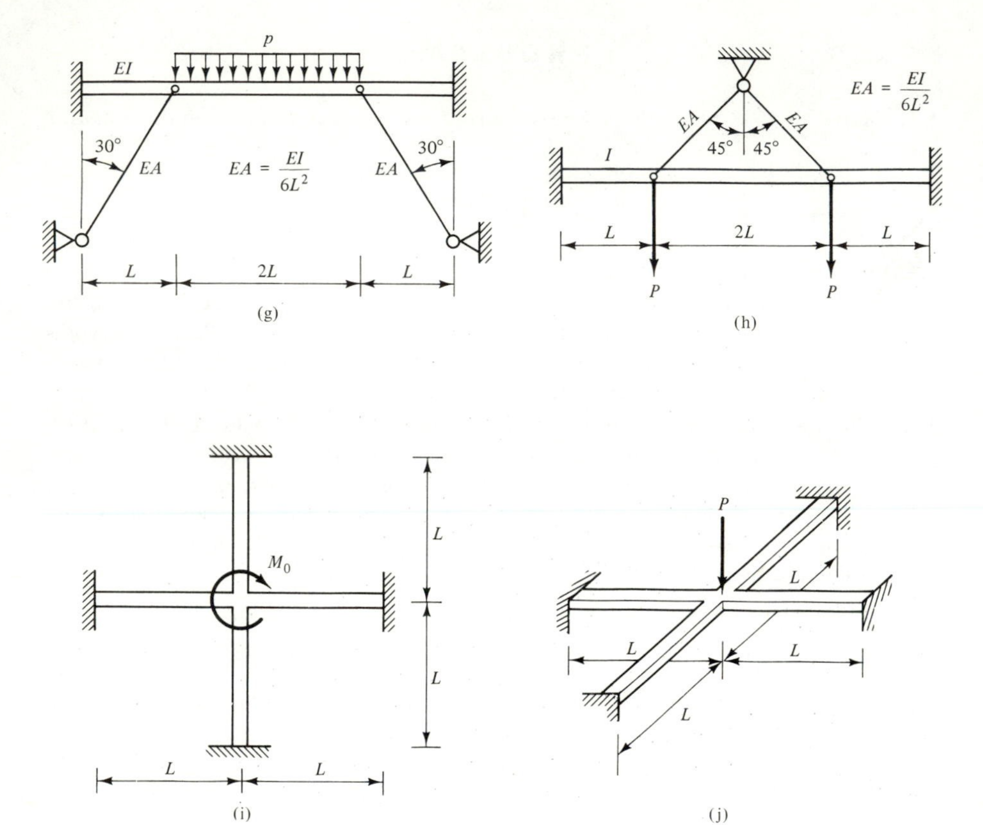

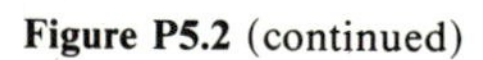

Figure P5.2 (continued)

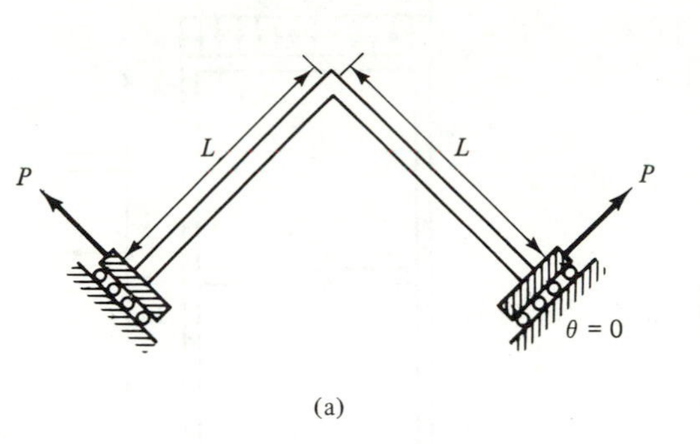

(a)

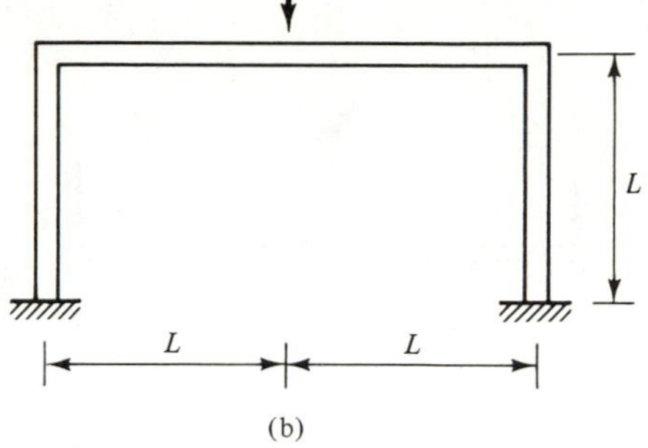

(b)

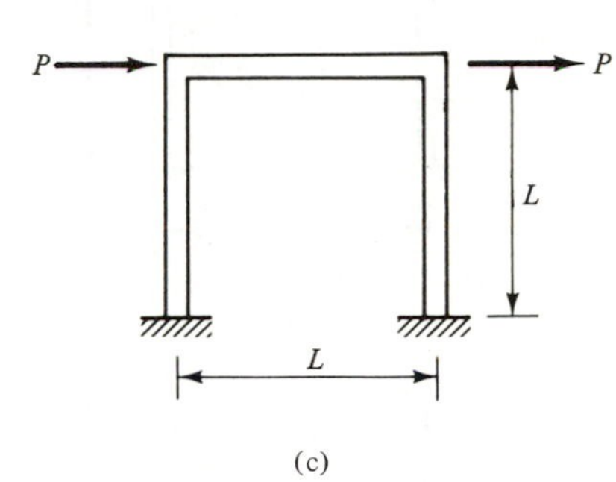

(c)

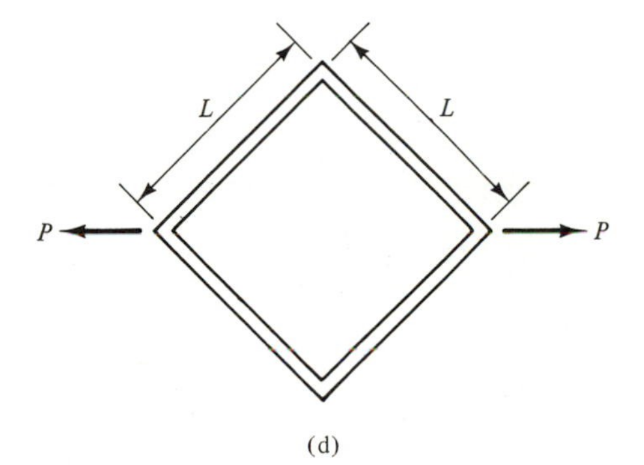

(d)

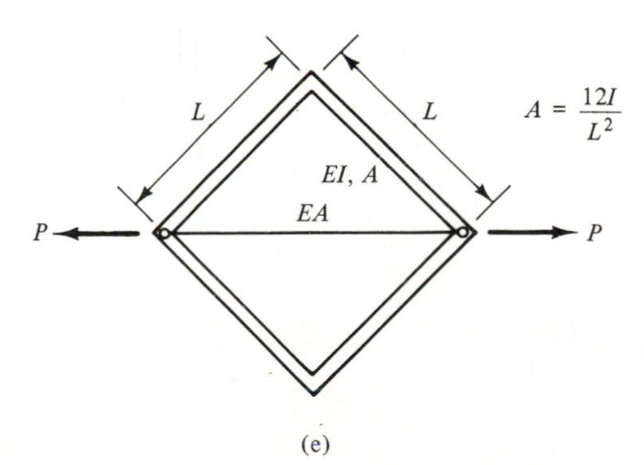

(e)

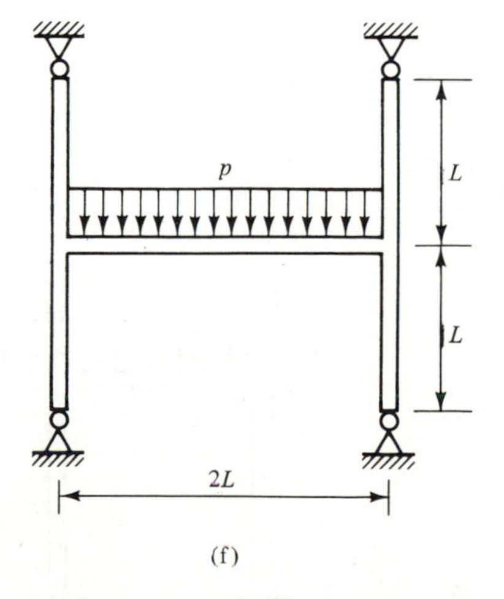

(f)

Figure P5.3

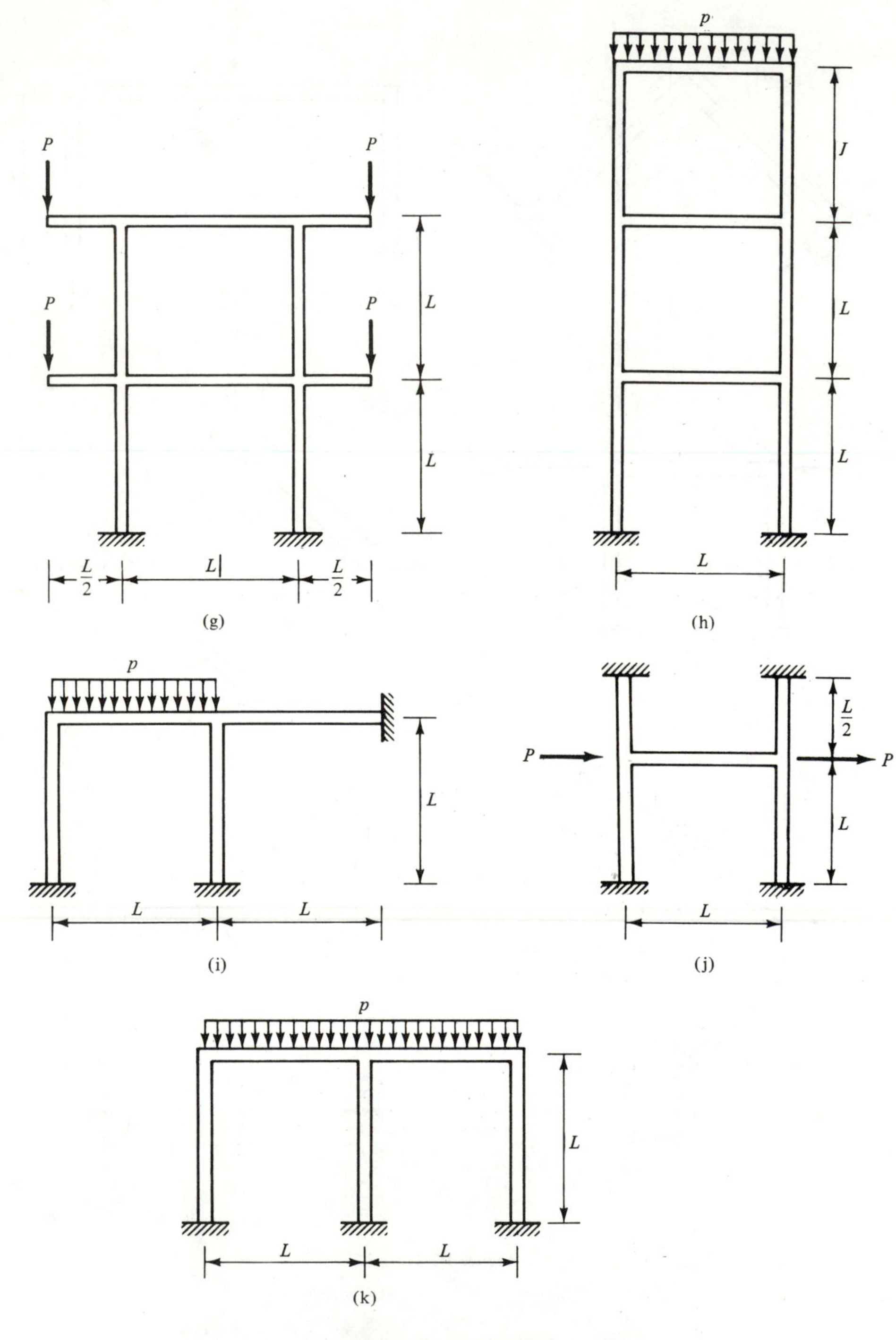

Figure P5.3 (continued)

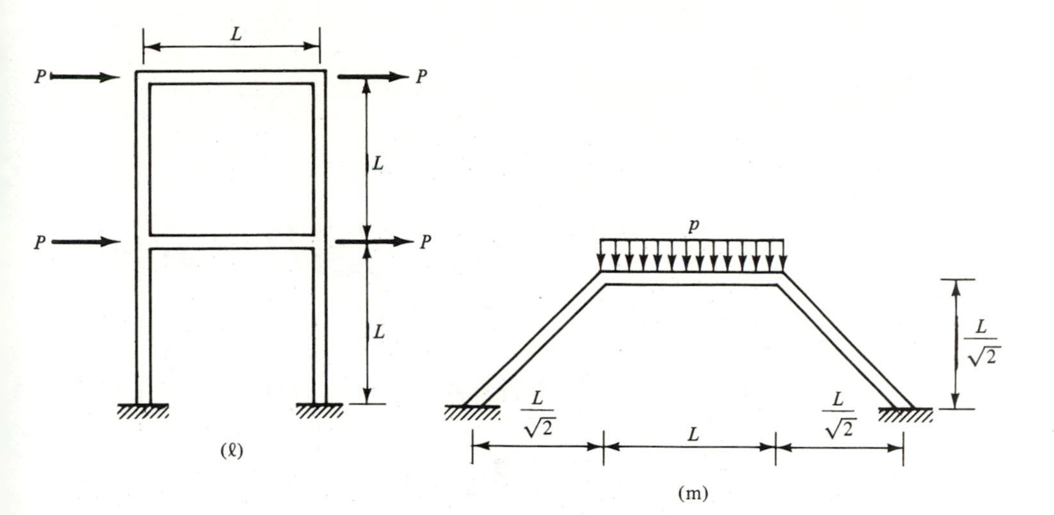

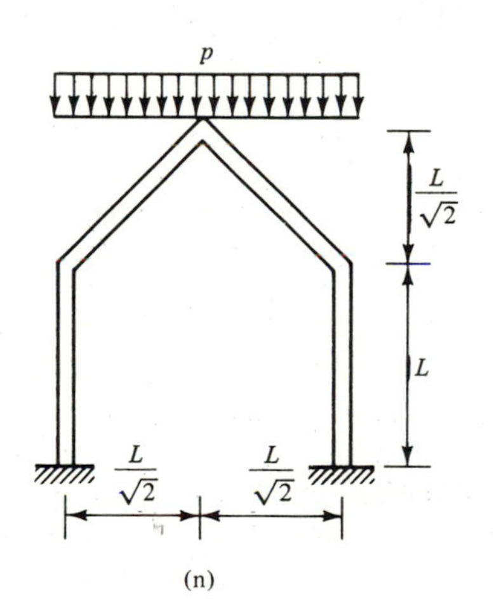

Figure P5.3 (continued)

CHAPTER 6

Nonuniform and Curved Beam Finite Elements

In structural design, we often encounter beams with nonuniform cross sections. When the weight or the cost of material is of major concern, the beams are designed with nonuniform cross sections. In civil engineering structural design, towers, arches, chimneys, and bridge girders are examples of tapered beams. In aerospace structural design, nearly all the beams are tapered. For example, wings are tapered for both weight and aerodynamic reasons. In automotive structural design, examples can be found in the suspension system where beams are designed in leaf-spring type. The shape of taper conforms to the moment diagram of a simply supported beam under a point load at midspan. Such a design can save material and produce the desirable structural damping.

Beams with curvatures are another form of structure that we often encounter in the structural design. In this chapter we limit our attention to beams curved and bent only in the plane of the curvature so that no torsion is involved. Examples for application of curved beams can be found in fuselage rings, reinforcing rings for cylindrical and conical shells, arches, curved bridge girders, hooks, and so on. It is common for curved beams to have nonuniform cross sections.

It is a common practice to model a tapered beam by a step representation using beam elements with a constant moment of inertia. It is also common to model a curved beam in a broken representation using straight beam elements. Such practices may not, however, give accurate results for a given problem unless a sufficient number of elements is used. On the other hand, it is a better practice to use straight elements with nonuniform cross section or curved elements, or even curved elements with nonuniform cross section so that accuracy can be achieved by using models with relatively fewer degrees of freedom.

In this chapter the stiffness matrix for a beam element with nonuniform moment of inertia is derived. The displacement function is based on the same cubic polynomial as that used for the uniform beam element. Superiority of the element is demonstrated by an example.

The stiffness matrix is also derived for a circularly curved beam element. Both the tangential and the radial displacement functions are based on cubic polynomials. The element is used to analyze a circular arch problem and is compared with the straight beam element. Developments in the study of curved beam finite elements are discussed.

6.1 NONUNIFORM STRAIGHT BEAM ELEMENT

In this section a beam element with nonuniform cross section as shown in Fig. 6.1 is considered. The element has the same degrees of freedom at each end as those assumed for the uniform beam element (i.e., a deflection v and a slope $\partial v/\partial x$ or θ). The element has modulus of elasticity E, length L, and a variable moment of inertia defined as

$$I(x) = I_0\left[1 + r\left(\frac{x}{L}\right)^{\alpha}\right] \tag{6.1}$$

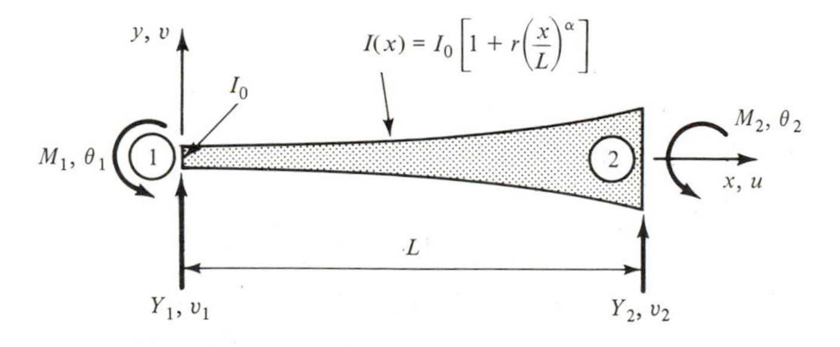

Figure 6.1 Nonuniform beam element.

where I_0 is the value of moment of inertia at nodal point 1 and r and α are parameters selected to define the approximation of the element geometry.

The displacement function for this element is assumed to be the same as that used for a uniform beam element,

$$v(x) = f_1(x)v_1 + f_2(x)\theta_1 + f_3(x)v_2 + f_4(x)\theta_2 \tag{6.2}$$

where the shape functions $f_i(x)$ are given in Eq. (5.7).

The strain energy for the element is

$$U = \frac{E}{2}\int_0^L I(x)\left(\frac{\partial^2 v}{\partial x^2}\right)^2 dx \tag{6.3}$$

where $I(x)$ is now inside the integral.

Substituting Eq. (6.2) into Eq. (6.3) and applying Castigliano's theorem, the stiffness coefficient is obtained in the form

$$k_{ij} = E\int_0^L I(x)f_i''(x)f_j''(x)\,dx \tag{6.4}$$

The only difference between Eqs. (6.4) and (5.13) is that $I(x)$ is inside the integral in Eq. (6.4).

The explicit form of the stiffness matrix is obtained by substituting Eqs. (6.1) and (5.7) into Eq. (6.4).

$$\begin{Bmatrix} Y_1 \\ M_1 \\ Y_2 \\ M_2 \end{Bmatrix} = \frac{EI_0}{L}\begin{bmatrix} \frac{12}{L^2}C_{11} & & \text{symmetric} & \\ \frac{6}{L}C_{21} & 4C_{22} & & \\ -\frac{12}{L^2}C_{11} & -\frac{6}{L}C_{21} & \frac{12}{L^2}C_{11} & \\ \frac{6}{L}C_{41} & 2C_{42} & -\frac{6}{L}C_{41} & 4C_{44} \end{bmatrix}\begin{Bmatrix} v_1 \\ \theta_1 \\ v_2 \\ \theta_2 \end{Bmatrix} \tag{6.5a}$$

where

$$\begin{aligned} C_{11} &= 1 + 3r\left(\frac{1}{\alpha+1} - \frac{4}{\alpha+2} + \frac{4}{\alpha+3}\right) \\ C_{21} &= 1 + 2r\left(\frac{2}{\alpha+1} - \frac{7}{\alpha+2} + \frac{6}{\alpha+3}\right) \\ C_{22} &= 1 + r\left(\frac{4}{\alpha+1} - \frac{12}{\alpha+2} + \frac{9}{\alpha+3}\right) \\ C_{41} &= 1 + 2r\left(\frac{1}{\alpha+1} - \frac{5}{\alpha+2} + \frac{6}{\alpha+3}\right) \\ C_{42} &= 1 + 2r\left(\frac{2}{\alpha+1} - \frac{9}{\alpha+2} + \frac{9}{\alpha+3}\right) \\ C_{44} &= 1 + r\left(\frac{1}{\alpha+1} - \frac{6}{\alpha+2} + \frac{9}{\alpha+3}\right) \end{aligned} \tag{6.5b}$$

Without the constants C_{ij}, Eq. (6.5a) is in precisely the same form as Eq. (5.14). These constants account for the effect of nonuniform moment of inertia. Equations (6.5a) and (6.5b) are available in Ref. 6.1.

For a more general case in which the beam elements are oriented arbitrarily in a two-dimensional plane, the stiffness matrix can be obtained by using the coordinate transformation procedure described in Section 5.2 or Eq. (5.31). In that case, however, the stiffness coefficients related to the two local axial displacement degrees of freedom must take into account the effect of nonuniform cross section. Let it be assumed that the cross-sectional areas are defined as

$$A(x) = A_0\left[1 + s\left(\frac{x}{L}\right)^{\beta}\right] \tag{6.6}$$

where A_0 is the cross-sectional area at nodal point 1 and s and β are geometric parameters. Recalling and employing Eq. (4.9) gives

$$k_{ij} = E\int_0^L A(x)f'_i f'_j(x)\,dx$$

We thus obtain the stiffness matrix related to the two local axial-displacement degrees of freedom:

$$\begin{Bmatrix} X_1 \\ X_2 \end{Bmatrix} = \frac{E}{L}A_0\left(1 + \frac{s}{\beta + 1}\right)\begin{bmatrix} 1 & -1 \\ -1 & 1 \end{bmatrix}\begin{Bmatrix} u_1 \\ u_2 \end{Bmatrix} \tag{6.7}$$

This stiffness matrix is the same as that for a truss bar element with uniform cross section with $A = A_0[1 + s/(\beta + 1)]$. For $\beta = 1$, $A(x)$ varies linearly with x, $A = A_0(1 + s/2)$, which is the value of A at midlength of the element. The taper of beams can, in general, vary differently from those defined in Eqs. (6.1) and (6.6). The definition of a completely general representation may, however, prohibit the explicit formulation of the stiffness coefficients. The more general case can still be accommodated by the use of numerical integration in the computation of the element stiffness coefficients.

6.2 APPLICATION TO TAPERED BEAM EXAMPLE

A tapered cantilever beam is shown in Fig. 6.2a. The moment of inertia is assumed as defined by Eq. (6.1) with $r = 8$ and $\alpha = 1$. Let it be desired to find the deflection and slope at the free end by using both nonuniform and uniform beam elements and compare the results with exact solution.

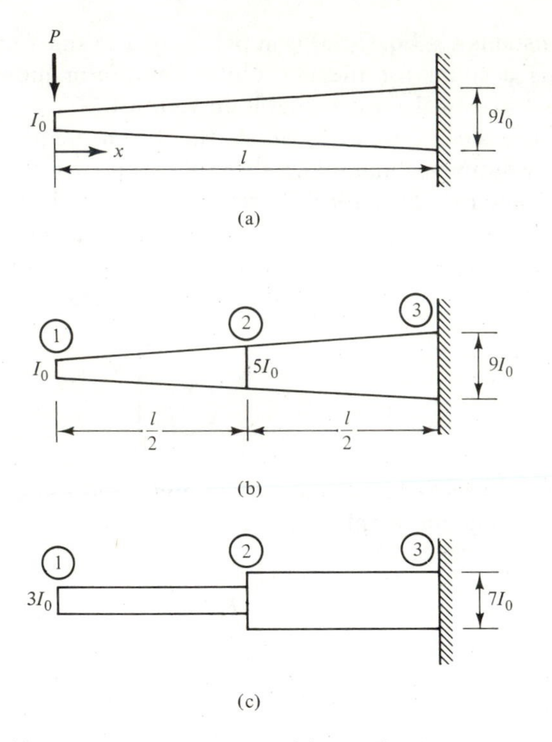

Figure 6.2 (a) Tapered cantilever beam; (b) modeling by two tapered elements; (c) modeling by two uniform elements.

6.2.1 Exact Solution

An exact solution for this problem can be obtained by direct integration of the differential equation for a beam,

$$EI(x)\frac{d^2v}{dx^2} = -Px \tag{6.8}$$

Integrating Eq. (6.8) once and twice yields θ and v, respectively. The two integration constants can be obtained based on the boundary conditions that $v = \theta = 0$ at $x = l$. The exact deflection and rotation at the free end are

$$\begin{aligned} v &= \frac{Pl^3}{r^2EI_0}\left[1 - \frac{r}{2} - \frac{1}{r}\log(1+r)\right] = -0.051166\frac{Pl^3}{EI_0} \\ \theta &= \frac{Pl^2}{r^2EI_0}[r - \log(1+r)] = 0.090668\frac{Pl^2}{EI_0} \end{aligned} \tag{6.9}$$

A derivation of Eq. (6.9) can be found in Ref. 6.2.

6.2.2 Nonuniform Beam Elements

Let us first model the tapered cantilever beam by using equal-length tapered beam elements.

If one element is used, we have only to form a 2×2 stiffness matrix due to the boundary conditions $v_2 = \theta_2 = 0$. Based on $r = 8$ and $\alpha = 1$, Eq. (6.5a) gives

$$\begin{Bmatrix} Y_1 = -P \\ M_1 = 0 \end{Bmatrix} = \frac{EI_0}{l}\begin{bmatrix} \dfrac{60}{l^2} & \dfrac{22}{l} \\ \dfrac{22}{l} & 12 \end{bmatrix}\begin{Bmatrix} v_1 \\ \theta_1 \end{Bmatrix}$$

$$\begin{Bmatrix} v_1 \\ \theta_1 \end{Bmatrix} = \frac{l^3}{118EI_0}\begin{bmatrix} 6 & -\dfrac{11}{l} \\ -\dfrac{11}{l} & \dfrac{30}{l^2} \end{bmatrix}\begin{Bmatrix} -P \\ 0 \end{Bmatrix} = \frac{Pl^2}{EI_0}\begin{Bmatrix} -0.05085l \\ 0.09322 \end{Bmatrix}$$

Compared with Eq. (6.9), the errors for v_1 and θ_1 are -0.62% and 2.82%, respectively.

A model of the beam using two tapered elements is shown in Fig. 6.2b. For the two elements, the moments of inertia are defined as

$$I_{1-2} = I_0\left[1 + 4\left(\frac{x}{L}\right)\right]$$

$$I_{2-3} = 5I_0\left[1 + 0.8\left(\frac{x}{L}\right)\right]$$

where $L = l/2$ is the half-length of the beam.

For element $1 - 2$, we obtain, for Eq. (6.5b),

$$C_{11} = 3 \qquad C_{21} = \tfrac{7}{3} \qquad C_{22} = 2$$

$$C_{41} = \tfrac{11}{3} \qquad C_{42} = 3 \qquad C_{44} = 4$$

For element $2 - 3$, we obtain, from Eq. (6.5b),

$$C_{11} = \tfrac{7}{5} \qquad C_{21} = \tfrac{19}{15} \qquad C_{22} = \tfrac{6}{5}$$

The boundary conditions are $v_3 = \theta_3 = 0$. We have a set of four stiffness equations:

$$\begin{Bmatrix} P_1 = -P \\ P_2 = 0 \\ P_3 = 0 \\ P_4 = 0 \end{Bmatrix} = \frac{EI_0}{L} \begin{bmatrix} \frac{36}{L^2} & \frac{14}{L} & -\frac{36}{L^2} & \frac{22}{L} \\ \frac{14}{L} & 8 & -\frac{14}{L} & 6 \\ -\frac{36}{L^2} & -\frac{14}{L} & \frac{36+84}{L^2} & -\frac{22-38}{L} \\ \frac{22}{L} & 6 & -\frac{22-38}{L} & 16-24 \end{bmatrix} \begin{Bmatrix} v_1 \\ \theta_1 \\ v_2 \\ \theta_2 \end{Bmatrix} \tag{6.10}$$

Multiplying out the last two rows of Eq. (6.10) and rearranging gives

$$\begin{Bmatrix} v_2 \\ \theta_2 \end{Bmatrix} = \frac{1}{284} \begin{bmatrix} 112 & 41L \\ -\frac{201}{L} & -59 \end{bmatrix} \begin{Bmatrix} v_1 \\ \theta_1 \end{Bmatrix} \tag{6.11}$$

Multiplying out the first two rows of Eq. (6.10) and employing the relation obtained in Eq. (6.11) gives

$$\begin{Bmatrix} -P \\ 0 \end{Bmatrix} = \frac{EI_0}{L} \begin{bmatrix} \frac{6.2324}{L^2} & \frac{4.2324}{L} \\ \frac{4.2324}{L} & 4.7324 \end{bmatrix} \begin{Bmatrix} v_1 \\ \theta_1 \end{Bmatrix}$$

$$\begin{Bmatrix} v_1 \\ \theta_1 \end{Bmatrix} = \frac{L^3}{11.581 EI_0} \begin{bmatrix} 4.7324 & -\frac{4.2324}{L} \\ -\frac{4.2324}{L} & \frac{6.2324}{L^2} \end{bmatrix} \begin{Bmatrix} -P \\ 0 \end{Bmatrix}.$$

Because $L = l/2$,

$$\begin{Bmatrix} v_1 \\ \theta_1 \end{Bmatrix} = \frac{PL^2}{EI_0} \begin{Bmatrix} -0.40864L \\ 0.36546 \end{Bmatrix} = \frac{Pl^2}{EI_0} \begin{Bmatrix} -0.051080l \\ 0.091365 \end{Bmatrix}$$

Compared with Eq. (6.9), the errors for v_1 and θ_1 are -0.17% and 0.77%, respectively. We thus find that at the two-element level, the errors for v_1 and θ_1 are already within 1%.

6.2.3 Uniform Beam Elements

Let us now model the tapered cantilever beam by stepped representation using equal-length uniform beam elements. Each element has a constant moment of inertia equal to its value at midlength. A two-element model is shown in Fig. 6.2c.

If one element is used, we have

$$\begin{Bmatrix} -P \\ 0 \end{Bmatrix} = \frac{E(5I_0)}{l} \begin{bmatrix} \frac{12}{l^2} & \frac{6}{l} \\ \frac{6}{l} & 4 \end{bmatrix} \begin{Bmatrix} v_1 \\ \theta_1 \end{Bmatrix}$$

$$\begin{Bmatrix} v_1 \\ \theta_1 \end{Bmatrix} = \frac{l^3}{30EI_0} \begin{bmatrix} 2 & -\frac{3}{l} \\ -\frac{3}{l} & \frac{6}{l^2} \end{bmatrix} \begin{Bmatrix} -P \\ 0 \end{Bmatrix} = \frac{Pl^2}{EI_0} \begin{Bmatrix} -\frac{l}{15} \\ \frac{1}{10} \end{Bmatrix}$$

Compared with Eq. (6.9), the errors in v_1 and θ_1 are 30.3% and 10.3%, respectively.

If two elements are used, we can find that the errors in v_1 and θ_1 become 8.58% and 5.04%, respectively. The errors in v_1 and θ_1 for different numbers of elements are shown in Table 6.1. It is seen that the errors are monotonically reduced as the number of elements is successively increased. The errors in v_1 and θ_1 become approximately 1% at the six-element level. The results obtained in Section 6.2.2 using tapered beam elements are also shown in Table 6.1 for comparison.

TABLE 6.1 Percentage Errors in Deflection and Slope at the Free End of a Tapered Cantilever Beam

Number of Elements	Uniform Elements		Tapered Elements	
	Error in Tip Deflection (%)	Error in Tip Slope (%)	Error in Tip Deflection (%)	Error in Tip Slope (%)
1	30.30	10.29	−0.62	2.82
2	8.58	5.04	−0.17	0.77
3	3.92	2.99		
4	2.22	1.97		
6	0.98	1.02		
8	0.55	0.62		
12	0.24	0.30		
16	0.14	0.17		

Table 6.1 shows that for this tapered cantilever beam problem, the tapered beam element is considerably more efficient than the uniform beam element. Such a conclusion is important when a tapered beam or frame problem requires

a large amount of computation. For example, computation of dynamic response of a structure consisting of tapered beams due to arbitrary forcing function, such as that caused by an earthquake, may involve formidable computational expenses. Such expenses can be reduced by minimizing the number of degrees of freedom using nonuniform beam elements.

6.3 CIRCULARLY CURVED BEAM FINITE ELEMENTS

Curved beams are a special form of curved shells. A study of curved beams is an important first step toward gaining insight into the more complex shells. Although there have been considerable developments in the curved shell finite elements, they are beyond the scope of this text. Only developments in curved beam elements are introduced in this chapter.

The basic difference between a curved and a straight beam is that, in the small deflection theory, axial and flexural behaviors are coupled in the curved beam but not in the straight beam. Furthermore, in the finite element formulation, the displacement functions for curved beam elements must be capable of representing three rigid-body displacements: two orthogonal displacements and a rotation, all in the plane of curvature of the element.

There are many circularly curved beam finite elements available. It is, however, not within the scope of this text to introduce all these elements in detail. For simplicity and clarity, we simply introduce seven types of circularly curved beam finite elements by describing the assumptions of displacement functions for tangential (u) and radial displacements (v) and the assumed degrees of freedom at each nodal point. To explain in depth, one element (element 3) will be formulated and evaluated in detail. Finally, an arch example will be used to compare all these elements.

Element 1: linear u and cubic v [6.3, 6.4]

$$u = a_1 + a_2 s$$

$$v = a_3 + a_4 s + a_5 s^2 + a_6 s^3 \tag{6.12}$$

$$\text{d.o.f.'s at each node} = u,\ v,\ \partial v/\partial s$$

where s is the curvilinear distance measured from nodal point 1.

Element 2: linear u and cubic v with rigid-body displacement terms [6.5]

$$u = a_1 \cos\phi - a_2 \sin\phi + a_3 R(\cos\beta \cos\phi - 1) + a_4 s$$

$$v = a_1 \sin\phi + a_2 \cos\phi + a_3 R \cos\beta \sin\phi + a_5 s^2 + a_6 s^3 \tag{6.13}$$

$$\text{d.o.f.'s at each node} = u,\ v,\ (\partial v/\partial s - u/R)$$

where R is the radius of curvature of the element, β is half of the subtending angle, ϕ is an angular variable measured from the bisecting line of the subtending angle, and $(\partial v/\partial s - u/R)$ is the rotation of the tangent.

Element 3: cubic u and cubic v [6.6]

$$u = a_1 + a_2 s + a_3 s^2 + a_4 s^3$$
$$v = a_5 + a_6 s + a_7 s^2 + a_8 s^3 \tag{6.14}$$
$$\text{d.o.f.'s at each node} = u, \partial u/\partial s, v, \partial v/\partial s$$

Element 4: cubic u and quintic v [6.7]

$$u = a_1 + a_2 s + a_3 s^2 + a_4 s^3$$
$$v = a_5 + a_6 s + a_7 s^2 + a_8 s^3 + a_9 s^4 + a_{10} s^5 \tag{6.15}$$
$$\text{d.o.f.'s at each node} = u, \partial u/\partial s, v, \partial v/\partial s, \partial^2 v/\partial s^2$$

Element 5: quintic u and cubic v [6.8]

$$u = a_1 + a_2 s + a_3 s^2 + a_4 s^3 + a_5 s^4 + a_6 s^5$$
$$v = a_7 + a_8 s + a_9 s^2 + a_{10} s^3 \tag{6.16}$$
$$\text{d.o.f.'s at each node} = u, \partial u/\partial s, \partial^2 u/\partial s^2, v, \partial v/\partial s$$

Element 6: quintic u and quintic v [6.8–6.10]

$$u = a_1 + a_2 s + a_3 s^2 + a_4 s^3 + a_5 s^4 + a_6 s^5$$
$$v = a_7 + a_8 s + a_9 s^2 + a_{10} s^3 + a_{11} s^4 + a_{12} s^5 \tag{6.17}$$
$$\text{d.o.f.'s at each node} = u, \partial u/\partial s, \partial^2 u/\partial s^2, v, \partial v/\partial s, \partial^2 v/\partial s^2$$

Element 7: constant strain and linear curvature [6.11]

$$u = -a_1 \sin \phi + a_2 \cos \phi + a_3 + a_5 \phi + \tfrac{1}{2} a_6 \phi^2$$
$$v = a_1 \cos \phi + a_2 \sin \phi + a_4 - a_6 \phi \tag{6.18}$$
$$\text{d.o.f.'s at each node} = u, v, (\partial v/\partial s - u/R)$$

where $\phi = s/R$, s is defined as the curvilinear distance measured from midpoint of the element, and $(\partial v/\partial s - u/R)$ is the rotation of tangent.

Because the curved beam finite elements are a special and simpler form of curved shell finite elements and because significant progress has been made in the developments of shell elements, the displacement functions for the curved beam finite elements are usually given in a reduced form of those for the equivalent shell elements. In the references listed above for elements 1 to 6, we can find the equivalent original unreduced displacement functions for the shell elements. Comprehensive surveys of curved beam finite elements can be found in Refs. 6.8 and 6.12.

Let us compare Eq. (6.13) for element 2 with Eq. (6.18) for element 7. For element 2, the tangential strain ϵ and the curvature κ of the middle surface are

$$\begin{aligned} \epsilon &= \frac{\partial u}{\partial s} + \frac{v}{R} = a_4 + \frac{1}{R}(a_5 s^2 + a_6 s^3) \\ \kappa &= \frac{1}{R}\frac{\partial u}{\partial s} - \frac{\partial^2 v}{\partial s^2} = \frac{1}{R}a_4 - 2a_5 - 6a_6 s \end{aligned} \tag{6.19}$$

For element 7,

$$\begin{aligned} \epsilon &= \frac{1}{R}(a_4 + a_5) \\ \kappa &= \frac{1}{R^2}(a_5 + a_6\phi) \end{aligned} \tag{6.20}$$

The terms a_1, a_2, and a_3 are absent in ϵ and κ in both equations. They represent the three rigid-body displacements (two orthogonal displacements and a rotation in the uv plane) which do not contribute to the strain energy. Such a representation appears simpler in Eq. (6.18) than in Eq. (6.13). The remaining terms (a_4, a_5, a_6) include, in Eq. (6.18), a coupling term a_6 which occurs in both u and v. We also observe that ϵ is constant in Eq. (6.20) but cubic in s in Eq. (6.19), while κ is linear in s in both equations.

Although the rigid-body displacement terms are not explicitly represented in the displacement functions for elements 3 to 6, it was shown in Ref. 6.6 that when the polynomial displacement functions are of sufficiently high order (cubic in this case), the necessary rigid-body displacements are implicitly included.

The stiffness matrices for elements 1, 2, and 3 were given explicitly in Ref. 6.13. In the free-vibration analysis of circular arches performed in Ref. 6.13, far better accuracy in frequency was obtained by using element 3 than elements 1 and 2. In a convergence analysis of a centrally loaded circular arch by using elements 1, 2, 3, and 7 performed in Ref. 6.12, it was shown that as the number of degrees of freedom increases, the solutions using the four different types of elements converge at quite different rates. Element 7 is superior to 3, element 3 is superior to 2, and element 2 is superior to 1. However, element 1 is reasonably good when the arch is inextensible ($EA \gg EI$) and not deep.

In order to demonstrate in depth the formulation and application of curved beam elements, we choose an element for which the stiffness matrix can be formulated explicitly and accurate results can be obtained for example application. Element 3 is chosen.

6.4 CUBIC–CUBIC CIRCULARLY CURVED BEAM FINITE ELEMENT

6.4.1 Element Description and Displacement Functions

A circularly curved beam finite element is shown in Fig. 6.3. The element has constant bending rigidity EI, axial rigidity EA, constant radius of curvature R, subtending angle β, and length L or $R\beta$. The angular variable ϕ and distance variable s or $R\phi$ are measured from nodal point 1. The element possesses four degrees of freedom at each nodal point: a tangential displacement u, a derivative of tangential displacement $\partial u/\partial s$ or u_s, a radial displacement v, and a derivative of radial displacement $\partial v/\partial s$ or v_s or slope θ.

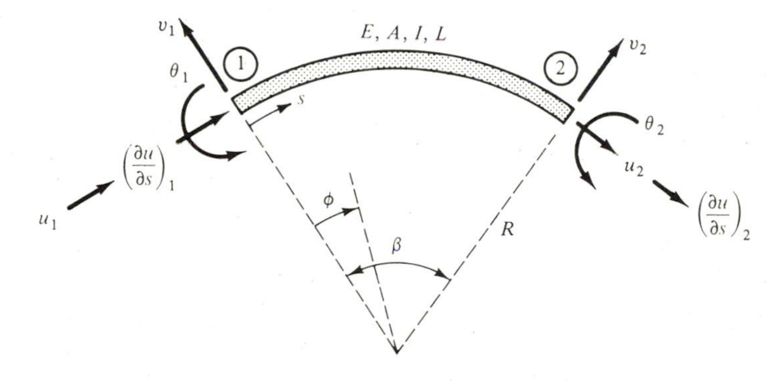

Figure 6.3 Eight-degree-of-freedom circular beam element.

The displacement functions for both the tangential and radial displacements for this element are of cubic polynomials in s as given in Eq. (6.14). The eight constants a are obtained by using the conditions of eight nodal degrees of freedom at both ends. After substituting the a's back into the displacement functions and factoring out each degree of freedom, we obtain the displacement functions in the form similar to that given in Eq. (5.6),

$$\begin{aligned} u &= f_1 u_1 + f_2 u_{s_1} + f_3 u_2 + f_4 u_{s_2} \\ v &= f_1 v_1 + f_2 v_{s_1} + f_3 v_2 + f_4 v_{s_2} \end{aligned} \tag{6.21a}$$

where the shape functions are

$$\begin{aligned} f_1 &= 1 - 3\xi^2 + 2\xi^3 \\ f_2 &= L(\xi - 2\xi^2 + \xi^3) \\ f_3 &= 3\xi^2 - 2\xi^3 \\ f_4 &= L(-\xi^2 + \xi^3) \end{aligned} \tag{6.21b}$$

and

$$\xi = \frac{s}{L} = \frac{R\phi}{R\beta} = \frac{\phi}{\beta} \tag{6.21c}$$

6.4.2 Strain Energy Expression

The strain energy expressions for general thin shells are well known (see, for example, Ref. 6.14). The strain energy expression for a curved beam is in a special reduced form of that for a thin shell,

$$U = \frac{EA}{2}\int \epsilon^2\, ds + \frac{EI}{2}\int \kappa^2\, ds \tag{6.22}$$

where ϵ and κ are the axial strain and curvature of the middle surface, respectively, with

$$\begin{aligned} \epsilon &= \frac{\partial u}{\partial s} + \frac{v}{R} = u' + \frac{v}{R} \\ \kappa &= \frac{1}{R}\frac{\partial u}{\partial s} - \frac{\partial^2 v}{\partial s^2} = \frac{1}{R}u' - v'' \end{aligned} \tag{6.23}$$

Substituting Eq. (6.23) into Eq. (6.22) gives

$$U = U_{uu} + U_{uv} + U_{vv} \tag{6.24a}$$

where

$$\begin{aligned} U_{uu} &= \frac{EA}{2}\int_0^L (u')^2\, ds + \frac{EI}{2R^2}\int_0^L (u')^2\, ds \\ U_{uv} &= \frac{EA}{R}\int_0^L u'v\, ds - \frac{EI}{R}\int_0^L u'v''\, ds \\ U_{vv} &= \frac{EA}{2R^2}\int_0^L v^2\, ds + \frac{EI}{2}\int_0^L (v'')^2\, ds \end{aligned} \tag{6.24b}$$

The energy expressions U_{uu}, U_{uv}, and U_{vv} are associated with axial, axial-flexural coupling, and flexural behaviors, respectively.

6.4.3 Stiffness Equations

Substituting the displacement functions for u and v (6.21a) into the energy expressions (6.24a) and (6.24b), and then performing partial differentiations of the strain energy with respect to each of the eight degrees of freedom, the 8 × 8 stiffness matrix equations for the element is obtained:

$$\begin{Bmatrix} X_1 \\ X_1' \\ X_2 \\ X_2' \\ \hdashline Y_1 \\ M_1 \\ Y_2 \\ M_2 \end{Bmatrix} = \begin{Bmatrix} \dfrac{\partial U}{\partial u_1} \\ \dfrac{\partial U}{\partial u_{s_1}} \\ \dfrac{\partial U}{\partial u_2} \\ \dfrac{\partial U}{\partial u_{s_2}} \\ \hdashline \dfrac{\partial U}{\partial v_1} \\ \dfrac{\partial U}{\partial \theta_1} \\ \dfrac{\partial U}{\partial v_2} \\ \dfrac{\partial U}{\partial \theta_2} \end{Bmatrix} = \left[\begin{array}{c:c} [\mathbf{k}_{uu}] & [\mathbf{k}_{uv}] \\ \hdashline [\mathbf{k}_{vu}] & [\mathbf{k}_{vv}] \end{array}\right] \begin{Bmatrix} u_1 \\ u_{s_1} \\ u_2 \\ u_{s_2} \\ \hdashline v_1 \\ \theta_1 \\ v_2 \\ \theta_2 \end{Bmatrix} \tag{6.25a}$$

where X_1' and X_2' are the counterpart forces in inch-pounds associated with the degrees of freedom u_{s_1} and u_{s_2}, respectively. The coefficients in the four 4 × 4 submatrices are obtained as

$$k_{uu_{ij}} = \int_0^L EA\left(1 + \frac{\alpha}{R^2}\right) f_i' f_j' \, ds$$

$$k_{uv_{ij}} = k_{vu_{ji}} = \int_0^L \frac{EA}{R}(f_i' f_j - \alpha f_i' f_j'') \, ds \tag{6.25b}$$

$$k_{vv_{ij}} = \int_0^L EA\left(\frac{f_i f_j}{R^2} + \alpha f_i'' f_j''\right) ds$$

where the primes indicate derivatives with respect to s and $\alpha = EI/EA$.

Explicitly, the submatrices are obtained as

$$[\mathbf{k}_{uu}] = \frac{EA}{5}\left(1+\frac{\alpha}{R^2}\right)\begin{bmatrix} \frac{6}{L} & & \text{symmetric} & \\ \frac{1}{2} & \frac{2L}{3} & & \\ -\frac{6}{L} & -\frac{1}{2} & \frac{6}{L} & \\ \frac{1}{2} & -\frac{L}{6} & -\frac{1}{2} & \frac{2L}{3} \end{bmatrix}$$

$$[\mathbf{k}_{uv}] = \frac{EA}{R}\begin{bmatrix} -\frac{1}{2}+0 & -\frac{L}{10}-\frac{\alpha}{L} & -\frac{1}{2}-0 & \frac{L}{10}+\frac{\alpha}{L} \\ \frac{L}{10}+\frac{\alpha}{L} & 0+\frac{\alpha}{2} & -\frac{L}{10}-\frac{\alpha}{L} & \frac{L^2}{60}+\frac{\alpha}{2} \\ \frac{1}{2}-0 & \frac{L}{10}+\frac{\alpha}{L} & \frac{1}{2}+0 & -\frac{L}{10}-\frac{\alpha}{L} \\ -\frac{L}{10}-\frac{\alpha}{L} & -\frac{L^2}{60}-\frac{\alpha}{2} & \frac{L}{10}+\frac{\alpha}{L} & 0-\frac{\alpha}{2} \end{bmatrix} \tag{6.25c}$$

$$[\mathbf{k}_{vv}] = \frac{EA}{L}\begin{bmatrix} \frac{13\beta^2}{35}+\frac{12\alpha}{L^2} & & \text{symmetric} & \\ \frac{11L\beta^2}{210}+\frac{6\alpha}{L} & \frac{L^2\beta^2}{105}+4\alpha & & \\ \frac{9\beta^2}{70}-\frac{12\alpha}{L^2} & \frac{13L\beta^2}{420}-\frac{6\alpha}{L} & \frac{13\beta^2}{35}+\frac{12\alpha}{L^2} & \\ -\frac{13L\beta^2}{420}+\frac{6\alpha}{L} & -\frac{L^2\beta^2}{140}+2\alpha & -\frac{11L\beta^2}{210}-\frac{6\alpha}{L} & \frac{L^2\beta^2}{105}+4\alpha \end{bmatrix}$$

Equations (6.25c) are available in Ref. 6.15, in which a vibration analysis of prestressed circular arches was performed. They were given earlier in Ref. 6.13 in a different form. It is seen in Eq. (6.25c) that for thin and shallow arches, the thickness/radius ratio is small, so that $\alpha/R^2 \propto (t/R)^2 \ll 1$; thus α/R^2 may be neglected from submatrix $[\mathbf{k}_{uu}]$.

For curved elements with nonuniform cross sections or variable radii of curvature, an explicit form of stiffness matrix may be difficult to obtain. The stiffness matrix may then be obtained through Eq. (6.25b) by using numerical integration.

6.5 APPLICATION TO CURVED BEAM EXAMPLE

In order to evaluate the performance of the eight-degree-of-freedom element (element 3), an example of a semicircular arch as shown in Fig. 6.4a is analyzed. The parameters are defined as $A = 1 \times 1 \text{ in.}^2$, $R = 17$ in., $P = 2000$ lb, and $E = 10^7$ psi.

Three different approaches are used to analyze this problem.

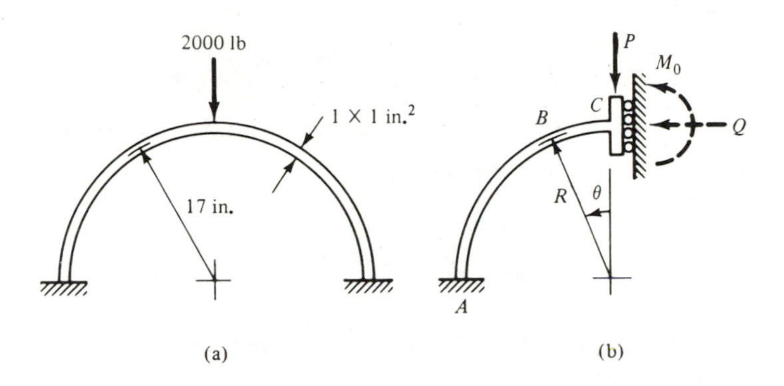

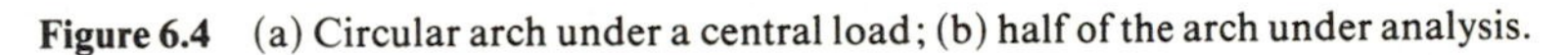

Figure 6.4 (a) Circular arch under a central load; (b) half of the arch under analysis.

6.5.1 Exact Solution by Castigliano's Theorem

Due to symmetry, only half of the arch as shown in Fig. 6.4b need be analyzed. At an arbitrary point B, the bending moment and axial force are, respectively,

$$M = PR \sin\theta - QR(1 - \cos\theta) - M_0$$

$$S = P \sin\theta + Q \cos\theta$$

The strain energy expressions are

$$U = \int_0^{\pi/2} \frac{M^2 R\, d\theta}{2EI} + \int_0^{\pi/2} \frac{S^2 R\, d\theta}{2EA}$$

Because of symmetry, the tangential displacement u_c and rotation θ_c at point C are both zeros. Hence

$$u_c = \frac{\partial U}{\partial Q} = \frac{R^2}{EI}\left[-\frac{PR}{2} + QR\left(\frac{3\pi}{4} - 2\right) + M_0\left(\frac{\pi}{2} - 1\right)\right] + \frac{R}{EA}\left(\frac{Q\pi}{4} + \frac{P}{2}\right) = 0$$

$$\theta_c = \frac{\partial U}{\partial M_0} = \frac{R}{EI}\left[\frac{M_0 \pi}{2} - PR + QR\left(\frac{\pi}{2} - 1\right)\right] = 0$$

Solving for the foregoing two equations simultaneously gives

$$Q = \frac{8R^2A(1 - \pi/4) - 2\pi I}{R^2A(\pi^2 - 8) + \pi^2 I}P$$

$$M_0 = \frac{2R}{\pi}\left[P - \left(\frac{\pi}{2} - 1\right)Q\right]$$

The radial deflection at C is obtained as

$$v_c = \frac{\partial U}{\partial P} = \frac{R^2}{EI}\left(\frac{PR\pi}{4} - \frac{QR}{2} - M_0\right) + \frac{R}{EA}\left(\frac{P\pi}{4} + \frac{Q}{2}\right)$$

For $R = 17$ in., $A = 1 \times 1$ in.^{2}, $I = 1/12$ in.^{4}, and $P = 1000$ lb,

$$Q = 0.91591372P \text{ lb}$$

$$M_0 = 0.30379484PR = 5.1645123P \text{ in.-lb}$$

$$v_c = 0.14152379 \text{ in.}$$

6.5.2 Using Eight-D.O.F. Curved Beam Elements (Element 3)

If one element is used to model half of the arch, the boundary conditions are

$$u_1 = v_1 = \left(\frac{\partial v}{\partial s}\right)_1 = u_2 = \left(\frac{\partial v}{\partial s}\right)_2 = 0$$

From Eqs. (6.25), the stiffness equations can be obtained as

$$\begin{Bmatrix} X_1' \\ X_2' \\ Y_2 \end{Bmatrix} = 10^5 \begin{bmatrix} 356.150 & -89.0375 & -15.7263 \\ -89.0375 & 356.150 & 15.7263 \\ -15.7263 & 15.7263 & 3.43724 \end{bmatrix} \begin{Bmatrix} u_{s_1} \\ u_{s_2} \\ v_2 \end{Bmatrix}$$

Inverting the matrix gives

$$\begin{Bmatrix} u_{s_1} \\ u_{s_2} \\ v_2 \end{Bmatrix} = 10^{-9} \begin{bmatrix} 35.3144 & 2.12302 & 151.860 \\ 2.12302 & 35.3144 & -151.860 \\ 151.860 & -151.860 & 4298.91 \end{bmatrix} \begin{Bmatrix} 0 \\ 0 \\ 1000 \end{Bmatrix}$$

which gives

$$v_2 = 0.004299 \text{ in.} \qquad \text{with } -97\% \text{ error}$$

The solution for the central radial deflection using 2, 3, 4, . . . , 12 equal-length elements is given in Table 6.2. The solution converges rapidly and monotonically as the number of elements increases. The error reduces to less than 1% at the four-element level (15 degrees of freedom). It is noted that better accuracy may be obtained if elements of unequal length are used (i.e., using smaller elements near the central load).

6.5.3 Using Six-D.O.F. Straight Beam Elements

It is of interest to test the simple straight beam elements for this problem. The results for the central radial deflection by using up to 16 equal-length elements are also given in Table 6.2 for comparison. It is seen that for nearly the same numbers of degrees of freedom (not differing by more than 1), the curved elements are better than the straight ones as number of d.o.f.'s >10.

TABLE 6.2 Convergence Study of the Two Types of Finite Elements for the Arch Problem

8-D.O.F. Curved Element				6-D.O.F. Straight Element			
Number of Elements	Number of D.O.F.	Center Deflection (in.)	Error (%)	Number of Elements	Number of D.O.F.	Center Deflection (in.)	Error (%)
1	3	0.004299	97.0	1	1	0.004800	96.6
2	7	0.114635	19.0	2	4	0.127804	9.69
3	11	0.137156	3.09	3	7	0.134743	4.79
4	15	0.140381	0.81	4	10	0.137483	2.86
5	19	0.141123	0.28	6	16	0.139643	1.33
6	23	0.141353	0.12	7	19	0.140127	0.99
7	27	0.141440	0.059	8	22	0.140447	0.76
8	31	0.141478	0.032	10	28	0.140829	0.49
9	35	0.141497	0.019	11	31	0.140948	0.41
10	39	0.141507	0.012	12	34	0.141039	0.34
11	43	0.141513	0.0076	14	40	0.141167	0.25
12	47	0.141516	0.0055	15	43	0.141213	0.22
				16	46	0.141250	0.19

Exact center deflection = 0.1415238 in.

This example has been analyzed in a review paper by Dawe [6.8] using elements 3, 4, 5, 6, and 7. Dawe's figure for the percentage error in center deflection vs. number of degrees of freedom is shown in Fig. 6.5. The present results by using straight beam elements are also plotted in Fig. 6.5. The curve for straight elements starts with quite good accuracy at a low number of degrees of freedom but converges very slowly as the number of degrees of freedom increases. Element 6 is by far the most superior; element 7 is quite superior.

This arch is considered as thick and deep. In Ref. 6.8, figures were also presented for the interesting cases of thin and deep ($t = 0.1$ in.), thick and shallow (subtending angle = 30°), and thin and shallow. When the section is relatively thin or EA/EI is relatively large, the arch is relatively inextensible and the flexural deformation is more dominant than the axial deformation in the strain energy. When the arch is relatively shallow, the flexural deformation is also more dominant in the strain energy.

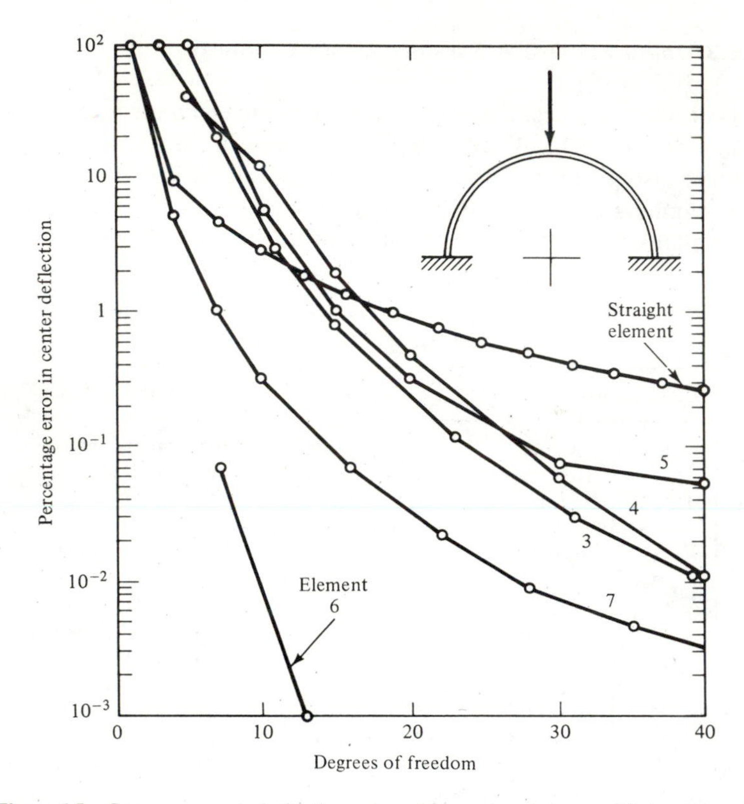

Figure 6.5 Convergence study for the arch problem using six types of finite elements.

In general, the axial-flexural behaviors are intricately coupled in the arch structures. It is recommended that we not be prejudiced against the axial displacement function u; that is, the degree of accuracy or order of polynomials assumed for the axial displacement function u should be comparable to that for the flexural displacement function v.

6.6 CONCLUDING REMARKS

A tapered beam element and a tapered axial force bar element based on assumed variations of moment of inertia and cross-sectional area, respectively, have been introduced. The two elements can be combined and oriented arbitrarily through coordinate transformation in two- and three-dimensional space for frame analysis.

A tapered cantilever beam example shows that the tapered beam element is far superior to the uniform beam element; that is, the former gives better accuracy than the latter for the same number of degrees of freedom. It is recommended that tapered beam elements be used for the problems of beams and frames with nonuniform cross sections. In the general case that the variation of cross section is such that explicit formulation of stiffness coefficients is prohibited, numerical integration is recommended.

Seven types of circularly curved beam finite elements, all based on assumed displacement functions, have been introduced. A review and comparison of curved beam finite elements are available in Refs. 6.8 and 6.12. One of the advantages of using displacement functions to formulate an element is that based on the energy principles, the effect of inertia force and initial axial force can be straightforwardly included for the analysis of dynamic, buckling, and finite deflection problems (see, for example, Refs. 6.13, 6.15, and 6.16).

It has been learned, from the development of shell finite elements, that because the axial and flexural behaviors are coupled in the curved beam finite elements, the displacement function for the tangential displacement has to be as sophisticated (of the same order) as that for the radial displacement.

We should note that numerical integration is a convenient and accurate tool with which to formulate the stiffness matrix when its explicit form is difficult to obtain. The difficulty arises when the curved element assumes an irregularly curved shape, irregular variation of cross section, or complicated displacement functions.

Because curved beam finite elements are a special and simpler form of shell finite elements, learning how to deal with curved beam elements is a logical first step toward an understanding of the shell elements.

REFERENCES

6.1. Gallagher, R. H., and Lee, C. H., "Matrix Dynamic and Instability Analysis with Nonuniform Elements," *International Journal for Numerical Methods in Engineering*, Vol. 2, No. 2, 1970, pp. 265–276.

6.2. Martin, H. C., *Introduction to Matrix Methods of Structural Analysis*, McGraw-Hill Book Company, New York, 1966, Chap. 5.

6.3. Connor, J., and Brebbia, C., "A Stiffness Matrix for a Shallow Rectangular Shell Element," *Journal of the Engineering Mechanics Division*, ASCE, Vol. 93, No. EM5, 1967, pp. 43–65.

6.4. Gallagher, R. H., "The Development and Evaluation of Matrix Methods for Thin Shell Structural Analysis," Ph.D. thesis, State University of New York at Buffalo, 1966.

6.5. Cantin, G., and Clough, R. W., "A Curved, Cylindrical Shell, Finite Element," *AIAA Journal*, Vol. 6, No. 6, 1968, pp. 1057–1062.

6.6. Bogner, F. K., Fox, R. L., and Schmit, L. A., "A Cylindrical Shell Discrete Element," *AIAA Journal*, Vol. 5, No. 4, 1967, pp. 745–750.

6.7. Cowper, G. R., Lindberg, G. M., and Olsen, M. D., "A Shallow Shell Finite Element of Triangular Shape," *International Journal of Solids and Structures*, Vol. 6, No. 8, 1970, pp. 1133–1156.

6.8. Dawe, D. J., "Some High-Order Elements for Arches and Shells," in *Finite Elements for Thin Shells and Curved Members*, ed. D. G. Ashwell and R. H. Gallagher, John Wiley & Sons, Inc., New York, 1976, pp. 131–142.

6.9. Argyris, J. H., and Scharpf, D. W., "The SHEBA family of Shell Elements for the Matrix Displacement Method," *Aeronautical Journal*, Vol. 72, No. 694, 1968, pp. 873–883.

6.10. Dupuis, G., and Goël, J. J., "A Curved Finite Element for Thin Elastic Shells," *International Journal of Solids and Structures*, Vol. 6, No. 11, 1970, pp. 1413–1428.

6.11. Ashwell, D. G., Sabir, A. B., and Roberts, T. M., "Further Studies in the Application of Curved Finite Elements to Circular Arches," *International Journal of Mechanical Science*, Vol. 13, No. 6, 1971, pp. 507–517.

6.12. Ashwell, D. G., "Strain Elements, with Applications to Arches, Rings, and Cylindrical Shells," in *Finite Elements for Thin Shells and Curved Members*, ed. D. G. Ashwell and R. H. Gallagher, John Wiley & Sons, Inc., New York, 1976, pp. 91–111.

6.13. Petyt, M., and Fleischer, C. C., "Free Vibration of a Curved Beam," *Journal of Sound and Vibration*, Vol. 18, No. 1, 1971, pp. 17–30.

6.14. Novozhilov, V. V., *Thin Shell Theory*, translated from 2nd Russian edition by Lowe, P. G.; ed. J. R. M. Radok, Wolters-Noordhoff BV, Groningen, The Netherlands, 1970, pp. 42–47.

6.15. Yang, T. Y., and Kim, H. W., "Vibration and Buckling of Shells under Initial Stress," *AIAA Journal*, Vol. 11, No. 11, 1973, pp. 1525–1531.

6.16. Dawe, D. J., "A Finite-Deflection Analysis of Shallow Arches by the Discrete Element Method," *International Journal for Numerical Methods in Engineering*, Vol. 3, No. 4, 1971, pp. 529–552.

PROBLEMS

6.1. Derive the 10 stiffness coefficients shown in Eq. (6.5a).

6.2. Show the solution for the deflection and slope at the free end of the cantilever beam as given in Eq. (6.9) by integrating the differential equation of beam and applying boundary conditions.

6.3. Using two uniform beam elements to model the tapered cantilever beam by stepped representation as shown in Fig. 6.2c, find the deflection and slope at the free end. *Hint*: The percentage errors are given in Table 6.1.

6.4. For the simply supported tapered beam shown in Fig. P6.4, find the slopes at both ends by using **(a)** exact integration of the differential equation of beam, **(b)** one uniform beam element, and **(c)** one tapered beam element. *Hint*: In this case, the first terms "1" in Eq. (6.5b) vanish.

6.5. Derive the stiffness matrix for a circular beam finite element with length L and with constant EI, EA, and R based on the displacement functions given in Eq. (6.12).

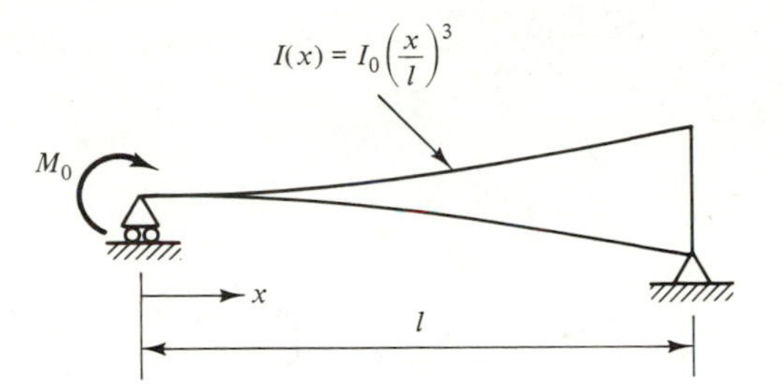

Figure P6.4

6.6. Equation (6.25a) for the eight-degree-of-freedom element can be arranged in the following sequence:

$$\begin{Bmatrix} F' \\ F \end{Bmatrix} = \begin{bmatrix} k_{11} & k_{12} \\ k_{21} & k_{22} \end{bmatrix} \begin{Bmatrix} q' \\ q \end{Bmatrix}$$

where

$$\lfloor \mathbf{F}' \rfloor = \lfloor X_1' \quad X_2' \rfloor \qquad \lfloor \mathbf{q}' \rfloor = \lfloor u_{s_1} \quad u_{s_2} \rfloor$$

$$\lfloor \mathbf{F} \rfloor = \lfloor X_1 \quad Y_1 \quad M_1 \quad X_2 \quad Y_2 \quad M_2 \rfloor$$

$$\lfloor \mathbf{q} \rfloor = \lfloor u_1 \quad v_1 \quad \theta_1 \quad u_2 \quad v_2 \quad \theta_2 \rfloor$$

Assuming that $X_1' = X_2' = 0$, the two degrees of freedom u_{s_1} and u_{s_2} can be written in terms of $\{\mathbf{q}\}$. We can then obtain

$$\underset{6\times 1}{\{\mathbf{F}\}} = \underset{6\times 6}{[\bar{\mathbf{k}}]} \underset{6\times 1}{\{\mathbf{q}\}}$$

Find $[\bar{\mathbf{k}}]$ in terms of the submatrices $[\mathbf{k}_{11}]$, $[\mathbf{k}_{12}]$, $[\mathbf{k}_{21}]$, and $[\mathbf{k}_{22}]$. What difference do you see between the original eight-d.o.f. and the new six-d.o.f. elements in terms of interelement compatibility at the common nodal point?

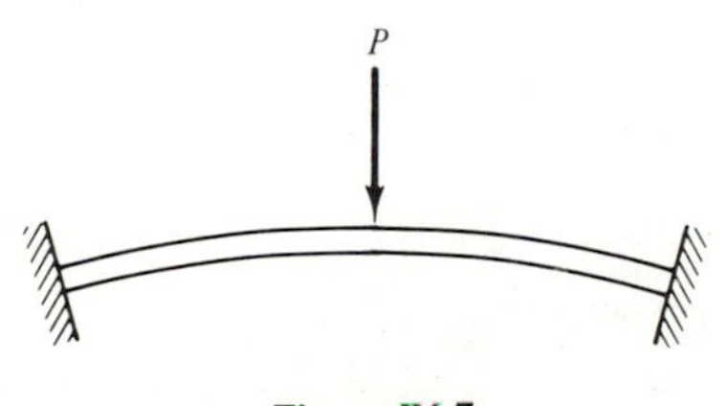

Figure P6.7

6.7. Figure P6.7 shows a circular arch with subtending angle = 30°, $R = 17$ in., $A = 1 \times 1$ in.^{2}, $E = 30 \times 10^6$ psi, and $P = 2 \times 10^5$ lb. Find the central deflection by using **(a)** Castigliano's theorem, **(b)** one eight-d.o.f. circular element, and **(c)** one straight beam element. *Hint*: The exact central deflection = 0.223073 in.

CHAPTER 7

Free Vibration of Truss Bar, Beam, and Plane Frame Finite Elements

7.1 ONE-D.O.F. SPRING–MASS SYSTEM

The simplest example of a free vibration system is a one-degree-of-freedom spring–mass system, as shown in Fig. 7.1. When the mass is vibrating at a distance x from the original static position, the equation of motion is

$$kx + m\ddot{x} = 0 \tag{7.1}$$

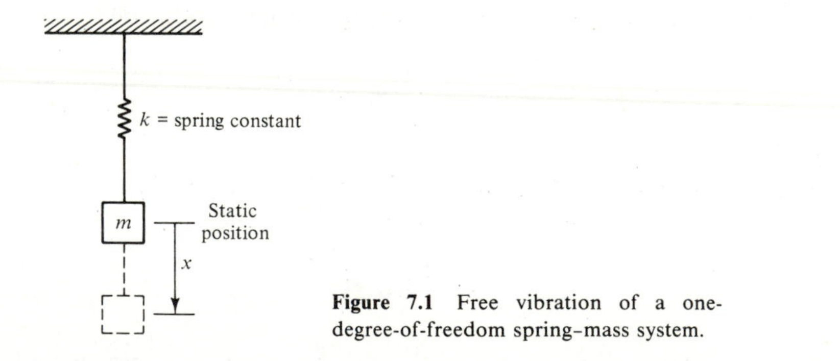

Figure 7.1 Free vibration of a one-degree-of-freedom spring–mass system.

where k is the stiffness of the spring and the dot represents derivative with respect to time. The term kx is the elastic force in the spring, while the term $m\ddot{x}$ is the inertia force of the mass. The inertia force always acts in the opposite direction of (resists) the spring force. At the static position, the acceleration is zero but the absolute value of velocity is maximum. At the extreme positions ($x = \pm$amplitude), the absolute value of the acceleration is maximum but the velocity is zero. The mass accelerates as the spring force increases in an absolute sense.

If we define

$$\omega = \sqrt{\frac{k}{m}} \tag{7.2}$$

Eq. (7.1) becomes

$$\ddot{x} + \omega^2 x = 0 \tag{7.3}$$

The solution to Eq. (7.3) is

$$x = A \cos \omega t + B \sin \omega t \tag{7.4}$$

Equation (7.4) states that the motion is harmonic with natural frequency ω. The constants A and B (amplitudes) in the equation are determined by the initial conditions.

If this spring-mass system is subjected to a forcing function $f(t)$, the equation of motion becomes

$$kx + m\ddot{x} = f(t) \tag{7.5}$$

In this chapter, the equations of motion for the truss bar, beam, and plane frame elements are to be formulated in a form similar to Eq. (7.5) but with more degrees of freedom:

$$[\mathbf{k}]\{\mathbf{x}\} + [\mathbf{m}]\{\ddot{\mathbf{x}}\} = \{\mathbf{f(t)}\} \tag{7.6}$$

where $[\mathbf{k}]$ is the stiffness matrix, $[\mathbf{m}]$ the mass matrix, $\{\mathbf{x}\}$ the vector of nodal displacements and rotations, $\{\ddot{\mathbf{x}}\}$ the vector of accelerations, and $\{\mathbf{f(t)}\}$ the vector of forcing functions.

7.2 AXIAL VIBRATION OF TWO-D.O.F. TRUSS BAR ELEMENTS

The stiffness equations for a truss bar element with two axial displacement degrees of freedom have been formulated in Chapter 4 for static analysis. A mass matrix is needed to extend the stiffness equations to equations of motion for dynamic analysis. There are two common ways to formulate such mass matrix.

7.2.1 Lumped Mass Formulation

Figure 7.2a shows a truss bar element with length L, modulus of elasticity E, cross-sectional area A, and mass density ρ. The simplest way to account

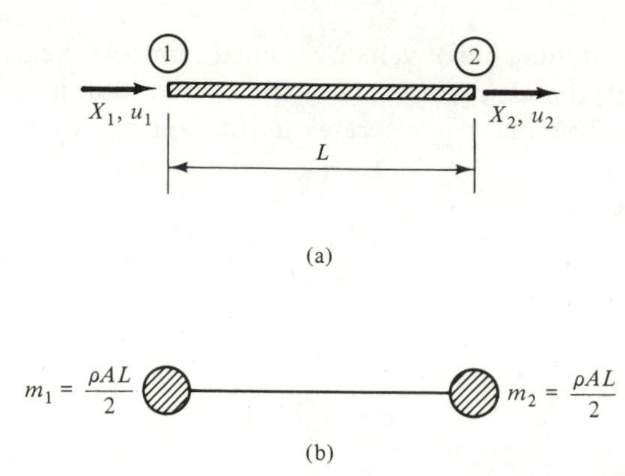

Figure 7.2 Two-degree-of-freedom axial force bar element with mass lumped at both ends.

for the effect of inertia force is to split the total mass ρAL and lump half of the total mass at each end like a dumb bell, as shown in Fig. 7.2b.

Including the effect of inertia forces at both ends, the equilibrium stiffness equations as given in Chapter 4 become

$$\begin{Bmatrix} X_1 - \dfrac{\rho AL}{2}\ddot{u}_1 \\ X_2 - \dfrac{\rho AL}{2}\ddot{u}_2 \end{Bmatrix} = \frac{EA}{L}\begin{bmatrix} 1 & -1 \\ -1 & 1 \end{bmatrix}\begin{Bmatrix} u_1 \\ u_2 \end{Bmatrix} \tag{7.7}$$

where the negative signs in front of the inertia forces $(\rho AL/2)\ddot{u}_1$ and $(\rho AL/2)\ddot{u}_2$ indicate that the inertia forces are acting in the opposite direction to (resist) the elastic forces X_1 and X_2.

Replacing the inertia forces to the right-hand side of Eq. (7.7), we obtain the equations of motion:

$$\begin{Bmatrix} X_1 \\ X_2 \end{Bmatrix} = \frac{EA}{L}\begin{bmatrix} 1 & -1 \\ -1 & 1 \end{bmatrix}\begin{Bmatrix} u_1 \\ u_2 \end{Bmatrix} + \frac{\rho AL}{2}\begin{bmatrix} 1 & 0 \\ 0 & 1 \end{bmatrix}\begin{Bmatrix} \ddot{u}_1 \\ \ddot{u}_2 \end{Bmatrix} \tag{7.8}$$

or symbolically,

$$\{\mathbf{X}\} = [\mathbf{k}]\{\mathbf{u}\} + [\mathbf{m}]\{\ddot{\mathbf{u}}\} \tag{7.8a}$$

It is seen that to formulate the mass matrix $[\mathbf{m}]$, we simply put the lumped nodal masses along the main diagonal.

For simple harmonic motion, we assume that the displacements vary sinusoidally with respect to time with a natural frequency ω,

$$\{\mathbf{u}\} = \{\mathbf{U}\}\sin \omega t \tag{7.9}$$

where $\{\mathbf{U}\}$ is the vector of amplitudes.

The accelerations are

$$\{\ddot{\mathbf{u}}\} = -\omega^2\{\mathbf{U}\}\sin\omega t$$
$$= -\omega^2\{\mathbf{u}\} \tag{7.10}$$

Substituting Eq. (7.10) into Eq. (7.8), we obtain

$$\begin{Bmatrix} X_1 \\ X_2 \end{Bmatrix} = \left[\frac{EA}{L}\begin{bmatrix} 1 & -1 \\ -1 & 1 \end{bmatrix} - \omega^2\frac{\rho AL}{2}\begin{bmatrix} 1 & 0 \\ 0 & 1 \end{bmatrix}\right]\begin{Bmatrix} u_1 \\ u_2 \end{Bmatrix} \tag{7.11}$$

For the case of free vibration, the forces X_1 and X_2 are zeros and Eqs. (7.11) describe an eigenvalue problem.

7.2.2 Consistent Mass Formulation

Alternatively, we can formulate the mass matrix by the use of Lagrange's formula with the strain energy and kinetic energy expressions written in terms of the displacement functions. If the displacement functions used in deriving the mass matrix are the same as those used in deriving the stiffness matrix, the mass matrix is called a "consistent" mass matrix [7.1].

It has been derived in Chapter 4 that the axial displacement function for this element is

$$u(x) = f_1(x)u_1 + f_2(x)u_2 \tag{7.12}$$

where

$$f_1(x) = 1 - \frac{x}{L}$$
$$f_2(x) = \frac{x}{L} \tag{7.12a}$$

The strain energy expression is in the form

$$U = \frac{EA}{2}\int_0^L \{u'(x)\}^2\,dx \tag{7.13}$$

where the prime indicates derivative with respect to x. The kinetic energy expression is in the form

$$T = \frac{\rho A}{2}\int_0^L \{\dot{u}(x)\}^2\,dx \tag{7.14}$$

The equations of motion can be obtained by the use of the well-known Lagrange's equation [7.2] as follows:

$$\frac{d}{dt}\left(\frac{\partial T}{\partial \dot{q}_i}\right) + \frac{\partial U}{\partial q_i} = F_i \tag{7.15}$$

where q_i represents the ith degree of freedom. It is seen in Eq. (7.15) that if the kinetic energy T is absent, it is precisely Castigliano's equation for deriving the stiffness matrix.

For the simplest case of a single mass m_i with velocity $\dot{q}_i$, the kinetic energy is known as

$$T = \frac{m_i \dot{q}_i^2}{2}$$

Substituting this T into Eq. (7.15) and assuming the absence of strain energy U gives

$$m_i \ddot{q}_i = F_i$$

which describes precisely Newton's first law.

Substituting the displacement function (7.12) into the strain energy and kinetic energy expressions (7.13) and (7.14) and then performing the differentiations as indicated in Eq. (7.15), we obtain

$$\begin{aligned}
X_1 &= \frac{d}{dt}\left(\frac{\partial T}{\partial \dot{u}_1}\right) + \frac{\partial U}{\partial u_1} \\
&= \frac{d}{dt}\left(\frac{\rho A}{2}\right) \int_0^L 2\{\dot{u}(x)\}\frac{\partial}{\partial \dot{u}_1}\{\dot{u}(x)\}\, dx + \frac{EA}{2}\int_0^L \{u'(x)\}\frac{\partial}{\partial u_1}\{u'(x)\}\, dx \\
&= \frac{d}{dt}(\rho A) \int_0^L \{f_1(x)\dot{u}_1 + f_2(x)\dot{u}_2\}\{f_1(x)\}\, dx \\
&\quad + EA \int_0^L \{f_1'(x)u_1 + f_2'(x)u_2\}\{f_1'(x)\}\, dx \\
&= \left\lfloor \rho A \int_0^L f_1(x) f_1(x)\, dx \quad \rho A \int_0^L f_1(x) f_2(x)\, dx \right\rfloor \begin{Bmatrix} \ddot{u}_1 \\ \ddot{u}_2 \end{Bmatrix} \\
&\quad + \left\lfloor EA \int_0^L f_1'(x) f_1'(x)\, dx \quad EA \int_0^L f_1'(x) f_2'(x)\, dx \right\rfloor \begin{Bmatrix} u_1 \\ u_2 \end{Bmatrix} \\
&= \lfloor m_{11} \quad m_{12} \rfloor \begin{Bmatrix} \ddot{u}_1 \\ \ddot{u}_2 \end{Bmatrix} + \lfloor k_{11} \quad k_{12} \rfloor \begin{Bmatrix} u_1 \\ u_2 \end{Bmatrix}
\end{aligned} \tag{7.16}$$

By the same token,

$$X_2 = \lfloor m_{21} \quad m_{22} \rfloor \begin{Bmatrix} \ddot{u}_1 \\ \ddot{u}_2 \end{Bmatrix} + \lfloor k_{21} \quad k_{22} \rfloor \begin{Bmatrix} u_1 \\ u_2 \end{Bmatrix} \tag{7.17}$$

where, in general form,

$$m_{ij} = \rho A \int_0^L f_i(x) f_j(x)\, dx \tag{7.18}$$

and

$$k_{ij} = EA \int_0^L f_i'(x) f_j'(x) \, dx \tag{7.19}$$

Equation (7.19) was derived in Chapter 4.

Carrying out the integrations as defined in Eqs. (7.18) and (7.19) for $i = 1, 2$ and $j = 1, 2$ and then assuming harmonic motion with natural frequency ω results in

$$\begin{Bmatrix} X_1 \\ X_2 \end{Bmatrix} = \left[\frac{EA}{L} \begin{bmatrix} 1 & -1 \\ -1 & 1 \end{bmatrix} - \omega^2 \frac{\rho A L}{6} \begin{bmatrix} 2 & 1 \\ 1 & 2 \end{bmatrix} \right] \begin{Bmatrix} u_1 \\ u_2 \end{Bmatrix} \tag{7.20}$$

Example 7.1

Let it be desired to find the natural frequencies for a cantilever bar vibrating freely in the axial direction by using one and two elements, respectively, as shown in Fig. 7.3. The exact solution is, from Ref. 7.3,

$$\omega = \frac{n\pi}{2L} \sqrt{\frac{E}{\rho}}$$

where $n = 1, 3, 5, \ldots$ is the mode number.

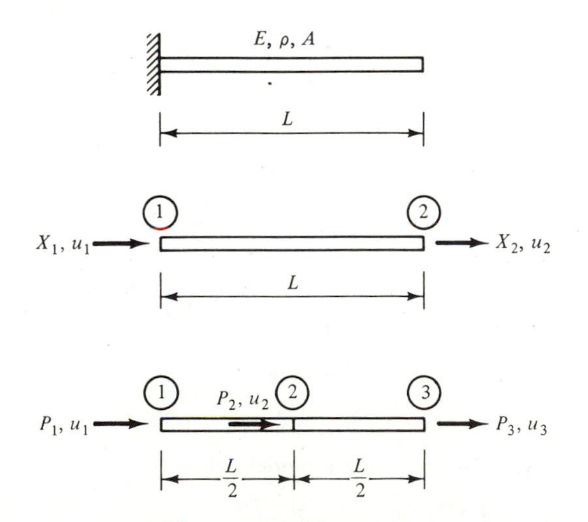

Figure 7.3 One- and two-element modelings of a cantilever bar.

One-element solution. If lumped mass formulation (7.11) is used, we have

$$\begin{Bmatrix} X_1 = ? \\ X_2 = 0 \end{Bmatrix} = \left[\frac{EA}{L} \begin{bmatrix} 1 & -1 \\ -1 & 1 \end{bmatrix} - \omega^2 \frac{\rho A L}{2} \begin{bmatrix} 1 & 0 \\ 0 & 1 \end{bmatrix} \right] \begin{Bmatrix} u_1 = 0 \\ u_2 = ? \end{Bmatrix}$$

It is necessary only to multiply out the second equation:

$$\{0\} = \left[\frac{EA}{L} - \omega^2\frac{\rho AL}{2}\right]\{\mathbf{u}_2\}$$

The first mode frequency is

$$\omega_1 = \frac{1.414}{L}\sqrt{\frac{E}{\rho}} \quad \text{vs.} \quad \frac{\pi}{2L}\sqrt{\frac{E}{\rho}} \rightarrow -10\% \text{ error}$$

If consistent mass formulation (7.20) is used, we have for the second equation,

$$\{\mathbf{0}\} = \left[\frac{EA}{L} - \omega^2\frac{\rho AL}{3}\right]\{\mathbf{u}_2\}$$

The first mode frequency is

$$\omega_1 = \frac{1.732}{L}\sqrt{\frac{E}{\rho}} \quad \text{vs.} \quad \frac{\pi}{2L}\sqrt{\frac{E}{\rho}} \rightarrow 10.3\% \text{ error}$$

Two-element solution. If lumped mass formulation (7.11) is used, we have

$$\begin{Bmatrix} P_1 = ? \\ P_2 = 0 \\ P_3 = 0 \end{Bmatrix} = \left[\frac{2EA}{L}\begin{bmatrix} 1 & -1 & 0 \\ -1 & 2 & -1 \\ 0 & -1 & 1 \end{bmatrix} - \omega^2\frac{\rho AL}{4}\begin{bmatrix} 1 & 0 & 0 \\ 0 & 2 & 0 \\ 0 & 0 & 1 \end{bmatrix}\right]\begin{Bmatrix} u_1 = 0 \\ u_2 = ? \\ u_3 = ? \end{Bmatrix}$$

Neglecting the first equation and following the eigenvalue solution procedure given in Chapter 2, we have

$$\det\left|\frac{2EA}{L}\begin{bmatrix} 2 & -1 \\ -1 & 1 \end{bmatrix} - \omega^2\frac{\rho AL}{4}\begin{bmatrix} 2 & 0 \\ 0 & 1 \end{bmatrix}\right| = 0$$

Let

$$\lambda = \frac{\omega^2(\rho AL/4)}{2EA/L} = \omega^2\frac{\rho L^2}{8E}$$

We have

$$\det\begin{vmatrix} 2 - 2\lambda & -1 \\ -1 & 1 - \lambda \end{vmatrix} = 0$$

$$2\lambda^2 - 4\lambda + 1 = 0$$

$$\lambda_1 = 0.293 \quad \text{and} \quad \lambda_2 = 1.707$$

The natural frequencies for the first two modes are

$$\omega_1 = \frac{1.531}{L}\sqrt{\frac{E}{\rho}} \quad \text{vs.} \quad \frac{\pi}{2L}\sqrt{\frac{E}{\rho}} \rightarrow -2.53\% \text{ error}$$

$$\omega_2 = \frac{3.695}{L}\sqrt{\frac{E}{\rho}} \quad \text{vs.} \quad \frac{3\pi}{2L}\sqrt{\frac{E}{\rho}} \rightarrow -21.6\% \text{ error}$$

If consistent mass formulation (7.20) is used, we have

$$\det \left| \frac{2EA}{L} \begin{bmatrix} 2 & -1 \\ -1 & 1 \end{bmatrix} - \omega^2 \frac{\rho AL}{12} \begin{bmatrix} 4 & 1 \\ 1 & 2 \end{bmatrix} \right| = 0$$

Let

$$\lambda = \omega^2 \frac{\rho L^2}{24E}$$

$$\det \begin{vmatrix} 2 - 4\lambda & -1 - \lambda \\ -1 - \lambda & 1 - 2\lambda \end{vmatrix} = 0$$

$$7\lambda^2 - 10\lambda + 1 = 0$$

$$\lambda_1 = 0.1082 \quad \text{and} \quad \lambda_2 = 1.320$$

The natural frequencies for the two modes are

$$\omega_1 = \frac{1.611}{L}\sqrt{\frac{E}{\rho}} \quad \text{vs.} \quad \frac{\pi}{2L}\sqrt{\frac{E}{\rho}} \rightarrow 2.56\% \text{ error}$$

$$\omega_2 = \frac{5.629}{L}\sqrt{\frac{E}{\rho}} \quad \text{vs.} \quad \frac{3\pi}{2L}\sqrt{\frac{E}{\rho}} \rightarrow 19.5\% \text{ error}$$

It is seen that for this problem the lumped mass solution for natural frequencies converges from lower values toward the upper-bound exact solution while the consistent mass solution converges from higher values toward the lower-bound exact solution. Both solutions give about the same level of accuracy.

The results for the natural frequencies for the first six modes by the use of up to 16 bar elements with consistent mass formulation are presented in Fig. 7.4.

7.3 AXIAL VIBRATION OF FOUR-D.O.F. TRUSS BAR ELEMENTS

The accuracy and efficiency of the two-d.o.f. truss bar elements in free-vibration analysis can be improved by adding a displacement derivative or axial strain degree of freedom ($\partial u/\partial x$) at each end. Thus the bar element has four degrees of freedom, as shown in Fig. 7.5.

The axial displacement function may be assumed as a cubic polynomical in x,

$$u(x) = a_1 + a_2 x + a_3 x^2 + a_4 x^3 \tag{7.21}$$

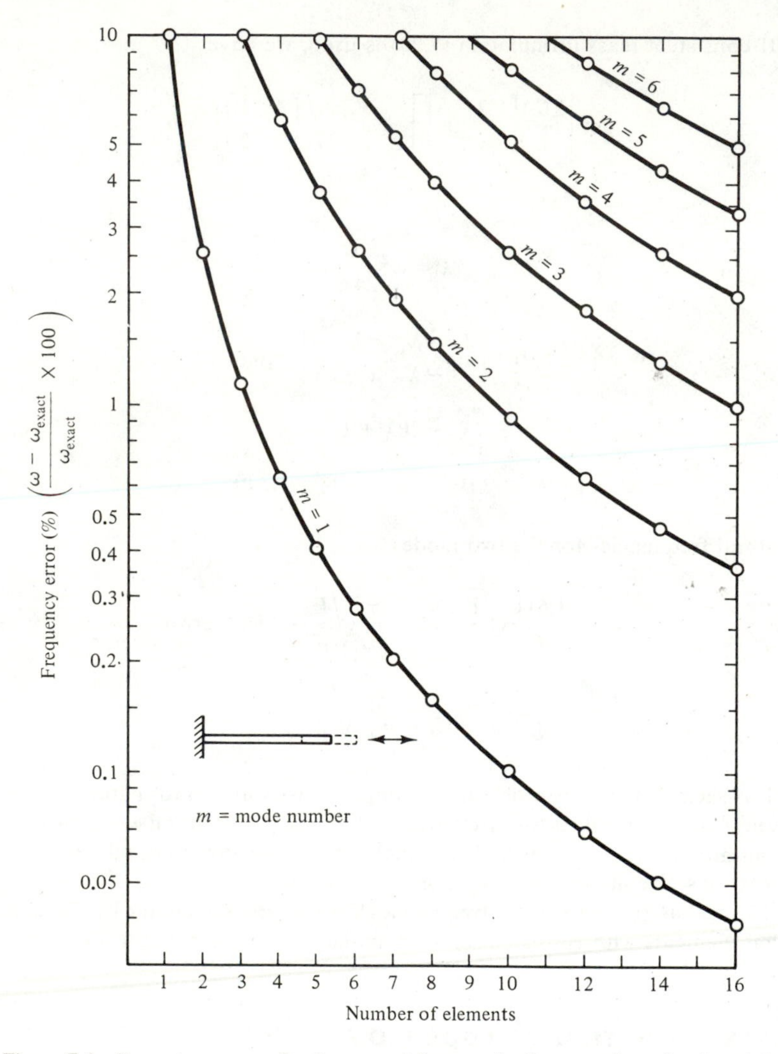

Figure 7.4 Percentage errors for the natural frequencies for a cantilever bar in axial vibration.

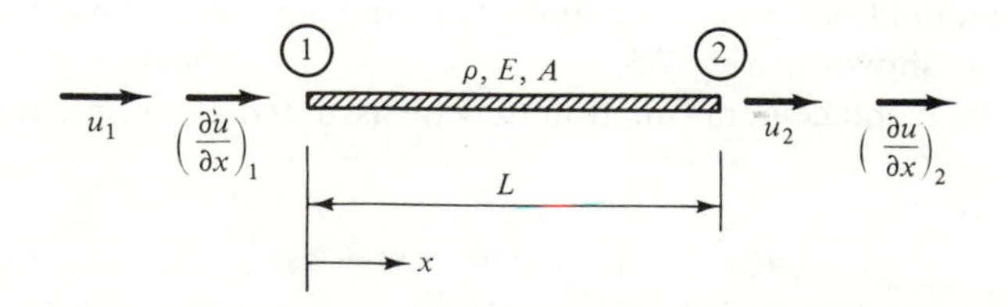

Figure 7.5 Four-degree-of-freedom axial force bar finite element.

The four constants are obtained by using the following boundary conditions:

$$u(0) = u_1$$

$$u(L) = u_2$$

$$\frac{\partial u(0)}{\partial x} = \left(\frac{\partial u}{\partial x}\right)_1$$

$$\frac{\partial u(L)}{\partial x} = \left(\frac{\partial u}{\partial x}\right)_2$$

Upon substitution of the solution of the four constants back into Eq. (7.21) and rearrangement of the terms, we can obtain the displacement function in the form

$$u(x) = f_1(x)u_1 + f_2(x)\left(\frac{\partial u}{\partial x}\right)_1 + f_3(x)u_2 + f_4(x)\left(\frac{\partial u}{\partial x}\right)_2 \tag{7.22}$$

where the four shape functions are in precisely the same form as those obtained for the beam element in Eq. (5.7).

The stiffness matrix and consistent mass matrix for this element can straightforwardly be derived by substituting the shape functions in Eq. (7.22) into Eqs. (7.18) and (7.19), respectively, and performing the integrations. The resulting equations of motion are in the following form:

$$\begin{Bmatrix} X_1 \\ X_1' \\ X_2 \\ X_2' \end{Bmatrix} = EA \begin{bmatrix} \frac{6}{5L} & & & \text{symmetric} \\ \frac{1}{10} & \frac{2L}{15} & & \\ -\frac{6}{5L} & -\frac{1}{10} & \frac{6}{5L} & \\ \frac{1}{10} & -\frac{L}{30} & -\frac{1}{10} & \frac{2L}{15} \end{bmatrix} \begin{Bmatrix} u_1 \\ \left(\frac{\partial u}{\partial x}\right)_1 \\ u_2 \\ \left(\frac{\partial u}{\partial x}\right)_2 \end{Bmatrix}$$

$$+ \frac{\rho A L}{420} \begin{bmatrix} 156 & & & \text{symmetric} \\ 22L & 4L^2 & & \\ 54 & 13L & 156 & \\ -13L & -3L^2 & -22L & 4L^2 \end{bmatrix} \begin{Bmatrix} \ddot{u}_1 \\ \left(\frac{\partial \ddot{u}}{\partial x}\right)_1 \\ \ddot{u}_2 \\ \left(\frac{\partial \ddot{u}}{\partial x}\right)_2 \end{Bmatrix} \tag{7.23}$$

or symbolically,

$$\{\mathbf{X}\} = [\mathbf{k}]\{\mathbf{u}\} + [\mathbf{m}]\{\ddot{\mathbf{u}}\} \tag{7.23a}$$

where X_1' and X_2' are the counterpart forces associated with the two strain degrees of freedom, respectively.

For the case of free vibration, the displacement vector may be assumed as sinusoidal function of time with circular frequency ω. Equation (7.23a) becomes a set of eigenvalue equations,

$$\{\mathbf{0}\} = [[\mathbf{k}] - \omega^2[\mathbf{m}]]\{\mathbf{u}\} \tag{7.24}$$

Equation (7.24) is used to find the natural frequencies of a cantilever bar in axial vibration. The results are given in Fig. 7.6 [7.4] on a semi-log scale. An alternative solution by Przemieniecki [7.5] using the two-degree-of-freedom bar elements but with quadratic equations of motion is also given in Fig. 7.6. Przemieniecki's solution is significantly more accurate and efficient than the

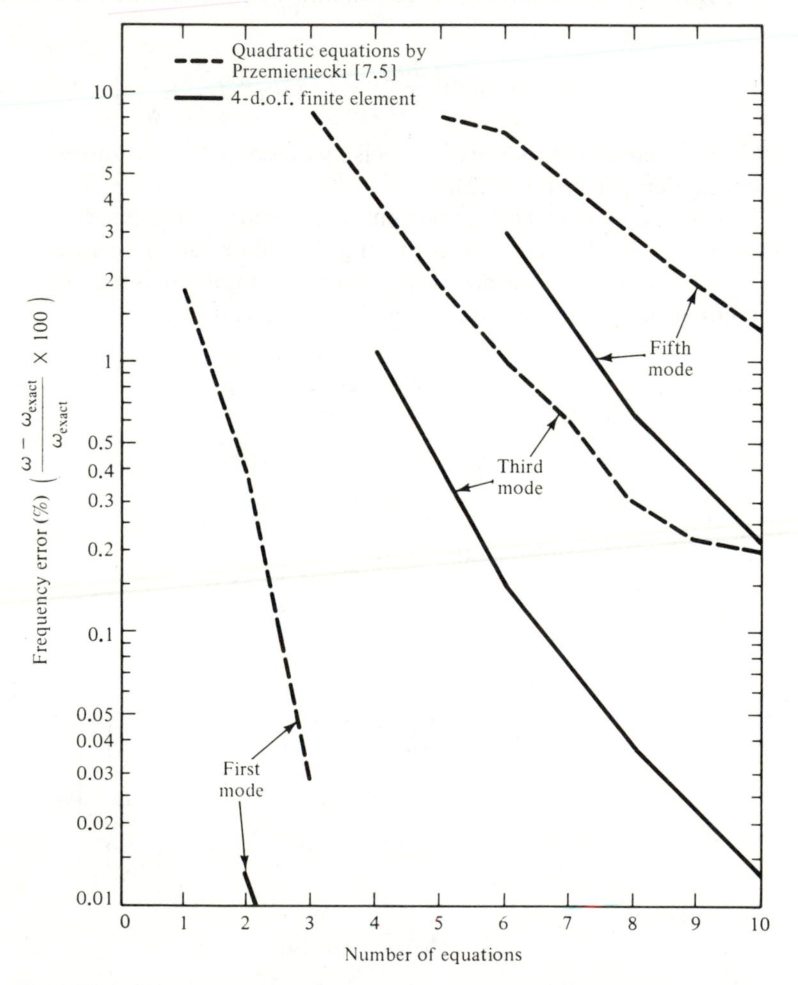

Figure 7.6 Comparison of the percentage errors in axial frequency for a cantilever bar.

solution given in Fig. 7.4 by using the two-degree-of-freedom bar elements with linear equations of motion. However, it is seen in Fig. 7.6 that with the use of the same number of degrees of freedom, the solution by the four-degree-of-freedom elements is order-of-magnitude improved in accuracy as compared to the solution by two-degree-of-freedom elements with quadratic equations of motion.

7.4 FLEXURAL VIBRATION OF BEAM ELEMENTS

Figure 7.7 shows a four-degree-of-freedom beam element for which the displacement function and stiffness matrix have been formulated in Chapter 5. It is necessary to derive a consistent mass matrix for this element.

The displacement function is in the form

$$v(x) = f_1(x)v_1 + f_2(x)\theta_1 + f_3(x)v_2 + f_4(x)\theta_2 \tag{7.25}$$

where the shape functions are defined in Eq. (5.7).

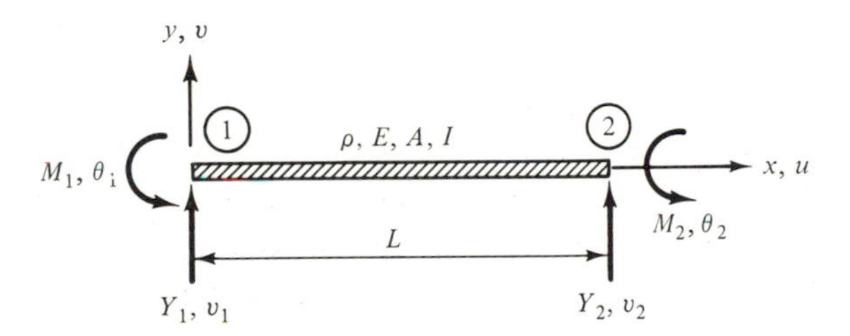

Figure 7.7 Four-degree-of-freedom beam finite element.

The kinetic energy for the beam element in flexural vibration is

$$T = \frac{\rho A}{2} \int_0^L \dot{v}(x)^2 \, dx \tag{7.26}$$

and the strain energy expression for the beam element in bending is

$$U = \frac{EI}{2} \int_0^L v''(x)^2 \, dx \tag{7.27}$$

where the prime indicates differentiation with respect to x.

Substituting the displacement function (7.25) into the kinetic energy expression (7.26) and the strain energy expression (7.27) and then performing the differentiations as indicated in the Lagrange's equation (7.15) with respect to each of the four degrees-of-freedom, we obtain

$$[\mathbf{m}]\{\ddot{\mathbf{q}}\} + [\mathbf{k}]\{\mathbf{q}\} = \{\mathbf{f(t)}\} \tag{7.28}$$

where the mass and stiffness terms are obtained in the similar way as those given in Eqs. (7.18) and (7.19), respectively:

$$m_{ij} = \rho A \int_0^L f_i(x) f_j(x)\, dx \tag{7.29}$$

$$k_{ij} = EI \int_0^L f_i''(x) f_j''(x)\, dx \tag{7.30}$$

The resulting equations of motion (7.28) in explicit form are as follows:

$$\begin{Bmatrix} Y_1 \\ M_1 \\ Y_2 \\ M_2 \end{Bmatrix} = \frac{EI}{L} \begin{bmatrix} \frac{12}{L^2} & \frac{6}{L} & -\frac{12}{L^2} & \frac{6}{L} \\ \frac{6}{L} & 4 & -\frac{6}{L} & 2 \\ -\frac{12}{L^2} & -\frac{6}{L} & \frac{12}{L^2} & -\frac{6}{L} \\ \frac{6}{L} & 2 & -\frac{6}{L} & 4 \end{bmatrix} \begin{Bmatrix} v_1 \\ \theta_1 \\ v_2 \\ \theta_2 \end{Bmatrix}$$

$$+ \frac{\rho A L}{420} \begin{bmatrix} 156 & 22L & 54 & -13L \\ 22L & 4L^2 & 13L & -3L^2 \\ 54 & 13L & 156 & -22L \\ -13L & -3L^2 & -22L & 4L^2 \end{bmatrix} \begin{Bmatrix} \ddot{v}_1 \\ \ddot{\theta}_1 \\ \ddot{v}_2 \\ \ddot{\theta}_2 \end{Bmatrix} \tag{7.31}$$

Equation (7.31) is due to Archer [7.1].

For the case of free vibration with natural frequency ω, Eq. (7.28) or (7.31) takes the following form:

$$\{\mathbf{0}\} = [[\mathbf{k}] - \omega^2[\mathbf{m}]]\{\mathbf{q}\} \tag{7.32}$$

which is suited for eigenvalue solutions.

The equations of motion given in Eq. (7.32) were used by Archer in Ref. 7.1 to perform free-vibration analysis for beam with free-free and simply supported boundary conditions. Results of percentage errors in natural frequencies obtained by using up to six elements are given in Table 7.1. The solutions by Fowler [7.6] based on three different kinds of mass lumping as explained in the footnote of Table 7.1 are also given for comparison. For the case studied, the improvement in accuracy by using consistent mass formulation is dramatic.

TABLE 7.1 Percentage Errors in Natural Frequencies for Beams

Number of Elements	Mode	Consistent Mass Matrix		Lumped Mass Matrix, Free-Free[a]		
		Free-Free	Simply Supported	A	B	C
1	1	20.00	11.00			
2	1	0.224	0.390		−38.07	
	2	13.77	11.00			
3	1	0.246	0.081	20.68	−23.67	−5.63
	2	0.520	1.182		−31.55	
	3	12.42	11.00			
4	1	0.107	0.026	10.79	−15.55	−3.18
	2	0.625	0.395	11.60	−32.17	−11.04
	3	0.795	1.83		−28.27	
	4	11.72	11.00			
5	1	0.048	0.011	6.64	−10.70	−2.14
	2	0.320	0.167	11.01	−16.19	−7.27
	3	0.969	0.792	9.87	−20.36	−15.58
	4	0.995	1.021		−26.58	
	5	11.28	11.00			
6	1	0.024	0.005	4.51	−7.76	−1.52
	2	0.170	0.082	7.63	−12.16	−5.13
	3	0.575	0.394	9.29	−15.64	−10.45
	4	1.262	1.183	4.55	−19.17	−16.97
	5	1.150	2.66		−25.87	
	6	10.96	11.00			

[a]A, mass lumped at center of gravity of the element; B, half the mass lumped at ends of the element; C, same as A with the inclusion of mass moment of inertia.
Source: Ref. 7.1.

Example 7.2

Using one element to idealize half of the beam as shown in Fig. 7.8, find the lowest four mode natural frequencies. Compare the results with the exact solution

$$\omega_n = \frac{(n\pi)^2}{l^2}\sqrt{\frac{EI}{\rho A}}$$

where $n = 1, 2, 3, \ldots$ is the mode number.

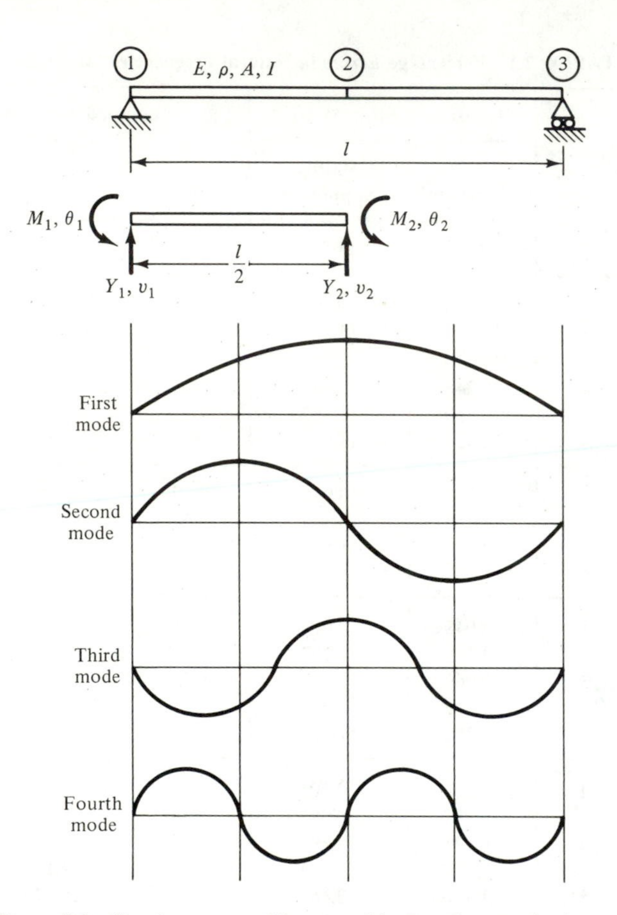

Figure 7.8 Simply supported beam and its first four mode shapes.

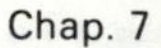

Solution. For the first and third modes as shown in Fig. 7.8, we have the same boundary conditions as for the one-element model,

$$v_1 = \theta_2 = 0$$

Thus we have a set of two equations of motion:

$$\begin{Bmatrix} 0 \\ 0 \end{Bmatrix} = \left[\frac{EI}{L} \begin{bmatrix} 4 & -\dfrac{6}{L} \\ -\dfrac{6}{L} & \dfrac{12}{L^2} \end{bmatrix} - \omega^2 \frac{\rho AL}{420} \begin{bmatrix} 4L^2 & 13L \\ 13L & 156 \end{bmatrix} \right] \begin{Bmatrix} \theta_1 \\ v_2 \end{Bmatrix}$$

Let

$$\lambda = \frac{\omega^2(\rho AL^4)}{420EI}$$

Then

$$\det \begin{vmatrix} 4-4\lambda & -\dfrac{6+13\lambda}{L} \\ -\dfrac{6+13\lambda}{L} & \dfrac{12-156\lambda}{L^2} \end{vmatrix} = 0$$

$$455\lambda^2 - 828\lambda + 12 = 0$$

$$\lambda = 0.01461 \quad \text{and} \quad 1.8052$$

Hence we have for the first mode,

$$\omega_1 = \frac{2.477}{L^2}\sqrt{\frac{EI}{\rho A}} = \frac{9.9085}{l^2}\sqrt{\frac{EI}{\rho A}} \qquad (0.39\% \text{ error})$$

and for the third mode,

$$\omega_3 = \frac{27.5}{L^2}\sqrt{\frac{EI}{\rho A}} = \frac{110.1}{l^2}\sqrt{\frac{EI}{\rho A}} \qquad (24\% \text{ error})$$

The second and fourth modes have the same boundary conditions for the one-element model,

$$v_1 = v_2 = 0$$

Thus we have a set of two equations of motion:

$$\begin{Bmatrix} 0 \\ 0 \end{Bmatrix} = \left[\frac{EI}{L}\begin{bmatrix} 4 & 2 \\ 2 & 4 \end{bmatrix} - \omega^2\frac{\rho AL}{420}\begin{bmatrix} 4L^2 & -3L^2 \\ -3L^2 & 4L^2 \end{bmatrix}\right]\begin{Bmatrix} \theta_1 \\ \theta_2 \end{Bmatrix}$$

$$\det \begin{vmatrix} 4-4\lambda & 2+3\lambda \\ 2+3\lambda & 4-4\lambda \end{vmatrix} = 0$$

$$7\lambda^2 - 44\lambda + 12 = 0$$

$$\lambda = 0.2857 \quad \text{and} \quad 6.0$$

Hence we have for the second mode,

$$\omega_2 = \frac{10.95}{L^2}\sqrt{\frac{EI}{\rho A}} = \frac{43.82}{l^2}\sqrt{\frac{EI}{\rho A}} \qquad (11\% \text{ error})$$

and for the fourth mode,

$$\omega_4 = \frac{50.2}{L^2}\sqrt{\frac{EI}{\rho A}} = \frac{200.8}{l^2}\sqrt{\frac{EI}{\rho A}} \qquad (27.2\% \text{ error})$$

The errors of 0.39% and 11% for the first- and second-mode frequencies, respectively, are the same as those shown in Table 7.1 at $N = 2$.

Example 7.3

Figure 7.9 shows a beam with both ends fixed. A lumped mass with $m = 4\rho AL/35$ and a massless elastic spring with constant $\alpha = 12EI/L^3$ are attached at the midspan.

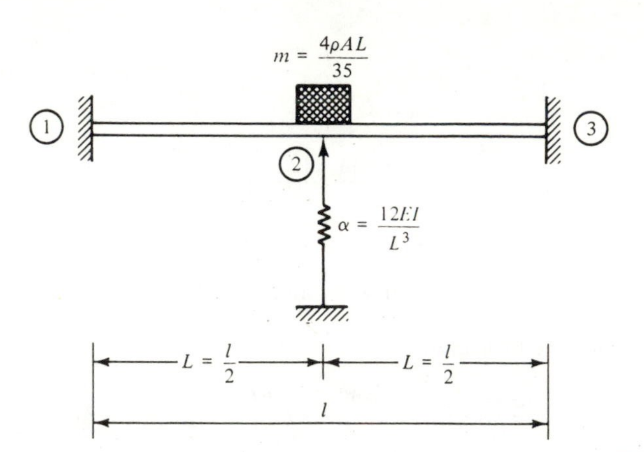

Figure 7.9 Fixed-end beam with a mass and a spring attached at midspan.

It is desired to find the approximate values of the first- and second-mode natural frequencies by using two beam elements to model the whole beam.

The boundary conditions are

$$v_1 = \theta_1 = v_3 = \theta_3 = 0$$

Thus we have a set of two equations of motion

$$\begin{Bmatrix} 0 \\ 0 \end{Bmatrix} = \left[\frac{EI}{L} \begin{bmatrix} \dfrac{24}{L^2} & 0 \\ 0 & 8 \end{bmatrix} - \omega^2 \frac{\rho AL}{420} \begin{bmatrix} 312 & 0 \\ 0 & 8L^2 \end{bmatrix} \right] \begin{Bmatrix} v_2 \\ \theta_2 \end{Bmatrix}$$

To include the effect of attached spring and lumped mass, we simply add α and m to the stiffness term k_{11} and mass term m_{11}, respectively.

$$\begin{Bmatrix} 0 \\ 0 \end{Bmatrix} = \left[\frac{EI}{L} \begin{bmatrix} \dfrac{36}{L^2} & 0 \\ 0 & 8 \end{bmatrix} - \omega^2 \frac{\rho AL}{420} \begin{bmatrix} 360 & 0 \\ 0 & 8L^2 \end{bmatrix} \right] \begin{Bmatrix} v_2 \\ \theta_2 \end{Bmatrix}$$

Let

$$\lambda = \frac{\omega^2(\rho AL^4)}{420EI}$$

$$\det \begin{vmatrix} \dfrac{36}{L}(1 - 10\lambda) & 0 \\ 0 & 8L(1 - \lambda) \end{vmatrix} = 0$$

$$288(1 - 10\lambda)(1 - \lambda) = 0$$

$$\lambda = 0.1 \text{ and } 1$$

So we have

$$\omega_1 = \frac{6.48}{L^2}\sqrt{\frac{EI}{\rho A}} = \frac{25.92}{l^2}\sqrt{\frac{EI}{\rho A}}$$

$$\omega_2 = \frac{20.5}{L^2}\sqrt{\frac{EI}{\rho A}} = \frac{82}{l^2}\sqrt{\frac{EI}{\rho A}}$$

Because only two elements are used, the two values obtained are only approximate. If more elements are used, it is advisable to use computer subroutines such as those suggested in Chapter 2 for solving for eigenvalues and eigenvectors.

If the massless elastic spring in Fig. 7.9 is replaced by a truss bar element with the same axial spring constant, we must include the effect due to the mass of the bar. We can either add $\rho AL/2$ (lumped mass) or $\rho AL/3$ (consistent mass) to the first diagonal term in the mass matrix above.

7.5 AXIAL-FLEXURAL VIBRATION OF FRAME ELEMENTS

Figure 7.10 shows an axial-flexural frame element oriented arbitrarily in a two-dimensional plane. The equations of motion for this element can be obtained by using the same coordinate transformation procedure as that used for obtaining the stiffness equations described in Chapter 5 [Eq. (5.32)].

$$\underset{6\times 6}{\{\mathbf{F}\}} = \underset{6\times 6}{[\mathbf{T}]^T}\left[\underset{6\times 6}{[\bar{\mathbf{k}}]} - \omega^2 \underset{6\times 6}{[\bar{\mathbf{m}}]}\right]\underset{6\times 6}{[\mathbf{T}]}\underset{6\times 1}{\{\mathbf{q}\}} \tag{7.33}$$

where [T] is the coordinate transformation matrix as derived in Eq. (5.24),

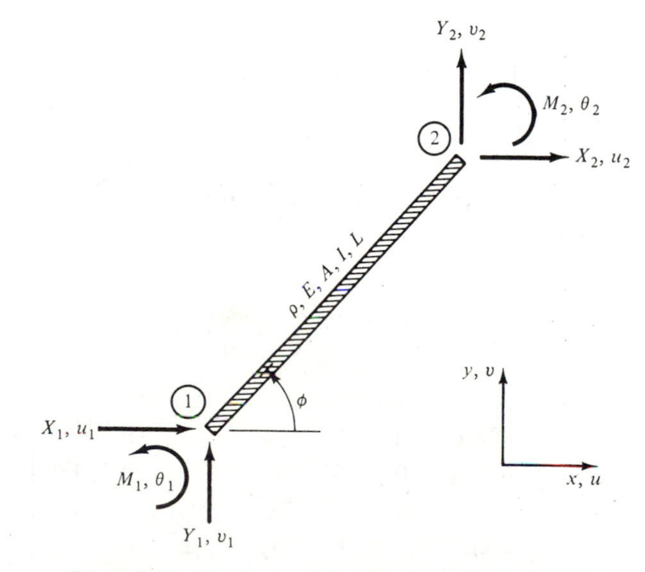

Figure 7.10 Six-degree-of-freedom frame finite element.

and the stiffness matrix $[\bar{\mathbf{k}}]$ and the mass matrix $[\bar{\mathbf{m}}]$ are with reference to the local coordinates.

In the following derivation, the equations of motion associated with axial motion and flexural motion are derived separately. Because the lumped mass matrices require no coordinate transformation, only consistent mass matrices are dealt with.

7.5.1 Axial Vibration

For an axial force bar (truss bar) element oriented arbitrarily in a two-dimensional plane, the augmented stiffness and mass matrices with reference to local coordinates are, from Eq. (7.20),

$$[\bar{\mathbf{k}}] = \frac{EA}{L}\begin{bmatrix} 1 & 0 & 0 & -1 & 0 & 0 \\ 0 & 0 & 0 & 0 & 0 & 0 \\ 0 & 0 & 0 & 0 & 0 & 0 \\ -1 & 0 & 0 & 1 & 0 & 0 \\ 0 & 0 & 0 & 0 & 0 & 0 \\ 0 & 0 & 0 & 0 & 0 & 0 \end{bmatrix}$$

$$[\bar{\mathbf{m}}] = \frac{\rho AL}{6}\begin{bmatrix} 2 & 0 & 0 & 1 & 0 & 0 \\ 0 & 0 & 0 & 0 & 0 & 0 \\ 0 & 0 & 0 & 0 & 0 & 0 \\ 1 & 0 & 0 & 2 & 0 & 0 \\ 0 & 0 & 0 & 0 & 0 & 0 \\ 0 & 0 & 0 & 0 & 0 & 0 \end{bmatrix} \tag{7.34}$$

Following the multiplications as indicated in Eq. (7.33), the equations of motion for the two-dimensional truss bar element are obtained as

$$\begin{Bmatrix} X_1 \\ Y_1 \\ M_1 \\ X_2 \\ Y_2 \\ M_2 \end{Bmatrix} = \left[\frac{EA}{L}\begin{bmatrix} \lambda^2 & \lambda\mu & 0 & -\lambda^2 & -\lambda\mu & 0 \\ \lambda\mu & \mu^2 & 0 & -\lambda\mu & -\mu^2 & 0 \\ 0 & 0 & 0 & 0 & 0 & 0 \\ -\lambda^2 & -\lambda\mu & 0 & \lambda^2 & \lambda\mu & 0 \\ -\lambda\mu & -\mu^2 & 0 & \lambda\mu & \mu^2 & 0 \\ 0 & 0 & 0 & 0 & 0 & 0 \end{bmatrix} \right.$$

$$\left. - \omega^2 \frac{\rho AL}{6}\begin{bmatrix} 2\lambda^2 & 2\lambda\mu & 0 & \lambda^2 & \lambda\mu & 0 \\ 2\lambda\mu & 2\mu & 0 & \lambda\mu & \mu^2 & 0 \\ 0 & 0 & 0 & 0 & 0 & 0 \\ \lambda^2 & \lambda\mu & 0 & 2\lambda^2 & 2\lambda\mu & 0 \\ \lambda\mu & \mu^2 & 0 & 2\lambda\mu & 2\mu^2 & 0 \\ 0 & 0 & 0 & 0 & 0 & 0 \end{bmatrix} \right] \begin{Bmatrix} u_1 \\ v_1 \\ \theta_1 \\ u_2 \\ v_2 \\ \theta_2 \end{Bmatrix} \tag{7.35}$$

7.5.2 Flexural Vibration

For a flexural beam element oriented arbitrarily in the two-dimensional plane, the augmented stiffness and mass matrices with reference to local coordinates are, from Eq. (7.31),

$$[\bar{\mathbf{k}}] = \frac{EI}{L}\begin{bmatrix} 0 & 0 & 0 & 0 & 0 & 0 \\ 0 & \dfrac{12}{L^2} & \dfrac{6}{L} & 0 & -\dfrac{12}{L^2} & \dfrac{6}{L} \\ 0 & \dfrac{6}{L} & 4 & 0 & -\dfrac{6}{L} & 2 \\ 0 & 0 & 0 & 0 & 0 & 0 \\ 0 & -\dfrac{12}{L^2} & -\dfrac{6}{L} & 0 & \dfrac{12}{L^2} & -\dfrac{6}{L} \\ 0 & \dfrac{6}{L} & 2 & 0 & -\dfrac{6}{L} & 4 \end{bmatrix}$$

$$[\bar{\mathbf{m}}] = \frac{\rho AL}{420}\begin{bmatrix} 0 & 0 & 0 & 0 & 0 & 0 \\ 0 & 156 & 22L & 0 & 54 & -13L \\ 0 & 22L & 4L^2 & 0 & 13L & -3L^2 \\ 0 & 0 & 0 & 0 & 0 & 0 \\ 0 & 54 & 13L & 0 & 156 & -22L \\ 0 & -13L & -3L^2 & 0 & -22L & 4L^2 \end{bmatrix} \tag{7.36}$$

Performing the multiplications as indicated in Eq. (7.33), the equations of motion for the two-dimensional flexural beam element are obtained:

$$\begin{Bmatrix} X_1 \\ Y_1 \\ M_1 \\ X_2 \\ Y_2 \\ M_2 \end{Bmatrix} = \left[\frac{EI}{L}\begin{bmatrix} \dfrac{12\mu^2}{L^2} & & & & & \\ -\dfrac{12\lambda\mu}{L^2} & \dfrac{12\lambda^2}{L^2} & & \text{symmetric} & & \\ -\dfrac{6\mu}{L} & \dfrac{6\lambda}{L} & 4 & & & \\ -\dfrac{12\mu^2}{L^2} & \dfrac{12\lambda\mu}{L^2} & \dfrac{6\mu}{L} & \dfrac{12\mu^2}{L^2} & & \\ \dfrac{12\lambda\mu}{L^2} & -\dfrac{12\lambda^2}{L^2} & -\dfrac{6\lambda}{L} & -\dfrac{12\lambda\mu}{L^2} & \dfrac{12\lambda^2}{L^2} & \\ -\dfrac{6\mu}{L} & \dfrac{6\lambda}{L} & 2 & \dfrac{6\mu}{L} & -\dfrac{6\lambda}{L} & 4 \end{bmatrix} \right. \tag{7.37}$$

$$
-\omega^2\frac{\rho AL}{420}\begin{bmatrix} 156\mu^2 & & & & & \\ -156\lambda\mu & 156\lambda^2 & & \text{symmetric} & & \\ -22L\mu & 22L\lambda & 4L^2 & & & \\ 54\mu^2 & -54\lambda\mu & -13L\mu & 156\mu^2 & & \\ -54\lambda\mu & 54\lambda^2 & 13L\lambda & -156\lambda\mu & 156\lambda^2 & \\ 13L\mu & -13L\lambda & -3L^2 & 22L\mu & -22L\lambda & 4L^2 \end{bmatrix}\Bigg]\begin{Bmatrix} u_1 \\ v_1 \\ \theta_1 \\ u_2 \\ v_2 \\ \theta_2 \end{Bmatrix}
$$

Equations (7.35) are used for the free-vibration analysis of plane truss structures. Equations (7.37) are used for free-vibration analysis of some uncommon plane frame and curved beam structures that are dominated by flexural motions only (with inextensible assumption). The combination of Eqs. (7.35) and (7.37) is used for the free-vibration analysis of common plane frame and curved beam structures where the axial and flexural motions are coupled.

Example 7.4

Figure 7.11 shows a two-bar truss. Both bars have the same E, ρ, and A. It is desired to find the approximate natural frequencies for the truss by using only two bar elements.

The boundary conditions are

$$u_2 = v_2 = u_3 = v_3 = 0$$

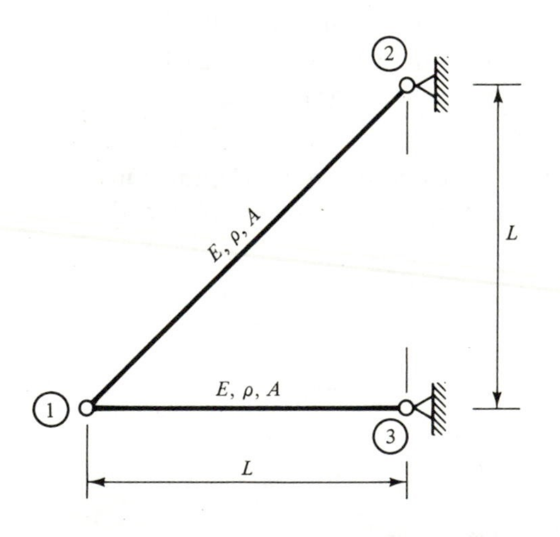

Figure 7.11 Two-bar truss for free-vibration analysis.

Thus we have a set of two equations of motion, from Eq. (7.35):

$$
\begin{Bmatrix} 0 \\ 0 \end{Bmatrix} = \left[\frac{EA}{L}\begin{bmatrix} 1+\dfrac{1}{2\sqrt{2}} & \dfrac{1}{2\sqrt{2}} \\ \dfrac{1}{2\sqrt{2}} & \dfrac{1}{2\sqrt{2}} \end{bmatrix} - \omega^2\frac{\rho AL}{6}\begin{bmatrix} 2+\sqrt{2} & \sqrt{2} \\ \sqrt{2} & \sqrt{2} \end{bmatrix}\right]\begin{Bmatrix} u_1 \\ v_1 \end{Bmatrix} \tag{7.38}
$$

Let

$$\lambda = \frac{\omega^2(\rho AL^2)}{6EA}$$

$$\det \left| \begin{array}{c|c} \left(1 + \dfrac{1}{2\sqrt{2}}\right) - (2+\sqrt{2})\lambda & \dfrac{1-4\lambda}{2\sqrt{2}} \\ \hline \dfrac{1-4\lambda}{2\sqrt{2}} & \dfrac{1-4\lambda}{2\sqrt{2}} \end{array} \right| = 0$$

$$8\lambda^2 - 6\lambda + 1 = 0$$

$$\lambda = \tfrac{1}{4} \quad \text{and} \quad \tfrac{1}{2}$$

Thus the natural frequencies are

$$\omega_1 = \frac{\sqrt{1.5}}{L}\sqrt{\frac{E}{\rho}} \qquad \text{and} \qquad \omega_2 = \frac{\sqrt{3}}{L}\sqrt{\frac{E}{\rho}}$$

Example 7.5

Figure 7.12 shows a fixed-end portal frame with the same E, ρ, A, I, and L for all three members. When such frame vibrates, the motion is predominantly flexural. It is customary to assume that the frame is inextensible. Find the approximate values of natural frequencies for the lowest (a) antisymmetrical and (b) symmetrical modes, using only three elements. The zero-displacement conditions are

$$u_1 = v_1 = \theta_1 = u_4 = v_4 = \theta_4 = 0 \qquad \text{for fixed ends}$$

$$v_2 = v_3 = 0 \qquad \text{for inextensibility}$$

Solution (a) *Lowest antisymmetrical mode*: Based on the combination use of Eqs. (7.35) and (7.37), the assembled equations of motion are

$$\begin{Bmatrix} 0 \\ 0 \\ 0 \\ 0 \end{Bmatrix} = \left[\frac{EI}{L} \begin{bmatrix} \dfrac{12}{L^2} + \dfrac{A}{I} & \dfrac{6}{L} & -\dfrac{A}{I} & 0 \\ \dfrac{6}{L} & 4+4 & 0 & 2 \\ -\dfrac{A}{I} & 0 & \dfrac{12}{L^2} + \dfrac{A}{I} & \dfrac{6}{L} \\ 0 & 2 & \dfrac{6}{L} & 4+4 \end{bmatrix} \right.$$

$$\left. - \omega^2 \frac{\rho AL}{420} \begin{bmatrix} 156+140 & 22L & 70 & 0 \\ 22L & 4L^2+4L^2 & 0 & -3L^2 \\ 70 & 0 & 156+140 & 22L \\ 0 & -3L^2 & 22L & 4L^2+4L^2 \end{bmatrix} \right] \begin{Bmatrix} u_2 \\ \theta_2 \\ u_3 \\ \theta_3 \end{Bmatrix} \tag{7.39}$$

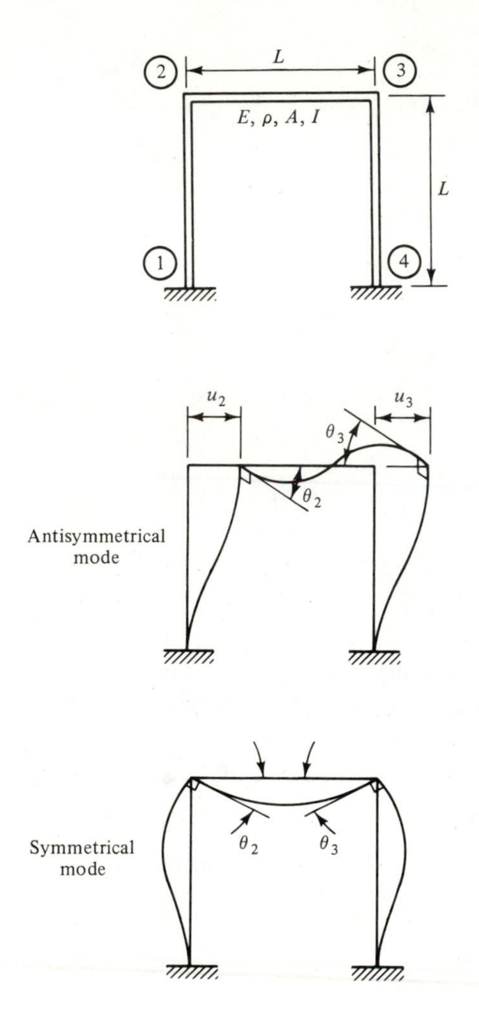

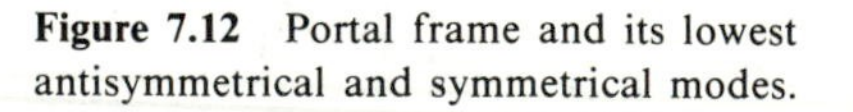
Figure 7.12 Portal frame and its lowest antisymmetrical and symmetrical modes.

Imposing the antisymmetrical conditions

$$u_2 = u_3$$
$$\theta_2 = \theta_3$$

the first (or second) two equations in Eq. (7.39) become

$$\begin{Bmatrix} 0 \\ 0 \end{Bmatrix} = \left[\frac{EI}{L} \begin{bmatrix} \dfrac{12}{L^2} & \dfrac{6}{L} \\ \dfrac{6}{L} & 10 \end{bmatrix} - \omega^2 \frac{\rho AL}{420} \begin{bmatrix} 366 & 22L \\ 22L & 5L^2 \end{bmatrix} \right] \begin{Bmatrix} u_2 \\ \theta_2 \end{Bmatrix} \tag{7.40}$$

Let

$$\lambda = \frac{\omega^2(\rho A L^4)}{420EI}$$

$$\det \begin{vmatrix} 12 - 366\lambda & (6 - 22\lambda)L \\ (6 - 22\lambda)L & (10 - 5\lambda)L^2 \end{vmatrix} = 0 \tag{7.41}$$

$$1346\lambda^2 - 3456\lambda + 84 = 0$$

$$\lambda = 0.02454 \text{ and } 2.543$$

Then we have

$$\omega_1 = \frac{3.21}{L^2}\sqrt{\frac{EI}{\rho A}} \qquad \text{and} \qquad \omega_2 = \frac{32.68}{L^2}\sqrt{\frac{EI}{\rho A}}$$

As compared with the closed-form inextensible solution from Ref. 7.7, the error for the first mode frequency is approximately 0.2%.

If lumped mass matrix instead of consistent mass matrix is used, the problem becomes a massless portal frame with a mass of ρAL lumped at joints 2 and 3. In that case, the equation equivalent to Eq. (7.41) is

$$\det \begin{vmatrix} \dfrac{12EI}{L^3} - \omega^2 \rho AL & \dfrac{6EI}{L^2} \\ \dfrac{6EI}{L^2} & \dfrac{10EI}{L} \end{vmatrix} = 0$$

$$\omega_1 = \frac{2.90}{L^2}\sqrt{\frac{EI}{\rho A}}$$

The error is −9.5%.

The results for the lowest antisymmetrical natural frequencies obtained by using three, six, and nine elements, respectively, are shown in Fig. 7.13 for both consistent mass and lumped mass formulations. The accuracies for both curves are good. One curve converges from above the exact solution and one converges from below.

(b) *Lowest symmetrical mode*: For the symmetrical mode shown in Fig. 7.12, there exist only two-degrees-of-freedom. The equations of motion are

$$\begin{Bmatrix} 0 \\ 0 \end{Bmatrix} = \left[\frac{EI}{L}\begin{bmatrix} 2+4 & 2 \\ 2 & 4+2 \end{bmatrix} - \omega^2 \frac{\rho AL}{420}\begin{bmatrix} 4L^2 + 4L^2 & -3L^2 \\ -3L^2 & 4L^2 + 4L^2 \end{bmatrix}\right]\begin{Bmatrix} \theta_2 \\ \theta_3 \end{Bmatrix}$$

The symmetrical condition provides

$$\theta_2 = -\theta_3$$

Thus we have

$$\{\mathbf{0}\} = \left[\frac{EI}{L}(4) - \omega^2 \frac{\rho AL}{420}(11L^2)\right]\{\boldsymbol{\theta}_2\}$$

$$\omega_1 = \frac{12.36}{L^2}\sqrt{\frac{EI}{\rho A}}$$

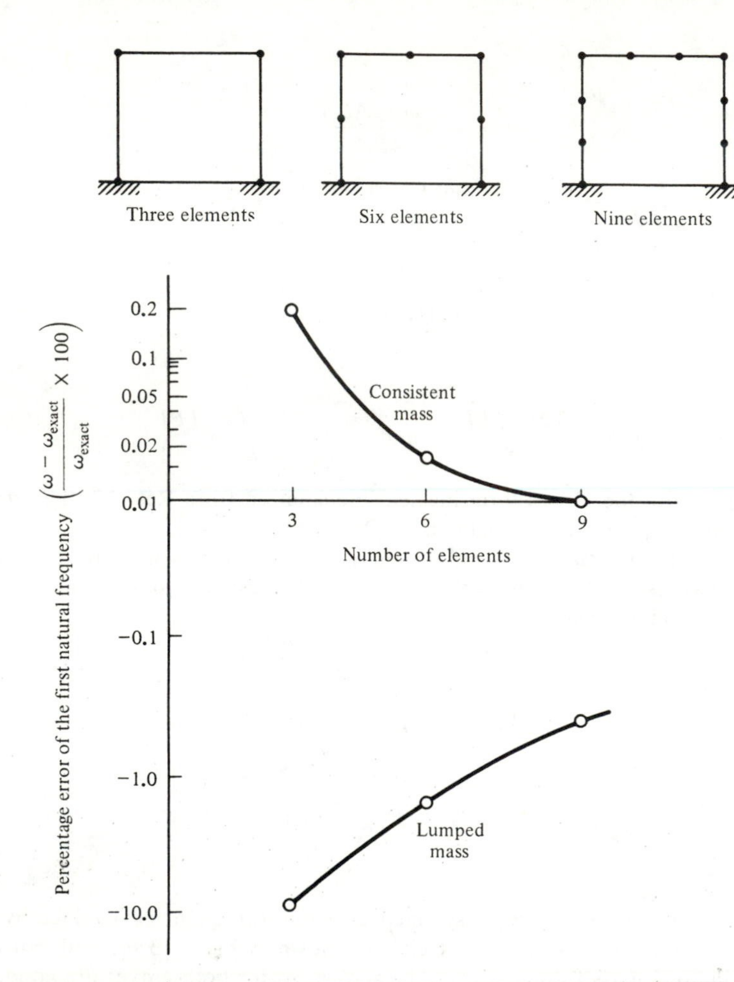

Figure 7.13 Percentage errors of the lowest antisymmetrical natural frequency for a square portal frame.

which is considerably higher than the first natural frequency for the antisymmetrical mode.

Example 7.6

Figure 7.14 shows an H frame with four ends fixed and with the same E, ρ, A, I, and L for all five members. Let it be desired to find the approximate values of the lowest natural frequencies for (a) the antisymmetrical and (b) symmetrical modes as shown in the figure by the use of only five frame elements.

Solution (a) *Lowest antisymmetrical mode*: Due to the conditions of fixed ends and antisymmetrical mode, we have only two-degrees-of-freedom. The equations of motion

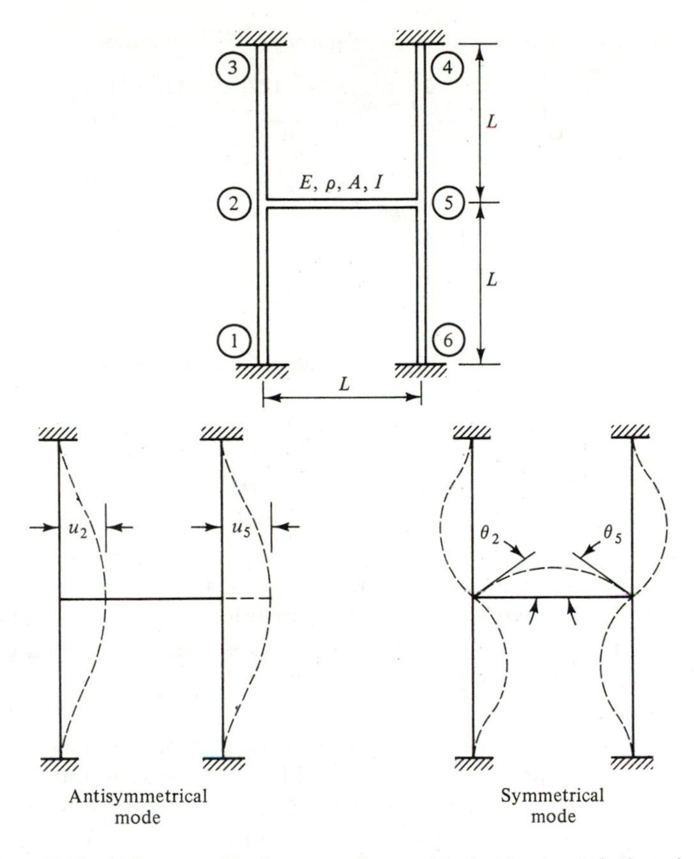

Figure 7.14 H frame and its lowest antisymmetrical and symmetrical modes.

are

$$\begin{Bmatrix} 0 \\ 0 \end{Bmatrix} = \left[\frac{EI}{L} \begin{bmatrix} \dfrac{24}{L^2} + \dfrac{A}{I} & -\dfrac{A}{I} \\ -\dfrac{A}{I} & \dfrac{24}{L^2} + \dfrac{A}{I} \end{bmatrix} - \omega^2 \frac{\rho A L}{420} \begin{bmatrix} 312 + 140 & 70 \\ 70 & 312 + 140 \end{bmatrix} \right] \begin{Bmatrix} u_2 \\ u_5 \end{Bmatrix}$$

Because $u_2 = u_5$,

$$\{0\} = \left[\frac{24EI}{L^3} - \omega^2 \frac{\rho A L}{420}(522) \right] \{\mathbf{u}_2\}$$

Thus

$$\omega_1 = \frac{4.39}{L^2} \sqrt{\frac{EI}{\rho A}}$$

(b) *Lowest symmetrical mode*: The equations of motion are

$$\begin{Bmatrix} 0 \\ 0 \end{Bmatrix} = \left[\frac{EI}{L} \begin{bmatrix} 12 & 2 \\ 2 & 12 \end{bmatrix} - \omega^2 \frac{\rho AL}{420} \begin{bmatrix} 12L^2 & -3L^2 \\ -3L^2 & 12L^2 \end{bmatrix} \right] \begin{Bmatrix} \theta_2 \\ \theta_5 \end{Bmatrix}$$

Because $\theta_2 = -\theta_5$,

$$\{0\} = \left[\frac{10EI}{L} - \omega^2 \frac{\rho AL}{420} (15L^2) \right] \{\theta_2\}$$

$$\omega_1 = \frac{16.73}{L^2} \sqrt{\frac{EI}{\rho A}}$$

which is considerably higher than the first natural frequency for the antisymmetrical mode.

7.6 AXIAL-FLEXURAL COUPLING EFFECT IN FRAME VIBRATION

When the equations of motion given in Eqs. (7.35) and (7.37) are combined, they can be used to analyze the vibration problems where the members are extensible and the axial and flexural motions are coupled. Such a capability is general, practical, and useful.

In the conventional frame analysis (see, for example, Ref. 7.7), the frames are usually assumed as inextensible and the axial motions are usually assumed as independent of flexural motions. The validity of such assumptions depends, of course, on the degree of coupling between the axial and flexural motions. The axial-flexural coupling effects were studied in Ref. 7.8 for a single-story frame with one and two bays and a single bay frame with two and four stories by the use of Eqs. (7.35) and (7.37). In each example studied, the cross-sectional area was varied and its effect on the natural frequencies and mode shapes was studied.

The example of a two- and four-story single bay frame studied in Ref. 7.8 is shown in Fig. 7.15. The floors are assumed to contain 10 times the mass density as the columns. The results of the natural frequencies for the first symmetrical mode are plotted against the nondimensional cross-sectional area (AL^2/I) in Fig. 7.15. The subscripts "*af*" and "*f*" for ω denote axial-flexural and flexural frequencies, respectively. The steel wide-flange columns used in the common designs fall within the range of the AL^2/I values studied. It is seen that the effect of axial motion on the first symmetrical mode natural frequency reduces as the cross-sectional area (or the degree of inextensibility) increases. The results obviously show that the axial motion is not at all negligible in the analysis of such type of frames. Even for the relatively inextensible case with $AL^2/I = 1000$, the effect of axial motion on the first symmetrical mode natural frequency for the four-story frame is as much as 18.6%.

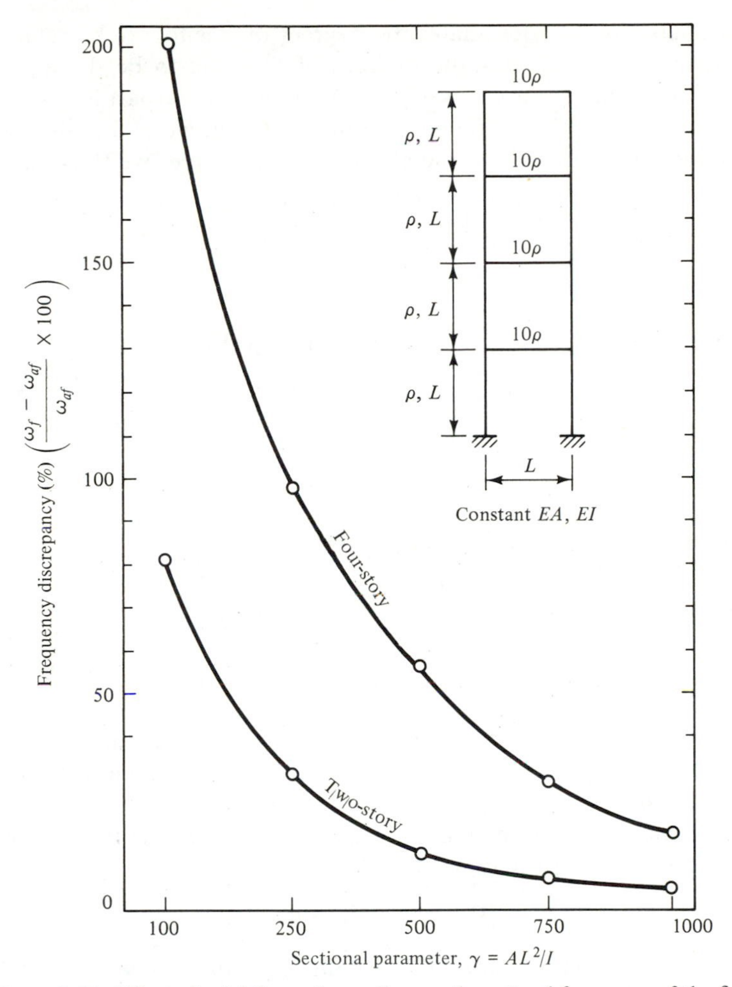

Figure 7.15 Effect of axial-flexural coupling on the natural frequency of the first symmetrical mode of a four-story frame: ω_f = flexural frequency; ω_{af} = axial-flexural frequency.

7.7 REDUCTION OF LARGE EIGENVALUE PROBLEM

The computation of eigenvalues and eigenvectors for a set of equations of motion is far more expensive than the computation of displacements for a set of linear stiffness equations of the same size. Thus the reduction of the size of the eigenvalue equations of motion in the free vibration analysis is far more essential than the reduction of the stiffness equations in the static analysis.

Let a set of equations of motion be arranged and partitioned as

$$\begin{Bmatrix} P_1 \\ P_2 = 0 \end{Bmatrix} = \left[\begin{bmatrix} K_{11} & K_{12} \\ K_{21} & K_{22} \end{bmatrix} - \begin{bmatrix} M_{11} & M_{12} \\ M_{21} & M_{22} \end{bmatrix}\right]\begin{Bmatrix} Q_1 \\ Q_2 \end{Bmatrix} \tag{7.42}$$

where the subscript "1" designates the degrees of freedom to be retained and the subscript "2" designates the degrees of freedom to be discarded. The degrees of freedom $\{Q_2\}$ are so selected that their counterpart loads are zeros.

To treat Eq. (7.42) by the method of reduction, let us first disregard the mass matrix $[\mathbf{M}]$, or disregard the effect of inertia force; we thus have

$$\begin{Bmatrix} P_1 \\ 0 \end{Bmatrix} = \begin{bmatrix} K_{11} & K_{12} \\ K_{21} & K_{22} \end{bmatrix} \begin{Bmatrix} Q_1 \\ Q_2 \end{Bmatrix} \tag{7.43}$$

Multiplying out Eq. (7.43) gives

$$\{\mathbf{P}_1\} = [\mathbf{K}_{11}]\{Q_1\} + [\mathbf{K}_{12}]\{Q_2\} \tag{7.44}$$

$$\{\mathbf{0}\} = [\mathbf{K}_{12}]\{Q_1\} + [\mathbf{K}_{22}]\{Q_2\} \tag{7.45}$$

From Eq. (7.45),

$$\{Q_2\} = -[\mathbf{K}_{22}]^{-1}[\mathbf{K}_{21}]\{Q_1\} \tag{7.46}$$

Substituting Eq. (7.46) into Eq. (7.44) gives

$$\{\mathbf{P}_1\} = [\bar{\mathbf{K}}]\{Q_1\} \tag{7.47}$$

where

$$[\bar{\mathbf{K}}] = [\mathbf{K}_{11}] - [\mathbf{K}_{12}][\mathbf{K}_{22}]^{-1}[\mathbf{K}_{21}] \tag{7.47a}$$

Equation (7.47) describes the standard reduction procedure of the static problem without inertia effect. This procedure has been introduced in Chapters 4 and 5.

Equation (7.47) may be considered as to have been arrived at by means of the congruent transformation

$$\{\mathbf{P}_1\} = [\mathbf{T}]^T[\mathbf{K}][\mathbf{T}]\{Q_1\} \tag{7.48}$$

where the transformation matrix $[\mathbf{T}]$ is defined by the following relationship:

$$\begin{Bmatrix} Q_1 \\ Q_2 \end{Bmatrix} = \begin{bmatrix} [\mathbf{I}] \\ -[\mathbf{K}_{22}]^{-1}[\mathbf{K}_{21}] \end{bmatrix} \{Q_1\} \tag{7.49}$$

To explain Eqs. (7.48) and (7.49), we have only to perform the following multiplications:

$$\begin{aligned} [\mathbf{T}]^T[\mathbf{K}][\mathbf{T}] &= [\mathbf{T}]^T \begin{bmatrix} \mathbf{K}_{11} & \mathbf{K}_{12} \\ \mathbf{K}_{21} & \mathbf{K}_{22} \end{bmatrix} \begin{bmatrix} [\mathbf{I}] \\ -[\mathbf{K}_{22}]^{-1}[\mathbf{K}_{21}] \end{bmatrix} \\ &= \lfloor [\mathbf{I}] \quad -[\mathbf{K}_{22}]^{-1}[\mathbf{K}_{21}] \rfloor \begin{bmatrix} [\mathbf{K}_{11}] - [\mathbf{K}_{12}][\mathbf{K}_{22}]^{-1}[\mathbf{K}_{21}] \\ [\mathbf{0}] \end{bmatrix} \\ &= [\mathbf{K}_{11}] - [\mathbf{K}_{12}][\mathbf{K}_{22}]^{-1}[\mathbf{K}_{21}] \end{aligned} \tag{7.50}$$

which has precisely the same definition as that given in Eq. (7.47a).

For the vibration problem, Eq. (7.48) represents a desirable transformation of the degrees of freedom for both the stiffness matrix [**K**] and the mass matrix [**M**]. The transformation is accomplished by application of the strain energy and the kinetic energy expressions in the following forms. For the strain energy, we have

$$U = \tfrac{1}{2}\lfloor Q_1 \quad Q_2 \rfloor [\mathbf{K}] \begin{Bmatrix} Q_1 \\ Q_2 \end{Bmatrix}$$

$$= \tfrac{1}{2}\{Q_1\}^T [\mathbf{T}]^T [\mathbf{K}][\mathbf{T}]\{Q_1\} \tag{7.51}$$

which results in the definition of the reduced stiffness matrix as given in Eq. (7.47a). For the kinetic energy, we have

$$[\mathbf{T}] = \tfrac{1}{2}\lfloor \dot{Q}_1 \quad \dot{Q}_2 \rfloor [\mathbf{M}] \begin{Bmatrix} \dot{Q}_1 \\ \dot{Q}_2 \end{Bmatrix}$$

$$= \tfrac{1}{2}\{\dot{Q}_1\}^T [\mathbf{T}]^T [\mathbf{M}][\mathbf{T}]\{\dot{Q}_1\} \tag{7.52}$$

so that the reduced mass matrix is obtained as

$$[\bar{\mathbf{M}}] = [\mathbf{T}]^T [\mathbf{M}][\mathbf{T}] \tag{7.53}$$

This reduction method is due to Guyan [7.9]. A more rigorous mathematical treatment of such method is given in Ref. 7.10.

Example 7.7

Find the lowest eigenvalues for the following equations by using the reduction method with the matrix partitions as indicated by dashed lines:

$$\left[\left[\begin{array}{cc|c} 3 & 1 & 0 \\ 1 & 4 & 1 \\ \hline 0 & -1 & 1 \end{array}\right] - \lambda \left[\begin{array}{cc|c} 1 & 0 & 0 \\ 0 & 1 & 0 \\ \hline 0 & 0 & 1 \end{array}\right]\right] \begin{Bmatrix} x_1 \\ x_2 \\ x_3 \end{Bmatrix} = \begin{Bmatrix} 0 \\ 0 \\ 0 \end{Bmatrix} \tag{7.54}$$

Solution

$$-[\mathbf{K}_{22}]^{-1}[\mathbf{K}_{21}] = -[1]^{-1}\lfloor 0 \quad -1 \rfloor = \lfloor 0 \quad 1 \rfloor$$

$$[\mathbf{T}] = \begin{bmatrix} [\mathbf{I}] \\ -[\mathbf{K}_{22}]^{-1}[\mathbf{K}_{21}] \end{bmatrix} = \begin{bmatrix} 1 & 0 \\ 0 & 1 \\ 0 & 1 \end{bmatrix}$$

$$[\mathbf{T}]^T [\mathbf{K}][\mathbf{T}] = \begin{bmatrix} 3 & 1 \\ 1 & 5 \end{bmatrix}$$

$$[\mathbf{T}]^T [\mathbf{M}][\mathbf{T}] = \begin{bmatrix} 1 & 0 \\ 0 & 2 \end{bmatrix}$$

Equation (7.54) is thus reduced to

$$\det \left| \begin{bmatrix} 3 & 1 \\ 1 & 5 \end{bmatrix} - \lambda \begin{bmatrix} 1 & 0 \\ 0 & 2 \end{bmatrix} \right| = 0$$

$$2\lambda^2 - 11\lambda + 14 = 0$$

$$\lambda = 2 \text{ and } 3.5$$

whereas the solution without using the reduction method is

$$\lambda = 1.58 \text{ and } 2.0$$

It is seen that the eigenvalues obtained by using the reduction method are not accurate. It must be realized that in realistic structural vibration problems, the numbers of degrees of freedom involved may well be several hundred. For practical purposes, we sometimes need only several, say, 10 to 50, lowest eigenvalues (natural frequencies) and the corresponding eigenvectors (mode shapes) that dominate the dynamic response behavior. In that case, the reduction method is very economical and it could be very accurate for the dominating lower frequencies and modes. This reduction method is widely used and has been adopted in the NASTRAN program.

7.8 CONCLUDING REMARKS

The equations of motion formulated here are suitable for general free-vibration analysis of two-dimensional trusses, beams, and plane frames. The coupling effect of axial and flexural motions are included. A general free-vibration analysis Fortran program, a user's manual, and two examples with input and output data are provided in Section 13.2.

The present formulations are general. They can be straightforwardly extended to be incorporated in the modal method and time-integration method for dynamic response analysis.

The present equations of motion are approximate formulations. Unless a sufficient number of elements are used, the solutions for natural frequencies and mode shapes are not exact. In the examples demonstrated in this chapter, close to the minimum number of elements is used. The solutions are only approximate.

Solving for eigenvalues and eigenvectors is a tedious numerical process. A simple iteration method, called the power method, for finding the highest eigenvalue and the corresponding eigenvector has been given in Chapter 2. An easy way to solve eigenvalue problems is to use existing computer subroutines. Such an approach was given in detail in Chapter 2. One of the popular simple methods for solving eigenvalue problems is Newton's method of tangents [7.11], provided that the eigenvalue equations are transformed into high-order polynomial equations. The reduction method is powerful and popular for practical purpose. Because the lumped mass method is easy to use and its accuracy is acceptable for practical purpose, it is not usually less popular than the consistent mass method. For example, in the SAP IV program, only the lumped mass method is used.

REFERENCES

7.1. Archer, J. S., "Consistent Mass Matrix for Distributed Mass System," *Journal of the Structural Division, Proceedings*, ASCE, Vol. 89, No. ST4, Aug. 1963, pp. 161–178.

7.2. Bisplinghoff, R. L., Ashley, H., and Halfman, R. L., *Aeroelasticity*, Addison-Wesley Publishing Co., Inc., Reading, Mass., 1955.

7.3. Timoshenko, S. P., and Young, D. H., *Vibration Problems in Engineering*, Van Nostrand Co., Inc., Princeton, N. J., 1956.

7.4. Yang, T. Y., and Sun, C. T., "Finite Elements for the Vibration of Framed Shear Walls," *Journal of Sound and Vibration*, Vol. 27, No. 3, 1973, pp. 297–311.

7.5. Przemieniecki, J. S., "Quadratic Matrix Equations for Determining Vibration Modes and Frequencies of Continuous Elastic Systems," *Proceedings, Conference on Matrix Methods in Structural Mechanics*, Air Force Flight Dynamics Laboratory, TR-66-80, 1965, pp. 779–802.

7.6. Fowler, J. R., "Accuracy of Lumped System Approximations for Lateral Vibrations of Free-Free Uniform Beam," Report EM 9-12, Engineering Mechanics Laboratory, Space Technology Laboratories, Inc., Los Angeles, July 1959.

7.7. Rieger, N. F., and McCallion, H., "The Natural Frequency of Portal Frames, I and II," *International Journal of Mechanical Sciences*, Vol. 7, 1965, pp. 253–276.

7.8. Yang, T. Y., and Sun, C. T., "Axial-Flexural Vibration of Frameworks Using Finite Element Approach," *Journal of the Acoustical Society of America*, Vol. 53, No. 1, 1973, pp. 137–146.

7.9. Guyan, R. J., "Reduction of Stiffness and Mass Matrices," *AIAA Journal*, Vol. 3, No. 2, Feb. 1965, p. 380.

7.10. Wright, G. C., and Miles, G. A., "An Economical Method for Determining the Smallest Eigenvalues of Large Linear Systems," *International Journal for Numerical Methods in Engineering*, Vol. 3, 1971, pp. 25–33.

7.11. Salvadori, M. G., and Baron, M. L., *Numerical Methods in Engineering*, Prentice-Hall, Inc., Englewood Cliffs, N.J., 1961, pp. 5 and 19.

PROBLEMS

7.1. Use Eq. (7.18) to derive the consistent mass matrix for the two-d.o.f. axial force bar element.

7.2. Use Eq. (7.18) to derive the consistent mass matrix for the four-d.o.f. axial force bar element.

7.3. Use Eq. (7.29) to derive the consistent mass matrix for the four-d.o.f. flexural beam element.

7.4. Using the least possible number of truss bar elements, find the natural frequencies for the truss problems as given in Figs. P4.1 to P4.12. Assume ρ as mass density and neglect all the loads in the figures. For Fig. P4.2, assume $\rho = 0.00073\ \text{lb-sec}^2/\text{in}^4$. For Fig. P4.3, assume $\rho = 0.00025\ \text{lb-sec}^2/\text{in}^4$. Use both lumped and consistent mass formulations.

7.5. Using the least possible number of truss bar, beam, and frame elements, find the natural frequencies for the problems given in Figs. P5.2 and P5.3. Use both lumped and consistent mass formulations. Assume inextensibility for beams and frames. Assume ρA as the mass per unit length of all members and neglect all the loads in the figures.

CHAPTER 8

Buckling and Large Deflection of Column and Plane Frame Finite Elements

A simple way to describe the buckling phenomenon is to use an example of an ideally straight bar with uniform and axisymmetrical cross section subjected to a compressive force along the center axis of the bar. Under such a force, the bar will be slightly shortened but remain straight with no bending. If a small lateral force such as a breeze is applied, the beam will be bent infinitesimally but return to its original straight form when the breeze disappears. If the axial force is gradually increased, a condition will be reached in which a small lateral force will cause a deflection which remains when the lateral force disappears. Such an unstable phenomenon is called *buckling* and the critical force is called *buckling load* or *Euler load.* Buckling usually occurs when the compressive stress is well below the material stress limit.

Elastic buckling problems of compressed members were first solved by Euler in 1744 [8.1]. Euler's theoretical solution long remained without practical applications. Only with extensive construction of truss railway bridges did buckling problems become of practical importance. Due to the advances in high-strength-material technology, the structural members used have become increasingly thinner and lighter and thus buckling problems have become of increasing concern.

Buckling can happen to structures in many forms, such as columns, truss members, components of thin-walled beams and plate girders, walls, arches, and shell roofs. Buckling can also happen to torispherical shells under internal pressure. In aerospace structures, minimum-weight design is an important criterion so that the structures are made of skins and thin members. The buckling problem is a predominant one.

In this chapter primary concern is focused on column problems; the buckling of plates and shells is beyond the scope of this book. The buckling formulations for column and frame elements can be used straightforwardly to treat large deflection problems. Such applications are also discussed.

8.1 GOVERNING DIFFERENTIAL EQUATIONS FOR A BEAM ELEMENT WITH THE EFFECT OF AXIAL FORCE

The differential equations for the analysis of beam-columns can be derived by considering Fig. 8.1a. The beam is subjected to an axial compressive force P and to a distributed lateral load $w(x)$. A differential segment between two cross sections taken normal to the original undeflected axis of the beam is shown in Fig. 8.1b. The lateral load may be considered as having a constant value of w over the distance dx. The axial force P, distributed load $w(x)$, deflection v, axes x and y, shearing force V, and bending moment M are all assumed as being positive in the directions shown.

Taking the moment about point i for the beam segment and assuming that the angle $\partial v/\partial x$ between the axis of the beam and the horizontal axis is small, yields:

$$M - \left(M + \frac{\partial M}{\partial x}dx\right) + w\frac{(dx)^2}{2} + \left(V + \frac{\partial V}{\partial x}dx\right)dx + P\left(\frac{\partial v}{\partial x}\right)dx = 0 \tag{8.1}$$

Neglecting terms in dx^2 which are small and then differentiating each term with respect to x, we obtain

$$\frac{\partial^2 M}{\partial x^2} - \frac{\partial V}{\partial x} - P\frac{\partial^2 v}{\partial x^2} = 0 \tag{8.2}$$

Considering equilibrium in the y direction gives

$$\frac{\partial V}{\partial x} = -w \tag{8.3}$$

If the effects of shearing deformation and shortening of the beam axis are neglected, the basic beam theory provides

$$M = -EI\frac{\partial^2 v}{\partial x^2} \tag{8.4}$$

Substituting Eqs. (8.3) and (8.4) into Eq. (8.2) and assuming a beam of uniform cross section, we finally obtain

$$EI\frac{\partial^4 v}{\partial x^4} = w - P\frac{\partial^2 v}{\partial x^2} \tag{8.5}$$

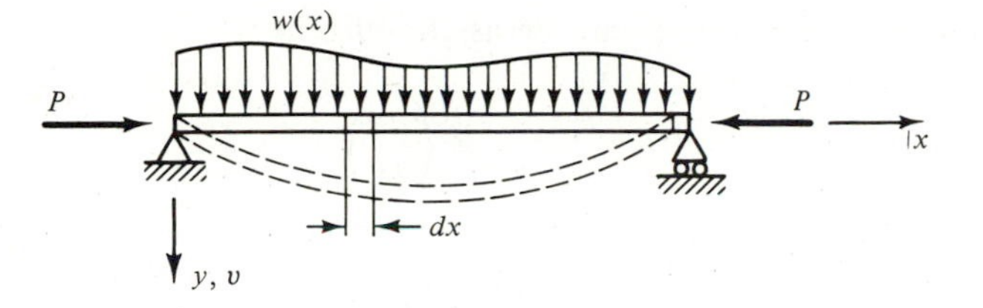

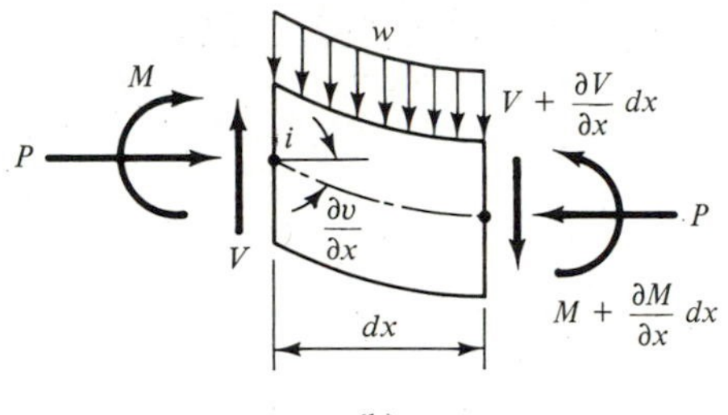

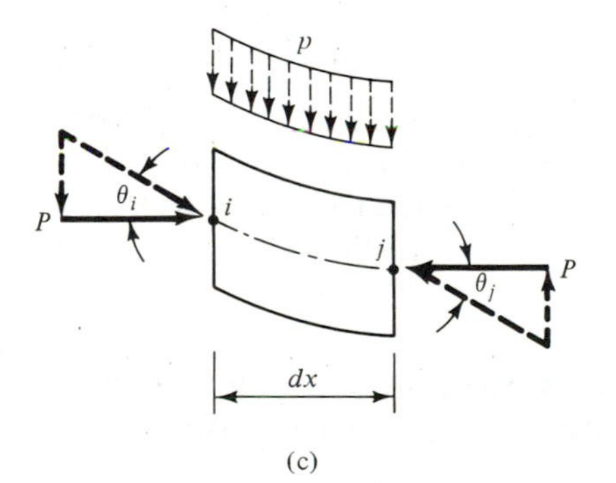

Figure 8.1 (a) Simply supported beam column; (b) differential segment; (c) effect of axial force P.

Equation (8.5) shows that to include the effect of an axial force P is simply to add a term $-P(\partial^2 v/\partial x^2)$ to the fundamental differential equation of beam in bending. This term can be regarded as an additional lateral loading, which may also be obtained from the following physical interpretation.

Let us consider the equilibrium of a beam segment subjected to an axial force P and a uniform lateral pressure p (Fig. 8.1c). The purpose is to find the p value equivalent to the effect of axial force P in the y direction.

$$p(dx) = P \tan \theta_j - P \tan \theta_i$$

For a small deflection, it can be assumed that

$$\theta_i \approx \tan \theta_i = \left(\frac{\partial v}{\partial x}\right)_i$$

$$\theta_j \approx \tan \theta_j = \left(\frac{\partial v}{\partial x}\right)_j$$

Hence

$$p = \frac{P(\partial v/\partial x)_j - P(\partial v/\partial x)_i}{dx}$$

In the limit, we have

$$p = P\left(\frac{\partial^2 v}{\partial x^2}\right) \tag{8.6}$$

It is seen that the effect of the axial force P is equivalent to an upward uniform pressure with a magnitude of $P(\partial^2 v/\partial x^2)$. This term is the product of the axial force and the curvature of the beam. It is independent of the bending rigidity EI.

The stiffness matrix for a beam element corresponding to the first two terms in Eq. (8.5) has been derived in Eq. (5.14). We are now to formulate an additional matrix equivalent to the third term in Eq. (8.5) for buckling and large deflection analyses. The formulation will be based on the energy method.

8.2 FORMULATION OF A UNIFORM BEAM FINITE ELEMENT WITH CONSTANT AXIAL FORCE

The formulation of a beam finite element with the effect of axial force can be carried out using Castigliano's first theorem, introduced in Sec. 3.4.2. It is, however, necessary to reconsider the theorem due to the presence of an axial force.

If a beam is subjected to a set of n external forces, $F_1, F_2, \ldots, F_i, \ldots, F_n$ but regarded as fixed against all displacements q except q_i, which occurs at point i and in the direction of the single force F_i, it follows from the definition of virtual work that [see Eq. (3.17a)]

$$F_i = \frac{\partial W}{\partial q_i} = \frac{\partial U}{\partial q_i} \tag{8.7}$$

where W and U are the work and strain energy, respectively, in the absence of axial force and $W = U$.

In the presence of axial force, however, we have the following relationship:

$$W = W_b + W_n = U_b \tag{8.8}$$

where W_b is the work done by the lateral forces $F_1, F_2, \ldots, F_i, \ldots, F_n$ during deflection; W_n is the work done by the axial force during deflection of the beam caused by the lateral forces $F_1, F_2, \ldots, F_i, \ldots, F_n$; and U_b is the strain energy stored during bending deflection.

If the restrained condition employed in the formulation of Eq. (8.7) is considered, it follows that

$$\frac{\partial W}{\partial q_i} = \frac{\partial W_b}{\partial q_i} \tag{8.9}$$

Consequently, from Eqs. (8.7) to (8.9), we obtain

$$F_i = \frac{\partial}{\partial q_i}(U_b - W_n) \tag{8.10}$$

This equation will be used to formulate a beam finite element with the effect of axial force. Such formulation requires the expressions for U_b, W_n, and a deflection function for the beam element.

8.2.1 Energy Expressions

The strain energy for a beam with uniform cross section has been given in Chapter 3 as

$$U_b = \frac{EI}{2}\int_0^L \left(\frac{\partial^2 v}{\partial x^2}\right)^2 dx \tag{8.11}$$

The work done by an axial force P due to bending of the beam can be derived by considering the beam shown in Fig. 8.2. Due to lateral deflection of the beam from its initial straight form, end B is displaced to the left by a small amount. This displacement is equal to the difference between the length

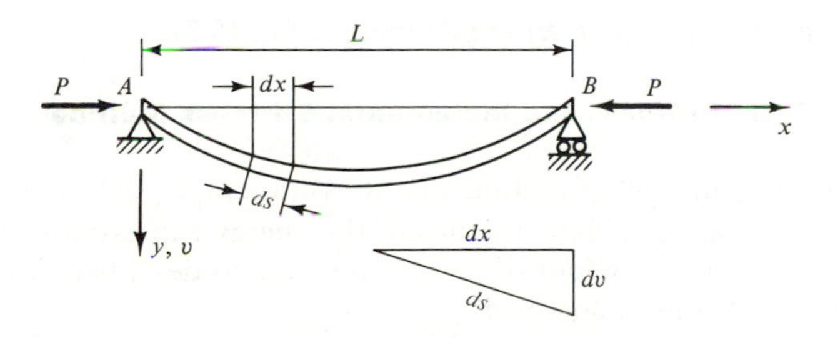

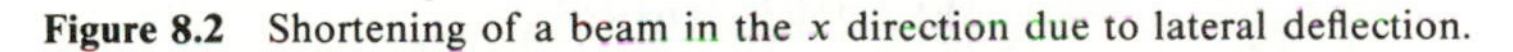
Figure 8.2 Shortening of a beam in the x direction due to lateral deflection.

of the deflected curve and the length of the chord AB if we consider the beam as inextensible.

We first consider the difference between the length of an element ds of the curve and the corresponding element dx of the chord,

$$\begin{aligned} ds - dx &= (dx^2 + dv^2)^{1/2} - dx \\ &= dx\left[1 + \frac{1}{2}\left(\frac{dv}{dx}\right)^2 - \frac{1}{4}\left(\frac{dv}{dx}\right)^4 + \cdots\right] - dx \\ &\approx \frac{1}{2}\left(\frac{dv}{dx}\right)^2 dx \end{aligned} \tag{8.12}$$

The value of the slope (dv/dx) is in radians. Its fourth order is certainly too small to be included. The horizontal displacement of B is the integration of $(ds - dx)$ through the beam length,

$$\Delta = \frac{1}{2}\int_0^L \left(\frac{dv}{dx}\right)^2 dx \tag{8.13}$$

Thus the work done by the axial force P due to the end displacement of B is

$$W_n = \frac{P}{2}\int_0^L \left(\frac{\partial v}{\partial x}\right)^2 dx \tag{8.14}$$

where the axial force P is positive when in compression.

For thin plates, the expressions for U_b and W_n are available in, for example, Sec. 8.4 of Ref. 8.2. For thin shells, the expressions for strain energy and work due to the effect of middle-surface forces are available in, for example, Sec. 9 of Chapter 1 of Ref. 8.3.

8.2.2 Deflection Function

The same deflection function for the beam finite element in bending is used:

$$v(x) = f_1(x)v_1 + f_2(x)\theta_1 + f_3(x)v_2 + f_4(x)\theta_2 \tag{8.15}$$

where the shape functions $f_i(x)$ are defined in Eq. (5.7).

8.2.3 Basic Stiffness and Incremental Stiffness Matrices

Substituting the deflection function (8.15) into Eqs. (8.11) and (8.14) and then performing partial differentiations of the energy expressions U_b and W_n with respect to each of the four degrees of freedom as described in Eq. (8.10), a set of four equations is obtained:

$$\{\mathbf{F}\} = [[\mathbf{k}] - [\mathbf{n}]]\{\mathbf{q}\} \tag{8.16}$$

where [**k**] is the basic stiffness matrix associated with bending deflection; [**n**] is called the *incremental stiffness matrix* associated with the effect of axial force on bending deflection. Equation (8.16) is equivalent to the differential equation (8.5) with {**F**}, [**k**]{**q**}, and [**n**]{**q**} corresponding to w, $EI(\partial^4 v/\partial x^4)$, and $P(\partial^2 v/\partial x^2)$, respectively.

The coefficients in [**k**] have been given in Eq. (5.13) as

$$k_{ij} = EI \int_0^L f_i''(x) f_j''(x)\, dx \tag{8.17}$$

The coefficients in [**n**] can be derived as

$$n_{ij} = P \int_0^L f_i'(x) f_j'(x)\, dx \tag{8.18}$$

Substituting the shape functions (5.7) into Eqs. (8.17) and (8.18), explicit expressions for Eq. (8.16) are obtained:

$$\begin{Bmatrix} Y_1 \\ M_1 \\ Y_2 \\ M_2 \end{Bmatrix} = \left[\frac{EI}{L^2} \begin{bmatrix} \frac{12}{L} & 6 & -\frac{12}{L} & 6 \\ 6 & 4L & -6 & 2L \\ -\frac{12}{L} & -6 & \frac{12}{L} & -6 \\ 6 & 2L & -6 & 4L \end{bmatrix} - \frac{P}{10} \begin{bmatrix} \frac{12}{L} & 1 & -\frac{12}{L} & 1 \\ 1 & \frac{4L}{3} & -1 & -\frac{L}{3} \\ -\frac{12}{L} & -1 & \frac{12}{L} & -1 \\ 1 & -\frac{L}{3} & -1 & \frac{4L}{3} \end{bmatrix} \right] \begin{Bmatrix} v_1 \\ \theta_1 \\ v_2 \\ \theta_2 \end{Bmatrix} \tag{8.19}$$

Equation (8.19) is written in a form such that if $PL^2/10EI$ is defined as the eigenvalues, the determinant becomes nondimensional.

Because the incremental stiffness matrix [**n**] contains P, it is often referred to as the *initial stress matrix.* Because this matrix contains L but no EI, it is sometimes referred to as the *geometric stiffness matrix.*

Derivation of Eq. (8.19) is due to Gallagher and Padlog [8.4]. For a more detailed discussion, the reader is referred to Ref. 8.5.

8.3 CONSISTENT INCREMENTAL STIFFNESS MATRIX FOR DISTRIBUTED AXIAL FORCE

The effect of axial force on a beam in bending depends upon the distribution of axial force $p(x)$. In this section we derive an incremental stiffness matrix [**n**] for distributed axial force but with the use of the same shape functions (5.7) for deriving the stiffness matrix [**k**].

Let the intensity of the distributed load be represented as

$$p(x) = p_0 \left[1 + r \left(\frac{x}{L} \right)^{\alpha} \right] \qquad \text{lb/in.} \tag{8.20}$$

where x is measured from node 1, and $p(x)$ acts in the direction from node 2 to node 1. The axial force at x is

$$P(x) = P_0 + P_d(x) \qquad \text{pounds} \tag{8.21}$$

where P_0 is the axial force at node 2 and

$$\begin{aligned} P_d(x) &= \int_x^L p(\xi)\, d\xi \\ &= p_0 L \left\{ 1 - \frac{x}{L} + \frac{r}{1+\alpha}\left[1 - \left(\frac{x}{L}\right)^{\alpha+1}\right]\right\} \end{aligned} \tag{8.22}$$

Following the same procedure as that used in deriving [**n**] for constant axial force and keeping $P(x)$ inside the integrals, it can be shown that

$$n_{ij} = \int_0^L P(x) f_i'(x) f_j'(x)\, dx \tag{8.23}$$

The coefficients in the symmetrical matrix [**n**] are obtained explicitly as

$$\begin{aligned}
n_{11} &= n_{33} = -n_{31} \\
&= \frac{6P_0}{5L} + p_0\left[\frac{3}{5} + \frac{r}{1+\alpha}\left(\frac{6}{5} - \frac{36}{4+\alpha} + \frac{72}{5+\alpha} - \frac{36}{6+\alpha}\right)\right] \\
n_{21} &= -n_{32} = \frac{P_0}{10} + \frac{p_0 r L}{1+\alpha}\left(\frac{1}{10} + \frac{6}{3+\alpha} - \frac{30}{4+\alpha} + \frac{42}{5+\alpha} - \frac{18}{6+\alpha}\right) \\
n_{41} &= -n_{43} = \frac{P_0}{10} + p_0 L\left[\frac{1}{10} + \frac{r}{1+\alpha}\left(\frac{1}{10} - \frac{12}{4+\alpha} + \frac{30}{5+\alpha} - \frac{18}{6+\alpha}\right)\right] \\
n_{22} &= \frac{2P_0 L}{15} + p_0 L^2\left[\frac{1}{10} + \frac{r}{1+\alpha}\left(\frac{2}{15} - \frac{1}{2+\alpha} + \frac{8}{3+\alpha} - \frac{22}{4+\alpha}\right.\right. \\
&\quad \left.\left. + \frac{24}{5+\alpha} - \frac{9}{6+\alpha}\right)\right] \\
n_{42} &= -\frac{P_0 L}{30} - p_0 L^2\left[\frac{1}{60} + \frac{r}{1+\alpha}\left(\frac{1}{30} - \frac{2}{3+\alpha} + \frac{11}{4+\alpha} - \frac{18}{5+\alpha} + \frac{9}{6+\alpha}\right)\right] \\
n_{44} &= \frac{2P_0 L}{15} + p_0 L^2\left[\frac{1}{30} + \frac{r}{1+\alpha}\left(\frac{2}{15} - \frac{4}{4+\alpha} + \frac{12}{5+\alpha} - \frac{9}{6+\alpha}\right)\right]
\end{aligned} \tag{8.24}$$

Because the shape functions used in deriving [**n**] are the same as those used in deriving [**k**], matrix [**n**] may be called the *consistent incremental stiffness matrix.*

Example 8.1 Buckling of a clamped-clamped column

A clamped-clamped column with bending rigidity EI and length l subjected to an axial compressive force P is shown in Fig. 8.3a. Because of the symmetrical buckling shape, only half of the column need be analyzed.

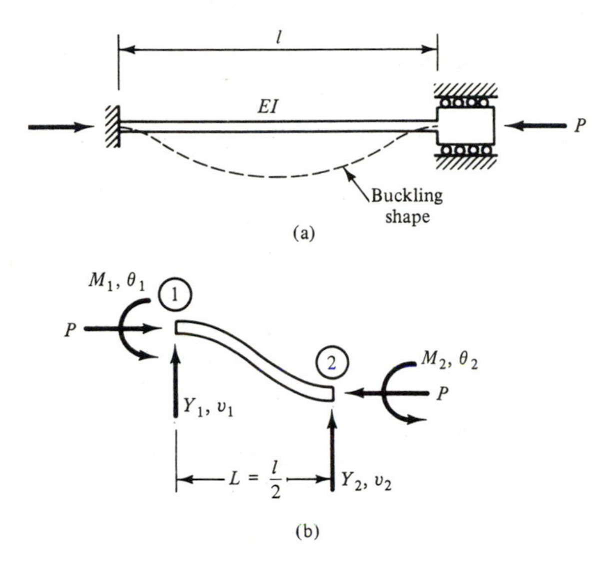

Figure 8.3 (a) Buckling of a clamped-clamped column; (b) one-element modeling for half of the column.

One-element solution. For the one-element model shown in Fig. 8.3b, there are three zero degrees of freedom,

$$v_1 = \theta_1 = \theta_2 = 0$$

By considering only the third equation in Eqs. (8.19), we have

$$\{Y_2 = 0\} = \left[\frac{EI}{L^2}\left(\frac{12}{L}\right) - \frac{P}{10}\left(\frac{12}{L}\right)\right]\{\mathbf{v}_2\}$$

Substituting the element length of $l/2$ for L, the critical (buckling) load is found as

$$P_{cr} = \frac{40EI}{l^2} \quad \text{vs.} \quad \frac{4\pi^2 EI}{l^2}\,(\text{exact}) \qquad \text{error} = 1.32\%$$

Two-element solution. For a two-element model, the zero degrees of freedom are

$$v_1 = \theta_1 = \theta_3 = 0$$

We have a set of three equations:

$$\begin{Bmatrix} 0 \\ 0 \\ 0 \end{Bmatrix} = \left[\frac{EI}{L^2} \begin{bmatrix} \frac{24}{L} & 0 & -\frac{12}{L} \\ 0 & 8L & -6 \\ -\frac{12}{L} & -6 & \frac{12}{L} \end{bmatrix} - \frac{P}{10} \begin{bmatrix} \frac{24}{L} & 0 & -\frac{12}{L} \\ 0 & \frac{8L}{3} & -1 \\ -\frac{12}{L} & -1 & \frac{12}{L} \end{bmatrix} \right] \begin{Bmatrix} v_2 \\ \theta_2 \\ v_3 \end{Bmatrix}$$

Letting $\lambda = PL^2/10EI$ and setting the determinant of the above matrix equal to zero, we have

$$(1 - \lambda)(15\lambda^2 - 52\lambda + 12) = 0$$

$$\lambda = 0.2486, 1.0, \text{ and } 3.2181$$

Substituting the element length of $l/4$ for L, the critical (buckling) loads are found:

First mode:

$$P_{cr} = \frac{39.78EI}{l^2} \quad \text{vs.} \quad \frac{4\pi^2 EI}{l^2} \text{ (exact)} \qquad \text{error} = 0.8\%$$

Third mode:

$$P_{cr} = \frac{160EI}{l^2} \quad \text{vs.} \quad \frac{36\pi^2 EI}{l^2} \text{ (exact)} \qquad \text{error} = -55\%$$

Fifth mode:

$$P_{cr} = \frac{515EI}{l^2} \quad \text{vs.} \quad \frac{100\pi^2 EI}{l^2} \text{ (exact)} \qquad \text{error} = -48\%$$

If we change the boundary conditions from $\theta_3 = 0$ to $v_3 = 0$, we can obtain the critical (buckling) loads for the second, fourth, and sixth modes.

Example 8.2 Buckling of a tapered column under its own weight

A tapered column with rectangular cross section under its own weight is shown in Fig. 8.4. Also shown is a step modeling using uniform elements. The critical (buckling) value for q_0 is given on page 131 of Ref. [8.2] as 13.0 EI_0/l^2.

The column is modeled using uniform elements and then using tapered elements. For each type of element, the axial load is first assumed as constant by using **[n]** as given in Eq. (8.19) and then assumed as distributed by using **[n]** as given in Eq. (8.24). The results for percentage error in critical value of q_0 vs. number of elements are plotted in Fig. 8.4. Although it has been shown in Table 6.1 that more accurate representation in geometry by using tapered elements is essential in bending problems, it is seen that more accurate representation in distributed axial load by using Eq. (8.24) is essential in this buckling problem.

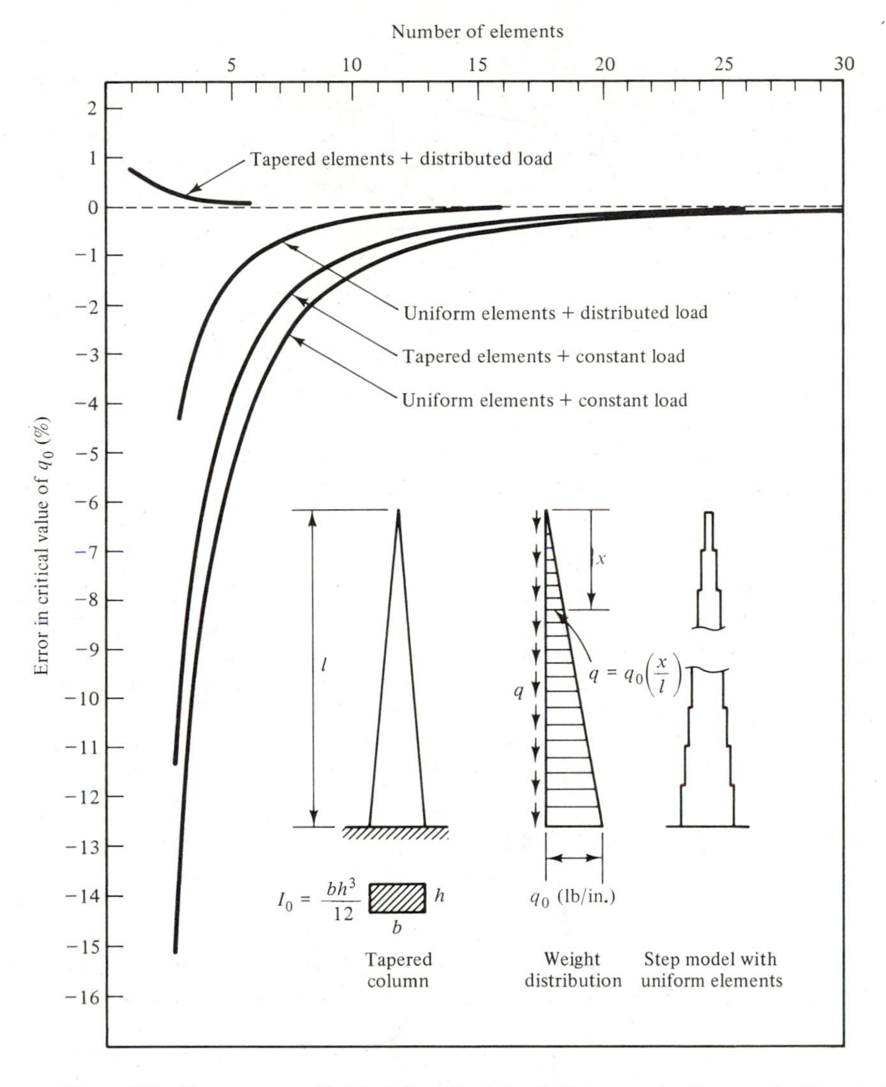

Figure 8.4 Convergence of critical (buckling) load of a tapered column under its own weight.

8.4 BUCKLING OF COLUMN ON ELASTIC FOUNDATION

If a column lies on an elastic foundation as shown in Fig. 8.5, it is common practice to assume the elastic foundation as a Winkler foundation with an infinite number of independent direct springs with elastic constant β (lb/in./in.). Such an assumption and its development are explained in Chapter 1 of Ref. 8.6. The differential equation (8.5) for a column on elastic foundation

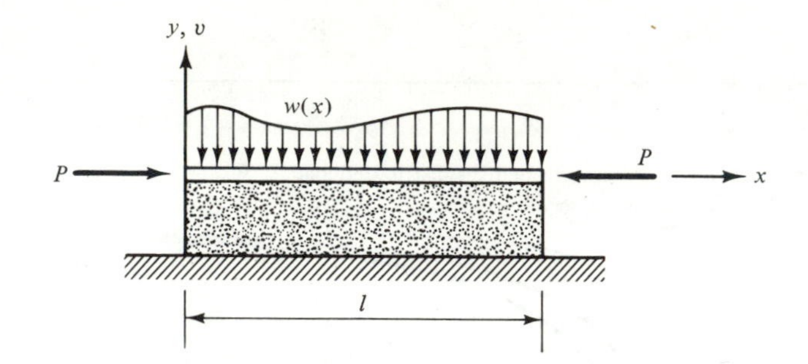

Figure 8.5 Beam-column on an elastic foundation.

then becomes

$$EI\frac{\partial^4 v}{\partial x^4} = w - P\frac{\partial^2 v}{\partial x^2} - \beta v \tag{8.25}$$

The term βv is a distributed reaction force proportional at every point to the deflection of the beam at that point. Compared to the differential equation for a beam in lateral vibration, the reaction force βv is analogous to the inertia force $m\ddot{v}$. It is apparent that the consistent mass matrix as derived in Eq. (7.31) or the lumped mass matrix can be used directly as an elastic foundation matrix by simply replacing ρA by β.

Example 8.3 Buckling of a simply supported column on elastic foundation

Depending on the value of the spring constant β, the buckling mode of a simply supported column with length l and bending rigidity EI may have different numbers of half-waves. It was shown in Ref. [8.6] that when $\beta \le 4\pi^2 EI/l^2$, the buckling mode is a single half-wave. Let it be desired to find the buckling load for $\beta = 336\ EI/l^4$ using one element to model half of the column.

Based on the conditions $v_1 = \theta_2 = 0$ and based on Eq. (8.19) and the modified form of Eq. (7.31), we have

$$\begin{Bmatrix} 0 \\ 0 \end{Bmatrix} = \left[\frac{EI}{L^2}\begin{bmatrix} 4L & -6 \\ -6 & \dfrac{12}{L} \end{bmatrix} - \frac{P}{10}\begin{bmatrix} \dfrac{4L}{3} & -1 \\ -1 & \dfrac{12}{L} \end{bmatrix} + \frac{\beta L^2}{420}\begin{bmatrix} 4L & 13 \\ 13 & \dfrac{156}{L} \end{bmatrix} \right] \begin{Bmatrix} \theta_1 \\ v_2 \end{Bmatrix}$$

where $\beta = 336\ EI/l^4 = 21\ EI/L^4$ for $L = l/2$. Letting $\lambda = PL^2/10EI$ and setting the determinant of the matrix above equal to zero, we have

$$15\lambda^2 - 66.1\lambda + 54.5375 = 0$$

$$\lambda_{cr} = 1.0993 \qquad \text{or} \qquad P_{cr} = \frac{43.97EI}{l^2}$$

Compared with the analytical solution given on page 145 of Ref. 8.6,

$$P_{cr} = \pi^2 \frac{EI}{l^2} + \frac{\beta l^2}{\pi^2} = 43.91 \frac{EI}{l^2}$$

The error is 0.14%.

8.5 REDUCTION METHOD FOR BUCKLING PROBLEM

A numerical approximate method given by Guyan [7.9] for reducing large eigenvalue equations of motion was described in detail in Chapter 7. This method can be applied to buckling problems as well by simply replacing the mass matrix [**m**] in Eq. (7.53) by the incremental stiffness matrix [**n**].

Example 8.4 Column buckling analysis using the reduction method

An example for buckling of a clamped-clamped column is chosen to evaluate the reduction method. Plots of the critical (buckling) values vs. number of elements used to model half of the column are shown in Fig. 8.6 [8.7]. It is seen that, although using the reduction method results in more error than without using it, the results eventually converge to the exact value as the number of elements increases.

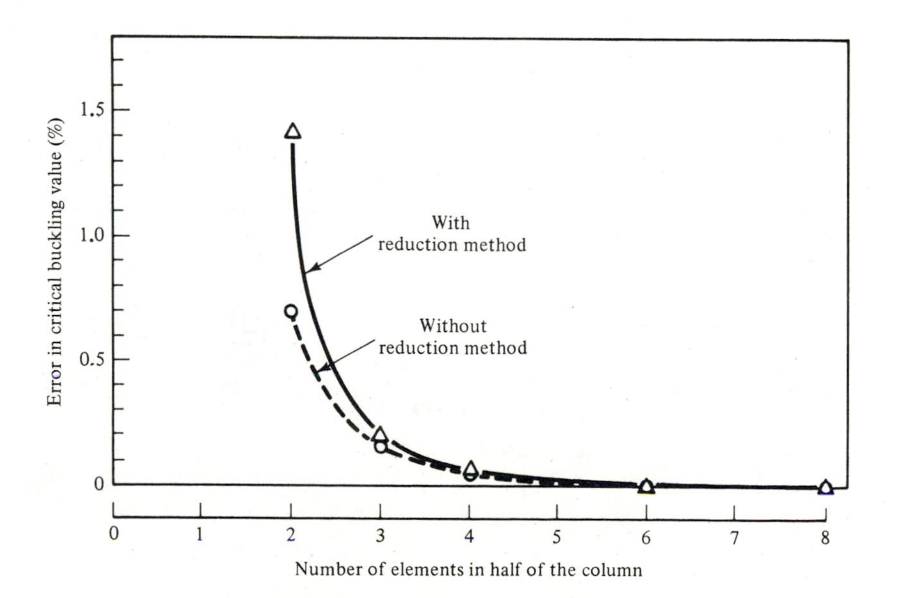

Figure 8.6 Buckling analysis of a clamped-clamped column.

8.6 BUCKLING OF PLANE FRAME

Buckling formulation for a plane frame element oriented arbitrarily in a two-dimensional plane can be obtained from Eq. (8.19) by using the coordinate transformation procedure as described in Sec. 5.2.2 for the stiffness formulation.

An axial-flexural beam element with reference to a set of local coordinates $(\bar{x}, \bar{y})$ and a set of global or reference coordinates (x, y) is shown in Fig. 5.5. It has been shown during the derivation of Eq. (5.31) that the stiffness matrix for the element with reference to global coordinates can be obtained by the following congruent transformation formula:

$$\{\mathbf{F}\} = [\mathbf{T}]^T[\bar{\mathbf{k}}][\mathbf{T}]\{\mathbf{q}\} \tag{8.26}$$

where $\{\mathbf{F}\}$ and $\{\mathbf{q}\}$ are the vectors for the six nodal forces and displacements, respectively; $[\mathbf{T}]$ is the coordinate transformation matrix given in Eq. (5.24); and $[\bar{\mathbf{k}}]$ is the augmented 6×6 stiffness matrix given in Eq. (5.22) or (5.23).

If the 4×4 matrix $[\mathbf{n}]$ as defined in Eq. (8.19) is augmented to a 6×6 matrix $[\bar{\mathbf{n}}]$ with the first row and column and fourth row and column containing zeros, the stiffness and incremental stiffness matrix formulations for a plane frame element can be obtained as

$$\{\mathbf{F}\} = [\mathbf{T}]^T\Big[[\bar{\mathbf{k}}] - [\bar{\mathbf{n}}]\Big][\mathbf{T}]\{\mathbf{q}\} \tag{8.27}$$

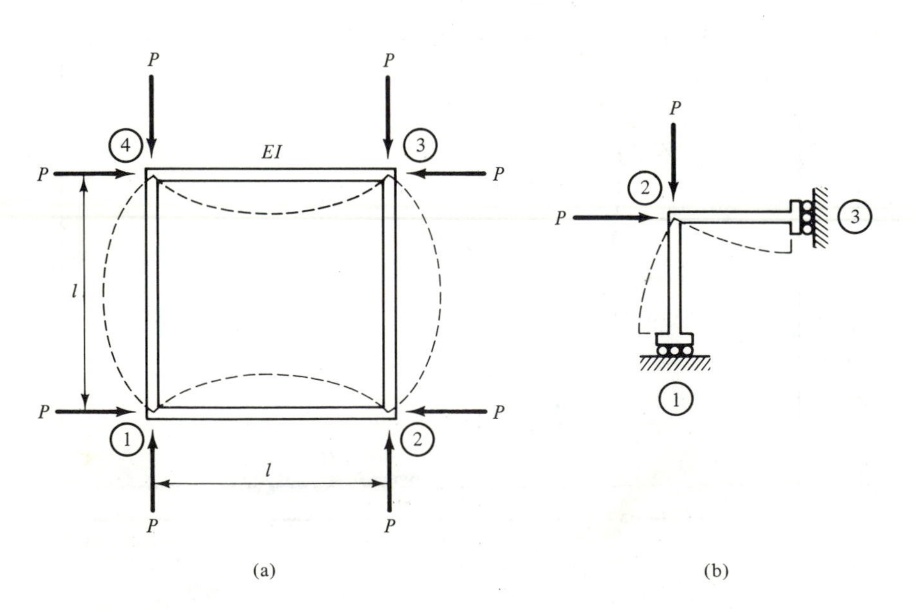

Figure 8.7 (a) Symmetrical buckling of a square frame; (b) two-element modeling for a quadrant of the frame.

or explicitly,

$$
\begin{Bmatrix} X_1 \\ Y_1 \\ M_1 \\ X_2 \\ Y_2 \\ M_2 \end{Bmatrix} = \left[\frac{E}{L^2} \begin{bmatrix}
AL\lambda^2 + \frac{12I\mu^2}{L} & & & & & \text{symmetric} \\
\left(AL - \frac{12I}{L}\right)\lambda\mu & AL\mu^2 + \frac{12I\lambda^2}{L} & & & & \\
-6I\mu & 6I\lambda & 4IL & & & \\
-\left(AL\lambda^2 + \frac{12I\mu^2}{L}\right) & -\left(AL - \frac{12I}{L}\right)\lambda\mu & 6I\mu & AL\lambda^2 + \frac{12I\mu^2}{L} & & \\
-\left(AL - \frac{12I}{L}\right)\lambda\mu & -\left(AL\mu^2 + \frac{12I\lambda^2}{L}\right) & -6I\lambda & \left(AL - \frac{12I}{L}\right)\lambda\mu & AL\mu^2 + \frac{12I\lambda^2}{L} & \\
-6I\mu & 6I\lambda & 2IL & 6I\mu & -6I\lambda & 4IL
\end{bmatrix} \right.
$$

$$
\left. - \frac{P}{10} \begin{bmatrix}
\frac{12\mu^2}{L} & & & & & \text{symmmetric} \\
\frac{-12\lambda\mu}{L} & \frac{12\lambda^2}{L} & & & & \\
-\mu & \lambda & \frac{4L}{3} & & & \\
\frac{-12\mu^2}{L} & \frac{12\lambda\mu}{L} & \mu & \frac{12\mu^2}{L} & & \\
\frac{12\lambda\mu}{L} & \frac{-12\lambda^2}{L} & -\lambda & \frac{-12\lambda\mu}{L} & \frac{12\lambda^2}{L} & \\
-\mu & \lambda & \frac{-L}{3} & \mu & -\lambda & \frac{4L}{3}
\end{bmatrix} \right] \begin{Bmatrix} u_1 \\ v_1 \\ \theta_1 \\ u_2 \\ v_2 \\ \theta_2 \end{Bmatrix} \tag{8.28}
$$

Example 8.5 Symmetrical buckling of a square frame

A square frame with a symmetrical buckling mode shape is shown in Fig. 8.7a. An analytic solution for this problem can be found in Ref. [8.8] with $P_{cr} = \pi^2 EI/l^2$. The case of antisymmetrical buckling was also considered in Ref. [8.8]. Let it be desired to find P_{cr} for symmetrical buckling mode.

Four-element solution. Using four elements based on Eq. (8.28) and using the conditions that u and v equal to zero at all four joints, we obtain

$$
\begin{Bmatrix} 0 \\ 0 \\ 0 \\ 0 \end{Bmatrix} = \left[\frac{EI}{L} \begin{bmatrix} 8 & 2 & 0 & 2 \\ 2 & 8 & 2 & 0 \\ 0 & 2 & 8 & 2 \\ 2 & 0 & 2 & 8 \end{bmatrix} - \frac{PL}{30} \begin{bmatrix} 8 & -1 & 0 & -1 \\ -1 & 8 & -1 & 0 \\ 0 & -1 & 8 & -1 \\ -1 & 0 & -1 & 8 \end{bmatrix} \right] \begin{Bmatrix} \theta_1 \\ \theta_2 \\ \theta_3 \\ \theta_4 \end{Bmatrix}
$$

Using the symmetrical conditions that $\theta_1 = -\theta_2 = \theta_3 = -\theta_4$ and letting $L = l$, the equations above can be reduced to

$$
\frac{4EI}{l} - \frac{Pl}{3} = 0
$$

which gives

$$P_{cr} = \frac{12EI}{l^2} \quad \text{vs.} \quad \frac{\pi^2 EI}{l^2}\text{(exact)} \qquad \text{error} = 21.6\%$$

Eight-element solution. Due to the symmetrical nature of the buckling mode shape, an eight-element model is in reality a two-element model, as shown in Fig. 8.7b. For this model, there are only three degrees of freedom and the equations are

$$\begin{Bmatrix} 0 \\ 0 \\ 0 \end{Bmatrix} = \left[\frac{EI}{L^2} \begin{bmatrix} \frac{12}{L} & -6 & 0 \\ -6 & 8L & -6 \\ 0 & -6 & \frac{12}{L} \end{bmatrix} - \frac{P}{10} \begin{bmatrix} \frac{12}{L} & -1 & 0 \\ -1 & \frac{8L}{3} & -1 \\ 0 & -1 & \frac{12}{L} \end{bmatrix} \right] \begin{Bmatrix} u_1 \\ \theta_2 \\ v_3 \end{Bmatrix}$$

Using the symmetrical conditions that $u_1 = v_3$, we have

$$\begin{Bmatrix} 0 \\ 0 \end{Bmatrix} = \left[\frac{EI}{L^2} \begin{bmatrix} \frac{12}{L} & -6 \\ -12 & 8L \end{bmatrix} - \frac{P}{10} \begin{bmatrix} \frac{12}{L} & -1 \\ -2 & \frac{8L}{3} \end{bmatrix} \right] \begin{Bmatrix} u_1 \\ \theta_2 \end{Bmatrix}$$

Letting $\lambda = PL^2/10EI$ and setting the determinant equal to zero gives

$$15\lambda^2 - 52\lambda + 12 = 0$$

$$\lambda_{cr} = 0.2486$$

Substituting $l/2$ for L gives

$$P_{cr} = 9.944\frac{EI}{l^2} \quad \text{vs.} \quad \frac{\pi^2 EI}{l^2}\text{(exact)} \qquad \text{error} = 0.8\%$$

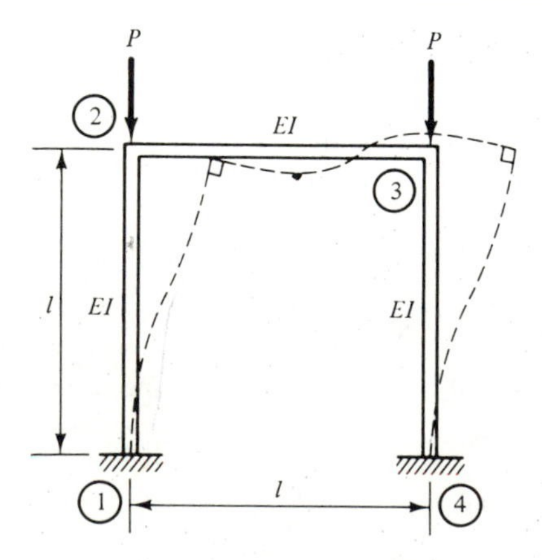

Figure 8.8 Antisymmetrical buckling of a clamped inextensible portal frame.

Due to double symmetry of the buckling shape, the model above can, in fact, be reduced to a one-element model for an octant of the frame. One purpose of performing the two-element solution above is to demonstrate the use of Eq. (8.28).

Example 8.6 Antisymmetrical buckling of a portal frame

An inextensible portal frame with antisymmetrical buckling mode shape is shown in Fig. 8.8. An exact solution is available in Ref. [8.8] with $P_{cr} = 7.344EI/l^2$. Let it be desired to find the value of P_{cr} by using a three-element model.

The frame has only four degrees of freedom and the assembled equations are, from Eq. (8.28),

$$\begin{Bmatrix} 0 \\ 0 \\ 0 \\ 0 \end{Bmatrix} = \left[\frac{E}{L^2} \begin{bmatrix} \dfrac{12I}{L} + AL & 6I & -AL & 0 \\ 6I & 8IL & 0 & 2IL \\ -AL & 0 & \dfrac{12I}{L} + AL & 6I \\ 0 & 2IL & 6I & 8IL \end{bmatrix} - \frac{P}{10} \begin{bmatrix} \dfrac{12}{L} & 1 & 0 & 0 \\ 1 & \dfrac{4L}{3} & 0 & 0 \\ 0 & 0 & \dfrac{12}{L} & 1 \\ 0 & 0 & 1 & \dfrac{4L}{3} \end{bmatrix} \right] \begin{Bmatrix} u_2 \\ \theta_2 \\ u_3 \\ \theta_3 \end{Bmatrix}$$

Using the antisymmetrical conditions that $u_2 = u_3$ and $\theta_2 = \theta_3$, we have

$$\begin{Bmatrix} 0 \\ 0 \end{Bmatrix} = \left[\frac{EI}{L^2} \begin{bmatrix} \dfrac{12}{L} & 6 \\ 6 & 10L \end{bmatrix} - \frac{P}{10} \begin{bmatrix} \dfrac{12}{L} & 1 \\ 1 & \dfrac{4L}{3} \end{bmatrix} \right] \begin{Bmatrix} u_2 \\ \theta_2 \end{Bmatrix}$$

Letting $\lambda = PL^2/10EI$, $L = l$, and setting the determinant for the matrix above equal to zero, we have

$$15\lambda^2 - 124\lambda + 84 = 0$$

$$\lambda_{cr} = 0.7445$$

$$P_{cr} = 7.445\frac{EI}{l^2} \quad \text{vs.} \quad 7.344\frac{EI}{l^2}\,(\text{exact}) \qquad \text{error} = 1.4\%$$

8.7 LARGE DEFLECTION OF BEAMS AND PLANE FRAMES

If the axial force P is assumed as positive when in tension, Eq. (8.28) can be written symbolically as

$$\{\mathbf{F}\} = [[\mathbf{K}] + [\mathbf{N}]]\{\mathbf{Q}\} \tag{8.29}$$

where the capital letters are used to symbolize assembled matrices for the overall finite element model. The positive sign in front of [**N**] shows that the structure becomes stiffer when P is in tension.

Because matrix [**N**] is a function of P, which, in turn, is a function of displacements {**Q**}, Eq. (8.29) is nonlinear. A simple way to solve this nonlinear equation is to use a step-by-step linear incremental procedure. The linearized incremental formulation can be obtained by applying an incremental operator Δ to Eq. (8.29),

$$\begin{aligned}\{\Delta\mathbf{F}\}_i &= [[\mathbf{K}]_{i-1} + [\mathbf{N}]_{i-1}]\{\Delta\mathbf{Q}\}_i \\ \{\Delta\mathbf{Q}\}_i &= [[\mathbf{K}]_{i-1} + [\mathbf{N}]_{i-1}]^{-1}\{\Delta\mathbf{F}\}_i\end{aligned} \tag{8.30}$$

which gives the vector of displacement increments at the end of step i. For the first step, $[\mathbf{N}]_0$ may be neglected.

In the solution of large deflection problems of beams and frames, axial force P and nodal point locations for each finite element change at every step. The axial force for each step is obtained as the resultant of the horizontal component X and the vertical component Y,

$$P = X\cos\theta + Y\sin\theta \tag{8.31}$$

with

$$\theta = \tan^{-1}\frac{y_2 - y_1}{x_2 - x_1}$$

where x and y are the global coordinates of the two nodes at the beginning of the current step. With an update of P, λ, and μ for every step, Eq. (8.30) can be used to predict the nonlinear load-displacement path for every degree of freedom.

The equilibrium equation (8.29) can also be solved by using iterative procedures, such as Newton-Raphson technique [8.9]. The solution can also be performed by combined incremental and iterative procedures.

Example 8.7 Large deflection of a cantilever beam

The nonlinear load-displacement curves for a cantilever beam subjected to a lateral load at the tip predicted by using four beam elements are shown in Fig. 8.9 [8.10]. The size of the load increment is progressively enlarged in accordance with a geometric series. The results from an analytic solution in terms of elliptic integrals are also shown [8.11]. It is seen that the simple finite element formulation (8.28) and the straightforward linear incremental procedure [Eq. (8.30)] are both highly accurate and efficient. Examples of a cantilever beam under distributed load and two plane frames under concentrated loads are also given in Ref. 8.10.

8.8 CONCLUDING REMARKS

The stiffness formulation for a beam finite element has been extended to include the effect of axial force, through the additional formulation of an incremental stiffness matrix using the energy method. The formulation can

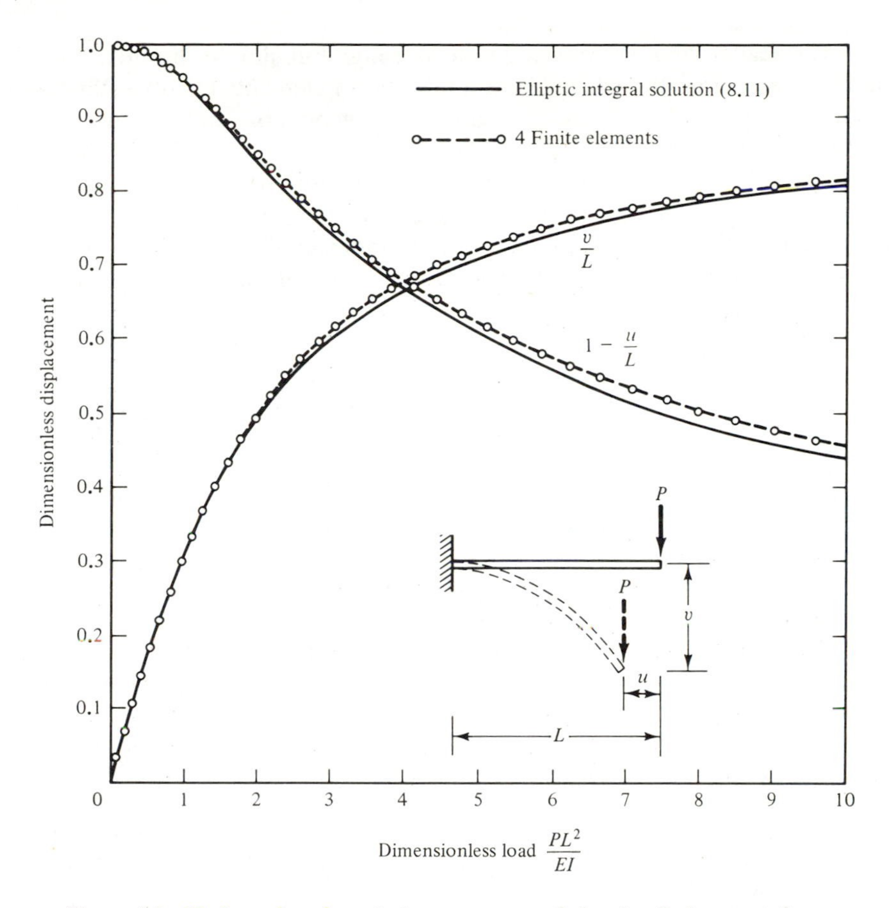

Figure 8.9 Horizontal and vertical components of the tip displacement for a cantilever beam under a concentrated load.

also be transformed to that for two- or three-dimensional frame elements. The formulations have been shown to be highly accurate for buckling analyses of columns and plane frames.

For columns with distributed axial force, a consistent incremental stiffness matrix has been formulated. The advantage of this matrix has been demonstrated in an example of buckling of a tapered column under its own weight. For columns on elastic foundation, it has been shown that an elastic foundation matrix analogous to the consistent mass matrix for dynamic problems can be used for buckling analysis.

Equation (8.28) and the simple linear incremental procedure [Eq. (8.30)] have been shown to be highly accurate and efficient in large deflection analyses of beams and frames. They can also be straightforwardly extended to large displacement problems of viscoelastic frames [8.12] and snap buckling problems of nonlinear elastic frames [8.13].

The reader who is interested in a thorough but simple formulation and numerical solution procedure using simple beam elements to solve static and dynamic response problems of beams with geometric and material nonlinearities is advised to see Ref. [8.14].

With a thorough understanding of the theories and procedures given in this chapter for beam and frame finite elements, the reader is equipped with the fundamental background for further understanding of more complex problems, such as buckling and large displacement of plates and shells using more sophisticated finite elements.

REFERENCES

8.1. Euler, Leonard, *Elastic Curves, Des Curvie Elasticis,* Lausanne and Geneva, 1744, translated and annotated by W. A. Oldfather, C. A. Ellis, and D. M. Brown, ISIS, Vol. XX, No. 1, Nov. 1933.

8.2. Timoshenko, S. P., and Gere, J. M., *Theory of Elastic Stability,* 2nd ed., McGraw-Hill Book Company, New York, 1961, Sec. 8.4, p. 131.

8.3. Novozhilov, V. V., *Thin Shell Theory,* trans. from the 2nd Russian edition by P. G. Lowe; ed. J. R. M. Radok, Wolters-Noordhoff BV, Groningen, The Netherlands, 1971.

8.4. Gallagher, R. H., and Padlog, J., "Discrete Element Approach to Structural Instability Analysis," *AIAA Journal,* Vol. 1, No. 6, June 1963, pp. 1437–1439.

8.5. Martin, H. C., "On the Derivation of Stiffness Matrices for the Analysis of Large Deflection and Stability Problems," *Proceedings,* Conference on Matrix Methods in Structural Mechanics, Air Force Flight Dynamics Laboratory, Report TR-66-80, 1966, pp. 697–716.

8.6. Hetényi, M., *Beams on Elastic Foundation,* University of Michigan Press, Ann Arbor, 1946, pp. 1 and 144–146.

8.7. Gallagher, R. H., and Yang, T. Y., "Elastic Instability Predictions for Doubly Curved Shells," *Proceedings,* Second Conference on Matrix Methods in Structural Mechanics, Air Force Flight Dynamics Laboratory, Report TR-68-150, 1968, pp. 711–739.

8.8. Simitses, G. J., *An Introduction to the Elastic Stability of Structures,* Prentice-Hall, Inc., Englewood Cliffs, N.J., 1976, Chap. 4 and p. 86.

8.9. Carnahan, B., Luther, H. A., and Wilkes, J. O., *Applied Numerical Methods,* John Wiley & Sons, Inc., New York, 1969.

8.10. Yang, T. Y., "Matrix Displacement Solution to Elastica Problems of Beams and Frames," *International Journal of Solids and Structures,* Vol. 9, 1973, pp. 829–842.

8.11. Bisshopp, K. E., and Drucker, D. C., "Large Deflection of Cantilever Beams," *Quarterly of Applied Mathematics,* Vol. 3, 1945, pp. 272–275.

8.12. Yang, T. Y., and Lianis, G., "Large Displacement Analysis of Viscoelastic Beams and Frames by the Finite Element Method," *Journal of Applied Mechanics,* Vol. 41, No. 3, 1974, pp. 635–640.

8.13. Yang, T. Y., and Wagner, R. J., "Snap Buckling of Nonlinearly-Elastic Finite Element Bars," *Computers and Structures*, Vol. 3, 1973, pp. 1473–1481.

8.14. Yang, T. Y., and Saigal, S., "A Simple Element for Static and Dynamic Response of Beams with Material and Geometric Nonlinearities," *International Journal for Numerical Methods in Engineering*, Vol. 20, 1984, pp. 851–867.

PROBLEMS

8.1. Using two elements, find the critical buckling load P_{cr} for the column shown in Fig. P8.1.

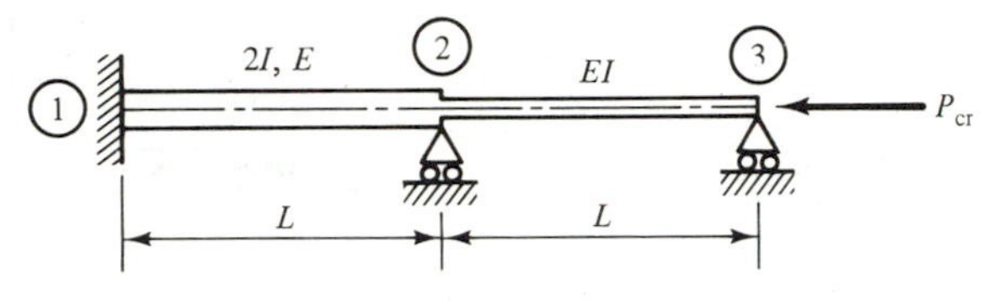

Figure P8.1

8.2. Find the critical buckling load P_{cr} for a column with both ends restrained from rotation, as shown in Fig. P8.2. Use one and two elements, respectively, to model half of the column. Compare the two solutions with the exact answer $P_{cr} = 4\pi^2 EI/L^2$. For the two-element solution, use the Guyan reduction method by eliminating the d.o.f. θ_2.

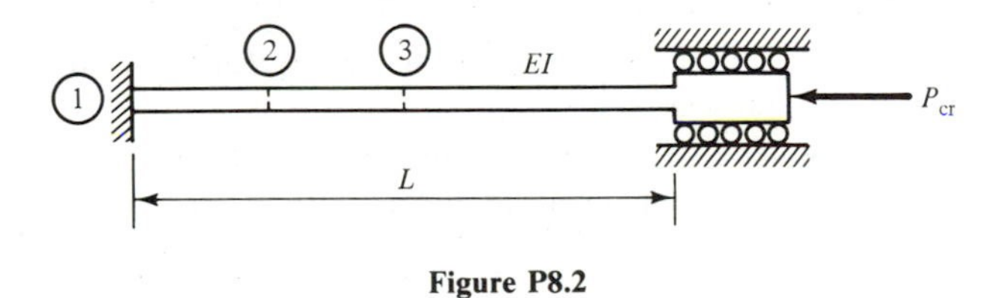

Figure P8.2

8.3. Using two elements to model the entire column shown in Fig. P8.3, find the lowest value of the spring constant $\alpha EI/L^3$ that will cause the column to buckle into the second mode as shown. An exact solution for α is given in Ref. 8.8 as $2\pi^2$.

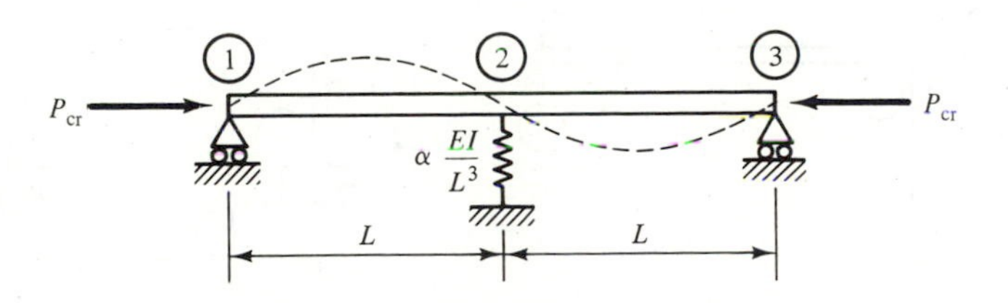

Figure P8.3

8.4. Figure P8.4 shows two identical columns joined rigidly to a rigid but weightless plate at the tops and fixed at the bases. Each column shares half of the weight W. Using one element to model half a column, find the critical buckling weight W. $E = 10^7$ psi. Exact solution: $W_{cr} = 2\pi^2 EI/L^2$ or 18.28 lb.

8.5. Using four elements, find the critical buckling load P_{cr} for the inextensible frame shown in Fig. P8.5.

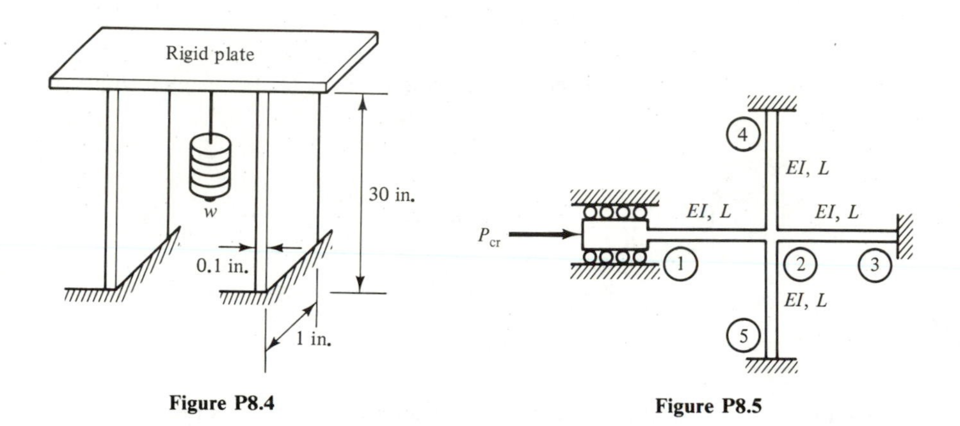

Figure P8.4

Figure P8.5

8.6. Using two elements, find the angle θ that can produce the minimum possible critical buckling loads P_{cr} for the inextensible frame shown in Fig. P8.6.

8.7. Using three elements, find the critical buckling load P_{cr} for the inextensible portal frame shown in Fig. P8.7.

8.8. Using four elements, find the critical buckling load P_{cr} for the inextensible square frame with antisymmetrical mode shape shown in Fig. P8.8.

8.9. First use four elements to model the entire inextensible square frame shown in Fig. P8.9. Then use two elements to model a quadrant of the frame to find the

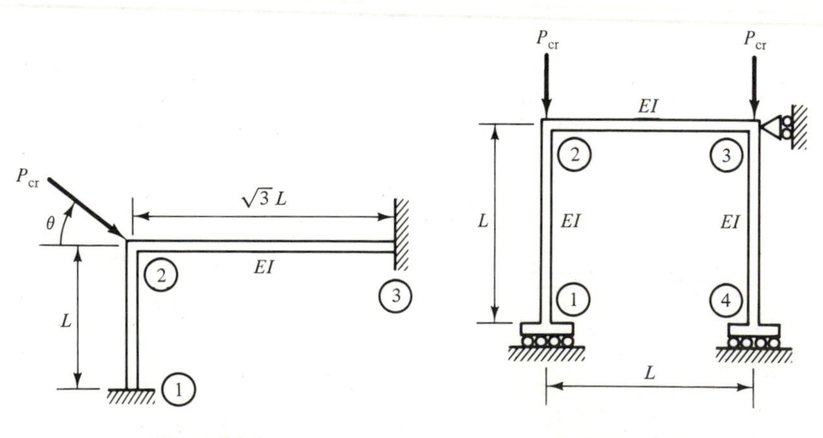

Figure P8.6

Figure P8.7

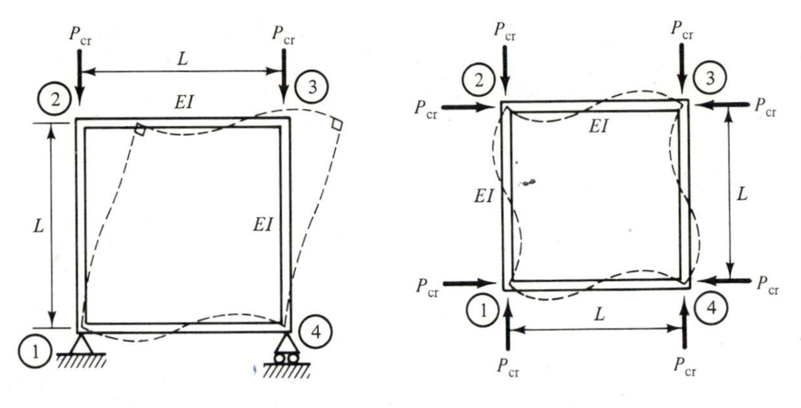

Figure P8.8

Figure P8.9

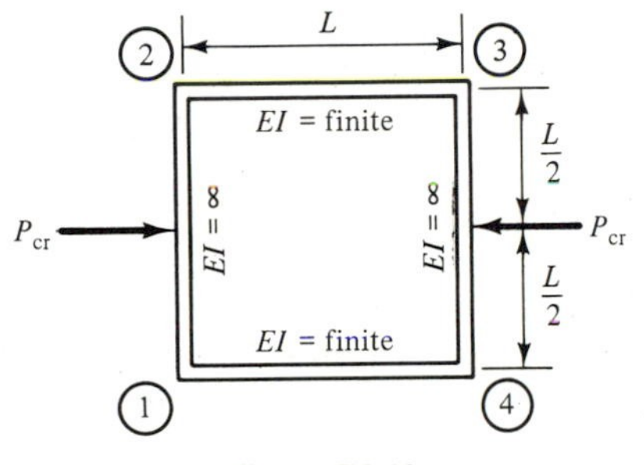

Figure P8.10

lowest critical buckling load P_{cr} for the antisymmetrical mode shape shown. Compare this P_{cr} value with that for the symmetrical mode as obtained in Example 8.5.

8.10. Using the least possible number of elements, find the critical buckling load P_{cr} for the inextensible square frame shown in Fig. P8.10. *Hint*: This problem is similar to Problem 8.4.

CHAPTER 9

Plane Stress and Plane Strain Finite Elements

The finite element method is a general method of structural analysis in which a continuous structure is replaced by a finite number of elements interconnected at a finite number of "nodal" points. Such an idealization is inherent in the conventional analysis of trusses and frames described in Chapters 4–8.

In this chapter the finite element "displacement" method is used to determine the stresses and displacements in two-dimensional elastic structures of arbitrary geometric and material properties as well as complex loading and boundary conditions. An assemblage of two-dimensional finite elements is used to represent the continuous structure. Forces acting on the actual structure are replaced by statically equivalent or work equivalent concentrated forces acting at the nodal points of the finite element system. The resulting large set of linear simultaneous equations is solved by pertinent numerical techniques using computers.

A brief review of the basic equations for two-dimensional elasticity is first given. The basic finite element formulative methods and numerical solution procedures are then demonstrated in great detail for two simplest plane stress and plane strain finite elements: an eight-degree-of-freedom rectangular element and a six-degree-of-freedom triangular element. Numerical examples are given. A computer program for the six-d.o.f. triangular element together with a user's manual and input and output data for sample problems are given in Chapter 13.

More sophisticated higher-order finite elements can normally be achieved in two ways: (1) including higher-order displacement derivatives as nodal degrees of freedom or (2) increasing the number of nodal points by placing them on the sides or inside the element. A state-of-the-art review of the higher-order plane stress and plane strain finite element displacement models of both rectangular and triangular shapes are presented in a self-contained fashion. The concept of natural coordinates for convenient integration of polynomials over a triangle is introduced.

Finally, the elements described in this chapter are evaluated through their performance on two popular examples: (1) a parabolically loaded square plate and (2) a cantilever plate subjected to parabolically distributed shear load at the end. Discussions and concluding remarks are provided.

9.1 TWO-DIMENSIONAL ELASTIC EQUATIONS

9.1.1 Equilibrium Equations for Two-Dimensional Stresses

Stress is defined as the amount of force ΔF applied on an area ΔA as ΔA approaches zero:

$$\sigma = \lim_{\Delta A \to 0} \frac{\Delta F}{\Delta A} \tag{9.1}$$

Figure 9.1 shows a differential cubic element with three direct stress components $\sigma_x, \sigma_y, \sigma_z$ in the x, y, z directions, respectively, and three shearing stress components τ_{xy} $(= \tau_{yx})$, τ_{yz} $(= \tau_{zy})$, and τ_{zx} $(= \tau_{xz})$. The shearing stress τ_{xy} indicates that the stress acts in the xy plane and in the y direction.

Figure 9.2 shows a plane state of stress. All the stress components not acting in the xy plane vanish; consequently,

$$\sigma_z = \tau_{xz} = \tau_{zx} = \tau_{yz} = \tau_{zy} = 0 \tag{9.2}$$

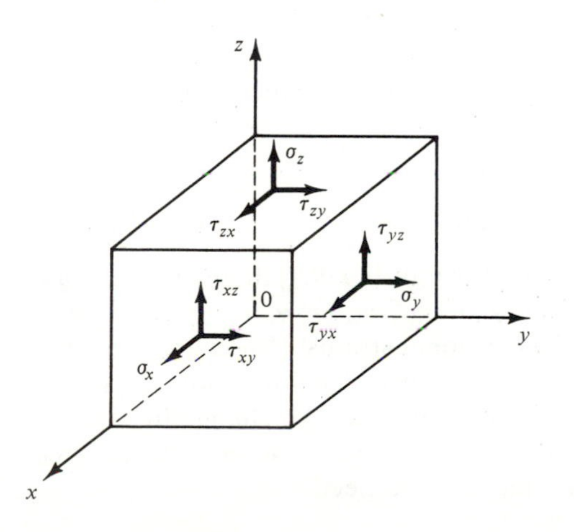

Figure 9.1 Three-dimensional stress components acting on a differential cube.

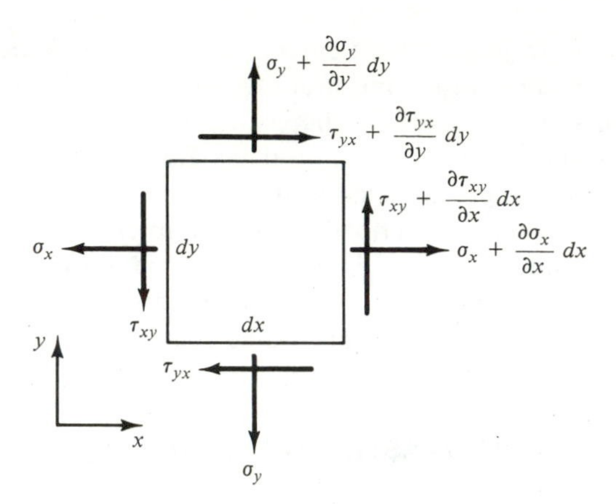

Figure 9.2 Two-dimensional stress components acting on a differential square.

The equilibrium equations for two-dimensional stresses can be obtained by first setting the total forces in the x direction to be zero:

$$\left(\sigma_x + \frac{\partial \sigma_x}{\partial x} dx\right) dy - \sigma_x\, dy + \left(\tau_{yx} + \frac{\partial \tau_{yx}}{\partial y} dy\right) dx - \tau_{yx}\, dx = 0$$

Canceling out terms yields

$$\frac{\partial \sigma_x}{\partial x} + \frac{\partial \tau_{yx}}{\partial y} = 0 \tag{9.3a}$$

Similarly, setting the equilibrium of forces in the y direction yields

$$\frac{\partial \sigma_y}{\partial y} + \frac{\partial \tau_{xy}}{\partial x} = 0 \tag{9.3b}$$

9.1.2 Two-Dimensional Strain–Displacement Relations

Strain is defined as the amount of elongation of a fiber, $\Delta\delta$, divided by its original length, ΔL, as ΔL approaches zero,

$$\epsilon = \lim_{\Delta L \to 0} \frac{\Delta \delta}{\Delta L} \tag{9.4}$$

Corresponding to the stress components shown in Fig. 9.1, there are the strain components: ϵ_x, ϵ_y, ϵ_z, γ_{xy} $(= \gamma_{yx})$, γ_{yz} $(= \gamma_{zy})$, and γ_{zx} $(= \gamma_{xz})$.

Figure 9.3 shows a state of two-dimensional strains. There are two types of strains: longitudinal strains and shearing strains. The longitudinal strain is defined in Eq. (9.4). The shearing strain is defined as the change in value of an originally right angle in an unstrained state. In Fig. 9.3, u and v are defined as the displacements in the x and y directions, respectively.

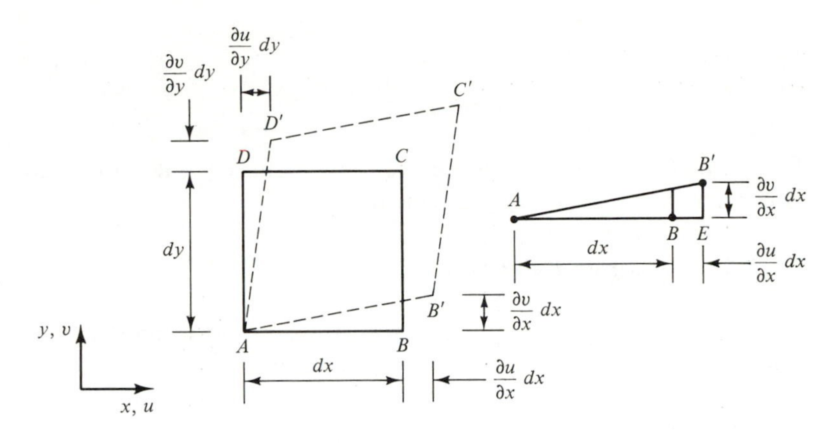

Figure 9.3 Deformation of a two-dimensional differential area.

From the definition of longitudinal strain, we have

$$\epsilon_x = \frac{\overline{AB'} - \overline{AB}}{\overline{AB}}$$

$$\overline{AB'} = \overline{AB}(1 + \epsilon_x) \tag{9.5a}$$

$$\overline{AB'}^2 = \overline{AB}^2(1 + 2\epsilon_x + \epsilon_x^2)$$

Following the rule of right triangles of Pythagoras (see triangle AEB' in Fig. 9.3), we have

$$\overline{AB'}^2 = \left(dx + \frac{\partial u}{\partial x}dx\right)^2 + \left(\frac{\partial v}{\partial x}dx\right)^2$$

$$= dx^2\left[1 + 2\left(\frac{\partial u}{\partial x}\right) + \left(\frac{\partial u}{\partial x}\right)^2 + \left(\frac{\partial v}{\partial x}\right)^2\right] \tag{9.5b}$$

Equating Eqs. (9.5a) and (9.5b) and neglecting the second-order terms of small strains yields

$$\epsilon_x = \frac{\partial u}{\partial x} \tag{9.6a}$$

Similarly,

$$\epsilon_y = \frac{\partial v}{\partial y} \tag{9.6b}$$

For shearing strains in the xy plane,

$$\gamma_{xy} = \gamma_{yx} = \text{sum of changes in angle} = \frac{\dfrac{\partial v}{\partial x}dx}{dx} + \frac{\dfrac{\partial u}{\partial y}dy}{dy}$$

Thus

$$\gamma_{xy} = \gamma_{yx} = \frac{\partial u}{\partial y} + \frac{\partial v}{\partial x} \tag{9.6c}$$

Equations (9.6a) to (9.6c) are useful in the subsequent formulation of plane stress and plane strain finite elements.

9.1.3 Stress–Strain Relations

According to Hooke's law, the uniaxial stress σ_x and strain ϵ_x are related by the modulus of elasticity or Young's modulus E as

$$\sigma_x = E\epsilon_x \tag{9.7}$$

Although E is normally a constant, it can be a function of strain for nonlinearly elastic and inelastic materials. It can also be a function of time for viscoelastic materials.

From experiment it is observed that extension of an element in the x direction is accompanied by the lateral contraction in the y and z directions, respectively,

$$\epsilon_y = \epsilon_z = -\nu\epsilon_x = -\nu\frac{\sigma_x}{E} \tag{9.8}$$

where ν is called *Poisson's ratio.* For common metals such as steel and aluminum, the value of ν is approximately 0.3. For concrete, it is roughly 0.16.

The shearing stress τ_{xy} and the shearing strain γ_{xy} are related by the modulus of shear G as

$$\tau_{xy} = G\gamma_{xy} \tag{9.9}$$

with

$$G = \frac{E}{2(1+\nu)} \tag{9.9a}$$

The value of Poisson's ratio ν can be obtained by first measuring G from a torsion test and then using Eq. (9.9a).

In terms of these engineering constants, the generalized Hooke's law may be written as

$$\epsilon_x = \frac{1}{E}[\sigma_x - \nu(\sigma_y + \sigma_z)]$$

$$\epsilon_y = \frac{1}{E}[\sigma_y - \nu(\sigma_z + \sigma_x)] \tag{9.10}$$

$$\epsilon_z = \frac{1}{E}[\sigma_z - \nu(\sigma_x + \sigma_y)]$$

$$\gamma_{xy} = \frac{2(1+\nu)}{E}\tau_{xy}$$

$$\gamma_{yz} = \frac{2(1+\nu)}{E}\tau_{yz}$$

$$\gamma_{zx} = \frac{2(1+\nu)}{E}\tau_{zx}$$

Or, in terms of strain components, Eqs. (9.10) become

$$\begin{aligned}
\sigma_x &= \frac{\nu E}{(1+\nu)(1-2\nu)}e + \frac{E}{1+\nu}\epsilon_x \\
\sigma_y &= \frac{\nu E}{(1+\nu)(1-2\nu)}e + \frac{E}{1+\nu}\epsilon_y \\
\sigma_z &= \frac{\nu E}{(1+\nu)(1-2\nu)}e + \frac{E}{1+\nu}\epsilon_z \\
\tau_{xy} &= \frac{E}{2(1+\nu)}\gamma_{xy} \\
\tau_{yz} &= \frac{E}{2(1+\nu)}\gamma_{yz} \\
\tau_{zx} &= \frac{E}{2(1+\nu)}\gamma_{zx}
\end{aligned} \tag{9.11}$$

where

$$e = \epsilon_x + \epsilon_y + \epsilon_z \tag{9.11a}$$

9.1.4 Stress–Strain Relations for Problems of Plane Stress and Plane Strain

Plane stress. For the case of plane stress, all the stresses related to the z axis vanish;

$$\sigma_z = \tau_{yz} = \tau_{zx} = 0 \tag{9.12}$$

Following Eqs. (9.10) and (9.12), the strain–stress relations can be expressed as

$$
\begin{aligned}
\epsilon_x &= \frac{1}{E}(\sigma_x - \nu\sigma_y) \\
\epsilon_y &= \frac{1}{E}(\sigma_y - \nu\sigma_x) \\
\epsilon_z &= \frac{-\nu}{E}(\sigma_x + \sigma_y) \\
\gamma_{xy} &= \frac{\tau_{xy}}{G}
\end{aligned}
\tag{9.13}
$$

Solving Eqs. (9.13) for stresses gives the stress-strain relations:

$$
\begin{aligned}
\sigma_x &= \frac{E}{1-\nu^2}(\epsilon_x + \nu\epsilon_y) \\
\sigma_y &= \frac{E}{1-\nu^2}(\epsilon_y + \nu\epsilon_x) \\
\tau_{xy} &= G\gamma_{xy}
\end{aligned}
\tag{9.14}
$$

Equations (9.14) can also be obtained by setting σ_z equal to zero in Eqs. (9.11) and substituting the resulting expression for ϵ_z into the equations for σ_x and σ_y in Eqs. (9.11).

There are many problems of plane stress—skins of a shear panel such as that shown in Fig. 3.11, skins of a wing box, shear walls of tall buildings, and so on.

Plane strain. For the case of plane strain, all the strains related to the z axis vanish:

$$\epsilon_z = \gamma_{yz} = \gamma_{zx} = 0 \tag{9.15}$$

Following Eqs. (9.11) and (9.15), the stress-strain relations can be expressed as

$$
\begin{aligned}
\sigma_x &= \frac{E}{(1+\nu)(1-2\nu)}[(1-\nu)\epsilon_x + \nu\epsilon_y] \\
\sigma_y &= \frac{E}{(1+\nu)(1-2\nu)}[\nu\epsilon_x + (1-\nu)\epsilon_y] \\
\sigma_z &= \frac{\nu E}{(1+\nu)(1-2\nu)}(\epsilon_x + \epsilon_y) = \nu(\sigma_x + \sigma_y) \\
\tau_{xy} &= G\gamma_{xy}
\end{aligned}
\tag{9.16}
$$

Solving Eqs. (9.16) for strains gives the strain–stress relations:

$$\epsilon_x = \frac{1+\nu}{E}[(1-\nu)\sigma_x - \nu\sigma_y]$$

$$\epsilon_y = \frac{1+\nu}{E}[(1-\nu)\sigma_y - \nu\sigma_x] \tag{9.17}$$

$$\gamma_{xy} = \frac{\tau_{xy}}{G}$$

Equations (9.17) can also be obtained by setting ϵ_z equal to zero in Eqs. (9.10) and substituting the resulting expressions for σ_z into the equations for ϵ_x and ϵ_y in Eqs. (9.10).

There are many problems of plane strain: a gravity dam, a retaining wall, a culvert or tunnel, a pipe under internal or external pressure, and a long solid rocket, to name a few. For a more detailed description of plane stress and plane strain, the reader is referred to Ref. 9.1.

9.1.5 Strain Energy for Plane Stress

The strain energy expression in three-dimensional stress may be written as

$$U = \tfrac{1}{2}\int_V \{\boldsymbol{\epsilon}\}^T\{\boldsymbol{\sigma}\}\,dV \tag{9.18}$$

where $\{\boldsymbol{\sigma}\}$ is the vector for the six stress components shown in Fig. 9.1 and $\{\boldsymbol{\epsilon}\}$ contains the six corresponding strain components.

By means of Hooke's law [Eq. (9.10)] the strain energy can be expressed in terms of the stress components:

$$U = \int_V \left[\frac{1}{2E}(\sigma_x^2 + \sigma_y^2 + \sigma_z^2) - \frac{\nu}{E}(\sigma_x\sigma_y + \sigma_y\sigma_z + \sigma_z\sigma_x) + \frac{1}{2G}(\tau_{xy}^2 + \tau_{yz}^2 + \tau_{zx}^2)\right] dV \tag{9.19}$$

For the case of plane stress, $\sigma_z = \tau_{xz} = \tau_{yz} = 0$, Eq. (9.19) can be reduced to

$$U = \int_V \left[\frac{1}{2E}(\sigma_x^2 + \sigma_y^2) - \frac{\nu}{E}\sigma_x\sigma_y + \frac{1}{2G}\tau_{xy}^2\right] dV \tag{9.20}$$

Using the stress–strain equations (9.14) and then using the strain–displacement equations (9.6) gives

$$U = \frac{E}{2(1-\nu^2)} \int_A \left[\left(\frac{\partial u}{\partial x}\right)^2 + 2\nu \left(\frac{\partial u}{\partial x}\right)\left(\frac{\partial v}{\partial y}\right) + \left(\frac{\partial v}{\partial y}\right)^2 + \frac{1-\nu}{2}\left(\frac{\partial u}{\partial y} + \frac{\partial v}{\partial x}\right)^2 \right] t\, dA \tag{9.21}$$

which can readily be used for derivation of the stiffness matrix for the plane stress finite element displacement model. As will be explained in Sec. 9.2.3, the plane stress stiffness matrix is readily convertible into plane strain stiffness matrix.

9.2 EIGHT-D.O.F. RECTANGULAR PLANE STRESS AND PLANE STRAIN FINITE ELEMENT

The simplest plane stress finite element is a rectangular element as shown in Fig. 9.4. The element has a length a, width b, and constant thickness t. Each of the four corners is assumed to possess two degrees of freedom: displacements u and v in the x and y directions, respectively. The four corners are usually called *nodal points*. Thus this element possesses eight nodal forces (four pairs of F_x and F_y) and eight nodal displacements or nodal degrees of freedom (four pairs of u and v).

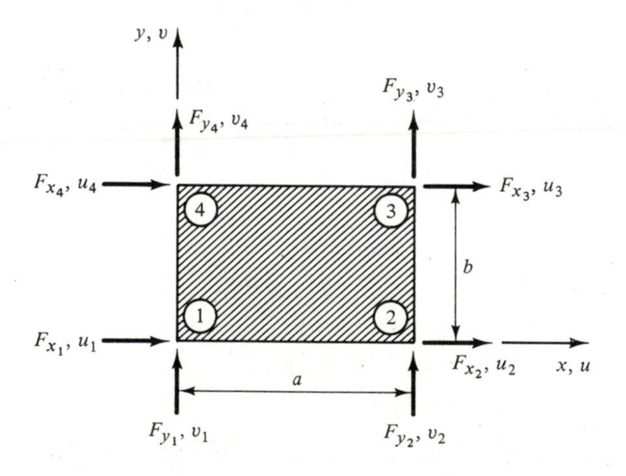

Figure 9.4 Eight-degree-of-freedom rectangular plane stress or plane strain finite element.

In the matrix displacement formulation for a finite element, an important step is to assume the pertinent displacement interpolation functions, usually in polynomial form.

9.2.1 Displacement Functions

The interpolation functions for the displacements u and v can be assumed in the following simple form [9.2]:

$$\begin{aligned} u(x, y) &= c_1 + c_2 x + c_3 y + c_4 xy \\ v(x, y) &= c_5 + c_6 x + c_7 y + c_8 xy \end{aligned} \tag{9.22}$$

One of the reasons for assuming such displacement functions is that the eight constants c can be determined uniquely by using the eight nodal displacement conditions

$$\begin{aligned} &u = u_1 \quad \text{and} \quad v = v_1 \quad \text{at } (0, 0) \\ &u = u_2 \quad \text{and} \quad v = v_2 \quad \text{at } (a, 0) \\ &u = u_3 \quad \text{and} \quad v = v_3 \quad \text{at } (a, b) \\ &u = u_4 \quad \text{and} \quad v = v_4 \quad \text{at } (0, b) \end{aligned} \tag{9.23}$$

Such displacement functions result in the following assumptions:

1. The boundary lines of each element remain straight after deformation.
2. The displacement distribution is represented by a second-degree surface, where for constant values of x (or y) the variation of displacement in the direction of y (or x) is linear.
3. The strain ϵ_x (or ϵ_y) is independent of x (or y) but varies linearly with y (or x). The shearing strain γ_{xy} varies linearly with both x and y.
4. The element stresses can be replaced by statically equivalent or work equivalent stress resultants which act at the corners of the element (nodal forces).

Substituting the eight nodal point conditions of Eq. (9.23) into Eq. (9.22), the eight constants c can be solved in terms of element dimensions a, b, and the eight nodal displacements. Then substituting the eight constants back into the displacement functions (9.22) and rearranging the terms yields

$$\begin{aligned} u(x, y) &= f_1(x, y)u_1 + f_2(x, y)u_2 + f_3(x, y)u_3 + f_4(x, y)u_4 \\ v(x, y) &= f_1(x, y)v_1 + f_2(x, y)v_2 + f_3(x, y)v_3 + f_4(x, y)v_4 \end{aligned} \tag{9.24}$$

where the four functions $f(x, y)$ can be called *shape functions* and are obtained as

$$
\begin{aligned}
f_1(x, y) &= \left(1 - \frac{x}{a}\right)\left(1 - \frac{y}{b}\right) \\
f_2(x, y) &= \frac{x}{a}\left(1 - \frac{y}{b}\right) \\
f_3(x, y) &= \frac{xy}{ab} \\
f_4(x, y) &= \left(1 - \frac{x}{a}\right)\frac{y}{b}
\end{aligned}
\tag{9.24a}
$$

Examining Eqs. (9.24) and (9.24a) reveals that the distribution of the u and v displacements along any edge is linear and that it depends only on the displacements at the two corner nodal points connected by the edge. Thus these displacement functions ensure that the compatibility of displacements on the boundaries of adjacent elements is satisfied.

Equations (9.24) provide the displacement values at any point (x, y) within the element. This is done by simply substituting the eight nodal displacement values at the four corners and the coordinates (x, y) of the point of interest into Eqs. (9.24). If derivatives of u and v functions (9.24) with respect to x and y are performed following Eqs. (9.6), the strains for any point within the element can also be obtained.

9.2.2 Strain–Displacement Relations

In the state of plane stress, we have three strain components. They can be written in terms of displacement derivatives in the following form:

$$
\begin{aligned}
\epsilon_x &= \frac{\partial u}{\partial x} \\
\epsilon_y &= \frac{\partial v}{\partial y} \\
\gamma_{xy} &= \frac{\partial u}{\partial y} + \frac{\partial v}{\partial x}
\end{aligned}
\tag{9.6}
$$

Substituting the displacement equations (9.24) into the strain–displacement equations (9.6), we obtain

$$\begin{Bmatrix} \epsilon_x \\ \epsilon_y \\ \gamma_{xy} \end{Bmatrix} = \begin{bmatrix} -\frac{1}{a}\left(1-\frac{y}{b}\right) & 0 & \frac{1}{a}\left(1-\frac{y}{b}\right) & 0 & \frac{y}{ab} & 0 & -\frac{y}{ab} & 0 \\ 0 & -\frac{1}{b}\left(1-\frac{x}{a}\right) & 0 & -\frac{x}{ab} & 0 & \frac{x}{ab} & 0 & \frac{1}{b}\left(1-\frac{x}{a}\right) \\ -\frac{1}{b}\left(1-\frac{x}{a}\right) & -\frac{1}{a}\left(1-\frac{y}{b}\right) & -\frac{x}{ab} & \frac{1}{a}\left(1-\frac{y}{b}\right) & \frac{x}{ab} & \frac{y}{ab} & \frac{1}{b}\left(1-\frac{x}{a}\right) & -\frac{y}{ab} \end{bmatrix} \begin{Bmatrix} u_1 \\ v_1 \\ u_2 \\ v_2 \\ u_3 \\ v_3 \\ u_4 \\ v_4 \end{Bmatrix} \tag{9.25}$$

or symbolically,

$$\{\boldsymbol{\epsilon}\} = [\mathbf{A}]\{\mathbf{q}\} \tag{9.25a}$$

It is seen in Eqs. (9.25) that the strains ϵ_x are constant in the x direction and vary linearly in the y direction. The shearing strains γ_{xy} vary linearly with both the x and y coordinates.

9.2.3 Stress–Strain Relations

One important advantage of the finite element method in two-dimensional elasticity is that structures with orthotropic or anisotropic material properties can be considered. In general, the stress-strain relations are of the form

$$\begin{Bmatrix} \sigma_x \\ \sigma_y \\ \tau_{xy} \end{Bmatrix} = \begin{bmatrix} c_{11} & c_{12} & c_{13} \\ c_{21} & c_{22} & c_{23} \\ c_{31} & c_{32} & c_{33} \end{bmatrix} \begin{Bmatrix} \epsilon_x \\ \epsilon_y \\ \gamma_{xy} \end{Bmatrix} \tag{9.26}$$

For the case of isotropic material, the stress-strain relations in the state of plane stress are of the form

$$\begin{Bmatrix} \sigma_x \\ \sigma_y \\ \tau_{xy} \end{Bmatrix} = \frac{E}{1-\nu^2} \begin{bmatrix} 1 & \nu & 0 \\ \nu & 1 & 0 \\ 0 & 0 & \frac{1-\nu}{2} \end{bmatrix} \begin{Bmatrix} \epsilon_x \\ \epsilon_y \\ \gamma_{xy} \end{Bmatrix} \tag{9.26a}$$

or in the state of plane strain

$$\begin{Bmatrix} \sigma_x \\ \sigma_y \\ \tau_{xy} \end{Bmatrix} = \frac{E}{(1+\nu)(1-2\nu)} \begin{bmatrix} 1-\nu & \nu & 0 \\ \nu & 1-\nu & 0 \\ 0 & 0 & \frac{1-2\nu}{2} \end{bmatrix} \begin{Bmatrix} \epsilon_x \\ \epsilon_y \\ \gamma_{xy} \end{Bmatrix} \tag{9.26b}$$

If E and ν in the plane stress equations (9.26a) are replaced by $E/(1-\nu^2)$ and $\nu/(1-\nu)$, respectively, Eqs. (9.26a) become exactly the same form as the plane strain equations (9.26b). Symbolically, Eqs. (9.26a) or (9.26b) can be written as

$$\{\boldsymbol{\sigma}\} = [\mathbf{C}]\{\boldsymbol{\epsilon}\} \tag{9.27}$$

9.2.4 Derivation of Stiffness Matrix by the Energy Method

The element stiffness matrix can be derived by first formulating the strain energy for the element and then performing partial differentiation of the strain energy with respect to each degree of freedom following Castigliano's theorem.

The strain energy in the subject finite element is in the form

$$U = \frac{1}{2}\int_0^a \int_0^b \int_0^t \{\boldsymbol{\epsilon}\}^T\{\boldsymbol{\sigma}\}\, dz\, dy\, dx \tag{9.28}$$

where

$$\{\boldsymbol{\epsilon}\} = \begin{Bmatrix} \epsilon_x \\ \epsilon_y \\ \gamma_{xy} \end{Bmatrix} \qquad \text{and} \qquad \{\boldsymbol{\sigma}\} = \begin{Bmatrix} \sigma_x \\ \sigma_y \\ \tau_{xy} \end{Bmatrix} \tag{9.28a}$$

Transposing the strain–displacement equations (9.25a) gives

$$\{\boldsymbol{\epsilon}\}^T = \{\mathbf{q}\}^T[\mathbf{A}]^T \tag{9.29}$$

From Eqs. (9.27) and (9.25a),

$$\{\boldsymbol{\sigma}\} = [\mathbf{C}]\{\boldsymbol{\epsilon}\} = [\mathbf{C}][\mathbf{A}]\{\mathbf{q}\} \tag{9.30}$$

Substituting Eqs. (9.29) and (9.30) into Eq. (9.28) and assuming constant thickness t, we have

$$U = \frac{t}{2}\{\mathbf{q}\}^T\left[\int_0^b \int_0^a [\mathbf{A}]^T[\mathbf{C}][\mathbf{A}]\, dx\, dy\right]\{\mathbf{q}\} \tag{9.31}$$

Applying Castigliano's first theorem,

$$F_i = \frac{\partial U}{\partial q_i} \tag{3.17a}$$

By performing partial differentiation of the strain energy with respect to each of the eight degrees of freedom of the finite element, we finally obtain

$$\{\mathbf{F}\} = [\mathbf{k}]\{\mathbf{q}\} \tag{9.32}$$

where

$$\begin{aligned} \lfloor \mathbf{F} \rfloor &= \lfloor F_{x_1} \quad F_{y_1} \quad F_{x_2} \quad F_{y_2} \quad F_{x_3} \quad F_{y_3} \quad F_{x_4} \quad F_{y_4} \rfloor \\ \lfloor \mathbf{q} \rfloor &= \lfloor u_1 \quad v_1 \quad u_2 \quad v_2 \quad u_3 \quad v_3 \quad u_4 \quad v_4 \rfloor \end{aligned} \tag{9.32a}$$

and

$$[\mathbf{k}] = t \iint_{\text{Area}} [\mathbf{A}]^T [\mathbf{C}][\mathbf{A}]\, dx\, dy \tag{9.33}$$

Equation (9.33) is in a popular form suitable for the derivation of stiffness matrix for finite elements with properly assumed displacement functions.

Alternatively, the stiffness coefficients for a plane stress element can be obtained by using strain energy equation (9.21), displacement functions, and Castigliano's first theorem,

$$[\mathbf{k}] = \begin{bmatrix} k_{uu} & k_{uv} \\ k_{vu} & k_{vv} \end{bmatrix} \tag{9.34}$$

where the coefficients are obtained as

$$\begin{aligned} (k_{uu})_{ij} &= \frac{E}{1-\nu^2} \int_A \left[\frac{\partial f_i}{\partial x} \left(\frac{\partial f_j}{\partial x} \right) + \frac{1-\nu}{2} \left(\frac{\partial f_i}{\partial y} \right) \frac{\partial f_j}{\partial y} \right] t\, dA \\ (k_{vv})_{ij} &= \frac{E}{1-\nu^2} \int_A \left[\frac{\partial f_i}{\partial y} \left(\frac{\partial f_j}{\partial y} \right) + \frac{1-\nu}{2} \left(\frac{\partial f_i}{\partial x} \right) \frac{\partial f_j}{\partial x} \right] t\, dA \\ (k_{uv})_{ij} &= (k_{vu})_{ji} \\ &= \frac{E}{1-\nu^2} \int_A \left[\nu \left(\frac{\partial f_i}{\partial x} \right) \frac{\partial f_j}{\partial y} + \frac{1-\nu}{2} \left(\frac{\partial f_i}{\partial y} \right) \frac{\partial f_j}{\partial x} \right] t\, dA \end{aligned} \tag{9.35}$$

in which $f_i(x, y)$ is the shape function associated with u_i or v_i, such as that defined in Eq. (9.24a).

9.2.5 Stiffness Matrix Equations

Using Eq. (9.33) or (9.35), the stiffness equations for the eight-d.o.f. rectangular element (Fig. 9.4) in the state of plane stress can be obtained as follows:

$$
\begin{Bmatrix} F_{x_1} \\ F_{y_1} \\ F_{x_2} \\ F_{y_2} \\ F_{x_3} \\ F_{y_3} \\ F_{x_4} \\ F_{y_4} \end{Bmatrix} =
\begin{bmatrix}
C_1 & & & & & & & \\
C_2 & C_3 & & & \text{symmetric} & & & \\
C_4 & -C_5 & C_1 & & & & & \\
C_5 & C_6 & -C_2 & C_3 & & & & \\
-\dfrac{C_1}{2} & -C_2 & C_7 & -C_5 & C_1 & & & \\
-C_2 & -\dfrac{C_3}{2} & C_5 & C_8 & C_2 & C_3 & & \\
C_7 & C_5 & -\dfrac{C_1}{2} & C_2 & C_4 & -C_5 & C_1 & \\
-C_5 & C_8 & C_2 & -\dfrac{C_3}{2} & C_5 & C_6 & -C_2 & C_3
\end{bmatrix}
\begin{Bmatrix} u_1 \\ v_1 \\ u_2 \\ v_2 \\ u_3 \\ v_3 \\ u_4 \\ v_4 \end{Bmatrix} \tag{9.36}
$$

where

$$
\begin{aligned}
C_1 &= \left(\frac{b}{3a} + \frac{1-\nu}{6}\frac{a}{b}\right)\frac{Et}{1-\nu^2} \\
C_2 &= \left(\frac{\nu}{4} + \frac{1-\nu}{8}\right)\frac{Et}{1-\nu^2} \\
C_3 &= \left(\frac{a}{3b} + \frac{1-\nu}{6}\frac{b}{a}\right)\frac{Et}{1-\nu^2} \\
C_4 &= \left(-\frac{b}{3a} + \frac{1-\nu}{12}\frac{a}{b}\right)\frac{Et}{1-\nu^2} \\
C_5 &= \left(\frac{\nu}{4} - \frac{1-\nu}{8}\right)\frac{Et}{1-\nu^2} \\
C_6 &= \left(\frac{a}{6b} - \frac{1-\nu}{6}\frac{b}{a}\right)\frac{Et}{1-\nu^2} \\
C_7 &= \left(\frac{b}{6a} - \frac{1-\nu}{6}\frac{a}{b}\right)\frac{Et}{1-\nu^2} \\
C_8 &= \left(-\frac{a}{3b} + \frac{1-\nu}{12}\frac{b}{a}\right)\frac{Et}{1-\nu^2}
\end{aligned} \tag{9.36a}
$$

To demonstrate how to derive this stiffness matrix, let it be desired to derive the stiffness coefficient k_{84} which is defined as the amount of force F_{y_4} produced by a unit displacement v_2. This coefficient can be obtained by substituting the eighth and fourth columns in [**A**] of Eq. (9.25), and matrix [**C**] of Eq. (9.26a) into Eq. (9.33) for $[\mathbf{A}]^T$, [**A**], and [**C**], respectively:

$$k_{84} = t\int_0^a\int_0^b \left[0 \quad \frac{1}{b}\left(1-\frac{x}{a}\right) \quad -\frac{y}{ab}\right]\frac{E}{1-\nu^2}\begin{bmatrix}1 & \nu & 0\\ \nu & 1 & 0\\ 0 & 0 & \dfrac{1-\nu}{2}\end{bmatrix}\begin{Bmatrix}0\\ -\dfrac{x}{ab}\\ \dfrac{1}{a}\left(1-\dfrac{y}{b}\right)\end{Bmatrix}dx\,dy$$

$$= -\frac{Et}{1-\nu^2}\left(\frac{a}{6b}+\frac{1-\nu}{12}\frac{b}{a}\right) \tag{9.37}$$

Alternatively, Eq. (9.35) can be used:

$$k_{84} = (k_{vv})_{42}$$

$$= \frac{Et}{1-\nu^2}\int_0^b\int_0^a\left[\frac{\partial f_4}{\partial y}\left(\frac{\partial f_2}{\partial y}\right)+\frac{1-\nu}{2}\left(\frac{\partial f_4}{\partial x}\right)\frac{\partial f_2}{\partial x}\right]dx\,dy$$

$$= \frac{Et}{1-\nu^2}\int_0^b\int_0^a\left[\left(\frac{1}{b}-\frac{x}{ab}\right)\frac{-x}{ab}+\frac{1-\nu}{2}\left(\frac{-y}{ab}\right)\left(\frac{1}{a}-\frac{y}{ab}\right)\right]dx\,dy$$

$$= -\frac{Et}{1-\nu^2}\left(\frac{a}{6b}+\frac{1-\nu}{12}\frac{b}{a}\right) \tag{9.38}$$

which is the same as that obtained in Eq. (9.37).

If Eq. (9.26b) instead of Eq. (9.26a) is used for the stress–strain matrix [**C**] as contained in Eq. (9.33), the stiffness equations for plane strain can be obtained. Thus Eqs. (9.36) become valid for plane strain if E and ν are replaced by $E/(1-\nu^2)$ and $\nu/(1-\nu)$, respectively.

9.2.6 Application of the Rectangular Plane Stress Element

Figure 9.5 shows an example of a square plate under a parabolically distributed stress σ_x acting on two opposite edges:

$$\sigma_x = \sigma_0\left[1-\left(\frac{y}{16}\right)^2\right] \tag{9.39}$$

The length, width, and thickness of the plate are 32, 32, and 0.1 in., respectively. It is assumed that $E = 10^7$ psi, $\nu = 0.3$, and $\sigma_0 = 1000$ psi. The stress distribution of this plane stress problem can be obtained by assuming stress function in the form of a series and then minimizing the strain energy [9.1].

This problem was solved by using the eight-d.o.f. elements [9.3]. Because of symmetry, only a quadrant of the plate need be modeled and 2×2, 4×4, and 8×8 meshes as shown in Fig. 9.5 were used. The concentrated loads

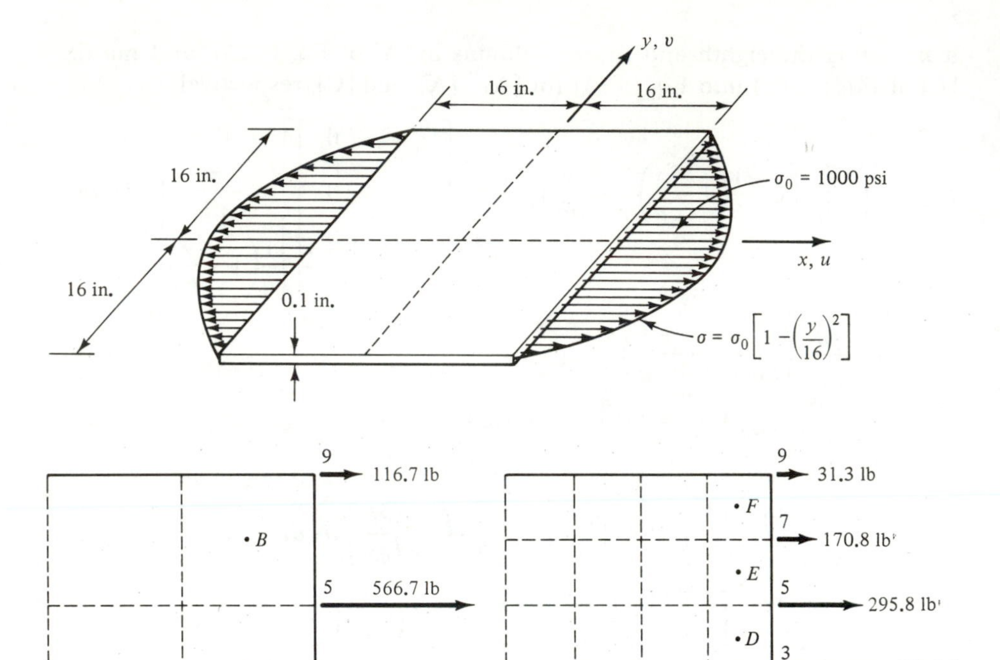

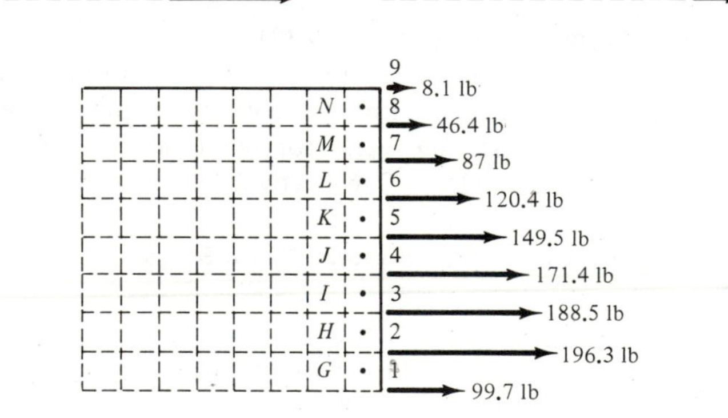

Figure 9.5 Parabolically loaded plate and three different element meshes modeling a quadrant.

shown were the work-equivalent loads. They were obtained using the following derivations.

Assuming that an element is under a distributed edge stress $\sigma(y)$ in the x direction acting on edge 2-3 of the element shown in Fig. 9.4, the work done is

$$W = \int_0^b \sigma(y)u(x, y)t\,dy \tag{9.40}$$

at $x = a$. From Eq. (9.24),

$$W = \sum_{i=1}^{4} \left[\int_0^b \sigma(y) f_i(a, y) t \, dy \right] u_i \tag{9.41}$$

The work can also be considered as done by four work-equivalent forces acting at the four corner nodes,

$$W = \sum_{i=1}^{4} F_{x_i} u_i \tag{9.42}$$

Equating the two work expressions in Eqs. (9.41) and (9.42) yields the expression for the work-equivalent load,

$$F_{x_i} = \int_0^b \sigma(y) f_i(a, y) t \, dy \tag{9.43}$$

As an example, let it be desired to find the work-equivalent load at point 1 of the 4×4 mesh in Fig. 9.5,

$$F_{x_1} = \int_0^4 \sigma(y) f_2(a, y) t \, dy = \int_0^4 100 \left[1 - \left(\frac{y}{16} \right)^2 \right] \left(1 - \frac{y}{4} \right) dy = 197.9 \text{ lb} \tag{9.44}$$

Table 9.1 shows the comparison of results for u displacement and stresses σ_x and σ_y at various locations between the solution of Ref. 9.1 and the 2×2, 4×4, and 8×8 finite element solutions, respectively [9.3].

Although the finite element models have changed the nature of the problem by changing the distributed load to concentrated loads, the results for u displacement are quite good even at the four-element level. This is because the displacement functions (9.24) are of very low order, modeling a smooth displaced curve well, such as in the case of a parabolically distributed load. However, if this plate is under concentrated loads, displacement functions of higher order may be desirable to produce the wavy distribution of displacements near the loads. The lack of agreement in u displacement at point 9 is obviously due to the concentrated load applied there, where in reality no load exists.

Because the strains are derivatives of the displacements as defined in Eqs. (9.6), their accuracy is expected to be not as good as the displacements. As indicated in Table 9.1, finer meshes are needed to obtain accurate stresses, especially near point N.

Although the 2×2 mesh provides the u displacements only at points 1, 5, and 9 and stresses at points A and B, the displacements and stresses at any other point of interest can be interpolated using the displacement functions and their derivatives.

TABLE 9.1 Comparison of Results in u Displacement and Stresses σ_x and σ_y between the 8-D.O.F. Rectangular Element and Analytical Solutions for the Problem Shown in Fig. 9.5

a. u displacement (10^{-3} in.)

Point Location	Analytic Solution (Ref. 9.1)	4 elements	16 elements	64 elements
1	1.4764	1.4729	1.4757	1.4769
2				1.4550
3	1.3895		1.3892	1.3899
4				1.2842
5	1.1428	1.1490	1.1442	1.1425
6				0.9712
7	0.7786		0.7873	0.7795
8				0.5972
9	0.3647	0.4956	0.4159	0.3862

b. Stress (psi)

Point Location	σ_x (Ref. 9.1)	σ_x (8×8 elements)	σ_y (Ref. 9.1)	σ_y (8×8 elements)
A	909.6	854.1	78.7	102.4
B	456.2	479.3	18.8	32.9
C	975.0	960.1	217.8	224.6
D	852.6	837.9	171.6	309.3
E	609.4	597.3	91.9	95.5
F	249.2	271.4	14.9	22.1
G	993.5	989.7	310.2	309.3
H	962.5	958.5	293.3	293.6
I	900.0	896.3	261.4	262.8
J	806.9	803.1	213.8	218.9
K	683.1	679.2	157.4	163.5
L	528.1	525.4	97.0	101.1
M	342.6	341.9	41.9	43.78
N	126.6	139.3	5.6	8.12

9.3 SIX-D.O.F. TRIANGULAR PLANE STRESS AND PLANE STRAIN FINITE ELEMENT

Due to the double symmetry of the geometry, finite elements with rectangular shape have some advantages over those with triangular shape. It is easier to assume more satisfactory displacement functions to meet the nodal point conditions and edge compatibility and it is also easier to perform area integration. In general, rectangular elements are more efficient and accurate than the triangular ones for plates with straight edges intersecting with 90° angles. However, for plates with curved edges or straight edges with other than 90° angles, elements of triangular shape are often preferable.

In this section a 6-d.o.f. triangular plane stress and plane strain finite element is formulated. The numerical solution procedures and a computer program are discussed in great detail. The work is based primarily on that developed in Refs. [4.1], [9.4], and [9.5].

9.3.1 Basic Assumptions

1. Within each element, lines initially straight remain straight in their displaced positions.
2. The strains ϵ_x, ϵ_y, and γ_{xy} are assumed to be constant within each element. Hence the stresses are also constant.
3. The element stresses are replaced by stress resultants which act at the corners of the element.

9.3.2 Displacement Functions

A typical finite element is shown in Fig. 9.6. The element has three nodal points i, j, and k. The coordinates for the three nodal points are $(0, 0)$, (a_j, b_j), and (a_k, b_k), respectively. The displaced shape of the element is also shown in the figure with nodal points i, j, and k displaced by (u_i, v_i), (u_j, v_j), and (u_k, v_k), respectively.

Based on the linear displacement assumption, the displacements at any point (x, y) can be described by the following linear functions:

$$\begin{aligned} u(x, y) &= u_i + c_1 x + c_2 y \\ v(x, y) &= v_i + c_3 x + c_4 y \end{aligned} \tag{9.45}$$

The four constants can be obtained in terms of the nodal point displacements and the geometry of the element by using the nodal point displacement conditions as follows:

At node j:

$$\begin{cases} x = a_j \\ y = b_j \end{cases} \text{and} \begin{cases} u_j = u_i + c_1 a_j + c_2 b_j \\ v_j = v_i + c_3 a_j + c_4 b_j \end{cases}$$

At node k:

$$\begin{cases} x = a_k \\ y = b_k \end{cases} \text{and} \begin{cases} u_k = u_i + c_1 a_k + c_2 b_k \\ v_k = v_i + c_3 a_k + c_4 b_k \end{cases} \tag{9.46}$$

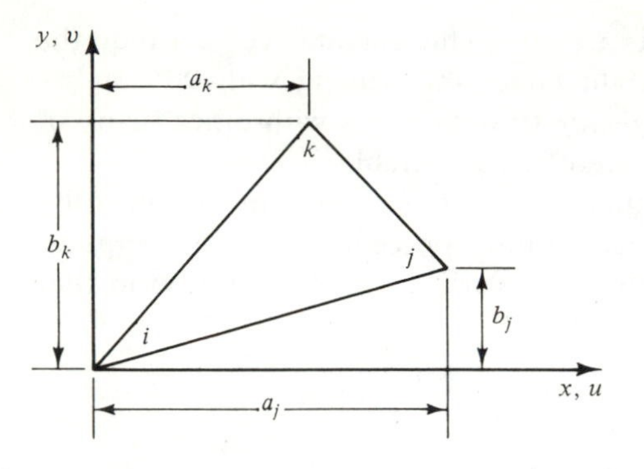

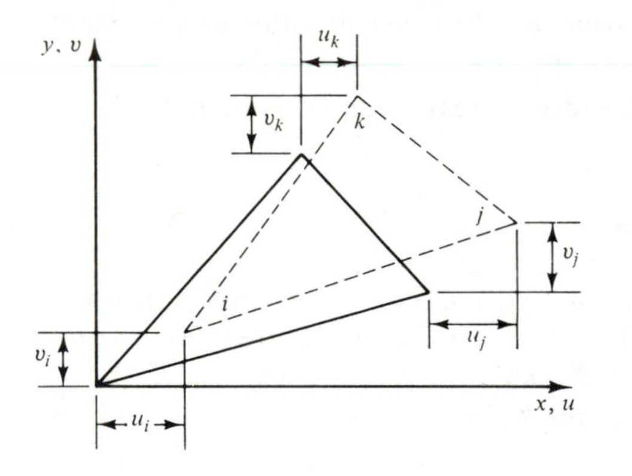

Figure 9.6 Geometry and assumed displacement shapes of the triangular plane stress and plane strain element.

Solving Eqs. (9.46) gives

$$\begin{Bmatrix} c_1 \\ c_2 \\ c_3 \\ c_4 \end{Bmatrix} = \frac{1}{a_j b_k - a_k b_j} \begin{bmatrix} b_j - b_k & 0 & b_k & 0 & -b_j & 0 \\ a_k - a_j & 0 & -a_k & 0 & a_j & 0 \\ 0 & b_j - b_k & 0 & b_k & 0 & -b_j \\ 0 & a_k - a_j & 0 & -a_k & 0 & a_j \end{bmatrix} \begin{Bmatrix} u_i \\ v_i \\ u_j \\ v_j \\ u_k \\ v_k \end{Bmatrix} \tag{9.47}$$

9.3.3 Strain–Displacement Equations

The three strain components within the element can be obtained by taking derivatives of the displacement functions (9.45) with respect to x and y as indicated in Eqs. (9.6):

$$\epsilon_x = \frac{\partial u}{\partial x} = c_1$$

$$\epsilon_y = \frac{\partial v}{\partial y} = c_4 \tag{9.48}$$

$$\gamma_{xy} = \frac{\partial u}{\partial y} + \frac{\partial v}{\partial x} = c_2 + c_3$$

If Eqs. (9.47) and (9.48) are combined, the three strain components can be obtained in terms of the six nodal displacement degrees of freedom,

$$\begin{Bmatrix} \epsilon_x \\ \epsilon_y \\ \gamma_{xy} \end{Bmatrix} = \frac{1}{a_j b_k - a_k b_j} \begin{bmatrix} b_j - b_k & 0 & b_k & 0 & -b_j & 0 \\ 0 & a_k - a_j & 0 & -a_k & 0 & a_j \\ a_k - a_j & b_j - b_k & -a_k & b_k & a_j & -b_j \end{bmatrix} \begin{Bmatrix} u_i \\ v_i \\ u_j \\ v_j \\ u_k \\ v_k \end{Bmatrix} \tag{9.49}$$

or in symbolic form,

$$\{\boldsymbol{\epsilon}\} = [\mathbf{A}]\{\mathbf{q}\} \tag{9.49a}$$

9.3.4 Stress-Strain Equations

The stress-strain relationship can be written for a general anisotropic material,

$$\begin{Bmatrix} \sigma_x \\ \sigma_y \\ \tau_{xy} \end{Bmatrix} = \begin{bmatrix} c_{11} & c_{12} & c_{13} \\ c_{21} & c_{22} & c_{23} \\ c_{31} & c_{32} & c_{33} \end{bmatrix} \begin{Bmatrix} \epsilon_x \\ \epsilon_y \\ \gamma_{xy} \end{Bmatrix} \tag{9.26}$$

For isotropic material and for the case of plane stress and plane strain, the stress-strain relations are given in Eqs. (9.26a) and (9.26b), respectively, or in symbolic form,

$$\{\boldsymbol{\sigma}\} = [\mathbf{C}]\{\boldsymbol{\epsilon}\} \tag{9.50}$$

Combining Eqs. (9.49a) and (9.50) yields

$$\{\boldsymbol{\sigma}\} = [\mathbf{C}][\mathbf{A}]\{\mathbf{q}\} \tag{9.51}$$

9.3.5 Nodal Point Force-Stress Equations

It is assumed that the uniform stresses acting on the edges of the finite element can be replaced by a set of static equivalent corner forces as shown in Fig. 9.7. This is done by dividing each of the three edges at its midpoint, splitting the stresses acting on the edge equally, and assigning each half to the adjacent corners. For example, the resultant force in the x direction at corner k is

$$F_x^k = \sigma_x \frac{b_k - b_j}{2} - \sigma_x \frac{b_k}{2} = \frac{-b_j}{2}\sigma_x \tag{9.52}$$

The corner nodal forces expressed in terms of the three stress components are thus obtained as

$$\begin{Bmatrix} F_x^i \\ F_y^i \\ F_x^j \\ F_y^j \\ F_x^k \\ F_y^k \end{Bmatrix} = \frac{1}{2} \begin{bmatrix} b_j - b_k & 0 & a_k - a_j \\ 0 & a_k - a_j & b_j - b_k \\ b_k & 0 & -a_k \\ 0 & -a_k & b_k \\ -b_j & 0 & a_j \\ 0 & a_j & -b_j \end{bmatrix} \begin{Bmatrix} \sigma_x \\ \sigma_y \\ \tau_{xy} \end{Bmatrix} \tag{9.53}$$

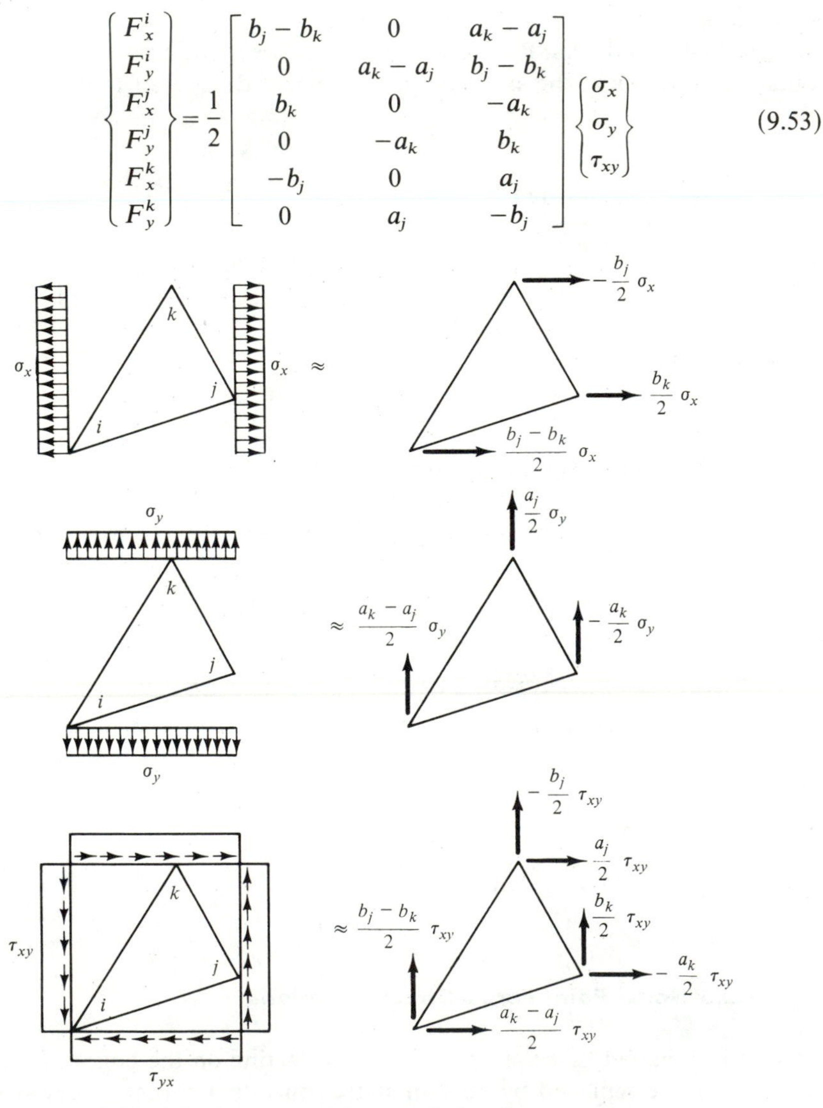

Figure 9.7 Corner nodal forces replacing element stresses.

or symbolically,

$$\{\mathbf{F}\} = [\mathbf{B}]\{\boldsymbol{\sigma}\} \tag{9.53a}$$

9.3.6 Element Stiffness Matrix Equations

The stiffness matrix equations for this element are finally obtained by substituting Eq. (9.51) into Eq. (9.53a):

$$\underset{6\times 6}{\{\mathbf{F}\}} = \underset{6\times 6}{[\mathbf{k}]}\ \underset{6\times 1}{\{\mathbf{q}\}} \tag{9.54}$$

with

$$[\mathbf{k}] = [\mathbf{B}][\mathbf{C}][\mathbf{A}] \tag{9.54a}$$

where $\{\mathbf{F}\}$ is the vector of the six nodal forces, $\{\mathbf{q}\}$ the vector of six nodal displacements, and $[\mathbf{k}]$ the stiffness matrix.

It should be remembered that the stiffness matrix can also be obtained by using the energy principle as described in Eq. (9.33) or (9.35). In the former case,

$$[\mathbf{k}] = \iint_{\text{Area}} t[\mathbf{A}]^T[\mathbf{C}][\mathbf{A}]\, dx\, dy \tag{9.55}$$

To perform an area integration for a triangle is not as easy as that for a rectangle. Fortunately, both $[\mathbf{A}]$ and $[\mathbf{C}]$ are constant matrices. Thus Eq. (9.55) can be simplified as

$$[\mathbf{k}] = t[\mathbf{A}]^T[\mathbf{C}][\mathbf{A}](\text{Area}) \tag{9.56}$$

where the area of the triangle is

$$\text{Area} = \tfrac{1}{2}\begin{bmatrix} 1 & 1 & 1 \\ a_i & a_j & a_k \\ b_i & b_j & b_k \end{bmatrix} \tag{9.57}$$

9.3.7 Stiffness Equation for Complete Structure (Assemblage)

The equilibrium of the complete system of elements, which is an expression for nodal point loads in terms of nodal point displacements, can be expressed as

$$\{\mathbf{P}\} = [\mathbf{K}]\{\mathbf{Q}\} \tag{9.58}$$

where the assembled or total stiffness matrix $[\mathbf{K}]$ can be found by a systematic addition of the stiffnesses of all elements in the system. This addition can best be illustrated if Eq. (9.54) is rewritten in terms of a typical element g:

$$\begin{Bmatrix} F_i^g \\ F_j^g \\ F_k^g \end{Bmatrix} = \begin{bmatrix} k_{ii}^g & k_{ij}^g & k_{ik}^g \\ k_{ji}^g & k_{jj}^g & k_{jk}^g \\ k_{ki}^g & k_{kj}^g & k_{kk}^g \end{bmatrix} \begin{Bmatrix} q_i \\ q_j \\ q_k \end{Bmatrix} \tag{9.59}$$

where, in terms of arbitrary nodal points l and m, F_l^g and q_m are vectors of the form

$$F_l^g = \begin{Bmatrix} F_x \\ F_y \end{Bmatrix}_l^g \quad \text{and} \quad q_m^g = \begin{Bmatrix} u \\ v \end{Bmatrix}_m^g = q_m \tag{9.60}$$

and the stiffness coefficient k_{lm}^g is a 2×2 submatrix of the form

$$k_{lm}^g = \begin{bmatrix} k_{xx} & k_{xy} \\ k_{yx} & k_{yy} \end{bmatrix}_{lm}^g \tag{9.61}$$

The coefficient k_{lm}^g is defined as the forces caused at nodal point l of element g due to a unit displacement at nodal point m. Thus the assembled stiffness coefficient, K_{lm} for the overall structure, which is the sum of forces acting at nodal point l of all surrounding elements due to the unit displacement at m, is obtained in the following summation form:

$$K_{lm} = \sum_{g=1}^{N} k_{lm}^g \tag{9.62}$$

where N is the total number of elements. It should be pointed out that K_{lm} exists only if l equals m or if l and m are adjacent nodal points sharing at least one common element.

9.3.8 Solution of the Total Stiffness Equations

For a practical structure, finite element modeling can easily involve several hundred nodal points or possibly close to twice the number of equations (d.o.f.'s). In solving such a large set of linear simultaneous equations, we often encounter the following major difficulties:

1. The storage required by the large stiffness matrix for the overall structure is equal to M^2, where M is the number of equations.
2. The time required for solution is approximately proportional to M^3.
3. The accuracy of the solution can be a serious problem.

In view of these difficulties, two elementary methods of solution are commonly used: the iteration method and the band matrix direct elimination method.

9.3.9 Iteration Procedure

The specific iteration method used is a modification of the well-known Gauss–Seidel iteration procedure. When applied to the total system of stiffness equations, this method involves repeated calculation of new displacements from the equation

$$Q_n^{(S+1)} = K_{nn}^{-1}\left[P_n - \sum_{i=1}^{n-1} K_{ni}Q_i^{(S+1)} - \sum_{i=n+1}^{N} K_{ni}Q_i^{(S)}\right] \tag{9.63}$$

where n is the number of the unknown displacement vector and S is the cycle of iteration. It is noted that Q_n contains the x and y components of displacement and the stiffness coefficients may be expressed in the 2×2 submatrix form of Eq. (9.61). It is also noted that the stiffness matrix is positive definite; therefore, the method will always converge.

An easy way to explain Eq. (9.63) is to use it to solve the following example:

$$\begin{bmatrix} 2 & 1 & 1 \\ 2 & 2 & 1 \\ 1 & 1 & 3 \end{bmatrix}\begin{Bmatrix} x_1 \\ x_2 \\ x_3 \end{Bmatrix} = \begin{Bmatrix} 4 \\ 5 \\ 5 \end{Bmatrix}$$

Following Eq. (9.63),

$$(1) \qquad x_1^{(S+1)} = 2^{-1}[4 - x_2^{(S)} - x_3^{(S)}]$$

$$(2) \qquad x_2^{(S+1)} = 2^{-1}[5 - 2x_1^{(S+1)} - x_3^{(S)}]$$

$$(3) \qquad x_3^{(S+1)} = 3^{-1}[5 - x_1^{(S+1)} - x_2^{(S+1)}]$$

First cycle: assuming that $x_2 = x_3 = 0$:

From (1), $x_1 = 2$
From (2), $x_2 = 2^{-1}(5 - 2 \times 2 - 0) = 0.5$
From (3), $x_3 = 3^{-1}(5 - 2 - 0.5) = 0.833$

Second cycle: $x_1 = 1.334;\ x_2 = 0.750;\ x_3 = 0.972$
Third cycle: $x_1 = 1.139;\ x_2 = 0.875;\ x_3 = 0.995$
⋮
Eighth cycle: $x_1 = 1.004;\ x_2 = 0.996;\ x_3 = 1.000$
⋮
Twelfth cycle: $x_1 = 1.000;\ x_2 = 1.000;\ x_3 = 1.000$

One of the major advantages of this method is that it requires very little computer storage. Only one equation need be stored at a time, and as iteration proceeds, each equation is subsequently destroyed for storage of the next equation. One of the major disadvantages of this method is that it converges slowly for certain matrices such as those with a narrow bandwidth and those with irregular or spotty populations. The former may occur when the structure is thin and long. The latter may occur when a structure is of irregular shape, with irregular cutouts; when the numbering of the nodal points fails to follow a regular pattern; and so on.

Overrelaxation factor. The rate of convergence of the Gauss–Seidel iterative procedure can be greatly increased by the use of an overrelaxation factor. However, to apply this factor, it is first necessary to calculate the change in the displacement of nodal point n between cycles of iteration:

$$\Delta Q_n^{(S)} = Q_n^{(S+1)} - Q_n^{(S)} = Q_n^{(S+1)} - K_{nn}^{-1} K_{nn} Q_n^{(S)} \tag{9.64}$$

Substitution of Eq. (9.63) into Eq. (9.64) yields for the change in displacement:

$$\Delta Q_n^{(S)} = K_{nn}^{-1}\left[P_n - \sum_{i=1}^{n-1} K_{ni} Q_i^{(S+1)} - \sum_{i=n}^{N} K_{ni} Q_i^{(S)}\right] \tag{9.65}$$

The new displacement of nodal point n is then determined from the following equation:

$$Q_n^{(S+1)} = Q_n^{(S)} + \beta\, \Delta Q_n^{(S)} \tag{9.66}$$

where β is the overrelaxation factor.

The selection of an overrelaxation factor, which gives the best convergence, depends on the characteristics of the particular problem. However, experience has indicated that for most two-dimensional structures the optimum overrelaxation factor is between 1.8 and 1.95 [9.5]. A discussion of theoretical determination of an optimum relaxation factor is given in Ref. 9.6.

Group relaxation. Following the concept of block and group relaxation by Southwell [9.7] and simulating Rayleigh's energy method [9.8] for calculating buckling loads and natural frequencies, Wilson [9.5] developed the following group relaxation method. After S cycles of iteration, it is assumed that $\alpha\{\mathbf{Q}^{(S)}\}$ represents a good approximation of the final displacements of the structure. In order to solve for α it is necessary to consider the energy of the system when subjected to this deformation pattern. The energy supplied externally to the system is given by

$$U = \tfrac{1}{2}\alpha \lfloor \mathbf{Q}^{(S)} \rfloor \{\mathbf{P}\} \tag{9.67}$$

From Eq. (9.58), we have

$$\{\mathbf{P}\} = [\mathbf{K}]\{\mathbf{Q}\} = [\mathbf{K}]\alpha\{\mathbf{Q}^{(S)}\} \tag{9.68}$$

The energy stored elastically within the elements of the system will be

$$U = \tfrac{1}{2}\alpha^2 \lfloor \mathbf{Q}^{(S)} \rfloor [\mathbf{K}]\{\mathbf{Q}^{(S)}\} \tag{9.69}$$

If the internal and external energy is equated, α is found to be

$$\alpha = \frac{\lfloor \mathbf{Q}^{(S)} \rfloor \{\mathbf{P}\}}{\lfloor \mathbf{Q}^{(S)} \rfloor [\mathbf{K}] \{\mathbf{Q}^{(S)}\}} \tag{9.70}$$

Therefore, before the start of the next cycle of iteration, the displacements may be modified as follows:

$$\{\mathbf{Q}^{(S)}\} = \alpha \{\mathbf{Q}^{(S)}\} \tag{9.71}$$

The determination of α involves approximately the same number of numerical operations as one cycle of Gauss-Seidel iteration. However, this group relaxation need only be applied once every several cycles and the same α is used for every degree of freedom.

9.3.10 Direct Elimination of Band Matrix

For two-dimensional structures, the stiffness matrix is commonly in the form of a band matrix. Closed-loop types of structures are exceptions. If the nodal points are numbered properly, the bandwidth of the stiffness matrix may be narrowed. Thus the storage problems can be reduced by storing only the coefficients in the band into the computer and the equations can be solved using the method of direct elimination instead of iteration. The convergence problem can then be avoided.

The bandwidth for an assembled set of finite elements can be found from a general equation,

$$\text{bandwidth} = 2[mn + (n-1)] + 1 \tag{9.72}$$

where m is the maximum numerical difference between any two nodal point numbers sharing a common finite element, and n is the number of degrees of freedom assumed for each nodal point.

As an example, a finite element system is shown in Fig. 9.8. For the present finite element, each nodal point has two d.o.f.'s. The numbering of both nodal points and d.o.f.'s is shown in the figure. In this case, $m = 4$ and $n = 2$;

$$\text{bandwidth} = 2[4 \times 2 + (2-1)] + 1 = 19$$

The order of the total stiffness matrix is 24. If the nonzero coefficients are designated by circles, a band matrix is shown graphically in Fig. 9.8. Equation (9.72) can readily be understood if the reader practices to construct such a band matrix and then observes a row in the band. The last term, "1," in Eq. (9.72) represents the diagonal coefficient and the "2" means that there may be equal number of coefficients on both sides of the diagonal coefficient.

If the nodal points in Fig. 9.8 are numbered horizontally instead of vertically, m becomes 5 and the bandwidth becomes 23.

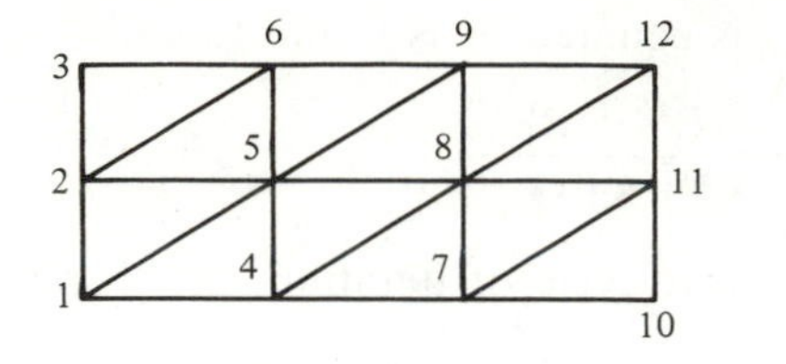

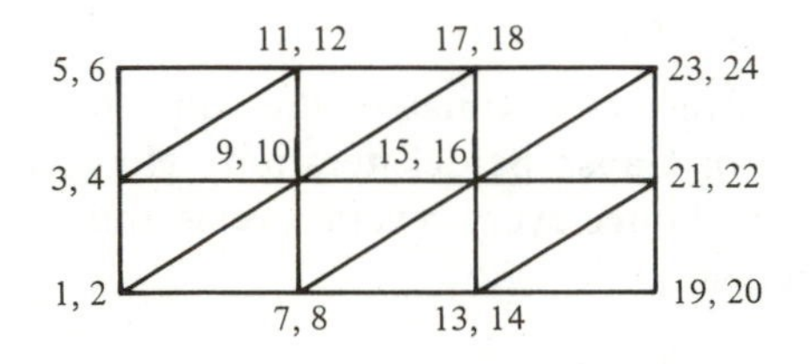

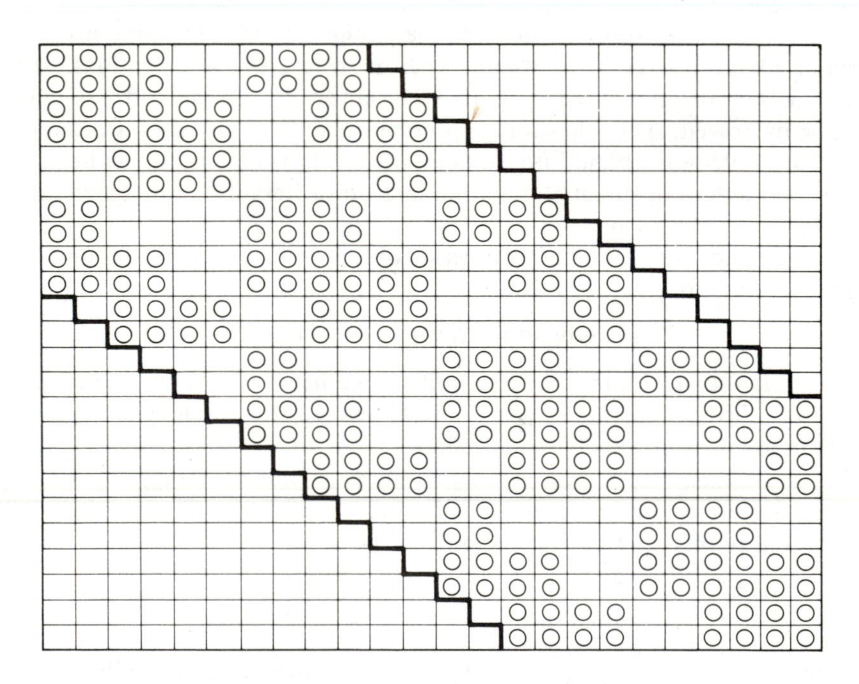

Figure 9.8 Numbering of nodal points, numbering of degrees of freedom, and the band matrix of a two-dimensional finite element system.

9.3.11 Some Physical Interpretation of the Iterative Methods

In Eq. (9.65) the term $[\mathbf{K}_{nn}]^{-1}$ is the flexibility of nodal point n. This represents the nodal displacements produced by unit nodal forces and can be written in the form of a submatrix:

$$[\mathbf{K}_{nn}]^{-1} = \begin{bmatrix} f_{xx} & f_{xy} \\ f_{yx} & f_{yy} \end{bmatrix} \tag{9.73}$$

where f is called the *flexibility coefficient.*

The two summation terms in Eq. (9.65) represent the elastic forces acting at nodal point n due to the deformations of the elements:

$$F_n^{(S+1)} = \sum_{i=1}^{n-1} K_{ni} Q_i^{(S+1)} + \sum_{i=n}^{N} K_{ni} Q_i^{(S)} \tag{9.74}$$

The difference between these elastic forces and the applied loads is the total unbalanced force, which in submatrix form may be written as

$$\begin{Bmatrix} X \\ Y \end{Bmatrix}_n^{(S+1)} = \begin{Bmatrix} P_x \\ P_y \end{Bmatrix}_n - \begin{Bmatrix} F_x \\ F_y \end{Bmatrix}_n^{(S+1)} \tag{9.75}$$

The amount of unbalanced force can be used to judge the degree of convergence.

Equation (9.66), which gives the new displacement of nodal point n, may now be rewritten in the following submatrix form:

$$\begin{Bmatrix} Q_x \\ Q_y \end{Bmatrix}^{(S+1)} = \begin{Bmatrix} Q_x \\ Q_y \end{Bmatrix}^{(S)} + \beta \begin{bmatrix} f_{xx} & f_{xy} \\ f_{yx} & f_{yy} \end{bmatrix} \begin{Bmatrix} X \\ Y \end{Bmatrix}^{(S+1)} \tag{9.76}$$

With $\beta = 1$, the application of this equation is physically equivalent to releasing nodal point n and permitting it to move freely to a new equilibrium position. With β greater than 1, the nodal point is moved beyond its equilibrium position before proceeding to the next point.

Any assumed nodal displacements $\{\mathbf{Q}_n\}^{(0)}$ may be used for the first cycle of iteration. Proper assumption will expedite the convergence of the solution. It is, however, common to start the iteration from zero displacements. A means to judge the convergence is to print out the total unbalanced force every, say, 10 or 20 cycles. The convergence is normally achieved if the total unbalanced force drops by, say, three orders of magnitude.

9.3.12 Boundary Conditions

Iterative method. Let us now consider a two-dimensional finite element system as shown in Fig. 9.9a. Equations (9.76) are valid for all but the two boundary nodal points, which are not free to move in both x and y directions. For the two boundary points, the flexibility coefficients must be modified to account for the specific restrained conditions.

Let us consider nodal point 1, which is free to slide along a straight boundary inclined with an angle ϕ with the x axis, where ϕ is positive when measured counterclockwise from the positive x axis. The unknown reaction

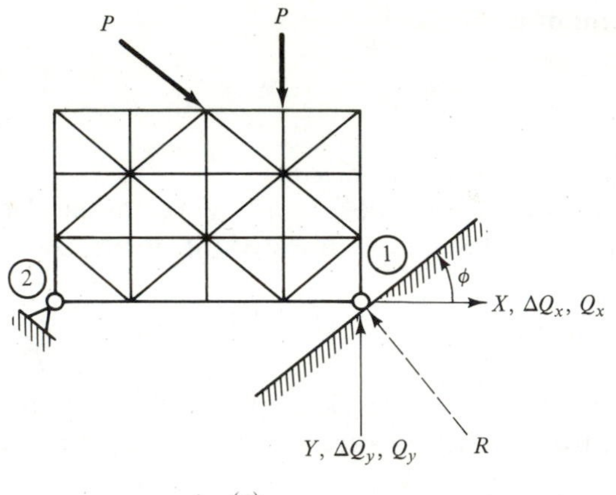

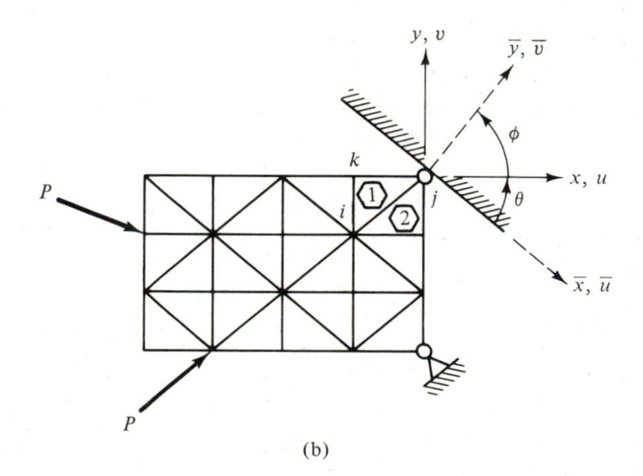

Figure 9.9 Boundary nodal point treated by (a) the iterative method and (b) the direct elimination method.

is represented by R. The unbalanced forces X and Y are determined from Eq. (9.75). The displacement components Q_x and Q_y and their increments ΔQ_x and ΔQ_y are determined from Eqs. (9.66) and (9.65), respectively.

Equation (9.65) can be rewritten in the following form:

$$\begin{Bmatrix} \Delta Q_x \\ \Delta Q_y \end{Bmatrix} = \begin{bmatrix} f_{xx} & f_{xy} \\ f_{yx} & f_{yy} \end{bmatrix} \begin{Bmatrix} X \\ Y \end{Bmatrix} \tag{9.77}$$

Considering R as an external force, Eq. (9.77) becomes

$$\begin{Bmatrix} \Delta Q_x \\ \Delta Q_y \end{Bmatrix} = \begin{bmatrix} f_{xx} & f_{xy} \\ f_{yx} & f_{yy} \end{bmatrix} \begin{Bmatrix} X - R \sin \phi \\ Y + R \cos \phi \end{Bmatrix} \tag{9.78}$$

Eliminating the unknown reaction R yields

$$\Delta Q_x = \frac{f_{xx} - Cf_{yx}}{1 - C \tan \phi} X + \frac{f_{xy} - Cf_{yy}}{1 - C \tan \phi} Y \tag{9.79}$$

where

$$C = \frac{f_{xx} \tan \phi - f_{xy}}{f_{xy} \tan \phi - f_{yy}} \tag{9.79a}$$

By definition of the slope shown in Fig. 9.9a,

$$\Delta Q_y = \Delta Q_x \tan \phi \tag{9.80}$$

Combining Eqs. (9.79) and (9.80) produces equations similar to Eq. (9.77):

$$\begin{Bmatrix} \Delta Q_x \\ \Delta Q_y \end{Bmatrix} = \begin{bmatrix} f^*_{xx} & f^*_{xy} \\ f^*_{yx} & f^*_{yy} \end{bmatrix} \begin{Bmatrix} X \\ Y \end{Bmatrix} \tag{9.81}$$

where the modified flexibility coefficients are defined as

$$\begin{aligned} f^*_{xx} &= \frac{f_{xx} - Cf_{xy}}{1 - C \tan \phi} \\ f^*_{xy} &= \frac{f_{xy} - Cf_{yy}}{1 - C \tan \phi} \\ f^*_{yx} &= \tan \phi f^*_{xx} \\ f^*_{yy} &= \tan \phi f^*_{xy} \end{aligned} \tag{9.81a}$$

For nodal points such as point 2 which are restrained from moving in both the x and y directions, ΔQ_x and $\Delta Q_y = 0$, all four coefficients in Eq. (9.81) are set equal to zero.

Direct elimination method. If the direct elimination method is used to take advantage of the band matrix, the boundary conditions for a certain nodal point can be treated using the coordinate transformation matrix.

Figure 9.9b shows a boundary point j considered by two coordinate systems: global coordinates (x, y) and boundary coordinates $(\bar{x}, \bar{y})$. Corresponding to the two sets of coordinates are the displacements (u_j, v_j) and $(\bar{u}_j, \bar{v}_j)$, respectively. Nodal point j is free to slide along but not to separate from the slope shown. Thus the two sets of displacements can be related as

$$\begin{Bmatrix} u_j \\ v_j \end{Bmatrix} = [\mathbf{t}] \begin{Bmatrix} \bar{u}_j \\ \bar{v}_j \end{Bmatrix} \tag{9.82}$$

where

$$[\mathbf{t}] = \left[\begin{array}{c|c} \cos \theta & \sin \theta \\ \hline -\sin \theta & \cos \theta \end{array}\right] \tag{9.82a}$$

Following the derivations given in Eqs. (4.19) to (4.21), the element stiffness matrix given in Eq. (9.54) can be transformed to that corresponding to the boundary coordinates $\bar{x}$ and $\bar{y}$ as

$$\{\mathbf{F}\} = [\mathbf{T}]^T[\mathbf{k}][\mathbf{T}]\{\bar{\mathbf{q}}\} \tag{9.83}$$

with

$$\underset{6\times 6}{[\mathbf{T}]} = \begin{bmatrix} I & & \\ & t & \\ & & I \end{bmatrix} \tag{9.83a}$$

This transformation process has to be applied to every element that shares the boundary nodal point j. In the case shown in Fig. 9.9b, only the stiffness matrices [**k**] for elements 1 and 2 need be modified.

After such a coordinate transformation, we can set $\bar{v}_j = 0$ and solve the assembled stiffness matrix equations. This can be done either by reducing the assembled stiffness matrix by eliminating the row and column corresponding to $\bar{v}_j$, or by adding a relatively very large stiffness coefficient to the diagonal term corresponding to $\bar{v}_j$. Thus displacement $\bar{u}_j$ can be obtained, and its components u_j and v_j can be found using Eq. (9.82) (i.e., $u_j = \bar{u}_j \cos\theta$ and $v_j = -\bar{u}_j \sin\theta$).

Physically, a slope boundary condition for a nodal point can also be achieved by adding a truss bar element perpendicular to the slope line, with one end of the bar connected to the nodal point and the other end fixed. By setting the axial rigidity of this bar to be very large (infinite), the nodal displacement normal to the slope is prevented. Mathematically, this is equivalent to performing the aforementioned coordinate transformation and then adding a very large coefficient to the diagonal stiffness term corresponding to $\bar{v}_j$.

9.3.13 Gravity, Distributed, and Thermal Loads

Three types of nodal point loads are discussed here.

1. *Gravity loads:* The weight of each element is obtained as the product of its unit weight and its volume. It is a common practice that the element weight be divided by three and distributed equally among the three nodal points.
2. *Distributed loads:* For distributed loads acting on the edge of an element, the two nodal point loads can be obtained by considering the edge as a simply supported beam and the support reaction forces found become the nodal point loads, adjusted to the opposite direction. Another way of converting distributed loads to nodal point loads is to use the work-equivalent loads described in Eqs. (9.40) through (9.44).

3. *Thermal loads:* To treat thermal loads, the analysis method is divided into two steps. First, assuming that all nodal points are restrained, the stresses developed within all elements due to temperature changes are found. From Eqs. (9.14) and (9.16), the uniform stresses for a typical element are given by

$$\sigma_x = \sigma_y = \sigma_t = \frac{E\alpha\,\Delta T}{1-\nu} \qquad \text{(for plane stress)}$$

$$= \frac{E\alpha\,\Delta T}{(1+\nu)(1-2\nu)} \qquad \text{(for plane strain)} \tag{9.84}$$

where α is the thermal coefficient of expansion and ΔT is the rise in temperature.

Setting $\sigma_x = \sigma_y = \sigma_t$ and $\tau_{xy} = 0$ in Eq. (9.53), the element corner forces that are necessary to maintain these stresses are obtained:

$$\begin{Bmatrix} S_x^i \\ S_y^i \\ S_x^j \\ S_y^j \\ S_x^k \\ S_y^k \end{Bmatrix} = -\frac{1}{2} \begin{Bmatrix} b_k - b_j \\ a_j - a_k \\ -b_k \\ a_k \\ b_j \\ -a_j \end{Bmatrix} \sigma_t \tag{9.85}$$

Second, to eliminate these forces the system is analyzed for nodal point loads which are equal in magnitude but opposite in sign to these restraining forces. The final thermal stress distribution is the sum of stresses due to these thermal loads and the initial stresses in the restrained system.

9.3.14 Element Stresses

After the nodal displacements are found by solving the assembled stiffness matrix, they can be converted to the coordinate stresses σ_x, σ_y, and τ_{xy} using Eq. (9.51) for each element. Since these stresses are constant within each element, a reasonably fine mesh must be used in the region that has relatively sharp stress gradient. It is a common practice to assume these stresses as occurring near the centroid of each triangle, as long as the mesh is reasonably fine.

For practical analysis and design, it is of interest to find the principal maximum and minimum stresses σ_1 and σ_2, respectively, and their directions. Such stresses can be obtained through the use of equations based on the method of the Mohr circle. It is common practice to construct contour lines for σ_1 and σ_2, respectively (isostress curves), which give us an overview of the stress distributions.

9.3.15 Computer Programs

Based on the Gauss–Seidel iterative method, Wilson [9.5] presented a computer program with many examples of plane stress and plane strain problems: a thick-walled cylinder, an infinite plate with elliptic hole, a gravity dam, and a cantilever beam.

Following the program developed basically by Wilson, the author and Guruswamy have presented a computer program in Sec. 13.3. Instead of using the Gauss–Seidel iterative technique, methods of band matrix storage and Gaussian elimination are used. Furthermore, the slope boundary conditions are treated using the coordinate transformation method described in Sec. 9.3.12. A program description, user's manual, and the input and output data for two sample examples are given in Chapter 13.

9.3.16 Example

To demonstrate the performance of this element and the program given in Sec. 13.3, an example of a rectangular aluminum plate with a circular hole subjected to uniform edge tensile stress as shown in Fig. 9.10 is analyzed. It is assumed that $E = 10^7$ psi and $\nu = 0.3$.

Due to double symmetry, only a quadrant need be analyzed. A finite element modeling with 202 elements and 122 nodal points is shown in Fig. 9.10. The roller boundary conditions are also shown.

The results for deformed configuration, normal stresses on two edges, and isostress curves for maximum and minimum principal stresses are shown in Fig. 9.11. For normal edge stress σ_x, as plotted in Fig. 9.11b, an approximate elasticity solution is available in Ref. [9.1] for comparison. The agreement is good. The stress concentration factor for σ_x is found to be approximately 3.

Let us examine the results by checking the equilibrium of the quadrant shown in Fig. 9.11b as a free body. The unbalanced forces are found to be 1,835 and 216 lb in the x and y directions, respectively. These magnitudes are insignificant as compared to the total loads ($5000 \times 20 = 100{,}000$ lb).

9.4 HIGHER-ORDER RECTANGULAR ELEMENTS

In Secs. 9.2 and 9.3 we introduced two simplest (lowest-order) plane stress and plane strain elements in great detail. We will now introduce the more sophisticated higher-order elements. We will first introduce the rectangular elements in this section, then the triangular elements in the next section, and finally compare the performance of these elements through numerical example analysis in the last section.

The higher-order elements are normally created either by increasing the number of nodal points by placing them on or inside the boundaries or by

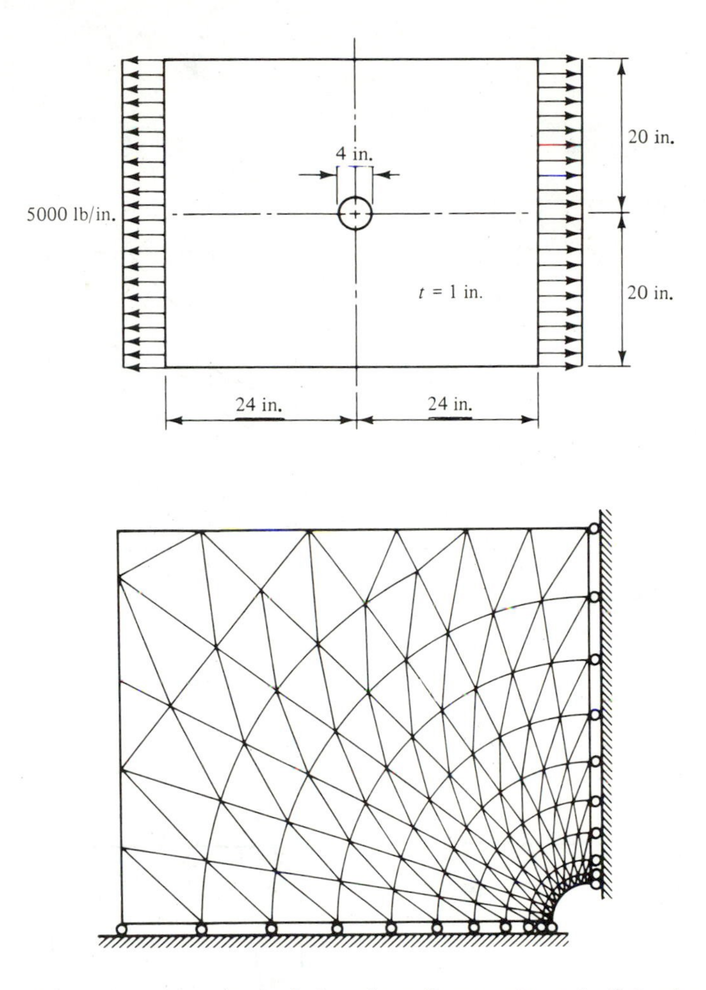

Figure 9.10 Plate with a circular hole under uniform tension and a finite element model for a quadrant.

including the higher-order derivative terms of displacements as degrees of freedom at nodal points. An advantage of some effective higher-order elements over lower-order elements is that the results of the same accuracy may be obtainable using not only fewer higher-order elements but also fewer total degrees of freedom. Such an advantage is most important for problems with highly concentrated stresses or sharp stress gradients, as it is possible that reasonable results may not be obtainable using a reasonable or manageable number of lower-order elements.

Early convergence study of plane stress elements can be attributed to, among others, Argyris [9.2], Melosh [9.9], and Fraeijs de Veubeke [9.10]. References [9.2] and [9.10] showed that if displacements are assumed which

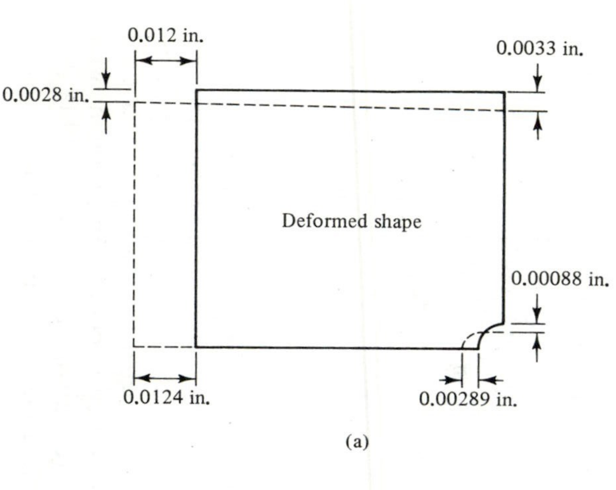

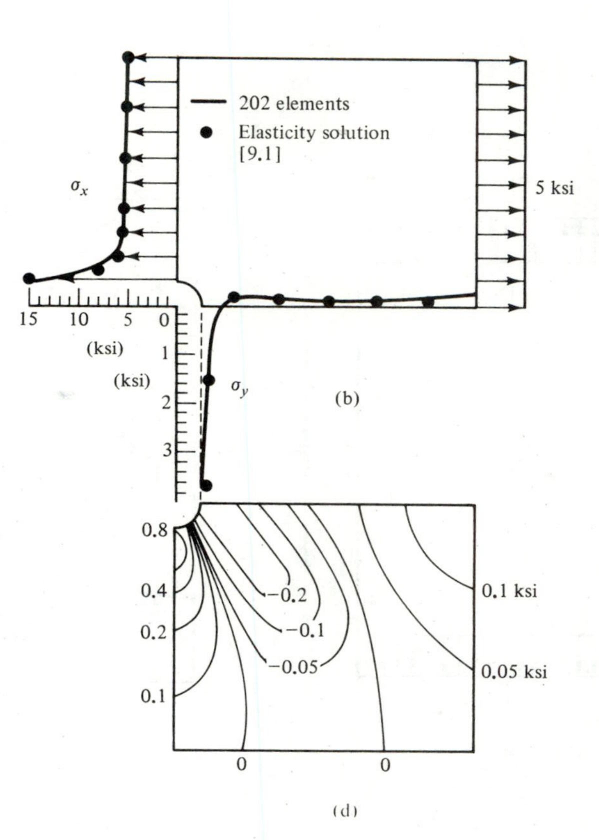

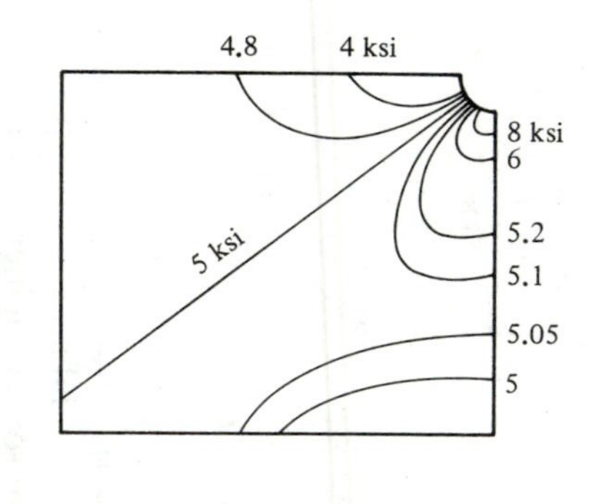

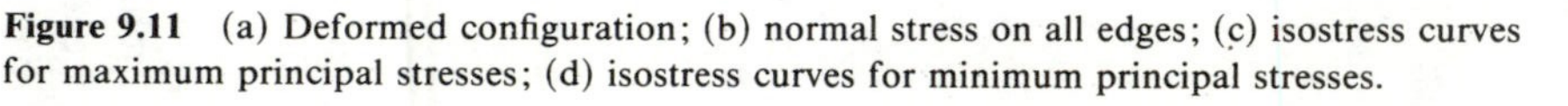

Figure 9.11 (a) Deformed configuration; (b) normal stress on all edges; (c) isostress curves for maximum principal stresses; (d) isostress curves for minimum principal stresses.

result in interelement compatibility, a direct stiffness coefficient (displacement at and in the direction of a unit load) will be underestimated and approaches a lower bound. If stresses are assumed which result in interelement equilibrium, an upper bound to a direct flexibility coefficient will be found.

9.4.1 Rectangular Elements with Additional Side Nodes

Higher-order elements may be achieved by adding one node at the midpoint of each side or adding two nodes at the one- and two-third points of each side, as shown in Fig. 9.12. For both elements, only u- and v-displacement d.o.f.'s are assumed at each node.

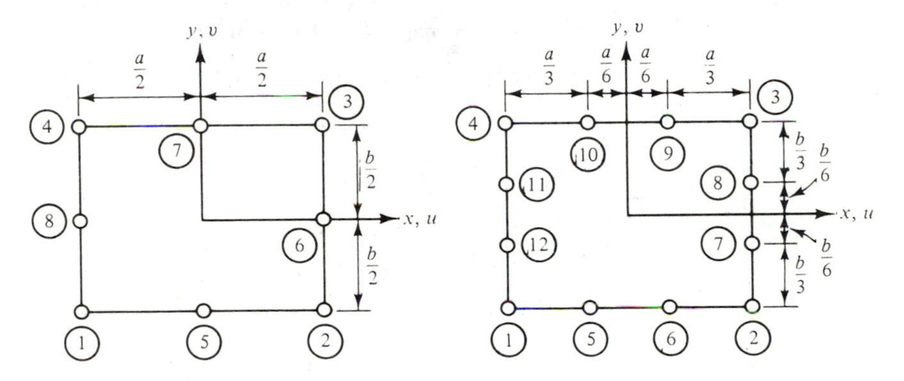

Figure 9.12 Eight-node 16-d.o.f. and 12-node 24-d.o.f. rectangular plane stress and plane strain finite elements.

For the eight-node 16-d.o.f. rectangular element, the u- or v-displacement function can be assumed as

$$u(\xi, \eta) = \alpha_1 + \alpha_2\xi + \alpha_3\eta + \alpha_4\xi^2 + \alpha_5\xi\eta + \alpha_6\eta^2 + \alpha_7\xi^2\eta + \alpha_8\xi\eta^2 \tag{9.86}$$

or

$$\begin{aligned} u(\xi, \eta) = &-\tfrac{1}{4}(1-\xi)(1-\eta)(1+\xi+\eta)u_1 \\ &-\tfrac{1}{4}(1+\xi)(1-\eta)(1-\xi+\eta)u_2 \\ &-\tfrac{1}{4}(1+\xi)(1+\eta)(1-\xi-\eta)u_3 \\ &-\tfrac{1}{4}(1-\xi)(1+\eta)(1+\xi-\eta)u_4 \\ &+\tfrac{1}{2}(1-\xi^2)(1-\eta)u_5 \\ &+\tfrac{1}{2}(1-\eta^2)(1+\xi)u_6 \\ &+\tfrac{1}{2}(1-\xi^2)(1+\eta)u_7 \\ &+\tfrac{1}{2}(1-\eta^2)(1-\xi)u_8 \end{aligned} \tag{9.87}$$

where $\xi = 2x/a$ and $\eta = 2y/b$.

For the 12-node 24-d.o.f. element, the u- or v-displacement function can be assumed as

$$u(\xi, \eta) = \alpha_1 + \alpha_2\xi + \alpha_3\eta + \alpha_4\xi^2 + \alpha_5\xi\eta + \alpha_6\eta^2 + \alpha_7\xi^3 + \alpha_8\xi^2\eta + \alpha_9\xi\eta^2 + \alpha_{10}\eta^3 + \alpha_{11}\xi^3\eta + \alpha_{12}\xi\eta^3 \tag{9.88}$$

or

$$\begin{aligned} u(\xi, \eta) = {} & \tfrac{1}{32}(1-\xi)(1-\eta)[-10+9(\xi^2+\eta^2)]u_1 \\ & + \tfrac{1}{32}(1+\xi)(1-\eta)[-10+9(\xi^2+\eta^2)]u_2 \\ & + \tfrac{1}{32}(1+\xi)(1+\eta)[-10+9(\xi^2+\eta^2)]u_3 \\ & + \tfrac{1}{32}(1-\xi)(1+\eta)[-10+9(\xi^2+\eta^2)]u_4 \\ & + \tfrac{9}{32}(1-\xi^2)(1-\eta)(1-3\xi)u_5 \\ & + \tfrac{9}{32}(1-\xi^2)(1-\eta)(1+3\xi)u_6 \\ & + \tfrac{9}{32}(1-\eta^2)(1+\xi)(1-3\eta)u_7 \\ & + \tfrac{9}{32}(1-\eta^2)(1+\xi)(1+3\eta)u_8 \\ & + \tfrac{9}{32}(1-\xi^2)(1+\eta)(1+3\xi)u_9 \\ & + \tfrac{9}{32}(1-\xi^2)(1+\eta)(1-3\xi)u_{10} \\ & + \tfrac{9}{32}(1-\eta^2)(1-\xi)(1+3\eta)u_{11} \\ & + \tfrac{9}{32}(1-\eta^2)(1-\xi)(1-3\eta)u_{12} \end{aligned} \tag{9.89}$$

Using the forms of displacement functions given in Eqs. (9.87) and (9.89), the stiffness matrix may be formulated using exact numerical integration.

It is noted that for elements with arbitrary quadrilateral shape, side nodes, interior nodes, and higher-order displacement functions, the stiffness matrix can be obtained using the concept of isoparametric finite element and various exact numerical integration methods. Details on this topic are given in Chapter 11.

9.4.2 Four-Corner-Node 12-D.O.F. Rectangular Element

Without increasing the number of nodal points, a first step to increase the order of the 8-d.o.f. rectangular element appears to assign one more degree of freedom to each of the four corner nodes. MacLeod [9.11] added an in-plane rotational d.o.f. $\partial v/\partial x$ and $-\partial u/\partial y$ at alternate corner nodes and analyzed shear walls using such elements in connection with beam elements. An advantage of such element is that the extra in-plane rotational d.o.f.'s can be used in connection with other types of elements possessing the same kind of nodal d.o.f.'s, such as a beam element.

For this 12-d.o.f. element, the displacement function for u or v may be assumed as, using the coordinates shown in Fig. 9.4,

$$u(x, y) = a_1 + a_2x + a_3y + a_4x^2 + a_5xy + a_6y^2 \tag{9.90}$$

where the 12 constants for both u and v functions can be determined using the 12 nodal d.o.f.'s.

9.4.3 Four-Corner-Node 24-D.O.F. Rectangular Element

For a four-corner-node 24-d.o.f. rectangular element, each nodal point is assumed to have six d.o.f.'s: u, $\partial u/\partial x$, $\partial u/\partial y$, v, $\partial v/\partial x$, and $\partial v/\partial y$. The displacement function for u or v may be assumed as, using the coordinates shown in Fig. 9.4,

$$\begin{aligned} u(x, y) = a_1 + a_2x + a_3y + a_4x^2 + a_5xy + a_6y^2 + a_7x^3 \\ + a_8x^2y + a_9xy^2 + a_{10}y^3 + a_{11}x^3y + a_{12}xy^3 \end{aligned} \tag{9.91}$$

For both u and v functions, there are a total of 24 constants, which can be determined by evaluating the 24-nodal d.o.f. values at the four corner nodes.

9.4.4 Four-Corner-Node 32 (or 24)-D.O.F. Rectangular Element

As a special case of a 48-d.o.f. rectangular shell finite element [9.12], a 32-d.o.f. rectangular plane stress element [9.13] was used in a free-vibration analysis of a framed shear wall. For such element, each of the four corner nodes possesses eight d.o.f.'s: u, $\partial u/\partial x$, $\partial u/\partial y$, $\partial^2 u/\partial x\,\partial y$, and the same for v. The displacement functions for u or v may be assumed as

$$u = (a_1 + a_2x + a_3x^2 + a_4x^3)(b_1 + b_2y + b_3y^2 + b_4y^3) \tag{9.92}$$

which results in 32 constants for both the u and v functions. Alternatively, Eq. (9.92) can also be expressed in the form of bicubic Hermitian polynomials [9.12], which will be explained in great detail in Chapter 12 in connection with rectangular plate elements in bending.

The 32-d.o.f. element can be reduced to a 24-d.o.f. element by writing the d.o.f.'s in $\partial^2 u/\partial x\,\partial y$ and $\partial^2 v/\partial x\,\partial y$ at the four nodes in terms of the rest of the d.o.f.'s. This can be done by assuming the counterpart forces for these eight d.o.f.'s to be zero and using the reduction method described in Sec. 7.7. Physically, such reduction means that the internodal compatibility of these eight d.o.f.'s is no longer imposed. Improvement of the performance on plane stress analysis due to such reduction was studied [9.14] and a numerical example is given in Sec. 9.6.3.

9.5 HIGHER-ORDER TRIANGULAR ELEMENTS

To formulate the stiffness matrix for a triangular element, Eq. (9.33) can be used. For a higher-order element, however, the strain–displacement matrix [**A**] is no longer a constant and the integration of $[\mathbf{A}]^T[\mathbf{C}][\mathbf{A}]$ over an arbitrary triangle becomes very cumbersome if rectangular coordinates are used. Such integration can be simplified considerably if we work with a natural coordinate system for an arbitrary triangle.

9.5.1 Natural Coordinate System for a Triangle

Figure 9.13a shows a triangle in both rectangular and triangular coordinates. Sides 1, 2, and 3 are identified by the opposite vertices 1, 2, and 3, respectively. The triangular coordinates ξ_i $(i = 1, 2, 3)$ for an interior point are defined as the ratios of the areas A_i to the total area A;

$$\xi_1 = \frac{A_1}{A} \qquad \xi_2 = \frac{A_2}{A} \qquad \xi_3 = \frac{A_3}{A} \tag{9.93}$$

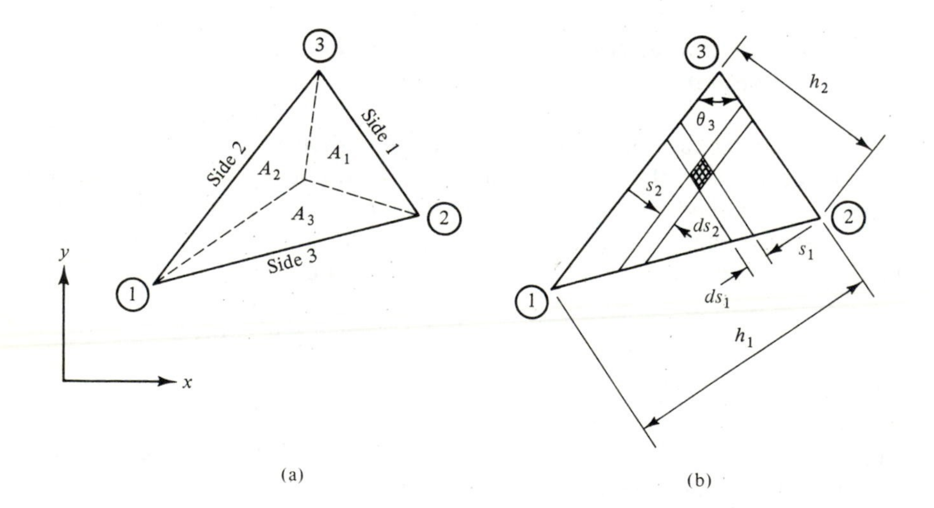

Figure 9.13 (a) Triangular coordinates; (b) integration domain.

Because

$$A_1 + A_2 + A_3 = A$$

then

$$\xi_1 + \xi_2 + \xi_3 = 1 \tag{9.94}$$

The rectangular coordinates, x and y, of this interior point may now be related to the triangular coordinates as

$$\begin{aligned} x &= \xi_1 x_1 + \xi_2 x_2 + \xi_3 x_3 \\ y &= \xi_1 y_1 + \xi_2 y_2 + \xi_3 y_3 \end{aligned} \tag{9.95}$$

These two equations can easily be interpreted with a few specific check points. For example, at centroid, $A_1 = A_2 = A_3 = A/3$, $x = (x_1 + x_2 + x_3)/3$, and $y = (y_1 + y_2 + y_3)/3$. At point 1, $A_1 = A$, $A_2 = A_3 = 0$, $x = x_1$, and $y = y_1$.

Collecting Eqs. (9.94) and (9.95) yields the relation between rectangular and triangular coordinates:

$$\begin{Bmatrix} 1 \\ x \\ y \end{Bmatrix} = \begin{bmatrix} 1 & 1 & 1 \\ x_1 & x_2 & x_3 \\ y_1 & y_2 & y_3 \end{bmatrix} \begin{Bmatrix} \xi_1 \\ \xi_2 \\ \xi_3 \end{Bmatrix} \tag{9.96}$$

and, after inverting,

$$\xi_i = \frac{1}{2A}(\alpha_i + \beta_i x + \gamma_i y) \qquad (i = 1, 2, 3) \tag{9.97}$$

where

$$\begin{aligned} A &= \tfrac{1}{2}(x_2 y_3 - x_3 y_2 + x_3 y_1 - x_1 y_3 + x_1 y_2 - x_2 y_1) \\ \alpha_i &= x_j y_k - x_k y_j \\ \beta_i &= y_j - y_k \\ \gamma_i &= x_k - x_j \end{aligned} \tag{9.97a}$$

with i, j, k being cyclic permutations among 1, 2, 3. Equations (9.97a) indicate that

$$\alpha_1 + \alpha_2 + \alpha_3 = 2A \tag{9.97b}$$

The integration may be evaluated by means of the parallelogram area element shown in Fig. 9.13b. The differential area element is given by

$$\begin{aligned} dA &= (ds_1)\csc(\theta_3)(ds_2) \\ &= (h_1\, d\xi_1)\csc(\theta_3)(h_2\, d\xi_2) \\ &= 2A\, d\xi_1\, d\xi_2 \end{aligned} \tag{9.98}$$

Thus for a polynomial term, the integral is in the general form

$$\int_A \xi_1^a \xi_2^b \xi_3^c\, dA = 2A \int_0^1 \left[\int_0^{1-\xi_1} \xi_1^a \xi_2^b (1 - \xi_1 - \xi_2)^c\, d\xi_2 \right] d\xi_1 \tag{9.99}$$

The substitution of $\xi_2 = t(1 - \xi_1)$ and $d\xi_2 = (1 - \xi_1)\,dt$ into the inner integral results in

$$\int_A \xi_1^a \xi_2^b \xi_3^c \, dA = 2A \int_0^1 \xi_1^a (1 - \xi_1)^{b+c+1} \, d\xi_1 \int_0^1 t^b (1 - t)^c \, dt \tag{9.100}$$

Each of the integrals on the right-hand side is of the form of the beta function [9.15]:

$$B(z, w) = \int_0^1 t^{z-1}(1 - t)^{w-1} \, dt = \frac{\Gamma(z)\Gamma(w)}{\Gamma(z + w)} \tag{9.101}$$

where Γ denotes the gamma function, which satisfies $\Gamma(n + 1) = n!$ for integers $n \geq 0$. Thus

$$\int_A \xi_1^a \xi_2^b \xi_3^c \, dA = 2A \frac{\Gamma(a + 1)\Gamma(b + 1)\Gamma(c + 1)}{\Gamma(a + b + c + 3)} \tag{9.102}$$

for complex numbers a, b, and c with real parts greater than -1.

Imposing restrictions that a, b, and c be nonnegative integers yields

$$\int_A \xi_1^a \xi_2^b \xi_3^c \, dA = \frac{a!b!c!}{(a + b + c + 2)!} 2A \tag{9.103}$$

Although this equation has been extensively quoted, the derivation above is due to Eisenberg and Malvern [9.16]. A comprehensive introduction of the triangular coordinates and the displacement functions can be found in Gallagher [9.17].

9.5.2 Triangular Elements with Additional Side and Interior Nodes

Higher-order elements may be achieved by (1) adding one node at the midpoint of each side or (2) adding two nodes at the one- and two-third points of each side plus an interior node at the centroid as shown in Fig. 9.14. The coordinates for each nodal point are also shown. The next-higher-order element would be a 15-node triangle, with three nodes equally spaced between the two corner nodes on each side and three interior nodes. The three interior nodes can be located at the intersecting points between lines, each connecting two nodes on two sides and parallel to the third side.

The six-node 12-d.o.f. triangular plane stress element, often called the *linear strain triangle*, was suggested by Fraejis de Veubeke [9.10] using rectangular coordinates. It was formulated by Argyris [9.18] in terms of natural strains and by Felippa [9.19] using triangular coordinates.

The displacement functions for u or v in triangular coordinates can be assumed in the form

$$u(\xi_1, \xi_2) = \alpha_1 + \alpha_2\xi_1 + \alpha_3\xi_2 + \alpha_4\xi_1^2 + \alpha_5\xi_1\xi_2 + \alpha_6\xi_2^2 \tag{9.104}$$

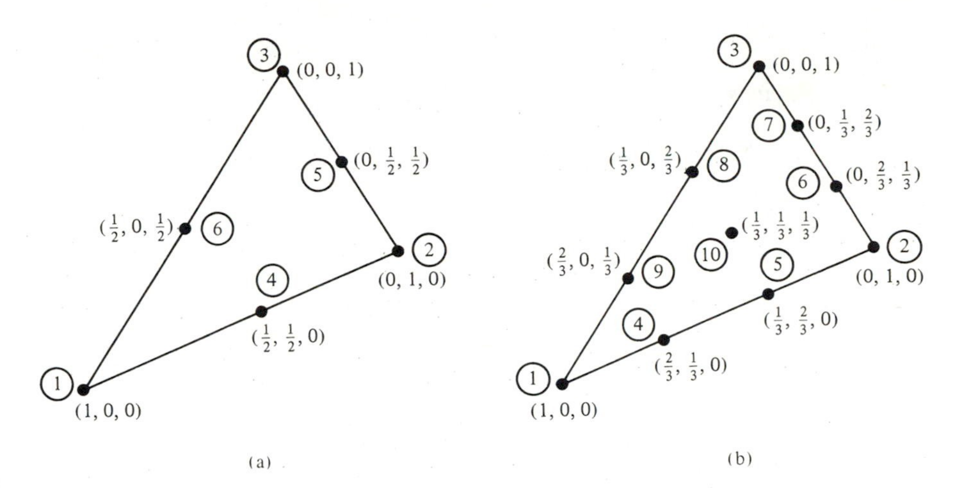

Figure 9.14 (a) Six-node 12-d.o.f. linear strain triangular element; (b) 10-node 20-d.o.f. quadratic strain triangular element.

For both u and v functions, there are a total of twelve constants which can be determined by the six u and v displacements at the six nodal points. Thus

$$u(\xi_1, \xi_2, \xi_3) = \xi_1(2\xi_1 - 1)u_1 + \xi_2(2\xi_2 - 1)u_2 + \xi_3(2\xi_3 - 1)u_3 + 4\xi_1\xi_2 u_4$$
$$+ 4\xi_2\xi_3 u_5 + 4\xi_3\xi_1 u_6 = \sum_{i=1}^{6} N_i(\xi_1, \xi_2, \xi_3)u_i \tag{9.105}$$

where the triangular coordinates for the six nodes are shown in Fig. 9.14a and N_i is called the *shape function.* The function for v holds the same form.

Using the chain rule for partial differentiation to express the strain-displacement equations (9.6) and the triangular-rectangular coordinate relation (9.97) gives

$$\begin{Bmatrix} \epsilon_x \\ \epsilon_y \\ \gamma_{xy} \end{Bmatrix} = \begin{Bmatrix} \dfrac{\partial u}{\partial x} \\ \dfrac{\partial v}{\partial y} \\ \dfrac{\partial u}{\partial y} + \dfrac{\partial v}{\partial x} \end{Bmatrix} = \sum_{i=1}^{3} \begin{Bmatrix} \dfrac{\partial u}{\partial \xi_i}\dfrac{\partial \xi_i}{\partial x} \\ \dfrac{\partial v}{\partial \xi_i}\dfrac{\partial \xi_i}{\partial y} \\ \dfrac{\partial u}{\partial \xi_i}\dfrac{\partial \xi_i}{\partial y} + \dfrac{\partial v}{\partial \xi_i}\dfrac{\partial \xi_i}{\partial x} \end{Bmatrix} = \frac{1}{2A}\sum_{i=1}^{3} \begin{Bmatrix} \beta_i \dfrac{\partial u}{\partial \xi_i} \\ \gamma_i \dfrac{\partial v}{\partial \xi_i} \\ \gamma_i \dfrac{\partial u}{\partial \xi_i} + \beta_i \dfrac{\partial v}{\partial \xi_i} \end{Bmatrix} \tag{9.106}$$

Substituting the displacement functions for u and v (9.105) into Eqs. (9.106) yields

$$\begin{Bmatrix} \epsilon_x \\ \epsilon_y \\ \gamma_{xy} \end{Bmatrix} = \begin{bmatrix} \mathbf{N}_x & 0 \\ 0 & \mathbf{N}_y \\ \mathbf{N}_y & \mathbf{N}_x \end{bmatrix} \begin{Bmatrix} u \\ v \end{Bmatrix} = \underset{3 \times 12}{[\mathbf{A}]} \underset{12 \times 1}{\begin{Bmatrix} u \\ v \end{Bmatrix}} \tag{9.107}$$

where

$$\lfloor u \rfloor = \lfloor u_1 \quad u_2 \quad u_3 \quad u_4 \quad u_5 \quad u_6 \rfloor \tag{9.107a}$$

$$\lfloor v \rfloor = \lfloor v_1 \quad v_2 \quad v_3 \quad v_4 \quad v_5 \quad v_6 \rfloor \tag{9.107b}$$

$$\lfloor \mathbf{N}_x \rfloor = \frac{1}{2A} \lfloor \beta_1 \quad \beta_2 \quad \beta_3 \rfloor [\mathbf{N}_\xi] \tag{9.107c}$$

$$\lfloor \mathbf{N}_y \rfloor = \frac{1}{2A} \lfloor \gamma_1 \quad \gamma_2 \quad \gamma_3 \rfloor [\mathbf{N}_\xi] \tag{9.107d}$$

$$[\mathbf{N}_\xi] = \begin{bmatrix} \frac{\partial N_1}{\partial \xi_1} & \frac{\partial N_2}{\partial \xi_1} & \frac{\partial N_3}{\partial \xi_1} & \frac{\partial N_4}{\partial \xi_1} & \frac{\partial N_5}{\partial \xi_1} & \frac{\partial N_6}{\partial \xi_1} \\ \frac{\partial N_1}{\partial \xi_2} & \frac{\partial N_2}{\partial \xi_2} & \frac{\partial N_3}{\partial \xi_2} & \frac{\partial N_4}{\partial \xi_2} & \frac{\partial N_5}{\partial \xi_2} & \frac{\partial N_6}{\partial \xi_2} \\ \frac{\partial N_1}{\partial \xi_3} & \frac{\partial N_2}{\partial \xi_3} & \frac{\partial N_3}{\partial \xi_3} & \frac{\partial N_4}{\partial \xi_3} & \frac{\partial N_5}{\partial \xi_3} & \frac{\partial N_6}{\partial \xi_3} \end{bmatrix}$$

$$= \begin{bmatrix} 4\xi_1 - 1 & 0 & 0 & 4\xi_2 & 0 & 4\xi_3 \\ 0 & 4\xi_2 - 1 & 0 & 4\xi_1 & 4\xi_3 & 0 \\ 0 & 0 & 4\xi_3 - 1 & 0 & 4\xi_2 & 4\xi_1 \end{bmatrix} \tag{9.107e}$$

or

$$\lfloor \mathbf{N}_x \rfloor = \frac{1}{2A} \lfloor \beta_1(4\xi_1 - 1) \quad \beta_2(4\xi_2 - 1) \quad \beta_3(4\xi_3 - 1) \quad 4(\beta_1\xi_2 + \beta_2\xi_1)$$
$$4(\beta_2\xi_3 + \beta_3\xi_2) \quad 4(\beta_1\xi_3 + \beta_3\xi_1) \rfloor \tag{9.108}$$

and $[\mathbf{N}_y]$ is of the same form as $[\mathbf{N}_x]$ with β_i's replaced by γ_i's.

The stiffness matrix can now be obtained using Eq. (9.33),

$$t \iint_A [\mathbf{A}]^T [\mathbf{C}][\mathbf{A}]\, dA \tag{9.109}$$

and formula (9.103) is readily available.

For the 10-node 20-d.o.f. triangular plane stress element, often called *quadratic strain triangle*, the displacement function for u or v in triangular coordinates can be assumed in the form of a complete cubic polynomial,

$$u(\xi_1, \xi_2) = \alpha_1 + \alpha_2\xi_1 + \alpha_3\xi_2 + \alpha_4\xi_1^2 + \alpha_5\xi_1\xi_2 + \alpha_6\xi_2^2 + \alpha_7\xi_1^3 + \alpha_8\xi_1^2\xi_2$$
$$+ \alpha_9\xi_1\xi_2^2 + \alpha_{10}\xi_2^3 \tag{9.110}$$

For both u and v functions, there are a total of 20 constants which can be determined by evaluating the 10 u and 10 v displacements at the 10 nodal points. Thus

$$
\begin{aligned}
u(\xi_1, \xi_2, \xi_3) = {} & \tfrac{1}{2}\xi_1(3\xi_1 - 1)(3\xi_1 - 2)u_1 + \tfrac{1}{2}\xi_2(3\xi_2 - 1)(3\xi_2 - 2)u_2 \\
& + \tfrac{1}{2}\xi_3(3\xi_3 - 1)(3\xi_3 - 2)u_3 + \tfrac{9}{2}\xi_1\xi_2(3\xi_1 - 1)u_4 \\
& + \tfrac{9}{2}\xi_1\xi_2(3\xi_2 - 1)u_5 + \tfrac{9}{2}\xi_2\xi_3(3\xi_2 - 1)u_6 \\
& + \tfrac{9}{2}\xi_2\xi_3(3\xi_3 - 1)u_7 + \tfrac{9}{2}\xi_3\xi_1(3\xi_3 - 1)u_8 \\
& + \tfrac{9}{2}\xi_3\xi_1(3\xi_1 - 1)u_9 + 27\xi_1\xi_2\xi_3u_{10}
\end{aligned} \tag{9.111}
$$

Following the procedure described in Eqs. (9.106) to (9.109) and using Eq. (9.103), the stiffness matrix for a quadratic strain triangular element can also be obtained.

9.5.3 Three-Corner-Node 18-D.O.F. Triangular Element

Because of the lesson learned during the development of curved shell finite elements—that membrane displacement functions u and v should not be looked upon with prejudice and should be assumed to be closer to or on the same level of sophistication as the transverse displacement function w—it is a common practice that not just w, but also u and v displacement functions, be assumed as higher-order polynomials. Thus a sophisticated plane stress finite element can be obtained by simply eliminating the curvatures and discarding the w-displacement-related d.o.f.'s of an existing higher-order shell finite element. It has become routine that in the process of evaluating the performance of a shell element, its membrane or plane stress behavior first be investigated.

Cowper et al. [9.20] developed a three-corner-node 36-d.o.f. triangular shell element, for which the function for membrane displacement u or v was assumed as a complete cubic polynomial,

$$
\begin{aligned}
u(\xi_1, \xi_2) = {} & \alpha_1 + \alpha_2\xi_1 + \alpha_3\xi_2 + \alpha_4\xi_1^2 + \alpha_5\xi_1\xi_2 + \alpha_6\xi_2^2 + \alpha_7\xi_1^3 + \alpha_8\xi_1^2\xi_2 \\
& + \alpha_9\xi_1\xi_2^2 + \alpha_{10}\xi_2^3
\end{aligned} \tag{9.112}
$$

As shown in Fig. 9.15, a rectangular coordinate system was so chosen that the ξ_1 axis coincides with one edge of the triangle and the ξ_2 axis passes through the vertex opposite to the edge. Thus the triangle is divided into two subtriangles, each with ξ_1 and ξ_2 axes not only coinciding with two edges but also at a 90° angle. For such a special case, if nondimensionalized, ξ_1 and ξ_2 fit the definition of natural coordinates as described in connection with Fig. 9.13 and the closed-form integration formula (9.103) can be used for evaluating the integrals necessary for formulating the stiffness matrix. For both u and v functions, there are a total of 20 constants which can be determined by evaluating the 20 nodal d.o.f.'s: u, $\partial u/\partial\xi_1$, $\partial u/\partial\xi_2$, v, $\partial v/\partial\xi_1$, and $\partial v/\partial\xi_2$ at the three vertices and u_c and v_c at the centroid. The 20 d.o.f.'s are reduced to 18

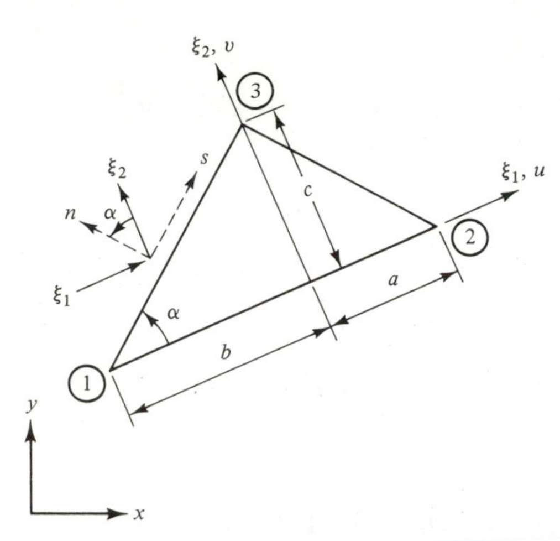

Figure 9.15 Coordinate systems.

by expressing u_c and v_c in terms of the rest of the d.o.f.'s through matrix partitioning and reduction.

The stiffness coefficients obtained in such local natural coordinates and the associated coordinate transformation matrix were given explicitly in Ref. 9.20. Numerical results for membrane behavior were also provided. More details of the formulation of this stiffness matrix are given in Chapter 12 in connection with plate elements in bending.

Alternative ways of forming three-node 18-d.o.f. triangular elements were also developed. One way is to divide the triangle into three subtriangles and assume the displacement function for each subtriangle such that $\partial u/\partial Y$ and $\partial v/\partial Y$ vary linearly along the outer edge, so as to ensure their full interelement compatibility, where the Y axis is normal to the outer edge [9.21]. This process will be given in more detail in Chapter 12 in connection with HCT plate element in bending. Another way is to choose a complete third-order 10-term polynomial displacement function for u or v, supplemented with a minimum energy requirement to eliminate the two superfluous polynomial coefficients [9.22, 9.23].

9.5.4 Three-Corner-Node 36-D.O.F. Triangular Element

A three-corner-node 54-d.o.f. triangular element was developed by Dawe [9.24] for shell analysis. The u, v, and w displacement functions are assumed in the same form as a complete fifth-order polynomial minus the $\xi_1^4\xi_2$ term; for example,

$$\begin{aligned} u(\xi_1, \xi_2) = {} & \alpha_1 + \alpha_2\xi_1 + \alpha_3\xi_2 + \alpha_4\xi_1^2 + \alpha_5\xi_1\xi_2 + \alpha_6\xi_2^2 + \alpha_7\xi_1^3 + \alpha_8\xi_1^2\xi_2 + \alpha_9\xi_1\xi_2^2 \\ & + \alpha_{10}\xi_2^3 + \alpha_{11}\xi_1^4 + \alpha_{12}\xi_1^3\xi_2 + \alpha_{13}\xi_1^2\xi_2^2 + \alpha_{14}\xi_1\xi_2^3 + \alpha_{15}\xi_2^4 \\ & + \alpha_{16}\xi_1^5 + \alpha_{17}\xi_1^3\xi_2^2 + \alpha_{18}\xi_1^2\xi_2^3 + \alpha_{19}\xi_1\xi_2^4 + \alpha_{20}\xi_2^5 \end{aligned} \tag{9.113}$$

The coordinates are chosen in the same way as those shown in Fig. 9.15. The term $\xi_1^4\xi_2$ is omitted to ensure that the in-plane rotation term $\partial u/\partial\xi_2$ be reduced to a third-order function in ξ_1 along edge 1-2 ($\xi_2 = 0$); thus

$$\frac{\partial u}{\partial \xi_2} = c_1 + c_2\xi_1 + c_3\xi_1^2 + c_4\xi_1^3 \tag{9.114}$$

where the four constants can be determined uniquely based on the four-nodal d.o.f. values for $\partial u/\partial\xi_2$ and $\partial^2 u/\partial\xi_2\,\partial\xi_1$ at nodes 1 and 2, respectively.

To solve for the 20 constants in Eq. (9.113), 20 equations are needed. Eighteen equations are obtained by evaluating the 18 nodal d.o.f.'s: u, $\partial u/\partial\xi_1$, $\partial u/\partial\xi_2$, $\partial^2 u/\partial\xi_1^2$, $\partial^2 u/\partial\xi_1\,\partial\xi_2$, and $\partial^2 u/\partial\xi_2^2$ at each of the three vertices. One of the two remaining equations is obtained by imposing the condition that the in-plane rotational d.o.f. $\partial u/\partial n$ varies as a cubic function of s along edge 1-3. This is done by eliminating all the coefficients associated with the fourth-order terms of s in the $\partial u/\partial n$ expression, which come from the fifth-order terms or the last five terms in Eq. (9.113). As shown in Fig. 9.15, the s axis is assumed to be parallel with edge 1-3 and the n axis is the outward normal to edge 1-3. The two sets of axes result in the following relations:

$$\begin{gathered}\frac{\partial u}{\partial n} = -\frac{\partial u}{\partial \xi_1}\sin\alpha + \frac{\partial u}{\partial \xi_2}\cos\alpha \\ \xi_1 = s\cos\alpha \quad \text{and} \quad \xi_2 = s\sin\alpha \\ \cos\alpha = \frac{b}{\sqrt{b^2+c^2}} \quad \text{and} \quad \sin\alpha = \frac{c}{\sqrt{b^2+c^2}}\end{gathered} \tag{9.115}$$

The condition for cubic variation of $\partial u/\partial n$ along edge 1-3 then is [9.25]

$$\begin{aligned}5b^4c\alpha_{16} + (3b^2c^3 - 2b^4c)\alpha_{17} + (2bc^4 - 3b^3c^2)\alpha_{18} + (c^5 - 4b^2c^3)\alpha_{19} \\ - 5bc^4\alpha_{20} = 0\end{aligned} \tag{9.116a}$$

Similarly, the condition for cubic variation of $\partial u/\partial n$ along edge 2-3 gives the last equation needed,

$$\begin{aligned}5a^4c\alpha_{16} + (3a^2c^3 - 2a^4c)\alpha_{17} + (-2ac^4 + 3a^3c^2)\alpha_{18} + (c^5 - 4a^2c^3)\alpha_{19} \\ + 5ac^4\alpha_{20} = 0\end{aligned} \tag{9.116b}$$

It is noted that for this 36-d.o.f. element, the six d.o.f.'s in $\partial^2 u/\partial\xi_1\,\partial\xi_2$ and $\partial^2 v/\partial\xi_1\,\partial\xi_2$ can be eliminated by assuming their counterpart forces equal to zero and then expressing them in terms of the rest of the d.o.f.'s through matrix partitioning and substitution. Thus the internodal compatibility in such degrees of freedom is relaxed and the element d.o.f.'s are reduced to 30. More details of the formulation of this stiffness matrix are given in Sec. 12.4.5 in connection with an 18-d.o.f. plate element in bending.

9.6 COMPARISON OF RESULTS

It is of interest to compare the performance of the various plane stress elements introduced in this chapter. Two types of examples are most commonly used in such evaluation: (1) a parabolically loaded square plate and (2) a cantilever plate loaded at the free end. Two cases are chosen for each of the two types of examples: (1) a case treated using rectangular elements and (2) a case treated using triangular elements.

9.6.1 Parabolically Loaded Plate Treated Using Rectangular Elements

This example is shown in Fig. 9.5. It is analyzed using the simplest eight d.o.f. rectangular elements with three different modelings and using the work-equivalent loads as shown in Fig. 9.5. The results in displacements and stresses are shown in Table 9.1. Since the results are relatively accurate for relatively coarse mesh, there seems to be no need to use higher-order rectangular elements for this type of example.

9.6.2 Parabolically Loaded Plate Treated Using Triangular Elements

Results for the u displacement at point A for the parabolically loaded plate with a quadrant modeled using four different types of triangular elements with various numbers of d.o.f.'s are shown in Fig. 9.16. The six-node 12-d.o.f. linear strain elements are seen to be much more efficient than the three-node 6-d.o.f. constant-strain elements. The two curves obtained using these two types of elements are taken from Fig. 9.10 of Ref. [9.17]. The linear strain elements are very widely used.

The solutions obtained using the three-node 18-d.o.f. and three-node 36-d.o.f. elements are seen to be highly accurate even at the two-element level (1×1 mesh). To give more insight into the performance of these two types of elements, the percentage errors in displacements and stresses at various points are plotted for different mesh sizes in Fig. 9.17 [9.20, 9.26].

9.6.3 Cantilever Plate Subjected to End Load Treated Using Rectangular Elements

Figure 9.18 shows a cantilever plate with unit thickness t and aspect ratio 3:1 under parabolically distributed shear stress at the end with total value equal to P. The distributed shear stress is assigned as work-equivalent loads to each nodal point as described in Eq. (9.43). Results for displacement v_A obtained using five different types of rectangular elements with various

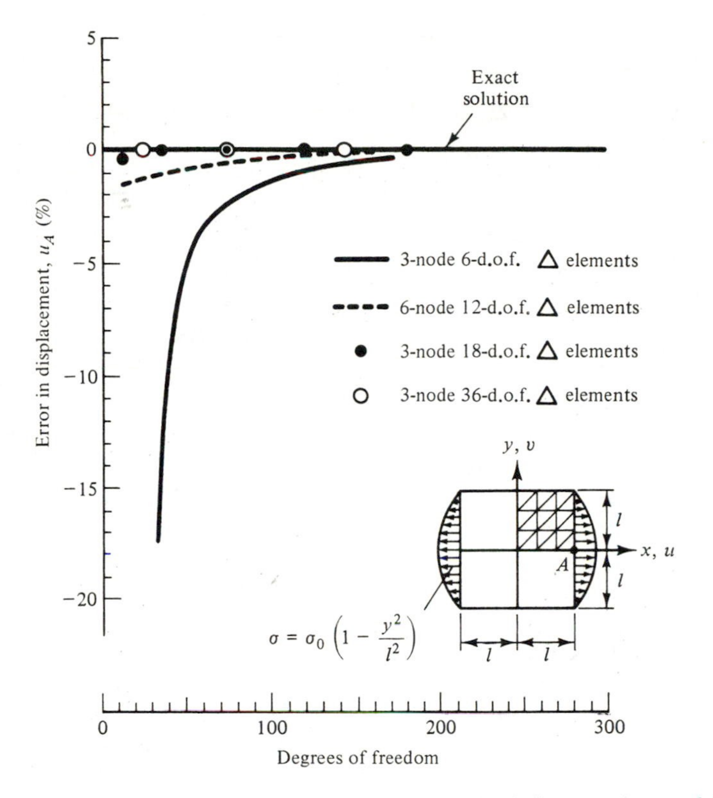

Figure 9.16 Comparison of results for a parabolically loaded square plate modeled by triangular elements.

numbers of d.o.f.'s are shown in Fig. 9.18. The comparison solution was obtained in Ref. [9.27].

The eight-node 16-d.o.f. elements are seen to be much more efficient than the four-node 8-d.o.f. elements, with the result in $v_A Et/P$ converging to 116.0 for a 198-d.o.f. model, the same value as that given in Ref. [9.27].

The four-node 32-d.o.f. elements are seen to be as efficient as the eight-node 16-d.o.f. elements. The 32-d.o.f. element can be reduced to a 24-d.o.f. element by writing the d.o.f.'s $\partial^2 u/\partial x\, \partial y$ and $\partial^2 v/\partial x\, \partial y$ at the four corners to be in terms of the rest of the d.o.f.'s. Physically, the interelement compatibilities for such d.o.f.'s are relaxed and the plate is modeled more flexibly. The results in Fig. 9.18 show that the 32-d.o.f. element is improved by such reduction.

It is of interest to examine the results obtained using a four-node 8-d.o.f. rectangular assumed stress hybrid element. The results are taken from Ref. [9.17]. This element is quite efficient. A description of this element is given in Sec. 9.6.5.

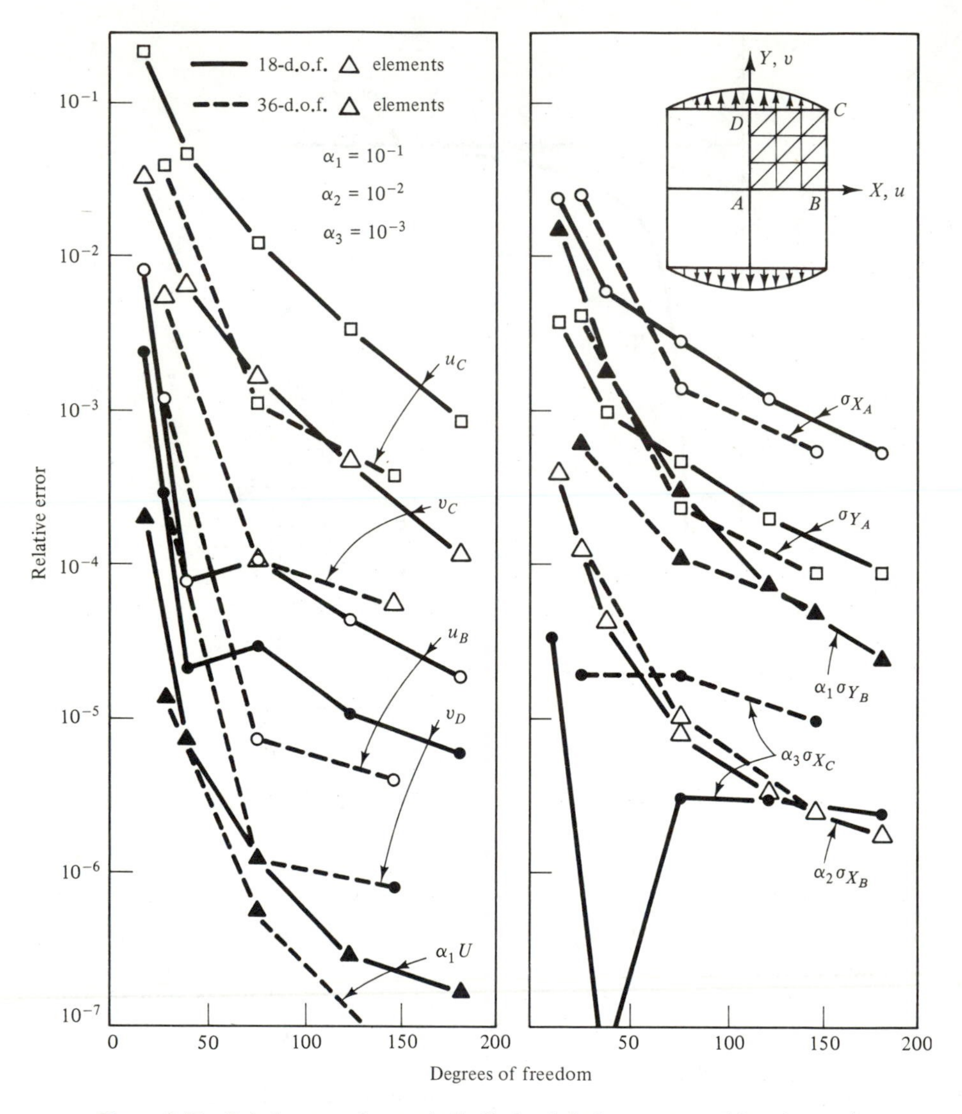

Figure 9.17 Relative error for parabolically loaded plane stress problem ($\nu = 0.3$).

9.6.4 Cantilever Plate Subjected to End Load Treated Using Triangular Elements

Results for the vertical v displacement at point A for the cantilever plate subjected to parabolically distributed shear stress at the end are modeled using three different types of triangular elements with various d.o.f.'s as shown in Fig. 9.19. Although the six-node 12-d.o.f. linear strain elements show considerable improvement over the three-node 6-d.o.f. constant strain elements, the

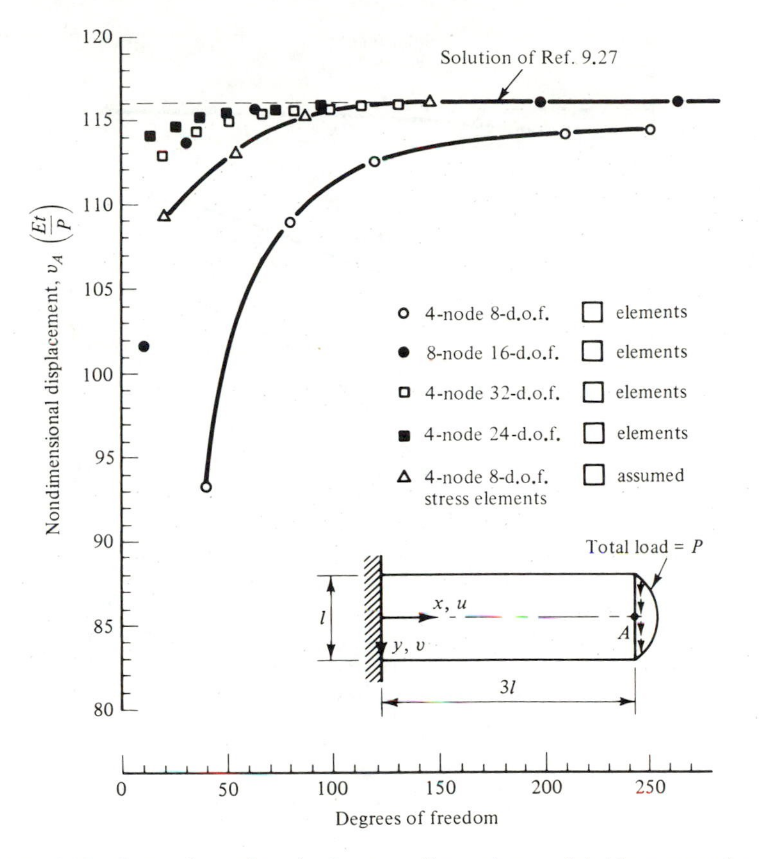

Figure 9.18 Comparison of results for a cantilever plate modeled by rectangular elements ($\nu = 0.2$).

results in v_A obtained using the 12-d.o.f. elements show approximately a 5 to 6% discrepancy at the 200-d.o.f. level compared to that given in Ref. [9.27]. These two curves are taken from Fig. 9.11 of Ref. [9.17].

The results obtained using the three-node 36-d.o.f. elements are seen to be highly accurate, showing 2% discrepancy at the 32-d.o.f. level (1×1 mesh). Further improved results may be obtained if the 36-d.o.f. element is reduced to 30 d.o.f. by writing the d.o.f.'s $\partial^2 u/\partial x\, \partial y$ and $\partial^2 v/\partial x\, \partial y$ in terms of the rest of the d.o.f.'s.

9.6.5 Concluding Remarks

The examples shown in Secs 9.6.1 through 9.6.4 provided an indication of the comparative performance of the various elements introduced in this chapter. There is a trade-off between rectangular and triangular elements. For approximately the same order of displacement functions, rectangular elements

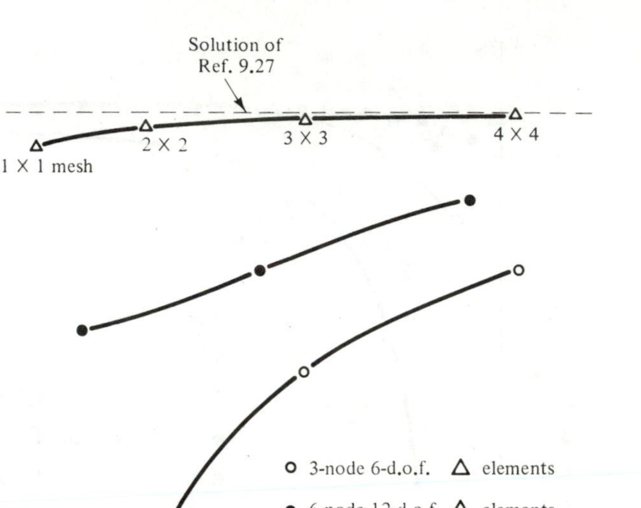
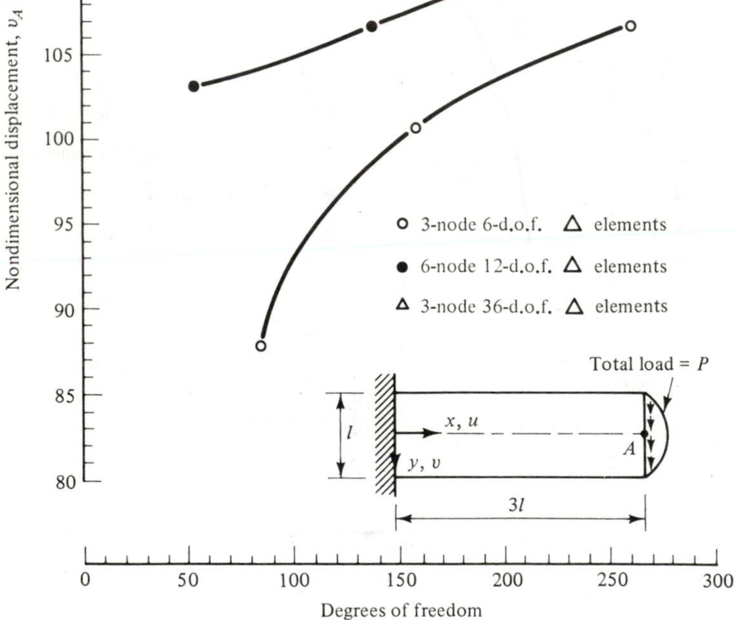

Figure 9.19 Comparison of results for a cantilever plate modeled by triangular elements ($\nu = 0.2$).

usually perform better than triangular elements. On the other hand, triangular elements are more versatile than rectangular elements in the geometric modeling of practical structures.

There is also a trade-off between lower- and higher-order elements. Lower-order elements are easy to use and practical in modeling geometry, loading, and boundary conditions. However, for problems with high stress concentration or sharp stress gradients, reasonable results may not be obtainable using reasonable or manageable number of lower-order elements, and higher-order ones may be needed. On the other hand, too high an order may render the element impractical due to the increase in formulative effort and the loss in modeling capability for geometry, loading, and boundary conditions caused by using fewer elements.

It is noted that researchers working on plane stress elements have also used approaches other than the displacement formulations. One of the earliest developments of the rectangular element stiffness matrix is based on the assumption of linear stress distribution for σ_x and σ_y and constant stress for τ_{xy}. These stresses can be used to form strain-stress equations by inverting the stress-strain relations and the displacement functions can be obtained by integrating the strains [4.1].

Pian later derived the element stiffness matrices for a rectangular hybrid model using assumed stress distributions [9.28]. The stiffness matrix for the special case turns out to be the same as that obtained using the displacement functions of Ref. [4.1]. A study of the basic convergence criteria for the general finite element methods was later given by Pian and Tong [9.29]. The basic concept of the hybrid stress approach and its usefulness were discussed by Gallagher [9.17].

A complementary energy formulation of a plane stress element involving a functional expressed in terms of second derivatives when the Airy stress function was used as nodal degrees of freedom was given by Gallagher and Dhalla [9.30]. A summary discussion of the various finite element models for plane stress analysis including compatible, hybrid displacement; equilibrium; hybrid stress; mixed Reissner; and a mixed variational principle is given in Ref. [9.31].

REFERENCES

9.1. Timoshenko, S. P., and Goodier, J. N., *Theory of Elasticity*, 2nd ed., McGraw-Hill Book Company, New York, 1951, pp. 1-13, 78-84, and 167-171.

9.2. Argyris, J. H., "Energy Theorems and Structural Analysis, Part 1: General Theory," *Aircraft Engineering*, Vol. 26, 1954, pp. 137 and 383, and Vol. 28, 1955, pp. 42, 80, and 145.

9.3. Gallagher, R. H., "The Development and Evaluation of Matrix Methods for Thin Shell Structural Analysis," Ph.D. thesis, State University of New York at Buffalo, June 1966.

9.4. Clough, R. W., "The Finite Element Method in Plane Stress Analysis," *Proceedings*, 2nd ASCE Conference on Electronic Computation, Pittsburgh, Pa., Sept. 1960.

9.5. Wilson, E. L., "Finite Element Analysis of Two-Dimensional Structures," Report No. 63-2, Department of Civil Engineering, University of California at Berkeley, June 1963.

9.6. Varga, R. S., *Matrix Iterative Analysis*, Prentice-Hall, Inc., Englewood Cliffs, N.J., 1962, Chap. 4.

9.7. Southwell, R. V., *Relaxation Methods in Engineering Science*, Oxford University Press, Oxford, 1940.

9.8. Temple, G., and Beckley, W. G., *Rayleigh's Principle and Its Application to Engineering*, Oxford University Press, Oxford, 1933.

9.9. Melosh, R. J., "Basis for Derivation of Matrices for the Direct Stiffness Method," *AIAA Journal*, Vol. 1, No. 7, 1963, pp. 1631-1637.

9.10. Fraeijs de Veubeke, B., "Displacement and Equilibrium Models in the Finite Element Method," in *Stress Analysis*, ed. O. C. Zienkiewicz and G. Hollister, John Wiley & Sons, Ltd., London, 1965, Chap. 9.

9.11. MacLeod, J. A., "New Rectangular Finite Element for Shear Wall Analysis," *Journal of the Structural Division*, ASCE, Vol. 95, No. ST3, 1969, pp. 399–409.

9.12. Bogner, F. K., Fox, R. L., and Schmit, L. A., Jr., "A Cylindrical Shell Discrete Element," *AIAA Journal*, Vol. 5, No. 4, 1967, pp. 745–750.

9.13. Yang, T. Y., and Sun, C. T., "Finite Elements for the Free Vibration of Framed Shear Walls," *Journal of Sound and Vibration*, Vol. 27, No. 3, 1973, p. 297–311.

9.14. Wu, T. M., "An Analytical Study of Shear Walls with Openings Including Effects of Floor Slabs," Ph.D. thesis, Department of Civil Engineering, Purdue University, West Lafayette, Ind., 1973.

9.15. Abramowitz, M., and Stegun, I. A., *Handbook of Mathematical Functions*, National Bureau of Standards, Washington, D.C., 1964, p. 258.

9.16. Eisenberg, M. A., and Malvern, L. E., "On Finite Element Integration in Natural Co-ordinates," *International Journal for Numerical Methods in Engineering*, Vol. 7, 1973, pp. 574–575.

9.17. Gallagher, R. H., *Finite Element Analysis Fundamentals*, Prentice-Hall, Inc., Englewood Cliffs, N.J., 1975, Chap. 8.

9.18. Argyris, J. H., "Triangular Elements with Linearly Varying Strain for the Matrix Displacement Method," *Journal of the Royal Aeronautical Society*, Vol. 69, 1965, pp. 711–713.

9.19. Felippa, C. A., "Refined Finite Element Analysis of Linear and Nonlinear Two-Dimensional Structures," Report 66-22, Department of Civil Engineering, University of California at Berkeley, Oct. 1966.

9.20. Cowper, G. R., Lindberg, G. M., and Olson, M. D., "A Shallow Shell Finite Element of Triangular Shape," *International Journal of Solids and Structures*, Vol. 6, 1970, pp. 1133–1156.

9.21. Tocher, J. L., and Hartz, B. J., "Higher-Order Finite Element for Plane Stress," *Journal of the Engineering Mechanics Division*, ASCE, Vol. 93, No. EM4, 1967, pp. 149–172.

9.22. Holand, I., and Bergan, P. G., Discussion of "Higher-Order Finite Element for Plane Stress," *Journal of the Engineering Mechanics Division*, ASCE, Vol. 94, No. EM2, 1968, pp. 698–702.

9.23. Holand, I., "The Finite Element Method in Plane Stress Analysis," in *The Finite Element Method in Stress Analysis*, ed. I. Holand and K. Bell, Tapir Press, Trondheim, Norway, 1969, Chap. 2.

9.24. Dawe, D. J., "High-Order Triangular Finite Element for Shell Analysis," *International Journal of Solids and Structures*, Vol. 11, 1975, pp. 1097–1110.

9.25. Cowper, G. R., Kosko, E., Lindberg, G. M., and Olson, M. D., "A High Precision Triangular Plate-Bending Element," Aeronautical Report LR-514, National Research Council of Canada, 1968.

9.26. Yang, T. Y., and Han, A. D., "Buckled Plate Vibrations and Large Amplitude Vibrations Using High-Order Triangular Elements," *AIAA Journal*, Vol. 21, No. 5, 1983, pp. 758–766.

9.27. Hooley, R. F., and Hibbert, P. D., "Bounding Plane Stress Solutions by Finite Elements," *Journal of the Structural Division*, ASCE, Vol. 92, No. ST1, 1966, pp. 39-48.

9.28. Pian, T. H. H., "Derivation of Element Stiffness Matrices by Assumed Stress Distributions," *AIAA Journal*, Vol. 2, 1964, pp. 1333-1335.

9.29. Pian, T. H. H., and Tong, P., "Basis of Finite Element Methods for Solid Continua," *International Journal for Numerical Methods in Engineering*, Vol. 1, No. 1, 1969, pp. 3-28.

9.30. Gallagher, R. H., and Dhalla, A. K., "Direct Flexibility-Finite Element Elastoplastic Analysis," *Proceedings*, 1st International Conference on Structural Mechanics in Reactor Technology, Vol. 6, Part M, Berlin, 1971.

9.31. Day, M. L., and Yang, T. Y., "A Mixed Variational Principle for Finite Element Analysis," *International Journal for Numerical Methods in Engineering*, Vol. 18, 1982, pp. 1213-1230.

PROBLEMS

9.1. Find the stiffness coefficient k_{21} in Eq. (9.36) using both Eqs. (9.33) and (9.35).

9.2. Show that the stress-strain equations (9.26a) for plane stress can be changed to those for plane strain (9.26b) by replacing E and ν by $E/(1-\nu^2)$ and $\nu/(1-\nu)$, respectively.

9.3. Verify the work-equivalent loads for the 4×4 mesh shown in Fig. 9.5.

9.4. With the initial assumption that $x_1 = x_2 = x_3 = x_4 = 0$, find the solution by using Gauss-Seidel iterative method, first without using overrelaxation factor β and then using a β value of 1.4.

$$\begin{bmatrix} 4 & 3 & 2 & 1 \\ 3 & 4 & 2 & 1 \\ 2 & 2 & 6 & 2 \\ 1 & 1 & 2 & 5 \end{bmatrix} \begin{Bmatrix} x_1 \\ x_2 \\ x_3 \\ x_4 \end{Bmatrix} = \begin{Bmatrix} 10 \\ 10 \\ 12 \\ 9 \end{Bmatrix}$$

9.5. For a plate modeled using three 6-d.o.f. constant plane stress elements with thickness 0.1 in. as shown in Fig. P9.5, find σ_x, σ_y, and τ_{xy} for all three elements.

9.6. Figure P9.6 shows a plate modeled using four rectangular elements, defined by Eqs. (9.36), and four truss bar elements, defined by Eqs. (4.10). Assuming that $Et = 2EA/l$ and $\nu = \sqrt{0.1}$, find **(a)** displacements u at nodes 2 and 3, and **(b)** ϵ_x and ϵ_y at point A.

9.7. Using one six-d.o.f. plane stress finite element as shown in Fig. P9.7 and assuming that $t = 0.1$ in., $E/(1-\nu^2) = 10^4$ ksi, find the displacements u_1, v_1, and u_2.

9.8. The roller at node 1 is free to slide along a parabolic locus as shown in Fig. P9.8. The flexibility matrix for node 1 is

$$\begin{Bmatrix} Q_x \\ Q_y \end{Bmatrix} = \begin{bmatrix} 10 & 2 \\ 2 & 4 \end{bmatrix} \begin{Bmatrix} X \\ Y \end{Bmatrix}$$

where X and Y are the unbalanced forces as defined in Eq. (9.75). Modify the equation above to incorporate such boundary condition.

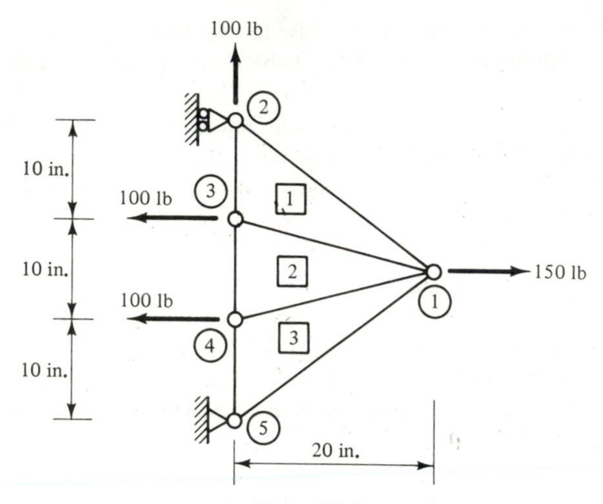

Figure P9.5

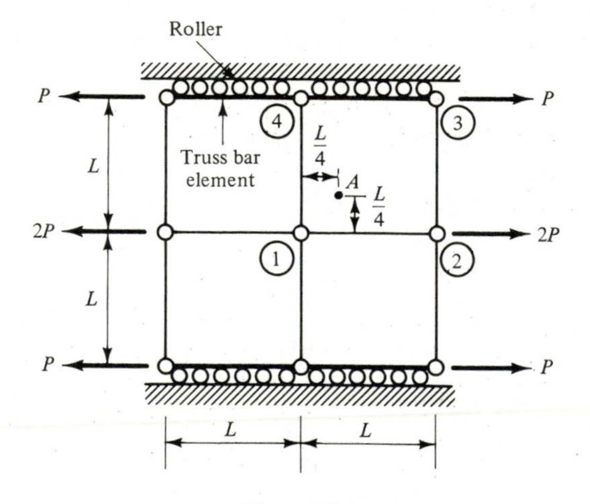

Figure P9.6

9.9. If a rectangular plate is modeled by a 4 × 8 mesh of six types of rectangular plane stress elements: **(a)** four-node eight-d.o.f., **(b)** eight-node 16-d.o.f., **(c)** 12-node 24-d.o.f., **(d)** four-node 12-d.o.f., **(e)** four-node 24-d.o.f., and **(f)** four-node 32-d.o.f. elements, find the minimum possible bandwidth for each type of assembled stiffness matrix.

9.10. Use the six-d.o.f. triangular plane stress finite element program given in Chapter 13 to perform a stress analysis for the panel with a cutout as shown in Fig. P9.10. Assume that $E = 10^7$ psi, $\nu = 0.3$, and $t = 0.1$ in. Due to double symmetry, only a quadrant need be modeled. Do not use more than 200 elements. **(a)** Plot the displacement configuration (magnify the nodal displacements by a very large

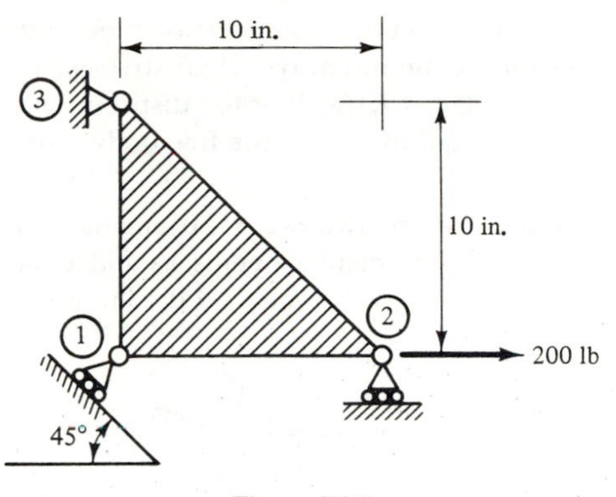

Figure P9.7

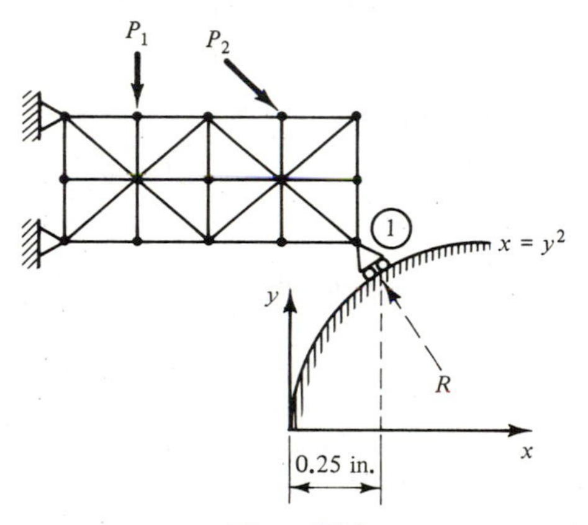

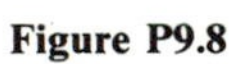
Figure P9.8

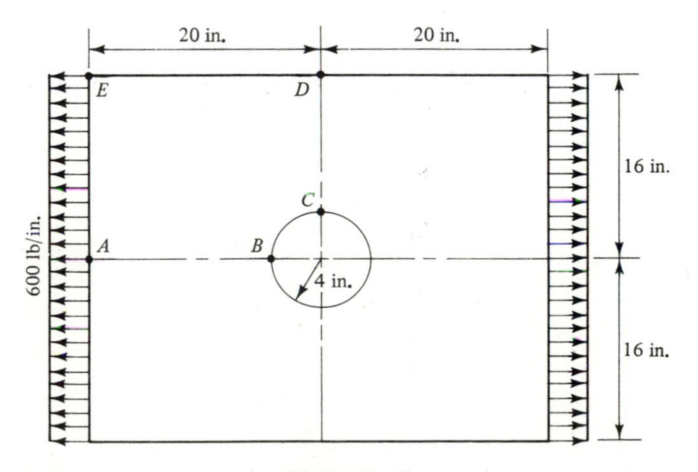

Figure P9.10

factor). **(b)** Plot the isostress curves for the maximum principal stresses. **(c)** Plot the isostress curves for the minimum principal stresses. **(d)** Plot the distribution of normal stress along edge $\overline{AB}$. **(e)** Plot the distribution of normal stress along edge $\overline{CD}$. **(f)** Check the equilibrium of the free body *ABCDE* in both the *x* and *y* directions.

9.11. The temperature rises ΔT in the two 6-d.o.f. triangular plane stress elements are shown in Fig. P9.11. Find the displacements u_2 and v_2 at joint 2. Assume that $E = 30 \times 10^6$ psi, $\alpha = 6 \times 10^{-6}$ in./in./°F, $t = 0.1$ in., and $\nu = \sqrt{0.1}$.

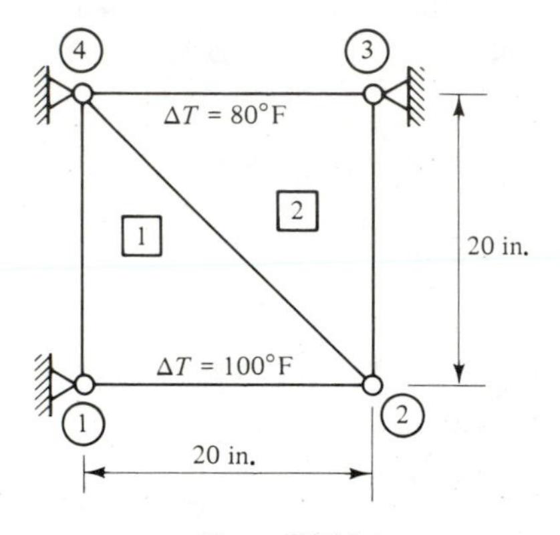

Figure P9.11

9.12. For a four-corner-node 8-d.o.f. rectangular plane stress element, the stress field can be assumed as $\sigma_x = c_1 + c_2 y$, $\sigma_y = c_3 + c_4 x$, and $\tau_{xy} = c_5$. Find the displacement function u and v by integration of the strain–displacement equations (9.6).

9.13. Derive in explicit closed form the stiffness coefficient k_{11} relating F_{x_1} to u_1 for the eight-node 16-d.o.f. rectangular plane stress element shown in Fig. 9.12a.

9.14. Derive in explicit closed form the stiffness coefficient k_{11} relating F_{x_1} to u_1 for the six-node 12-d.o.f. triangular plane stress element shown in Fig. 9.14a.

9.15. Show a step-by-step derivation of Eqs. (9.116a) and (9.116b).

9.16. For a rectangular cantilever plate subjected to a parabolically distributed end load and modeled using a 4 × 4 mesh of four-node 36-d.o.f. rectangular elements, give the zero d.o.f. boundary conditions.

CHAPTER 10

Axisymmetric and General Solid Finite Elements

Axisymmetric solid subjected to axisymmetric load is a special case of a general solid (i.e., a two-dimensional plane strain system). All the plane strain finite elements developed in Chapter 9 can be extended to become axisymmetric solid finite elements, simply by replacing the integration along the thickness to be that along the circumference. Because there are no additional degrees of freedom involved in such an extension for the case of axisymmetric loads, each version of the axisymmetric solid elements is as practical as its plane strain version. For the case of nonaxisymmetric loads, however, an additional displacement component in the circumferential direction is involved. The loads and the solutions are approximated by a Fourier series with a finite number (n) of terms. As compared to the case of axisymmetric load, the nonaxisymmetric solution involves the sum of a set of n solutions, each with approximately 50% more d.o.f.'s.

In this chapter an elementary axisymmetric solid element, with cross section being a constant strain triangle, is formulated in detail. Both axisymmetric and nonaxisymmetric loading conditions are included. Numerical examples are given.

Theoretically, we can extend the formulations introduced in Chapter 9 for all the triangular and rectangular two-dimensional elements to suitable formulations for tetrahedronal and rectangular hexahedronal elements, respectively. Practically, we encounter some difficulties in three-dimensional problems in addition to the formulation. The difficulties include (1) suitable element discretization and numbering of the nodes, (2) enormous increase in the total number of d.o.f.'s and bandwidth, and (3) displaying the results.

Let us consider the problem of the increase in computational effort due to the increase in dimension. For example, a 9×9 mesh of two-dimensional four-node 8-d.o.f. rectangular elements has one hundred nodes and two hundred d.o.f.'s with a minimum bandwidth of 47, as defined in Eq. (9.72). For a $9 \times 9 \times 9$ mesh of three-dimensional eight-node 24-d.o.f. "brick" elements, the number of nodes, d.o.f.'s, and minimum bandwidth becomes 1000, 3000, and 671, respectively.

All these difficulties have motivated the developments of automatic mesh generators, plotting routines, more efficient storage and computing methods, and more sophisticated and higher-order element formulations. These difficulties have been gradually overcome, not only due to these developments but also to the advances in computer graphics and vectorized supercomputers.

In this chapter we introduce two types of solid elements: tetrahedra and rectangular hexahedra, with low and higher orders. These solid elements can be generalized to have curved faces and the rectangular hexahedra can be generalized to be nonrectangular if we also use curvilinear coordinates and map the element shapes between the curvilinear and rectangular coordinates. This extension will be introduced in Chapter 11 for isoparametric elements.

10.1 AXISYMMETRIC SOLID ELEMENT UNDER SYMMETRIC LOADS

The axisymmetric solid element to be introduced here is in the shape of a circular ring with triangular cross section as shown in Fig. 10.1 [10.1, 10.2]. Also shown in the figure is an example of a closed-base thick-walled cylinder modeled using such elements. Because of axisymmetry, no tangential degree of freedom is assumed. Thus all the derivations are independent of the circumferential angular variable θ. Although axisymmetric solid elements with higher orders are available [10.3, 10.4], we will concentrate on this simplest element, which is a three-dimensional extension of the 6-d.o.f. plane strain triangular element introduced in Chapter 9. In this element, as shown in Fig. 10.1, each nodal circle is assumed to have two degrees of freedom: displacement u and w in the r and z directions, respectively. Such an element is not limited to axisymmetric loads, it can be extended to cope with nonaxisymmetric loads by using summations of trigonometric functions to represent the loads and displacements.

The axisymmetric elements have a wide range of applications in various fields of engineering. For example, in aerospace engineering, they have been applied to perform stress analysis of solid propellant grains, rocket nozzles and cases, spacecraft bulkheads and heat shields, and so on. In mechanical engineering, they have been applied to thick-walled pipes, tanks, pressure vessels, pistons, rotors, and so on. In civil engineering, they have been applied to tunnels, thick-walled pipes in soil, half-space, cylindrical and spherical

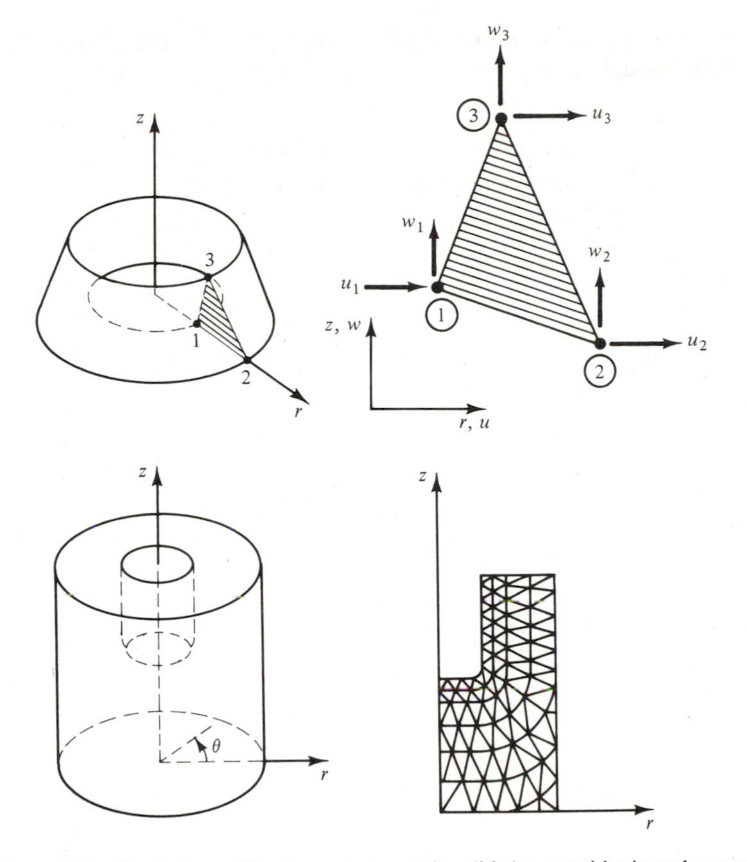

Figure 10.1 Six-degree-of-freedom axisymmetric solid element with triangular cross section and the modeling of a closed-base cylinder.

types of structures, and so on. In agricultural engineering, they have been applied to the stress analysis of soybeans, corncobs, and potatoes under the pressure of their piles. In bioengineering, they have been applied to the stress analysis of femurs and of teeth with fillings subjected to temperature change.

10.1.1 Displacement Functions

The cross section of this axisymmetric solid element is shown in Fig. 10.1. The element has three nodal circles, 1, 2, and 3. The cylindrical coordinates for the three nodes are (r_1, z_1), (r_2, z_2), and (r_3, z_3), respectively. Following the basic assumptions that edges remain straight after deformation and strains are constant throughout the triangle, the displacement functions are assumed as

$$\begin{aligned} u(r, z) &= a_1 + a_2 r + a_3 z \\ w(r, z) &= a_4 + a_5 r + a_6 z \end{aligned} \tag{10.1}$$

where the six constants can be obtained by evaluating the values of u and w at the three nodal circles,

$$\begin{Bmatrix} u_1 \\ u_2 \\ u_3 \\ w_1 \\ w_2 \\ w_3 \end{Bmatrix} = \begin{bmatrix} 1 & r_1 & z_1 & 0 & 0 & 0 \\ 1 & r_2 & z_2 & 0 & 0 & 0 \\ 1 & r_3 & z_3 & 0 & 0 & 0 \\ 0 & 0 & 0 & 1 & r_1 & z_1 \\ 0 & 0 & 0 & 1 & r_2 & z_2 \\ 0 & 0 & 0 & 1 & r_3 & z_3 \end{bmatrix} \begin{Bmatrix} a_1 \\ a_2 \\ a_3 \\ a_4 \\ a_5 \\ a_6 \end{Bmatrix} \tag{10.2}$$

or symbolically,

$$\{\mathbf{q}\} = [\mathbf{B}]\{\mathbf{a}\} \tag{10.2a}$$

Thus

$$\{\mathbf{a}\} = [\mathbf{B}]^{-1}\{\mathbf{q}\} \tag{10.3}$$

where

$$[\mathbf{B}]^{-1} = \frac{1}{\lambda}\begin{bmatrix} r_2 z_3 - r_3 z_2 & r_3 z_1 - r_1 z_3 & r_1 z_2 - r_2 z_1 & 0 & 0 & 0 \\ z_{23} & z_{31} & z_{12} & 0 & 0 & 0 \\ r_{32} & r_{13} & r_{21} & 0 & 0 & 0 \\ 0 & 0 & 0 & r_2 z_3 - r_3 z_2 & r_3 z_1 - r_1 z_3 & r_1 z_2 - r_2 z_1 \\ 0 & 0 & 0 & z_{23} & z_{31} & z_{12} \\ 0 & 0 & 0 & r_{32} & r_{13} & r_{21} \end{bmatrix} \tag{10.3a}$$

with

$$\lambda = r_1 z_{23} + r_2 z_{31} + r_3 z_{12} \qquad z_{ij} = z_i - z_j \qquad r_{ij} = r_i - r_j \tag{10.3b}$$

10.1.2 Strain–Displacement Equations

Because we will formulate this element for nonaxisymmetric loading in Sec. 10.2, let us first derive the general strain–displacement equations in polar coordinates and then simplify them to the special case of axisymmetric deformation only [9.1].

Figure 10.2 shows a differential element *abcd* with area $r\,d\theta\,dr$ and its deformed shape. If u is the radial displacement of side *ad*, the radial displacement of side *bc* is $u + (\partial u/\partial r)\,dr$. Then the unit elongation of the element *abcd* in the radial direction is

$$\epsilon_r = \frac{\partial u}{\partial r} \tag{10.4a}$$

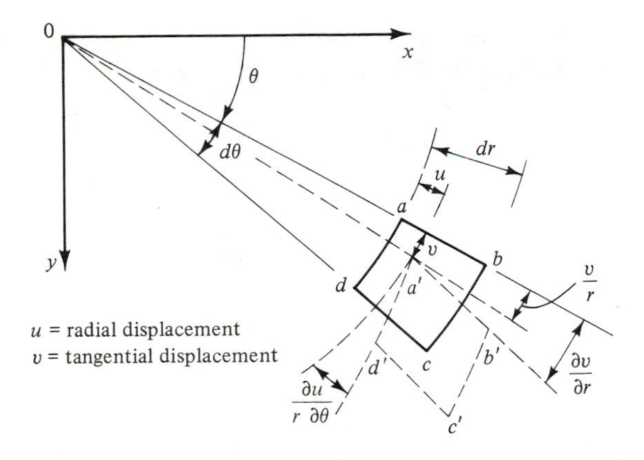

Figure 10.2 Strain-displacement definitions for polar coordinates.

The tangential strain depends not only on the tangential displacement v but also on the radial displacement u. Assuming that the points a and d displace only radially, the new length of the arc ad is $(r + u)\, d\theta$, and the tangential strain is

$$\frac{(r + u)\, d\theta - r\, d\theta}{r\, d\theta} = \frac{u}{r}$$

The difference in the tangential displacement of the sides ab and cd is $(\partial v / r\, \partial \theta) r\, d\theta$, and the tangential strain due to displacement v is $\partial v / r\, \partial \theta$. The total tangential strain then is

$$\epsilon_\theta = \frac{u}{r} + \frac{\partial v}{r\, \partial \theta} \tag{10.4b}$$

The shearing strain is the change in angle dab. The side da rotates by an angle of $\partial u / r\, \partial \theta$ to the position of $d'a'$. Although the side ab rotates by an angle of $\partial v / \partial r$ to the position of $a'b'$, the part of the angle v/r is caused by the rotation of the element $abcd$ as a rigid body about the axis through O. Thus the shearing strain is

$$\gamma_{r\theta} = \frac{\partial u}{r\, \partial \theta} + \frac{\partial v}{\partial r} - \frac{v}{r} \tag{10.4c}$$

The strains related to the rz plane are the same as those obtained in Eqs. (9.6) for rectangular coordinates.

For the special case of axisymmetric deformation, v displacement vanishes; we thus have, from Eqs. (10.4) and (9.6),

$$\begin{aligned}
\epsilon_r &= \frac{\partial u}{\partial r} = a_2 \\
\epsilon_\theta &= \frac{u}{r} = \frac{a_1}{r} + a_2 + \frac{a_3 z}{r} \\
\epsilon_z &= \frac{\partial w}{\partial z} = a_6 \\
\gamma_{rz} &= \frac{\partial u}{\partial z} + \frac{\partial w}{\partial r} = a_3 + a_5 \\
\gamma_{r\theta} &= \gamma_{\theta z} = 0
\end{aligned} \tag{10.5}$$

where the tangential strain is often called *hoop strain* and γ_{rz} is the shearing strain in planes through the axis of revolution. Equation (10.5) becomes

$$\begin{Bmatrix} \epsilon_r \\ \epsilon_\theta \\ \epsilon_z \\ \gamma_{rz} \end{Bmatrix} = \begin{bmatrix} 0 & 1 & 0 & 0 & 0 & 0 \\ \frac{1}{r} & 1 & \frac{z}{r} & 0 & 0 & 0 \\ 0 & 0 & 0 & 0 & 0 & 1 \\ 0 & 0 & 1 & 0 & 1 & 0 \end{bmatrix} \begin{Bmatrix} a_1 \\ a_2 \\ a_3 \\ a_4 \\ a_5 \\ a_6 \end{Bmatrix} \tag{10.6}$$

or symbolically,

$$\{\boldsymbol{\epsilon}\} = [\mathbf{G}]\{\mathbf{a}\} \tag{10.6a}$$

Equations (10.5) and (10.6) describe a case similar to plane strain, but with ϵ_θ being a function of $1/r$ and z/r, not a constant over the element. Moreover, ϵ_θ approaches infinity as r approaches zero. The ways to circumvent this difficulty will be discussed in Sec. 10.1.5.

Substituting Eq. (10.3) into Eq. (10.6a) gives

$$\{\boldsymbol{\epsilon}\} = [\mathbf{A}]\{\mathbf{q}\} \tag{10.7}$$

where

$$[\mathbf{A}] = [\mathbf{G}][\mathbf{B}]^{-1} \tag{10.7a}$$

10.1.3 Stress–Strain Equations

The stress-strain relationships can have a general form such as for an anisotropic material. For the case of an isotropic material, we have

$$\begin{Bmatrix} \sigma_r \\ \sigma_\theta \\ \sigma_z \\ \tau_{rz} \end{Bmatrix} = \frac{E}{(1+\nu)(1-2\nu)} \begin{bmatrix} 1-\nu & \nu & \nu & 0 \\ \nu & 1-\nu & \nu & 0 \\ \nu & \nu & 1-\nu & 0 \\ 0 & 0 & 0 & \frac{1}{2}-\nu \end{bmatrix} \left\{ \begin{Bmatrix} \epsilon_r \\ \epsilon_\theta \\ \epsilon_z \\ \gamma_{rz} \end{Bmatrix} - \begin{Bmatrix} \alpha\,\Delta T \\ \alpha\,\Delta T \\ \alpha\,\Delta T \\ 0 \end{Bmatrix} \right\} \tag{10.8}$$

or symbolically,

$$\{\sigma\} = [\mathbf{C}][\{\boldsymbol{\epsilon}\} - \{\boldsymbol{\epsilon}^i\}] \tag{10.8a}$$

where $\{\boldsymbol{\epsilon}^i\}$ is the vector of initial strains. In this case it contains thermal strains due to temperature rise ΔT and thermal coefficient of expansion α.

If the second row and second column, associated with σ_θ and ϵ_θ, are omitted, Eqs. (10.8) become the same as those for plane strain [Eqs. (9.26b)]. Combining with the strain–displacement expression (10.7), we have

$$\{\boldsymbol{\sigma}\} = [\mathbf{C}][\mathbf{A}]\{\mathbf{q}\} - [\mathbf{C}]\{\boldsymbol{\epsilon}^i\} \tag{10.9}$$

10.1.4 Stiffness Equations

The nodal force vs. displacement relations can be obtained using Castigliano's theorem, with an equation similar to Eq. (9.33). Thus we have

$$\{\mathbf{F}\} = [\mathbf{k}]\{\mathbf{q}\} - \{\mathbf{F}^i\} \tag{10.10}$$

where the stiffness matrix is obtained as

$$[\mathbf{k}] = \int_V [\mathbf{A}]^T[\mathbf{C}][\mathbf{A}]\, dV \tag{10.11}$$

Substituting Eq. (10.7a) into Eq. (10.11) yields

$$[\mathbf{k}] = ([\mathbf{B}]^{-1})^T \left[\int_V [\mathbf{G}]^T[\mathbf{C}][\mathbf{G}]\, dV \right] [\mathbf{B}]^{-1} \tag{10.12}$$

Also,

$$\{\mathbf{F}^i\} = ([\mathbf{B}]^{-1})^T \int_V [\mathbf{G}]^T[\mathbf{C}]\{\boldsymbol{\epsilon}^i\}\, dV \tag{10.13}$$

and the volume integration in polar coordinates is

$$V = \iiint dV = \iiint r\, d\theta\, dz\, dr = 2\pi \iint r\, dr\, dz \tag{10.14}$$

The explicit form of the stiffness equation (10.12) can now be obtained by performing volume integration of the following expression:

$$[\mathbf{G}]^T[\mathbf{C}][\mathbf{G}] = \frac{E}{(1+\nu)(1-2\nu)} \begin{bmatrix} \frac{1-\nu}{r^2} & & & \text{symmetric} & & \\ \frac{1}{r} & 2 & & & & \\ \frac{z(1-\nu)}{r^2} & \frac{z}{r} & \frac{z^2(1-\nu)}{r^2} + \frac{1}{2} - \nu & & & \\ 0 & 0 & 0 & 0 & & \\ 0 & 0 & \frac{1}{2} - \nu & 0 & \frac{1}{2} - \nu & \\ \frac{\nu}{r} & 2\nu & \frac{z\nu}{r} & 0 & 0 & 1-\nu \end{bmatrix} \tag{10.15}$$

The integration of Eqs. (10.15) involves the following six types of integrals [10.5, 10.6]:

$$I_1 = \iint dr\,dz = \frac{1}{2} \begin{vmatrix} 1 & 1 & 1 \\ r_1 & r_2 & r_3 \\ z_1 & z_2 & z_3 \end{vmatrix} \tag{10.16a}$$

$$I_2 = \iint r\,dr\,dz = \frac{r_1 + r_2 + r_3}{3} I_1 \tag{10.16b}$$

$$I_3 = \iint z\,dr\,dz = \frac{z_1 + z_2 + z_3}{3} I_1 \tag{10.16c}$$

$$I_4 = \iint \frac{dr\,dz}{r} = C_{12} + C_{23} + C_{31} \tag{10.16d}$$

where

$$C_{ij} = \frac{r_i z_j - r_j z_i}{r_i - r_j} \ln \frac{r_i}{r_j} \tag{10.17a}$$

$$I_5 = \iint \frac{z}{r}\,dr\,dz = D_{12} + D_{23} + D_{31} \tag{10.16e}$$

where

$$D_{ij} = \frac{z_j - z_i}{4(r_i - r_j)}[z_i(3r_j - r_i) - z_j(3r_i - r_j)] + \frac{1}{2}\left(\frac{r_i z_j - r_j z_i}{r_i - r_j}\right)^2 \ln \frac{r_i}{r_j} \tag{10.17b}$$

$$I_6 = \iint \frac{z^2}{r}\, dr\, dz = E_{12} + E_{23} + E_{31} \tag{10.16f}$$

where

$$E_{ij} = \frac{z_i - z_j}{18(r_i - r_j)^2}[z_j^2(11r_i^2 - 7r_i r_j + 2r_j^2) + z_i z_j(5r_i^2 - 22r_i r_j + 5r_j^2)$$

$$+ z_i^2(11r_j^2 - 7r_i r_j + 2r_i^2)] + \frac{1}{3}\left(\frac{r_i z_j - r_j z_i}{r_i - r_j}\right)^3 \ln \frac{r_i}{r_j} \tag{10.17c}$$

10.1.5 Treatment of Core Elements

If one or two nodal circles of an element are on the axis of revolution, or one of the sides of an element is parallel to the axis of revolution, r_i, r_i and r_j, or $r_i - r_j$ will vanish and C_{ij}, D_{ij}, and E_{ij} become indeterminate. Let us deal with such special cases.

Special case 1: When $r_i = 0$, r_j, $r_k \neq 0$, we can simply substitute zero value for r_i into Eqs. (10.16d–f) to obtain simplified expressions of I_4, I_5, and I_6:

$$I_4 = C_{jk} - z_i \ln \frac{r_j}{r_k}$$

$$I_5 = D_{jk} - \tfrac{1}{4}(z_j - z_i)(3z_i + z_j) - \tfrac{1}{4}(z_i - z_k)(3z_i + z_k) - \tfrac{1}{2}z_i^2 \ln \frac{r_j}{r_k}$$

$$I_6 = E_{jk} - \tfrac{1}{18}(z_j - z_i)(11z_i^2 + 5z_i z_j + 2z_j^2)$$
$$- \tfrac{1}{18}(z_i - z_k)(11z_i^2 + 5z_i z_k + 2z_k^2) - \tfrac{1}{3}z_i^3 \ln \frac{r_j}{r_k}$$

For this special case, $u_i = 0$, the element stiffness matrix can be reduced to the size of 5×5.

Special case 2: When $r_i = r_j = 0$, $r_k \neq 0$, u_i and u_j both vanish. The element stiffness matrix can be reduced to the size of 4×4. The derivation in Ref. 10.6 showed that the terms I_4, I_5, and I_6 do not appear in the element stiffness matrix.

Special case 3: When the side ij is parallel to the z axis, $r_i - r_j = 0$, which appears in the denominator of C_{ij}, D_{ij}, and E_{ij}. This difficulty can be circumvented by using L'Hospital's rule. Letting $t = r_i - r_j$, $r_j = r_i - t$, we have

$$C_{ij} = \frac{r_i z_j - r_i z_i + t z_i}{t}[\ln(r_i) - \ln(r_i - t)]$$

Differentiating both the numerator and denominator with respect to t gives

$$C_{ij} = z_i \ln r_i - z_i \ln (r_i - t) + \frac{r_i z_j - r_i z_i + t z_i}{r_i - t}$$

Setting $t = 0$ gives

$$C_{ij} = z_j - z_i \tag{10.18a}$$

By the same token it can be shown that [10.6]

$$D_{ij} = E_{ij} = 0 \tag{10.18b}$$

Special case 4: It is practical to use a core element with rectangular cross section to avoid the difficulty when $r_i = r_j = 0$. In this case, we simply assume that the element has only two nodes, k and l, both not at $r = 0$. We then use the displacement functions $u = a_1 r + a_2 z$ and $w = a_3 r + a_4 z$. The circumferential strain ϵ_θ may be assumed as $\epsilon_\theta = (u_k/r_k + u_l/r_l)/2$. The derivation of the stiffness matrix follows the same procedure as described above.

10.2 AXISYMMETRIC SOLID ELEMENT UNDER NONAXISYMMETRIC LOADS

If the loads are not axisymmetric, but are symmetric about a plane containing the axis of symmetry, such as the static equivalent wind load shown in Fig. 10.3, there are three displacement components at each nodal circle: u, v, and w. These displacements can be expanded in Fourier series, and due to the

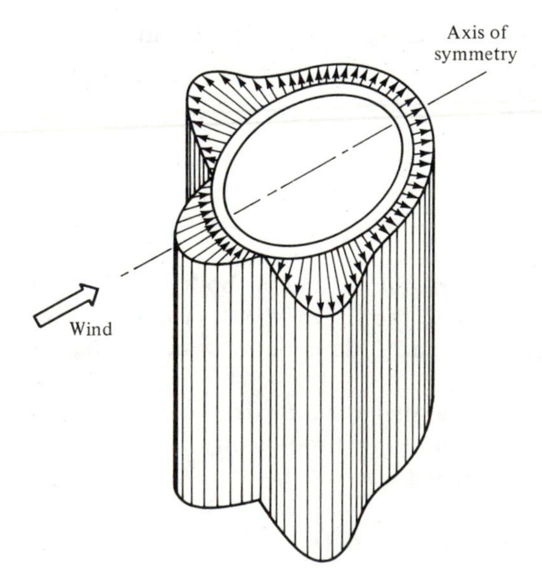

Figure 10.3 Nonaxisymmetric static equivalent wind load on a cylinder.

orthogonality of harmonic functions, the three-dimensional solution is the sum of a series of uncoupled two-dimensional solutions, in which the displacement amplitudes are the unknowns.

10.2.1 Stiffness Formulation

Expanding in Fourier series, the temperature distribution T and nodal circle loads P are in the form

$$\begin{aligned} T &= \sum T_n(r, z) \cos n\theta \\ P_r &= \sum P_{rn}(r, z) \cos n\theta \\ P_\theta &= \sum P_{\theta n}(r, z) \sin n\theta \\ P_z &= \sum P_{zn}(r, z) \cos n\theta \end{aligned} \tag{10.19}$$

and the nodal circle displacements are in the form

$$\begin{aligned} u &= \sum u_n(r, z) \cos n\theta \\ v &= \sum v_n(r, z) \sin n\theta \\ w &= \sum w_n(r, z) \cos n\theta \end{aligned} \tag{10.20}$$

where the amplitudes in the three displacement functions are assumed to be linear function in the rz plane within each element:

$$\begin{aligned} u_n(r, z) &= a_{1n} + a_{2n} r + a_{3n} z \\ v_n(r, z) &= a_{4n} + a_{5n} r + a_{6n} z \\ w_n(r, z) &= a_{7n} + a_{8n} r + a_{9n} z \end{aligned} \tag{10.20a}$$

It is observed that when $n = 0$, the temperature T, radial load P_r, axial load P_z, radial displacement u, and axial displacement w become independent of θ, and the circumferential load P_θ and circumferential displacement v vanish. Equations (10.19) and (10.20) reduce to the case of axisymmetric deformation.

Let us recall the strain–displacement relations derived in Eqs. (9.6) for the rz plane and those in Eqs. (10.4) for the $r\theta$ plane:

$$\begin{aligned} \epsilon_r &= \frac{\partial u}{\partial r} = \sum \epsilon_{rn} \cos n\theta \\ \epsilon_z &= \frac{\partial w}{\partial z} = \sum \epsilon_{zn} \cos n\theta \\ \epsilon_\theta &= \frac{\partial v}{r\,\partial \theta} + \frac{u}{r} = \sum \epsilon_{\theta n} \cos n\theta \\ \gamma_{r\theta} &= \frac{\partial u}{r\,\partial \theta} + \frac{\partial v}{\partial r} - \frac{v}{r} = \sum \gamma_{r\theta n} \sin n\theta \end{aligned} \tag{10.21}$$

$$\gamma_{z\theta} = \frac{\partial v}{\partial z} + \frac{\partial w}{r\,\partial \theta} = \sum \gamma_{z\theta n} \sin n\theta$$

$$\gamma_{rz} = \frac{\partial u}{\partial z} + \frac{\partial w}{\partial r} = \sum \gamma_{rzn} \cos n\theta$$

For a typical harmonic n, we have

$$\begin{Bmatrix} \epsilon_{rn} \\ \epsilon_{zn} \\ \epsilon_{\theta n} \\ \gamma_{r\theta n} \\ \gamma_{z\theta n} \\ \gamma_{rzn} \end{Bmatrix} = \begin{bmatrix} 0 & 1 & 0 & 0 & 0 & 0 & 0 & 0 & 0 \\ 0 & 0 & 0 & 0 & 0 & 0 & 0 & 0 & 1 \\ \frac{1}{r} & 1 & \frac{z}{r} & \frac{n}{r} & n & \frac{nz}{r} & 0 & 0 & 0 \\ -\frac{n}{r} & -n & -\frac{nz}{r} & -\frac{1}{r} & 0 & -\frac{z}{r} & 0 & 0 & 0 \\ 0 & 0 & 0 & 0 & 0 & 1 & -\frac{n}{r} & -n & -\frac{nz}{r} \\ 0 & 0 & 1 & 0 & 0 & 0 & 0 & 1 & 0 \end{bmatrix} \begin{Bmatrix} a_{1n} \\ a_{2n} \\ a_{3n} \\ \vdots \\ a_{9n} \end{Bmatrix} \tag{10.22}$$

or symbolically,

$$\{\boldsymbol{\epsilon}_n\} = [\mathbf{G}_n]\{\mathbf{a}_n\} \tag{10.22a}$$

The rest of the derivation follows the same pattern as described in the preceding section. Thus we have a set of nine stiffness equations for each harmonic n,

$$\underset{9\times 1}{\{\mathbf{F}\}} = \underset{9\times 9}{[\mathbf{k}]}\ \underset{9\times 1}{\{\mathbf{q}\}} \tag{10.23}$$

with the stiffness matrix defined as

$$\underset{9\times 9}{[\mathbf{k}]} = \underset{9\times 9}{[\mathbf{B}]^{-1}} \left[\int_V \underset{9\times 6}{[\mathbf{G}]^T}\ \underset{6\times 6}{[\mathbf{C}]}\ \underset{6\times 9}{[\mathbf{G}]}\ dV \right] \underset{9\times 9}{[\mathbf{B}]^{-1}} \tag{10.23a}$$

More details on these derivations are given in Ref. 10.7.

10.2.2 Determination of Fourier Coefficients

Let us now become familiar with the means to determine the Fourier coefficients. Assume that a pressure distribution along the circumference can be expressed in Fourier series as

$$p(\theta) = \sum_{n=0}^{N} a_n \cos n\theta \tag{10.24}$$

where a number of N coefficients, a_n's, are to be determined based on, say, a given function or a set of tabulated data of surface pressure.

Multiplying both sides of Eq. (10.24) by cos $m\theta$ and integrating along the circumference gives

$$\int_0^{2\pi} p(\theta) \cos m\theta \, d\theta = \int_0^{2\pi} \sum_{n=0}^{N} a_n \cos n\theta \cos m\theta \, d\theta \tag{10.25}$$

Using the orthogonality property of the trigonometric functions that

$$\begin{aligned} \int_0^{2\pi} \cos m\theta \cos n\theta \, d\theta &= 2\pi \qquad \text{for } m = n = 0 \\ &= \pi \qquad \text{for } m = n \neq 0 \\ &= 0 \qquad \text{for } m \neq n \end{aligned} \tag{10.26}$$

we have

$$\begin{aligned} a_0 &= \frac{1}{2\pi} \int_0^{2\pi} p(\theta) \, d\theta \\ a_n &= \frac{1}{\pi} \int_0^{2\pi} p(\theta) \cos n\theta \, d\theta \qquad \text{with } n = 1, 2, 3, \ldots, N \end{aligned} \tag{10.27}$$

A means to perform the integration above is by numerical method. The simplest methods are the trapezoidal, Simpson's $\frac{1}{3}$ and $\frac{3}{8}$ rules, and similar methods. Details of the numerical integration methods are given in Chapter 11. The total number of coefficients N can be determined from a practical point of view, that is, when the plotted curve for $p(\theta)$ appears close to the real distribution of $p(\theta)$.

10.3 COMPUTER PROGRAMS AND SAMPLE ANALYSIS

Computer programs using the triangular ring element described above are available in the common general-purpose finite element programs. The reader is referred to a special canned program developed by Dunham and Nickell [10.7], available from the U.S. National Technical Information Service. The program features a ring element with quadrilateral cross section, which is composed of four triangles sharing a common interior node with averaged coordinate values of the four corners. The equations on the interior node are eliminated by the standard reduction method described in Sec. 4.6.2.

Example 10.1 *Infinitely long cylinder under external pressure*

An infinitely long cylinder with an inner radius a and outer radius $2a$ under external pressure p_0 (psi) is analyzed. This is a plane strain problem, so only a slice as shown in Fig. 10.4 need be analyzed, and twenty triangular ring elements with gradually decreasing sizes toward the loading zone are used. The model is restrained from deformation in the axial z direction.

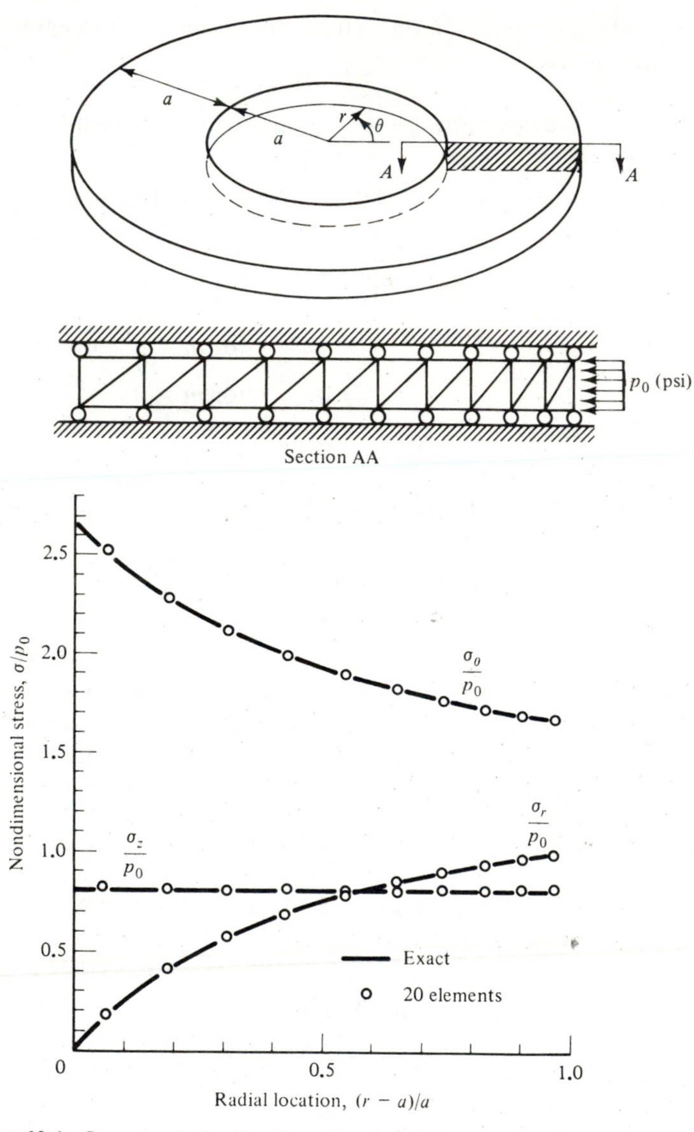

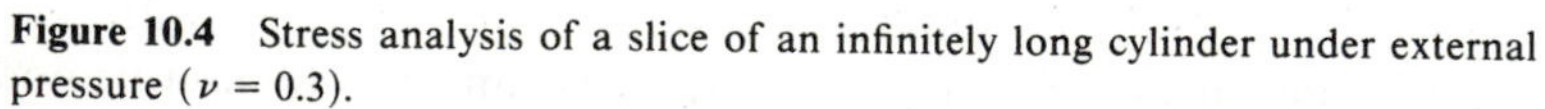

Figure 10.4 Stress analysis of a slice of an infinitely long cylinder under external pressure ($\nu = 0.3$).

The results for nondimensional radial, circumferential, and axial stresses are plotted in Fig. 10.4 and are compared with the exact solution [9.1]. The accuracy and practicality of this element are evident.

Example 10.2 *Cylindrical pressure vessel*

Figure 10.5 shows the cross section of a hypothetical concrete cylindrical pressure vessel subjected to an internal pressure of 100 psi. It is assumed that $E = 5 \times 10^6$ psi

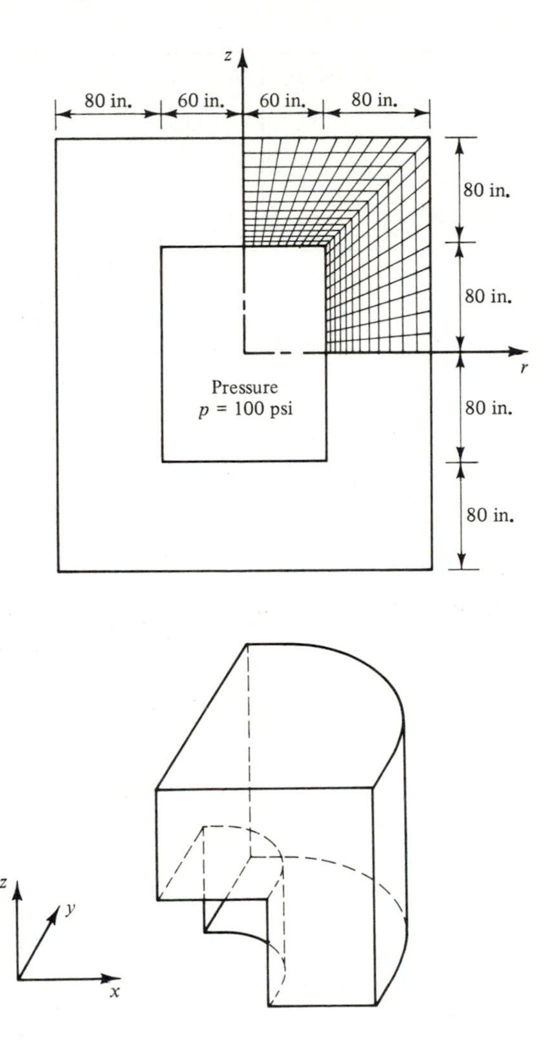

Figure 10.5 Cylindrical pressure vessel, its finite element modeling, and an octant for equilibrium check.

and $\nu = 0.16$. Due to double symmetry, only half of the vessel need be modeled and 240 axisymmetric solid elements with quadrilateral cross sections are used. The bandwidth is 59.

When preparing the input data, it is convenient that we input only the data for nodes and elements at the boundaries and let the computer interpolate the data in between. Because of the regular pattern of the mesh shown in Fig. 10.5, programming for the automatic mesh generation is particularly simple.

The isostress curves obtained for σ_r, σ_θ, σ_z and τ_{rz} are shown in Fig. 10.6. It is of interest to validate the results by equilibrium check of the free body shown in Fig.

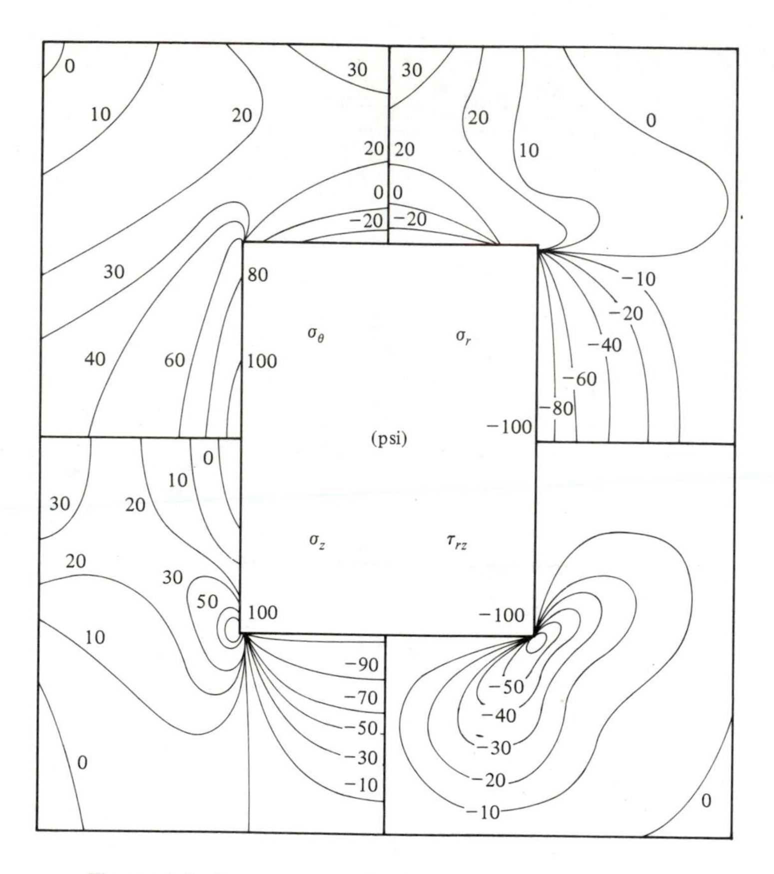

Figure 10.6 Isostress curves for the cylindrical pressure vessel.

10.5 in the x, y, and z directions, respectively,

$$\sum F_x = \sum F_y = \sum_{i=1}^{240} \sigma_{\theta_i} A_i - p(80 \times 60) = 480{,}762 - 480{,}000 = 762 \text{ lb}$$

$$= 0.16\% \text{ of total load in } x \text{ or } y \text{ direction}$$

where A_i is the cross-sectional area of element i.

$$\sum F_z = \sum_{i=1}^{12} \sigma_{z_i} \frac{\pi}{4}(r_2^2 - r_1^2)_i - p\left(60^2 \frac{\pi}{4}\right) = 282{,}192 - 282{,}743 = -551 \text{ lb}$$

$$= 0.20\% \text{ of total load in } z \text{ direction}$$

where $\pi(r_2^2 - r_1^2)$ is the base area at $z = 0$ for element i across the wall thickness.

Example 10.3 *Concentrated load on an infinite solid cylinder* [10.7]

To demonstrate the capability of the axisymmetric solid element in treating a nonaxisymmetric load, an infinitely long solid cylinder under a pair of diametrical, concentrated line loads as shown in Fig. 10.7 is considered. Although this problem was treated earlier by others, an analytical solution is given by Muskhelishvili [10.8].

This is a plane strain problem. We can consider only a slice with unit thickness (a solid disk). Results are plotted in Fig. 10.7 for radial displacements under the load and 90° apart, using twenty axisymmetric solid elements with quadrilateral cross sections and eight Fourier coefficients. The exact solution is also shown. The agreement is very good except as the radius approaches a. From a practical point of view, this solution using only eight Fourier terms is acceptable.

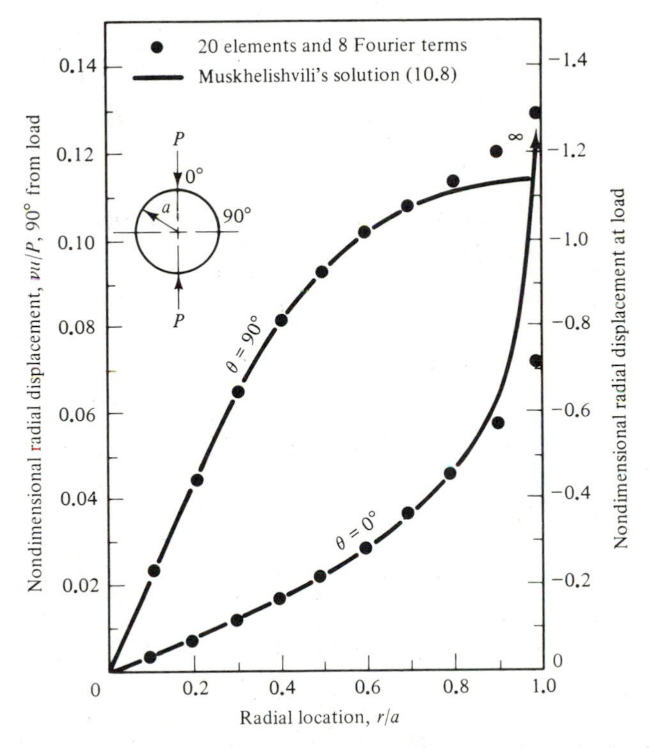

Figure 10.7 Radial displacement at the load and at 90° from the load for a diametrically loaded infinite solid cylinder ($\nu = 0.25$).

10.4 TETRAHEDRAL ELEMENTS

The first formulation of a simple tetrahedral element was by Gallagher et al. [10.9]. Earlier elaborations were by Melosh [10.10], Argyris [10.11], and Clough [10.12], among others. One of the early extensive numerical applications was due to Rashid [10.13].

Similar to the formulation of the two-dimensional triangular element, we will use natural coordinates to formulate the tetrahedral elements.

10.4.1 Natural Coordinate System for Tetrahedron

Figure 10.8 shows a tetrahedron in both rectangular and tetrahedral coordinates. Faces 1, 2, 3, and 4 are identified by the opposite vertices 1, 2, 3, and 4 and have tetrahedral coordinates of $\xi_1 = 0$, $\xi_2 = 0$, $\xi_3 = 0$, and $\xi_4 = 0$, respectively. The tetrahedral coordinates ξ_i ($i = 1, 2, 3,$ and 4) for an interior point are identified as the ratio of the volumes V_i to the total volume V, with V_i defined as the volume of the subtetrahedron bounded by the interior point and face i; thus

$$\xi_1 = \frac{V_1}{V} \qquad \xi_2 = \frac{V_2}{V} \qquad \xi_3 = \frac{V_3}{V} \qquad \xi_4 = \frac{V_4}{V} \tag{10.28}$$

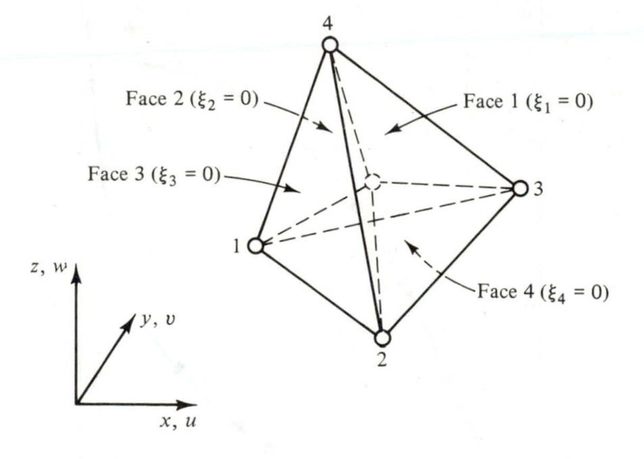

Figure 10.8 Tetrahedral coordinates (ξ_1, ξ_2, ξ_3, and ξ_4).

Because

$$V_1 + V_2 + V_3 + V_4 = V$$

then

$$\xi_1 + \xi_2 + \xi_3 + \xi_4 = 1 \tag{10.29}$$

Combining Eq. (10.29) and the equations relating the rectangular coordinates of the interior point to the tetrahedral coordinates, we have

$$\begin{Bmatrix} 1 \\ x \\ y \\ z \end{Bmatrix} = \begin{bmatrix} 1 & 1 & 1 & 1 \\ x_1 & x_2 & x_3 & x_4 \\ y_1 & y_2 & y_3 & y_4 \\ z_1 & z_2 & z_3 & z_4 \end{bmatrix} \begin{Bmatrix} \xi_1 \\ \xi_2 \\ \xi_3 \\ \xi_3 \end{Bmatrix} \tag{10.30}$$

Using the adjoint method of inverse as given in Eq. (2.39) yields

$$\begin{Bmatrix} \xi_1 \\ \xi_2 \\ \xi_3 \\ \xi_4 \end{Bmatrix} = \frac{1}{6V} \begin{bmatrix} \alpha_1 & \alpha_2 & \alpha_3 & \alpha_4 \\ \beta_1 & \beta_2 & \beta_3 & \beta_4 \\ \gamma_1 & \gamma_2 & \gamma_3 & \gamma_4 \\ \delta_1 & \delta_2 & \delta_3 & \delta_4 \end{bmatrix}^T \begin{Bmatrix} 1 \\ x \\ y \\ z \end{Bmatrix} \tag{10.31}$$

where $6V$ is the determinant of the 4×4 matrix and α_i, β_i, γ_i, and δ_i are the cofactors of the ith coefficients in row 1, 2, 3, and 4, respectively. The definition of cofactor is given in Sec. 2.12.1.

Equation (10.31) may now be written in a general form

$$\xi_i = \frac{1}{6V}(\alpha_i + \beta_i x + \gamma_i y + \delta_i z) \qquad (i = 1, 2, 3, 4) \tag{10.32}$$

According to Laplace expansion formula (2.37) for the determinant,

$$\alpha_1 + \alpha_2 + \alpha_3 + \alpha_4 = 6V \tag{10.33}$$

The volume integration of polynomials in tetrahedral coordinates may be performed following the procedure outlined in the derivation of the area integration formula (9.103) [9.16]:

$$\int_{\text{Vol}} \xi_1^a \xi_2^b \xi_3^c \xi_4^d \, dV = \frac{a!\,b!\,c!\,d!}{(a + b + c + d + 3)!} 6V \tag{10.34}$$

10.4.2 Four-Corner-Node 12-D.O.F. Constant Strain Tetrahedron [10.9]

The simplest tetrahedral element is the one with four corner nodes, each with three d.o.f.'s (u, v, and w), as shown in Fig. 10.9. These displacement components may be assumed to be linear functions of the tetrahedral coordinates:

$$\begin{aligned} u &= \xi_1 u_1 + \xi_2 u_2 + \xi_3 u_3 + \xi_4 u_4 = \sum_{i=1}^{4} N_i u_i \\ v &= \xi_1 v_1 + \xi_2 v_2 + \xi_3 v_3 + \xi_4 v_4 = \sum_{i=1}^{4} N_i v_i \\ w &= \xi_1 w_1 + \xi_2 w_2 + \xi_3 w_3 + \xi_4 w_4 = \sum_{i=1}^{4} N_i w_i \end{aligned} \tag{10.35}$$

where u_i, v_i, and w_i are the three displacement components at the vertex i with $i = 1, 2, 3$, and 4, respectively, and N_i is the shape function. Based on Eq. (10.32) for tetrahedral coordinates, the standard strain-displacement equations can be obtained in the form

$$\begin{Bmatrix} \epsilon_x \\ \epsilon_y \\ \epsilon_z \\ \gamma_{xy} \\ \gamma_{yz} \\ \gamma_{zx} \end{Bmatrix} = \begin{Bmatrix} \dfrac{\partial u}{\partial x} \\ \dfrac{\partial v}{\partial y} \\ \dfrac{\partial w}{\partial z} \\ \dfrac{\partial u}{\partial y} + \dfrac{\partial v}{\partial x} \\ \dfrac{\partial v}{\partial z} + \dfrac{\partial w}{\partial y} \\ \dfrac{\partial w}{\partial x} + \dfrac{\partial u}{\partial z} \end{Bmatrix} = \sum_{i=1}^{4} \begin{Bmatrix} \dfrac{\partial u}{\partial \xi_i}\dfrac{\partial \xi_i}{\partial x} \\ \dfrac{\partial v}{\partial \xi_i}\dfrac{\partial \xi_i}{\partial y} \\ \dfrac{\partial w}{\partial \xi_i}\dfrac{\partial \xi_i}{\partial z} \\ \dfrac{\partial u}{\partial \xi_i}\dfrac{\partial \xi_i}{\partial y} + \dfrac{\partial v}{\partial \xi_i}\dfrac{\partial \xi_i}{\partial x} \\ \dfrac{\partial v}{\partial \xi_i}\dfrac{\partial \xi_i}{\partial z} + \dfrac{\partial w}{\partial \xi_i}\dfrac{\partial \xi_i}{\partial y} \\ \dfrac{\partial w}{\partial \xi_i}\dfrac{\partial \xi_i}{\partial x} + \dfrac{\partial u}{\partial \xi_i}\dfrac{\partial \xi_i}{\partial z} \end{Bmatrix} = \frac{1}{6V}\sum_{i=1}^{4} \begin{Bmatrix} \beta_i \dfrac{\partial u}{\partial \xi_i} \\ \gamma_i \dfrac{\partial v}{\partial \xi_i} \\ \delta_i \dfrac{\partial w}{\partial \xi_i} \\ \gamma_i \dfrac{\partial u}{\partial \xi_i} + \beta_i \dfrac{\partial v}{\partial \xi_i} \\ \delta_i \dfrac{\partial v}{\partial \xi_i} + \gamma_i \dfrac{\partial w}{\partial \xi_i} \\ \beta_i \dfrac{\partial w}{\partial \xi_i} + \delta_i \dfrac{\partial u}{\partial \xi_i} \end{Bmatrix} \tag{10.36}$$

Substituting Eqs. (10.35) for displacement functions into Eqs. (10.36), we have

$$\underset{6\times 1}{\begin{Bmatrix} \epsilon_x \\ \epsilon_y \\ \epsilon_z \\ \gamma_{xy} \\ \gamma_{yz} \\ \gamma_{zx} \end{Bmatrix}} = \underset{6\times 3n}{\begin{bmatrix} N_x & 0 & 0 \\ 0 & N_y & 0 \\ 0 & 0 & N_z \\ N_y & N_x & 0 \\ 0 & N_z & N_y \\ N_z & 0 & N_x \end{bmatrix}} \underset{3n\times 1}{\begin{Bmatrix} u \\ v \\ w \end{Bmatrix}} = [\mathbf{A}]\{\mathbf{q}\} \tag{10.37}$$

where

$$\underset{1\times 4}{\lfloor \mathbf{N}_x \rfloor} = \frac{1}{6V} \lfloor \beta_1 \quad \beta_2 \quad \beta_3 \quad \beta_4 \rfloor \begin{bmatrix} \dfrac{\partial N_1}{\partial \xi_1} & \dfrac{\partial N_2}{\partial \xi_1} & \dfrac{\partial N_3}{\partial \xi_1} & \dfrac{\partial N_4}{\partial \xi_1} \\ \dfrac{\partial N_1}{\partial \xi_2} & \dfrac{\partial N_2}{\partial \xi_2} & \dfrac{\partial N_3}{\partial \xi_2} & \dfrac{\partial N_4}{\partial \xi_2} \\ \dfrac{\partial N_1}{\partial \xi_3} & \dfrac{\partial N_2}{\partial \xi_3} & \dfrac{\partial N_3}{\partial \xi_3} & \dfrac{\partial N_4}{\partial \xi_3} \\ \dfrac{\partial N_1}{\partial \xi_4} & \dfrac{\partial N_2}{\partial \xi_4} & \dfrac{\partial N_3}{\partial \xi_4} & \dfrac{\partial N_4}{\partial \xi_4} \end{bmatrix} \tag{10.37a}$$

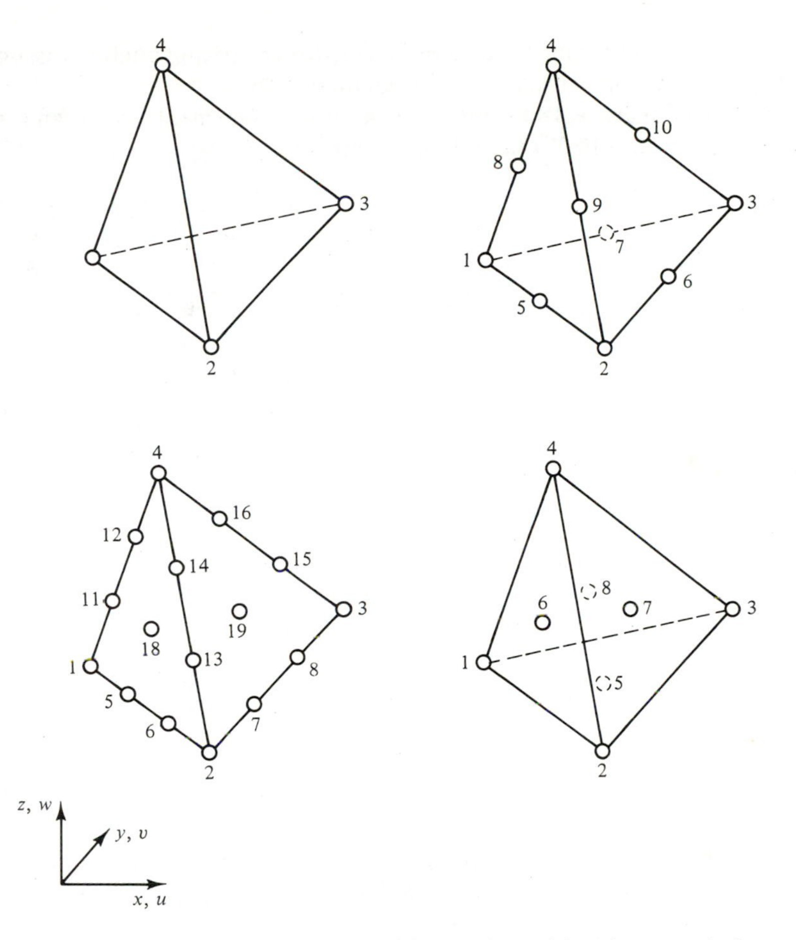

Figure 10.9 Four tetrahedral elements with 4, 10, 20, and 8 nodes, respectively.

Now we can write the three submatrices in Eq. (10.37) for the derivatives of the shape functions with respect to x, y, and z, in the following general form:

$$\underset{1\times 4}{\lfloor \mathbf{N}_x \rfloor} = \frac{1}{6V}\lfloor \beta_1 \quad \beta_2 \quad \beta_3 \quad \beta_4 \rfloor \underset{4\times n}{[\mathbf{N}_\xi]}$$

$$\underset{1\times 4}{\lfloor \mathbf{N}_y \rfloor} = \frac{1}{6V}\lfloor \gamma_1 \quad \gamma_2 \quad \gamma_3 \quad \gamma_4 \rfloor \underset{4\times n}{[\mathbf{N}_\xi]} \tag{10.38}$$

$$\underset{1\times 4}{\lfloor \mathbf{N}_z \rfloor} = \frac{1}{6V}\lfloor \delta_1 \quad \delta_2 \quad \delta_3 \quad \delta_4 \rfloor \underset{4\times n}{[\mathbf{N}_\xi]}$$

where n is the total number of degrees of freedom or shape functions associated with u, v, or w, respectively. For this element, $n = 4$.

For the linear displacement or constant strain element, $[\mathbf{N}_\xi]$ is an identity matrix. Equations (10.37) take the following form:

$$\begin{Bmatrix} \epsilon_x \\ \epsilon_y \\ \epsilon_z \\ \gamma_{xy} \\ \gamma_{yz} \\ \gamma_{zx} \end{Bmatrix} = \frac{1}{6V} \begin{bmatrix} \beta_1 & \beta_2 & \beta_3 & \beta_4 & 0 & 0 & 0 & 0 & 0 & 0 & 0 & 0 \\ 0 & 0 & 0 & 0 & \gamma_1 & \gamma_2 & \gamma_3 & \gamma_4 & 0 & 0 & 0 & 0 \\ 0 & 0 & 0 & 0 & 0 & 0 & 0 & 0 & \delta_1 & \delta_2 & \delta_3 & \delta_4 \\ \gamma_1 & \gamma_2 & \gamma_3 & \gamma_4 & \beta_1 & \beta_2 & \beta_3 & \beta_4 & 0 & 0 & 0 & 0 \\ 0 & 0 & 0 & 0 & \delta_1 & \delta_2 & \delta_3 & \delta_4 & \gamma_1 & \gamma_2 & \gamma_3 & \gamma_4 \\ \delta_1 & \delta_2 & \delta_3 & \delta_4 & 0 & 0 & 0 & 0 & \beta_1 & \beta_2 & \beta_3 & \beta_4 \end{bmatrix} \begin{Bmatrix} u_1 \\ u_2 \\ u_3 \\ u_4 \\ v_1 \\ v_2 \\ v_3 \\ v_4 \\ w_1 \\ w_2 \\ w_3 \\ w_4 \end{Bmatrix} \tag{10.39}$$

or symbolically,

$$\underset{6 \times 1}{\{\boldsymbol{\epsilon}\}} = \underset{6 \times 12}{[\mathbf{A}]} \; \underset{12 \times 1}{\{\mathbf{q}\}} \tag{10.39a}$$

It is seen that the six strain components are all constant within an element.

The three-dimensional stress–strain relations for an isotropic material have been derived in Eq. (9.11) as

$$\begin{Bmatrix} \sigma_x \\ \sigma_y \\ \sigma_z \\ \tau_{xy} \\ \tau_{yz} \\ \tau_{zx} \end{Bmatrix} = \frac{E}{(1+\nu)(1-2\nu)} \begin{bmatrix} 1-\nu & \nu & \nu & 0 & 0 & 0 \\ \nu & 1-\nu & \nu & 0 & 0 & 0 \\ \nu & \nu & 1-\nu & 0 & 0 & 0 \\ 0 & 0 & 0 & \dfrac{1-2\nu}{2} & 0 & 0 \\ 0 & 0 & 0 & 0 & \dfrac{1-2\nu}{2} & 0 \\ 0 & 0 & 0 & 0 & 0 & \dfrac{1-2\nu}{2} \end{bmatrix} \begin{Bmatrix} \epsilon_x \\ \epsilon_y \\ \epsilon_z \\ \gamma_{xy} \\ \gamma_{yz} \\ \gamma_{zx} \end{Bmatrix} \tag{10.40}$$

or symbolically,

$$\{\boldsymbol{\sigma}\} = [\mathbf{C}]\{\boldsymbol{\epsilon}\} \tag{10.40a}$$

Finally, the stiffness equations are obtained using the three-dimensional version of Eq. (9.33):

$$[\mathbf{k}] = \iiint_{\text{volume}} [\mathbf{A}]^T[\mathbf{C}][\mathbf{A}]\, dV \tag{10.41}$$

Equation (10.34) is readily available for performing this volume integration. Because $[\mathbf{A}]$ is a constant matrix for this constant strain element,

$$[\mathbf{k}] = [\mathbf{A}]^T[\mathbf{C}][\mathbf{A}]V \tag{10.42}$$

10.4.3 Ten-Node 30-D.O.F. Linear Strain Tetrahedron [10.12]

If we add one node at the midpoint of each of the six sides of a tetrahedron and assume three degrees of freedom (u, v, w) at each node, we have a ten-node 30-d.o.f. tetrahedral element, as shown in Fig. 10.9.

The displacement function for u, v, or w may be assumed as a complete 10-term quadratic polynomial in tetrahedral coordinates ξ_1, ξ_2, and ξ_3, with $\xi_4 = 1 - \xi_1 - \xi_2 - \xi_3$.

$$\begin{aligned} u(\xi_1, \xi_2, \xi_3) = {} & \alpha_1 + \alpha_2\xi_1 + \alpha_3\xi_2 + \alpha_4\xi_3 + \alpha_5\xi_1^2 \\ & + \alpha_6\xi_2^2 + \alpha_7\xi_3^2 + \alpha_8\xi_1\xi_2 + \alpha_9\xi_2\xi_3 + \alpha_{10}\xi_3\xi_1 \end{aligned} \tag{10.43}$$

For u, v, and w functions, there are a total of thirty constants which can be determined by evaluating the thirty displacement values at the ten nodes. The tetrahedral coordinates for the ten nodes are given in Table 10.1.

TABLE 10.1 Coordinate Values for the Ten Nodes of a Tetrahedron

	Node									
	1	2	3	4	5	6	7	8	9	10
ξ_1	1	0	0	0	$\frac{1}{2}$	0	$\frac{1}{2}$	$\frac{1}{2}$	0	0
ξ_2	0	1	0	0	$\frac{1}{2}$	$\frac{1}{2}$	0	0	$\frac{1}{2}$	0
ξ_3	0	0	1	0	0	$\frac{1}{2}$	$\frac{1}{2}$	0	0	$\frac{1}{2}$
ξ_4	0	0	0	1	0	0	0	$\frac{1}{2}$	$\frac{1}{2}$	$\frac{1}{2}$

Thus we can write for u, v, or w an alternative form containing shape functions,

$$\begin{aligned} u(\xi_1, \xi_2, \xi_3, \xi_4) &= \xi_1(2\xi_1 - 1)u_1 + \xi_2(2\xi_2 - 1)u_2 + \xi_3(2\xi_3 - 1)u_3 \\ &\quad + \xi_4(2\xi_4 - 1)u_4 + 4\xi_1\xi_2 u_5 + 4\xi_2\xi_3 u_6 + 4\xi_3\xi_1 u_7 \\ &\quad + 4\xi_1\xi_4 u_8 + 4\xi_2\xi_4 u_9 + 4\xi_3\xi_4 u_{10} \\ &= \sum_{i=1}^{10} N_i(\xi_1, \xi_2, \xi_3, \xi_4)u_i \end{aligned} \tag{10.44}$$

We readily see that the u-displacement functions given in Eqs. (10.43) and (10.44) are similar in form to those given in Eqs. (9.104) and (9.105), respectively.

The strain–displacement equations can be obtained in the same form as that given in Eqs. (10.37) and (10.38) with $n = 10$:

$$\underset{4\times 10}{[\mathbf{N}_\xi]} = \begin{bmatrix} \frac{\partial N_1}{\partial \xi_1} & \frac{\partial N_2}{\partial \xi_1} & \cdots & \frac{\partial N_{10}}{\partial \xi_1} \\ \frac{\partial N_1}{\partial \xi_2} & \frac{\partial N_2}{\partial \xi_2} & \cdots & \frac{\partial N_{10}}{\partial \xi_2} \\ \frac{\partial N_1}{\partial \xi_3} & \frac{\partial N_2}{\partial \xi_3} & \cdots & \frac{\partial N_{10}}{\partial \xi_3} \\ \frac{\partial N_1}{\partial \xi_4} & \frac{\partial N_2}{\partial \xi_4} & \cdots & \frac{\partial N_{10}}{\partial \xi_4} \end{bmatrix} \tag{10.45a}$$

where the coefficient at the ith row and jth column is

$$n_{\xi_{ij}} = \frac{\partial N_j}{\partial \xi_i} \tag{10.45b}$$

$$[\mathbf{N}_\xi] = \begin{bmatrix} 4\xi_1 - 1 & 0 & 0 & 0 & 4\xi_2 & 0 & 4\xi_3 & 4\xi_4 & 0 & 0 \\ 0 & 4\xi_2 - 1 & 0 & 0 & 4\xi_1 & 4\xi_3 & 0 & 0 & 4\xi_4 & 0 \\ 0 & 0 & 4\xi_3 - 1 & 0 & 0 & 4\xi_2 & 4\xi_1 & 0 & 0 & 4\xi_4 \\ 0 & 0 & 0 & 4\xi_4 - 1 & 0 & 0 & 0 & 4\xi_1 & 4\xi_2 & 4\xi_3 \end{bmatrix} \tag{10.45c}$$

Substituting Eqs. (10.45c) into Eq. (10.38) and then in Eqs. (10.37), strain–displacement matrix [**A**] is obtained. The stiffness matrix [**k**] is readily obtained using Eqs. (10.41) and (10.34).

10.4.4 Twenty-Node 60-D.O.F. Quadratic Strain Tetrahedron [10.14]

If we add two nodes at the one- and two-third points of each of the six sides of a tetrahedron and also add one node at the centroid of each of the four triangular faces, we have a 20-node tetrahedron as shown in Fig. 10.9. The element can be assumed as having sixty d.o.f.'s, three at each node.

The displacement function may be assumed as a complete twenty-term cubic polynomial in tetrahedral coordinates,

$$
\begin{aligned}
u(\xi_1, \xi_2, \xi_3) = \alpha_1 &+ \alpha_2 \xi_1 + \alpha_3 \xi_2 + \alpha_4 \xi_3 + \alpha_5 \xi_1^2 + \alpha_6 \xi_2^2 + \alpha_7 \xi_3^2 + \alpha_8 \xi_1 \xi_2 \\
&+ \alpha_9 \xi_2 \xi_3 + \alpha_{10} \xi_3 \xi_1 + \alpha_{11} \xi_1^3 + \alpha_{12} \xi_2^3 + \alpha_{13} \xi_3^3 \\
&+ \alpha_{14} \xi_1^2 \xi_2 + \alpha_{15} \xi_1^2 \xi_3 + \alpha_{16} \xi_2^2 \xi_1 \\
&+ \alpha_{17} \xi_2^2 \xi_3 + \alpha_{18} \xi_3^2 \xi_1 + \alpha_{19} \xi_3^2 \xi_2 + \alpha_{20} \xi_1 \xi_2 \xi_3 \qquad (10.46)
\end{aligned}
$$

The twenty constants can be determined by evaluating the u values at the twenty nodes. Thus we can write for u, v, or w an alternative form containing shape functions,

$$
\begin{aligned}
u(\xi_1, \xi_2, \xi_3, \xi_4) &= \sum_{i=1}^{20} N_i(\xi_1, \xi_2, \xi_3, \xi_4) u_i \\
&= \tfrac{1}{2}\xi_1(3\xi_1 - 1)(3\xi_1 - 2)u_1 + \tfrac{1}{2}\xi_2(3\xi_2 - 1)(3\xi_2 - 2)u_2 \\
&\quad + \tfrac{1}{2}\xi_3(3\xi_3 - 1)(3\xi_3 - 2)u_3 + \tfrac{1}{2}\xi_4(3\xi_4 - 1)(3\xi_4 - 2)u_4 \\
&\quad + \tfrac{9}{2}\xi_1 \xi_2(3\xi_1 - 1)u_5 + \tfrac{9}{2}\xi_1 \xi_2(3\xi_2 - 1)u_6 \\
&\quad + \tfrac{9}{2}\xi_2 \xi_3(3\xi_2 - 1)u_7 + \tfrac{9}{2}\xi_2 \xi_3(3\xi_3 - 1)u_8 \\
&\quad + \tfrac{9}{2}\xi_3 \xi_1(3\xi_3 - 1)u_9 + \tfrac{9}{2}\xi_3 \xi_1(3\xi_1 - 1)u_{10} \\
&\quad + \tfrac{9}{2}\xi_1 \xi_4(3\xi_1 - 1)u_{11} + \tfrac{9}{2}\xi_1 \xi_4(3\xi_4 - 1)u_{12} \\
&\quad + \tfrac{9}{2}\xi_2 \xi_4(3\xi_2 - 1)u_{13} + \tfrac{9}{2}\xi_2 \xi_4(3\xi_4 - 1)u_{14} \\
&\quad + \tfrac{9}{2}\xi_3 \xi_4(3\xi_3 - 1)u_{15} + \tfrac{9}{2}\xi_3 \xi_4(3\xi_4 - 1)u_{16} \\
&\quad + 27\xi_1 \xi_2 \xi_3 u_{17} + 27\xi_1 \xi_2 \xi_4 u_{18} \\
&\quad + 27\xi_2 \xi_3 \xi_4 u_{19} + 27\xi_1 \xi_3 \xi_4 u_{20} \qquad (10.47)
\end{aligned}
$$

Equations (10.46) and (10.47) are the three-dimensional extension of the two-dimensional equations (9.110) and (9.111) for a ten-node triangle.

Substituting Eq. (10.47) into Eq. (10.45b), $[\mathbf{N}_\xi]$ is obtained. Thus using Eqs. (10.38) and (10.37), $[\mathbf{A}]$ is obtained. Finally, using Eqs. (10.41) and (10.34), $[\mathbf{k}]$ is obtained.

10.4.5 Sixteen-Node 48-D.O.F. Tetrahedron [10.14, 10.15]

If the nodal points at the centroid of the faces are omitted, we have a 16-node 48-d.o.f. element. The displacement functions may be assumed the same as those given in Eq. (10.47) minus the last four terms.

10.4.6 Eight-Node 60-D.O.F. Tetrahedron [10.14]

If a node is placed at the centroid of each of the four faces in addition to the four vertices, we have an eight-node tetrahedron as shown in Fig. 10.9. If each of the three displacement functions u, v, and w is assumed as a complete third-order polynomial in ξ_1, ξ_2, and ξ_3 as shown in Eq. (10.46), sixty d.o.f.'s are needed to define the 60 constants. Each of the four vertices can then be assumed as having 12 d.o.f.'s: u, $\partial u/\partial x$, $\partial u/\partial y$, $\partial u/\partial z$, and the same for v and w. Each of the four centroids on the four faces can be assumed as having three d.o.f.'s: u, v, and w.

When formulating the stiffness matrix, we do not have to express the displacements in the form consisting of shape functions as in Eq. (10.47) and thus we do not have to express strain–displacement functions in the same form as in Eqs. (10.37) and (10.38). There is an alternative procedure to form the stiffness matrix.

Upon evaluation of the sixty nodal d.o.f. values using the displacement functions (10.46) for u, v, and w, we can obtain $\{\mathbf{q}\} = [\mathbf{B}]\{\boldsymbol{\alpha}\}$ and the 60 constants become $\{\boldsymbol{\alpha}\} = [\mathbf{B}]^{-1}\{\mathbf{q}\}$. The strains can then be written as $\{\boldsymbol{\epsilon}\} = [\mathbf{G}]\{\boldsymbol{\alpha}\} = [\mathbf{G}][\mathbf{B}]^{-1}\{\mathbf{q}\}$. Thus the stiffness matrix can be obtained using Eq. (10.12) based on the integration formula (10.34).

10.4.7 Four-Node 48-D.O.F. Tetrahedron [10.11, 10.15–10.17]

If the four centroidal nodes on the four faces are omitted from the eight-node 60-d.o.f. tetrahedron, we have a four-node 48-d.o.f. tetrahedron, which includes the displacements u, v, w, and their first derivatives with respect to x, y, and z, respectively, as nodal d.o.f.'s.

The displacement function for u, v, or w may be assumed as an incomplete cubic polynomial [10.11]:

$$\begin{aligned} u(\xi_1, \xi_2, \xi_3, \xi_4) = {} & \alpha_1 \xi_1 + \alpha_2 \xi_2 + \alpha_3 \xi_3 + \alpha_4 \xi_4 + \alpha_5 \xi_1 \xi_2 + \alpha_6 \xi_3 \xi_4 + \alpha_7 \xi_4 \xi_1 \\ & + \alpha_8 \xi_3 \xi_2 + \alpha_9 \xi_2 \xi_4 + \alpha_{10} \xi_3 \xi_1 + \alpha_{11}(\xi_1^2 \xi_2 - \xi_1 \xi_2^2) \\ & + \alpha_{12}(\xi_3^2 \xi_4 - \xi_3 \xi_4^2) + \alpha_{13}(\xi_4^2 \xi_1 - \xi_4 \xi_1^2) + \alpha_{14}(\xi_3^2 \xi_2 - \xi_3 \xi_2^2) \\ & + \alpha_{15}(\xi_2^2 \xi_4 - \xi_2 \xi_4^2) + \alpha_{16}(\xi_3^2 \xi_1 - \xi_3 \xi_1^2) \end{aligned} \tag{10.48}$$

The stiffness matrix can be formulated in the same manner as described for the eight-node 60-d.o.f. tetrahedron in Sec. 10.4.6.

Further elaborations on this element, including the methods of formulation, can be found in Refs. 10.15 and 10.16. Due to the inclusion of displacement derivatives as d.o.f.'s, this four-node tetrahedron results in the smallest bandwidth as compared with a similar assemblage of other tetrahedra with more nodes but comparable numbers of element d.o.f.'s (try Problem 10.12).

This is the most advantageous tetrahedron introduced. For comparative studies on the efficiency of various tetrahedral elements, the reader is referred to Refs. 10.10, 10.12, 10.15, and 10.17.

10.4.8 Mesh Generation

It is difficult manually to model a three-dimensional solid with tetrahedra without having mixed-up or missing pieces inside. If, however, tetrahedra are first assembled to form a hexahedron, the modeling of three-dimensional solids becomes more practical. The construction of a solid hexahedron by five tetrahedra using computer graphics is shown in Fig. 10.10. To give the reader a three-dimensional (stereo) view [10.18] of the hexahedron and the five decomposed tetrahedra, a stereo pair is shown in Fig. 10.11. A stereo picture can be seen if the reader uses a pair of stereo glasses or puts a piece of white paper in between two eyes with proper viewing distance.

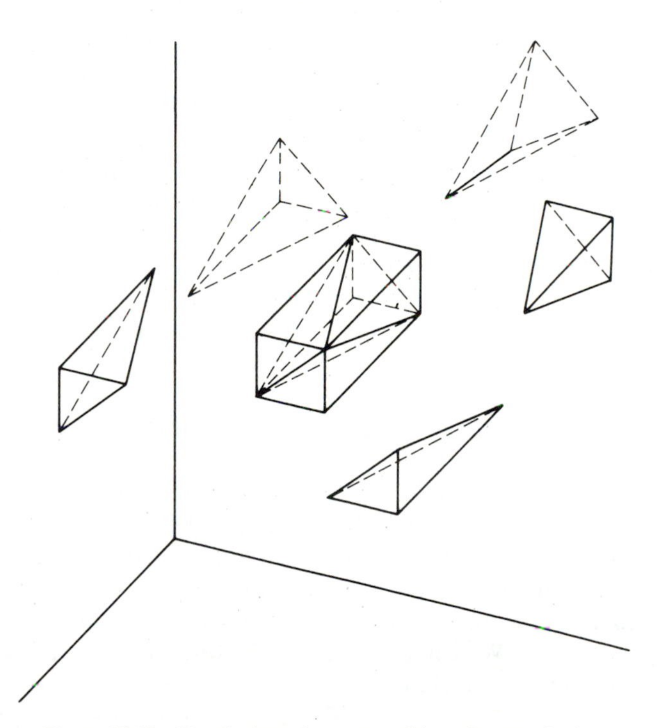

Figure 10.10 Hexahedron decomposed into five tetrahedra.

With the advances in computer graphics, the feasibility of using solid elements has been enhanced. Automatic mesh generation routines have been increasingly developed. For example, a module concept has been used [10.19]. First, a set of modules is generated, each with its own mesh inside. Then the

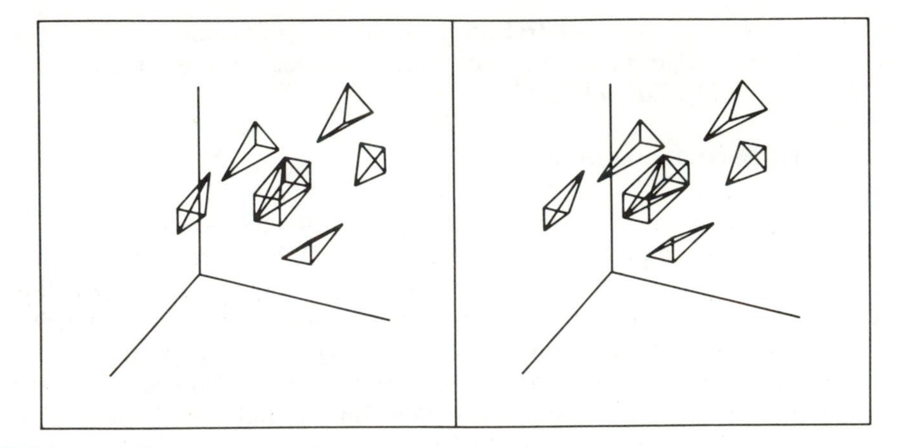

Figure 10.11 Stereo pair of a hexahedron decomposed into five tetrahedra.

modules are placed in contact, and a matrix describing the connectivity of the entire mesh is determined. Finally, the geometry of the mesh is specified and all nodal coordinates are determined by the computer. The user's effort required to specify a mesh representing a certain physical configuration resembles the effort involved in specifying the configuration to the model machinist in the shop. Another similar concept, for example, is based on the construction of network structures [10.20], each of which consists of a number of tetrahedra. The number and the coordinates of the nodes are input manually. The mesh of the constructed network structure can be refined automatically by splitting every tetrahedron into, say, eight smaller ones.

10.5 RECTANGULAR HEXAHEDRONAL ELEMENTS

Because the faces and the sides of the rectangular hexahedronal elements are orthogonal to one another, such elements can be formulated using non-dimensional local coordinates. The most straightforward way to construct a rectangular hexahedral element is to assume only three d.o.f.'s (u, v, w) at each node and to increase the number of nodes from 8 to 20, 32, and 64, respectively, as shown in Fig. 10.12. We can also include the displacement derivative terms as d.o.f.'s to the eight-node element.

10.5.1 Eight-Node 24-D.O.F. Linear Displacement Rectangular Hexahedron [10.10, 10.12]

An eight-node 24-d.o.f. rectangular hexahedronal element with dimensions $2a$, $2b$, and $2c$ is shown in Fig. 10.12. Each node has three d.o.f.'s

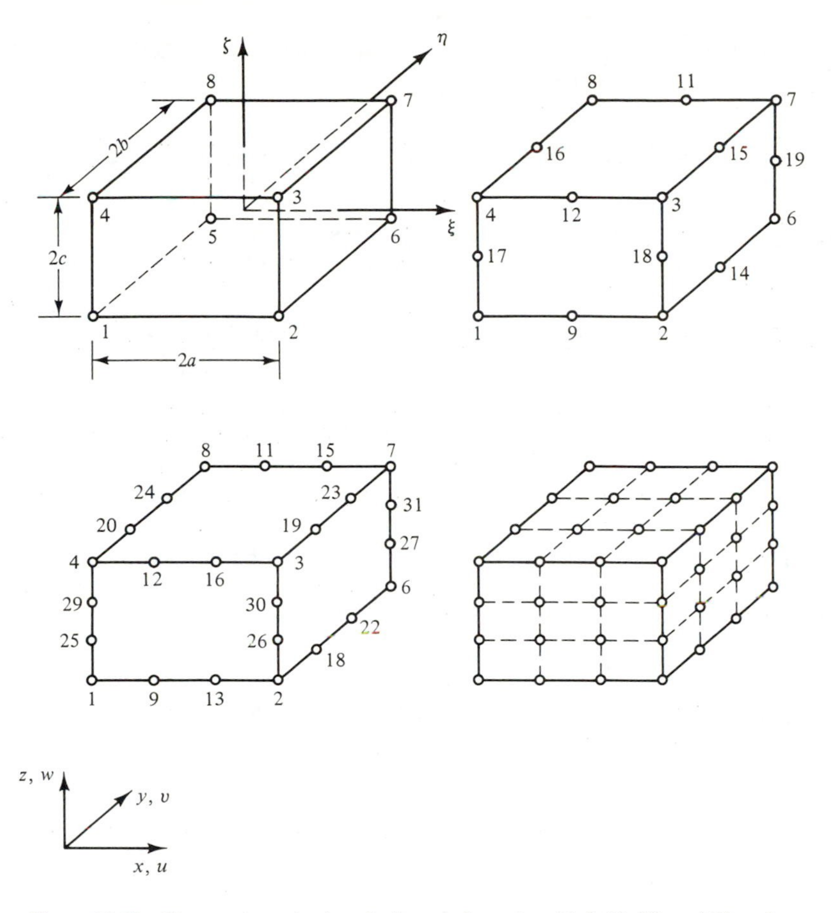

Figure 10.12 Four rectangular hexahedronal elements with 8, 20, 32, and 64 nodes, respectively.

(u, v, w). It is convenient now to use dimensionless local coordinates (ξ, η, ζ) with origin at the centroid, whose global coordinates are x_0, y_0, and z_0. Thus

$$\xi = \frac{x - x_0}{a} \qquad \eta = \frac{y - y_0}{b} \qquad \zeta = \frac{z - z_0}{c} \tag{10.49}$$

where each coordinate axis varies from -1 to 1.

The displacement function for u, v, or w may be assumed to be of the following form, linear along a side:

$$u(\xi, \eta, \zeta) = \alpha_1 + \alpha_2 \xi + \alpha_3 \eta + \alpha_4 \zeta + \alpha_5 \xi\eta + \alpha_6 \eta\zeta + \alpha_7 \zeta\xi + \alpha_8 \xi\eta\zeta \tag{10.50}$$

where the eight constants are obtained by evaluating the u values at the eight nodes.

The displacement function for u, v, or w may also be written in a form consisting of shape functions,

$$u(\xi, \eta, \zeta) = \sum_{i=1}^{8} N_i(\xi, \eta, \zeta)u_i = \sum_{i=1}^{8} \tfrac{1}{8}(1 + \xi_i\xi)(1 + \eta_i\eta)(1 + \zeta_i\zeta)u_i \quad (10.51)$$

where ξ_i, η_i, and ζ_i are the coordinate values of node i, with $i = 1, 2, \ldots, 8$. The stiffness matrix for this element is given explicitly in Ref. [10.10].

10.5.2 Twenty-Node 60-D.O.F. Quadratic Displacement Rectangular Hexahedron [10.12, 10.21–10.23]

If one node is added at the midpoints of the twelve sides of a rectangular hexahedron in addition to the eight corner nodes, and three d.o.f.'s (u, v, w) are assumed at each node, we have a 20-node 60-d.o.f. element, as shown in Fig. 10.12.

The displacement function for u, v, or w may be assumed to be

$$\begin{aligned} u(\xi, \eta, \zeta) = {} & \alpha_1 + \alpha_2\xi + \alpha_3\eta + \alpha_4\zeta + \alpha_5\xi^2 + \alpha_6\eta^2 + \alpha_7\zeta^2 + \alpha_8\xi\eta + \alpha_9\eta\zeta \\ & + \alpha_{10}\xi\zeta + \alpha_{11}\xi^2\eta + \alpha_{12}\xi^2\zeta + \alpha_{13}\eta^2\xi + \alpha_{14}\eta^2\zeta \\ & + \alpha_{15}\zeta^2\xi + \alpha_{16}\zeta^2\eta + \alpha_{17}\xi\eta\zeta + \alpha_{18}\xi^2\eta\zeta \\ & + \alpha_{19}\eta^2\xi\zeta + \alpha_{20}\zeta^2\xi\eta \end{aligned} \quad (10.52)$$

which varies as a quadratic function along any side.

After the twenty constants are obtained by evaluating the twenty nodal d.o.f. values for u, we have

$$\begin{aligned} u(\xi, \eta, \zeta) &= \sum_{i=1}^{20} N_i(\xi, \eta, \zeta)u_i \\ &= \sum_{i=1}^{8} \tfrac{1}{8}(1 + \xi_i\xi)(1 + \eta_i\eta)(1 + \zeta_i\zeta)(\xi_i\xi + \eta_i\eta + \zeta_i\zeta - 2)u_i \\ &\quad + \sum_{i=9}^{12} \tfrac{1}{4}(1 - \xi^2)(1 + \eta_i\eta)(1 + \zeta_i\zeta)u_i \\ &\quad + \sum_{i=13}^{16} \tfrac{1}{4}(1 - \eta^2)(1 + \xi_i\xi)(1 + \zeta_i\zeta)u_i \\ &\quad + \sum_{i=17}^{20} \tfrac{1}{4}(1 - \zeta^2)(1 + \xi_i\xi)(1 + \eta_i\eta)u_i \end{aligned} \quad (10.53)$$

Equation (10.53) is the three-dimensional version of Eq. (9.87).

10.5.3 Thirty-two-Node 96-D.O.F. Cubic Displacement Rectangular Hexahedron [10.12, 10.24]

If two nodes are added, in addition to the eight corner nodes, at the $\frac{1}{3}$ and $\frac{2}{3}$ points on the twelve sides of a rectangular hexahedron, and three d.o.f.'s are assumed at each node, we have a 32-node 96-d.o.f. element, as shown in Fig. 10.12.

The displacement function for u, v, or w may be assumed to be

$$\begin{aligned} u(\xi,\eta,\zeta) = {} & \alpha_1 + \alpha_2\xi + \alpha_3\eta + \alpha_4\zeta + \alpha_5\xi^2 + \alpha_6\eta^2 + \alpha_7\zeta^2 + \alpha_8\xi\eta + \alpha_9\eta\zeta \\ & + \alpha_{10}\xi\zeta + \alpha_{11}\xi^3 + \alpha_{12}\eta^3 + \alpha_{13}\zeta^3 + \alpha_{14}\xi^2\eta + \alpha_{15}\xi^2\zeta + \alpha_{16}\eta^2\xi \\ & + \alpha_{17}\eta^2\zeta + \alpha_{18}\zeta^2\xi + \alpha_{19}\zeta^2\eta + \alpha_{20}\xi\eta\zeta + \alpha_{21}\xi^3\eta + \alpha_{22}\xi^3\zeta \\ & + \alpha_{23}\eta^3\zeta + \alpha_{24}\eta^3\xi + \alpha_{25}\zeta^3\xi + \alpha_{26}\zeta^3\eta + \alpha_{27}\xi^2\eta\zeta + \alpha_{28}\eta^2\xi\zeta \\ & + \alpha_{29}\zeta^2\xi\eta + \alpha_{30}\xi^3\eta\zeta + \alpha_{31}\eta^3\xi\zeta + \alpha_{32}\zeta^3\xi\eta \end{aligned} \tag{10.54}$$

Alternatively, we can write

$$\begin{aligned} u(\xi,\eta,\zeta) &= \sum_{i=1}^{32} N_i(\xi,\eta,\zeta)u_i \\ &= \sum_{i=1}^{8} \tfrac{1}{64}(1+\xi_i\xi)(1+\eta_i\eta)(1+\zeta_i\zeta)[-19+9(\xi^2+\eta^2+\zeta^2)]u_i \\ &\quad + \sum_{i=9}^{16} \tfrac{9}{64}(1-\xi^2)(1+\eta_i\eta)(1+\zeta_i\zeta)(1+9\xi_i\xi)u_i \\ &\quad + \sum_{i=17}^{24} \tfrac{9}{64}(1-\eta^2)(1+\xi_i\xi)(1+\zeta_i\zeta)(1+9\eta_i\eta)u_i \\ &\quad + \sum_{i=25}^{32} \tfrac{9}{64}(1-\zeta^2)(1+\xi_i\xi)(1+\eta_i\eta)(1+9\zeta_i\zeta)u_i \end{aligned} \tag{10.55}$$

Equation (10.55) is the three-dimensional version of Eq. (9.89).

10.5.4 Other Higher-Order Rectangular Hexahedra

As shown in the fourth hexahedron in Fig. 10.12, we can place sixteen nodes on the face at $\xi = 1$. If we move this face to $\xi = \frac{1}{3}$, $-\frac{1}{3}$, and -1, respectively, we can have sixty-four nodes. If three d.o.f.'s (u, v, w) are assumed at each node, we have a 64-node 192-d.o.f. hexahedron. Details of this element were given in Ref. [10.25]. For such an element, the displacement function u, v, or w may be assumed as the product of three complete third-order polynomials in ξ, η, and ζ, respectively:

$$u(\xi, \eta, \zeta) = (a_1 + a_2\xi + a_3\xi^2 + a_4\xi^3)(b_1 + b_2\eta + b_3\eta^2 + b_4\eta^3) \times (c_1 + c_2\zeta + c_3\zeta^2 + c_4\zeta^3) \tag{10.56}$$

which contains 64 constants to be evaluated using the 64 u-related d.o.f.'s.

We can include not only u, v, and w, but also their first derivatives with respect to x, y, and z. We can then have 12 d.o.f.'s at each node. The simplest possible hexahedronal element of this kind thus has eight nodes and ninety-six d.o.f.'s.

For such an element, the displacement functions can be assumed to be in the form of Hermitian polynomials [10.26], which are more commonly used in plate and shell finite elements. They can also be assumed as incomplete fifth-order polynomials [10.27].

10.5.5 Numerical Results for General Solid Elements

Because of the enormous computational effort needed in three-dimensional finite element analysis, numerical results for the comparative studies of the various kinds of three-dimensional elements have been spotty, far from complete compared to those for the two-dimensional elements. Notable numerical studies include Refs. [10.10], [10.12], [10.16], [10.17], [10.28], and [10.29], with Ref. [10.12] containing the most useful information.

To give the reader a rough indication of the performance of the various kinds of solid elements, let us consider two examples of three-dimensional cantilever beams: a short beam under an end moment and a slender beam under an end shear force. The results for the end deflection obtained using different meshes formed by different kinds of elements are given in Table 10.2.

For the short cantilever beam [10.12], four types of solid elements are used. The two types of tetrahedra are used by first forming a hexahedron using five elements, as shown in Fig. 10.10. The results indicate that the directly formed general hexahedra are more efficient than the tetrahedra. For the slender cantilever beam [10.28], highly accurate results are obtained using the eight-node 96-d.o.f. hexahedra, which include displacement derivatives as d.o.f.'s.

When comparing the eight-node 24-d.o.f. hexahedron and the 20-node 60-d.o.f. hexahedron, it is noted that the connectivity of the 20-node hexahedra is much more complex and the resulting bandwidth of the stiffness matrix is much larger than a similar assemblage of eight-node hexahedra. From a practical point of view, it is of interest to compare the efficiency (accuracy achieved per unit of computing time) of the two hexahedra. Let us now consider two examples, a slender and a square cantilever plate, both under an in-plane load at the free end. The results for the tip deflection ratios and the solution times using an IBM 7094 computer for four series of successively refined meshes are given in Table 10.3.

TABLE 10.2 Numerical Results for Three-Dimensional Cantilever Beams Using Various Solid Elements

(Length) × (Depth) × (Width)	Element Description				Assemblage		Error of Deflection at Center of Section at Tip (%)	Remark	Reference
	Load	Shape	Node	D.O.F.	Mesh	D.O.F.			
$21 \times 6 \times 4$ (in.^{3})	End moment		4	12	$7 \times 3 \times 2 \times 5 = 210$	$84 \times 3 = 252$	−25.8	Five tetrahedra used to form a hexahedron	10.12
			8	24	$7 \times 3 \times 2 = 42$	$84 \times 3 = 252$	−10.2		
			10	30	$7 \times 3 \times 2 \times 5 = 210$	$448 \times 3 = 1{,}344$	≈ Exact	Five tetrahedra used to form a hexahedron	
			20	60	$7 \times 3 \times 2 = 42$	$287 \times 3 = 861$	≈ Exact		
$20 \times 2 \times 2$ (in.^{3})	End shear force		8	96	$2 \times 1 = 2$	100	−0.081		10.28
					$4 \times 2 \times 2 = 16$	441	0.011		

TABLE 10.3 Study of Efficiency (Accuracy Achieved per Unit of Computing Time) for Cantilever Plate under Load at Free End

(Length) × (Height) × (Width)	Element Description			Mesh Size $(m \times n)$[a]	Ratio of Tip Deflection[b]		IBM 7094 Solution Time (seconds)	Reference
	Shape	Node	D.O.F.		Δ/Δ_a	Δ/Δ_b		
100 × 10 × 1 (in.^{3})		8	24	10 × 3	0.37		9.8	10.12
				20 × 3	0.71		18.4	
				30 × 3	0.88		31.9	
				40 × 3	0.97		50.2	
		20	60	1 × 1	0.71		7.2	
				3 × 1	0.97		18.0	
				4 × 1	1.0		24.2	
				5 × 1	1.0		28.9	
10 × 10 × 1 (in.^{3})		8	24	5 × 3		0.75	10.0	
				10 × 3		0.94	22.3	
				15 × 3		1.00	52.0	
		20	60	1 × 2		0.74	15.9	
				2 × 2		0.80	30.8	
				3 × 2		0.84	47.2	
				4 × 2		0.86	64.1	
		32	96	1 × 1	$\Delta/\Delta_c = 1.0$		Not given	10.29

[a] m, number of elements along plate length; n, number of elements along plate height.

[b] Δ_a, tip deflection in case (a) obtained using the 5 × 1 mesh of 20-node 60-d.o.f. elements; Δ_b, tip deflection in case (b) obtained using the 15 × 3 mesh of eight-node 24-d.o.f. elements; Δ_c, tip deflection obtained using beam theory, including shear deformation ($\nu = 0$).

For the slender cantilever plate where a state of bending stress is dominating, the 20-node hexahedron is superior to and more efficient than the eight-node hexahedron. For the square cantilever plate where a state of more general stresses is present, however, the eight-node hexahedron is more efficient than the 20-node hexahedron. The square plate problem was also tackled using a 32-node 96-d.o.f. hexahedron [10.29].

In order to be conclusive on the comparisons of the various kinds of solid elements, many computational experiments need to be done. With the growth in computer graphics and vectorized supercomputers, it is expected that much numerical work will be conducted in mesh generation, efficient algorithms, results display, and more conclusive comparisons will be made among the various kinds of three-dimensional elements.

10.6 CONCLUDING REMARKS

We have introduced in detail an elementary axisymmetric solid element with triangular cross section for both axisymmetric and nonaxisymmetric (still symmetric about a diameter) loadings. Although higher-order elements are available [10.3, 10.4], examples on both symmetric and nonsymmetric loading cases indicate that such elementary element is practical.

We have introduced the formulations for a series of tetrahedral elements using tetrahedral coordinates and a series of rectangular hexahedronal elements using nondimensionalized rectangular coordinates. Arbitrary hexahedronal elements can usually be decomposed into a fixed number of tetrahedronal elements, thus easing the difficulty in modeling using tetrahedronal elements.

Because the faces of the rectangular hexahedronal elements are flat and orthogonal to one another, such elements are not practical in geometric modeling, especially when their orders are high. This impracticality can be eased if we use the displacement functions of variable orders to describe a distorted hexahedron with curved faces and formulate the stiffness matrix using numerical integration based on curvilinear coordinates. This procedure will be described in Chapter 11 for isoparametric curved elements.

A practical usage of the hexahedronal elements is in treating the thick as well as thin plates and shells (if the concept of curved isoparametric elements is used). This can be done by reducing the order of numerical integration applied to certain terms to avoid the excessive bending stiffness [10.22, 10.23].

Numerical experimentations on the performance of the various solid elements are far from sufficient. Difficulties lie in the mesh generation, computer storage, computing time, and results interpretation. Advances in computer graphics, vectorized supercomputers, efficient pre- and postprocessing routines, and efficient computational methods will ease such difficulties.

It should be noted that an alternative to the displacement models is the hybrid-stress model, in which compatible displacements and equilibrating intraelement stresses are independently interpolated. Stress parameters are eliminated on the element level, and a stiffness matrix is obtained [10.30]. Some three-dimensional hybrid-stress elements have been developed: eight-node elements [10.31, 10.32], 20-node elements [10.33], special-purpose three-dimensional elements for thick plate analysis [10.34], 20-node quadratic displacement three-dimensional isoparametric elements [10.35], and an eight-node solid element [10.36].

REFERENCES

10.1. Clough, R. W., and Rashid, Y., "Finite Element Analysis of Axi-symmetric Solids," *Journal of the Engineering Mechanics Division*, ASCE, Vol. 91, No. EMl, 1965, pp. 71–85.

10.2. Wilson, E. L., "Structural Analysis of Axisymmetric Solids," *AIAA Journal*, Vol. 3, No. 12, 1965, pp. 2269–2274.

10.3. Chacour, S., "A High Precision Axisymmetric Triangular Element Used in the Analysis of Hydraulic Turbine Components," *Journal of Basic Engineering*, ASME, Vol. 92, 1970, pp. 819–826.

10.4. Silvester, P., and Konrad, A., "Axisymmetric Triangular Finite Elements for the Scalar Helmholtz Equation," *International Journal for Numerical Methods in Engineering*, Vol. 5, 1973, pp. 481–498.

10.5. Zienkiewicz, O. C., *The Finite Element Method in Structural and Continuum Mechanics*, McGraw-Hill Publishing Company, New York, 1967, pp. 53–55.

10.6. Utku, S., "Explicit Expressions for Triangular Torus Element Stiffness Matrix," *AIAA Journal*, 1968, pp. 1174–1176.

10.7. Dunham, R. S., and Nickell, R. E., "Finite Element Analysis of Axisymmetric Solids with Arbitrary Loadings," Report AD 655 253, National Technical Information Service, Springfield, Va., June 1967.

10.8. Muskhelishvili, N. I., *Some Basic Problems in the Mathematical Theory of Elasticity*, Wolters-Noordhoff BV, Groningen, The Netherlands, 1953, pp. 324–328.

10.9. Gallagher, R. H., Padlog, J., and Bijlaard, P. P., "Stress Analysis of Heated Complex Shapes," *Journal of the American Rocket Society*, Vol. 32, No. 5, 1962, pp. 700–707.

10.10. Melosh, R. J., "Structural Analysis of Solids," *Journal of the Structural Division*, ASCE, Vol. 89, No. ST4, 1963, pp. 205–223.

10.11. Argyris, J. H., "Matrix Analysis of Three-Dimensional Elastic Media Small and Large Displacements," *AIAA Journal*, Vol. 3, No. 1, 1965, pp. 45–51.

10.12. Clough, R. W., "Comparison of Three-Dimensional Finite Elements," *Proceedings*, Symposium on Application of Finite Element Methods in Civil Engineering, Vanderbilt University, Nashville, Tenn., 1969 (published by ASCE), pp. 1–26.

10.13. Rashid, Y. R., "Three-Dimensional Analysis of Elastic Solids: I. Analysis Procedure; II. The Computational Problems," *International Journal of Solids and Structures*, Vol. 5, 1969, pp. 1311–1331, and Vol. 6, 1970, pp. 195–207.

10.14. Argyris, J. H., Fried, I., and Scharpf, D. W., "The TET 20 and TEA 8 Elements for the Matrix Displacement Method," *The Aeronautical Journal of the Royal Aeronautical Society*, Vol. 72, No. 691, 1968, pp. 618–623.

10.15. Rashid, Y. R., Smith, P. D., and Prince, N., "On Further Application of Finite Element Method to Three-Dimensional Elastic Analysis," *Proceedings*, Symposium on High Speed Computing of Elastic Structures, University of Liège, Belgium, Vol. 2, 1970, pp. 433–454.

10.16. Hughes, J. R., and Allik, H., "Finite Elements for Compressible and Incompressible Continua," *Proceedings*, Symposium on Application of Finite Element Methods in Civil Engineering, Vanderbilt University, Nashville, Tenn., 1969 (published by ASCE), pp. 27–62.

10.17. Fjeld, S. A., "Three Dimensional Theory of Elasticity," in *Finite Element Methods in Stress Analysis*, Tapir Press, Trondheim, Norway, 1969, pp. 333–364.

10.18. Pammer, Z., and Szabo, L., "Stereo Decomposition Subroutines for Three-Dimensional Plotter Programs," *International Journal for Numerical Methods in Engineering*, Vol. 17, 1981, pp. 1571–1575.

10.19. Pissanetzky, S., "Kubik: An Automatic Three-Dimensional Finite Element Mesh Generator," *International Journal for Numerical Methods in Engineering*, Vol. 17, 1981, pp. 255–269.

10.20. Nguyen, V. P., "Automatic Mesh Generation with Tetrahedron Elements," *International Journal for Numerical Methods in Engineering*, Vol. 18, 1982, pp. 273–289.

10.21. Rigby, G. L., and McNeice, G. M., "A Strain Energy Basis for Studies of Element Stiffness Matrices," *AIAA Journal*, Vol. 10, No. 11, 1972, pp. 1490–1493.

10.22. Zienkiewicz, O. C., Taylor, R. L. and Too, J. M., "Reduced Integration Technique in General Analysis of Plates and Shells," *International Journal for Numerical Methods in Engineering*, Vol. 3, 1971, pp. 275–290.

10.23. Pawsey, S. F., and Clough, R. W., "Improved Numerical Integration of Thick Shell Finite Elements," *International Journal for Numerical Methods in Engineering*, Vol. 3, 1971, pp. 575–586.

10.24. Ergatoudis, J., Irons, B. M., and Zienkiewicz, O. C., "Three Dimensional Analysis of Arch Dams and Their Foundations," Symposium on Arch Dams at the Institution of Civil Engineers, London, 1968.

10.25. Argyris, J. H., and Fried, I., "The LUMINA Element for the Matrix Displacement Method," *The Aeronautical Journal of the Royal Aeronautical Society*, Vol. 72, No. 690, 1968, pp. 514–517.

10.26. Argyris, J. H., Fried, I., and Scharpf, D. W., "The Hermes 8 Element for the Matrix Displacement Method," *The Aeronautical Journal of the Royal Aeronautical Society*, Vol. 72, No. 691, pp. 613–617.

10.27. Zienkiewicz, O. C., Irons, B. M., Scott, F. C., and Campbell, J. S., "Three Dimensional Stress Analysis," *Proceedings*, Symposium on High Speed Computing of Elastic Structures, University of Liège, Belgium, Vol. 1, 1970, pp. 413–432.

10.28. Chacour, S., "'DANUTA', A Three-Dimensional Finite Element Program Used in the Analysis of Turbomachinery," *Journal of Basic Engineering*, ASME, Vol. 94, 1972, pp. 71–77.

10.29. Ferguson, G. H., and Clark, R. D., "A Variable Thickness, Curved Beam and Shell Stiffening Element with Shear Deformations," *International Journal for Numerical Methods in Engineering*, Vol. 14, 1979, pp. 581–592.

10.30. Pian, T. H. H., "Hybrid Models," in *Numerical Methods and Computer Methods in Applied Mechanics*, ed. S. J. Fenves, N. Perrone, A. R. Robinson, and W. C. Schnobrich, Academic Press, Inc., New York, 1973.

10.31. Irons, B. M., "An Assumed Stress Version of the Wilson 8-Node Element," University of Wales, Computer Report CNME/CR/56, 1972.

10.32. Lee, S. W., "An Assumed Stress Hybrid Finite Element for Three Dimensional Elastic Structural Analysis," MIT, ASRL-TR-170-3; also AFOSR-TR-75-0087, 1974.

10.33. Ahmad, S., and Irons, B. M., "An Assumed Stress Approach to Refined Isoparametric Finite Elements in Three Dimensions," in *Finite Element Methods in Engineering*, University of New South Wales, 1974, pp. 85–100.

10.34. Spilker, R. L., "High-Order Three-Dimensional Hybrid-Stress Elements for Thick Plate Analysis," *International Journal for Numerical Methods in Engineering*, Vol. 17, 1981, pp. 53–69.

10.35. Spilker, R. L., "Three-Dimensional Hybrid-Stress Isoparametric Quadratic Displacement Elements," *International Journal for Numerical Methods in Engineering*, Vol. 18, 1982, pp. 445–465.

10.36. Bretl, J. L., and Cook, R. D., "A New Eight-Node Solid Element," *International Journal for Numerical Methods in Engineering*, Vol. 14, 1979, pp. 593–615.

PROBLEMS

10.1. Derive the expression for the stiffness coefficient k_{12} relating the radial force F_{r_1} to the axial displacement w_1 at nodal circle 1 for the axisymmetric solid element shown in Fig. 10.1.

10.2. Figure P10.2 shows the rectangular cross section of an eight-d.o.f. axisymmetric solid element. Based on the displacement functions of Eq. (9.24), first derive the 4×8 strain displacement matrix and then the stiffness coefficient k_{12} relating the radial force F_{r_1} to the axial displacement w_1 at nodal circle 1.

10.3. Verify the derivation for I_5 given by Eq. (10.16e).

10.4. Using L'Hospital's rule, show that both D_{ij} and E_{ij} defined in Eqs. (10.17b and c) vanish when $r_i - r_j = 0$.

10.5. Derive the expression for the stiffness coefficient k_{12} relating the radial force F_{r_1} to the axial displacement w_1 for the harmonic n for the axisymmetric solid element shown in Fig. 10.1 subjected to nonsymmetric loads.

10.6. Figure P10.6 shows the cross section of a cylinder subjected to a uniform radial load of 100 lb/in. extended over one-sixth of the circumference. Let this load be represented as

$$p(\theta) = \sum_{n=0}^{9} a_n \cos n\theta$$

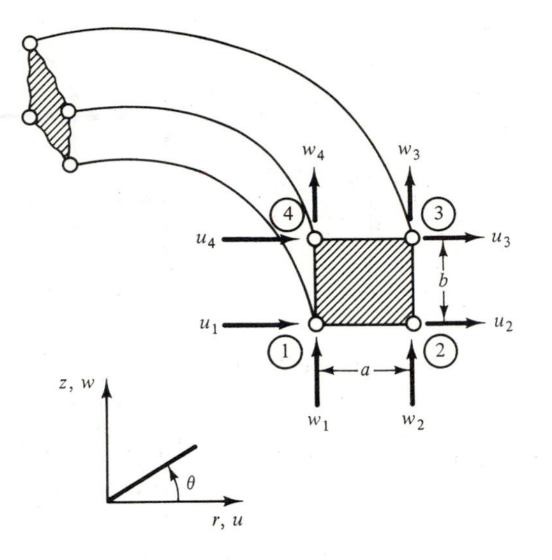

Figure P10.2

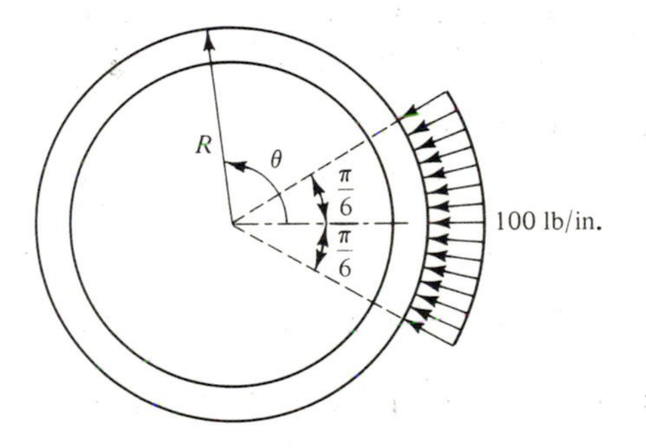

Figure P10.6

Find the values of the 10 coefficients a_n. Also find all the resulting $p(\theta)$ values at an interval of $\Delta\theta = 3°$.

10.7. Derive in explicit closed form the stiffness coefficient k_{11} relating F_{x_1} to u_1 for the four-node 12-d.o.f. tetrahedron shown in Fig. 10.9.

10.8. Derive in explicit closed form the stiffness coefficient k_{11} relating F_{x_1} to u_1 for the 10-node 30-d.o.f. tetrahedron shown in Fig. 10.9.

10.9. Derive in explicit closed form the stiffness coefficient k_{11} relating F_{x_1} to u_1 for the eight-node 24-d.o.f. hexahedron shown in Fig. 10.12.

10.10. Decompose an eight-node hexahedron diagrammatically into six tetrahedra without creating new nodes.

10.11. Figure P10.11 shows four 3×3 meshes of different kinds of two-dimensional elements, all with comparable numbers of element d.o.f.'s. Each rectangle in

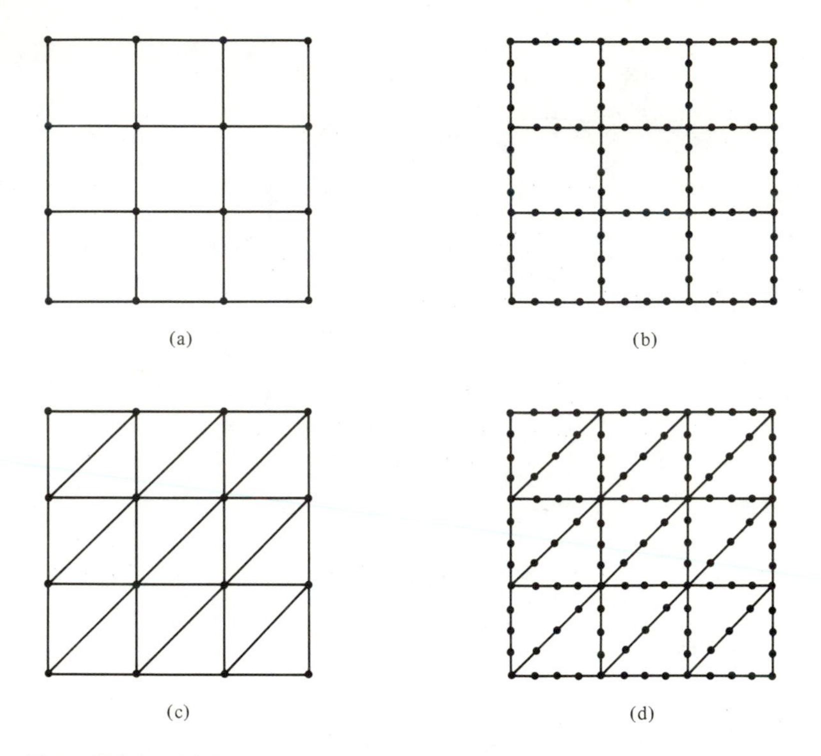

Figure P10.11 (a) Four-node 32-d.o.f. rectangles; (b) 16-node 32-d.o.f. rectangles; (c) three-node 24-d.o.f. triangles; (d) 12-node 24-d.o.f. triangles.

meshes (c) and (d) is formed by two triangles. Find the total number of d.o.f.'s and the minimum bandwidth of the assembled stiffness matrix for each of the four meshes.

10.12. This is a continuation of Problem 10.11. Figure P10.12 shows four $3 \times 3 \times 3$ meshes of different kinds of three-dimensional elements, all with comparable numbers of element d.o.f.'s. Each hexahedron in meshes (c) and (d) is formed by five tetrahedra, as shown in Fig. 10.10. Find the total number of d.o.f.'s and the minimum bandwidth of the assembled stiffness matrix for each of the four meshes.

10.13. The analysis of the stress distribution in a semi-infinite elastic solid (half-space) subjected to a concentrated load at the flat surface is known in classical elasticity as the *Bøussinesq problem.* By the use of not more than 200 axisymmetric solid elements with quadrilateral cross sections, perform a stress analysis for a clay foundation (half-space) subjected to a concentrated load of 10^6 lb. Assume that, for St. Louis Clay, $E = 2500$ psi and $\nu = 0.35$. The computer program is available in Ref. 10.7. Based on the principle of Saint-Venant, the nodal points at a distance sufficiently far from the load may be assumed as fixed. The complete work should include:

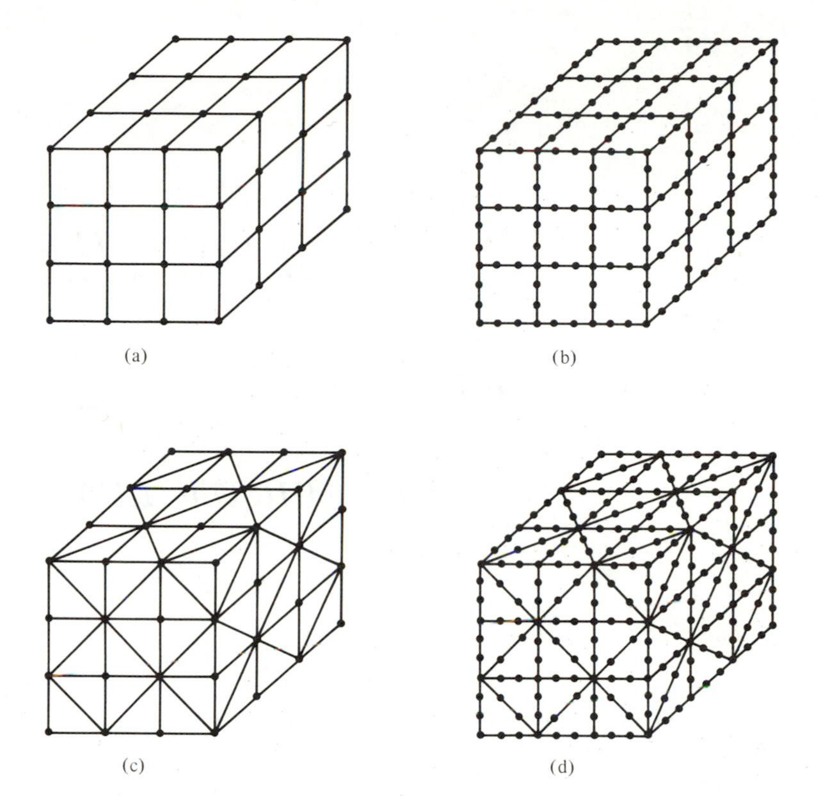

Figure P10.12 (a) Eight-node 96-d.o.f. hexahedra; (b) 32-node 96-d.o.f. hexahedra; (c) four-node 48-d.o.f. tetrahedra; (d) 16-node 48-d.o.f. tetrahedra.

1. A layout of the element mesh
2. The computer output
3. A plot of the displaced configuration
4. The isostress plots for axial stress σ_z, radial stress σ_r, circumferential stress σ_θ, and shearing stress τ_{rz}, respectively
5. An equilibrium check of a finite quadrant of the half-space as a free body
6. Comparison of all the isostress plots with those from elasticity solution (pp. 362–366 of Ref. 9.1)

A finite element solution for the Boussinesq problem was first given in Ref. 10.1.

CHAPTER 11

Numerical Integration and Curved Isoparametric Elements

In previous chapters we have derived the stiffness coefficients by performing closed-form integrations throughout the element areas or volumes. By the use of natural coordinates for triangles and tetrahedra and by the nondimensionalization of Cartesian coordinates for rectangles and rectangular hexahedra, the integration limits for each coordinate variable over an element area or volume can be converted to be constants. Thus these integrations may be obtained numerically and even exactly, not necessarily approximately. In this chapter we first introduce the methods of numerical integration.

From an economic point of view, the trend in the formulation of finite element matrices is to use more numerical integrations instead of manual integrations. This is due partly to the fact that computer operations become cheaper and faster all the time, but not necessarily so for engineering manpower.

In previous chapters we have formulated the stiffness matrices for the elements of the shapes of triangle, rectangle, tetrahedron, and rectangular hexahedron, all with straight edges and flat surfaces. In the modeling of a realistic structure, a substantial number of such elements may be needed merely for the purpose of geometric modeling for the curved edges or surfaces. Thus it is desirable that elements with curved edges and surfaces be used, so that a substantial amount of degrees of freedom can be saved. In fact, all the elements with straight edges and flat faces formulated in previous chapters can be distorted into elements with curved edges and surfaces if the concept of isoparametric elements is used. For such curved elements, the formulation is based on the mapping between Cartesian and curvilinear coordinates and the use of numerical integration.

The main scope of this chapter is to introduce the formulation of the isoparametric form of elements. Although the formulation for the curved shell elements is beyond the scope of this text, the reader is reminded that a form of plate and shell elements can be deduced from the three-dimensional isoparametric element through reduced integration [10.22, 10.23].

11.1 NUMERICAL INTEGRATION

11.1.1 One-Dimensional Numerical Integration

The evaluation of

$$I = \int_a^b f(x)\, dx \tag{11.1}$$

is termed *quadrature.* There are many methods of numerical integration. A few of the simplest and most popular methods are the trapezoidal rule and Simpson's $\frac{1}{3}$ and $\frac{3}{8}$ rules. For numerical integration of a function of one variable, we can use one of the two general and popular methods.

Newton–Cotes quadrature. Newton–Cotes quadrature is a numerical integration method which is exact for a single variable polynomial of degree n, where $n + 1$ is the number of function evaluations,

$$I = \int_a^b f(x)\, dx = \sum_{i=0}^{n} C_{ni} f(x_i) \tag{11.2}$$

The abscissas x_i of the points at which the function is to be evaluated are predetermined, usually equally spaced. The method to obtain the weights C_{ni} is illustrated by an example of a four-point quadrature of a polynomial function $f(x)$.

Let us assume that $f(x)$ has been evaluated at four distinct points, x_0, x_1, x_2, and x_3, at equal intervals h to obtain $f(x_0), f(x_1), f(x_2)$, and $f(x_3)$, respectively, and that a polynomial $g(x)$ is to pass through these data, with

$$g(x) = c_1 + c_2 x + c_3 x^2 + c_4 x^3 \tag{11.3}$$

The function $g(x)$ can be easily obtained in the form of Lagrangian interpolation functions $l(x)$,

$$g(x) = f(x_0) l_0(x) + f(x_1) l_1(x) + f(x_2) l_2(x) + f(x_3) l_3(x) \tag{11.4}$$

where

$$l_0(x) = \frac{(x - x_1)(x - x_2)(x - x_3)}{(x_0 - x_1)(x_0 - x_2)(x_0 - x_3)} \qquad l_1(x) = \frac{(x - x_0)(x - x_2)(x - x_3)}{(x_1 - x_0)(x_1 - x_2)(x_1 - x_3)}$$
$$l_2(x) = \frac{(x - x_0)(x - x_1)(x - x_3)}{(x_2 - x_0)(x_2 - x_1)(x_2 - x_3)} \qquad l_3(x) = \frac{(x - x_0)(x - x_1)(x - x_2)}{(x_3 - x_0)(x_3 - x_1)(x_3 - x_2)} \tag{11.5}$$

These functions can easily be understood by a simple inspection, for example, $l_0(x) = 1$ and $l_1(x) = l_2(x) = l_3(x) = 0$ when $x = x_0$. Thus $g(x) = f(x_0)$, $f(x_1), f(x_2)$, and $f(x_3)$ as $x = x_0, x_1, x_2$, and x_3, respectively.

The numerical integration for $f(x)$ can now be approximated using the interpolation polynomial $g(x)$ with four sampling points: $x_0 = a$, $x_1 = a + h$, $x_2 = a + 2h$, and $x_3 = a + 3h$ where the interval $h = (b - a)/3$

$$\int_a^b f(x)\,dx = \sum_{i=0}^{3} \left[\int_a^b l_i(x)\,dx\right] f(x_i) + E_3 = \sum_{i=0}^{3} C_{3i} f(x_i) + E_3 \tag{11.6}$$

where E_3 is the error term and the four Newton–Cotes numbers are obtained by

$$C_{3i} = \int_a^b l_i(x)\,dx \qquad \text{with } i = 0, 1, 2, 3 \tag{11.6a}$$

which results in $C_{30} = C_{33} = 3h/8$ and $C_{31} = C_{32} = 9h/8$.

For the four-point quadrature, the result is exact for a polynomial of third degree or less. We can now straightforwardly generalize the Newton–Cotes quadrature formula for a polynomial of degree n. The Newton–Cotes number C_{ni} and the coefficient of $h^{k+1}f^{(k)}(x)$ in the error term for $n = 1$ to 8 are given in Table 11.1 [11.1] with the definition that

$$\begin{aligned} c_{ni} &= hAW_i \\ f^{(k)}(x) &= \frac{d^k f(x)}{dx^k} \\ k &= n + 1 \qquad \text{if } n \text{ is odd number} \\ k &= n + 2 \qquad \text{if } n \text{ is even number} \end{aligned} \tag{11.7}$$

Example 11.1

Evaluate $I = \int_1^4 x^2\,dx$.

Solution. For $n = 1$, we have $h = (4 - 1)/1 = 3$, $A = \frac{1}{2}$, and

$$C_0 = hAW_0 = (3)(\tfrac{1}{2})(1) = 1.5$$
$$C_1 = hAW_1 = (3)(\tfrac{1}{2})(1) = 1.5$$

TABLE 11.1 Weights and Error-Term Coefficients for Newton–Cotes Quadrature Formula

n	A	W_0	W_1	W_2	W_3	W_4	Error
1	$\frac{1}{2}$	1	1				$-\frac{1}{12}$
2	$\frac{1}{3}$	1	4	1			$-\frac{1}{90}$
3	$\frac{3}{8}$	1	3	3	1		$-\frac{3}{80}$
4	$\frac{2}{45}$	7	32	12	32	7	$-\frac{8}{945}$
5	$\frac{5}{288}$	19	75	50	50	75	$-\frac{275}{12{,}096}$
6	$\frac{1}{140}$	41	216	27	272	27	$-\frac{9}{1{,}400}$
7	$\frac{7}{17{,}280}$	751	3,577	1,323	2,989	2,989	$-\frac{8{,}183}{518{,}400}$
8	$\frac{4}{14{,}175}$	989	5,888	−928	10,946	−4,540	$-\frac{2{,}368}{467{,}775}$

Thus

$$\begin{aligned} I &= C_0 f(1) + C_1 f(4) \\ &= (1.5)(1) + (1.5)(16) \\ &= 25.5 \neq 21 \text{ (exact value)} \end{aligned}$$

This is the *trapezoidal rule*, which is exact only for linear functions.

For $n = 2$, we have $h = (4 - 1)/2 = 1.5$, $A = \frac{1}{3}$, and

$$\begin{aligned} C_0 &= hAW_0 = (1.5)(\tfrac{1}{3})(1) = 0.5 \\ C_1 &= hAW_1 = (1.5)(\tfrac{1}{3})(4) = 2.0 \\ C_2 &= hAW_2 = (1.5)(\tfrac{1}{3})(1) = 0.5 \end{aligned}$$

Thus

$$\begin{aligned} I &= C_0 f(1) + C_1 f(2.5) + C_2 f(4) \\ &= (0.5)(1) + (2.0)(2.5)^2 + (0.5)(4)^2 \\ &= 21 \text{ (exact!)} \end{aligned}$$

This is *Simpson's* $\frac{1}{3}$ *rule*, which is exact for quadratic and linear functions.

For $n = 3$, we have $h = (4 - 1)/3 = 1.0$, $A = \frac{3}{8}$, $C_0 = C_3 = \frac{3}{8}$, and $C_1 = C_2 = \frac{9}{8}$.

Thus

$$\begin{aligned} I &= C_0 f(1) + C_1 f(2) + C_2 f(3) + C_3 f(4) \\ &= (\tfrac{3}{8})(1) + (\tfrac{9}{8})(4) + (\tfrac{9}{8})(9) + (\tfrac{3}{8})(16) \\ &= 21 \text{ (exact!)} \end{aligned}$$

This is *Simpson's* $\frac{3}{8}$ *rule*, which is exact for functions with cubic order and less.

Example 11.2

For a beam element as shown in Fig. 5.1, the displacement function can be expressed as

$$v(x) = f_1(x)v_1 + f_2(x)\theta_1 + f_3(x)v_2 + f_4(x)\theta_2 \tag{5.6}$$

where

$$\begin{aligned} f_1(x) &= 1 - 3\left(\frac{x}{L}\right)^2 + 2\left(\frac{x}{L}\right)^3 \\ f_2(x) &= x - 2\left(\frac{x^2}{L}\right) + \left(\frac{x^3}{L^2}\right) \\ f_3(x) &= 3\left(\frac{x}{L}\right)^2 - 2\left(\frac{x}{L}\right)^3 \\ f_4(x) &= -\left(\frac{x^2}{L}\right) + \left(\frac{x^3}{L^2}\right) \end{aligned} \tag{5.7}$$

Let it be desired to derive the stiffness matrix by using Newton-Cotes quadrature with $n = 2$.

Solution. The strain energy expression is

$$\begin{aligned} U &= \frac{1}{2}\int_0^L EI\left(\frac{\partial^2 v}{\partial x^2}\right)^2 dx \\ &= \tfrac{1}{2}\lfloor v_1 \quad \theta_1 \quad v_2 \quad \theta_2 \rfloor \int_0^L EI \begin{Bmatrix} f_1''(x) \\ f_2''(x) \\ f_3''(x) \\ f_4''(x) \end{Bmatrix} \\ &\quad \times \lfloor f_1''(x) \quad f_2''(x) \quad f_3''(x) \quad f_4''(x) \rfloor \, dx \begin{Bmatrix} v_1 \\ \theta_1 \\ v_2 \\ \theta_2 \end{Bmatrix} \\ &= \tfrac{1}{2}\lfloor v_1 \quad \theta_1 \quad v_2 \quad \theta_2 \rfloor [\mathbf{k}] \begin{Bmatrix} v_1 \\ \theta_1 \\ v_2 \\ \theta_2 \end{Bmatrix}. \end{aligned}$$

where

$$f_1''(x) = -\frac{6}{L^2} + \frac{12x}{L^3} \qquad f_2''(x) = -\frac{4}{L} + \frac{6x}{L^2}$$

$$f_3''(x) = \frac{6}{L^2} - \frac{12x}{L^3} \qquad f_4''(x) = -\frac{2}{L} + \frac{6x}{L^2}$$

For $n = 2$, we have $h = L/2$, $A = 1/3$, $C_0 = C_2 = L/6$, $C_1 = 2L/3$, and

$$k_{ij} = I = C_0 f(0) + C_1 f\left(\frac{L}{2}\right) + C_2 f(L)$$

or

$$[\mathbf{k}] = \frac{EIL}{6}\begin{Bmatrix} -\frac{6}{L^2} \\ -\frac{4}{L} \\ \frac{6}{L^2} \\ -\frac{2}{L} \end{Bmatrix}\left\lfloor -\frac{6}{L^2} \quad -\frac{4}{L} \quad \frac{6}{L^2} \quad -\frac{2}{L} \right\rfloor + \frac{2EIL}{3}\begin{Bmatrix} 0 \\ -\frac{1}{L} \\ 0 \\ \frac{1}{L} \end{Bmatrix}\left\lfloor 0 \quad -\frac{1}{L} \quad 0 \quad \frac{1}{L} \right\rfloor$$

$$+ \frac{EIL}{6}\begin{Bmatrix} \frac{6}{L^2} \\ \frac{2}{L} \\ -\frac{6}{L^2} \\ \frac{4}{L} \end{Bmatrix}\left\lfloor \frac{6}{L^2} \quad \frac{2}{L} \quad \frac{-6}{L^2} \quad \frac{4}{L} \right\rfloor$$

$$= \frac{EI}{L}\begin{bmatrix} \frac{12}{L^2} & & \text{symmetrical} & \\ \frac{6}{L} & 4 & & \\ -\frac{12}{L^2} & -\frac{6}{L} & \frac{12}{L^2} & \\ \frac{6}{L} & 2 & -\frac{6}{L} & 4 \end{bmatrix}$$

which is identical to that obtained analytically in Eq. (5.14). This example demonstrates that numerical integration can be analytically accurate without numerical approximation.

Gaussian quadrature. In the Newton–Cotes quadrature, the positions of the n sampling points for function evaluations of a polynomial of degree $n-1$ are predetermined, usually equally spaced. It is, however, perceivable that for n sampling points there are $2n$ unknowns (n abscissas and n functional values) which could be used to construct a polynomial of degree $2n-1$ for exact integration. Such a method obviously requires the most efficient selection of function evaluations which is achieved by optimizing both the positions of the sampling points and the weights. Such a superior numerical integration method is Gaussian quadrature. The Gaussian quadrature is exact for polynomials of degree $2n-1$, where n is the number of function evaluations,

$$I = \int_{-1}^{1} f(x)\,dx = \sum_{i=1}^{n} H_i f(x_i) \tag{11.8}$$

where H_i are the weights and x_i are the abscissas of the sampling points. The method to obtain the optimized abscissas of the sampling points and the weights can be illustrated using an example of a three-point Gaussian quadrature of a polynomial function $f(x)$.

Let us assume that $f(x)$ has been evaluated at three distinct points x_1, x_2, and x_3 to obtain $f(x_1)$, $f(x_2)$, and $f(x_3)$, respectively, and that a polynomial $g(x)$ is to pass through these data, with

$$g(x) = c_1 + c_2 x + c_3 x^2 \tag{11.9}$$

This function can be easily obtained in the form of Lagrangian interpolation functions $l(x)$,

$$g(x) = f(x_1)l_1(x) + f(x_2)l_2(x) + f(x_3)l_3(x) \tag{11.10}$$

where

$$l_1(x) = \frac{(x-x_2)(x-x_3)}{(x_1-x_2)(x_1-x_3)} \qquad l_2(x) = \frac{(x-x_1)(x-x_3)}{(x_2-x_1)(x_2-x_3)}$$
$$l_3(x) = \frac{(x-x_1)(x-x_2)}{(x_3-x_1)(x_3-x_2)} \tag{11.11}$$

Thus we see that as $x = x_1, x_2$, and x_3, $g(x) = f(x_1), f(x_2)$, and $f(x_3)$, respectively.

Let us now define another fictitious polynomial function of third order,

$$h(x) = (x-x_1)(x-x_2)(x-x_3) \tag{11.12}$$

which vanishes at points x_1, x_2, and x_3, respectively. Thus we can approximate the original function $f(x)$ as

$$f(x) = g(x) + h(x)(a_0 + a_1 x + a_2 x^2) \tag{11.13}$$

The function $f(x)$ now becomes a fifth-order polynomial which still passes through the three distinct data points $f(x_1)$, $f(x_2)$, and $f(x_3)$ at the three yet unknown sampling abscissas x_1, x_2, and x_3, respectively.

Integrating Eq. (11.13) gives

$$\int_{-1}^{1} f(x)\,dx = \sum_{i=1}^{3} \left[\int_{-1}^{1} l_i(x)\,dx\right] f(x_i) + \sum_{j=0}^{2} a_j \int_{-1}^{1} h(x)x^j\,dx \quad (11.14)$$

where the first term on the right of Eq. (11.14) contains the integration of functions of order 2 and less and the second term contains integration of functions of order 3 to 5.

The originally intended form of Gaussian quadrature formula (11.8) for a three-point integration can be written as

$$I = \int_{-1}^{1} f(x)\,dx = \sum_{i=1}^{3} H_i f(x_i) \quad (11.15)$$

The conditions for determining the three unknown abscissas and the weights can now be obtained by equating the expressions on the right of Eq. (11.14) to that of Eq. (11.15)

For abscissas:

$$\int_{-1}^{1} h(x)x^j\,dx = 0 \qquad \text{with } j = 0, 1, 2 \quad (11.16a)$$

For weights:

$$H_i = \int_{-1}^{1} l_i(x)\,dx \qquad \text{with } i = 1, 2, 3 \quad (11.16b)$$

Equation (11.16a) results in a set of three nonlinear equations:

$$\int_{-1}^{1} h(x)\,dx = -\tfrac{2}{3}(x_1 + x_2 + x_3) - 2x_1x_2x_3 = 0 \quad (11.17a)$$

$$\int_{-1}^{1} h(x)x\,dx = \tfrac{2}{5} + \tfrac{2}{3}(x_1x_2 + x_2x_3 + x_1x_3) = 0 \quad (11.17b)$$

$$\int_{-1}^{1} h(x)x^2\,dx = -\tfrac{2}{5}(x_1 + x_2 + x_3) - \tfrac{2}{3}x_1x_2x_3 = 0 \quad (11.17c)$$

Solution for Eqs. (11.17) gives the values of the three abscissas,

$$x_1 = \sqrt{\frac{3}{5}} \qquad x_2 = -\sqrt{\frac{3}{5}} \qquad x_3 = 0 \quad (11.18)$$

It is of interest to note that these three roots are also the roots for the equation based on the third-order Legendre polynomial

$$P_3(x) = \tfrac{1}{2}(5x^3 - 3x) = 0 \quad (11.19)$$

With the values of the three abscissas known, Eqs. (11.16b) can be used to obtain the three weights:

$$
\begin{aligned}
H_1(x) &= \int_{-1}^{1} l_1(x)\,dx = \tfrac{5}{9} \\
H_2(x) &= \int_{-1}^{1} l_2(x)\,dx = \tfrac{5}{9} \\
H_3(x) &= \int_{-1}^{1} l_3(x)\,dx = \tfrac{8}{9}
\end{aligned}
\tag{11.20}
$$

For this three-point Gaussian quadrature, the result is exact for the polynomial of fifth degree and less. We can now straightforwardly generalize the Gaussian quadrature formula to that for polynomial of degree n. The abscissas and the weight coefficients for $n = 1$ to 6 are given in Table 11.2 [11.2]. A general derivation of the Gaussian quadrature formula can be found in Ref. 11.3.

TABLE 11.2 Abscissas and Weights in Gaussian Quadrature

n	Abscissa, x_i	Weight, H_i
1	0	2
2	$\pm \frac{1}{\sqrt{3}}$	1
3	0	$\frac{8}{9}$
	$\pm \sqrt{0.6}$	$\frac{5}{9}$
4	±0.33998 10435 84856	0.65214 51548 62546
	±0.86113 63115 94053	0.34785 48451 37454
5	0	0.56888 88888 88889
	±0.53846 93101 05683	0.47862 86704 99366
	±0.90617 98459 38664	0.23692 68850 56189
6	±0.23861 91860 83197	0.46791 39345 72691
	±0.66120 93864 66265	0.36076 15730 48139
	±0.93246 95142 03152	0.17132 44923 79170

It is noted that the n abscissa values solved from the generalized set of n nonlinear equations (11.16a) are the same as the roots for the equation resulting from setting the nth-order Legendre polynomial to zero, $P_n(x) = 0$. Thus this integration method is frequently referred to as the Gauss–Legendre quadrature.

Before the advent of computers, Gaussian quadrature formulas were seldom used in practice. This was because the use of simple numbers such as integers and rational numbers in the common quadrature formulas (see, for example, Table 11.1) is much more convenient on calculators than the non-simple numbers in the Gaussian quadrature formula (Table 11.2). However, because Gaussian quadrature requires the least number of functional evaluations as compared with other methods of quadrature, this method is ideally suited for computers and has been exclusively used in the finite element formulations.

In place of Eq. (11.8), the Gaussian quadrature formula can be generalized to the form

$$I = \int_{-1}^{1} w(x)f(x)\,dx = \sum_{i=1}^{n} H_i f(x_i) \tag{11.21}$$

with certain degree of accuracy for some weight functions $w(x)$ which are practically and mathematically meaningful. A summary of some weight functions, abscissas, weights, and errors are given in Table 4.4 in Ref. 11.1.

Example 11.3

Examine the limits of exactness of Gaussian quadrature formula with (a) $n = 3$ and (b) $n = 4$.

Solution. (a) *Gauss three-point rule*:

$$I = \int_{-1}^{1} f(x)\,dx = \tfrac{8}{9}f(0) + \tfrac{5}{9}f(\sqrt{0.6}) + \tfrac{5}{9}f(-\sqrt{0.6})$$

is exact for a polynomial of degree up to $2n - 1$ or 5. Let

$f(x) = 1$: we have $I = \frac{8}{9} + \frac{5}{9} + \frac{5}{9} = 2$ (exact!)

$f(x) = x$: we have $I = \frac{5}{9}(\sqrt{0.6}) + \frac{5}{9}(-\sqrt{0.6}) = 0$ (exact!)

$f(x) = x^2$: we have $I = \frac{5}{9}(0.6) + \frac{5}{9}(0.6) = \frac{2}{3}$ (exact!)

$f(x) = x^3$: we have $I = \frac{5}{9}(\sqrt{0.6})^3 - \frac{5}{9}(\sqrt{0.6})^3 = 0$ (exact!)

$f(x) = x^4$: we have $I = \frac{5}{9}(0.36) + \frac{5}{9}(0.36) = \frac{2}{5}$ (exact!)

$f(x) = x^5$: we have $I = \frac{5}{9}(\sqrt{0.6})^5 - \frac{5}{9}(\sqrt{0.6})^5 = 0$ (exact!)

$f(x) = x^6$: we have $I = \frac{5}{9}(0.216) + \frac{5}{9}(0.216)$

$= 0.24 \neq \frac{2}{7}$ or 0.2857 (no longer exact!)

(b) *Gauss four-point rule*:

$$\begin{aligned} I &= \int_{-1}^{1} f(x)\,dx \\ &= 0.65214\,5155\{f(0.33998\,1044) + f(-0.33998\,1044)\} \\ &\quad + 0.34785\,4845\{f(0.86113\,6312) + f(-0.86113\,6312)\} \end{aligned}$$

is exact for a polynomial of degree up to $2n - 1$ or 7. We retain only nine digits after the decimal points.

Let $f(x) = x^6$.

$$I = 0.65214\,5155[2 \times (0.33998\,1044)^6] + 0.34785\,4845[2 \times (0.86113\,6312)^6]$$
$$= 0.28571\,4286 = 2/7 \qquad \text{(exact!)}$$

Let $f(x) = x^7$:

$$I = 0 \qquad \text{(exact!)}$$

Let $f(x) = x^8$:

$$I = 0.65214\,5155[2 \times (0.33998\,1044)^8] + 0.34785\,4845[2 \times (0.86113\,6312)^8]$$
$$= 0.21061\,2246 \neq 2/9 \qquad \text{(no longer exact!)}$$

Example 11.4

Evaluate $\int_1^3 dx/x$ using Gaussian three-point formula.

Solution. Since the integration limits for Gaussian quadrature are from -1 to 1, we change variables first. Let $y = x - 2$; then

$$I = \int_1^3 \frac{dx}{x} = \int_{-1}^1 \frac{dy}{y+2}$$
$$= \left(\frac{5}{9}\right)\frac{1}{2-\sqrt{0.6}} + \left(\frac{8}{9}\right)\frac{1}{2} + \left(\frac{5}{9}\right)\frac{1}{2+\sqrt{0.6}}$$
$$= 1.09803\,9216$$

whereas the true value of the integral is $\ln 3 = 1.09861\,2289$. The error is -0.052%. For the upper and lower bounds of this error, the reader is referred to Ref. 11.1.

11.1.2 Numerical Integration Over Rectangles or Rectangular Hexahedra

In finite element formulations, we often have to evaluate the double integral

$$I = \int_{-1}^1 \int_{-1}^1 f(\xi, \eta)\, d\xi\, d\eta \tag{11.22}$$

where ξ and η can be the nondimensionalized Cartesian coordinates, or curvilinear coordinates.

This integration can obviously be done by first obtaining the inner integral numerically holding η constant, and then obtaining the outer integral varying η.

$$I = \int_{-1}^{1} \sum_{j=1}^{n} H_j f(\xi_j, \eta)\, d\eta = \int_{-1}^{1} g(\eta)\, d\eta$$

$$= \sum_{i=1}^{m} H_i g(\eta_i) = \sum_{i=1}^{m} \sum_{j=1}^{n} H_i H_j f(\xi_j, \eta_i) \tag{11.23}$$

This equation can be generalized to the three-dimensional case,

$$I = \int_{-1}^{1} \int_{-1}^{1} \int_{-1}^{1} f(\xi, \eta, \zeta)\, d\xi\, d\eta\, d\zeta$$

$$= \sum_{i=1}^{l} \sum_{j=1}^{m} \sum_{k=1}^{n} H_i H_j H_k f(\xi_k, \eta_j, \zeta_i) \tag{11.24}$$

In most cases, there appears to be no obvious advantage to use different numbers of integrating points in each direction; we usually assume that $l = m = n$. If the function $f(\xi, \eta, \zeta)$ is integrated with, say, $l = m = n = 3$, the triple summation is equivalent to a single summation over $3 \times 3 \times 3$ points, which is exact for a polynomial of order 5 in each direction.

Alternatively and more efficiently, we could integrate Eq. (11.24) in single-summation form,

$$I = \int_{-1}^{1} \int_{-1}^{1} \int_{-1}^{1} f(\xi, \eta, \zeta)\, d\xi\, d\eta\, d\zeta = \sum_{i=1}^{n} w_i f(\xi_i, \eta_i, \zeta_i) \tag{11.25}$$

Such formulas have been derived [11.4] and applied [11.5].

Example 11.5

Let us illustrate a 14-point rule [11.6]:

$$I = \int_{-1}^{1} \int_{-1}^{1} \int_{-1}^{1} f(\xi, \eta, \zeta)\, d\xi\, d\eta\, d\zeta$$

$$= w_1[f(-b, 0, 0) + f(b, 0, 0) + f(0, -b, 0) + f(0, b, 0) + f(0, 0, -b) + f(0, 0, b)]$$

$$+ w_2[f(-c, -c, -c) + f(c, -c, -c) + f(-c, c, -c) + f(-c, -c, c)$$

$$+ f(c, c, c) + f(-c, c, c) + f(c, -c, c) + f(c, c, -c)]$$

with

$$w_1 = 0.88642\,6593 \qquad w_2 = 0.33518\,0055 \qquad b = 0.79582\,2426$$
$$c = 0.75878\,6911 \tag{11.26}$$

Equation (11.26) is exact for a complete quintic polynomial of 56 terms:

$$1$$

$$\xi \quad \eta \quad \zeta$$

$$\xi^2 \quad \eta^2 \quad \zeta^2 \quad \xi\eta \quad \eta\zeta \quad \xi\zeta$$

$$\xi^3 \quad \eta^3 \quad \zeta^3 \quad \xi^2\eta \quad \xi^2\zeta \quad \eta^2\xi \quad \eta^2\zeta \quad \zeta^2\xi \quad \zeta^2\eta \quad \xi\eta\zeta$$

$$\xi^4 \quad \eta^4 \quad \zeta^4 \quad \xi^3\eta \quad \xi^3\zeta \quad \eta^3\xi \quad \eta^3\zeta \quad \zeta^3\xi \quad \zeta^3\eta \quad \xi^2\eta\zeta \quad \eta^2\xi\zeta \quad \zeta^2\xi\eta \quad \xi^2\eta^2 \quad \eta^2\zeta^2 \quad \zeta^2\xi^2$$

$$\xi^5 \quad \eta^5 \quad \zeta^5 \quad \xi^4\eta \quad \xi^4\zeta \quad \eta^4\xi \quad \eta^4\zeta \quad \zeta^4\xi \quad \zeta^4\eta \quad \xi^3\eta^2 \quad \xi^3\zeta^2 \quad \eta^3\xi^2 \quad \eta^3\zeta^2 \quad \zeta^3\xi^2 \quad \zeta^3\eta^2$$

$$\xi^3\eta\zeta \quad \eta^3\xi\zeta \quad \zeta^3\xi\eta \quad \xi^2\eta^2\zeta \quad \eta^2\zeta^2\xi \quad \zeta^2\xi^2\eta$$

1: $I = w_1(1 \times 6) + w_2(1 \times 8) = 8$ (exact!)

ξ, η, ζ: $I = 0$ (exact!)

ξ^2: $I = w_1(b^2 \times 2) + w_2(c^2 \times 8) = \frac{8}{3}$ (exact!)

$\xi\eta$: $I = 0$ (exact!)

$\xi^3, \xi^2\eta, \xi\eta\zeta$: $I = 0$ (exact!)

ξ^4: $I = w_1(b^4 \times 2) + w_2(c^4 \times 8) = \frac{8}{5}$ (exact!)

$\xi^3\eta, \xi^2\eta\zeta$: $I = 0$ (exact!)

$\xi^2\eta^2$: $I = w_2(c^4 \times 8) = \frac{8}{9}$ (exact!)

$\xi^5, \xi^4\eta, \xi^3\eta^2, \xi^3\eta\zeta, \xi^2\eta^2\zeta$: $I = 0$ (exact!)

ξ^6: $I = w_1(b^6 \times 2) + w_2(c^6 \times 8)$

$= 0.96215\,4884 \neq \frac{8}{7}$ (no longer exact!)

$\xi^4\eta^2, \xi^2\eta^2\zeta^2$: $I = w_2(c^6 \times 8) = 0.51178\,4513$

$\neq \frac{8}{15}$ and $\frac{8}{27}$ (no longer exact!)

For three-dimensional integration, we have four unknowns (ξ_i, η_i, ζ_i, and f_i) for each sampling point. For the 56-term quintic polynomial above, it is obvious that fourteen sampling points provide the exact integration.

11.1.3 Numerical Integration Over Triangles and Tetrahedra

As we derived earlier in Fig. 9.13 and Eqs. (9.98) and (9.99), the integrals over a triangle in terms of the area coordinates are of the form

$$I = \int_0^1 \int_0^{1-\xi_1} f(\xi_1, \xi_2, \xi_3)\, d\xi_2\, d\xi_1$$

$$= \sum_{i=1}^{n} H_i f(\xi_{1_i}, \xi_{2_i}, \xi_{3_i}) \tag{11.27}$$

Here the limits of integral involve the variable ξ_1. Some suitable quadrature formulas have been derived [11.7–11.11] and compiled [11.12, 11.13]. From the viewpoint of both numerical efficiency and aesthetics, the best formulas appear to be those derived by Hammer et al. [11.9] with additions given by Cowper [11.10] and Dunavant [11.11]. A series of triangular coordinates of sampling points and the weights for Gaussian quadrature formulas (11.27) for polynomials exact to the orders from 1 to 7 is given in Table 11.3 [11.10]. Since $\xi_1 + \xi_2 + \xi_3 = 1$, it is seen in Table 11.3 that $a + b + b = c + d + d = g + e + f = 1$ for all n's. Quadrature rules of degree up to 20 for the triangle together with data and a Fortran program were given in Ref. 11.11.

Example 11.6

Using both (a) the closed-form integral formula (9.103) and (b) the four-point Gaussian quadrature formula (Table 11.3), integrate the representative terms $(1, x, x^2, xy, x^3, x^2y)$ in a complete third-order polynomial over an arbitrary triangle as shown in Fig. 11.1.

Solution. (a) *Closed-form formula (9.103)*: Based on the coordinates of the vertices in Fig. 11.1, we can use Eqs. (9.95) to relate the Cartesian coordinates to the area coordinates:

$$x = \xi_1 x_1 + \xi_2 x_2 + \xi_3 x_3 = \xi_1 + 3\xi_2 + 2\xi_3$$

$$y = \xi_1 y_1 + \xi_2 y_2 + \xi_3 y_3 = \xi_1 + 2\xi_2 + 3\xi_3$$

$$\iint (1)\, dA = \frac{2A}{2!} = A = \frac{1}{2}\begin{vmatrix} 1 & 1 & 1 \\ 1 & 3 & 2 \\ 1 & 2 & 3 \end{vmatrix} = \frac{3}{2}$$

$$\iint x\, dA = \iint (\xi_1 + 3\xi_2 + 2\xi_3)\, dA = \frac{2A}{3!}(1 + 3 + 2) = 3$$

$$\iint x^2\, dA = \iint (\xi_1^2 + 9\xi_2^2 + 4\xi_3^2 + 6\xi_1\xi_2 + 12\xi_2\xi_3 + 4\xi_1\xi_3)\, dA$$

$$= \frac{2A}{4!}(2 + 9 \times 2 + 4 \times 2 + 6 + 12 + 4) = \frac{25}{4}$$

$$\iint xy\, dA = \iint (\xi_1^2 + 6\xi_2^2 + 6\xi_3^2 + 5\xi_1\xi_2 + 13\xi_2\xi_3 + 5\xi_1\xi_3)\, dA$$

$$= \frac{2A}{4!}(2 + 6 \times 2 + 6 \times 2 + 5 + 13 + 5) = \frac{49}{8}$$

$$\iint x^3\, dA = \iint (\xi_1^3 + 27\xi_2^3 + 8\xi_3^3 + 9\xi_1^2\xi_2 + 27\xi_1\xi_2^2 + 54\xi_2^2\xi_3 + 36\xi_2\xi_3^2$$

$$+ 6\xi_1^2\xi_3 + 12\xi_1\xi_3^2 + 36\xi_1\xi_2\xi_3)\, dA = \frac{27}{2}$$

$$\iint x^2 y\, dA = \iint (\xi_1^3 + 18\xi_2^3 + 12\xi_3^3 + 8\xi_1^2\xi_2 + 21\xi_1\xi_2^2 + 51\xi_2^2\xi_3 + 44\xi_2\xi_3^2$$

$$+ 7\xi_1^2\xi_3 + 16\xi_1\xi_3^2 + 38\xi_1\xi_2\xi_3)\, dA = \frac{259}{20}$$

TABLE 11.3 Gaussian Quadrature Formulas for Triangles

Number and Locations of Sampling Points	Triangular Coordinates of Sampling Points	Weights (H_i/A)	Values of Coefficients	Degree of Accuracy
$n = 1$	$\left(\frac{1}{3}, \frac{1}{3}, \frac{1}{3}\right)$	1		1
$n = 3$	$\left(\frac{1}{2}, \frac{1}{2}, 0\right)$ $\left(0, \frac{1}{2}, \frac{1}{2}\right)$ $\left(\frac{1}{2}, 0, \frac{1}{2}\right)$	$\frac{1}{3}$		2
$n = 4$	$\left(\frac{1}{3}, \frac{1}{3}, \frac{1}{3}\right)$	$-\frac{27}{48}$	$a = \frac{3}{5}, \quad b = \frac{1}{5}$	3
	$(a, b, b); (b, a, b); (b, b, a)$	$\frac{25}{48}$		
$n = 6$	$(a, b, b); (b, a, b); (b, b, a)$	0.10995 17436 55322	$a = 0.81684\ 75729\ 80459$ $b = 0.09157\ 62135\ 09771$	4
	$(c, d, d); (d, c, d); (d, d, c)$	0.22338 15896 78011	$c = 0.10810\ 30181\ 68070$ $d = 0.44594\ 84909\ 15965$	

$n = 7$	$\left(\frac{1}{3}, \frac{1}{3}, \frac{1}{3}\right)$	0.22500 00000 00000		5
	$(a, b, b); (b, a, b); (b, b, a)$	0.12593 91805 44827	$a = 0.79742\ 69853\ 53087$ $b = 0.10128\ 65073\ 23456$	
	$(c, d, d); (d, c, d); (d, d, c)$	0.13239 41527 88506	$c = 0.05971\ 58717\ 89770$ $d = 0.47014\ 20641\ 05115$	
$n = 12$	$(a, b, b); (b, a, b); (b, b, a)$	0.05084 49063 70207	$a = 0.87382\ 19710\ 16996$ $b = 0.06308\ 90144\ 91502$	6
	$(c, d, d); (d, c, d); (d, d, c)$	0.11678 62757 26379	$c = 0.50142\ 65096\ 58179$ $d = 0.24928\ 67451\ 70910$	
	$(e, f, g); (e, g, f); (f, g, e);$ $(f, e, g); (g, e, f); (g, f, e)$	0.08285 10756 18374	$e = 0.63650\ 24991\ 21399$ $f = 0.31035\ 24510\ 33785$ $g = 0.05314\ 50498\ 44816$	
$n = 13$	$\left(\frac{1}{3}, \frac{1}{3}, \frac{1}{3}\right)$	−0.14957 00444 67670		7
	$(a, b, b); (b, a, b); (b, b, a)$	0.17561 52574 33204	$a = 0.47930\ 80678\ 41923$ $b = 0.26034\ 59660\ 79038$	
	$(c, d, d); (d, c, d); (d, d, c)$	0.05334 72356 08839	$c = 0.86973\ 97941\ 95568$ $d = 0.06513\ 01029\ 02216$	
	$(e, f, g); (e, g, f); (f, g, e);$ $(f, e, g); (g, e, f); (g, f, e)$	0.07711 37608 90257	$e = 0.63844\ 41885\ 69809$ $f = 0.31286\ 54960\ 04875$ $g = 0.04869\ 03154\ 25316$	

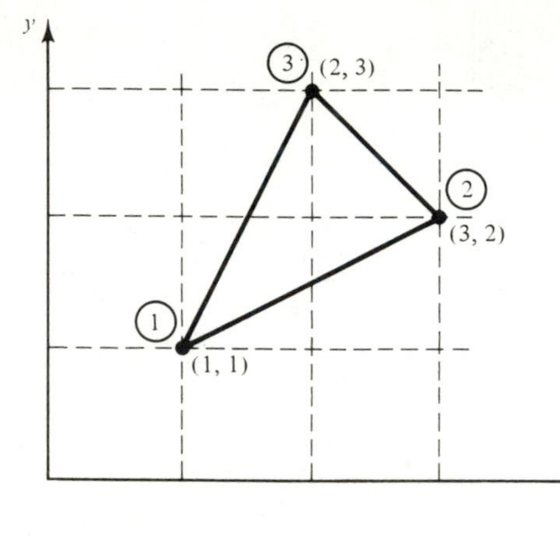

Figure 11.1 Arbitrary triangle for numerical integration.

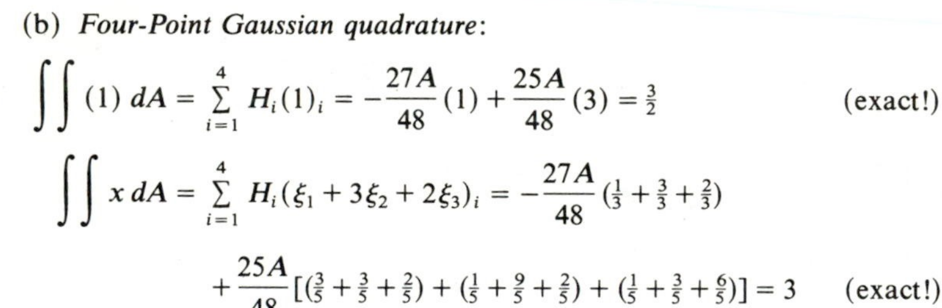

(b) *Four-Point Gaussian quadrature*:

$$\iint (1)\, dA = \sum_{i=1}^{4} H_i(1)_i = -\frac{27A}{48}(1) + \frac{25A}{48}(3) = \tfrac{3}{2} \qquad \text{(exact!)}$$

$$\iint x\, dA = \sum_{i=1}^{4} H_i(\xi_1 + 3\xi_2 + 2\xi_3)_i = -\frac{27A}{48}(\tfrac{1}{3} + \tfrac{3}{3} + \tfrac{2}{3})$$

$$+\frac{25A}{48}[(\tfrac{3}{5} + \tfrac{3}{5} + \tfrac{2}{5}) + (\tfrac{1}{5} + \tfrac{9}{5} + \tfrac{2}{5}) + (\tfrac{1}{5} + \tfrac{3}{5} + \tfrac{6}{5})] = 3 \qquad \text{(exact!)}$$

$$\iint x^2\, dA = -\frac{27A}{48}(2)^2 + \frac{25A}{48}[(\tfrac{8}{5})^2 + (\tfrac{12}{5})^2 + (\tfrac{10}{5})^2] = \tfrac{25}{4} \qquad \text{(exact!)}$$

$$\iint xy\, dA = -\frac{27A}{48}(2)^2 + \frac{25A}{48}[(\tfrac{8}{5})^2 + (\tfrac{12}{5})(\tfrac{10}{5})(2)] = \tfrac{49}{8} \qquad \text{(exact!)}$$

$$\iint x^3\, dA = -\frac{27A}{48}(2)^3 + \frac{25A}{48}[(\tfrac{8}{5})^3 + (\tfrac{12}{5})^3 + (\tfrac{10}{5})^3] = \tfrac{27}{2} \qquad \text{(exact!)}$$

$$\iint x^2y\, dA = -\frac{27A}{48}(2)^3 + \frac{25A}{48}[(\tfrac{8}{5})^3 + (\tfrac{12}{5})^2(\tfrac{10}{5}) + (\tfrac{12}{5})(\tfrac{10}{5})^2] = \tfrac{259}{20} \qquad \text{(exact!)}$$

Example 11.7

Using the seven-point Gaussian quadrature formula (Table 11.3), integrate the fourth- and fifth-order terms ($x^4, x^3y, x^2y^2, x^5, x^4y, x^3y^2$) over a simplest possible triangle as shown in Fig. 11.2 for *w*.

From Fig. 11.2 we have

$$x = \xi_1 x_1 + \xi_2 x_2 + \xi_3 x_3 = \xi_2$$

$$y = \xi_1 y_1 + \xi_2 y_2 + \xi_3 y_3 = \xi_3$$

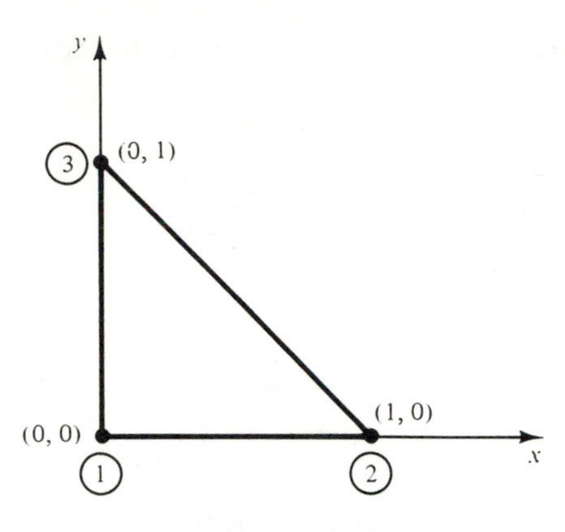

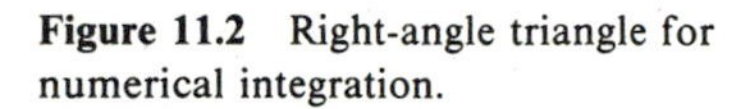
Figure 11.2 Right-angle triangle for numerical integration.

Let $w_1 = H_1$, $w_2 = H_2 = H_3 = H_4$, $w_3 = H_5 = H_6 = H_7$.

$$\iint x^4\, dA = \sum_{i=1}^{7} H_i(\xi_2)_i^4 = w_1(\tfrac{1}{3})^4 + w_2(a^4 + 2b^4) + w_3(c^4 + 2d^4)$$

$$= 0.06666\,66669\,27916A = 0.03333\,33334\,63958 \qquad \text{(exact!)}$$

$$\iint x^3 y\, dA = w_1(\tfrac{1}{3})^4 + w_2(a^3 b + b^3 a + b^4) + w_3(c^3 d + d^3 c + d^4)$$

$$= 0.08333\,33340\,49011 \qquad \text{(exact!)}$$

$$\iint x^2 y^2\, dA = w_1(\tfrac{1}{3})^4 + w_2(2a^2 b^2 + b^4) + w_3(2c^2 d^2 + d^4)$$

$$= 0.00555\,55556\,21922 \qquad \text{(exact!)}$$

$$\iint x^5\, dA = w_1(\tfrac{1}{3})^5 + w_2(a^5 + 2b^5) + w_3(c^5 + 2d^5)$$

$$= 0.02380\,95238\,86289 \qquad \text{(exact!)}$$

$$\iint x^4 y\, dA = w_1(\tfrac{1}{3})^5 + w_2\,(a^4 b + b^4 a + b^5) + w_3(c^4 d + d^4 c + d^5)$$

$$= 0.00476\,19048\,04190 \qquad \text{(exact!)}$$

$$\iint x^3 y^2\, dA = w_1(\tfrac{1}{3})^5 + w_2(a^3 b^2 + b^3 a^2 + b^5) + w_3(c^3 d^2 + c^2 d^3 + d^5)$$

$$= 0.00238\,09524\,19738 \qquad \text{(exact!)}$$

Examples 11.6 and 11.7 demonstrated that for the closed-form integration formula (9.103), the cumbersome expansion and product of polynomials in triangular coordinates have to be done first, whereas in Gaussian quadrature such analytical process is avoided but traded for numerical process. For certain formulations, such as those for crack-tip singular elements and curved shell elements, the integrand may include polynomials with noninteger powers. In that case, formula (9.103) is no longer valid but Gaussian quadrature formulas still apply.

Extension of the Gaussian quadrature formulas for triangles to those for tetrahedra could be done in the form

$$I = \int_0^1 \int_0^{1-\xi_1} \int_0^{1-\xi_1-\xi_2} f(\xi_1, \xi_2, \xi_3, \xi_4)\, d\xi_3\, d\xi_2\, d\xi_1$$

$$= \sum_{i=1}^{n} H_i f(\xi_{1_i}, \xi_{2_i}, \xi_{3_i}, \xi_{4_i}) \tag{11.28}$$

Some such formulas, again in an affine symmetry fashion, are given in Table 11.4 [11.9].

TABLE 11.4 Gaussian Quadrature Formulas for Tetrahedra

Number and Locations of Sampling Points	Tetrahedronal Coordinates of Sampling Points	Weights (H_i/V)	Values of Coefficients	Degree of Accuracy
$n = 1$	$(\frac{1}{4}, \frac{1}{4}, \frac{1}{4}, \frac{1}{4})$	1		1
$n = 4$	(a, b, b, b) (b, a, b, b) (b, b, a, b) (b, b, b, a)	$\frac{1}{4}$	$a = \frac{1}{4}\left(1 + \frac{3}{\sqrt{5}}\right)$ $= 0.58541\,01970\,00000$ $b = \frac{1}{4}\left(1 - \frac{1}{\sqrt{5}}\right)$ $= 0.13819\,66010\,00000$	2
$n = 5$	$(\frac{1}{4}, \frac{1}{4}, \frac{1}{4}, \frac{1}{4})$	$-\frac{4}{5}$		3
	(a, b, b, b) (b, a, b, b) (b, b, a, b) (b, b, b, a)	$\frac{9}{20}$	$a = \frac{1}{2}$ $b = \frac{1}{6}$	

Example 11.8

Using both (a) the closed-form integral formula (10.34) and (b) the five-point Gaussian quadrature formula (Table 11.4), integrate the representative terms $(1, x, x^2, xy, x^3, x^2y, xyz)$ in a complete third-order polynomial over an arbitrary tetrahedron as shown in Fig. 11.3.

Solution. From Fig. 11.3 we have

$$x = 3\xi_2 + \xi_3 + \xi_4$$
$$y = 3\xi_3 + \xi_4$$
$$z = \xi_2 + 4\xi_4$$

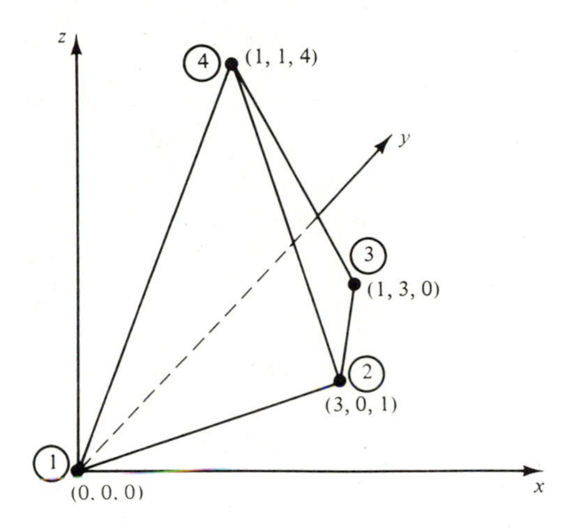

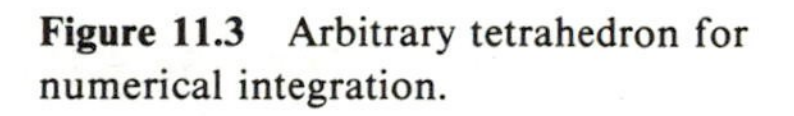
Figure 11.3 Arbitrary tetrahedron for numerical integration.

(a) *closed-form formula (10.34)*:

$$\int_{\text{vol}} (1)\, dV = V = \frac{1}{6}\begin{vmatrix} 1 & 1 & 1 & 1 \\ 0 & 3 & 1 & 1 \\ 0 & 0 & 3 & 1 \\ 0 & 1 & 0 & 4 \end{vmatrix} = \frac{17}{3}$$

$$\int_{\text{vol}} x\, dV = \int_{\text{vol}} (3\xi_2 + \xi_3 + \xi_4)\, dV = \frac{6V}{4!}(3 + 1 + 1) = \frac{85}{12}$$

$$\int_{\text{vol}} x^2\, dV = \int_{\text{vol}} (9\xi_2^2 + \xi_3^2 + \xi_4^2 + 6\xi_2\xi_3 + 2\xi_3\xi_4 + 6\xi_2\xi_4)\, dV$$

$$= \frac{6V}{5!}(9 \times 2 + 2 + 2 + 6 + 2 + 6) = \frac{51}{5}$$

$$\int_{\text{vol}} xy\, dV = \frac{34}{5}$$

$$\int_{\text{vol}} x^3\, dV = \int_{\text{vol}} (27\xi_2^3 + \xi_3^3 + \xi_4^3 + 27\xi_2^2\xi_3 + 27\xi_2^2\xi_4 + 9\xi_3^2\xi_2 + 3\xi_3^2\xi_4$$

$$+ 9\xi_4^2\xi_2 + 3\xi_4^2\xi_3 + 18\xi_2\xi_3\xi_4)\, dV$$

$$= \frac{6V}{6!}[(27 + 1 + 1)(3!) + (27 + 27 + 9 + 3 + 9 + 3)(2) + 18]$$

$$= \frac{493}{30}$$

$$\int_{\text{vol}} x^2 y \, dV = \frac{136}{15}$$

$$\int_{\text{vol}} xyz \, dV = \frac{374}{45}$$

(b) *Five-point Gaussian quadrature*:

$$\int_{\text{vol}} (1) \, dV = \sum_{i=1}^{5} H_i(1)_i = -\frac{4V}{5}(1) + \frac{9V}{20}(1 \times 4) = \frac{17}{3} \qquad \text{(exact!)}$$

$$\int_{\text{vol}} x \, dV = \sum_{i=1}^{5} H_i(3\xi_2 + \xi_3 + \xi_4)_i = -\frac{4V}{5}\left(\frac{3}{4} + \frac{1}{4} + \frac{1}{4}\right)$$

$$+ \frac{9V}{20}[(5b) + (3a + 2b) + 2(4b + a)] = \frac{85}{12} \qquad \text{(exact!)}$$

$$\int_{\text{vol}} x^2 \, dV = -\frac{4V}{5}\left(\frac{5}{4}\right)^2 + \frac{9V}{20}[(5b)^2 + (3a + 2b)^2 + 2(4b + a)^2] = \frac{51}{5} \qquad \text{(exact!)}$$

$$\int_{\text{vol}} xy \, dV = -\frac{4V}{5}\left(\frac{5}{4}\right)\left(\frac{4}{4}\right) + \frac{9V}{20}[20b^2 + (3a + 2b)(4b)$$

$$+ (4b + a)(3a + b) + (4b + a)(3b + a)] = \frac{34}{5} \qquad \text{(exact!)}$$

$$\int_{\text{vol}} x^3 \, dV = -\frac{4V}{5}\left(\frac{5}{4}\right)^3 + \frac{9V}{20}[(5b)^3 + (3a + 2b)^3 + 2(4b + a)^3] = \frac{493}{30} \qquad \text{(exact!)}$$

$$\int_{\text{vol}} x^2 y \, dV = -\frac{4V}{5}\left(\frac{5}{4}\right)^2 + \frac{9V}{20}[100b^3 + (3a + 2b)^2(4b) + (4b + a)^2(3a + b)$$

$$+ (4b + a)^2(3b + a)] = \frac{136}{15} \qquad \text{(exact!)}$$

$$\int_{\text{vol}} xyz \, dV = -\frac{4V}{5}\left(\frac{5}{4}\right)^2 + \frac{9V}{20}[100b^3 + (3a + 2b)(4b)(a + 4b) + (4b + a)(3a + b)$$

$$\times (5b) + (4b + a)(3b + a)(b + 4a)] = \frac{374}{45} \qquad \text{(exact!)}$$

11.2 CURVED ISOPARAMETRIC ELEMENTS [11.14]

Isoparametric elements are those for which the displacement functions are also used to describe the element geometry. Isoparametric elements usually are described by the *mapping* of nondimensionalized elements of regular shapes (i.e., rectangles, triangles, axisymmetric solids, tetrahedra, bricks, and curved surfaces with rectangular or triangular projections, etc.) into actual elements of distorted shapes (i.e., irregular nodal locations and curved edges and surfaces). The formulation of isoparametric elements is based on nondimensional local curvilinear, triangular, or tetrahedral coordinates so that numerical integration can be used.

For an isoparametric element, the geometric functions and the displacement functions are of equal orders. This is not necessary and on occasion it may be advantageous to use geometric and displacement functions with unequal orders. If the orders of the geometrtic functions are relatively higher, the element is called *superparametric.* If the orders of the geometric functions are relatively lower, the element is called *subparametric.* Intuitively, it appears that for some structures with very complex geometry but simple loadings (e.g., a pinched cylindrical shell [11.15]), superparametric elements might be of an advantage; the opposite might be true for subparametric elements.

Obviously, if the displacement functions for two adjacent elements satisfy compatibility along the common boundary, the geometric functions for the two distorted isoparametric elements will satisfy continuity along the common boundary with neither gaps nor overlaps.

The basic concept and detailed formulative procedures for isoparametric elements can best be described using two-dimensional quadrilateral elements as follows. The linear version of such type of elements was established in Ref. 11.16. Earlier fundamental works for variable-order two dimensional quadrilaterals and generalizations to other types of elements were given in Refs. 11.8, 11.17, and 11.18.

11.2.1 Linear Two-Dimensional Isoparametric Quadrilateral

For an 8-d.o.f. rectangular linear element as shown in Fig. 11.4a, the displacement functions are

$$\begin{aligned} u(x, y) &= c_1 + c_2 x + c_3 y + c_4 xy \\ v(x, y) &= c_1 + c_2 x + c_3 y + c_4 xy \end{aligned} \tag{11.29}$$

The x and y coordinates are so nondimensionalized that the four corner nodes have the coordinate values of ± 1 and the four constants in $u(x, y)$ are the same as those in $v(x, y)$.

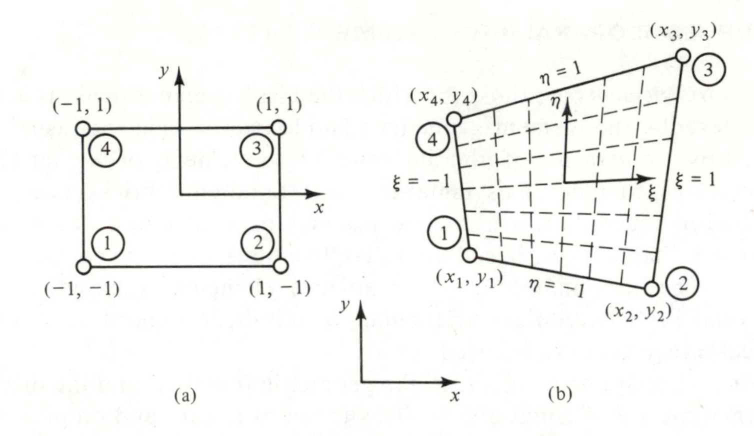

Figure 11.4 Linear two-dimensional isoparametric quadrilateral element mapped from a rectangular element: (a) rectangular element; (b) distorted element.

Substituting in the nodal values u_i and the appropriate coordinates, c_i's can be found. Through rearranging, we have

$$u = \sum_{i=1}^{4} N_i(x, y)u_i = \lfloor N_1 \quad N_2 \quad N_3 \quad N_4 \rfloor \begin{Bmatrix} u_1 \\ u_2 \\ u_3 \\ u_4 \end{Bmatrix}$$

$$v = \sum_{i=1}^{4} N_i(x, y)v_i = \lfloor N_1 \quad N_2 \quad N_3 \quad N_4 \rfloor \begin{Bmatrix} v_1 \\ v_2 \\ v_3 \\ v_4 \end{Bmatrix} \tag{11.30}$$

where

$$\begin{aligned} N_1(x, y) &= \tfrac{1}{4}(1 - x)(1 - y) \\ N_2(x, y) &= \tfrac{1}{4}(1 + x)(1 - y) \\ N_3(x, y) &= \tfrac{1}{4}(1 + x)(1 + y) \\ N_4(x, y) &= \tfrac{1}{4}(1 - x)(1 + y) \end{aligned} \tag{11.31}$$

where the nondimensional shape function N_i equals unity at point i and zero elsewhere.

Having the nondimensional rectangular element defined, we can now turn our attention to an actual quadrilateral isoparametric element as shown in Fig. 11.4b. While the four corners are defined by the global Cartesian coordinates (x, y), the element is formulated using curvilinear coordinates (ξ, η).

The shape functions $N_i(x, y)$ that define the displacement distributions of the rectangular element are now used in the same form but in curvilinear coordinates as $N_i(\xi, \eta)$ to describe the geometry of the distorted quadrilateral element, for which ξ and η coordinates take unit value on the element edges, the same as x and y coordinates do on the rectangular element.

Thus for the distorted element, the displacement functions are

$$u = \sum_{i=1}^{4} N_i(\xi, \eta) u_i$$
$$v = \sum_{i=1}^{4} N_i(\xi, \eta) v_i \tag{11.32}$$

and the geometric functions are

$$x = \sum_{i=1}^{4} N_i(\xi, \eta) x_i$$
$$y = \sum_{i=1}^{4} N_i(\xi, \eta) y_i \tag{11.33}$$

where

$$\begin{aligned} N_1(\xi, \eta) &= \tfrac{1}{4}(1 - \xi)(1 - \eta) \\ N_2(\xi, \eta) &= \tfrac{1}{4}(1 + \xi)(1 - \eta) \\ N_3(\xi, \eta) &= \tfrac{1}{4}(1 + \xi)(1 + \eta) \\ N_4(\xi, \eta) &= \tfrac{1}{4}(1 - \xi)(1 + \eta) \end{aligned} \tag{11.34}$$

Again, the function $N_i(\xi, \eta)$ takes the value of unity at point i and zero elsewhere.

Using the energy approach, the stiffness matrix for the subject rectangular element can be derived as, similar to Eq. (9.33),

$$[\mathbf{k}] = t \iint_{\text{Area}} [\mathbf{A}]^T[\mathbf{C}][\mathbf{A}]\, dx\, dy \tag{11.35}$$

where t is the thickness, the elasticity matrix $[\mathbf{C}]$ defines the stresses in terms of strains, and the strain matrix $[\mathbf{A}]$ relates strains to nodal displacements.

The elasticity matrices $[\mathbf{C}]$ were given in Eqs. (9.26a) and (9.26b) for plane stress and plane strain, respectively. For example, for the case of plane stress,

$$\begin{Bmatrix} \sigma_x \\ \sigma_y \\ \tau_{xy} \end{Bmatrix} = \frac{E}{1 - \nu^2} \begin{bmatrix} 1 & \nu & 0 \\ \nu & 1 & 0 \\ 0 & 0 & \dfrac{1 - \nu}{2} \end{bmatrix} \begin{Bmatrix} \epsilon_x \\ \epsilon_y \\ \gamma_{xy} \end{Bmatrix} \tag{11.36}$$

or symbolically,

$$\{\boldsymbol{\sigma}\} = [\mathbf{C}]\{\boldsymbol{\epsilon}\} \tag{11.36a}$$

The strain matrix is in the form

$$\begin{Bmatrix} \epsilon_x \\ \epsilon_y \\ \gamma_{xy} \end{Bmatrix} = \begin{Bmatrix} \dfrac{\partial u}{\partial x} \\ \dfrac{\partial v}{\partial y} \\ \dfrac{\partial u}{\partial y} + \dfrac{\partial v}{\partial x} \end{Bmatrix} = \begin{bmatrix} \dfrac{\partial N_1}{\partial x} & \dfrac{\partial N_2}{\partial x} & \dfrac{\partial N_3}{\partial x} & \dfrac{\partial N_4}{\partial x} & 0 & 0 & 0 & 0 \\ 0 & 0 & 0 & 0 & \dfrac{\partial N_1}{\partial y} & \dfrac{\partial N_2}{\partial y} & \dfrac{\partial N_3}{\partial y} & \dfrac{\partial N_4}{\partial y} \\ \dfrac{\partial N_1}{\partial y} & \dfrac{\partial N_2}{\partial y} & \dfrac{\partial N_3}{\partial y} & \dfrac{\partial N_4}{\partial y} & \dfrac{\partial N_1}{\partial x} & \dfrac{\partial N_2}{\partial x} & \dfrac{\partial N_3}{\partial x} & \dfrac{\partial N_4}{\partial x} \end{bmatrix} \begin{Bmatrix} u_1 \\ u_2 \\ u_3 \\ u_4 \\ v_1 \\ v_2 \\ v_3 \\ v_4 \end{Bmatrix} \tag{11.37}$$

or

$$\{\boldsymbol{\epsilon}\} = \begin{bmatrix} \dfrac{\partial N_i}{\partial x} & 0 \\ 0 & \dfrac{\partial N_i}{\partial y} \\ \dfrac{\partial N_i}{\partial y} & \dfrac{\partial N_i}{\partial x} \end{bmatrix} \begin{Bmatrix} u \\ v \end{Bmatrix} = [\mathbf{A}]\{\mathbf{q}\} \tag{11.37a}$$

As N_i is defined in terms of ξ and η in Eq. (11.34), it is necessary to change the derivatives from $\partial/\partial x$ and $\partial/\partial y$ to $\partial/\partial \xi$ and $\partial/\partial \eta$ using the chain rule of partial differentiation,

$$\begin{Bmatrix} \dfrac{\partial N_i}{\partial \xi} \\ \dfrac{\partial N_i}{\partial \eta} \end{Bmatrix} = \begin{bmatrix} \dfrac{\partial x}{\partial \xi} & \dfrac{\partial y}{\partial \xi} \\ \dfrac{\partial x}{\partial \eta} & \dfrac{\partial y}{\partial \eta} \end{bmatrix} \begin{Bmatrix} \dfrac{\partial N_i}{\partial x} \\ \dfrac{\partial N_i}{\partial y} \end{Bmatrix} = [\mathbf{J}] \begin{Bmatrix} \dfrac{\partial N_i}{\partial x} \\ \dfrac{\partial N_i}{\partial y} \end{Bmatrix} \tag{11.38}$$

where the Jacobian matrix $[\mathbf{J}]$ serves to transform the derivatives from (ξ, η) to (x, y) coordinates, and vice versa:

$$\underset{2\times 2}{[\mathbf{J}]} = \begin{bmatrix} \dfrac{\partial N_1}{\partial \xi} & \dfrac{\partial N_2}{\partial \xi} & \dfrac{\partial N_3}{\partial \xi} & \dfrac{\partial N_4}{\partial \xi} \\ \dfrac{\partial N_1}{\partial \eta} & \dfrac{\partial N_2}{\partial \eta} & \dfrac{\partial N_3}{\partial \eta} & \dfrac{\partial N_4}{\partial \eta} \end{bmatrix} \begin{bmatrix} x_1 & y_1 \\ x_2 & y_2 \\ x_3 & y_3 \\ x_4 & y_4 \end{bmatrix}$$

$$= \frac{1}{4} \begin{bmatrix} -(1-\eta) & 1-\eta & 1+\eta & -(1+\eta) \\ -(1-\xi) & -(1+\xi) & 1+\xi & 1-\xi \end{bmatrix} \begin{bmatrix} x_1 & y_1 \\ x_2 & y_2 \\ x_3 & y_3 \\ x_4 & y_4 \end{bmatrix} \tag{11.39}$$

As an example, let it be assumed that the four nodes of the distorted element are at (1, 1), (5, 2), (4, 5), and (2, 4), respectively. Then

$$[\mathbf{J}] = \frac{1}{4}\begin{bmatrix} 6 - 2\eta & 2 \\ -2\xi & 6 \end{bmatrix} \tag{11.40}$$

Equation (11.38) can now be written in inverted form:

$$\begin{Bmatrix} \dfrac{\partial N_i}{\partial x} \\ \dfrac{\partial N_i}{\partial y} \end{Bmatrix} = [\mathbf{J}]^{-1}\begin{Bmatrix} \dfrac{\partial N_i}{\partial \xi} \\ \dfrac{\partial N_i}{\partial \eta} \end{Bmatrix} = \frac{1}{9 + \xi - 3\eta}\begin{bmatrix} 6 & -2 \\ 2\xi & 6 - 2\eta \end{bmatrix}\begin{Bmatrix} \dfrac{\partial N_i}{\partial \xi} \\ \dfrac{\partial N_i}{\partial \eta} \end{Bmatrix} \tag{11.41}$$

$$[\mathbf{A}] = \frac{1}{8|J|}\begin{bmatrix} -2 - \xi + 3\eta & 4 + \xi - 3\eta & 2 - \xi + 3\eta & -4 + \xi - 3\eta & 0 & 0 & 0 & 0 \\ 0 & 0 & 0 & 0 & -3 + 2\xi + \eta & -3 - 2\xi + \eta & 3 + 4\xi - \eta & 3 - 4\xi - \eta \\ -3 + 2\xi + \eta & -3 - 2\xi + \eta & 3 + 4\xi - \eta & 3 - 4\xi - \eta & -2 - \xi + 3\eta & 4 + \xi - 3\eta & 2 - \xi + 3\eta & -4 + \xi - 3\eta \end{bmatrix} \tag{11.42}$$

Using the theorem of transformation of double integrals, we have [11.19]

$$dx\,dy = |J|\,d\xi\,d\eta = \tfrac{1}{4}(9 + \xi - 3\eta)\,d\xi\,d\eta \tag{11.43}$$

The determinant of the Jacobian matrix is referred to in mathematics as the *Jacobian* of x, y with respect to ξ, η and written as $\partial(x, y)/\partial(\xi, \eta)$.

It is obvious that if the distorted element is a rectangle, the Jacobian matrix **[J]** is a diagonal matrix populated with constants only (try Problem 11.8a). If the distorted element is a parallelogram, **[J]** is populated with constants only (try Problem 11.8b).

The stiffness matrix can now be obtained as

$$[\mathbf{k}] = t\int_{-1}^{1}\int_{-1}^{1} [\mathbf{A}]^T[\mathbf{C}][\mathbf{A}]|J|\,d\xi\,d\eta \tag{11.44}$$

Assuming that $\nu = 0.3$ and $Et = 10^4$, substituting Eqs. (11.36) and (11.42) in Eq. (11.44), and performing Gaussian quadrature with $n = 3$ yields the stiffness matrix as follows (only four consecutive digits or less are retained here):

$$[\mathbf{k}] = \begin{bmatrix} 3058 & -2106 & -807 & -146 & 1054 & -635 & -1405 & 986 \\ -2106 & 6292 & 498 & -4684 & -361 & -2451 & 611 & 2200 \\ -807 & 498 & 4345 & -4036 & -1405 & 886 & 854 & -335 \\ -146 & -4684 & -4036 & 8866 & 711 & 2200 & -60 & -2851 \\ 1054 & -361 & -1405 & 711 & 3991 & 1309 & -1892 & -3409 \\ -635 & -2451 & 886 & 2200 & 1309 & 4873 & -2809 & -3374 \\ -1405 & 611 & 854 & -60 & -1892 & -2809 & 5191 & -490 \\ 986 & 2200 & -335 & -2851 & -3409 & -3374 & -490 & 7273 \end{bmatrix} \quad (11.45)$$

As an example, we can find k_{11} as

$$\begin{aligned} k_{11} &= t \sum_{i=1}^{n} \sum_{j=1}^{n} H_i H_j f(\xi_i, \eta_j) \\ &= t \sum_{i=1}^{n} \sum_{j=1}^{n} H_i H_j \left(\frac{1}{64|J|}\right) \frac{E}{1-\nu^2} \\ &\quad \times \begin{Bmatrix} -2 - \xi_i + 3\eta_j \\ 0 \\ -3 + 2\xi_i + \eta_j \end{Bmatrix}^T \begin{bmatrix} 1 & \nu & 0 \\ \nu & 1 & 0 \\ 0 & 0 & \dfrac{1-\nu}{2} \end{bmatrix} \begin{Bmatrix} -2 - \xi_i + 3\eta_j \\ 0 \\ -3 + 2\xi_i + \eta_j \end{Bmatrix} \\ &= t \sum_{i=1}^{n} \sum_{j=1}^{n} \frac{H_i H_j E}{16 \times 0.91 \times (9 + \xi_i - 3\eta_j)} \\ &\quad \times (7.15 - 0.2\xi_i - 14.1\eta_j + 2.4\xi_i^2 - 4.6\xi_i\eta_j + 9.35\eta_j^2) \\ &= 0.30371795Et \qquad \text{for } n = 2 \\ &= 0.30581455Et \qquad \text{for } n = 3 \\ &= 0.30587903Et \qquad \text{for } n = 4 \end{aligned}$$

Although in this example both the inverse of [J] and the products of $[\mathbf{A}]^T[\mathbf{C}][\mathbf{A}]$ are formed analytically and explicitly, they do not have to be and are generally not formed explicitly in numerical integration.

11.2.2 Higher-Order Two-Dimensional Isoparametric Quadrilaterals

Four two-dimensional isoparametric quadrilateral elements with curved boundaries of quadratic and cubic orders are shown in Table 11.5. The rectangular version of these elements was presented in Chapter 9. Also given in the table are the geometric functions in the form of polynomials as well as

TABLE 11.5 Higher-Order Two-Dimensional Isoparametric Quadrilaterals

Number of Nodes n and Element Description	Geometric Functions: $x = \sum_1^n N_i(\xi, \eta)x_i$; $y = \sum_1^n N_i(\xi, \eta)y_i$	
	In Polynomial	Shape Function $N_i(\xi, \eta)$
$n = 8$ 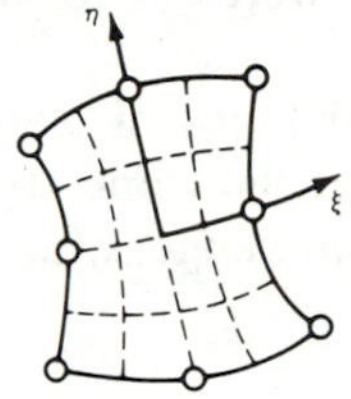	1 $\xi \quad \eta$ $\xi^2 \quad \xi\eta \quad \eta^2$ $\xi^2\eta \quad \xi\eta^2$	Corner nodes at $\xi_i = \pm 1$; $\eta_i = \pm 1$: $\frac{1}{4}(1 + \xi\xi_i)(1 + \eta\eta_i) - \frac{1}{4}(1 - \xi^2)(1 + \eta\eta_i)$ $-\frac{1}{4}(1 + \xi\xi_i)(1 - \eta^2)$ Side nodes at $\xi_i = 0$; $\eta_i = \pm 1$: $\frac{1}{2}(1 - \xi^2)(1 + \eta\eta_i)$; Side nodes at $\xi_i = \pm 1$; $\eta_i = \upsilon$ $\frac{1}{2}(1 - \eta^2)(1 + \xi\xi_i)$
$n = 9$ 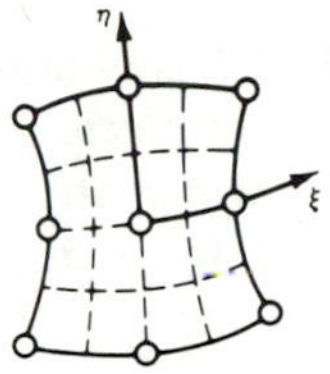	1 $\xi \quad \eta$ $\xi^2 \quad \xi\eta \quad \eta^2$ $\xi^2\eta \quad \xi\eta^2$ $\xi^2\eta^2$	Corner nodes at $\xi_i = \pm 1$; $\eta_i = \pm 1$: $\frac{1}{4}\xi\xi_i\eta\eta_i(1 + \xi\xi_i)(1 + \eta\eta_i)$ Nodes at $\xi_i = 0$; $\eta_i = \pm 1$: $\frac{1}{2}\eta\eta_i(1 + \eta\eta_i)(1 - \xi^2)$ Nodes at $\xi_i = \pm 1$; $\eta_i = 0$: $\frac{1}{2}\xi\xi_i(1 + \xi\xi_i)(1 - \eta^2)$ Nodes at origin: $(1 - \xi^2)(1 - \eta^2)$
$n = 12$ 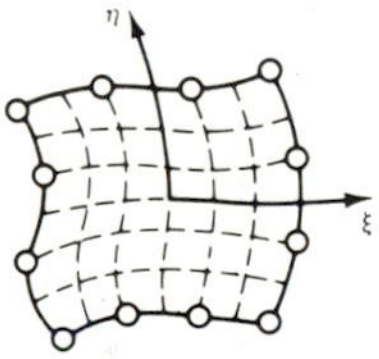	1 $\xi \quad \eta$ $\xi^2 \quad \xi\eta \quad \eta^2$ $\xi^3 \quad \xi^2\eta \quad \xi\eta^2 \quad \eta^3$ $\xi^3\eta \quad \xi\eta^3$	Corner nodes at $\xi_i = \pm 1$; $\eta_i = \pm 1$: $\frac{1}{32}(1 + \xi\xi_i)(1 + \eta\eta_i)[-10 + 9(\xi^2 + \eta^2)]$ Nodes at $\xi_i = \pm 1/3$; $\eta_i = \pm 1$: $\frac{9}{32}(1 + 9\xi\xi_i)(1 + \eta\eta_i)(1 - \xi^2)$ Nodes at $\xi_i = \pm 1$; $\eta_i = \pm 1/3$: $\frac{9}{32}(1 + \xi\xi_i)(1 + 9\eta\eta_i)(1 - \eta^2)$
$n = 16$ 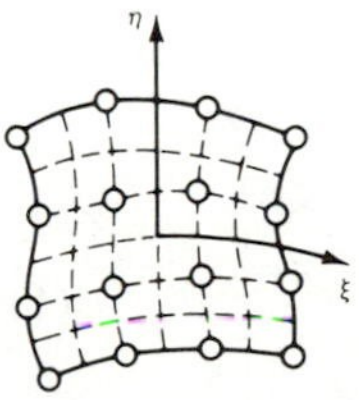	1 $\xi \quad \eta$ $\xi^2 \quad \xi\eta \quad \eta^2$ $\xi^3 \quad \xi^2\eta \quad \xi\eta^2 \quad \eta^3$ $\xi^3\eta \quad \xi^2\eta^2 \quad \xi\eta^3$ $\xi^3\eta^2 \quad \xi^2\eta^3$ $\xi^3\eta^3$	$N_i(\xi, \eta) = L_j(\xi)L_k(\eta)$ where $\left\{ \begin{matrix} j = 1 \\ j = 2 \\ k = 1 \\ k = 2 \end{matrix} \right\}$ for nodes at $\left\{ \begin{matrix} \xi_i = \pm 1 \\ \xi_i = \pm 1/3 \\ \eta_i = \pm 1 \\ \eta_i = \pm 1/3 \end{matrix} \right\}$ $L_1(\alpha) = (1 + \alpha\alpha_i)(9\alpha^2 - 1) \div 16$ $L_2(\alpha) = 9(1 - \alpha^2)(1 + 9\alpha\alpha_i) \div 16$

shape functions. All the shape functions can, of course, be obtained by evaluating the element polynomial function at the nodes and solving for the constant coefficients. For each element, the curvilinear coordinates ξ and η take unit value on the boundaries. The shape function N_i takes unit value at node i and vanishes at the rest of the nodes.

The geometric functions for the first three elements were suggested by Ergatoudis et al. [11.18] and those for the 16-node element were due to Argyris [10.25].

For practical purposes, we could combine the four types of elements into one. For example, it is possible to have an element with one straight edge, two quadratically curved edges, and a cubically curved edge, with some or no interior nodes.

11.2.3 Isoparametric Hexahedra

The stiffness formulation for the isoparametric hexahedra is obtained by straightforward extension from those derived for the isoparametric quadrilaterals. In the three-dimensional case, the strain-displacement matrix [A] is defined in Eq. (10.37) instead of Eq. (11.37). The definition of Jacobian matrix [J] now becomes, instead of Eqs. (11.38) and (11.39),

$$\begin{Bmatrix} \dfrac{\partial N_i}{\partial \xi} \\ \dfrac{\partial N_i}{\partial \eta} \\ \dfrac{\partial N_i}{\partial \zeta} \end{Bmatrix} = \begin{bmatrix} \dfrac{\partial x}{\partial \xi} & \dfrac{\partial y}{\partial \xi} & \dfrac{\partial z}{\partial \xi} \\ \dfrac{\partial x}{\partial \eta} & \dfrac{\partial y}{\partial \eta} & \dfrac{\partial z}{\partial \eta} \\ \dfrac{\partial x}{\partial \zeta} & \dfrac{\partial y}{\partial \zeta} & \dfrac{\partial z}{\partial \zeta} \end{bmatrix} \begin{Bmatrix} \dfrac{\partial N_i}{\partial x} \\ \dfrac{\partial N_i}{\partial y} \\ \dfrac{\partial N_i}{\partial z} \end{Bmatrix} = [\mathbf{J}] \begin{Bmatrix} \dfrac{\partial N_i}{\partial x} \\ \dfrac{\partial N_i}{\partial y} \\ \dfrac{\partial N_i}{\partial z} \end{Bmatrix} \tag{11.46}$$

where

$$\underset{3\times 3}{[\mathbf{J}]} = \begin{bmatrix} \dfrac{\partial N_1}{\partial \xi} & \dfrac{\partial N_2}{\partial \xi} & \dfrac{\partial N_3}{\partial \xi} & \cdots \\ \dfrac{\partial N_1}{\partial \eta} & \dfrac{\partial N_2}{\partial \eta} & \dfrac{\partial N_3}{\partial \eta} & \cdots \\ \dfrac{\partial N_1}{\partial \zeta} & \dfrac{\partial N_2}{\partial \zeta} & \dfrac{\partial N_3}{\partial \zeta} & \cdots \end{bmatrix} \begin{bmatrix} x_1 & y_1 & z_1 \\ x_2 & y_2 & z_2 \\ x_3 & y_3 & z_3 \\ \vdots & \vdots & \vdots \end{bmatrix} \tag{11.47}$$

The stiffness matrix is obtained as

$$[\mathbf{k}] = \int_{-1}^{1}\int_{-1}^{1}\int_{-1}^{1} [\mathbf{A}]^T[\mathbf{C}][\mathbf{A}]|J|\, d\xi\, d\eta\, d\zeta \tag{11.48}$$

Four isoparametric hexahedronal elements with 8 [10.10, 10.12], 20 [10.12, 10.21–10.23], 32 [10.12, 10.24], and 64 [10.25] nodes, respectively, are shown in Table 11.6. Each node has three degrees of freedom: u, v, and w. The rectangular hexahedronal versions of these elements were given in Chapter 10. Also shown in the table are the geometric functions in the form of polynomials as well as shape functions.

A significant application of the curved isoparametric hexahedra is for thick shell analysis [11.20, 11.21]. For example, the quadratically curved 20-node 60-d.o.f. isoparametric hexahedron could be used as a thick shell element by using quadratic shape functions in the surface coordinates and linear shape functions through the thickness, as shown in Fig. 11.5a. This element could be converted to an equivalent eight-node thick shell element as shown in Fig. 11.5b by transforming the six degrees of freedom of each surface node pair to the five d.o.f.'s at each midsurface node: three orthogonal displacements and two rotations about two surface coordinates, respectively.

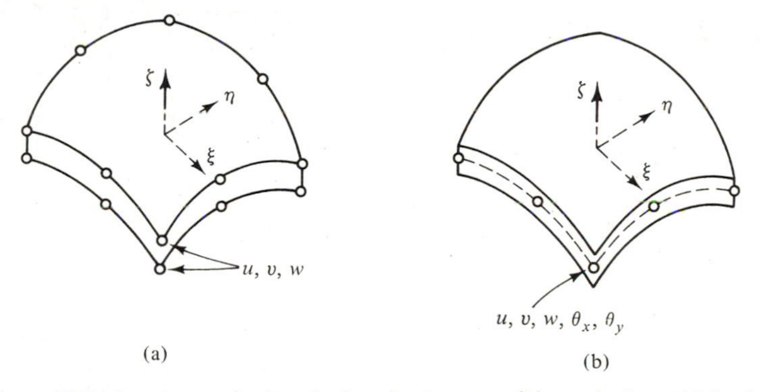

Figure 11.5 Isoparametric hexahedronal element; (b) equivalent thick shell element.

Elements of such type have been proven very successful for thick shell analysis due to their simplicity and ability to account for shear deformation. But they are too stiff for thin shell analysis due to the spurious shear strain energy. It was shown that [10.22, 10.23] this excessive bending stiffness can be avoided by reducing the order of numerical integration applied to certain terms related to shear strain energy. Thus improved performance of this type of element could be obtained for thin plate and shell analyses.

11.2.4 Isoparametric Quadrilaterals and Hexahedra with Displacement Derivatives as Degrees of Freedom

In Chapters 9 and 10 we introduced rectangular and rectangular hexahedronal elements with nodes only at the corners but with displacement derivatives also as degrees of freedom. Due to their relatively narrower resulting

TABLE 11.6 Some Isoparametric Hexahedra

Number of Nodes n and Element Description	Geometric Functions $x = \sum_1^n N_i(\xi, \eta, \zeta)x_i;\ y = \sum_1^n N_i(\xi, \eta, \zeta)y_i;\ z = \sum_1^n N_i(\xi, \eta, \zeta)z_i$	
	In Polynomial	Shape Functions $N_i(\xi, \eta, \zeta)$
$n = 8$ 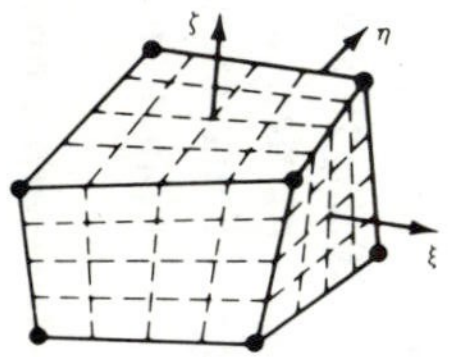	8 terms: 1 $\xi \quad \eta \quad \zeta$ $\xi\eta \quad \eta\zeta \quad \zeta\xi$ $\xi\eta\zeta$	8 corners at $\xi_i = \pm 1,\ \eta_i = \pm 1,\ \zeta_i = \pm 1$: $\frac{1}{8}(1 + \xi\xi_i)(1 + \eta\eta_i)(1 + \zeta\zeta_i)$
$n = 20$ 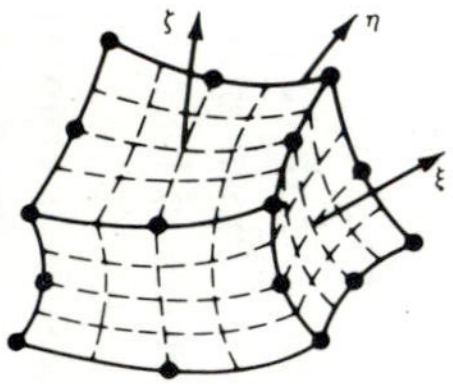	20 terms: 1 $\xi \quad \eta \quad \zeta$ $\xi^2 \quad \eta^2 \quad \zeta^2 \quad \xi\eta \quad \eta\zeta \quad \zeta\xi$ $\xi^2\eta \quad \xi^2\zeta \quad \eta^2\xi \quad \eta^2\zeta$ $\zeta^2\xi \quad \zeta^2\eta \quad \xi\eta\zeta$ $\xi^2\eta\zeta \quad \eta^2\xi\zeta \quad \zeta^2\xi\eta$	8 corners at $\xi_i = \pm 1;\ \eta_i = \pm 1;\ \zeta_i = \pm 1$: $\frac{1}{8}(1 + \xi\xi_i)(1 + \eta\eta_i)(1 + \zeta\zeta_i)(\xi\xi_i + \eta\eta_i + \zeta\zeta_i - 2)$ 4 side nodes at $\xi_i = 0;\ \eta_i = \zeta_i = \pm 1$: $\frac{1}{4}(1 - \xi^2)(1 + \eta\eta_i)(1 + \zeta\zeta_i)$ 4 side nodes at $\eta_i = 0;\ \xi_i = \zeta_i = \pm 1$: $\frac{1}{4}(1 - \eta^2)(1 + \xi\xi_i)(1 + \zeta\zeta_i)$ 4 side nodes at $\zeta_i = 0;\ \xi_i = \eta_i = \pm 1$: $\frac{1}{4}(1 - \zeta^2)(1 + \xi\xi_i)(1 + \eta\eta_i)$

$n = 32$	Same 20 terms for the 20-node element plus 12 terms: $\xi^3 \quad \eta^3 \quad \zeta^3$ $\xi^3\eta \quad \xi^3\zeta \quad \eta^3\xi$ $\eta^3\zeta \quad \zeta^3\xi \quad \zeta^3\eta$ $\xi^3\eta\zeta \quad \eta^3\zeta\xi \quad \zeta^3\xi\eta$	8 corners at $\xi_i = \pm 1;\ \eta_i = \pm 1;\ \zeta_i = \pm 1$: $(1 + \xi\xi_i)(1 + \eta\eta_i)(1 + \zeta\zeta_i)$ $\times[-19 + 9(\xi^2 + \eta^2 + \zeta^2)] \div 64$ 8 side nodes at $\xi_i = \pm 1/3;\ \eta_i = \pm 1;\ \zeta_i = \pm 1$: $9(1 - \xi^2)(1 + \eta\eta_i)(1 + \zeta\zeta_i)(1 + 9\xi\xi_i) \div 64$ 8 side nodes at $\eta_i = \pm 1/3;\ \xi_i = \pm 1;\ \zeta_i = \pm 1$: $9(1 - \eta^2)(1 + \xi\xi_i)(1 + \zeta\zeta_i)(1 + 9\eta\eta_i) \div 64$ 8 side nodes at $\zeta_i = \pm 1/3;\ \xi_i = \pm 1;\ \eta_i = \pm 1$: $9(1 - \zeta^2)(1 + \xi\xi_i)(1 + \eta\eta_i)(1 + 9\zeta\zeta_i) \div 64$
$n = 64$	64 terms resulted from the product of: $(a_1 + a_2\xi + a_3\xi^2 + a_4\xi^3)$ $(b_1 + b_2\eta + b_3\eta^2 + b_4\eta^3)$ $(c_1 + c_2\zeta + c_3\zeta^2 + c_4\zeta^3)$	$N_i(\xi, \eta, \zeta) = L_j(\xi)L_k(\eta)L_l(\zeta)$ where $\begin{Bmatrix} j = 1 \\ j = 2 \\ k = 1 \\ k = 2 \\ l = 1 \\ l = 2 \end{Bmatrix}$ for nodes at $\begin{Bmatrix} \xi_i = \pm 1 \\ \xi_i = \pm 1/3 \\ \eta_i = \pm 1 \\ \eta_i = \pm 1/3 \\ \zeta_i = \pm 1 \\ \zeta_i = \pm 1/3 \end{Bmatrix}$ $L_1(\alpha) = (1 + \alpha\alpha_i)(9\alpha^2 - 1) \div 16$ $L_2(\alpha) = 9(1 - \alpha^2)(1 + 9\alpha\alpha_i) \div 16$

bandwidth and relatively higher order displacement functions, such elements could be relatively more efficient. Such efficiency could be further enhanced if the elements are distorted in isoparametric form for more practical geometric modeling.

Two very sophisticated and efficient isoparametric elements of such type, a four-node 32-d.o.f. quadrilateral and an eight-node 96-d.o.f. hexahedron, are described in Table 11.7. As seen in the third column of the table, the geometric functions now involve not only the global coordinates x_i, y_i, and z_i at the node i, but also the derivatives of these coordinates with respect to the local curvilinear coordinates ξ, η, and ζ.

11.2.5 Isoparametric Triangles and Tetrahedra

Two higher-order isoparametric triangles and two higher-order isoparametric tetrahedra are described in Table 11.8. Also shown in the table are the assumptions for nodal degrees of freedom and the geometric functions. The original elements with straight edges and flat faces were described in Chapters 9 and 10, respectively.

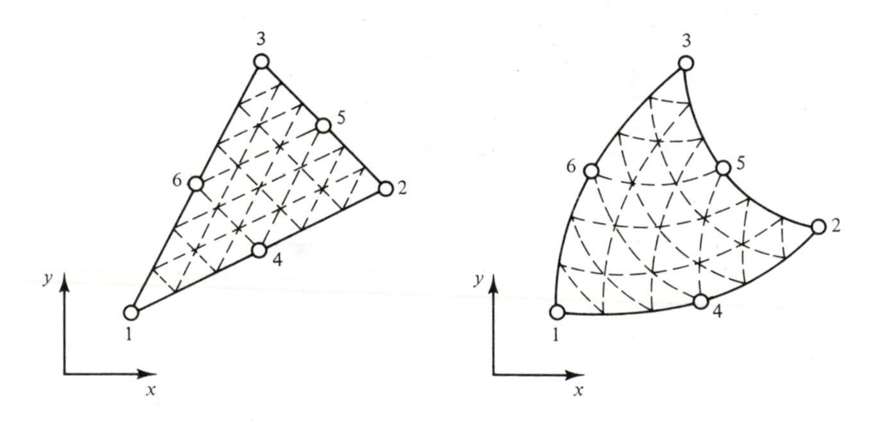

Figure 11.6 Six-node 12-d.o.f. plane stress triangle and a distorted isoparametric triangle.

To illustrate the derivation of the isoparametric triangles and tetrahedra, let us derive the stiffness coefficient k_{11} relating F_{x_1} to u_1 for a six-node 12-d.o.f. plane stress triangle as shown in Fig. 11.6.

TABLE 11.7 Some Isoparametric Elements with Displacement Derivatives at Nodes

Number of Nodes n and Element Description	Degrees of Freedom at Each Node	Geometric Functions for x, y, and z
$n = 4$ 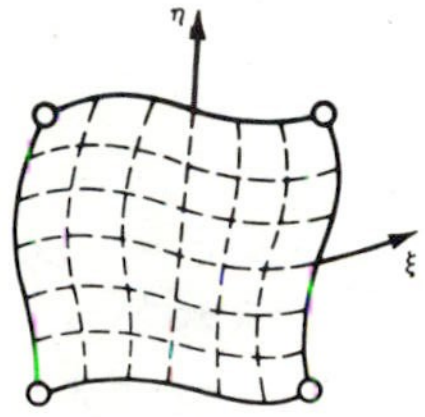	u, u_x, u_y, u_{xy} v, v_x, v_y, v_{xy}	$x(\xi, \eta) = (a_1 + a_2\xi + a_3\xi^2 + a_4\xi^3)(b_1 + b_2\eta + b_3\eta^2 + b_4\eta^3)$ $= \sum_1^4 \left[H(\xi)H(\eta)x_i + G(\xi)H(\eta)\left(\frac{\partial x}{\partial \xi}\right)_i + H(\xi)G(\eta)\left(\frac{\partial x}{\partial \eta}\right)_i + G(\xi)G(\eta)\left(\frac{\partial^2 x}{\partial \xi \partial \eta}\right)_i \right]$ where $H(\alpha) = (2 + 3\alpha\alpha_i - \alpha^3\alpha_i) \div 4$ $G(\alpha) = (-\alpha_i - \alpha + \alpha^2\alpha_i + \alpha^3) \div 4$
$n = 8$ 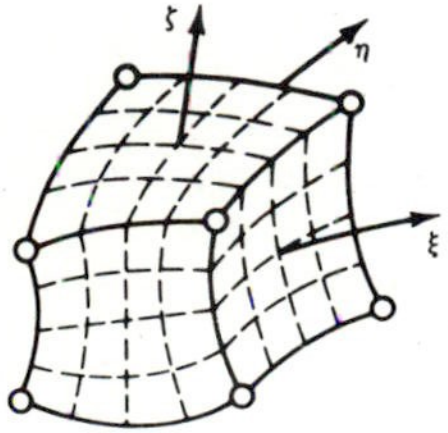	u, u_x, u_y, u_z v, v_x, v_y, v_z w, w_x, w_y, w_z	$x(\xi, \eta, \zeta) = \sum_1^8 \left[H(\xi)H(\eta)H(\zeta)x_i + G(\xi)H(\eta)H(\zeta)\left(\frac{\partial x}{\partial \xi}\right)_i + H(\xi)G(\eta)H(\zeta)\left(\frac{\partial x}{\partial \eta}\right)_i + H(\xi)H(\eta)G(\zeta)\left(\frac{\partial x}{\partial \zeta}\right)_i \right]$ where $H(\alpha) = (2 + 3\alpha\alpha_i - \alpha^3\alpha_i) \div 4$ $G(\alpha) = (-\alpha_i - \alpha + \alpha^2\alpha_i + \alpha^3) \div 4$

TABLE 11.8 Some Isoparametric Triangles and Tetrahedra

Number of Nodes n and Element Description	Degrees of Freedom at Each Node	Geometric Functions x, y, or z
$n = 10$ 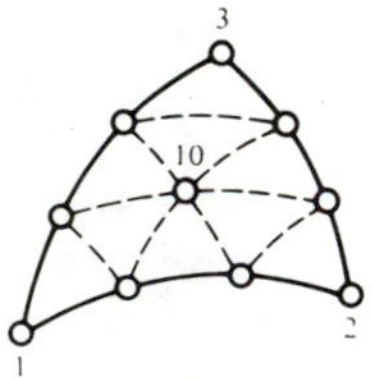	u, v	$x = \sum_{i=1}^{3} \frac{1}{2}\xi_i(3\xi_i - 1)(3\xi_i - 2)x_i$ $+ \sum_{4}^{9} N_i x_i + 27\xi_1\xi_2\xi_3 x_{10}$ where $N_i = \frac{9}{2}\xi_j\xi_k(3\xi_j - 1)$ for node i at 1/3 distance along edge jk from node j
$n = 4$ 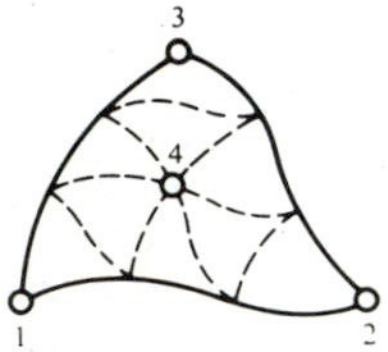	u, u_x, u_y v, v_x, v_y For Node 4: u, v only	$x = \sum_{i=1}^{3} \left\{ (\xi_i^3 + 3\xi_i^2\xi_j + 3\xi_i^2\xi_k - 7\xi_i\xi_j\xi_k)x_i + (\xi_{ji}C_{ij} + \xi_{ki}C_{ik})\left(\frac{\partial x}{\partial \xi}\right)_i + (\eta_{ji}C_{ij} + \eta_{ki}C_{ik})\left(\frac{\partial x}{\partial \eta}\right)_i \right\} + 27\xi_1\xi_2\xi_3 x_4$ where $\xi_{ij} = (\xi)_i - (\xi)_j$; $\eta_{ij} = (\eta)_i - (\eta)_j$; $C_{ij} = \xi_i^2\xi_j - \xi_i\xi_j\xi_k$ with i, j, k in cyclic permutation

$n = 16$ 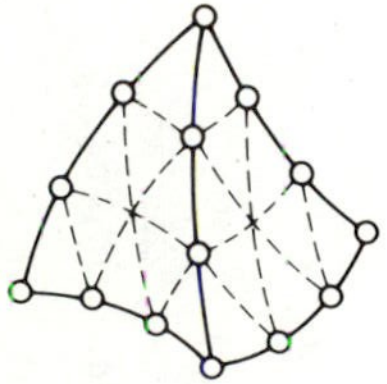	u, v, w	$x = \sum_{1}^{4} \frac{1}{2}\xi_i(3\xi_i - 1)(3\xi_i - 2)x_i + \sum_{5}^{16} N_i x_i$ where $N_i = \frac{9}{2}\xi_j\xi_k(3\xi_j - 1)$ for node i at $\frac{1}{3}$ distance along edge jk from node j
$n = 4$ 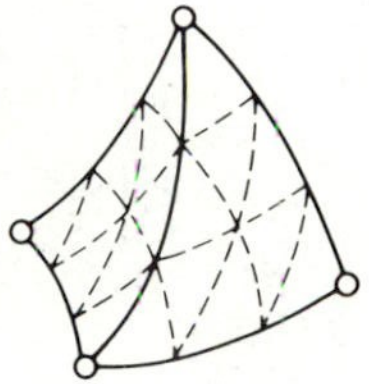	u, u_x, u_y, u_z v, v_x, v_y, v_x w, w_x, w_y, w_z	$x = \sum_{i=1}^{4} \Big\{ [\xi_i + \xi_i^2 - \xi_i^3 - \xi_i(\xi_j^2 + \xi_k^2 + \xi_l^2)]x_i$ $+ \frac{1}{2}(\xi_{ij}C_{ij} + \xi_{ik}C_{ik} + \xi_{il}C_{il})\left(\frac{\partial x}{\partial \xi}\right)_i$ $+ \frac{1}{2}(\eta_{ij}C_{ij} + \eta_{ik}C_{ik} + \eta_{il}C_{il})\left(\frac{\partial x}{\partial \eta}\right)_i$ $+ \frac{1}{2}(\zeta_{ij}C_{ij} + \zeta_{ik}C_{ik} + \zeta_{il}C_{il})\left(\frac{\partial x}{\partial \zeta}\right)_i \Big\}$ where $\xi_{ij} = (\xi)_i - (\xi)_j$; $\eta_{ij} = (\eta)_i - (\eta)_j$; $\zeta_{ij} = (\zeta)_i - (\zeta)_j$; $C_{ij} = \xi_i\xi_j^2 - \xi_i^2\xi_j - \xi_i\xi_j$ with i, j, k in cyclic permutation

The displacement or the geometric functions for the element are, from Eq. (9.105),

$$\begin{aligned} u(\xi_1, \xi_2, \xi_3) &= \sum_{1}^{6} N_i(\xi_1, \xi_2, \xi_3) u_i \\ &= \xi_1(2\xi_1 - 1)u_1 + \xi_2(2\xi_2 - 1)u_2 + \xi_3(2\xi_3 - 1)u_3 \\ &\quad + 4\xi_1\xi_2 u_4 + 4\xi_2\xi_3 u_5 + 4\xi_3\xi_1 u_6 \end{aligned} \tag{11.49}$$

and the same for v.

The strain displacement relations are

$$\begin{Bmatrix} \epsilon_x \\ \epsilon_y \\ \gamma_{xy} \end{Bmatrix} = \begin{Bmatrix} \dfrac{\partial u}{\partial x} \\ \dfrac{\partial v}{\partial y} \\ \dfrac{\partial u}{\partial y} + \dfrac{\partial v}{\partial x} \end{Bmatrix} = \sum_{i=1}^{6} \begin{Bmatrix} \dfrac{\partial N_i}{\partial x} u_i \\ \dfrac{\partial N_i}{\partial y} v_i \\ \dfrac{\partial N_i}{\partial y} u_i + \dfrac{\partial N_i}{\partial x} v_i \end{Bmatrix} \tag{11.50}$$

The transformation relations are the same as Eq. (11.38), with the Jacobian matrix defined as, similar to Eq. (11.39),

$$[\mathbf{J}] = \begin{bmatrix} \dfrac{\partial N_1}{\partial \xi} & \dfrac{\partial N_2}{\partial \xi} & \cdots & \dfrac{\partial N_6}{\partial \xi} \\ \dfrac{\partial N_1}{\partial \eta} & \dfrac{\partial N_2}{\partial \eta} & \cdots & \dfrac{\partial N_6}{\partial \eta} \end{bmatrix} \begin{bmatrix} x_1 & y_1 \\ x_2 & y_2 \\ \vdots & \vdots \\ x_6 & y_6 \end{bmatrix} \tag{11.51}$$

Let us now relate one triangular coordinate in terms of the other two,

$$\begin{aligned} \xi_1 &= \xi \\ \xi_2 &= \eta \\ \xi_3 &= 1 - \xi - \eta \end{aligned} \tag{11.52}$$

Then

$$\begin{aligned} \frac{\partial N_i}{\partial \xi} &= \frac{\partial N_i}{\partial \xi_1}\frac{\partial \xi_1}{\partial \xi} + \frac{\partial N_i}{\partial \xi_2}\frac{\partial \xi_2}{\partial \xi} + \frac{\partial N_i}{\partial \xi_3}\frac{\partial \xi_3}{\partial \xi} \\ &= \frac{\partial N_i}{\partial \xi_1} - \frac{\partial N_i}{\partial \xi_3} \end{aligned} \tag{11.53a}$$

$$\frac{\partial N_i}{\partial \eta} = \frac{\partial N_i}{\partial \xi_2} - \frac{\partial N_i}{\partial \xi_3} \tag{11.53b}$$

For simplicity, let us assume that this isoparametric element has straight edges and that the six nodes be given some specific values,

$$[\mathbf{J}] = \begin{bmatrix} 4\xi_1 - 1 & 0 & -4\xi_3 + 1 & 4\xi_2 & -4\xi_2 & 4\xi_3 - 4\xi_1 \\ 0 & 4\xi_2 - 1 & -4\xi_3 + 1 & 4\xi_1 & 4\xi_3 - 4\xi_2 & -4\xi_1 \end{bmatrix} \begin{bmatrix} 1 & 1 \\ 5 & 3 \\ 3 & 5 \\ 3 & 2 \\ 4 & 4 \\ 2 & 3 \end{bmatrix}$$

$$= \begin{bmatrix} -2 & -4 \\ 2 & -2 \end{bmatrix} \tag{11.54}$$

Substituting Eq. (11.54) in Eq. (11.38) or (11.41) gives

$$\begin{Bmatrix} \dfrac{\partial N_1}{\partial x} \\ \dfrac{\partial N_1}{\partial y} \end{Bmatrix} = \frac{1}{12}\begin{bmatrix} -2 & 4 \\ -2 & -2 \end{bmatrix} \begin{Bmatrix} \dfrac{\partial N_1}{\partial \xi} \\ \dfrac{\partial N_1}{\partial \eta} \end{Bmatrix} = \frac{1}{6}\begin{Bmatrix} 1 - 4\xi_1 \\ 1 - 4\xi_1 \end{Bmatrix} \tag{11.55}$$

We finally have, from Eq. (11.35),

$$k_{11} = \iint_A \left\lfloor \frac{\partial N_1}{\partial x} \quad 0 \quad \frac{\partial N_1}{\partial y} \right\rfloor \left(\frac{Et}{1 - \nu^2}\right) \begin{bmatrix} 1 & \nu & 0 \\ \nu & 1 & 0 \\ 0 & 0 & \dfrac{1 - \nu}{2} \end{bmatrix} \begin{Bmatrix} \dfrac{\partial N_1}{\partial x} \\ 0 \\ \dfrac{\partial N_1}{\partial y} \end{Bmatrix} dx\,dy$$

$$= \frac{Et}{72(1 - \nu^2)} \int_0^1 \int_0^{1-\xi_2} (1 - 4\xi_1)^2 (3 - \nu)|J|\, d\xi_1\, d\xi_2$$

$$= \frac{Et(3 - \nu)}{6(1 - \nu^2)} \sum_{i=1}^{n} H_i (1 - 4\xi_1)_i^2$$

Using the three-point Gaussian quadrature formula, from Table 11.3,

$$k_{11} = \frac{Et(3 - \nu)}{6(1 - \nu^2)} [\tfrac{1}{3}(1 - 4 \times \tfrac{1}{2})^2 + \tfrac{1}{3}(1)^2 + \tfrac{1}{3}(1 - 4 \times \tfrac{1}{2})^2](A)$$

For the triangle defined in the transformed coordinates ξ and η, the area is $\frac{1}{2}$; then

$$k_{11} = \frac{Et(3 - \nu)}{12(1 - \nu^2)} \tag{11.56}$$

This answer can be verified using an alternative procedure outlined in Eqs. (9.105) to (9.109) plus the closed-form integration formula (9.103). In fact, this was given as Problem 9.14.

The foregoing procedure can be straightforwardly extended to the case of isoparametric tetrahedra. Basically, Eq. (11.52) becomes

$$\begin{aligned} \xi_1 &= \xi \\ \xi_2 &= \eta \\ \xi_3 &= \zeta \\ \xi_4 &= 1 - \xi - \eta - \zeta \end{aligned} \tag{11.57}$$

Equation (11.53a) becomes

$$\frac{\partial N_i}{\partial \xi} = \frac{\partial N_i}{\partial \xi_1} - \frac{\partial N_i}{\partial \xi_4} \tag{11.58}$$

The stiffness matrix is obtained in the form

$$[\mathbf{k}] = \int_0^1 \int_0^{1-\xi_3} \int_0^{1-\xi_2-\xi_3} [\mathbf{A}]^T[\mathbf{C}][\mathbf{A}]|J|\, d\xi_1\, d\xi_2\, d\xi_3 \tag{11.59}$$

11.3 CONCLUDING REMARKS

In this chapter we have introduced the fundamental method of numerical integration in one, two, and three dimensions. For the two-dimensional case, we have treated rectangles and triangles. For the three-dimensional case, we have dealt with rectangular hexahedra and tetrahedra. Associated with each method of numerical integration, formulas and coefficients are given and examples are illustrated in detail.

Numerical integrations and isoparametric elements are synonymous. We have introduced the basic methods for the formulations of curved isoparametric elements in the forms of triangles, quadrilaterals, hexahedra, and tetrahedra. The assumptions for nodes and geometric (and displacement) functions for the various isoparametric elements are described in Tables 11.5 through 11.8. Examples for deriving the explicit stiffness coefficients including the process of numerical integration are given for a four-node eight-d.o.f. quadrilateral and a six-node 12-d.o.f. triangle.

Since the computer operations become cheaper and faster all the time, the trend in finite element formulations is to use more numerical integrations instead of manual integrations. Because of its accuracy and efficiency in terms of minimum number of integration points, and also because of its ability to handle variables with noninteger powers in the stiffness coefficients such as those for curved shell elements and crack-tip singular elements, Gaussian quadrature is exclusively recommended.

It has been established that by reducing the orders of numerical integrations of certain terms related to shear strain energy, curved isoparametric hexahedra can be used effectively for plate and shell analyses [10.22, 10.23].

With the advent of computer graphics, it is possible that the geometric functions and the displacement functions be assumed separately so that the geometric data base stored in the computer-aided design graphics system for realistic structures can be used to fit the geometric functions. Such a concept has been successfully demonstrated using rational B-spline geometric surface fit for an efficient four-node 48-d.o.f. quadrilateral shell element based on thin shell theory [11.15].

A detailed discussion on minimum order of numerical integration required for convergence, order of integration for no loss of convergence, and matrix singularity due to numerical integration is given in the text by Zienkiewicz [11.13]. Some applications of isoparametric elements in two- and three-dimensional stress analysis with extensive graphical illustrations are also given in the same text [11.13].

REFERENCES

11.1. Ralston, A., *A First Course in Numerical Analysis*, McGraw-Hill Book Company, New York, 1965, pp. 114–117, 90, and 103.

11.2. Loxan, A. N., Davis, N., and Levenson, A., "Table of Zeros of the Legendre Polynomials of Order 1–16 and the Weight Coefficients for Gauss' Mechanical Quadrature Formula," *Bulletin of the American Mathematical Society*, Vol. 48, 1942, pp. 739–743.

11.3. Bathe, K.-J., and Wilson, E. L., *Numerical Methods in Finite Element Analysis*, Prentice-Hall, Inc., Englewood Cliffs, N.J., 1976, pp. 151–162.

11.4. Irons, B. M., "Quadrature Rules for Brick Based Finite Elements," *International Journal for Numerical Methods in Engineering*, Vol. 3, 1971, pp. 293–294.

11.5. Hellen, T. K., "Effective Quadrature Rules for Quadratic Solid Isoparametric Finite Elements," *Internal Journal for Numerical Methods in Engineering*, Vol. 4, 1972, pp. 597–600.

11.6. Hammer, P. C., and Stroud, A. H., "Numerical Evaluation of Multiple Integral II," *Mathematical Tables and Other Aids to Computation*, Vol. 12, 1958, pp. 272–280.

11.7. Silvester, P., "Newton–Cotes Quadrature Formulae for *N*-Dimensional Simplexes," *Proceedings*, 2nd Canadian Congress on Applied Mechanics, Waterloo, Canada, 1969.

11.8. Irons, B. M., "Engineering Applications of Numerical Integration in Stiffness Methods," *AIAA Journal*, Vol. 4, No. 11, Nov. 1966, pp. 2035–2037.

11.9. Hammer, P. C., Marlowe, O. J., and Stroud, A. H., "Numerical Integration over Simplexes and Cones," *Mathematical Tables and Other Aids to Computation*, Vol. X, No. 54, Apr. 1956, pp. 130–137.

11.10. Cowper, G. R., "Gaussian Quadrature Formulas for Triangles," *International Journal for Numerical Methods in Engineering*, Vol. 7, 1973, pp. 405–408.

11.11. Dunavant, D. A., "High Degree Efficient Symmetrical Gaussian Quadrature Rules for the Triangle," *International Journal for Numerical Methods in Engineering*, Vol. 21, No. 7, 1985.

11.12. Felippa, C. A., "Refined Finite Element Analysis of Linear and Nonlinear Two-Dimensional Structures," Report 66-22, Department of Civil Engineering, University of California at Berkeley, 1966.

11.13. Zienkiewicz, O. C., *The Finite Element Method*, McGraw-Hill Book Company (U.K.) Ltd., London, 1977, p. 201.

11.14. Zienkiewicz, O. C., "Isoparametric and Allied Numerically Integrated Elements—A Review," *ONR International Symposium on Numerical and Computer Methods in Structural Mechanics*, University of Illinois, Urbana, 1971.

11.15. Moore, C. J., Yang, T. Y., and Anderson, D. C., "A New 48 D.O.F. Quadrilateral Shell Element with Variable-Order Polynomial and Rational B-Spline Geometries with Rigid Body Modes," *International Journal for Numerical Methods in Engineering*, Vol. 20, No. 8, 1984, pp. 2121–2141.

11.16. Taig, I. C., "Structural Analysis by the Matrix Displacement Method," England Electric Aviation Report S017, 1961.

11.17. Irons, B. M., "Numerical Integration Applied to Finite Element Methods," *Conference on Use of Digital Computers in Structural Engineering*, University of Newcastle, England, 1966.

11.18. Ergatoudis, I., Irons, B. M., and Zienkiewicz, O. C., "Curved, Isoparametric, 'Quadrilateral' Elements for Finite Element Analysis," *International Journal of Solids and Structures*, Vol. 4, No. 1, Jan. 1968, pp. 31–42.

11.19. Widder, D. V., *Advanced Calculus*, Prentice-Hall, Inc., Englewood Cliffs, N.J., 1961, pp. 28–31 and 241–243.

11.20. Ahmad, S., Irons, B. M., and Zienkiewicz, O. C., "Curved Thick Shell and Membrane Elements with Particular Reference to Axi-symmetric Problems," *Proceedings*, Second Conference on Matrix Methods in Structural Mechanics, TR-68-150, Air Force Flight Dynamics Laboratory, Fairborn, Ohio, 1968, pp. 539–572.

11.21. Ahmad, S., Irons, B. M., and Zienkiewicz, O. C., "Analysis of Thick and Thin Shell Structures by Curved Finite Elements," *International Journal for Numerical Methods in Engineering*, Vol. 2, 1970, pp. 419–451.

PROBLEMS

11.1. Derive the coefficients A and W_0 to W_4 for the five-point Newton–Cotes formula as given in Table 11.1 for $n = 4$.

11.2. Using Newton–Cotes quadrature with $n = 6$ and Eq. (7.29), derive the coefficient m_{13} in the consistent mass matrix (7.31) for the four-d.o.f. beam element (Fig. 5.1) with the displacement function (7.25),

11.3. Derive the abscissas x_i and the weights H_i for the Gaussian quadrature formula, as given in Table 11.2, for $n = 4$. Also check the four abscissa values x_i with the roots of the fourth-order Legendre polynomial equation $P_4(x) = 0$.

11.4. Using Gaussian quadrature with $n = 4$ and Eq. (7.29), derive the coefficient m_{13} in the consistent mass matrix (7.31) for the four-d.o.f. beam element (Fig. 5.1) with the displacement function (7.25).

11.5. Using Gaussian quadrature with $n = 2$ and Eq. (9.35), derive the coefficient k_{11} in the stiffness matrix (9.36) for the 8-d.o.f. rectangular plane stress element (Fig. 9.4) with the displacement functions (9.24).

11.6. Using Gaussian quadrature with $n = 3$, 4, and 5, respectively (Table 11.3), evaluate

$$I = \iint \frac{y^2}{x}\, dx\, dy$$

over the area of a triangle as defined in Fig. 11.1. Compare the result with that from the exact solution (10.17c).

11.7. Using Gaussian quadrature with $n = 4$ (Table 11.4) and Eq. (10.41), derive the stiffness coefficient k_{11} relating F_{x_1} to u_1 for the 10-node 30-d.o.f. tetrahedron shown in Fig. 10.9 with vertex coordinates defined in Fig. 11.3. The shape functions for N_i and matrix [**A**] are defined in Eqs. (10.44) and (10.37), respectively. Derivation of this coefficient was assigned as Problem 10.8.

11.8. Find the Jacobian matrix [**J**] for a four-node 8-d.o.f. isoparametric quadrilateral with coordinates as shown in Fig. 11.8a and Fig. 11.8b.

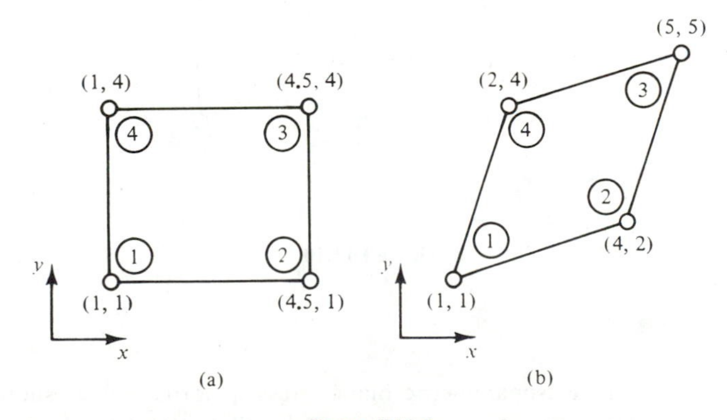

Figure P11.8

11.9. **(a)** Find the determinant of the Jacobian matrix [**J**] for a nine-node isoparametric quadrilateral as shown in Fig. P11.9. The geometric functions are given in Table 11.5 with $n = 9$. **(b)** Find $|J|$ for the same quadrilateral minus the central node. The geometric functions are given in Table 11.5 with $n = 8$.

11.10. Find the determinant of the Jacobian matrix [**J**] for an eight-node isoparametric quadrilateral as shown in Fig. P11.10. The geometric functions are given in Table 11.5 with $n = 8$.

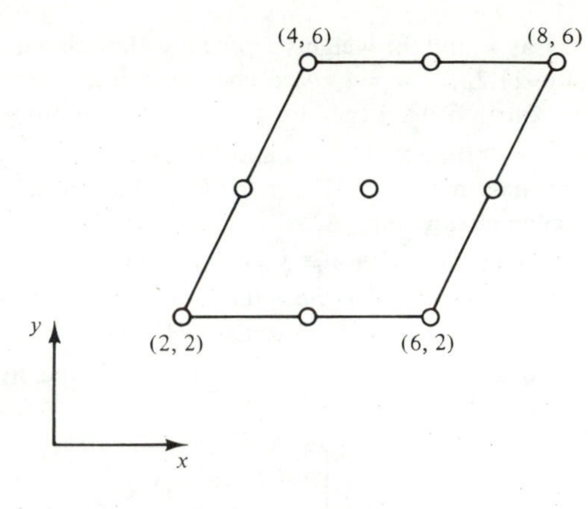

Figure P11.9

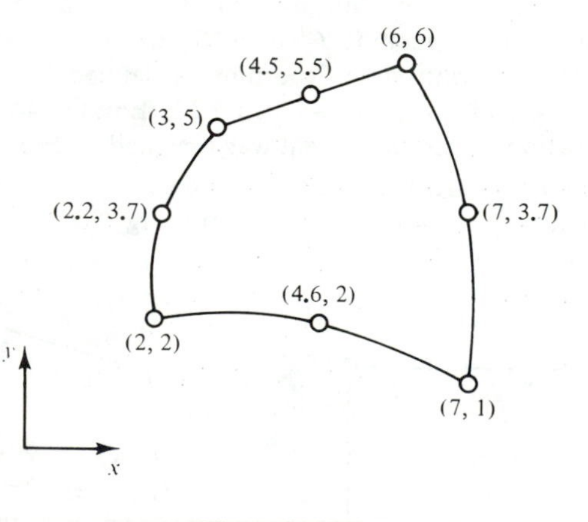

Figure P11.10

11.11. For an eight-node isoparametric plane stress quadrilateral as shown in Fig. P11.10, **(a)** derive the stiffness coefficient k_{11} that relates the force F_{x_1} to the displacement u_1 in closed form, and **(b)** find k_{11} using Gaussian quadrature with $n = 2$.

11.12. For a 12-node isoparametric plane stress quadrilateral as shown in Fig. P11.12, **(a)** find $|J|$, **(b)** derive the stiffness coefficient k_{11} relating the force F_{x_1} to the displacement u_1 in closed form, **(c)** find k_{11} using Gaussian quadrature with $n = 3$, and **(d)** find $|J|$ if side 2-3 is curved with two side-node coordinates at (6.4, 1) and (6.4, 2).

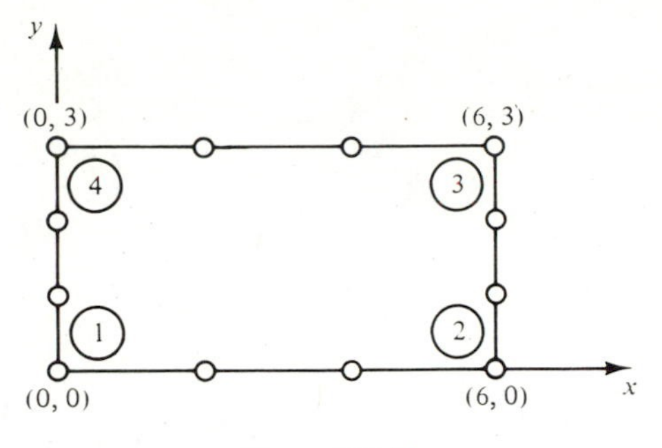

Figure P11.12

11.13. **(a)** Find $|J|$ for an eight-node isoparametric rectangular hexahedron with coordinates given as

	Node Number							
	1	2	3	4	5	6	7	8
x coordinate	0	4	4	0	0	4	4	0
y coordinate	0	0	0	0	4	4	4	4
z coordinate	0	0	4	4	0	0	4	4

The shape functions are given in Table 11.6 with $n = 8$. **(b)** Find $|J|$ if the x and y coordinates of nodes 3, 4, 7, and 8 are changed as follows:

	Node Number							
	1	2	3	4	5	6	7	8
x coordinate	0	4	6	−2	0	4	6	−2
y coordinate	0	0	−2	−2	4	4	6	6
z coordinate	0	0	4	4	0	0	4	4

11.14. A six-node isoparametric triangle is shown in Fig. P11.14 with its geometric functions given by Eq. (11.49). **(a)** Find the expression for $|J|$. **(b)** Show that $|J| = 2A$ when side 2-3 is a straight line. **(c)** Find the limiting value of a such that $|J| = 0$.

11.15. A 10-node isoparametric isosceles triangle is shown in Fig. P11.15 with its geometric functions given in Table 11.8. **(a)** Find $|J|$. **(b)** Derive the stiffness coefficient k_{11} that relates the force F_{x_1} to the displacement u_1 in the closed form. **(c)** Find k_{11} using the Gaussian quadrature formula given in Table 11.3 with $n = 6$.

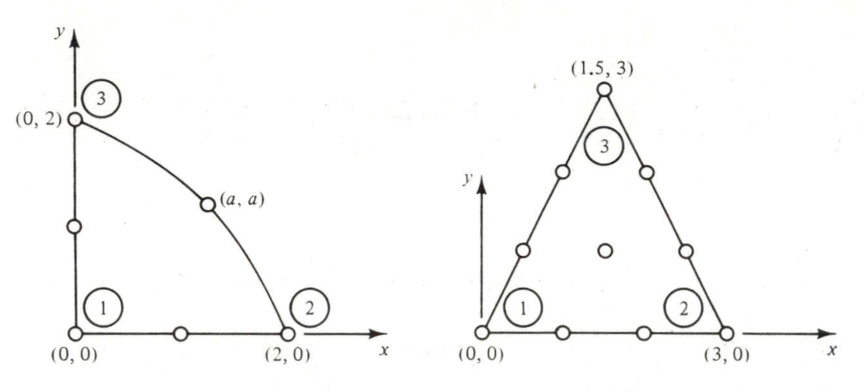

Figure P11.14 **Figure P11.15**

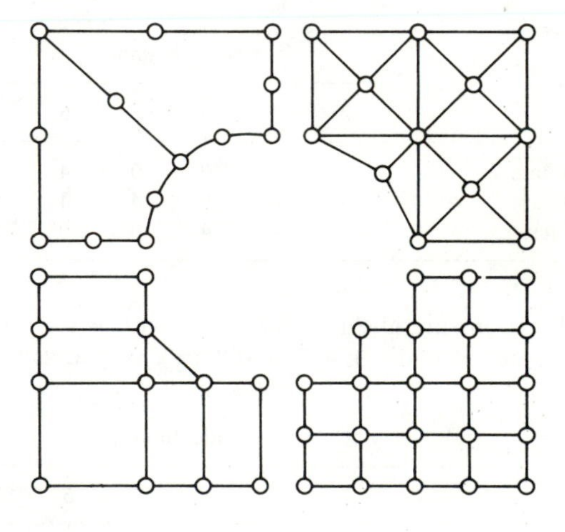

Figure P11.17

11.16. Repeat Problem 11.15 by changing side 2-3 from a straight line to a curve with the two side-node coordinates to be (2.8, 1.14) and (2.3, 2.15).

11.17. Figure P11.17 shows four different meshes that can be used to model a quadrant of a plate with a circular cutout. Assuming that there are two d.o.f.'s at each node (u and v), compare and discuss the four models in terms of **(a)** the total number of equations, **(b)** the minimum bandwidth, **(c)** the geometric representation, and **(d)** the strain or stress distributions.

CHAPTER 12

Plate Elements in Bending

A prime feature of this book is that it is self-contained. It is not necessary for the reader to be well exposed to the theory of thin plates beforehand. The starting point of this chapter will be to introduce the fundamental assumptions of the plate theory and the derivations of the stress–strain (moment–curvature) equations, equilibrium differential equations, and strain energy expression.

For plane stress, plane strain, and solid elements, the equilibrium differential equations are of the second order, whereas for plate bending elements they are of the fourth order. For plane and solid elements the strain energy expressions contain the first derivatives of displacements, whereas for plate bending elements they contain the second-order derivatives. Consequently, meeting the convergence criteria for plate bending elements is far more complicated than meeting them for plane and solid elements. It is therefore necessary that before formulating the plate elements in this chapter, an elementary but thorough explanation of the convergence criteria and the patch test be given.

Formulations of flat plate elements in bending have constituted a most exciting area in the development of finite element theory. The efforts in the 1960s are especially notable. These static bending formulations have been extended extensively to practical applications in the areas of vibration, dynamic response, impact, buckling and postbuckling, geometric and material nonlinearities, fluid–plate interactions, panel flutter, thermal effects, viscoelasticity and viscoplasticity, fracture, laminated plates, lattice plates for space structures, and so on. Most of these developments have now been well coded into commercial programs.

In this chapter we formulate each plate bending element in chronological order of its development. We first introduce the rectangular elements. The simplest 12-d.o.f. nonconforming rectangle is derived in great detail to demonstrate the basic formulative method, the convergence examination procedure, and the patch test technique. The 16-d.o.f. conforming rectangle is introduced next. Due to its simplicity and efficiency, it is used to demonstrate the applicability of plate bending elements to vibration, buckling, and large deflection analyses.

We then introduce the triangular elements. Basically, there are four types of triangles (displacement models) that meet the convergence criteria, depending on the variation in the normal slope $(\partial w/\partial n)$ along the interelement boundaries: linear, quadratic, cubic, and quartic variations. Both Cartesian and triangular coordinates are popular. In some formulations, only a single deflection function is used for the entire element, whereas in other formulations, an element is divided into subtriangles, each of which has its own deflection function. Quadrilateral elements can simply be considered as those composed of existing triangular elements or they can be formulated in the same fashion as for the triangular elements. A discussion of some quadrilateral elements is given.

A special but elementary section is devoted to the hybrid stress and mixed formulations for plate bending elements.

In addition to the discussion of how each element meets the convergence criteria, summary numerical comparisons of some representative elements are given using the example of a simply supported square plate under a central load.

After superposition with plane stress formulation and coordinate transformation, the flat plate elements can be used for curved shell analysis. Although the resulting convergence rate may not be as rapid as that for some curved shell elements, the assembled flat plate elements do possess rigid-body modes and have been routinely used for curved shell analysis.

As mentioned in Chapter 11, with some proper reduced-order integration techniques, isoparametric solid elements can be used for thin and thick plate analysis.

12.1 THEORY OF PLATES IN BENDING [12.1]

A plate is defined as thin when its thickness is small compared with its other dimensions. In this chapter we focus on the formulations of thin plate finite elements. First we introduce the basic assumptions of plate theory and the necessary equations.

12.1.1 Thin Plates with Small Deflection

For thin plates bending with small deflection, we have the following assumptions:

1. The middle plane of the plate remains neutral (undeformed) during bending.
2. Straight lines initially normal to the middle plane of the plate remain straight and normal to the middle surface. This assumption is equivalent to the neglect of the effect of transverse shear deformation.
3. The stresses normal to the middle surface are negligible.

12.1.2 Thin Plates with Large Deflection

In the case that the deflections w are no longer small compared with the thickness of the plate but are still small compared to the other dimensions, assumptions 2 and 3 for small deflections still hold, but assumption 1 no longer holds, with the exception of a plate bent into a developable surface. Instead of assumption 1, we have the following assumptions:

1. The boundary conditions include not only those related to the transverse deflection w but also in-plane displacements u and v.
2. The strains in the middle surface are functions of the first derivatives of u and v with respect to in-plane coordinates x and y as well as the quadratic terms of the first derivative of w with respect to x and y.

12.1.3 Thick Plates

In cases of plates with considerable thickness and in cases of highly concentrated loads, the approximate theory of thin plates no longer holds.

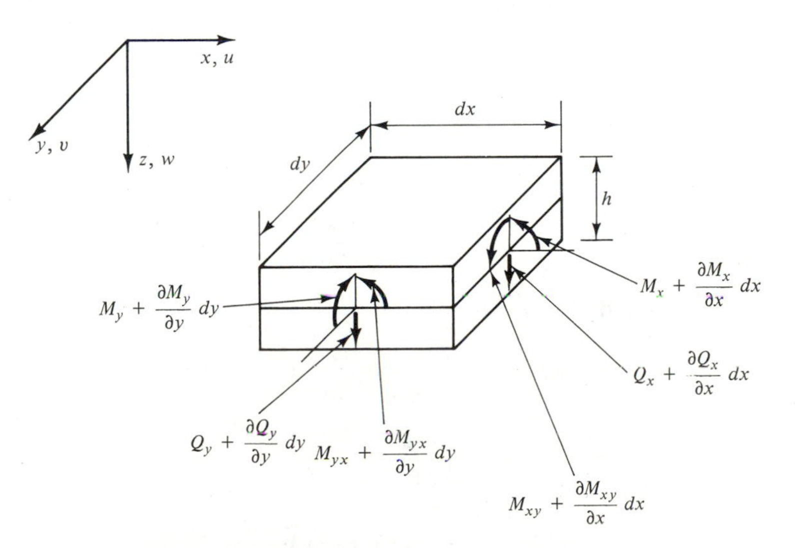

Figure 12.1 Differential element of a thin plate in bending.

Instead, the thick-plate theory based on the theory of three-dimensional elasticity should be used. The stress analysis of thick plates can be achieved for some particular cases by making corrections to that for the thin plates, such as by adding the effect of transverse shear deformation.

12.1.4 Differential Equations for Thin Plates with Small Deflection

A differential element of a plate in bending is shown in Fig. 12.1. Acting on this element are bending moments M_x and M_y, twisting moments M_{xy} and M_{yx}, shearing forces Q_x and Q_y, and the rates of change of these quantities along the x and y directions, respectively.

Physically, a thin plate can be considered as composed of numerous layers of plane stress membranes with the stresses varying linearly across the thickness. The bending and twisting moments per unit length can be obtained by integrating these plane stresses through the thickness.

$$\begin{aligned} M_x &= \int_{-h/2}^{h/2} \sigma_x z\, dz = \frac{E}{1-\nu^2}\int_{-h/2}^{h/2} (\epsilon_x + \nu\epsilon_y) z\, dz \\ M_y &= \int_{-h/2}^{h/2} \sigma_y z\, dz = \frac{E}{1-\nu^2}\int_{-h/2}^{h/2} (\epsilon_y + \nu\epsilon_x) z\, dz \\ M_{xy} &= -M_{yx} = -\int_{-h/2}^{h/2} \tau_{xy} z\, dz = \frac{-E}{2(1+\nu)}\int_{-h/2}^{h/2} \gamma_{xy} z\, dz \end{aligned} \tag{12.1}$$

The bending and twisting curvatures of the plate are the gradients of the strains, which vary linearly through the thickness:

$$\begin{aligned} \epsilon_x &= -z\frac{\partial^2 w}{\partial x^2} \\ \epsilon_y &= -z\frac{\partial^2 w}{\partial y^2} \\ \gamma_{xy} &= -2z\frac{\partial^2 w}{\partial x\,\partial y} \end{aligned} \tag{12.2}$$

From Eqs. (12.1) and (12.2), we have

$$\begin{aligned} M_x &= -D\left(\frac{\partial^2 w}{\partial x^2} + \nu\frac{\partial^2 w}{\partial y^2}\right) \\ M_y &= -D\left(\frac{\partial^2 w}{\partial y^2} + \nu\frac{\partial^2 w}{\partial x^2}\right) \\ M_{xy} &= -M_{yx} = D(1-\nu)\frac{\partial^2 w}{\partial x\,\partial y} \end{aligned} \tag{12.3}$$

where $D = Eh^3/12(1 - \nu^2)$ is the bending rigidity. In the absence of Poisson's ratio ν, Eqs. (12.3) reduce to those for the case of a beam.

We now assume that the differential element is under a downward load $p\,dx\,dy$ distributed over the upper surface. Considering the equilibrium of all the shearing forces, bending and twisting moments acting on the differential element, we have

$$\frac{\partial Q_x}{\partial x} + \frac{\partial Q_y}{\partial y} + p = 0 \tag{12.4a}$$

$$\frac{\partial M_{xy}}{\partial x} - \frac{\partial M_y}{\partial y} + Q_y = 0 \tag{12.4b}$$

$$\frac{\partial M_{yx}}{\partial y} + \frac{\partial M_x}{\partial x} - Q_x = 0 \tag{12.4c}$$

Let us eliminate the shearing forces Q_x and Q_y from these equations by obtaining them in terms of the moment derivatives from Eqs. (12.4b) and (12.4c) and substituting them in Eq. (12.4a):

$$\frac{\partial^2 M_x}{\partial x^2} - 2\frac{\partial^2 M_{xy}}{\partial x\,\partial y} + \frac{\partial^2 M_y}{\partial y^2} = -p \tag{12.5}$$

Substituting Eqs. (12.3) into (12.5) yields the well-known fourth-order differential equation of a plate,

$$\frac{\partial^4 w}{\partial x^4} + 2\frac{\partial^4 w}{\partial x^2\,\partial y^2} + \frac{\partial^4 w}{\partial y^4} = \frac{p}{D} \tag{12.6}$$

An interesting historical account of the developments of Eq. (12.6) was given in Ref. [12.2]. Among the investigators are Euler, Bernoulli, Chladni, Germain, Lagrange, Poisson, Navier, and Kirchhoff. In the absence of Poisson's ratio ν and the y coordinate, Eq. (12.6) reduces to that for the case of a beam.

12.1.5 Strain Energy for Thin Plates with Small Deflection

The strain energy expression for a differential plate element subjected to distributed bending moments M_x and M_y and twisting moments M_{xy} and M_{yx} is

$$\begin{aligned} dU = &-\tfrac{1}{2}(M_x\,dy)\left(\frac{\partial^2 w}{\partial x^2}\,dx\right) - \tfrac{1}{2}(M_y\,dx)\left(\frac{\partial^2 w}{\partial y^2}\,dy\right) \\ &- \tfrac{1}{2}(M_{xy}\,dy)\left(\frac{\partial^2 w}{\partial x\,\partial y}\,dx\right) - \tfrac{1}{2}(M_{yx}\,dx)\left(\frac{\partial^2 w}{\partial y\,\partial x}\,dy\right) \end{aligned} \tag{12.7}$$

This equation can be derived by physical inspection. The term $-(\partial^2 w/\partial x^2)$ represents the curvature of the plate in the xz plane. The angle corresponding to the bending moment $M_x\, dy$ is $-(\partial^2 w/\partial x^2)\, dx$ and the work done by this moment is

$$-\tfrac{1}{2}(M_x\, dy)\left(\frac{\partial^2 w}{\partial x^2}\, dx\right) \qquad \text{with units} \left(\frac{\text{in.} - \text{lb}}{\text{in.}}\right)(\text{in.})\left(\frac{1}{\text{in.}}\right)(\text{in.})$$

Corresponding to the twisting moment $M_{xy}\, dy$, the angle of twist is equal to the rate of change of the slope $\partial w/\partial y$, as x varies, multiplied by dx. The work done is

$$\tfrac{1}{2}(M_{xy}\, dy)\frac{\partial}{\partial x}\left(\frac{\partial w}{\partial y}\right) dx$$

Substituting Eqs. (12.3) in Eq. (12.7) and performing integration of the differential energy dU throughout the volume of the whole plate, the strain energy expression is obtained.

$$U = \iint_{\text{area}} \frac{D}{2}\left[\left(\frac{\partial^2 w}{\partial x^2}\right)^2 + \left(\frac{\partial^2 w}{\partial y^2}\right)^2 + 2\nu\left(\frac{\partial^2 w}{\partial x^2}\right)\left(\frac{\partial^2 w}{\partial y^2}\right) + 2(1-\nu)\left(\frac{\partial^2 w}{\partial x\, \partial y}\right)^2\right] dx\, dy \tag{12.8}$$

12.1.6 Equations of Bending Stresses

The equations for bending stresses at a distance z from the middle plane are of the form

$$\begin{aligned} \sigma_x &= \frac{M_x z}{I_y} = \frac{12 M_x z}{h^3} \\ \sigma_y &= \frac{M_y z}{I_x} = \frac{12 M_y z}{h^3} \\ \tau_{xy} &= \tau_{yx} = \frac{12 M_{xy} z}{h^3} \end{aligned} \tag{12.9}$$

12.2 SELECTION OF DISPLACEMENT FUNCTIONS AND PATCH TEST

12.2.1 Criteria for Selection of Displacement Functions

Assumption of nodal degrees of freedom and selection of displacement functions are the key to the development of a successful finite element displacement model. The displacement functions must be chosen so that as the element mesh is successively refined, the resulting displacement solution converges to the correct answer. To ensure the convergence of the results with mesh refinement, the displacement approximations must satisfy certain criteria, known as the *convergence criteria.* These criteria are explained as follows.

Criterion 1: The displacement functions should be continuous within an element.

This criterion is easily met by polynomial displacement functions.

Criterion 2: The displacement functions should be so assumed that no straining of the element occurs when all nodal points are subjected to a rigid-body displacement.

This criterion means that when substituting the rigid-body nodal displacements in the strain–nodal displacement or the nodal force–displacement equations of an element, neither strains nor nodal forces exist. This criterion can also be tested by solving for a set of eigenvalue equations for the element,

$$\big[[\mathbf{K}] - \omega^2[\mathbf{I}]\big]\{\mathbf{q}\} = \{\mathbf{0}\} \tag{12.10}$$

where $[\mathbf{I}]$ is an identity matrix simulating the element mass matrix. The solution for the eigenvalues and eigenvectors should include all the rigid-body modes, that is, zero natural frequencies ω and the corresponding rigid-body displacements $\{\mathbf{q}\}$. For plate elements in bending, there are three rigid-body modes: a transverse displacement and two rotations about x and y axes, respectively. For solid and curved shell elements, there are six rigid-body modes: three displacements along and three rotations about x, y, and z axes, respectively. This criterion is extremely important in the formulation of curved shell finite elements (see, for example, Refs. [9.12], [11.15], [12.3], and [12.4]).

Criterion 3: The displacement functions should be so assumed that a state of constant strain is obtained when the nodal displacements correspond to a constant strain condition [12.5].

This criterion can be explained using the example of beam elements. Figure 12.2 shows a cantilever beam under uniform loads. It is modeled by 10 elements. Let the deflection function for element 3 be assumed as

$$w(x) = a_1 + a_2x + a_3x^2 + a_4x^3 \tag{12.11}$$

The axial strain in the upper extreme fiber is

$$\epsilon_x = \frac{h}{2}\frac{\partial^2 w}{\partial x^2} = \frac{h}{2}(2a_3 + 6a_4x) \tag{12.12}$$

which varies linearly with x as shown in Fig. 12.2. If x approaches zero as the element length approaches zero, ϵ_x approaches a constant value. Thus the ϵ_x curve is modeled smoothly like a stairway with near-zero step size.

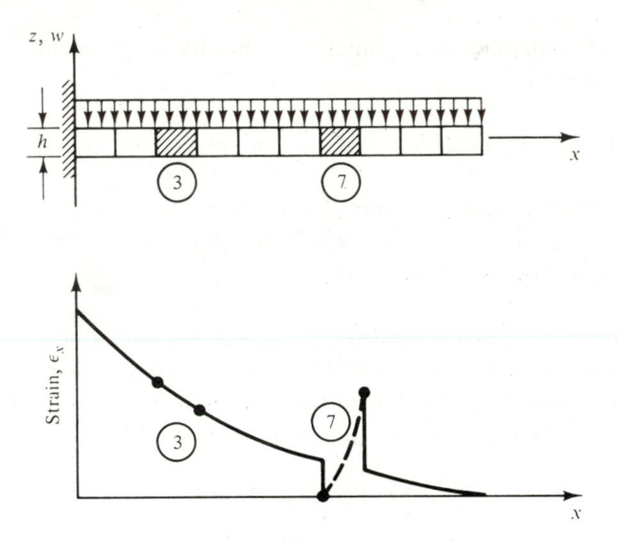

Figure 12.2 Variation of strains in upper extreme fiber for two elements: one with and one without the state of constant strain.

Now let the deflection function for element 7 be assumed as

$$w(x) = b_1 + b_2x + b_3x^3 + b_4x^4 \tag{12.13}$$

The axial strain in the upper extreme fiber is

$$\epsilon_x = \frac{h}{2}\frac{\partial^2 w}{\partial x^2} = \frac{h}{2}(6b_3x + 12b_4x^2) \tag{12.14}$$

which starts from a zero value and varies parabolically with x as shown in Fig. 12.2. As the element length approaches zero, ϵ_x varies like a spike. Thus convergence to a correct curve cannot be accomplished.

Actually, criterion 2 is a special case of criterion 3 as zero strain is a special case of constant strain. Rigorously speaking, both criteria need only be met when the element size approaches zero, not necessarily finite. It is seen from Eqs. (12.11) and (12.12) that for the rigid-body displacements w and $\partial w/\partial x$, the terms a_1 and a_2x are needed and for constant strain, the term a_3x^2 is needed. Thus criteria 2 and 3 suggest that the lower-order terms of the polynomial should be retained. This is part of the completeness requirement.

Criterion 4: The displacement functions should be so assumed that the element is geometrically invariant or isotropic, with no preferential directions.

This is a desirable but not a necessary criterion. Although for most elements, the displacement functions are polynomials with incomplete higher-order terms, they are geometrically isotropic. For instance, if the terms x^3y and x^3y^2 are included in the displacement function of a plate bending element, the terms xy^3 and x^2y^3 must also be included. The stiffness matrix for an element is normally formulated in local coordinates. It becomes geometrically invariant after transforming to that in global coordinates.

Criterion 5: The displacement functions must be so chosen that at the interelement boundaries, the displacements and their derivatives up to at least one order less than the highest-order derivative in the energy expression, must be continuous. In other words, the displacement derivatives with the same order must be finite.

This criterion can be explained using an example of plane stress or solid elements, for which the strain energy contains the products of only the terms of first-order displacement derivatives (i.e., $\partial u/\partial x$, $\partial v/\partial x$, $\partial w/\partial x$, $\partial u/\partial y, \ldots,$ $\partial w/\partial z$, etc.). Based on this criterion, it is sufficient to choose the displacement functions so that only the displacements be continuous at interelement boundaries, not necessarily the displacement derivatives. Figure 12.3 shows an example of the variation of displacement u and its derivative $\partial u/\partial x$ along a line crossing three such elements. Displacement u has the same value for two neighboring elements at joint A but not at joint B. As a result, the derivative or strain du/dx at joint A is finite:

$$\lim_{\Delta \to 0} \frac{u_2 - u_1}{\Delta} = \frac{0}{0} = \text{finite}$$

On the other hand, du/dx at joint B is infinite:

$$\lim_{\Delta \to 0} \frac{u_4 - u_3}{\Delta} = \text{infinite}$$

Thus criterion 5 is satisfied at joint A but not at joint B.

The elements that satisfy criterion 5 are called *conforming* elements. The elements that satisfy interelement compatibility in the zero-, first-, and second-order derivative terms of displacements are known as the elements with C_0, C_1, and C_2 *continuity*, respectively.

Criterion 5 certainly makes the selection of the displacement function w for a plate bending element more complicated than that for a plane stress or solid element, as the strain energy expression (12.8) for plates in bending

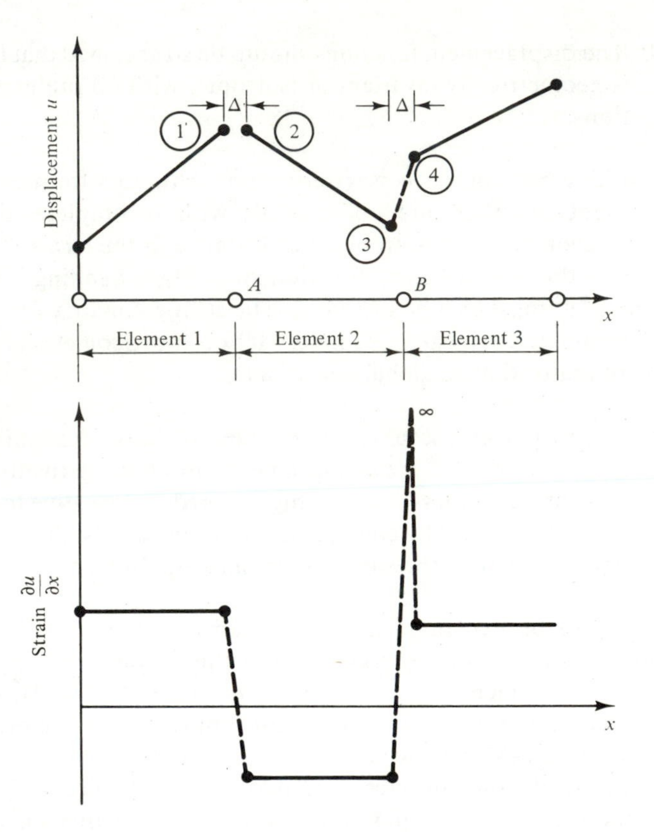

Figure 12.3 Interelement relations in displacement u and its derivative.

contains the products of the second-order derivative terms of w. The displacement function for w and its first-order derivatives $\partial w/\partial x$ and $\partial w/\partial y$ should be continuous along interelement boundaries, so as to ensure that the second-order derivative terms of w in the energy expression do not become infinite. Such elements are C_1 continuous.

There are plate bending elements that are *nonconforming* (i.e., not C_1 continuous). For such elements, although the deflection w is continuous along the interelement boundaries, the continuities of $\partial w/\partial x$ and $\partial w/\partial y$ are ensured only at the adjoining nodal points. If these elements satisfy criterion 3 for constant strain and pass the patch test (to be described in the next section), they are still convergent and often used. The convergence may, however, no longer be guaranteed to be monotonic and upperbound. Mathematical bases for the foregoing convergence criteria are given in, among others, Refs. [12.6] to [12.10].

12.2.2 Patch Test

The patch test is a simple numerical test to evaluate the capability of certain finite elements based on an assemblage or a patch of such elements. The patch should have at least one nodal point completely surrounded by elements. In a patch test, the nodal points at the outside boundaries of the patch are subjected to either displacements or work equivalent (consistent) loads which in an exact analysis result in a state of constant strain. If the finite element analysis shows that as the patch becomes infinitesimally small, the solution of nodal displacements results in a state of constant strain with exact values within the computer accuracy at any point within any element, the patch test is passed [11.17, 12.5, 12.11]. As an example, a simple test of a cantilever patch with five plane stress quadrilaterals subjected to a uniformly distributed displacement $l(\partial u/\partial x)$ (in.) or loads P_0 (lb/in.) at the free edge is shown in Fig. 12.4. In a patch test, the inner nodes should neither be loaded nor fixed.

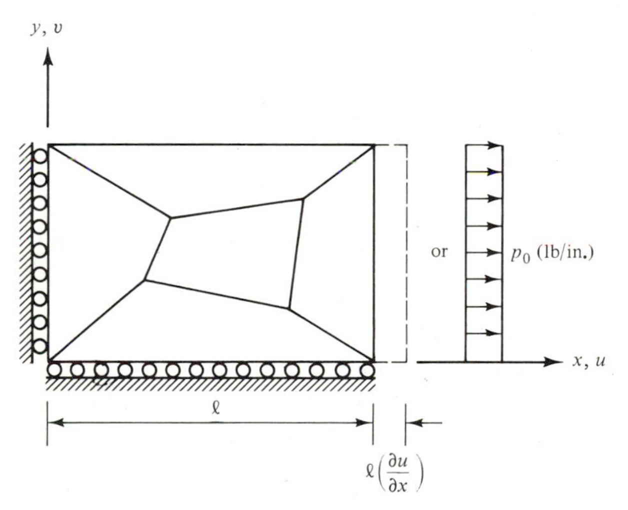

Figure 12.4 Patch test for plane stress elements.

Although the patch test need only be passed when the size of the element patch becomes infinitesimal, it has been quite common that passing the patch test be required for any element size. For many cases, however, such as those for the curved shell elements based on curvilinear coordinates, the patch test is still required to be passed in the limit but not for a finite size of patch.

In reality, the patch test checks the requirements for rigid-body modes (criterion 2), constant strains (criterion 3), and interelement compatibility (criterion 5). Rigid-body modes are a special case of constant strains with zero values. When element mesh is infinitely refined, constant strains and interelement compatibility exhibit simultaneously. Thus on physical grounds only, passing the patch test for infinitesimal patch sizes appears to be a necessary and sufficient condition for convergence [12.10–12.14].

It is possible that some elements may pass some specific patch tests but not others. Some quadrilateral elements may pass the patch test when used as rectangles or parallelograms but fail when used as general quadrilaterals. Some triangular elements may only pass the patch test for certain specifically regular geometry and mesh. Thus it is advisable that for a given element, the patch test be conducted for more than one patch configuration, mesh distribution, and strain type.

It is a common belief that an element that does not pass the patch test should not be trusted. On the other hand, passing the patch test does not guarantee satisfaction because the rate of convergence may be too slow. The original concept of the patch test does not yield information on the rate of convergence. However, the concept has been generalized to provide, in addition to a sufficient condition for convergence, an indicator of the expected degree of convergence [12.15].

12.2.3 Patch Test for Plate Bending Elements

For the patch test of plate bending elements, constant strains mean constant curvatures $\partial^2 w/\partial x^2$, $\partial^2 w/\partial y^2$, and $\partial^2 w/\partial x\,\partial y$, as defined in Eq. (12.2). The patch may be subjected to either a set of displacements (deflections and rotations) or boundary loads (transverse forces and moments) which correspond to a state of constant curvature.

Prescribed displacements. If nodal displacements are prescribed, the following form may be suggested [12.11]:

$$
\begin{aligned}
w &= \frac{C(x^2 + xy + y^2)}{2} \\
\frac{\partial w}{\partial x} &= C\left(x + \frac{y}{2}\right) \\
\frac{\partial w}{\partial y} &= C\left(\frac{x}{2} + y\right) \\
\frac{\partial^2 w}{\partial x\,\partial y} &= \frac{C}{2}
\end{aligned}
\tag{12.15}
$$

where C is a constant so assumed that $w \ll$ thickness. For an example patch of five elements as shown in Fig. 12.5a, the deflected surface based on Eq. (12.15) is plotted in Fig. 12.5b. In this patch test, the displacements may be prescribed in either of two ways:

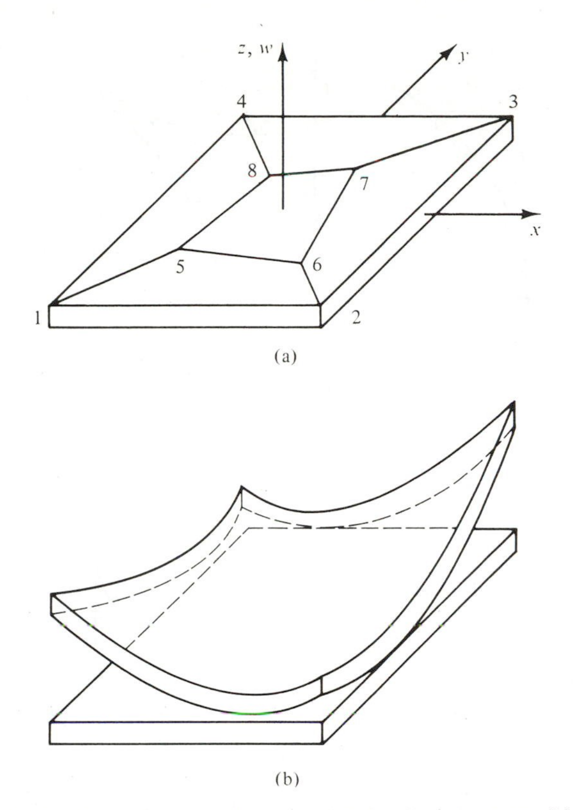

Figure 12.5 Patch test: (a) a patch of five plate bending elements; (b) prescribed deflection shape.

1. First prescribe the values of w, $\partial w/\partial x$, and $\partial w/\partial y$ at boundary nodes 1, 2, 3, and 4, respectively, and solve the assembled stiffness equations for the values of degrees of freedom at the *unloaded* inner nodes 5, 6, 7, and 8, respectively. Then check the state of constant strain or curvature in each element.
2. First prescribe the values of w, $\partial w/\partial x$, and $\partial w/\partial y$ at all the eight nodes. Such nodal values assure a state of constant strain or curvature in every element. Then compute the total lateral forces and bending moments at each of the four inner nodes using either the element stiffness equations or the assembled stiffness equations. The patch test is passed if the lateral forces and moments at each inner node, as summed up from those contributed by all the elements surrounding the node, vanish.

Applied boundary loads. If boundary loads are applied, the types of loading conditions as illustrated in Fig. 12.6 may provide an example of the patch test for (a) constant curvatures $\partial^2 w/\partial x^2$ and $\partial^2 w/\partial y^2$ and (b) constant

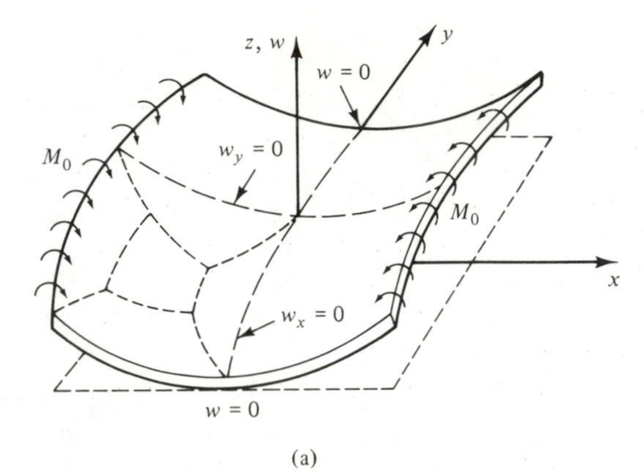

(a)

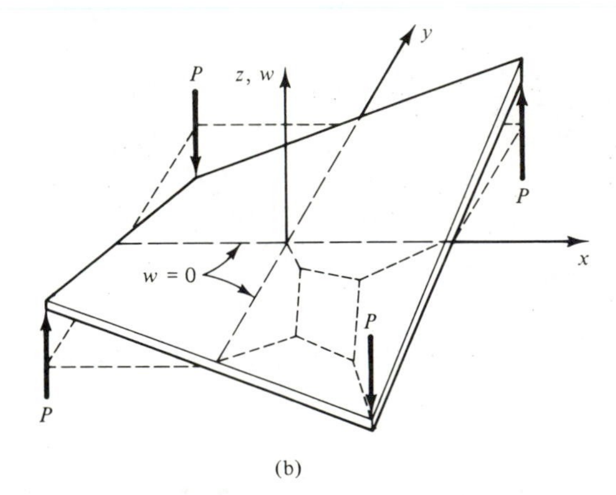

(b)

Figure 12.6 Patch test for plate bending elements: (a) constant curvature; (b) constant twist.

twist $\partial^2 w/\partial x\, \partial y$. Because of double symmetry and antisymmetry as shown for cases (a) and (b), respectively, only a quadrant of the flat plate need be tested and a patch of five arbitrary quadrilaterals may be suggested. The boundary conditions for each quadrant or patch are given in the figure.

For the constant curvature test as shown in Fig. 12.6a, $M_x = M_0$ (in.-lb/in.) and $M_y = M_{xy} = 0$ everywhere. Solving Eqs. (12.3) results in

$$\frac{\partial^2 w}{\partial y^2} = -\nu \frac{\partial^2 w}{\partial x^2} = -\frac{12\nu M_0}{Eh^3} \qquad \frac{\partial^2 w}{\partial x\, \partial y} = 0 \tag{12.16}$$

For the constant twist test as shown in Fig. 12.6b, let us first interpret the deflection surface physically. Due to double antisymmetry about x and y axes, the center and the four midside points of the deflected plate lie in the same plane with no deflection. An equilibrium check of any of the quadrants as a free body reveals that the quadrant is balanced by four corner loads in the same manner as that for the whole plate. The center and the four midside points of the quadrant again lie in the same plane, although deflected and not parallel to the xy plane. If this quadrant is subdivided again into four subquadrants, each subquadrant is under the same type of corner loads. Successive division in this way will lead to infinitesimally small quadrants, similar to a differential element $dx\,dy$, subjected to the same type of corner loads. Each of the four corner loads is now split into two halves to produce two twisting moments, $(P/2)\,dx$ and $(P/2)\,dy$. Thus we see that the plate is subjected to a state of constant twisting moments $(P\,dx/2)/dx$ and $(P\,dy/2)/dy$ (in.-lb/in.) everywhere and remains flat after deflection (anticlastic). Solving Eq. (12.3) gives

$$\begin{aligned} \frac{\partial^2 w}{\partial x^2} &= \frac{\partial^2 w}{\partial y^2} = 0 \qquad \text{for } M_x = M_y = 0 \\ \frac{\partial^2 w}{\partial x\,\partial y} &= \frac{6P(1+\nu)}{Eh^3} \qquad \text{for } M_{xy} = -M_{yx} = \frac{P}{2} \end{aligned} \tag{12.17}$$

Physical interpretation of this four-corner load twist problem was first given by Kelvin and Tait in 1883 and explained in detail in the text by Timoshenko [12.1].

As shown in Figs. 12.4 to 12.6, the outside configurations of the patches are chosen as rectangles so that the states of constant strain, curvature, or twist can be conveniently produced. It is, however, essential that each element be of irregular shape so as to ensure a sound patch test. Some patch tests of plate bending elements in the commercial finite element programs are available (see, for example, Ref. [12.16]). A set of problems for patch test was proposed in Ref. [12.17].

12.3 RECTANGULAR ELEMENTS

12.3.1 Four-Node 12-D.O.F. Nonconforming Rectangle: 12-Term Polynomial [12.18–12.21]

It is more convenient to formulate an element of rectangular shape than that of other shapes simply because the element sides are parallel to the Cartesian coordinates. A typical rectangular element with four corner nodal points is shown in Fig. 12.7. The element has length a, width b, and thickness h. Let us assume that the origin of the Cartesian coordinates is at nodal point 1 and that there are three degrees of freedom at each nodal point: deflection w; slope in the x direction, $\partial w/\partial x$ or w_x; and slope in the y direction, $\partial w/\partial y$ or w_y. We can derive the stiffness matrix by assuming the displacement function as a 12-term polynomial.

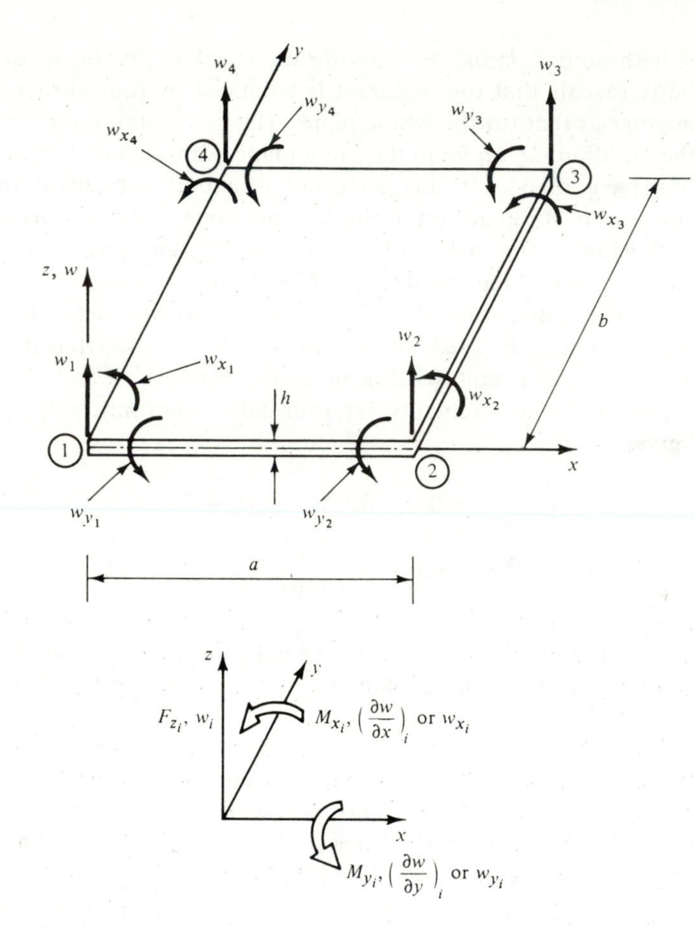

Figure 12.7 Four-node 12-d.o.f. rectangular plate bending element.

Displacement function. The displacement function is assumed as

$$w(x, y) = c_1 + c_2x + c_3y + c_4x^2 + c_5xy + c_6y^2 + c_7x^3 + c_8x^2y + c_9xy^2 + c_{10}y^3 + c_{11}x^3y + c_{12}xy^3 \tag{12.18}$$

It can be viewed more conveniently by using a Pascal's triangle:

$$\begin{array}{ccccccccc} & & & & 1 & & & & \\ & & & x & & y & & & \\ & & x^2 & & xy & & y^2 & & \\ & x^3 & & x^2y & & xy^2 & & y^3 & \\ \textcircled{x^4} & & x^3y & & \textcircled{x^2y^2} & & xy^3 & & \textcircled{y^4} \end{array}$$

It is an incomplete fourth-order polynomial in x and y with x^4, x^2y^2, and y^4 missing. A reason for selecting the two fourth-order terms x^3y and xy^3 is that it maintains geometric isotropy and also satisfies the differential equation (12.6) of a differential element in the unloaded region of the plate,

$$\frac{\partial^4 w}{\partial x^4} + 2\frac{\partial^4 w}{\partial x^2\,\partial y^2} + \frac{\partial^4 w}{\partial y^4} = \frac{p}{D} = 0 \tag{12.19}$$

The term x^2y^2 is not suitable because we cannot find another pairing term. The pair of terms x^4 and y^4 is not suitable because they do not satisfy interelement compatibility in displacement, whereas x^3y and xy^3 do.

Rigid-body modes, constant strains, and interelement compatibility. The plate element has three possible rigid-body movements: a deflection w and two rotations $\partial w/\partial x$ and $\partial w/\partial y$. In the displacement function (12.18), the three terms c_1, c_2x, and c_3y provide the needed constants c_1, c_2, and c_3 for the three respective rigid-body modes.

The plate element has three strain terms: $\partial^2 w/\partial x^2$, $\partial^2 w/\partial y^2$, and $\partial^2 w/\partial x\,\partial y$. The term c_4x^2, c_6y^2, and c_5xy provide the needed constants c_4, c_6, and c_5 for the three respective constant strains.

To check interelement compatibility, let us consider the displacement function along edge 2-3 of the element shown in Fig. 12.7. We have from Eq. (12.18) for $x = a$,

$$w(a, y) = a_1 + a_2y + a_3y^2 + a_4y^3 \tag{12.20a}$$

$$\frac{\partial w}{\partial y}(a, y) = a_2 + 2a_3y + 3a_4y^2 \tag{12.20b}$$

The four constants a_1, a_2, a_3, and a_4 are uniquely defined by the four d.o.f. values at nodes 2 and 3:

$$\begin{aligned} w(a, 0) &= a_1 = w_2 \\ w(a, b) &= a_1 + a_2b + a_3b^2 + a_4b^3 = w_3 \\ \frac{\partial w}{\partial y}(a, 0) &= a_2 = \left(\frac{\partial w}{\partial y}\right)_2 \\ \frac{\partial w}{\partial y}(a, b) &= a_2 + 2a_3b + 3a_4b^2 = \left(\frac{\partial w}{\partial y}\right)_3 \end{aligned} \tag{12.21}$$

Because the same four nodal d.o.f. values are shared by the two neighboring elements with common edge 2-3, both elements share the same cubic deflection function (12.20a) along this edge. The interelement compatibility of deflection along edge 2-3 (in the y direction) means that the slopes $\partial w/\partial y$ are also fully compatible.

Let us now consider the normal slope along edge 2-3 of the element. We have from Eq. (12.18) for $x = a$,

$$\frac{\partial w}{\partial x}(a, y) = a_5 + a_6 y + a_7 y^2 + a_8 y^3 \tag{12.22}$$

The four constants cannot be uniquely determined by only two nodal values of $\partial w/\partial x$ at joints 2 and 3. Thus the two neighboring elements will have two undefined functions for $\partial w/\partial x$ and the interelement compatibility of the normal slope along edge 2-3 of the two elements will not be satisfied.

This element does not satisfy convergence criterion 5 and the solution does not result in minimum strain energy. It is, however, of interest to note that this element was proven theoretically to be convergent [12.22] and it passes a patch test later in this section.

Stiffness formulation. To form the stiffness matrix, we first evaluate the values of the 12 nodal degrees of freedom based on Eq. (12.18).

$$\begin{Bmatrix} w_1 \\ w_{x_1} \\ w_{y_1} \\ w_2 \\ w_{x_2} \\ w_{y_2} \\ w_3 \\ w_{x_3} \\ w_{y_3} \\ w_4 \\ w_{x_4} \\ w_{y_4} \end{Bmatrix} = \begin{bmatrix} 1 & 0 & 0 & 0 & 0 & 0 & 0 & 0 & 0 & 0 & 0 & 0 \\ 0 & 1 & 0 & 0 & 0 & 0 & 0 & 0 & 0 & 0 & 0 & 0 \\ 0 & 0 & 1 & 0 & 0 & 0 & 0 & 0 & 0 & 0 & 0 & 0 \\ 1 & a & 0 & a^2 & 0 & 0 & a^3 & 0 & 0 & 0 & 0 & 0 \\ 0 & 1 & 0 & 2a & 0 & 0 & 3a^2 & 0 & 0 & 0 & 0 & 0 \\ 0 & 0 & 1 & 0 & a & 0 & 0 & a^2 & 0 & 0 & a^3 & 0 \\ 1 & a & b & a^2 & ab & b^2 & a^3 & a^2b & ab^2 & b^3 & a^3b & ab^3 \\ 0 & 1 & 0 & 2a & b & 0 & 3a^2 & 2ab & b^2 & 0 & 3a^2b & b^3 \\ 0 & 0 & 1 & 0 & a & 2b & 0 & a^2 & 2ab & 3b^2 & a^3 & 3ab^2 \\ 1 & 0 & b & 0 & 0 & b^2 & 0 & 0 & 0 & b^3 & 0 & 0 \\ 0 & 1 & 0 & 0 & b & 0 & 0 & 0 & b^2 & 0 & 0 & b^3 \\ 0 & 0 & 1 & 0 & 0 & 2b & 0 & 0 & 0 & 3b^2 & 0 & 0 \end{bmatrix} \begin{Bmatrix} c_1 \\ c_2 \\ c_3 \\ c_4 \\ c_5 \\ c_6 \\ c_7 \\ c_8 \\ c_9 \\ c_{10} \\ c_{11} \\ c_{12} \end{Bmatrix} \tag{12.23a}$$

or symbolically,

$$\{\mathbf{q}\} = [\mathbf{B}]\{\mathbf{c}\} \tag{12.23b}$$

Then

$$\{\mathbf{c}\} = [\mathbf{B}]^{-1}\{\mathbf{q}\} = [\mathbf{T}]\{\mathbf{q}\} \tag{12.24}$$

The strain-displacement relations are obtained by substituting Eq. (12.18) into (12.2):

$$\begin{Bmatrix} \epsilon_x \\ \epsilon_y \\ \gamma_{xy} \end{Bmatrix} = -z \begin{Bmatrix} \dfrac{\partial^2 w}{\partial x^2} \\ \dfrac{\partial^2 w}{\partial y^2} \\ 2\dfrac{\partial^2 w}{\partial x\, \partial y} \end{Bmatrix} = -z \begin{bmatrix} 2 & 0 & 0 & 6x & 2y & 0 & 0 & 6xy & 0 \\ 0 & 0 & 2 & 0 & 0 & 2x & 6y & 0 & 6xy \\ 0 & 2 & 0 & 0 & 4x & 4y & 0 & 6x^2 & 6y^2 \end{bmatrix} \begin{Bmatrix} c_4 \\ c_5 \\ \vdots \\ c_{12} \end{Bmatrix}$$

$$= [\mathbf{G}]\{\mathbf{c}\} = [\mathbf{G}][\mathbf{T}]\{\mathbf{q}\} = [\mathbf{A}]\{\mathbf{q}\} \tag{12.25}$$

The state of plane stress is present in a thin plate. For an isotropic material the stress–strain equations can be written as

$$\begin{Bmatrix} \sigma_x \\ \sigma_y \\ \tau_{xy} \end{Bmatrix} = \frac{E}{1-\nu^2} \begin{bmatrix} 1 & \nu & 0 \\ \nu & 1 & 0 \\ 0 & 0 & \dfrac{1-\nu}{2} \end{bmatrix} \begin{Bmatrix} \epsilon_x \\ \epsilon_y \\ \gamma_{xy} \end{Bmatrix} = [\mathbf{C}]\{\boldsymbol{\epsilon}\} \tag{12.26}$$

The stiffness matrix can be obtained in the usual manner as described in Eq. (9.33):

$$[\mathbf{k}] = \int_0^a \int_0^b \int_{-h/2}^{h/2} [\mathbf{A}]^T[\mathbf{C}][\mathbf{A}]\, dz\, dy\, dx \tag{12.27}$$

There is a common term $z^2E/(1-\nu^2)$ in the integrand; it becomes the bending rigidity D after integrating through the thickness.

Alternatively, we can operate on constants $\{\mathbf{c}\}$ instead of nodal d.o.f.'s $\{\mathbf{q}\}$. We write the strain energy expression in terms of $\{\mathbf{c}\}$ by substituting Eq. (12.18) in Eq. (12.8):

$$U = \tfrac{1}{2}\{\mathbf{c}\}^T[\bar{\mathbf{k}}]\{\mathbf{c}\} \tag{12.28}$$

where the coefficients in the kernel stiffness matrix can be obtained by an analogous use of Castigliano's theorem,

$$\bar{k}_{ij} = \frac{\partial^2 U}{\partial c_i\, \partial c_j} \tag{12.29}$$

For example, we can obtain from Eqs. (12.8), (12.18), and (12.29),

$$\begin{aligned} \bar{k}_{44} &= \frac{\partial^2 U}{\partial c_4^2} = \int_0^b \int_0^a \frac{D}{2}(8)\, dx\, dy = 4abD \\ \bar{k}_{88} &= \frac{\partial^2 U}{\partial c_8^2} = \int_0^b \int_0^a \frac{D}{2}[8y^2 + 16(1-\nu)x^2]\, dx\, dy \\ &= \frac{4Dab}{3}[2(1-\nu)a^2 + b^2] \end{aligned} \tag{12.30}$$

The strain energy can be written in terms of $\{\mathbf{q}\}$, from Eqs. (12.24) and (12.28),

$$U = \tfrac{1}{2}\{\mathbf{q}\}^T[\mathbf{T}]^T[\bar{\mathbf{k}}][\mathbf{T}]\{\mathbf{q}\} \tag{12.31}$$

From this equation we see the stiffness matrix as

$$[\mathbf{k}] = [\mathbf{T}]^T[\bar{\mathbf{k}}][\mathbf{T}] \tag{12.32}$$

Alternatively, Eq. (12.23a) may be solved analytically so that the displacement function be expressed explicitly. It is easier to write it in nondimensional centroidal coordinates,

$$w(\xi, \eta) = \sum_{i=1}^{4} \frac{1}{8}(1 + \xi\xi_i)(1 + \eta\eta_i)(2 + \xi\xi_i + \eta\eta_i - \xi^2 - \eta^2)w_i$$

$$+ \sum_{i=1}^{4} \frac{a}{16}\xi_i(1 + \xi\xi_i)^2(1 + \eta\eta_i)(\xi\xi_i - 1)\left(\frac{\partial w}{\partial x}\right)_i$$

$$+ \sum_{i=1}^{4} \frac{b}{16}\eta_i(1 + \xi\xi_i)(1 + \eta\eta_i)^2(\eta\eta_i - 1)\left(\frac{\partial w}{\partial y}\right)_i$$

$$= \sum_{i=1}^{12} f_i(\xi, \eta)q_i \tag{12.33}$$

where $\xi = 2(x - x_0)/a$, $\eta = 2(y - y_0)/b$, (x_0, y_0) are the coordinates of the centroid of the plate, and the nondimensional coordinates for nodes 1, 2, 3, and 4 are $(-1, -1)$, $(1, -1)$, $(1, 1)$, and $(-1, 1)$, respectively.

If we substitute Eq. (12.33) into the strain energy expression (12.8) for thin plates and then perform partial differentiation of the strain energy with respect to each of the 12 d.o.f.'s following Castigliano's theorem, the stiffness matrix can be obtained with its coefficient in the form

$$k_{ij} = D\int_0^b \int_0^a \left\{\frac{\partial^2 f_i}{\partial x^2}\left(\frac{\partial^2 f_j}{\partial x^2}\right) + \frac{\partial^2 f_i}{\partial y^2}\left(\frac{\partial^2 f_j}{\partial y^2}\right)\right.$$

$$\left. + \nu\left[\frac{\partial^2 f_i}{\partial x^2}\left(\frac{\partial^2 f_j}{\partial y^2}\right) + \frac{\partial^2 f_j}{\partial x^2}\left(\frac{\partial^2 f_i}{\partial y^2}\right)\right] + 2(1 - \nu)\frac{\partial^2 f_i}{\partial x\, \partial y}\left(\frac{\partial^2 f_j}{\partial x\, \partial y}\right)\right\} dx\, dy \tag{12.34}$$

Example 12.1

Find the stiffness term k_{34} that relates F_{z_3} to w_4 for the 12-term polynomial rectangular plate bending element.

Solution

$$k_{34} = D\int_{-1}^{1}\int_{-1}^{1}\left\{\frac{4\partial^2 f_3}{a^2\,\partial\xi^2}\left(\frac{4\partial^2 f_4}{a^2\,\partial\xi^2}\right) + \frac{4\partial^2 f_3}{b^2\,\partial\eta^2}\left(\frac{4\partial^2 f_4}{b^2\,\partial\eta^2}\right) + \frac{16\nu}{a^2b^2}\left[\frac{\partial^2 f_3}{\partial\xi^2}\left(\frac{\partial^2 f_4}{\partial\eta^2}\right) + \frac{\partial^2 f_4}{\partial\xi^2}\left(\frac{\partial^2 f_3}{\partial\eta^2}\right)\right]\right.$$

$$\left. + \frac{32(1 - \nu)}{a^2b^2}\left(\frac{\partial^2 f_3}{\partial\xi\,\partial\eta}\right)\frac{\partial^2 f_4}{\partial\xi\,\partial\eta}\right\}\left(\frac{ab}{4}\right) d\xi\, d\eta \tag{12.35a}$$

From the displacement function (12.33), we have for $k = 3$ or 4,

$$f_k(\xi, \eta) = \tfrac{1}{8}(1 + \xi\xi_k)(1 + \eta\eta_k)(2 + \xi\xi_k + \eta\eta_k - \xi^2 - \eta^2)$$

$$\frac{\partial^2 f_k}{\partial\xi^2} = \tfrac{1}{4}(1 + \eta\eta_k)(\xi_k^2 - 3\xi\xi_k - 1)$$

$$\frac{\partial^2 f_k}{\partial\eta^2} = \tfrac{1}{4}(1 + \xi\xi_k)(\eta_k^2 - 3\eta\eta_k - 1)$$

$$\frac{\partial^2 f_k}{\partial\xi\,\partial\eta} = \tfrac{1}{8}[4\xi_k\eta_k - 2(\xi_k\eta + \xi\eta_k) + 2\xi_k\eta_k(\xi\xi_k + \eta\eta_k) - 3\xi_k\eta_k(\xi^2 + \eta^2)] \tag{12.35b}$$

where $(\xi_3, \eta_3) = (1, 1)$ and $(\xi_4, \eta_4) = (-1, 1)$. Substituting Eqs. (12.35b) in (12.35a) gives

$$k_{34} = D(I_1 + I_2 + I_3 + I_4 + I_5) \tag{12.36a}$$

where

$$\begin{aligned}
I_1 &= \frac{b}{4a^3}\int_{-1}^{1}\int_{-1}^{1}(1+\eta)^2(-9\xi^2)\,d\xi\,d\eta = -\frac{4b}{a^3}\\
I_2 &= \frac{a}{4b^3}\int_{-1}^{1}\int_{-1}^{1}(1-\xi^2)(9\eta^2)\,d\xi\,d\eta = \frac{2a}{b^3}\\
I_3 &= \frac{\nu}{4ab}\int_{-1}^{1}\int_{-1}^{1}(1+\eta)(1-\xi)(9\xi\eta)\,d\xi\,d\eta = -\frac{\nu}{ab}\\
I_4 &= \frac{\nu}{4ab}\int_{-1}^{1}\int_{-1}^{1}(1+\eta)(1+\xi)(-9\xi\eta)\,d\xi\,d\eta = -\frac{\nu}{ab}\\
I_5 &= \frac{1-\nu}{8ab}\int_{-1}^{1}\int_{-1}^{1}-[4-3(\xi^2+\eta^2)]^2\,d\xi\,d\eta = -\frac{14(1-\nu)}{5ab}
\end{aligned} \tag{12.36b}$$

Hence

$$k_{34} = \frac{D}{5ab}\left[10\left(\frac{a}{b}\right)^2 - 20\left(\frac{b}{a}\right)^2 + 4\nu - 14\right] \tag{12.36c}$$

A complete explicit presentation of this stiffness matrix is available in the texts by Zienkiewicz [12.23] and Gallagher [12.24]. The performance of this element in numerical examples will be discussed in Sec. 12.8.

Patch test. A patch of nine 12-d.o.f. plate bending elements is shown in Fig. 12.8. This assemblage is subjected to three patch tests as described in Sec. 12.2.3 and Figs. 12.5 and 12.6.

Prescribed displacements. The following deflection surface can be used to produce a state of constant curvatures $\partial^2 w/\partial x^2$ and $\partial^2 w/\partial y^2$ and constant twist $\partial^2 w/\partial x\,\partial y$:

$$w(x, y) = 5 \times 10^{-4}(x^2 + xy + y^2) \tag{12.37}$$

This deflection function results in, from Eq. (12.3),

$$M_x = M_y = 1.11111 \times 10^{-7}\ \text{in.-lb/in.} \tag{12.38a}$$

$$M_{xy} = -M_{yx} = 0.333333 \times 10^{-7}\ \text{in.-lb/in.} \tag{12.38b}$$

everywhere in the plate. The constant twisting moment M_{xy} corresponds to, from Eq. (12.17) and Fig. 12.6b,

$$P = 2M_{xy} = 0.666666 \times 10^{-7}\ \text{in.-lb/in.} \tag{12.38c}$$

Based on Eq. (12.37), nodal displacements w and rotations $\partial w/\partial x$ and $\partial w/\partial y$ are substituted into the stiffness equations of the patch and the results are shown in Table 12.1, which agree with the values given by Eqs. (12.38). Thus the patch test is passed as the prescribed state of displacement (constant curvatures and twist) is found to correspond to a state of constant bending moments and twisting moments (four corner loads) with correct values.

TABLE 12.1 Patch Test of 12-d.o.f. Rectangles Due to Prescribed Displacements

Node Number	Lateral Nodal Loads (10^{-7} lb)	Nodal Moment, M_x (10^{-8} in.-lb)	Nodal Moment, M_y (10^{-8} in.-lb)
1	0.666666	−0.333333	−0.222222
2	a	−0.555555	b
3	a	−0.333333	b
4	−0.666666	−0.111111	0.222222
5	a	b	−0.666666
6	a	b	b
7	a	b	b
8	a	b	0.666666
9	a	b	−1.111111
10	a	b	b
11	a	b	b
12	a	b	1.111111
13	−0.666666	0.333333	−0.666666
14	a	0.555555	b
15	a	0.333333	b
16	0.666666	0.111111	0.666666

[a] $<10^{-11}$ lb.
[b] $<10^{-13}$ in.-lb.

Uniform moments at two opposite edges. A patch test as described in Fig. 12.6a is conducted. Due to double symmetry, only the first quadrant is tested with a mesh as shown in Fig. 12.8. A uniformly distributed moment $M_x = 0.0001$ in.-lb/in. is applied along edge 13-16. The boundary conditions are that $w = 0$ at node 1, $\partial w/\partial x = 0$ along edge 1-4, and $\partial w/\partial y = 0$ along edge 1-13. As an alternative, we could specify $w = 0$ at node 4 instead of node 1, as shown in Fig. 12.6a.

The state of constant curvature caused by the constant loading conditions that $M_x = 0.0001$ in.-lb/in. and $M_y = 0$ can be obtained from Eqs. (12.3):

$$\frac{\partial^2 w}{\partial x^2} = \frac{12M_x}{Eh^3} = 1.2\left(\frac{1}{\text{in.}}\right)$$
$$\frac{\partial^2 w}{\partial y^2} = -\nu\frac{\partial^2 w}{\partial x^2} = -0.3\left(\frac{1}{\text{in.}}\right) \tag{12.39}$$

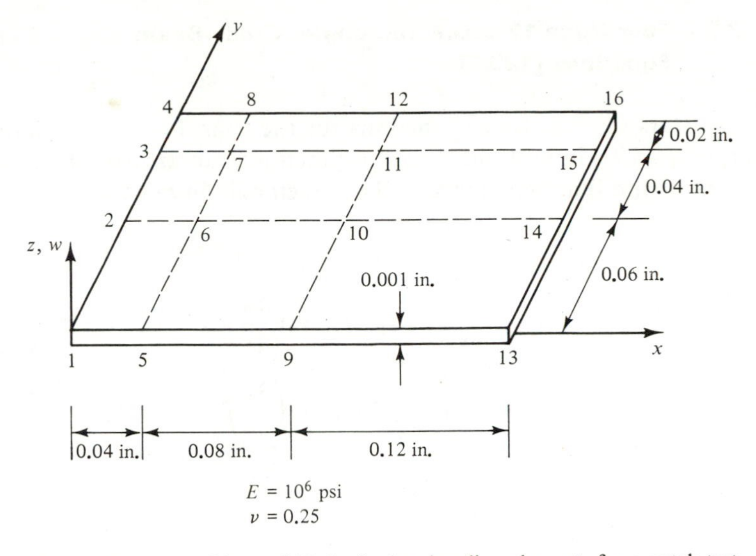

Figure 12.8 Assemblage of 12-d.o.f. plate bending elements for a patch test.

A patch test is conducted by first finding all the nodal deflections w and slopes $\partial w/\partial x$ and $\partial w/\partial y$. These values are found to agree with those values obtained through integrating Eqs. (12.39). These nodal values are then substituted in the strain-nodal displacement equations (12.25) for each element and a state of constant curvatures with values agreeing with those given in Eq. (12.39) are found.

Four corner loads producing constant twist. The patch shown in Fig. 12.8 is tested by applying four corner loads with $P = 10^{-6}$ lb as shown in Fig. 12.6b. These loads produce a constant twist, Eq. (12.17):

$$\frac{\partial^2 w}{\partial x\,\partial y} = \frac{6P(1+\nu)}{Eh^3} = 0.0075\left(\frac{1}{\text{in.}}\right) \tag{12.40}$$

Integrating gives

$$\frac{\partial w}{\partial x} = 0.0075y \qquad \frac{\partial w}{\partial y} = 0.0075x \qquad w = 0.0075xy \tag{12.41}$$

It is found that the patch test results in nodal displacements and slopes that agree totally with the values from Eq. (12.41). The nodal values are then substituted in the strain-displacement equations (12.25) for each element and indeed a state of constant twist with values of 0.0075 (1/in.) is obtained.

12.3.2 Four Node 12-D.O.F. rectangle: Cross-Beam Functions [12.25]

It appears that the shape functions for the four-d.o.f. elementary beam elements [Eq. (5.7)] in both the x and y directions can properly be combined to form the shape functions for a 12-d.o.f. rectangle in bending.

$$
\begin{aligned}
w(x, y) = {} & f_1(x)f_1(y)w_1 + f_2(x)f_1(y)w_2 + f_2(x)f_2(y)w_3 + f_1(x)f_2(y)w_4 \\
& + f_3(x)f_1(y)\left(\frac{\partial w}{\partial x}\right)_1 + f_4(x)f_1(y)\left(\frac{\partial w}{\partial x}\right)_2 + f_4(x)f_2(y)\left(\frac{\partial w}{\partial x}\right)_3 \\
& + f_3(x)f_2(y)\left(\frac{\partial w}{\partial x}\right)_4 + f_1(x)f_3(y)\left(\frac{\partial w}{\partial y}\right)_1 + f_2(x)f_3(y)\left(\frac{\partial w}{\partial y}\right)_2 \\
& + f_2(x)f_4(y)\left(\frac{\partial w}{\partial y}\right)_3 + f_1(x)f_4(y)\left(\frac{\partial w}{\partial y}\right)_4
\end{aligned}
\tag{12.42}
$$

where

$$
\begin{aligned}
f_1(x) &= 1 - 3\left(\frac{x}{a}\right)^2 + 2\left(\frac{x}{a}\right)^3 \\
f_2(x) &= 3\left(\frac{x}{a}\right)^2 - 2\left(\frac{x}{a}\right)^3 \\
f_3(x) &= x\left(\frac{x}{a} - 1\right)^2 \\
f_4(x) &= x\left[\left(\frac{x}{a}\right)^2 - \frac{x}{a}\right]
\end{aligned}
\tag{12.42a}
$$

and $f_i(y)$ is obtained by replacing x and a by y and b, respectively, in Eq. (12.42a).

Following the procedures of Sec. 12.3.1, we see that this displacement function satisfies the interelement compatibility in w, $\partial w/\partial x$, and $\partial w/\partial y$ along all four edges. If we form the products of the shape functions in Eq. (12.42), we see a polynomial in which the three terms c_1, c_2x, c_3y needed for the three rigid-body modes w, $\partial w/\partial x$, and $\partial w/\partial y$ are present. Also, the terms c_4x^2 and c_5y^2 needed for the two constant strains ϵ_x and ϵ_y (or curvatures) are present. But the term c_6xy for the constant shearing strain γ_{xy} or twist $\partial^2 w/\partial x\,\partial y$ is missing. This xy term is included in the 12-term polynomial rectangle [12.18–12.21] and it is also included in the following 16-d.o.f. conforming rectangle in bending.

12.3.3 Four-Node 16-D.O.F. Conforming Rectangle: 16-Term Polynomial [12.26]

By adding the second-order twist derivative terms $\partial^2 w/\partial x\, \partial y$ as the nodal d.o.f.'s to the 12-d.o.f. element as shown in Fig. 12.7, we have a 16-d.o.f. element. Thus each corner node now has four d.o.f.'s: w, $\partial w/\partial x$, $\partial w/\partial y$, and $\partial^2 w/\partial x\, \partial y$.

Displacement function. The 16-term polynomial displacement function is assumed as obtained from the product of two cross-beam displacements, each in cubic polynomial form:

$$w(x, y) = (a_1 + a_2x + a_3x + a_4x^3)(b_1 + b_2y + b_3y^2 + b_4y^3)$$

$$\begin{array}{ccccccc} & & & = \quad c_1 & & & \\ & & & c_2x \quad c_3y & & & \\ & & c_4x^2 & c_5xy & c_6y^2 & & \\ & c_7x^3 & c_8x^2y & & c_9xy^2 & c_{10}y^3 & \\ x^4 & c_{11}x^3y & & c_{12}x^2y^2 & & c_{13}xy^3 & y^4 \\ x^5 \quad x^4y & c_{14}x^3y^2 & & & c_{15}x^2y^3 & xy^4 \quad y^5 & \\ x^6 \quad x^5y \quad x^4y^2 & & & c_{16}x^3y^3 & & x^2y^4 \quad xy^5 \quad y^6 & \end{array} \tag{12.43}$$

where the sixteen constants can be determined by evaluating the sixteen nodal values.

This displacement function can also be written explicitly in terms of bicubic Hermitian polynomials in nondimensional centroidal coordinates ξ and η:

$$w(\xi, \eta) = \sum_{i=1}^{4} \left[G_i(\xi)G_i(\eta)w_i + \frac{a}{2}H_i(\xi)G_i(\eta)\left(\frac{\partial w}{\partial x}\right)_i \right.$$
$$\left. + \frac{b}{2}G_i(\xi)H_i(\eta)\left(\frac{\partial w}{\partial y}\right)_i + \frac{ab}{4}H_i(\xi)H_i(\eta)\left(\frac{\partial^2 w}{\partial x\, \partial y}\right)_i \right] \tag{12.44}$$

where

$$G_i(\xi) = \tfrac{1}{4}(-\xi_i\xi^3 + 3\xi_i\xi + 2)$$
$$H_i(\xi) = \tfrac{1}{4}(\xi^3 + \xi_i\xi^2 - \xi - \xi_i) \tag{12.44a}$$
$$\xi = \frac{x - (a/2)}{a/2} = 2\left(\frac{x}{a}\right) - 1 \qquad \eta = 2\left(\frac{y}{b}\right) - 1$$

The Hermitian polynomials (12.44a) and the beam displacement functions (12.42a) are the same functions;

$$\begin{aligned} G_1(\xi) &= G_4(\xi) = f_1(x) & G_1(\eta) &= G_2(\eta) = f_1(y) \\ G_2(\xi) &= G_3(\xi) = f_2(x) & G_3(\eta) &= G_4(\eta) = f_2(y) \\ H_1(\xi) &= H_4(\xi) = f_3(x) & H_1(\eta) &= H_2(\eta) = f_3(y) \\ H_2(\xi) &= H_3(\xi) = f_4(x) & H_3(\eta) &= H_4(\eta) = f_4(y) \end{aligned} \tag{12.45}$$

The first twelve terms in Eqs. (12.44) are precisely the same as those in cross-beam displacement functions (12.42). Equations (12.44) are in a form convenient for mathematical manipulations. However, it is of interest to write them in a form convenient for physical interpretation,

$$\begin{aligned} w(x, y) &= \sum_{i=1}^{16} f_i(x, y) q_i \\ &= \frac{1}{a^3 b^3}\Bigg[(a^3 + 2x^3 - 3ax^2)(b^3 + 2y^3 - 3by^2) w_1 \\ &\quad + (3ax^2 - 2x^3)(b^3 + 2y^3 - 3by^2) w_2 \\ &\quad + (3ax^2 - 2x^3)(3by^2 - 2y^3) w_3 \\ &\quad + (a^3 + 2x^3 - 3ax^2)(3by^2 - 2y^3) w_4 \\ &\quad + ax(x - a)^2 (b^3 + 2y^3 - 3by^2)\left(\frac{\partial w}{\partial x}\right)_1 \\ &\quad + a(x^3 - ax^2)(b^3 + 2y^3 - 3by^2)\left(\frac{\partial w}{\partial x}\right)_2 \\ &\quad + a(x^3 - ax^2)(3by^2 - 2y^3)\left(\frac{\partial w}{\partial x}\right)_3 \\ &\quad + ax(x - a)^2 (3by^2 - 2y^3)\left(\frac{\partial w}{\partial x}\right)_4 \\ &\quad + b(a^3 + 2x^3 - 3ax^2) y(y - b)^2 \left(\frac{\partial w}{\partial y}\right)_1 \\ &\quad + b(3ax^2 - 2x^3) y(y - b)^2 \left(\frac{\partial w}{\partial y}\right)_2 \\ &\quad + b(3ax^2 - 2x^3)(y^3 - by^2)\left(\frac{\partial w}{\partial y}\right)_3 \\ &\quad + b(a^3 + 2x^3 - 3ax^2)(y^3 - by^2)\left(\frac{\partial w}{\partial y}\right)_4 \end{aligned}$$

$$+ abxy(x-a)^2(y-b)^2\left(\frac{\partial^2 w}{\partial x\,\partial y}\right)_1$$

$$+ abxy(x^2-ax)(y-b)^2\left(\frac{\partial^2 w}{\partial x\,\partial y}\right)_2$$

$$+ abxy(x^2-ax)(y^2-by)\left(\frac{\partial^2 w}{\partial x\,\partial y}\right)_3$$

$$+ abxy(x-a)^2(y^2-by)\left(\frac{\partial^2 w}{\partial x\,\partial y}\right)_4\Bigg] \tag{12.46}$$

Let us interpret these 16 shape functions physically. For example, if we set the local coordinates to be those for node 1 (i.e., $x = y = 0$), the values of w, $\partial w/\partial x$, $\partial w/\partial y$, and $\partial^2 w/\partial x\,\partial y$ will equal those of the four corresponding d.o.f.'s at node 1. This means that at node 1 $f_1(x, y)$, $\partial f_5(x, y)/\partial x$, $\partial f_9(x, y)/\partial y$, and $\partial^2 f_{13}(x, y)/\partial x\,\partial y$ will equal 1 and the rest of the twelve shape functions and their derivatives will vanish. Plots of these four shape functions f_1, f_5, f_9, and f_{13} in Fig. 12.9 show that each shape function represents the deflection surface of the element when its corresponding d.o.f. is set to 1 and the rest of the fifteen d.o.f.'s are set to zero.

Rigid-body modes, constant strains, and interelement compatibility. From Eq. (12.43) it is seen that this displacement function contains the three rigid-body motion terms (c_1, c_2x, c_3y) and the three constant strain terms (c_4x^2, c_5xy, and c_6y^2).

Following Eqs. (12.20) and (12.21), it can be shown that this function satisfies interelement compatibility in w and $\partial w/\partial y$ along edge 2-3. To show interelement compatibility in the normal slope $\partial w/\partial y$ along edge 2-3 with $x = a$, we first obtain

$$\frac{\partial w}{\partial x}(a, y) = c(b_1 + b_2y + b_3y^2 + b_4y^3) \tag{12.47}$$

where the four constants can be uniquely determined by the four conditions

$$\begin{aligned} \frac{\partial w}{\partial x} &= \left(\frac{\partial w}{\partial x}\right)_2 \quad \text{and} \quad \frac{\partial^2 w}{\partial y\,\partial x} = \left(\frac{\partial^2 w}{\partial x\,\partial y}\right)_2 \quad \text{for } y = 0 \\ \frac{\partial w}{\partial x} &= \left(\frac{\partial w}{\partial x}\right)_3 \quad \text{and} \quad \frac{\partial^2 w}{\partial y\,\partial x} = \left(\frac{\partial^2 w}{\partial x\,\partial y}\right)_3 \quad \text{for } y = b \end{aligned} \tag{12.48}$$

Thus the normal slopes $\partial w/\partial x$ between the two adjacent elements are compatible along edge 2-3. The interelement compatibility in normal slopes along the other three edges can be shown in the same manner.

This displacement function fits all the criteria for monotonic convergence. The element is *conforming* and C_1 *continuous*. It passes the patch test.

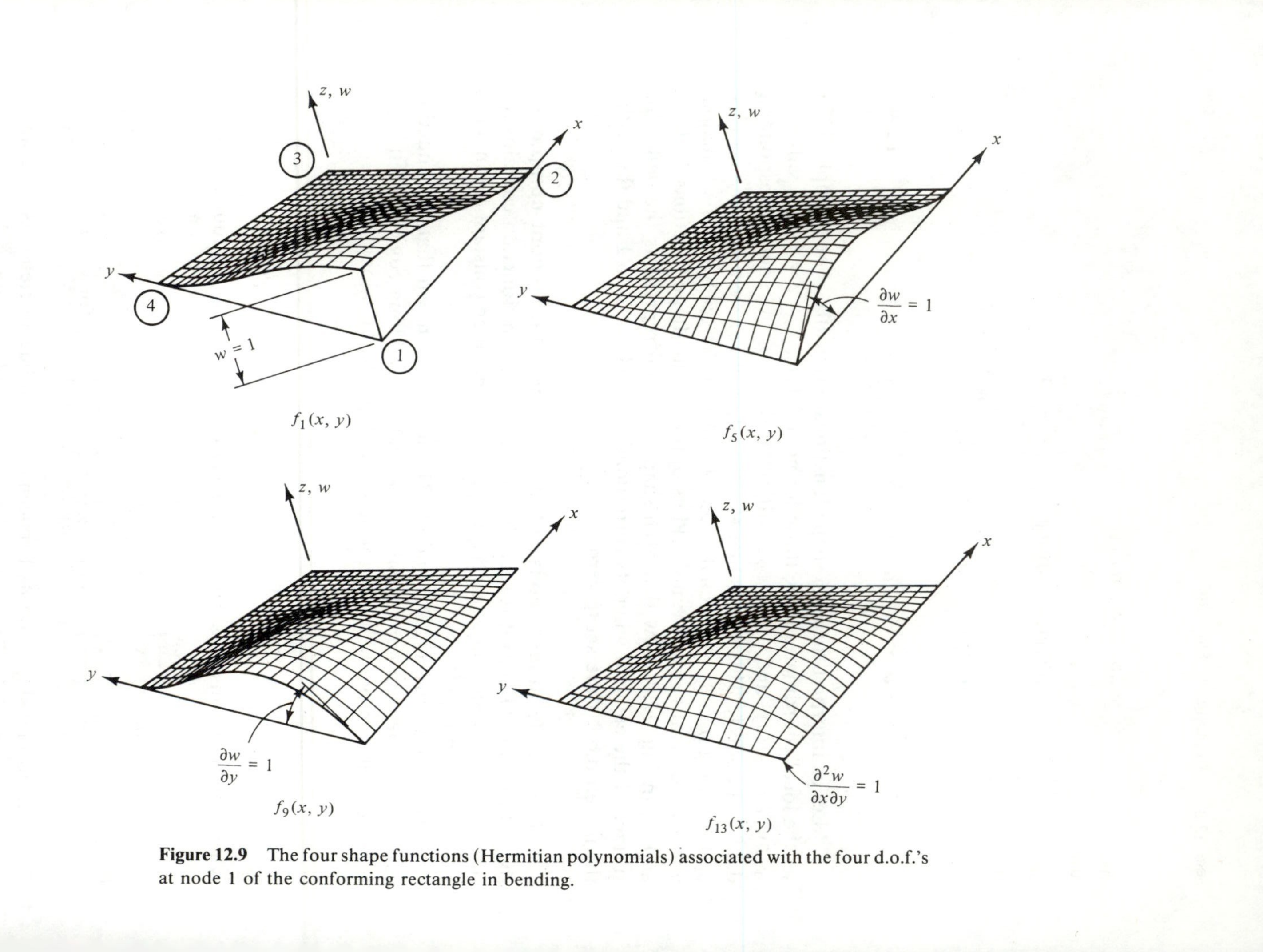

Figure 12.9 The four shape functions (Hermitian polynomials) associated with the four d.o.f.'s at node 1 of the conforming rectangle in bending.

Stiffness matrix. The stiffness matrix can be formulated using one of the three equations (12.27), (12.32), and (12.34). Since the displacement function (12.44) is available explicitly, Eq. (12.34) is preferred.

Example 12.2

Find the stiffness term k_{34} that relates f_{z_3} to w_4 for the 16-d.o.f. rectangular plate bending element.

Solution. Equation (12.35) can be used to derive k_{34}. It is necessary first to obtain the second derivatives of the shape function (12.44a);

$$\begin{aligned} &G_3(\xi) = \tfrac{1}{4}(-\xi^3 + 3\xi + 2) \qquad G_4(\xi) = \tfrac{1}{4}(\xi^3 - 3\xi + 2) \\ &G_3'(\xi) = -G_4'(\xi) = \tfrac{3}{4}(-\xi^2 + 1) \\ &G_3''(\xi) = -G_4''(\xi) = -\frac{3\xi}{2} \\ &G_3(\eta) = G_4(\eta) = \tfrac{1}{4}(-\eta^3 + 3\eta + 2) \\ &G_3'(\eta) = G_4'(\eta) = \tfrac{3}{4}(-\eta^2 + 1) \\ &G_3''(\eta) = G_4''(\eta) = -\frac{3\eta}{2} \end{aligned} \tag{12.49}$$

From Eq. (12.35),

$$k_{34} = D(I_1 + I_2 + I_3 + I_4 + I_5) \tag{12.50a}$$

where

$$\begin{aligned} I_1 &= \int_{-1}^{1}\int_{-1}^{1} \frac{4b}{a^3}(-\tfrac{9}{4}\xi^2)(\tfrac{1}{16})(-\eta^3 + 3\eta + 2)^2\, d\xi\, d\eta = -\frac{156}{35}\frac{b}{a^3} \\ I_2 &= \int_{-1}^{1}\int_{-1}^{1} \frac{4a}{b^3}\left(\frac{3\eta}{2}\right)^2 \frac{-\xi^3 + 3\xi + 2}{4}\left(\frac{\xi^3 - 3\xi + 2}{4}\right) d\xi\, d\eta = \frac{54}{35}\frac{a}{b^3} \\ I_3 &= \int_{-1}^{1}\int_{-1}^{1} \frac{4\nu}{ab}\left(\frac{-3\xi}{2}\right)\frac{-\eta^3 + 3\eta + 2}{4}\left(\frac{\xi^3 - 3\xi + 2}{4}\right)\left(-\frac{3\eta}{2}\right) d\xi\, d\eta = -\frac{36\nu}{25ab} \\ I_4 &= \int_{-1}^{1}\int_{-1}^{1} \frac{4\nu}{ab}\left(\frac{\xi^3 - 3\xi + 2}{4}\right)\frac{-3\xi}{2}\left(\frac{-3\eta}{2}\right)\frac{-\eta^3 + 3\eta + 2}{4}\, d\xi\, d\eta = \frac{-36\nu}{25ab} \\ I_5 &= \int_{-1}^{1}\int_{-1}^{1} \frac{8(1-\nu)}{ab}(\tfrac{3}{4})^4(-1)(-\xi^2 + 1)^2(-\eta^2 + 1)^2\, d\xi\, d\eta = -\frac{72(1-\nu)}{25ab} \end{aligned} \tag{12.50b}$$

Hence

$$k_{34} = D\left(-\frac{156}{35}\frac{b}{a^3} + \frac{54}{35}\frac{a}{b^3} - \frac{72}{25ab}\right) \tag{12.50c}$$

This 16×16 stiffness matrix is given explicitly in Table 12.2, in which the coefficient in ith row and jth column of the matrix $[\mathbf{k}]$ is defined as

$$k_{ij} = \frac{Eh^3}{12ab(1-\nu^2)}\left[\alpha_1\left(\frac{b}{a}\right)^2 + \alpha_2\left(\frac{a}{b}\right)^2 + \alpha_3 + \alpha_4\nu\right]a^{\alpha_5}b^{\alpha_6} \tag{12.51}$$

where the constants $\alpha_1, \alpha_2, \ldots, \alpha_6$ are given in Table 12.3. Although 136 terms are populated in the half of the matrix shown, they share only forty patterns with forty sets of constants, as given in Table 12.3.

A computer program for this element and a sample problem of a simply supported square plate under uniform loads is given in Sec. 13.4. Performance of this element in numerical example will be given in Sec. 12.8.

Mass matrix. Formulation of mass matrices for free-vibration analysis has been discussed in Chapter 7 for bar, beam, and frame elements. Extension to plate elements is straightforward.

For a lumped mass matrix, we simply divide the total mass of the element by 4 and put them along the diagonal of the matrix, with a sequence corresponding to that of w_1, w_2, w_3, and w_4.

For a consistent mass matrix, we use the same displacement function (12.44) or (12.46), and Lagrange's equation (7.15).

The kinetic energy for a vibrating plate in bending with small deflection is

$$T = \frac{\rho h}{2} \int_0^a \int_0^b (\dot{w})^2 \, dx \, dy \tag{12.52}$$

where ρ is the mass per unit volume, in lb-sec^2/in.^{4}, and h is the plate thickness.

Let us recall Lagrange's equation from Chapter 7:

$$F_i = \frac{\partial U}{\partial q_i} + \frac{d}{dt}\left(\frac{\partial T}{\partial \dot{q}_i}\right) \tag{7.15}$$

The first term on the right, $\partial U/\partial q_i$, produces the stiffness matrix with coefficients defined by Eq. (12.51). The second term produces the mass matrix.

Substituting the displacement function in the kinetic energy expression (12.52) and then performing differentiation of the kinetic energy as shown in Eq. (7.15), a mass matrix is obtained.

The equations of motion may now be written as

$$\underset{16 \times 1}{\{\mathbf{F}\}} = \underset{16 \times 16}{[\mathbf{k}]} \{\mathbf{q}\} + \underset{16 \times 16}{[\mathbf{m}]} \{\ddot{\mathbf{q}}\} \tag{12.53}$$

Assuming sinusoidal motion with natural frequency ω as described in Eqs. (7.9) and (7.10), we obtain a set of eigenvalue equations for free vibration,

$$\{\mathbf{0}\} = [[\mathbf{k}] - \omega^2[\mathbf{m}]]\{\mathbf{q}\} \tag{12.54}$$

with ω being the eigenvalues or natural frequencies and $\{\mathbf{q}\}$ the eigenvectors or mode shapes. The methods of solution for eigenvalue equations are given in detail in Chapter 2.

The mass matrix coefficient in Eq. (12.53) is given as

$$m_{ij} = \rho h \int_0^b \int_0^a f_i(x, y) f_j(x, y) \, dx \, dy \tag{12.55}$$

TABLE 12.2 Stiffness Matrix for the 16-d.o.f. Rectangle in Bending[a]

	w_1	w_2	w_3	w_4	w_{x_1}	w_{x_2}	w_{x_3}	w_{x_4}	w_{y_1}	w_{y_2}	w_{y_3}	w_{y_4}	w_{xy_1}	w_{xy_2}	w_{xy_3}	w_{xy_4}
w_1	$k_{1,1}$															
w_2	$k_{2,1}$	$k_{1,1}$								symmetric						
w_3	$k_{3,1}$	$k_{4,1}$	$k_{1,1}$													
w_4	$k_{4,1}$	$k_{3,1}$	$k_{2,1}$	$k_{1,1}$												
w_{x_1}	$k_{5,1}$	$-k_{6,1}$	$-k_{7,1}$	$k_{8,1}$	$k_{5,5}$											
w_{x_2}	$k_{6,1}$	$-k_{5,1}$	$-k_{8,1}$	$k_{7,1}$	$k_{6,5}$	$k_{5,5}$										
w_{x_3}	$k_{7,1}$	$-k_{8,1}$	$-k_{5,1}$	$k_{6,1}$	$k_{7,5}$	$k_{8,5}$	$k_{5,5}$									
w_{x_4}	$k_{8,1}$	$-k_{7,1}$	$-k_{6,1}$	$k_{5,1}$	$k_{8,5}$	$k_{7,5}$	$k_{6,5}$	$k_{5,5}$								
w_{y_1}	$k_{9,1}$	$k_{10,1}$	$-k_{11,1}$	$-k_{12,1}$	$k_{9,5}$	$-k_{10,5}$	$k_{11,5}$	$-k_{12,5}$	$k_{9,9}$							
w_{y_2}	$k_{10,1}$	$k_{9,1}$	$-k_{12,1}$	$-k_{11,1}$	$k_{10,5}$	$-k_{9,5}$	$k_{12,5}$	$-k_{11,5}$	$k_{10,9}$	$k_{9,9}$						
w_{y_3}	$k_{11,1}$	$k_{12,1}$	$-k_{9,1}$	$-k_{10,1}$	$k_{11,5}$	$-k_{12,5}$	$k_{9,5}$	$-k_{10,5}$	$k_{11,9}$	$k_{12,9}$	$k_{9,9}$					
w_{y_4}	$k_{12,1}$	$k_{11,1}$	$-k_{10,1}$	$-k_{9,1}$	$k_{12,5}$	$-k_{11,5}$	$k_{10,5}$	$-k_{9,5}$	$k_{12,9}$	$k_{11,9}$	$k_{10,9}$	$k_{9,9}$				
w_{xy_1}	$k_{13,1}$	$-k_{14,1}$	$k_{15,1}$	$-k_{16,1}$	$k_{13,5}$	$k_{14,5}$	$-k_{15,5}$	$-k_{16,5}$	$k_{13,9}$	$-k_{14,9}$	$-k_{15,9}$	$k_{16,9}$	$k_{13,13}$			
w_{xy_2}	$k_{14,1}$	$-k_{13,1}$	$k_{16,1}$	$-k_{15,1}$	$k_{14,5}$	$k_{13,5}$	$-k_{16,5}$	$-k_{15,5}$	$k_{14,9}$	$-k_{13,9}$	$-k_{16,9}$	$k_{15,9}$	$k_{14,13}$	$k_{13,13}$		
w_{xy_3}	$k_{15,1}$	$-k_{16,1}$	$k_{13,1}$	$-k_{14,1}$	$k_{15,5}$	$k_{16,5}$	$-k_{13,5}$	$-k_{14,5}$	$k_{15,9}$	$-k_{16,9}$	$-k_{13,9}$	$k_{14,9}$	$k_{15,13}$	$k_{16,13}$	$k_{13,13}$	
w_{xy_4}	$k_{16,1}$	$-k_{15,1}$	$k_{14,1}$	$-k_{13,1}$	$k_{16,5}$	$k_{15,5}$	$-k_{14,5}$	$-k_{13,5}$	$k_{16,9}$	$-k_{15,9}$	$-k_{14,9}$	$k_{13,9}$	$k_{16,13}$	$k_{15,13}$	$k_{14,13}$	$k_{13,13}$

[a] For a consistent mass matrix, replace k by m. Both k_{ij} and m_{ij} re given in Table 12.3.

TABLE 12.3 Constants for the Stiffness and Consistent Mass Matrix Coefficients for the 16-d.o.f. Rectangle in Bending[a]

Row i	Column j	α_1	α_2	α_3	α_4	α_5	α_6	α_7
1	1	$\frac{156}{35}$	$\frac{156}{35}$	$\frac{72}{25}$	0	0	0	169
2	1	$-\frac{156}{35}$	$\frac{54}{35}$	$-\frac{72}{25}$	0	0	0	$\frac{117}{2}$
3	1	$-\frac{54}{35}$	$-\frac{54}{35}$	$\frac{72}{25}$	0	0	0	$\frac{81}{4}$
4	1	$\frac{54}{35}$	$-\frac{156}{35}$	$-\frac{72}{25}$	0	0	0	$\frac{117}{2}$
5	1	$\frac{78}{35}$	$\frac{22}{35}$	$\frac{6}{25}$	$\frac{6}{5}$	1	0	$\frac{143}{6}$
6	1	$\frac{78}{35}$	$-\frac{13}{35}$	$\frac{6}{25}$	0	1	0	$-\frac{169}{12}$
7	1	$\frac{27}{35}$	$\frac{13}{35}$	$-\frac{6}{25}$	0	1	0	$-\frac{39}{8}$
8	1	$\frac{27}{35}$	$-\frac{22}{35}$	$-\frac{6}{25}$	$-\frac{6}{5}$	1	0	$\frac{33}{4}$
9	1	$\frac{22}{35}$	$\frac{78}{35}$	$\frac{6}{25}$	$\frac{6}{5}$	0	1	$\frac{143}{6}$
10	1	$-\frac{22}{35}$	$\frac{27}{35}$	$-\frac{6}{25}$	$-\frac{6}{5}$	0	1	$\frac{33}{4}$
11	1	$\frac{13}{35}$	$\frac{27}{35}$	$-\frac{6}{25}$	0	0	1	$-\frac{39}{8}$
12	1	$-\frac{13}{35}$	$\frac{78}{35}$	$\frac{6}{25}$	0	0	1	$-\frac{169}{12}$
13	1	$\frac{11}{35}$	$\frac{11}{35}$	$\frac{1}{50}$	$\frac{1}{5}$	1	1	$\frac{121}{36}$
14	1	$\frac{11}{35}$	$-\frac{13}{70}$	$\frac{1}{50}$	$\frac{1}{10}$	1	1	$-\frac{143}{72}$
15	1	$-\frac{13}{70}$	$-\frac{13}{70}$	$\frac{1}{50}$	0	1	1	$\frac{169}{144}$
16	1	$-\frac{13}{70}$	$\frac{11}{35}$	$\frac{1}{50}$	$\frac{1}{10}$	1	1	$-\frac{143}{72}$
5	5	$\frac{52}{35}$	$\frac{4}{35}$	$\frac{8}{25}$	0	2	0	$\frac{13}{3}$
6	5	$\frac{26}{35}$	$-\frac{3}{35}$	$-\frac{2}{25}$	0	2	0	$-\frac{13}{4}$
7	5	$\frac{9}{35}$	$\frac{3}{35}$	$\frac{2}{25}$	0	2	0	$-\frac{9}{8}$
8	5	$\frac{18}{35}$	$-\frac{4}{35}$	$-\frac{8}{25}$	0	2	0	$\frac{3}{2}$
9	5	$\frac{11}{35}$	$\frac{11}{35}$	$\frac{1}{50}$	$\frac{6}{5}$	1	1	$\frac{121}{36}$
10	5	$-\frac{11}{35}$	$\frac{13}{70}$	$-\frac{1}{50}$	$-\frac{1}{10}$	1	1	$\frac{143}{72}$
11	5	$\frac{13}{70}$	$\frac{13}{70}$	$-\frac{1}{50}$	0	1	1	$-\frac{169}{144}$
12	5	$-\frac{13}{70}$	$\frac{11}{35}$	$\frac{1}{50}$	$\frac{1}{10}$	1	1	$-\frac{143}{72}$
13	5	$\frac{22}{105}$	$\frac{2}{35}$	$\frac{2}{75}$	$\frac{2}{15}$	2	1	$\frac{11}{18}$
14	5	$\frac{11}{105}$	$-\frac{3}{70}$	$-\frac{1}{150}$	$-\frac{1}{30}$	2	1	$-\frac{11}{24}$
15	5	$-\frac{13}{210}$	$-\frac{3}{70}$	$-\frac{1}{150}$	0	2	1	$\frac{13}{48}$
16	5	$-\frac{13}{105}$	$\frac{2}{35}$	$\frac{2}{75}$	0	2	1	$-\frac{13}{36}$
9	9	$\frac{4}{35}$	$\frac{52}{35}$	$\frac{8}{25}$	0	0	2	$\frac{13}{3}$
10	9	$-\frac{4}{35}$	$\frac{18}{35}$	$-\frac{8}{25}$	0	0	2	$\frac{3}{2}$
11	9	$\frac{3}{35}$	$\frac{9}{35}$	$\frac{2}{25}$	0	0	2	$-\frac{9}{8}$
12	9	$-\frac{3}{35}$	$\frac{26}{35}$	$-\frac{2}{25}$	0	0	2	$-\frac{13}{4}$
13	9	$\frac{2}{35}$	$\frac{22}{105}$	$\frac{2}{75}$	$\frac{2}{15}$	1	2	$\frac{11}{18}$
14	9	$\frac{2}{35}$	$-\frac{13}{105}$	$\frac{2}{75}$	0	1	2	$-\frac{13}{36}$
15	9	$-\frac{3}{70}$	$-\frac{13}{210}$	$-\frac{1}{150}$	0	1	2	$\frac{13}{48}$
16	9	$-\frac{3}{70}$	$\frac{11}{105}$	$-\frac{1}{150}$	$-\frac{1}{30}$	1	2	$-\frac{11}{24}$
13	13	$\frac{4}{105}$	$\frac{4}{105}$	$\frac{8}{225}$	0	2	2	$\frac{1}{9}$
14	13	$\frac{2}{105}$	$-\frac{1}{35}$	$-\frac{2}{225}$	0	2	2	$-\frac{1}{12}$
15	13	$-\frac{1}{70}$	$-\frac{1}{70}$	$\frac{1}{450}$	0	2	2	$\frac{1}{16}$
16	13	$-\frac{1}{35}$	$\frac{2}{105}$	$-\frac{2}{225}$	0	2	2	$-\frac{1}{12}$

[a]See Table 12.2 for [**k**] and [**m**]).

$$k_{ij} = \frac{D}{ab}\left[\alpha_1\left(\frac{b}{a}\right)^2 + \alpha_2\left(\frac{a}{b}\right)^2 + \alpha_3 + \alpha_4\nu\right]a^{\alpha_5}b^{\alpha_6}; \qquad m_{ij} = \frac{\rho hab}{1225}\alpha_7 a^{\alpha_5}b^{\alpha_6}$$

If the shape functions used in deriving the mass matrix are the same as those used for deriving the stiffness matrix, the mass matrix may be referred to as a consistent mass matrix.

Example 12.3

Find the consistent mass matrix coefficient m_{34} that relates F_{z_3} to w_4 for the 16-d.o.f. rectangular plate bending element.

Solution. Let us try the displacement function (12.44) in an alternative form (12.46):

$$m_{34} = \rho h \int_0^b \int_0^a f_3(x, y) f_4(x, y)\, dx\, dy$$

$$= \rho h \int_0^b \int_0^a (3ax^2 - 2x^3)(a^3 + 2x^3 - 3ax^2)(3by^2 - 2y^3)^2\, dx\, dy \div a^6 b^6$$

$$m_{ij} = \tfrac{117}{2450} \rho hab \tag{12.56}$$

The consistent mass matrix is obtained explicitly as shown in Table 12.2 with k replaced by m. The coefficients are in the form

$$m_{ij} = \frac{\rho hab}{1225} \alpha_7 a^{\alpha_5} b^{\alpha_6} \tag{12.57}$$

where the constants α_5, α_6, and α_7 are as given in Table 12.3.

Both the lumped and consistent mass matrices of this element are used to test a free-vibration problem of a simply supported square plate. The results for the six fundamental natural frequencies together with the exact values are given in Table 12.4. Both formulations appear to be good tools for dynamic analysis.

TABLE 12.4 Natural Frequencies ω_{ij} (rad/sec) for a Simply Supported Square Plate[a]

Number of Elements	Number of D.O.F.'s	Mass Matrix Type	ω_{11}	ω_{12}, ω_{21}	ω_{22}	ω_{13}, ω_{31}	ω_{23}, ω_{32}	ω_{33}
4	16	Lumped	996					
		Consistent	1037	2777	4399	6222	7702	10745
16	64	Lumped	1033	2555	3983			
		Consistent	1035	2593	4147	5251	6791	9410
Exact solution[b]			1035	2587	4138	5173	6725	9311

[a] $l_1 = l_2 = 10$ in.; $h = 0.1$ in.; $E = 30 \times 10^6$ psi; $\nu = 0.3$; $\rho = 0.001$ lb-sec²/in.⁴

[b] $\omega_{mn} = \pi^2 \sqrt{\dfrac{D}{\rho h}} \left(\dfrac{m^2}{l_1^2} + \dfrac{n^2}{l_2^2} \right)$ with (m, n) being the mode number.

Incremental stiffness matrix. The formulation of incremental stiffness matrices for buckling and large deflection analyses was discussed in Chapter 8 for column and frame elements. Extension to plate elements is discussed here.

In formulating the plate elements for buckling analysis, the effect of in-plane compressive stresses is included and expressed in the form of incremental stiffness matrix. The strain energy or work done in a plate element due to the in-plane forces can be obtained by straightforward extension of that derived in Eq. (8.14) for the beam element [12.1]:

$$W = \frac{1}{2}\int_0^b \int_0^a \left[N_x\left(\frac{\partial w}{\partial x}\right)^2 + N_y\left(\frac{\partial w}{\partial y}\right)^2 + 2N_{xy}\left(\frac{\partial w}{\partial x}\right)\frac{\partial w}{\partial y}\right] dx\, dy \qquad (12.58)$$

where $N_x = \sigma_x h$, $N_y = \sigma_y h$, $N_{xy} = \tau_{xy} h$, and σ_x, σ_y, and τ_{xy} are the direct stresses in the x and y directions and the shearing stress, respectively (see Fig. 12.10). They do not have to be uniformly distributed as shown. They are assumed as negative when acting in the directions shown in Fig. 12.10.

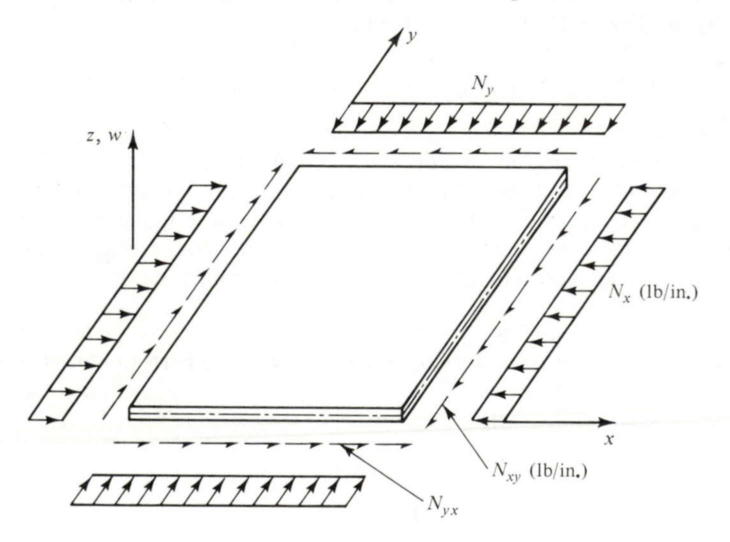

Figure 12.10 In-plane direct and shearing forces on an element.

When the potential energy of Eq. (12.58) is included in U in the Lagrange's equation (7.15), we obtain

$$\{\mathbf{F}\} = [[\mathbf{k}] + [\mathbf{n}] - \omega^2[\mathbf{m}]]\{\mathbf{q}\} \qquad (12.59)$$

where $[\mathbf{n}]$ can be called the *incremental stiffness matrix* and it can be obtained as

$$n_{ij} = \int_0^b \int_0^a \left\{ N_x\left(\frac{\partial f_i}{\partial x}\right)\frac{\partial f_j}{\partial x} + N_y\left(\frac{\partial f_i}{\partial y}\right)\frac{\partial f_j}{\partial y} + N_{xy}\left[\left(\frac{\partial f_i}{\partial x}\right)\frac{\partial f_j}{\partial y} + \left(\frac{\partial f_i}{\partial y}\right)\frac{\partial f_j}{\partial x}\right]\right\} dx\, dy \qquad (12.60)$$

Example 12.4

Find the incremental stiffness matrix coefficient n_{34} that relates F_{z_3} to w_4 for the 16-d.o.f. rectangular plate bending element.

Solution. Substituting Eq. (12.46) in (12.60), we have

$$\begin{aligned} n_{34} &= \int_0^b \int_0^a \{N_x(-36)(x^2 - ax)^2(3by^2 - 2y^3)^2 + N_y(3ax^2 - 2x^3)(a^3 + 2x^3 - 3ax^2) \\ &\quad \times (36)(by - y^2)^2 + N_{xy}[36(ax - x^2)(a^3 + 2x^3 - 3ax^2)(by - y^2)(3by^2 - 2y^3) \\ &\quad + 36(3ax^2 - 2x^3)(x^2 - ax)(by - y^2)(3by^2 - 2y^3)]\}\, dx\, dy / a^6 b^6 \\ &= -\frac{78}{175}\left(\frac{b}{a}\right) N_x + \frac{27}{175}\left(\frac{a}{b}\right) N_y \end{aligned} \tag{12.61}$$

The incremental stiffness matrix is obtained explicitly as shown in Table 12.5 with its coefficients defined as

$$n_{ij} = \left[\beta_1\left(\frac{b}{a}\right)N_x + \beta_2\left(\frac{a}{b}\right)N_y + \beta_3 N_{xy}\right] a^{\beta_4} b^{\beta_5} \tag{12.62}$$

where the constants β_1 to β_5 are given in Table 12.6. Although 136 terms are populated in the half of the matrix shown, they share only sixty-four patterns with those constants as given in Table 12.6.

Equation (12.59) can be used to solve several types of problems:

1. When the incremental stiffness matrix [**n**] is absent, it states a free-vibration problem. Furthermore, the natural frequencies and mode shapes obtained can be used to treat dynamic response problems of plates using the method of modal superposition. Alternatively, if it is in the form of Eq. (12.53), it can be used to solve dynamic response problems using the direct time-integration method.
2. When the mass matrix [**m**] is absent, it states a buckling problem. The eigenvalues N_x, N_y, and N_{xy} are the buckling loads and {**q**} the mode shapes. If N_x, N_y, and N_{xy} are derived as a function of {**q**}, the equations become geometrically nonlinear but still with small linear deformations. These nonlinear equations can be used to treat large deflection problems of a plate.
3. When both [**n**] and [**m**] are present, it states a free-vibration problem of plates with initial in-plane stresses. In reality, plates often vibrate with initial stresses. For example, wing panels vibrate while they are stressed due to bending of the wing. Shear walls vibrate while they are stressed due to bending of the building. If the initial stresses are expressed as functions of {**q**}, the equations can be used to treat large-amplitude vibration problems of plates.

This element is used to test a buckling problem of a simply supported square plate. The results for the critical buckling loads using different meshes are given in Table 12.7. It is seen that the Guyan reduction scheme as explained in Chapter 7 [7.9] for vibration analysis is also effective in buckling analysis. It is also seen that this element serves as a good tool for buckling analysis.

TABLE 12.5 Incremental Stiffness Matrix for the 16-d.o.f. Rectangle in Bending[a]

	w_1	w_2	w_3	w_4	w_{x_1}	w_{x_2}	w_{x_3}	w_{x_4}	w_{y_1}	w_{y_2}	w_{y_3}	w_{y_4}	w_{xy_1}	w_{xy_2}	w_{xy_3}	w_{xy_4}
w_1	$n_{1,1}$															
w_2	$n_{2,1}$	$n_{2,2}$														
w_3	$n_{3,1}$	$n_{3,2}$	$n_{1,1}$													
w_4	$n_{4,1}$	$n_{4,2}$	$n_{2,1}$	$n_{2,2}$					symmetric							
w_{x_1}	$n_{5,1}$	$-n_{6,1}$	$-n_{7,1}$	$-n_{7,2}$	$n_{5,5}$											
w_{x_2}	$n_{6,1}$	$-n_{5,1}$	$-n_{8,1}$	$-n_{8,2}$	$n_{6,5}$	$n_{5,5}$										
w_{x_3}	$n_{7,1}$	$n_{7,2}$	$-n_{5,1}$	$n_{6,1}$	$n_{7,5}$	$n_{7,6}$	$n_{5,5}$									
w_{x_4}	$n_{8,1}$	$n_{8,2}$	$-n_{6,1}$	$-n_{6,2}$	$n_{8,5}$	$n_{8,6}$	$n_{6,5}$	$n_{5,5}$								
w_{y_1}	$n_{9,1}$	$n_{9,2}$	$-n_{11,1}$	$-n_{12,1}$	$n_{9,5}$	$n_{9,6}$	$n_{11,5}$	$n_{11,6}$	$n_{9,9}$							
w_{y_2}	$n_{10,1}$	$n_{9,1}$	$-n_{12,1}$	$-n_{12,2}$	$n_{10,5}$	$n_{10,6}$	$n_{12,5}$	$n_{12,6}$	$n_{10,9}$	$n_{9,9}$						
w_{y_3}	$n_{11,1}$	$n_{12,1}$	$-n_{9,1}$	$-n_{9,2}$	$n_{11,5}$	$n_{11,6}$	$n_{9,5}$	$n_{9,6}$	$n_{11,9}$	$n_{12,9}$	$n_{9,9}$					
w_{y_4}	$n_{12,1}$	$n_{12,2}$	$-n_{10,1}$	$-n_{10,2}$	$n_{12,5}$	$n_{12,6}$	$n_{10,5}$	$n_{10,6}$	$n_{12,9}$	$n_{12,10}$	$n_{10,9}$	$n_{9,9}$				
w_{xy_1}	$n_{13,1}$	$n_{13,2}$	$n_{15,1}$	$n_{15,2}$	$n_{13,5}$	$n_{13,6}$	$-n_{15,5}$	$-n_{15,6}$	$n_{13,9}$	$-n_{14,9}$	$-n_{15,9}$	$-n_{15,10}$	$n_{13,13}$			
w_{xy_2}	$n_{14,1}$	$n_{14,2}$	$n_{16,1}$	$n_{16,2}$	$n_{14,5}$	$n_{13,5}$	$-n_{16,5}$	$-n_{16,6}$	$n_{14,9}$	$-n_{13,9}$	$-n_{16,9}$	$-n_{16,10}$	$n_{16,13}$	$n_{13,13}$		
w_{xy_3}	$n_{15,1}$	$n_{15,2}$	$n_{13,1}$	$n_{13,2}$	$n_{15,5}$	$n_{15,6}$	$-n_{13,5}$	$-n_{13,6}$	$n_{15,9}$	$n_{15,10}$	$-n_{13,9}$	$-n_{13,10}$	$n_{15,13}$	$n_{16,13}$	$n_{13,13}$	
w_{xy_4}	$n_{16,1}$	$n_{16,2}$	$n_{14,1}$	$n_{14,2}$	$n_{16,5}$	$n_{16,6}$	$-n_{14,5}$	$-n_{14,6}$	$n_{16,9}$	$n_{16,10}$	$-n_{14,8}$	$-n_{14,10}$	$n_{16,13}$	$n_{16,14}$	$n_{14,13}$	$n_{13,13}$

[a] n given in Table 12.6

TABLE 12.6 Constants for the Incremental Stiffness Matrix Coefficients for the 16-d.o.f. Rectangle in Bending[a]

i	j	β_1	β_2	β_3	β_4	β_5	i	j	β_1	β_2	β_3	β_4	β_5
1	1	$\frac{78}{175}$	$\frac{78}{175}$	$\frac{1}{2}$	0	0	10	5	$-\frac{11}{2100}$	$\frac{13}{4200}$	$-\frac{1}{50}$	1	1
2	1	$-\frac{78}{175}$	$\frac{27}{175}$	0	0	0	11	5	$\frac{13}{4200}$	$\frac{13}{4200}$	$\frac{1}{50}$	1	1
3	1	$-\frac{27}{175}$	$-\frac{27}{175}$	$-\frac{1}{2}$	0	0	12	5	$-\frac{13}{4200}$	$\frac{11}{2100}$	$-\frac{1}{50}$	1	1
4	1	$\frac{27}{175}$	$-\frac{78}{175}$	0	0	0	13	5	$\frac{11}{1575}$	$\frac{1}{1050}$	0	2	1
5	1	$\frac{13}{350}$	$\frac{11}{175}$	0	1	0	14	5	$-\frac{11}{6300}$	$-\frac{1}{1400}$	$\frac{1}{300}$	2	1
6	1	$\frac{13}{350}$	$-\frac{13}{350}$	0	1	0	15	5	$\frac{13}{12600}$	$-\frac{1}{1400}$	$-\frac{1}{300}$	2	1
7	1	$\frac{9}{700}$	$\frac{13}{350}$	$\frac{1}{10}$	1	0	16	5	$-\frac{13}{3150}$	$\frac{1}{1050}$	0	2	1
8	1	$-\frac{9}{700}$	$\frac{11}{175}$	$\frac{1}{10}$	1	0	7	6	$\frac{3}{175}$	$-\frac{2}{175}$	0	2	0
9	1	$\frac{11}{175}$	$\frac{13}{350}$	0	0	1	8	6	$-\frac{3}{700}$	$\frac{3}{350}$	$-\frac{1}{60}$	2	0
10	1	$-\frac{11}{175}$	$\frac{9}{700}$	$-\frac{1}{10}$	0	1	9	6	$\frac{11}{2100}$	$-\frac{13}{4200}$	$-\frac{1}{50}$	1	1
11	1	$\frac{13}{350}$	$\frac{9}{700}$	$\frac{1}{10}$	0	1	10	6	$-\frac{11}{2100}$	$-\frac{11}{2100}$	$\frac{1}{50}$	1	1
12	1	$-\frac{13}{350}$	$\frac{13}{350}$	0	0	1	11	6	$\frac{13}{4200}$	$-\frac{11}{2100}$	$-\frac{1}{50}$	1	1
13	1	$\frac{11}{2100}$	$\frac{11}{2100}$	$-\frac{1}{50}$	1	1	12	6	$-\frac{13}{4200}$	$-\frac{13}{4200}$	$\frac{1}{50}$	1	1
14	1	$\frac{11}{2100}$	$-\frac{13}{4200}$	$\frac{1}{50}$	1	1	13	6	$-\frac{11}{6300}$	$-\frac{1}{1400}$	$-\frac{1}{300}$	2	1
15	1	$-\frac{13}{4200}$	$-\frac{13}{4200}$	$-\frac{1}{50}$	1	1	15	6	$-\frac{13}{3150}$	$\frac{1}{1050}$	0	2	1
16	1	$-\frac{13}{4200}$	$\frac{11}{2100}$	$\frac{1}{50}$	1	1	16	6	$\frac{13}{12600}$	$-\frac{1}{1400}$	$\frac{1}{300}$	2	1
2	2	$\frac{78}{150}$	$\frac{78}{175}$	$-\frac{1}{2}$	0	0	9	9	$\frac{2}{175}$	$\frac{26}{525}$	0	0	2
3	2	$\frac{27}{150}$	$-\frac{78}{175}$	0	0	0	10	9	$-\frac{2}{175}$	$\frac{3}{175}$	0	0	2
4	2	$-\frac{27}{150}$	$-\frac{27}{175}$	$\frac{1}{2}$	0	0	11	9	$\frac{3}{350}$	$-\frac{3}{700}$	$\frac{1}{60}$	0	2
7	2	$-\frac{9}{700}$	$\frac{11}{175}$	$-\frac{1}{10}$	1	0	12	9	$-\frac{3}{350}$	$-\frac{13}{1050}$	0	0	2
8	2	$-\frac{9}{700}$	$\frac{13}{350}$	$\frac{1}{10}$	1	0	13	9	$\frac{1}{1050}$	$\frac{11}{1575}$	0	1	2
9	2	$-\frac{11}{175}$	$\frac{9}{700}$	$-\frac{1}{10}$	0	1	14	9	$\frac{1}{1050}$	$-\frac{13}{3150}$	0	1	2
12	2	$\frac{13}{350}$	$\frac{9}{700}$	$-\frac{1}{10}$	0	1	15	9	$-\frac{1}{1400}$	$\frac{13}{12600}$	$-\frac{1}{300}$	1	2
13	2	$-\frac{11}{2100}$	$\frac{13}{4200}$	$\frac{1}{50}$	1	1	16	9	$-\frac{1}{1400}$	$-\frac{11}{6300}$	$\frac{1}{300}$	1	2
14	2	$-\frac{11}{2100}$	$-\frac{11}{2100}$	$-\frac{1}{50}$	1	1	12	10	$\frac{3}{350}$	$-\frac{3}{700}$	$-\frac{1}{60}$	0	2
15	2	$\frac{13}{4200}$	$-\frac{11}{2100}$	$\frac{1}{50}$	1	1	15	10	$\frac{1}{1400}$	$\frac{11}{6300}$	$\frac{1}{300}$	1	2
16	2	$\frac{13}{4200}$	$\frac{13}{4200}$	$-\frac{1}{50}$	1	1	16	10	$\frac{1}{1400}$	$-\frac{13}{12600}$	$-\frac{1}{300}$	1	2
5	5	$\frac{26}{525}$	$\frac{2}{175}$	0	2	0	13	13	$\frac{2}{1575}$	$\frac{2}{1575}$	0	2	2
6	5	$-\frac{13}{1050}$	$-\frac{3}{350}$	0	2	0	14	13	$-\frac{1}{3150}$	$-\frac{1}{1050}$	0	2	2
7	5	$-\frac{3}{700}$	$\frac{3}{350}$	$\frac{1}{60}$	2	0	15	13	$\frac{1}{4200}$	$\frac{1}{4200}$	$-\frac{1}{1800}$	2	2
8	5	$\frac{3}{175}$	$-\frac{2}{175}$	0	2	0	16	13	$-\frac{1}{1050}$	$-\frac{1}{3150}$	0	2	2
9	5	$\frac{11}{2100}$	$\frac{11}{2100}$	$\frac{1}{50}$	1	1	16	14	$\frac{1}{4200}$	$\frac{1}{4200}$	$\frac{1}{1800}$	2	2

[a] See Table 12.5 for [**n**]. $n_{ij} = \left[\beta_1\left(\frac{b}{a}\right)N_x + \beta_2\left(\frac{a}{b}\right)N_y + \beta_3 N_{xy}\right]a^{\beta_4}b^{\beta_5}$.

TABLE 12.7 Percentage Error in Critical Buckling Load $\{[N_x/N_x(\text{exact}) - 1](100)\}$ for a Simply Supported Square Plate, with $N_x(\text{exact}) = 4\pi^2 D/l^2$

	Element Mesh for a Quadrant		
	1 × 1	2 × 2	4 × 4
Without Guyan reduction	0.394	0.025	−0.004
With Guyan reduction (w_x, w_y, w_{xy} d.o.f.'s removed)	4.68	0.099	0.012

This element has been used successfully to treat large deflection problems of plates using a linear incremental approach [12.27]. The example shown in Fig. 12.11 is self-explanatory.

This element has been used successfully to treat the vibration problems of plates with initial stresses. An example is shown in Fig. 12.12, which shows

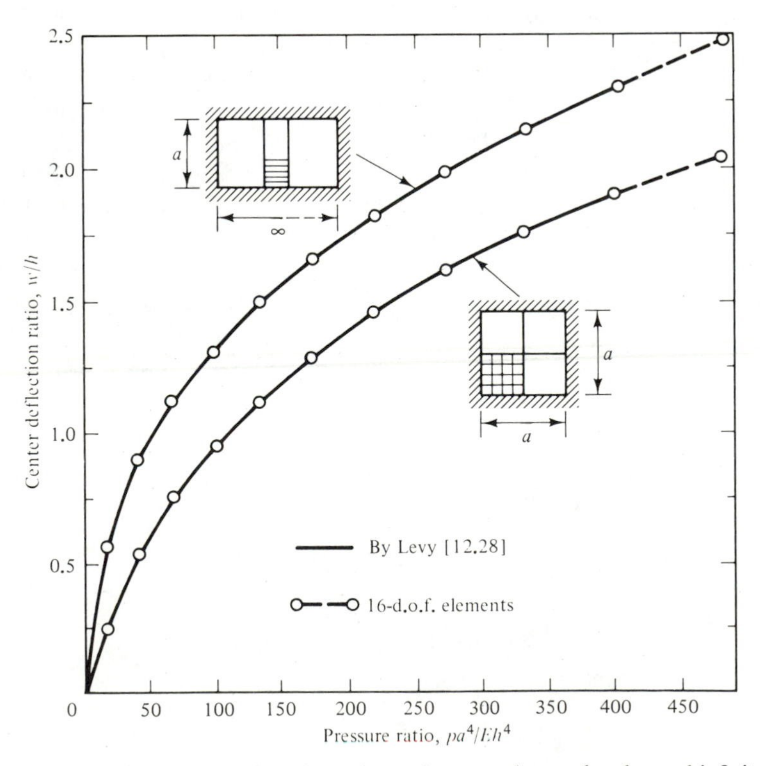

Figure 12.11 Large deflections for a clamped square plate and a clamped infinite strip ($\nu = 0.316$).

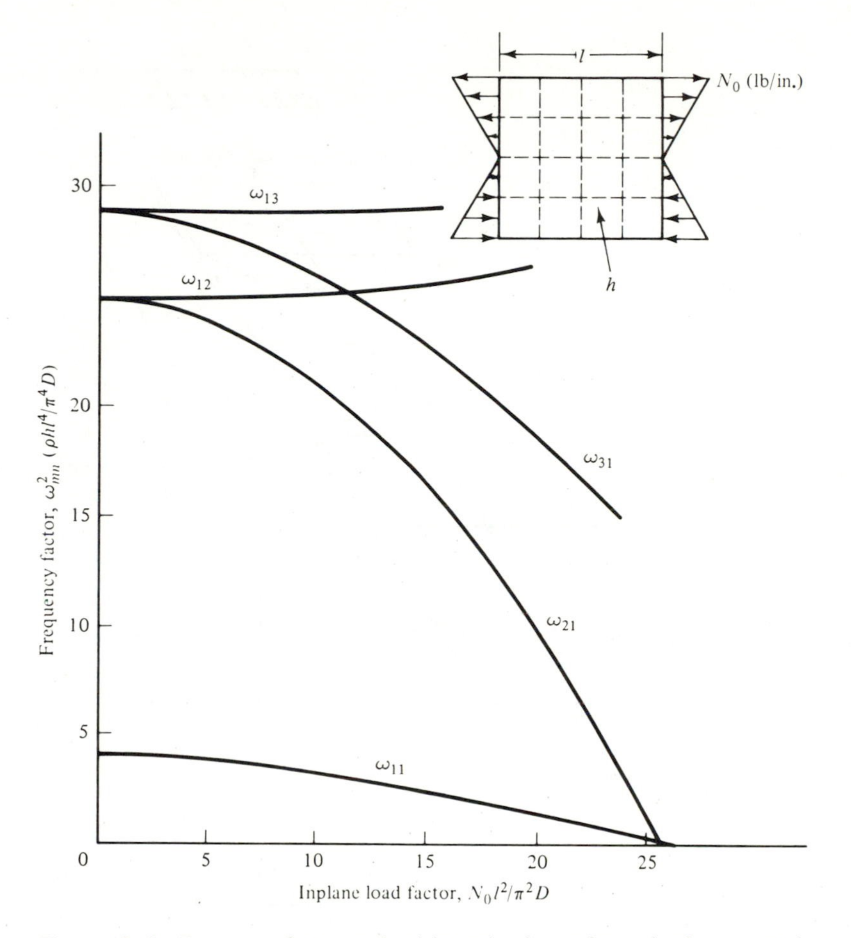

Figure 12.12 Frequency factor vs. load intensity factor for a simply supported square plate under in-plane bending.

that the squares of the frequencies for a mode can vary nonlinearly with certain types of in-plane loads [12.29].

Using a formulation similar to Eq. (12.59) based on a sophisticated 54-d.o.f. triangular plate bending element, it has been shown that finite elements can be used effectively to solve problems of large-amplitude vibration of plates, postbuckling of plates, vibration of buckled plates [9.26], and nonlinear panel flutter [12.30].

Boundary conditions. In a finite element analysis, it is necessary that all the zero boundary d.o.f.'s be identified and the matrix equations be reduced accordingly. It is also essential that all the zero interior d.o.f.'s be identified, as they can be of large number. Such identification needs the physical insight of how the plate is to be deflected. As an example, let us consider a simply

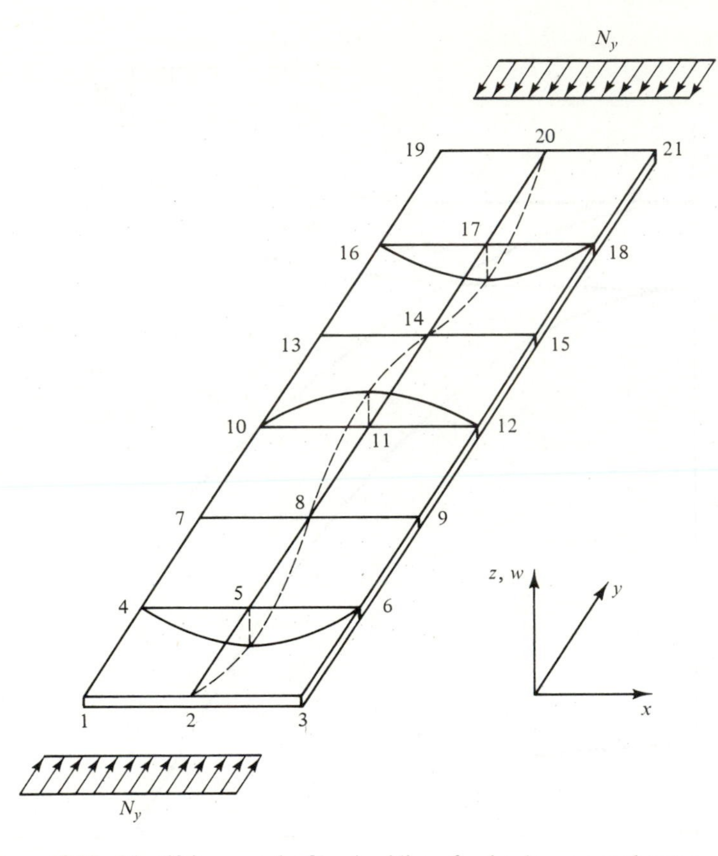

Figure 12.13 Identifying zero d.o.f.'s—buckling of a simply supported rectangular plate with an aspect ratio of 1:3 modeled by 12 equal-size elements.

supported rectangular plate of aspect ratio of 3:1 as shown in Fig. 12.13, which is modeled by a 2×6 mesh of 16-d.o.f. elements with equal size. Also shown is the critical buckling mode under the uniform in-plane compressive loads N_y, with lines 7-8-9 and 13-14-15 undeflected. As listed in Table 12.8, there are 63 zero and 21 nonzero d.o.f.'s. It is noted that the twist derivative d.o.f. can be interpreted as

$$\frac{\partial^2 w}{\partial x\, \partial y} = \frac{\partial^2 w}{\partial y\, \partial x} = \lim_{\Delta x \to 0} \frac{(\partial w/\partial y)_{+\Delta x} - (\partial w/\partial y)_{-\Delta x}}{2\,\Delta x}$$

$$\text{or} \quad \lim_{\Delta y \to 0} \frac{(\partial w/\partial x)_{+\Delta y} - (\partial w/\partial x)_{-\Delta y}}{2\,\Delta y} \tag{12.63}$$

TABLE 12.8 Zero (0) and Nonzero (×) d.o.f.'s for the Plate Buckling Example Shown in Fig. 12.13

Degrees of Freedom	Nodal Point																				
	1	2	3	4	5	6	7	8	9	10	11	12	13	14	15	16	17	18	19	20	21
w	0	0	0	0	×	0	0	0	0	0	×	0	0	0	0	0	×	0	0	0	0
$\partial w/\partial x$	0	0	0	×	0	×	0	0	0	×	0	×	0	0	0	×	0	×	0	0	0
$\partial w/\partial y$	0	×	0	0	0	0	0	×	0	0	0	0	0	×	0	0	0	0	0	×	0
$\partial^2 w/\partial x\,\partial y$	×	0	×	0	0	0	×	0	×	0	0	0	×	0	×	0	0	0	×	0	×

12.4 TRIANGULAR ELEMENTS

12.4.1 Three-Corner-Node Nine-D.O.F. Triangle with a Nine- or Ten-Term Polynomial

One of the earliest triangular plate elements in bending is the one shown in Fig. 12.14, with three nodes at the three vertices and three d.o.f.'s at each node: w, $\partial w/\partial x$, and $\partial w/\partial y$.

The displacement function was chosen by Adini [12.31] as a nine-term polynomial

$$w(x, y) = c_1 + c_2 x + c_3 y + c_4 x^2 + c_5 y^2 + c_6 x^3 + c_7 x^2 y + c_8 xy^2 + c_9 y^3 \quad (12.64)$$

which is a complete third-order polynomial minus the uniform twist term xy. Omitting the xy term has the advantage of maintaining geometric isotropy but has the defect of violating the constant twist requirement for convergence.

To accommodate all ten terms in a complete third-order polynomial, Tocher [12.32] combined two of the cubic terms to share a single coefficient c_8:

$$w(x, y) = c_1 + c_2 x + c_3 y + c_4 x^2 + c_5 xy + c_6 y^2 + c_7 x^3 + c_8(x^2 y + xy^2) + c_9 y^3 \quad (12.65)$$

These constants are obtained by evaluating the nine nodal d.o.f.'s such as in Eqs. (12.23). Through inverse of matrix [**B**], the constants are related to the d.o.f.'s by $[\mathbf{B}]^{-1}$. Thus, limiting the number of constants from 10 to 9 by equating two constants has an undesirable effect of imposing a restriction by relating among certain d.o.f.'s. Furthermore, for certain orientations of the triangle edges, matrix [**B**] is singular. For example, this happens when edge 1-2 is perpendicular to 1-3 with both edges parallel to the two triangular coordinates ξ_2 and ξ_3, respectively.

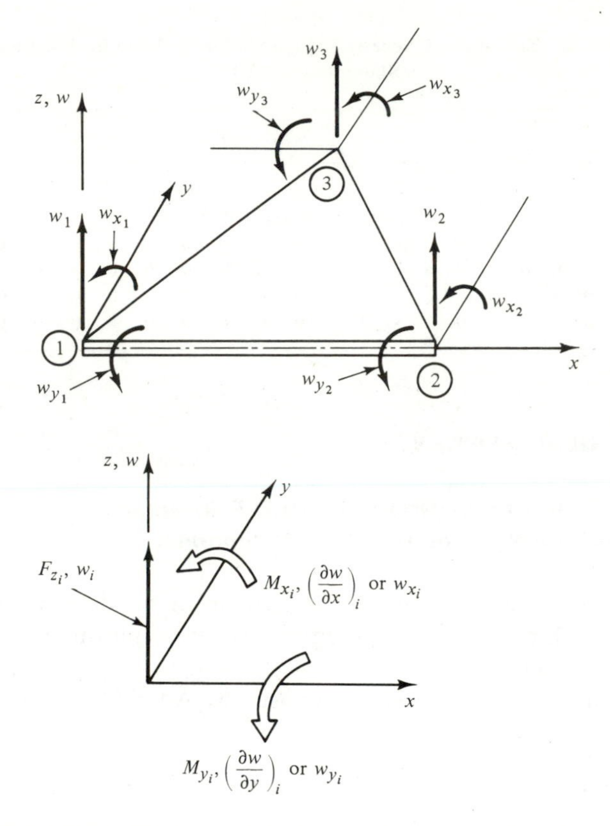

Figure 12.14 Three-node 9-d.o.f. triangular plate bending element.

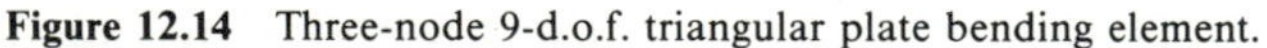

12.4.2 Three-Corner-Node Nine-D.O.F. (HCT) Triangle with a Nine-Term Polynomial in Each of the Three Subtriangles (*Normal Slope Varies Linearly along Edges*) [12.33]

For the two triangles introduced above, the displacement functions provide for interelement compatibility in displacement but not in normal slope $\partial w/\partial n$ along the common edges. It is somewhat awkward to achieve full slope and displacement compatibility in an assembly of triangular elements. However, this result can be achieved by dividing an element into three subtriangles. This concept was first suggested by Hsieh in correspondence with Clough, and the procedure for deriving the element stiffness was developed by Clough and Tocher [12.33].

Displacement function. Figure 12.15 shows the HCT triangular element. The interior point O may be located arbitrarily; it is located at the

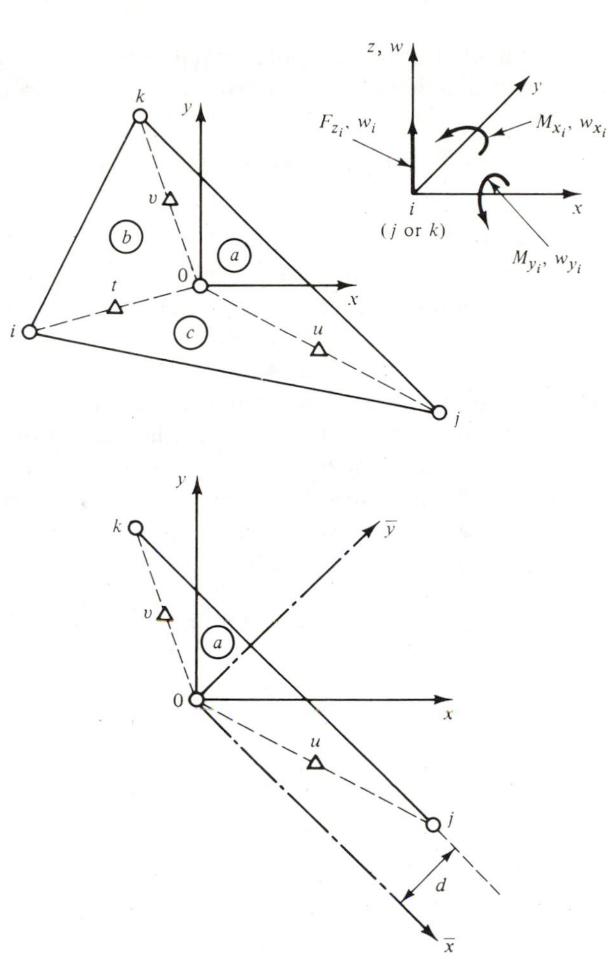

Figure 12.15 Three-corner-node 9-d.o.f. conforming triangle in bending (HCT) with three subtriangles.

centroid for convenience. The additional nodes t, u, and v are at the midpoints of the interior subtriangular boundaries. We assume three independent polynomial displacement functions for the three subtriangles.

For subtriangle a as shown in Fig. 12.15, a local coordinate system $(\bar{x}, \bar{y})$ is assumed. The $\bar{x}$ axis is parallel with the exterior edge of the subtriangle. A different coordinate system of this type applies to each subtriangle.

The displacement function for subtriangle a is assumed as

$$w_a(\bar{x}, \bar{y}) = \alpha_1 + \alpha_2\bar{x} + \alpha_3\bar{y} + \alpha_4\bar{x}^2 + \alpha_5\bar{x}\bar{y} + \alpha_6\bar{y}^2 + \alpha_7\bar{x}^3 + \alpha_8\bar{x}\bar{y}^2 + \alpha_9\bar{y}^3$$

$$= \underset{1\times 9}{\lfloor \mathbf{M}_a \rfloor} \underset{9\times 1}{\{\boldsymbol{\alpha}_a\}} \tag{12.66}$$

This function is a complete third-order polynomial minus $\bar{x}^2\bar{y}$. Had this term been included, the normal slope along the exterior edge $\overline{jk}$ ($\bar{y} = d$) would be

$$\frac{\partial w_a}{\partial n} = \frac{\partial w_a}{\partial \bar{y}} = c_1 + c_2\bar{x} + c_3\bar{x}^2 \tag{12.67}$$

The term $c_3\bar{x}^2$ results from the inclusion of $\bar{x}^2\bar{y}$ in Eq. (12.66). Omission of this term allows for the normal slope to vary only linearly along the exterior boundary. It is this assumption that ensures the interelement compatibility in normal slope.

Stiffness formulation. A displacement function similar to Eq. (12.66) is assumed for each subtriangle; thus the three displacement functions of the complete element involve a total of twenty-seven constants $\{\mathbf{c}\}$, which need be solved by twenty-seven equations. We can obtain eighteen equations by satisfying internal compatibility requirements between the adjacent subtriangles and the remaining nine equations by relating nine internal d.o.f.'s to the nine exterior corner d.o.f.'s which represent the entire element.

To establish the necessary twenty-seven equations, we first evaluate the corner displacements and slopes in each subtriangle based on the local coordinates $(\bar{x}, \bar{y})$ and transform them to those for the global coordinates (x, y). We then equate the displacements and slopes at the corners of each subtriangle to those of adjacent subtriangles as well as the whole triangle.

The displacement and slopes at corner i in subtriangle c may be written in local coordinates as

$$\begin{Bmatrix} w \\ \dfrac{\partial w}{\partial \bar{x}} \\ \dfrac{\partial w}{\partial \bar{y}} \end{Bmatrix}_i^c = \{\bar{\mathbf{q}}\}_i^c = [\bar{\mathbf{A}}]_i^c\{\boldsymbol{\alpha}\}_c$$

$$= \begin{bmatrix} 1 & \bar{x} & \bar{y} & \bar{x}^2 & \bar{x}\bar{y} & \bar{y}^2 & \bar{x}^3 & \bar{x}\bar{y}^2 & \bar{y}^3 \\ 0 & 1 & 0 & 2\bar{x} & \bar{y} & 0 & 3\bar{x}^2 & \bar{y}^2 & 0 \\ 0 & 0 & 1 & 0 & \bar{x} & 2\bar{y} & 0 & 2\bar{x}\bar{y} & 3\bar{y}^2 \end{bmatrix}_i^c \begin{Bmatrix} \alpha_1 \\ \alpha_2 \\ \vdots \\ \alpha_9 \end{Bmatrix} \tag{12.68}$$

where $\bar{x}$ and $\bar{y}$ take the local coordinate values at i in subtriangle c.

Denoting θ to be the angle between $\bar{x}$ and x axes measured counterclockwise from x as shown in Fig. 2.3, the local coordinates and the global coordinates are related in Eq. (2.33) as

$$\begin{Bmatrix} \bar{x} \\ \bar{y} \end{Bmatrix} = \begin{bmatrix} \cos\theta & \sin\theta \\ -\sin\theta & \cos\theta \end{bmatrix} \begin{Bmatrix} x \\ y \end{Bmatrix} \tag{2.33}$$

The displacement and slopes in global coordinates are

$$\begin{Bmatrix} w \\ \dfrac{\partial w}{\partial x} \\ \dfrac{\partial w}{\partial y} \end{Bmatrix}_i^c = \begin{bmatrix} 1 & 0 & 0 \\ 0 & \dfrac{\partial \bar{x}}{\partial x} & \dfrac{\partial \bar{y}}{\partial x} \\ 0 & \dfrac{\partial \bar{x}}{\partial y} & \dfrac{\partial \bar{y}}{\partial y} \end{bmatrix} \begin{Bmatrix} w \\ \dfrac{\partial w}{\partial \bar{x}} \\ \dfrac{\partial w}{\partial \bar{y}} \end{Bmatrix}_i^c = \begin{bmatrix} 1 & 0 & 0 \\ 0 & \cos\theta & -\sin\theta \\ 0 & \sin\theta & \cos\theta \end{bmatrix} \{\bar{\mathbf{q}}\}_i^c \tag{12.69}$$

or

$$\{\mathbf{q}\}_i^c = [\mathbf{T}]\{\bar{\mathbf{q}}\}_i^c = [\mathbf{T}][\bar{\mathbf{A}}]_i^c\{\boldsymbol{\alpha}\}_c = [\mathbf{A}]_i^c\{\boldsymbol{\alpha}\}_c \tag{12.70}$$

Similar expressions may be written for other corners and subtriangles. The following set of 24 compatibility equations may be established,

$$\begin{Bmatrix} q_i \\ q_j \\ q_k \\ \hline q_i^c \\ q_j^a \\ q_k^b \\ \hline q_0^a \\ q_0^b \end{Bmatrix} = \begin{Bmatrix} q_i^c \\ q_j^a \\ q_k^b \\ \hline q_i^b \\ q_j^c \\ q_k^a \\ \hline q_0^b \\ q_0^c \end{Bmatrix} \quad \begin{matrix} \text{nine equations between the complete element d.o.f.'s and the subtriangle d.o.f.'s} \\ \text{nine equations for the d.o.f.'s between adjacent subtriangles at corners } i, j, \text{ and } k \\ \text{six equations for the d.o.f.'s between adjacent subtriangles at interior node O} \end{matrix} \tag{12.71}$$

in which the superscripts refer to subtriangles and symbols without superscripts refer to nodal d.o.f.'s of the complete element. The first nine equations state that the nodal d.o.f.'s in the complete element correspond with the subtriangle d.o.f.'s at the appropriate corners. The remaining 15 equations equate the nodal d.o.f.'s between adjacent subtriangles at the corners i, j, k, and O, respectively.

To enforce normal slope compatibility along interior edges of the subtriangles, additional constraints must be imposed. The normal slope at point v of subtriangle b is written as

$$\left(\frac{\partial w}{\partial n}\right)_v^b = (w_n)_v^b = \left(\frac{\partial w}{\partial x}\cos\phi + \frac{\partial w}{\partial y}\sin\phi\right)_v^b \tag{12.72}$$

where the angle ϕ, shown in Fig. 12.16, defines the orientation of the outward normal axis n measured counterclockwise from the direction parallel to the x axis. Substituting $\partial w/\partial x$ and $\partial w/\partial y$ as defined in the form of Eqs. (12.69) and (12.68) in Eq. (12.72) gives

$$(w_n)_v^b = \lfloor 0 \mid e \mid f \mid 2e\bar{x} \mid f\bar{x} + e\bar{y} \mid 2f\bar{y} \mid 3e\bar{x}^2 \mid 2f\bar{x}\bar{y} + e\bar{y}^2 \mid 3f\bar{y}^2 \rfloor \begin{Bmatrix} \alpha_1 \\ \alpha_2 \\ \vdots \\ \alpha_9 \end{Bmatrix}$$

$$= \lfloor \mathbf{A} \rfloor_v^b \{\boldsymbol{\alpha}\}_b \tag{12.73}$$

where $e = \cos\theta\cos\phi + \sin\theta\sin\phi$, $f = \cos\theta\sin\phi - \sin\theta\cos\phi$, and $\bar{x}$ and $\bar{y}$ take the local coordinate values at v in subtriangle b.

Similar expressions may be written for the other interior points t and u, and for the other subtriangles. Thus we have three equations enforcing inter-subtriangle compatibility for a normal slope:

$$\begin{Bmatrix} (w_n)_t^c \\ (w_n)_u^a \\ (w_n)_v^b \end{Bmatrix} = \begin{Bmatrix} (w_n)_t^b \\ (w_n)_u^c \\ (w_n)_v^a \end{Bmatrix} \tag{12.74}$$

Making use of expressions in the form of Eqs. (12.70) and (12.73), the complete set of twenty seven compatibility equations (12.71) and (12.74) take the form

$$\begin{matrix} 9\times 1 \\ \\ 15\times 1 \\ \\ 3\times 1 \end{matrix}\begin{Bmatrix} q_i \\ q_j \\ q_k \\ \hline 0 \\ 0 \\ 0 \\ 0 \\ 0 \\ \hline 0 \\ 0 \\ 0 \end{Bmatrix} = \left[\begin{array}{c|c|c} 0 & 0 & A_i^c \\ A_j^a & 0 & 0 \\ 0 & A_k^b & 0 \\ \hline 0 & -A_i^b & A_i^c \\ A_j^a & 0 & -A_j^c \\ -A_k^a & A_k^b & 0 \\ 0 & A_0^b & -A_0^c \\ A_0^a & -A_0^b & 0 \\ \hline 0 & -A_t^b & A_t^c \\ A_u^a & 0 & -A_u^c \\ -A_v^a & A_v^b & 0 \end{array}\right] \begin{Bmatrix} \alpha_a \\ \alpha_b \\ \alpha_c \end{Bmatrix}_{27\times 1} \tag{12.75}$$

which may be partitioned as follows:

$$\begin{matrix} 9 \text{ rows} \rightarrow \\ 18 \text{ rows} \rightarrow \end{matrix} \begin{Bmatrix} q \\ 0 \end{Bmatrix} = \begin{bmatrix} A_{aa} & A_{a0} \\ A_{0a} & A_{00} \end{bmatrix} \begin{Bmatrix} \alpha_a \\ \alpha_0 \end{Bmatrix} \tag{12.75a}$$

Multiplying out the second set of equations gives

$$\{\boldsymbol{\alpha}_0\} = -[\mathbf{A}_{00}]^{-1}[\mathbf{A}_{0a}]\{\boldsymbol{\alpha}_a\} \tag{12.76}$$

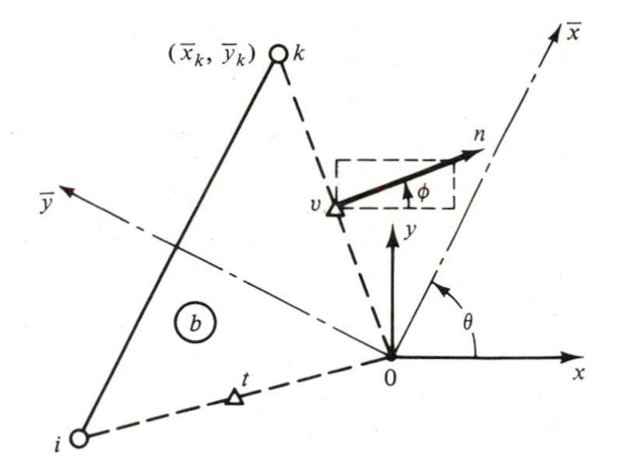

Figure 12.16 Coordinate transformation for w_n at v of subtriangle b in the nine-d.o.f. HCT triangle.

Multiplying out the first set of equations (12.75a) and incorporating the solution for $\{\boldsymbol{\alpha}_0\}$ from Eq. (12.76) gives

$$\{\mathbf{q}\} = [[\mathbf{A}_{aa}] - [\mathbf{A}_{a0}][\mathbf{A}_{00}]^{-1}[\mathbf{A}_{0a}]]\{\boldsymbol{\alpha}_a\} = [\bar{\mathbf{A}}]\{\boldsymbol{\alpha}_a\} \tag{12.77}$$

Inverting Eq. (12.77) for $\{\boldsymbol{\alpha}_a\}$ and substituting $\{\boldsymbol{\alpha}_a\}$ in Eq. (12.76) give

$$\begin{matrix} 9\times 1 \\ 18\times 1 \end{matrix}\begin{Bmatrix} \alpha_a \\ \alpha_0 \end{Bmatrix} = \begin{bmatrix} [\bar{\mathbf{A}}]^{-1} \\ -[\mathbf{A}_{00}]^{-1}[\mathbf{A}_{0a}][\bar{\mathbf{A}}]^{-1} \end{bmatrix}\{\mathbf{q}\} = \underset{27\times 9}{[\mathbf{T}]}\ \underset{9\times 1}{\{\mathbf{q}\}} \tag{12.78}$$

The total strain energy of the three subtriangles can be written in terms of the twenty seven constants as

$$U = \frac{1}{2}\begin{Bmatrix} \alpha_a \\ \alpha_0 \end{Bmatrix}^T [\bar{\mathbf{k}}]\begin{Bmatrix} \alpha_a \\ \alpha_0 \end{Bmatrix} = \frac{1}{2}\begin{Bmatrix} \alpha_a \\ \alpha_b \\ \alpha_c \end{Bmatrix}^T [\bar{\mathbf{k}}]\begin{Bmatrix} \alpha_a \\ \alpha_b \\ \alpha_c \end{Bmatrix} \tag{12.79}$$

Substituting Eq. (12.78) in Eq. (12.79) gives

$$U = \tfrac{1}{2}\{\mathbf{q}\}^T[\mathbf{T}]^T[\bar{\mathbf{k}}][\mathbf{T}]\{\mathbf{q}\} = \tfrac{1}{2}\{\mathbf{q}\}^T[\mathbf{k}]\{\mathbf{q}\} \tag{12.80}$$

The stiffness matrix for the complete element is finally obtained:

$$\underset{9\times 9}{[\mathbf{k}]} = \underset{9\times 27}{[\mathbf{T}]^T}\ \underset{27\times 27}{[\bar{\mathbf{k}}]}\ \underset{9\times 27}{[\mathbf{T}]} \tag{12.81}$$

Example 12.5

For the HCT element as shown in Fig. 12.15, find the coefficient $\bar{k}_{13\text{-}13}$ in the kernel stiffness matrix $[\bar{\mathbf{k}}]$ contained in Eq. (12.79).

Solution. The thirteenth constant is α_4 in subtriangle b. The displacement function in local coordinates is

$$w_b(\bar{x}, \bar{y}) = \alpha_1 + \alpha_2\bar{x} + \alpha_3\bar{y} + \alpha_4\bar{x}^2 + \alpha_5\bar{x}\bar{y} + \alpha_6\bar{y}^2 + a_7\bar{x}^3 + \alpha_8\bar{x}\bar{y}^2 + \alpha_9\bar{y}^3$$

and

$$\frac{\partial^2 w_b}{\partial \bar{x}^2} = 2\alpha_4 + 6\alpha_7 \bar{x}$$

$$\frac{\partial^2 w_b}{\partial \bar{y}^2} = 2\alpha_6 + 2\alpha_8 \bar{x} + 6\alpha_9 \bar{y} \tag{12.82}$$

$$\frac{\partial^2 w_b}{\partial \bar{x}\, \partial \bar{y}} = \alpha_5 + 2\alpha_8 \bar{y}$$

Substituting Eqs. (12.82) in the strain energy expression (12.8) and performing second partial differentiation of U with respect to α_4 gives, in local coordinates,

$$\begin{aligned}
\bar{k}_{13\text{-}13} &= \frac{\partial^2 U}{\partial \alpha_4\, \partial \alpha_4} \\
&= \frac{\partial^2}{\partial \alpha_4^2} \iint D/2\,[(2\alpha_4 + 6\alpha_7 \bar{x})^2 + (2\alpha_6 + 2\alpha_8 \bar{x} + 6\alpha_9 \bar{y})^2 \\
&\quad + 2\nu(2\alpha_4 + 6\alpha_7 \bar{x})(2\alpha_6 + 2\alpha_8 \bar{x} + 6\alpha_9 \bar{y}) \\
&\quad + 2(1-\nu)(\alpha_5 + 2\alpha_8 \bar{y})^2]\, d\bar{x}\, d\bar{y} \\
&= 4D \text{ (area of subtriangle } b)
\end{aligned} \tag{12.83}$$

12.4.3 Six-Node 12-D.O.F. Triangle with a 10-Term Polynomial in Each of the Three Subtriangles (*Normal Slope Varies Quadratically along Edges*)

As pointed out in Ref. 12.33, improvement in the HCT-compatible triangle is possible if a complete 10-term polynomial is assumed together with an additional normal-slope d.o.f. at the midpoint of each exterior edge. Such an element is shown in Fig. 12.17 without showing the interior sides and nodes. The normal slope along each exterior boundary of the element may then vary quadratically rather than linearly. The detailed formulation for such a six-node 12-d.o.f. element based on the 10-term polynomial for each subtriangle is available in the text by Gallagher [12.24].

12.4.4 Nine-D.O.F. Nonconforming and Conforming Triangles with Single Shape Functions in Triangular Coordinates (*Normal Slope Varies Linearly and Quadratically along Edges*) [12.5]

For a nine-d.o.f. triangular element as shown in Fig. 12.14, a set of nonconforming shape functions in triangular coordinates was proposed in

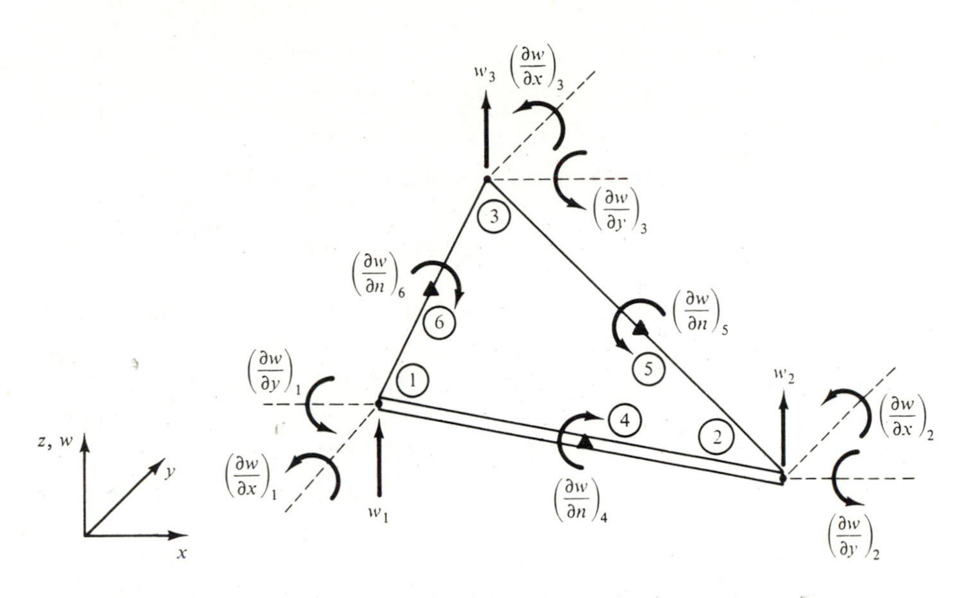

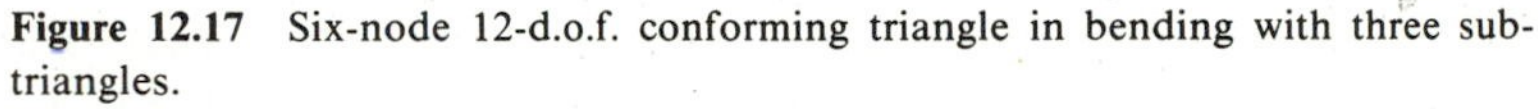

Figure 12.17 Six-node 12-d.o.f. conforming triangle in bending with three subtriangles.

Ref. [12.5] to represent the displacement:

$$
\begin{aligned}
W(\xi_1, \xi_2, \xi_3) &= \alpha_1 \xi_1 + \alpha_2 \xi_2 + \alpha_3 \xi_3 + \alpha_4(\xi_2^2 \xi_1 + \tfrac{1}{2}\xi_1 \xi_2 \xi_3) \\
&\quad + \alpha_5(\xi_2^2 \xi_3 + \tfrac{1}{2}\xi_1 \xi_2 \xi_3) + \cdots + \alpha_9(\xi_1^2 \xi_3 + \tfrac{1}{2}\xi_1 \xi_2 \xi_3) \\
&= \sum_{i=1}^{3} \left[f_i(\xi_1, \xi_2, \xi_3) W_i + g_i(\xi_1, \xi_2, \xi_3) \left(\frac{\partial W}{\partial x}\right)_i \right. \\
&\quad \left. + h_i(\xi_1, \xi_2, \xi_3) \left(\frac{\partial W}{\partial y}\right)_i \right]
\end{aligned}
\tag{12.84}
$$

where, for node 1, the shape functions are

$$
\begin{aligned}
f_1(\xi_1, \xi_2, \xi_3) &= \xi_1 + \xi_1^2 \xi_2 + \xi_1^2 \xi_3 - \xi_1 \xi_2^2 - \xi_1 \xi_3^2 \\
g_1(\xi_1, \xi_2, \xi_3) &= (x_2 - x_1)(\xi_1^2 \xi_2 + \tfrac{1}{2}\xi_1 \xi_2 \xi_3) + (x_3 - x_1)(\xi_3 \xi_1^2 + \tfrac{1}{2}\xi_1 \xi_2 \xi_3) \\
h_1(\xi_1, \xi_2, \xi_3) &= (y_2 - y_1)(\xi_1^2 \xi_2 + \tfrac{1}{2}\xi_1 \xi_2 \xi_3) + (y_3 - y_1)(\xi_3 \xi_1^2 + \tfrac{1}{2}\xi_1 \xi_2 \xi_3)
\end{aligned}
\tag{12.85}
$$

The other two sets of shape functions for nodes 2 and 3 are in the form obtained by cyclically permutating the three node numbers.

The function $f_1(\xi_1, \xi_2, \xi_3)$ takes the value of 1 at node 1 ($\xi_1 = 1, \xi_2 = \xi_3 = 0$) and zero at the other two nodes. Let us now examine $\partial g_1(\xi_1, \xi_2, \xi_3)/\partial x$

at node 1:

$$\frac{\partial g_1}{\partial x} = \frac{\partial g_1}{\partial \xi_1}\frac{\partial \xi_1}{\partial x} + \frac{\partial g_1}{\partial \xi_2}\frac{\partial \xi_2}{\partial x} + \frac{\partial g_1}{\partial \xi_3}\frac{\partial \xi_3}{\partial x} \tag{12.86}$$

where ξ_1, ξ_2, ξ_3 are written in terms of x and y in Eq. (9.97):

$$\frac{\partial g_1}{\partial x} = \frac{1}{2A}\left[(y_2 - y_3)\frac{\partial g_1}{\partial \xi_1} + (y_3 - y_1)\frac{\partial g_1}{\partial \xi_2} + (y_1 - y_2)\frac{\partial g_1}{\partial \xi_3}\right] \tag{12.87}$$

Substituting the expression (12.85) for g_1 in Eq. (12.87), we can show that

$$\begin{aligned} \frac{\partial g_1}{\partial x} &= 1 \qquad \text{at node 1 } (\xi_1 = 1, \xi_2 = \xi_3 = 0) \\ &= 0 \qquad \text{at the other two nodes} \end{aligned} \tag{12.88}$$

Similarly, we can show that $\partial h_1/\partial y$ equals 1 at node 1 and zero at the other two nodes.

If a normal-slope d.o.f. $\partial w/\partial n$ is added to each midside and new functions are superimposed on the function (12.84), a unique quadratic variation of the normal slope along interelement edges can be achieved. The result is again a six-node 12 d.o.f. conforming triangle as shown in Fig. 12.17. Let us now use capital letter W to represent the nonconforming displacement function (12.84) and a lowercase letter w for the new conforming function. Then

$$\begin{aligned} w(\xi_1, \xi_2, \xi_3) &= W(\xi_1, \xi_2, \xi_3) \quad [\text{Equation (12.84)}] \\ &\quad + \sum_{i=4}^{6} f_i(\xi_1, \xi_2, \xi_3)\left\{\left(\frac{\partial w}{\partial n}\right)_i - \left(\frac{\partial W}{\partial n}\right)_i\right\} \end{aligned} \tag{12.89}$$

where

$$\begin{aligned} f_4 &= \frac{8A}{l_{12}}\frac{\xi_1^2\xi_2^2\xi_3}{(\xi_3 + \xi_1)(\xi_3 + \xi_2)} \\ f_5 &= \frac{8A}{l_{23}}\frac{\xi_1\xi_2^2\xi_3^2}{(\xi_1 + \xi_2)(\xi_1 + \xi_3)} \\ f_6 &= \frac{8A}{l_{13}}\frac{\xi_1^2\xi_2\xi_3^2}{(\xi_2 + \xi_1)(\xi_2 + \xi_3)} \end{aligned} \tag{12.90}$$

in which l_{ij} is the length of side ij. This displacement function can be explained using Fig. 12.18 while noting the following properties of f_i (say, f_4):

1. It has zero values along all sides of the triangle.
2. $\partial f_4/\partial x = \partial f_4/\partial y = \partial f_4/\partial n = 0$ at three corner nodes.
3. $\partial f_4/\partial n = 0$ along the two sides opposite to node 4.
4. $\partial f_4/\partial n$ varies parabolically along edge 1-2 with its maximum value equal to 1 at midside node 4.

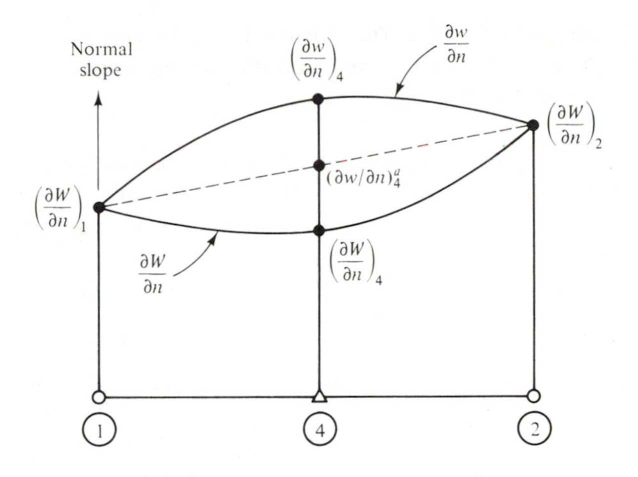

Figure 12.18 Variation of normal slope based on nonconforming displacement function W and the superimposed conforming function w along side 1-4-2.

For example, we can see that

$$\frac{\partial f_4}{\partial \xi_3} = \frac{8A}{l_{12}} \xi_1 \xi_2 \qquad \text{along } \xi_3 = 0 \tag{12.91}$$

which varies quadratically along with side 1-2. We see further that, at midside node 4,

$$\frac{\partial f_4}{\partial n} = \frac{\partial f_4}{h_3 \, \partial \xi_3} = 1 \qquad \text{at } \xi_1 = \xi_2 = \tfrac{1}{2}, \quad \xi_3 = 0 \tag{12.92}$$

where h_3 is the length of the normal from vertex 3 to side 1-2.

For convenience of assemblage, the midside normal slope can be constrained by setting it to be the average of the two normal slopes at the ends of the edge. This means that the normal slope varies linearly along the edge as shown by the dashed line in Fig. 12.18. Equation (12.89) is now modified by replacing $(\partial w/\partial n)_i$ by $(\partial w/\partial n)_i^a$, where the superscript indicates the average value. This concept results in a conforming triangle with a stiffness matrix the same as that for the HCT element. Since the displacement functions for such type of elements are continuously defined over the whole triangle, it is convenient to use numerical integration to form the stiffness matrix.

The foregoing concept was given in Ref. [12.5] and explained in Ref. [12.23]. An improvement of this type of formulation was achieved in Ref. [12.34] by smoothing out the second-order displacement derivatives through a least-square fit.

12.4.5 Three-Node 18-D.O.F. Conforming Triangle (*Normal Slope Varies Cubically along Edges*) [12.35–12.37]

The conforming triangles with normal slopes varying linearly (nine d.o.f.) and quadratically (12 d.o.f.) along the edges were presented in 1965 [12.5, 12.33]. Three to four years later, the conforming triangles with normal slopes varying cubically (18 d.o.f.) [12.35-12.37] and quartically (21 d.o.f.) [12.35, 12.37-12.40] along the edges were developed. Let us first explain in detail the 18-d.o.f. triangle of Ref. [12.36].

Displacement function. Figure 12.19 shows an 18-d.o.f. triangle with six d.o.f.'s at each vertex: w, w_x, w_y, w_{xx}, w_{xy}, and w_{yy}. The local coordinates are so chosen that the ξ axis coincides with edge 1-2 and the η axis passes through the vertex 3. The angle θ between the x and ξ axes and the dimensions a, b, and c are related to the coordinates of the vertices as

$$\begin{aligned}
\theta &= \tan^{-1}\frac{y_2 - y_1}{x_2 - x_1} \\
a &= (x_2 - x_3)\cos\theta - (y_3 - y_2)\sin\theta \\
b &= (x_3 - x_1)\cos\theta + (y_3 - y_1)\sin\theta \\
c &= (y_3 - y_1)\cos\theta - (x_3 - x_1)\sin\theta
\end{aligned} \tag{12.93}$$

The displacement function is assumed as a complete quintic polynomial of ξ and η minus the term $\xi^4\eta$:

$$\begin{aligned}
w(\xi, \eta) = \sum_{i=1}^{20} \alpha_i \xi^{m_i}\eta^{n_i} \\
= \quad & \alpha_1 \\
& \alpha_2\xi \quad \alpha_3\eta \\
& \alpha_4\xi^2 \quad \alpha_5\xi\eta \quad \alpha_6\eta^2 \\
& \alpha_7\xi^3 \quad \alpha_8\xi^2\eta \quad \alpha_9\xi\eta^2 \quad \alpha_{10}\eta^3 \\
& \alpha_{11}\xi^4 \quad \alpha_{12}\xi^3\eta \quad \alpha_{13}\xi^2\eta^2 \quad \alpha_{14}\xi\eta^3 \quad \alpha_{15}\eta^4 \\
& \alpha_{16}\xi^5 \quad \boxed{\xi^4\eta} \quad \alpha_{17}\xi^3\eta^2 \quad \alpha_{18}\xi^2\eta^3 \quad \alpha_{19}\xi\eta^4 \quad \alpha_{20}\eta^5
\end{aligned} \tag{12.94}$$

where

$$m_i = 0, 1, 0, 2, 1, 0, 3, 2, 1, 0, 4, 3, 2, 1, 0, 5, 3, 2, 1, 0$$

$$n_i = 0, 0, 1, 0, 1, 2, 0, 1, 2, 3, 0, 1, 2, 3, 4, 0, 2, 3, 4, 5$$

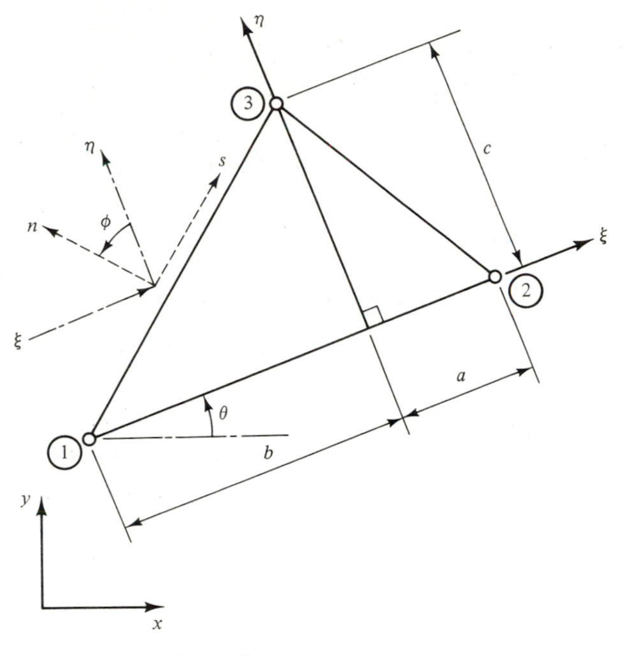

Figure 12.19 Local and global coordinates for an 18-d.o.f. triangle in bending.

Omitting the term $\xi^4\eta$ is to ensure that the slope normal to the edge $\eta = 0$ varies cubically along ξ,

$$\frac{\partial w}{\partial n} = \frac{\partial w}{\partial \eta} = c_1 + c_2\xi + c_3\xi^2 + c_4\xi^3 \tag{12.95}$$

where the four constants can be uniquely defined by evaluating the four d.o.f. values of $\partial w/\partial \eta$ and $\partial^2 w/\partial \xi\, \partial \eta$ at nodes 1 and 2, respectively.

To solve for the twenty constants in Eq. (12.94), eighteen equations are obtained by evaluating the 18-d.o.f. values at the three vertices. One of the two remaining equations is obtained by imposing the condition that the normal slope w_n varies cubically along edge 1-3. This is done by eliminating all the coefficients associated with the fourth-order terms in s from the w_n expression, which come from the fifth-order terms or the last five terms in Eq. (12.94). As shown in Fig. 12.19, the s axis is parallel with edge 1-3 and the n axis is the outward normal to edge 1-3. The two sets of axes result in the following relations:

$$\frac{\partial w}{\partial n} = -\frac{\partial w}{\partial \xi}\sin\phi + \frac{\partial w}{\partial \eta}\cos\phi$$

$$\xi = s\cos\phi \qquad \text{and} \qquad \eta = s\sin\phi \qquad \text{along edge 1-3} \qquad (12.96)$$

$$\tan\phi = \frac{c}{b}$$

The condition for cubic variation of $\partial w/\partial n$ along edge 1-3 then is

$$5b^4c\alpha_{16} + (3b^2c^3 - 2b^4c)\alpha_{17} + (2bc^4 - 3b^3c^2)\alpha_{18} + (c^5 - 4b^2c^3)\alpha_{19} - 5bc^4\alpha_{20} = 0 \qquad (12.97)$$

Similarly, the condition for cubic variation of $\partial w/\partial n$ along edge 2-3 gives the final equation that is needed:

$$5a^4c\alpha_{16} + (3a^2c^3 - 2a^4c)\alpha_{17} + (-2ac^4 + 3a^3c^2)\alpha_{18} + (c^5 - 4a^2c^3)\alpha_{19} + 5ac^4\alpha_{20} = 0 \qquad (12.98)$$

Thus twenty equations can be obtained to solve for the twenty constants in the displacement function (12.94):

$$\begin{matrix} 18\times 1 \\ 2\times 1 \end{matrix}\begin{Bmatrix} \bar{q} \\ 0 \end{Bmatrix} = \underset{20\times 20}{[\mathbf{B}]}\;\underset{20\times 1}{\{\boldsymbol{\alpha}\}} \qquad (12.99)$$

where $\lfloor\bar{\mathbf{q}}\rfloor = \lfloor w_1, w_{\xi_1}, w_{\eta_1}, w_{\xi\xi_1}, w_{\xi\eta_1}, w_{\eta\eta_1}, w_2, \ldots, w_{\eta\eta_4}\rfloor$ are the eighteen nodal d.o.f.'s in local coordinates and the matrix $[\mathbf{B}]$ is listed in Table 12.9.

Inverting $[\mathbf{B}]$ and discarding the last two columns gives

$$\underset{20\times 1}{\{\boldsymbol{\alpha}\}} = \underset{20\times 18}{[\mathbf{T}]}\;\underset{18\times 1}{\{\bar{\mathbf{q}}\}} \qquad (12.100)$$

Stiffness matrix. Substituting the displacement function (12.94) in the strain energy expression (12.8) in local coordinates and integrating over the areas of the two subtriangles gives

$$U = \frac{D}{2}\iint\left[w_{\xi\xi}^2 + w_{\eta\eta}^2 + 2\nu w_{\xi\xi}w_{\eta\eta} + 2(1-\nu)w_{\xi\eta}^2\right]d\xi\,d\eta$$

$$= \tfrac{1}{2}\{\boldsymbol{\alpha}\}[\bar{\mathbf{k}}]\{\boldsymbol{\alpha}\} \qquad (12.101)$$

where the "kernel" stiffness coefficient can be obtained as

$$\bar{k}_{ij} = \frac{\partial^2 U}{\partial\alpha_i\,\partial\alpha_j} \qquad (12.102)$$

TABLE 12.9 Matrix [B] in the 20 Equations Needed to Solve for the 20 Constants in the Displacement Function

$$
\begin{bmatrix}
1 & -b & 0 & b^2 & 0 & 0 & -b^3 & 0 & 0 & 0 & b^4 & 0 & 0 & 0 & 0 & -b^5 & 0 & 0 & 0 & 0 \\
0 & 1 & 0 & -2b & 0 & 0 & 3b^2 & 0 & 0 & 0 & -4b^3 & 0 & 0 & 0 & 0 & 5b^4 & 0 & 0 & 0 & 0 \\
0 & 0 & 1 & 0 & -b & 0 & 0 & b^2 & 0 & 0 & 0 & -b^3 & 0 & 0 & 0 & 0 & 0 & 0 & 0 & 0 \\
0 & 0 & 0 & 2 & 0 & 0 & -6b & 0 & 0 & 0 & 12b^2 & 0 & 0 & 0 & 0 & -20b^3 & 0 & 0 & 0 & 0 \\
0 & 0 & 0 & 0 & 1 & 0 & 0 & -2b & 0 & 0 & 0 & 3b^2 & 0 & 0 & 0 & 0 & 0 & 0 & 0 & 0 \\
0 & 0 & 0 & 0 & 0 & 2 & 0 & 0 & -2b & 0 & 0 & 0 & 2b^2 & 0 & 0 & 0 & -2b^3 & 0 & 0 & 0 \\
1 & a & 0 & a^2 & 0 & 0 & a^3 & 0 & 0 & 0 & a^4 & 0 & 0 & 0 & 0 & a^5 & 0 & 0 & 0 & 0 \\
0 & 1 & 0 & 2a & 0 & 0 & 3a^2 & 0 & 0 & 0 & 4a^3 & 0 & 0 & 0 & 0 & 5a^4 & 0 & 0 & 0 & 0 \\
0 & 0 & 1 & 0 & a & 0 & 0 & a^2 & 0 & 0 & 0 & a^3 & 0 & 0 & 0 & 0 & 0 & 0 & 0 & 0 \\
0 & 0 & 0 & 2 & 0 & 0 & 6a & 0 & 0 & 0 & 12a^2 & 0 & 0 & 0 & 0 & 20a^3 & 0 & 0 & 0 & 0 \\
0 & 0 & 0 & 0 & 1 & 0 & 0 & 2a & 0 & 0 & 0 & 3a^2 & 0 & 0 & 0 & 0 & 0 & 0 & 0 & 0 \\
0 & 0 & 0 & 0 & 0 & 2 & 0 & 0 & 2a & 0 & 0 & 0 & 2a^2 & 0 & 0 & 0 & 2a^3 & 0 & 0 & 0 \\
1 & 0 & c & 0 & 0 & c^2 & 0 & 0 & 0 & c^3 & 0 & 0 & 0 & 0 & c^4 & 0 & 0 & 0 & 0 & c^5 \\
0 & 1 & 0 & 0 & c & 0 & 0 & 0 & c^2 & 0 & 0 & 0 & 0 & c^3 & 0 & 0 & 0 & 0 & c^4 & 0 \\
0 & 0 & 1 & 0 & 0 & 2c & 0 & 0 & 0 & 3c^2 & 0 & 0 & 0 & 0 & 4c^3 & 0 & 0 & 0 & 0 & 5c^4 \\
0 & 0 & 0 & 2 & 0 & 0 & 0 & 2c & 0 & 0 & 0 & 0 & 2c^2 & 0 & 0 & 0 & 0 & 2c^3 & 0 & 0 \\
0 & 0 & 0 & 0 & 1 & 0 & 0 & 0 & 2c & 0 & 0 & 0 & 0 & 3c^2 & 0 & 0 & 0 & 0 & 4c^3 & 0 \\
0 & 0 & 0 & 0 & 0 & 2 & 0 & 0 & 0 & 6c & 0 & 0 & 0 & 0 & 12c^2 & 0 & 0 & 0 & 0 & 20c^3 \\
0 & 0 & 0 & 0 & 0 & 0 & 0 & 0 & 0 & 0 & 0 & 0 & 0 & 0 & 0 & 5a^4c & 3a^2c^3 - 2a^4c & -2ac^4 + 3a^3c^2 & c^5 - 4a^2c^3 & 5ac^4 \\
0 & 0 & 0 & 0 & 0 & 0 & 0 & 0 & 0 & 0 & 0 & 0 & 0 & 0 & 0 & 5b^4c & 3b^2c^3 - 2b^4c & 2bc^4 - 3b^3c^2 & c^5 - 4b^2c^3 & -5bc^4
\end{bmatrix}
$$

To derive $\bar{k}_{ij}$, let us first operate on the first term in Eq. (12.101):

$$w_{\xi\xi} = \sum_{i=1}^{20} \alpha_i m_i(m_i - 1)\xi^{m_i-2}\eta^{n_i}$$

$$w_{\xi\xi}^2 = \sum_{i=1}^{20}\sum_{j=1}^{20} \alpha_i\alpha_j m_i m_j(m_i - 1)(m_j - 1)\xi^{m_i-2}\xi^{m_j-2}\eta^{n_i}\eta^{n_j}$$

$$\frac{\partial^2 w_{\xi\xi}^2}{\partial\alpha_i\,\partial\alpha_j} = 2m_i m_j(m_i - 1)(m_j - 1)\xi^{m_i-2}\xi^{m_j-2}\eta^{n_i}\eta^{n_j}$$

$$\frac{\partial^2}{\partial\alpha_i\,\partial\alpha_j}\iint w_{\xi\xi}^2\,d\xi\,d\eta = 2m_i m_j(m_i - 1)(m_j - 1)\iint \xi^{m_i+m_j-4}\eta^{n_i+n_j}\,d\xi\,d\eta \tag{12.103}$$

Noting the relations between the triangular coordinates and the local coordinates in the two subtriangles as shown in Fig. 12.19—$\xi = a\xi_2$, $-\xi = b\xi_1$, and $\eta = c\xi_3$—and using the integral formula (9.103) for the two subtriangles, we obtain a formula for the integral:

$$\begin{aligned} F(m, n) &= \iint \xi^m\eta^n\,d\xi\,d\eta \\ &= \iint a^m c^n \xi_2^m \xi_3^n\,dA + \iint (-b)^m c^n \xi_1^m \xi_3^n\,dA \\ &= c^{n+1}[a^{m+1} - (-b)^{m+1}]\frac{m!\,n!}{(m+n+2)!} \end{aligned} \tag{12.104}$$

With Eq. (12.104) available for a typical integral, we can generalize from Eqs. (12.101) to (12.103) to obtain

$$\begin{aligned} \bar{k}_{ij} = D[&m_i m_j(m_i - 1)(m_j - 1)F(m_i + m_j - 4, n_i + n_j) \\ &+ n_i n_j(n_i - 1)(n_j - 1)F(m_i + m_j, n_i + n_j - 4) \\ &+ \{2(1 - \nu)m_i m_j n_i n_j + \nu m_i n_j(m_i - 1)(n_j - 1) \\ &+ \nu m_j n_i(m_j - 1)(n_i - 1)\}F(m_i + m_j - 2, n_i + n_j - 2)] \end{aligned} \tag{12.105}$$

Substituting Eq. (12.105) for $[\bar{\mathbf{k}}]$ and Eq. (12.100) for $\{\boldsymbol{\alpha}\}$ in Eq. (12.101), the strain energy expression becomes

$$U = \tfrac{1}{2}\{\bar{\mathbf{q}}\}^T[\mathbf{T}]^T[\bar{\mathbf{k}}][\mathbf{T}]\{\bar{\mathbf{q}}\} \tag{12.106}$$

The final step toward obtaining the stiffness matrix is to transfer $\{\bar{\mathbf{q}}\}$ in local coordinates to $\{\mathbf{q}\}$ in global coordinates, defined as

$$\{\mathbf{q}\}^T = \lfloor w_1 \quad w_{x_1} \quad w_{y_1} \quad w_{xx_1} \quad w_{xy_1} \quad w_{yy_1} \quad w_2 \quad \cdots \quad w_{yy_4}\rfloor \tag{12.107}$$

We can make use of the following relations:

$$\frac{\partial w}{\partial \xi} = \frac{\partial w}{\partial x}\left(\frac{\partial x}{\partial \xi}\right) + \frac{\partial w}{\partial y}\left(\frac{\partial y}{\partial \xi}\right) = w_x \cos\theta + w_y \sin\theta$$

$$\frac{\partial^2 w}{\partial \xi^2} = \frac{\partial}{\partial \xi}\left(\frac{\partial w}{\partial \xi}\right) = \left(\cos\theta\frac{\partial}{\partial x} + \sin\theta\frac{\partial}{\partial y}\right)\left(\cos\theta\frac{\partial w}{\partial x} + \sin\theta\frac{\partial w}{\partial y}\right)$$

$$= w_{xx}\cos^2\theta + w_{xy}\sin 2\theta + w_{yy}\sin^2\theta \tag{12.108}$$

Similar expressions can be obtained for w_η, $w_{\xi\eta}$, and $w_{\eta\eta}$. We can relate the d.o.f.'s in local coordinates at node 1 to those in global coordinates as

$$\begin{Bmatrix} w_1 \\ w_{\xi_1} \\ w_{\eta_1} \\ w_{\xi\xi_1} \\ w_{\xi\eta_1} \\ w_{\eta\eta_1} \end{Bmatrix} = \begin{bmatrix} 1 & 0 & 0 & 0 & 0 & 0 \\ 0 & \lambda & \mu & 0 & 0 & 0 \\ 0 & -\mu & \lambda & 0 & 0 & 0 \\ 0 & 0 & 0 & \lambda^2 & 2\lambda\mu & \mu^2 \\ 0 & 0 & 0 & -\lambda\mu & \lambda^2-\mu^2 & \lambda\mu \\ 0 & 0 & 0 & \mu^2 & -2\lambda\mu & \lambda^2 \end{bmatrix} \begin{Bmatrix} w_1 \\ w_{x_1} \\ w_{y_1} \\ w_{xx_1} \\ w_{xy_1} \\ w_{yy_1} \end{Bmatrix} \tag{12.109}$$

or symbolically,

$$\{\bar{\mathbf{q}}_1\} = [\mathbf{R}_1]\{\mathbf{q}_1\} \tag{12.109a}$$

where $\lambda = \cos\theta$ and $\mu = \sin\theta$.

Generalizing Eqs. (12.109) to relate the d.o.f.'s in local coordinates to those in global coordinates for the other two nodes, we have

$$\{\bar{\mathbf{q}}\} = \begin{Bmatrix} \bar{q}_1 \\ \bar{q}_2 \\ \bar{q}_3 \end{Bmatrix} = \begin{bmatrix} R_1 & 0 & 0 \\ 0 & R_1 & 0 \\ 0 & 0 & R_1 \end{bmatrix} \begin{Bmatrix} q_1 \\ q_2 \\ q_3 \end{Bmatrix} = [\mathbf{R}]\{\mathbf{q}\} \tag{12.110}$$

Substituting Eq. (12.110) in Eq. (12.106), we obtain in global coordinates,

$$U = \tfrac{1}{2}\{\mathbf{q}\}^T[\mathbf{k}]\{\mathbf{q}\} \tag{12.111}$$

where the stiffness matrix is defined as

$$[\mathbf{k}] = [\mathbf{R}]^T[\mathbf{T}]^T[\bar{\mathbf{k}}][\mathbf{T}][\mathbf{R}] \tag{12.112}$$

12.4.6 Six-Node 21-D.O.F. Conforming Triangle (*Normal Slope Varies Quartically along Edges*) [12.35, 12.37–12.40]

Figure 12.20 shows a six-node 21-d.o.f. triangular plate element with six d.o.f.'s (w, w_x, w_y, w_{xx}, w_{xy}, w_{yy}) at each of the three corners and one d.o.f. (w_n) at each of the three midside nodes. Thus we can assume the displacement function as a complete fifth-order polynomial in global coordinates x, y with 21 constants:

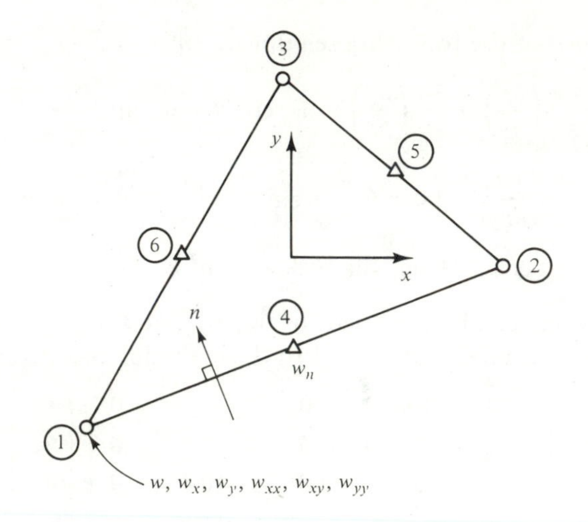

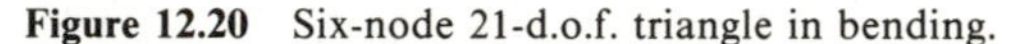

Figure 12.20 Six-node 21-d.o.f. triangle in bending.

$$
\begin{array}{l}
w(x, y) = \\
\qquad c_1 \\
\qquad c_2 x \quad c_3 y \\
\qquad c_4 x^2 \quad c_5 xy \quad c_6 y^2 \\
\qquad c_7 x^3 \quad c_8 x^2 y \quad c_9 xy^2 \quad c_{10} y^3 \\
\qquad c_{11} x^4 \quad c_{12} x^3 y \quad c_{13} x^2 y^2 \quad c_{14} xy^3 \quad c_{15} y^4 \\
\qquad c_{16} x^5 \quad c_{17} x^4 y \quad c_{18} x^3 y^2 \quad c_{19} x^2 y^3 \quad c_{20} xy^4 \quad c_{21} y^5
\end{array}
\tag{12.113}
$$

The displacement w describes a quintic curve along any interelement edge in, say, direction s and the normal slope w_n follows a quartic curve. The six constants in the quintic curve for the displacement w can be uniquely defined by the six nodal conditions (w, w_s, w_{ss} at each end node). The five constants in the quartic curve for the normal slope w_n can uniquely be defined by the five nodal conditions (w_n and w_{ns} at each end node and w_n at the midside node).

For convenience of assemblage, the midside node can be eliminated by imposing a unique cubic variation on the normal slope w_n along the interelement edge. This can be obtained by expressing w_n at each midside node in terms of w_n and w_{ns} at the two end nodes. Thus the element is reduced to the three-node 18-d.o.f. triangle described in the preceding section.

There is no obvious difference in numerical performance between the 18-d.o.f. and the 21-d.o.f. triangles. While the higher-order elements give very good accuracy, the practicing engineer may still prefer to use more but lower-order elements simply to achieve the convenience of geometric modeling and to avoid the difficulty of imposing the boundary conditions and interpreting the nodal forces associated with the second-order derivative d.o.f.'s.

12.4.7 Six-Node 6-D.O.F. Nonconforming Triangle with Constant Curvatures [12.41, 12.42]

Figure 12.21 shows a six-node six-d.o.f. triangle with deflection w at the three corners and normal slope w_n at the three midside nodes. We can assume a complete quadratic displacement function,

$$w(x, y) = c_1 + c_2 x + c_3 y + c_4 x^2 + c_5 xy + c_6 y^2 \tag{12.114}$$

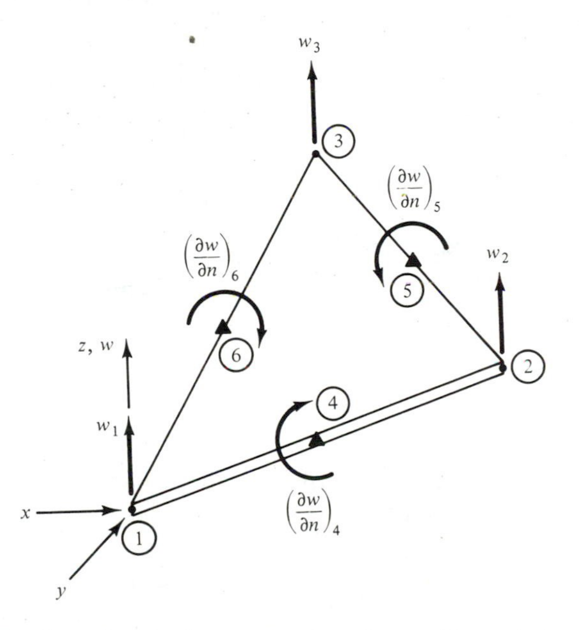

Figure 12.21 Six-node 6-d.o.f. nonconforming triangle with constant curvatures.

This displacement function results in constant curvatures and moments throughout the triangle. The element is equivalent to the constant strain triangle and is the simplest bending element possible.

Due to the constant strains, the integrands for deriving the stiffness matrix contain only constants, which considerably simplifies the derivation.

It was shown that [12.41, 12.42] although the element does not provide interelement continuity in deflection, it satisfies all conditions of equilibrium and it gives convergent results. The moments satisfy the conditions of internal equilibrium as well as interelement equilibrium.

12.4.8 Three-Node 9-D.O.F. Triangle with a Centroidal D.O.F. *w* [12.43]

Figure 12.22 shows a three-node 9-d.o.f. triangular element with an additional centroidal degree of freedom w_4. The presence of the centroidal node allows us to assume a complete 10-term polynomial displacement function,

$$w(x, y) = c_1 + c_2 x + c_3 y + c_4 x^2 + c_5 xy + c_6 y^2 + c_7 x^3 + c_8 x^2 y + c_9 xy^2 + c_{10} y^3 \tag{12.115}$$

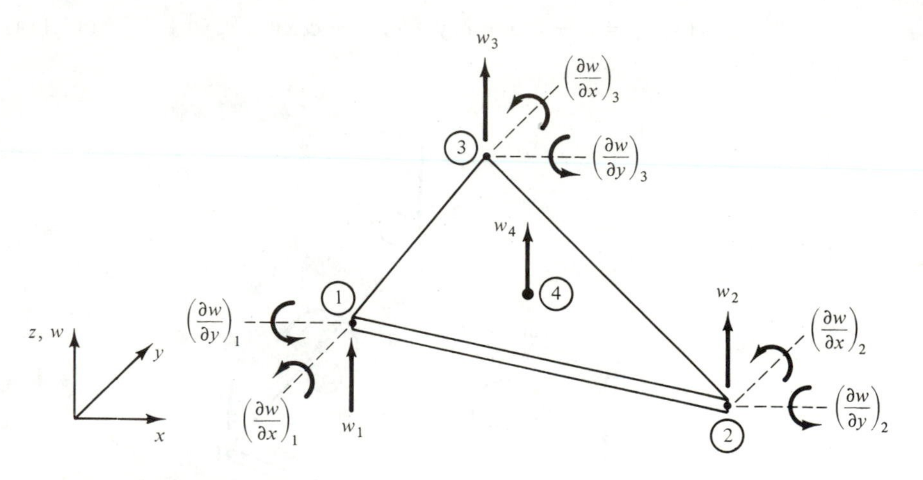

Figure 12.22 Three-node 9-d.o.f. triangle in bending with a centroidal d.o.f.

This displacement function can also be written in triangular coordinates in terms of the shape functions as

$$w(\xi_1, \xi_2, \xi_3) = \sum_{i=1}^{3} \left[f_i(\xi_1, \xi_2, \xi_3) w_i + g_i(\xi_1, \xi_2, \xi_3) \left(\frac{\partial w}{\partial x}\right)_i + h_i(\xi_1, \xi_2, \xi_3) \left(\frac{\partial w}{\partial y}\right)_i \right] + f_4(\xi_1, \xi_2, \xi_3) w_4 \tag{12.116}$$

where for node 1, the shape functions are

$$\begin{aligned} f_1(\xi_1, \xi_2, \xi_3) &= \xi_1^2(\xi_1 + 3\xi_2 + 3\xi_3) - 7\xi_1 \xi_2 \xi_3 \\ g_1(\xi_1, \xi_2, \xi_3) &= (x_2 - x_1)(\xi_1^2 \xi_2 - \xi_1 \xi_2 \xi_3) + (x_3 - x_1)(\xi_3 \xi_1^2 - \xi_1 \xi_2 \xi_3) \\ h_1(\xi_1, \xi_2, \xi_3) &= (y_2 - y_1)(\xi_1^2 \xi_2 - \xi_1 \xi_2 \xi_3) + (y_3 - y_1)(\xi_3 \xi_1^2 - \xi_1 \xi_2 \xi_3) \end{aligned} \tag{12.117}$$

The functions for nodes 2 and 3 are in the same form with cyclic permutation of the subscripts. Also,

$$f_4(\xi_1, \xi_2, \xi_3) = 27\xi_1 \xi_2 \xi_3 \tag{12.118}$$

The function $f_1(\xi_1, \xi_2, \xi_3)$ takes the value of 1 at vertex 1 ($\xi_1 = 1, \xi_2 = \xi_3 = 0$) and zero at the other two vertices and the centroidal node ($\xi_1 = \xi_2 = \xi_3 = \frac{1}{3}$). The derivatives $\partial f_1/\partial x$ and $\partial f_1/\partial y$ also vanish at the three vertices. As an example, let us examine $\partial g_1(\xi_1, \xi_2, \xi_3)/\partial x$ at node 1:

$$\begin{aligned}\frac{\partial g_1}{\partial x} &= \frac{\partial g_1}{\partial \xi_1}\frac{\partial \xi_1}{\partial x} + \frac{\partial g_1}{\partial \xi_2}\frac{\partial \xi_2}{\partial x} + \frac{\partial g_1}{\partial \xi_3}\frac{\partial \xi_3}{\partial x} \\ &= \frac{1}{2A}\left[(y_2 - y_3)\frac{\partial g_1}{\partial \xi_1} + (y_3 - y_1)\frac{\partial g_1}{\partial \xi_2} + (y_1 - y_2)\frac{\partial g_1}{\partial \xi_3}\right] \\ &= 1 \qquad \text{at node 1 } (\xi_1 = 1, \xi_2 = \xi_3 = 0) \\ &= 0 \qquad \text{at nodes 2 and 3}\end{aligned} \tag{12.119}$$

Similarly, we can show that $\partial h_1/\partial y$ is equal to 1 at vertex 1 and to zero at the other two vertices. Finally, f_4 takes the value 1 at the centroid and zero at the three vertices.

The function (12.115) or (12.116) allows the displacement w to vary cubically along an interelement edge in the direction, say s. The four constants are uniquely defined by the displacements w and the derivatives $\partial w/\partial s$ at the two end nodes. The normal slope $\partial w/\partial n$ varies parabolically along the edge direction s. However, the three constants cannot be uniquely determined by the two end node values $\partial w/\partial n$; thus the interelement compatibility requirement is violated. Steps to overcome this difficulty have been suggested [12.43] using the principle of generalized potential energy.

The first step is to form element and assembled equations. Because of the gaps induced by the incompatible normal slope along interelement boundaries, the strain energy for the assembled system is improperly defined (as explained in Figs. 12.2 and 12.3). This difficulty can be overcome by writing constraint equations in a way that enforces the normal slopes at the midside of each edge to be the same for adjacent elements. Thus a third equation can be created to define uniquely the quadratically varying normal slope.

Let us assume that A and B are two adjacent elements. The continuity of the normal slope is required so that at the common midside point,

$$\left(\frac{\partial w}{\partial n}\right)_A - \left(\frac{\partial w}{\partial n}\right)_B = 0 \tag{12.120}$$

Substituting the displacement functions (12.116) in Eq. (12.120) yields a single constraint equation of the form

$$\{\mathbf{C}_j\}^T\{\mathbf{q}\} = 0 \qquad (j = 1, 2, 3, \ldots, m) \tag{12.121}$$

where j defines the element interface number, $\{\mathbf{C}_j\}$ contains the normal derivatives of the shape functions, and $\{\mathbf{q}\}$ contains the d.o.f.'s of the whole assemblage. The collection of these conditions for all interfaces results in a set of m equations,

$$[\mathbf{C}]\{\mathbf{q}\} = \{\mathbf{0}\} \tag{12.122}$$

We shall now introduce the Lagrange multiplier technique. First, the potential energy π_p is augmented by adding to it the sum of the products of each constraint condition (12.121) and a corresponding Lagrange multiplier λ_j,

$$\begin{aligned}\pi_A &= \pi_p + \{\boldsymbol{\lambda}\}^T[\mathbf{C}]\{\mathbf{q}\} \\ &= \tfrac{1}{2}\{\mathbf{q}\}^T[\mathbf{K}]\{\mathbf{q}\} - \{\mathbf{F}\}^T\{\mathbf{q}\} + \{\boldsymbol{\lambda}\}^T[\mathbf{C}]\{\mathbf{q}\}\end{aligned} \tag{12.123}$$

where $[\mathbf{K}]$ is the assembled stiffness matrix and $\{\mathbf{F}\}$ contains the externally applied nodal forces.

Carrying out the principle of minimum generalized potential energy that

$$\frac{\partial \pi_A}{\partial q_i} = 0 \qquad \frac{\partial \pi_A}{\partial \lambda_i} = 0 \tag{12.124}$$

gives

$$[\mathbf{K}]\{\mathbf{q}\} - \{\mathbf{F}\} + [\mathbf{C}]^T\{\boldsymbol{\lambda}\} = \{\mathbf{0}\} \tag{12.125}$$

and

$$[\mathbf{C}]\{\mathbf{q}\} = \{\mathbf{0}\} \tag{12.126}$$

Combining Eqs. (12.125) and (12.126) gives

$$\begin{bmatrix} [\mathbf{K}] & [\mathbf{C}]^T \\ [\mathbf{C}] & 0 \end{bmatrix} \begin{Bmatrix} q \\ \lambda \end{Bmatrix} = \begin{Bmatrix} F \\ 0 \end{Bmatrix} \tag{12.127}$$

This expanded set of equations now guarantees interelement compatibility in the normal slope. If the d.o.f.'s and the Lagrange multipliers are properly ordered, the set of equations (12.127) can be banded and possible zero pivot terms (used in the Gaussian elimination method described in Chapter 2) can be avoided.

The vector $\{\boldsymbol{\lambda}\}$ of Lagrange multipliers may be interpreted physically as generalized constraint forces necessary to remove the slope mismatches.

The centroidal nodal d.o.f.'s w can be eliminated by writing them in terms of the other d.o.f.'s through matrix partitioning and substitution as outlined in Sec. 4.6.2. In such a condensation process, the centroidal nodal forces need not be zero.

12.5 QUADRILATERAL ELEMENTS

Formulation of quadrilateral elements in bending is as difficult as that for the triangular elements due to the complexity involved in satisfying the interelement compatibility requirement for the normal slopes along arbitrarily oriented edges. For this reason, a simple 12-d.o.f. quadrilateral displacement element using a single polynomial expansion with a C_1 continuity for w does not exist. A series of quadrilateral elements can be obtained by combining several conforming triangular elements just introduced. Some such quadrilateral elements are shown in Fig. 12.23. Quadrilateral *a* is composed of two three-node nine-d.o.f. (HCT) triangles. For quadrilateral *b*, the three (or six) d.o.f.'s at the internal node are eliminated by writing them in terms of the other 12 (or 24) d.o.f.'s through static condensation. For quadrilateral *c*, the three (or six) d.o.f.'s at the internal corner node and the eight normal-slope d.o.f.'s at the eight midside nodes are eliminated through static condensation.

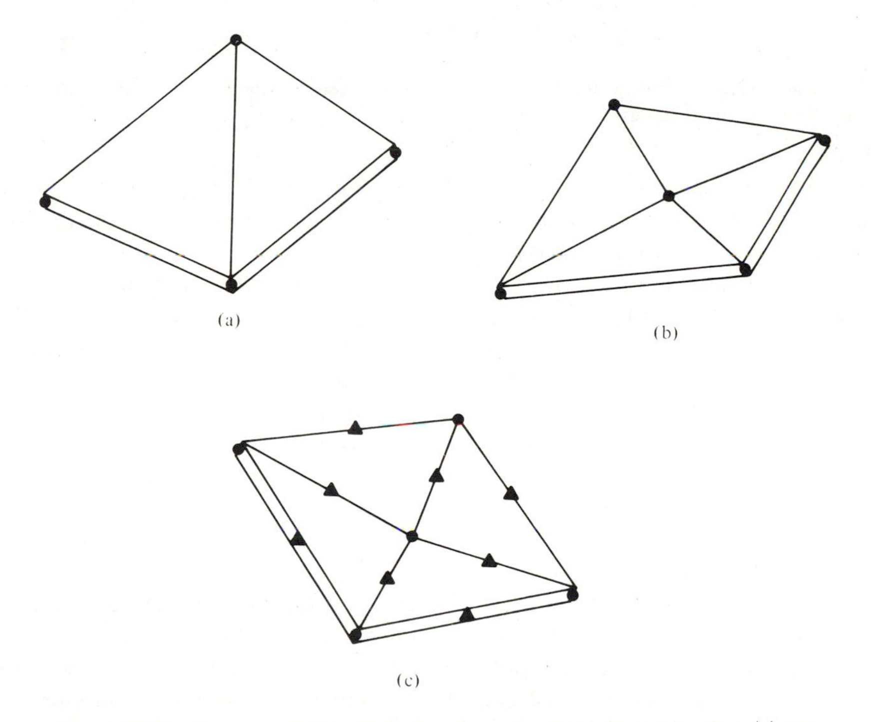

Figure 12.23 Some quadrilaterals in bending assembled from triangles: (a) two three-node 9-d.o.f. (HCT) triangles; (b) four three-node 9-d.o.f. (or 18-d.o.f.) triangles; (c) four six-node 12-d.o.f. (or 21-d.o.f.) triangles.

An interesting quadrilateral element was developed by Fraeijs de Veubeke [12.44]. The element is a 16-d.o.f. interelement-compatible quadrilateral. It is obtained from a combination of complete 10-term cubic deflection functions in the four respective triangles, bounded by a quadrilateral and its diagonals as shown in Fig. 12.24. The element possesses three conventional d.o.f.'s (w, w_x, w_y) at each of the four corner nodes and one d.o.f. in the form of a normal slope at the midside of each edge. The four midside normal slopes

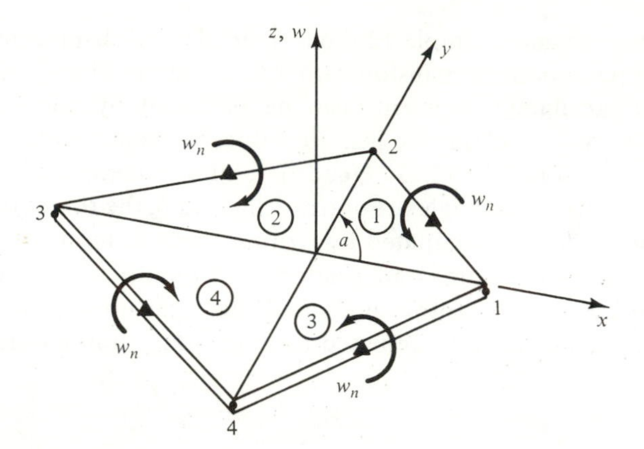

Figure 12.24 Four-node 16-d.o.f. quadrilateral in bending (with w, w_x, and w_y at the four corner nodes).

can be condensed out to yield a 12-d.o.f. quadrilateral element. It is noted that the element stiffnesses were formulated in terms of the oblique coordinates that lie along the two diagonals.

Clough and Felippa [12.45] presented an alternative formulation of a quadrilateral element based on the combination of four conforming triangles. Before assembling the triangles to form the quadrilateral, constraints are placed on sides that are to be exterior boundaries so as to remove the normal-slope d.o.f.'s at the midpoints of such sides. Thus the final quadrilateral element is a four-node 12-d.o.f. element and it satisfies the interelement compatibility.

12.6 HYBRID STRESS PLATE ELEMENTS

The element stiffness formulations for the various plate bending elements described in this chapter have been based on the displacement models and on the principle of minimum potential energy. The procedure involves the displacement representation of each finite element in terms of a polynomial having m independent, undetermined constants where m is equal to the number of nodal displacement d.o.f.'s. An important requirement for the plate displacement models is that the assumed functions maintain interelement compatibility in deflection and normal slope (C_1 continuity) along the common edges. However, while the displacement functions satisfy the compatibility requirements, the resulting stresses in the interior of the element may not be in equilibrium. The hybrid stress method was first introduced by Pian [12.46, 12.47] to circumvent this difficulty encountered in the development of pure displacement models.

In this method the derivation of the stiffness matrix is based on the use of assumed stress distributions and on the principle of minimum complementary energy. Instead of assuming a required continuous displacement function over the element, it is necessary to write down the boundary displacements that guarantee a complete displacement compatibility with the adjacent elements.

12.6.1 Outline of the Method

The basic approach is to assume an equilibrium stress field $\{\boldsymbol{\sigma}\}$ written within the element in terms of generalized parameters $\{\boldsymbol{\beta}\}$. Simultaneously, a set of displacement d.o.f.'s is chosen at the nodes and simple displacement functions along the edges are assigned so that the compatibility conditions within the neighboring elements can be satisfied. With the prescribed displacements at the boundary A_2, it is desirable to determine the stress distribution over the element. The problem can be solved by using the principle of minimum complementary energy, which may be stated as

$$\pi_c = U - \int_{A_2} \begin{Bmatrix} s_1 \\ s_2 \\ s_3 \end{Bmatrix}^T \begin{Bmatrix} u_1 \\ u_2 \\ u_3 \end{Bmatrix} dA_2 = \text{minimum} \tag{12.128}$$

where U is the strain energy in terms of stress components σ_{ij}; u_1, u_2, and u_3 are the components of the prescribed boundary displacements in the three orthogonal directions; and s_1, s_2, and s_3 are the components of surface forces (pounds per square inch) in the three orthogonal directions. These forces can be expressed as

$$\begin{Bmatrix} s_1 \\ s_2 \\ s_3 \end{Bmatrix} = \begin{bmatrix} \sigma_{11} & \sigma_{12} & \sigma_{13} \\ \sigma_{21} & \sigma_{22} & \sigma_{23} \\ \sigma_{31} & \sigma_{32} & \sigma_{33} \end{bmatrix} \begin{Bmatrix} n_1 \\ n_2 \\ n_3 \end{Bmatrix} \tag{12.129}$$

where σ_{ij} are the stress components on the edge surface and n_i are the direction cosines of the surface normals.

When applying the principle of minimum complementary energy, we begin by expressing the stress distribution $\{\boldsymbol{\sigma}\}$ in terms of m undetermined stress coefficients $\{\boldsymbol{\beta}\}$ as follows:

$$\{\boldsymbol{\sigma}\} = [\mathbf{P}]\{\boldsymbol{\beta}\} \tag{12.130}$$

where $\lfloor \boldsymbol{\sigma} \rfloor = \lfloor \sigma_{11}, \sigma_{12}, \sigma_{13}, \sigma_{22}, \sigma_{23}, \sigma_{33} \rfloor$ and the terms in the matrix $[\mathbf{P}]$ are functions of the coordinates in polynomial form. The number of terms in $\{\boldsymbol{\beta}\}$ is unlimited, but $[\mathbf{P}]$ and $\{\boldsymbol{\beta}\}$ must be chosen such that the stress distribution satisfies both the equations of equilibrium in the interior and any prescribed stresses at the boundary.

Substituting Eq. (12.130) and the strain–stress relations

$$\{\boldsymbol{\epsilon}\} = [\mathbf{N}]\{\boldsymbol{\sigma}\} \tag{12.131}$$

in the expression for internal strain energy gives

$$U = \frac{1}{2}\int_V \{\boldsymbol{\sigma}\}^T[\mathbf{N}]\{\boldsymbol{\sigma}\}\, dV$$
$$= \tfrac{1}{2}\{\boldsymbol{\beta}\}^T[\mathbf{H}]\{\boldsymbol{\beta}\} \tag{12.132}$$

where [**H**] is a symmetric matrix determined by

$$[\mathbf{H}] = \int_V [\mathbf{P}]^T[\mathbf{N}][\mathbf{P}]\, dV \tag{12.133}$$

The prescribed displacements at the boundary A_2 can be written in terms of the n nodal d.o.f.'s {**q**} in the form

$$\{\mathbf{u}\} = [\mathbf{L}]\{\mathbf{q}\} \tag{12.134}$$

where the terms in the matrix [**L**] contain coordinates of the edge surface.

The edge surface force {**S**} can be expressed in terms of the stresses $\{\boldsymbol{\sigma}\}$ by means of Eq. (12.129) and hence can be related to the undetermined stress coefficients $\{\boldsymbol{\beta}\}$ as, from Eq. (12.130),

$$\{\mathbf{S}\} = [\mathbf{R}]\{\boldsymbol{\beta}\} \tag{12.135}$$

where the terms in [**R**] also contain the coordinates of the edge surface.

The total complementary energy is obtained by substituting Eqs. (12.132), (12.134), and (12.135) in Eq. (12.128):

$$\pi_c = \tfrac{1}{2}\{\boldsymbol{\beta}\}^T[\mathbf{H}]\{\boldsymbol{\beta}\} - \{\boldsymbol{\beta}\}^T[\mathbf{T}]\{\mathbf{q}\} \tag{12.136}$$

where

$$[\mathbf{T}] = \int_{A_2} [\mathbf{R}]^T[\mathbf{L}]\, dA_2 \tag{12.137}$$

The principle of minimum complementary energy states that

$$\frac{\partial \pi_c}{\partial \beta_i} = 0 \qquad (i = 1, 2, \ldots, m) \tag{12.138}$$

which yields

$$[\mathbf{H}]\{\boldsymbol{\beta}\} = [\mathbf{T}]\{\mathbf{q}\} \tag{12.139a}$$

or

$$\{\boldsymbol{\beta}\} = [\mathbf{H}]^{-1}[\mathbf{T}]\{\mathbf{q}\} \tag{12.139b}$$

Substituting Eqs. (12.139) in (12.132) and noting that [H] is symmetric gives

$$U = \tfrac{1}{2}\{\mathbf{q}\}^T[\mathbf{T}]^T[\mathbf{H}]^{-1}[\mathbf{T}]\{\mathbf{q}\}$$
$$= \tfrac{1}{2}\{\mathbf{q}\}^T[\mathbf{K}]\{\mathbf{q}\} \tag{12.140}$$

which results in the desired stiffness matrix:

$$[\mathbf{K}] = [\mathbf{T}]^T[\mathbf{H}]^{-1}[\mathbf{T}] \tag{12.141}$$

The corresponding column of nodal forces {F} is given by

$$\{\mathbf{F}\} = [\mathbf{K}]\{\mathbf{q}\}$$
$$= [\mathbf{T}]^T[\mathbf{H}]^{-1}[\mathbf{T}]\{\mathbf{q}\} \tag{12.142}$$

$$\{\mathbf{F}\} = [\mathbf{T}]^T\{\boldsymbol{\beta}\} \tag{12.143}$$

The matrix $[\mathbf{T}]^T$ thus relates the equivalent nodal forces {F} and the assumed stress coefficient {β}.

The plate finite elements based on the foregoing concept have been developed by numerous authors. Pian [12.46], Severn and Taylor [12.48], and Neale et al. [12.49] developed a number of hybrid stress formulations for rectangular plate elements. They all differ by the number of internal stress parameters considered. Allwood and Cornes [12.50] used the foregoing approach to derive the stiffness matrix of a plate bending element of general polygonal shape having any number of nodes. These are based on a single field assumption. Two alternative subdomain formulations using hybrid stress concepts were presented by Cook [12.51]. Other quadrilateral elements based on hybrid stress models have been presented by Torbe and Church [12.52]. Triangular hybrid stress elements have been presented by, among others, Allman [12.53], Bartholomew [12.54], and Pian and Tong [12.55].

Recently, Spilker and Munir [12.56] examined the application of the assumed stress hybrid finite element model to the bending analysis of thin plates based on the hybrid stress functional by using a Mindlin-type displacement assumption (i.e., assuming independent transverse displacement and cross-section rotations) and including all components of stresses. The development and testing of some other successful Mindlin-type plate bending elements by the hybrid stress model, both with and without transverse shear effects included, were given by Mau and Witmer [12.57], Pian and Mau [12.58], Cook [12.59], and Cook and Ladkany [12.60]. In these studies [12.57–12.60], however, emphasis was not placed on studying the behavior of these elements in the thin plate limit. This was accomplished by Spilker and Munir [12.56].

12.7 MIXED PLATE BENDING ELEMENTS

Plate bending elements have also been derived based on the Hellinger–Reissner functional, which for the case of thin plates with homogeneous boundary conditions was given by Herrmann [12.61] and Bron and Dhatt [12.62]:

$$\pi_R = \int_A \Big[(M_{x,x} + M_{xy,y}) w_{,x} + (M_{xy,x} + M_{y,y}) w_{,y} \\ - \frac{D}{2} (M_x^2 + M_y^2 - 2\nu M_x M_y + 2(1+\nu) M_{xy}^2) \Big] \, dA \\ + \int M_{ns} w_{,s} \, ds - \int_A pw \, dA \tag{12.144}$$

where $M_{x,x}$, $M_{xy,y}, \ldots$ represent the derivatives of M_x, $M_{xy}, \ldots$ with respect to x and y, respectively; the twisting moment M_{ns} is defined in the orthogonal coordinate system (n, s) with s parallel to a side and n normal to the side; $D = 12/Eh^3$; and p is the distributed load.

The finite elements based on the stationary condition of the foregoing functional have as primary unknowns both force and displacement d.o.f.'s. Such elements are, therefore, called *mixed elements.* The moments M_x, M_y, and M_{xy} and the deflection w are approximated by assumed polynomials. Bron and Dhatt [12.62] developed a 32-d.o.f. rectangular element by approximating each of the three moments and also the deflection by an eight-term symmetrical polynomial

$$f(x, y) = a_1 + a_2 x + a_3 y + a_4 x^2 + a_5 xy + a_6 y^2 + a_7 x^2 y + a_8 xy^2 \tag{12.145}$$

The element has four corner nodes and four midside nodes.

They also developed a four-corner-node 16-d.o.f. rectangular element by introducing linear variations of the moments and the deflection.

In mixed elements, since neither the force nor the displacement is given preferential treatment, they may be computed with comparable accuracy.

12.8 NUMERICAL RESULTS

In the previous sections a thorough review of the rectangles, triangles, and quadrilaterals in bending has been given. A comparison of some of the representative elements is now given using an example of a simply supported square plate under a central load. The results are plotted as the percentage error in central deflection vs. the number of elements in a quadrant. The exact solution is available in Ref. [12.1].

Figure 12.25 shows the comparison of results for various rectangular and quadrilateral elements. The results for the nonconforming 12-d.o.f. rectangles

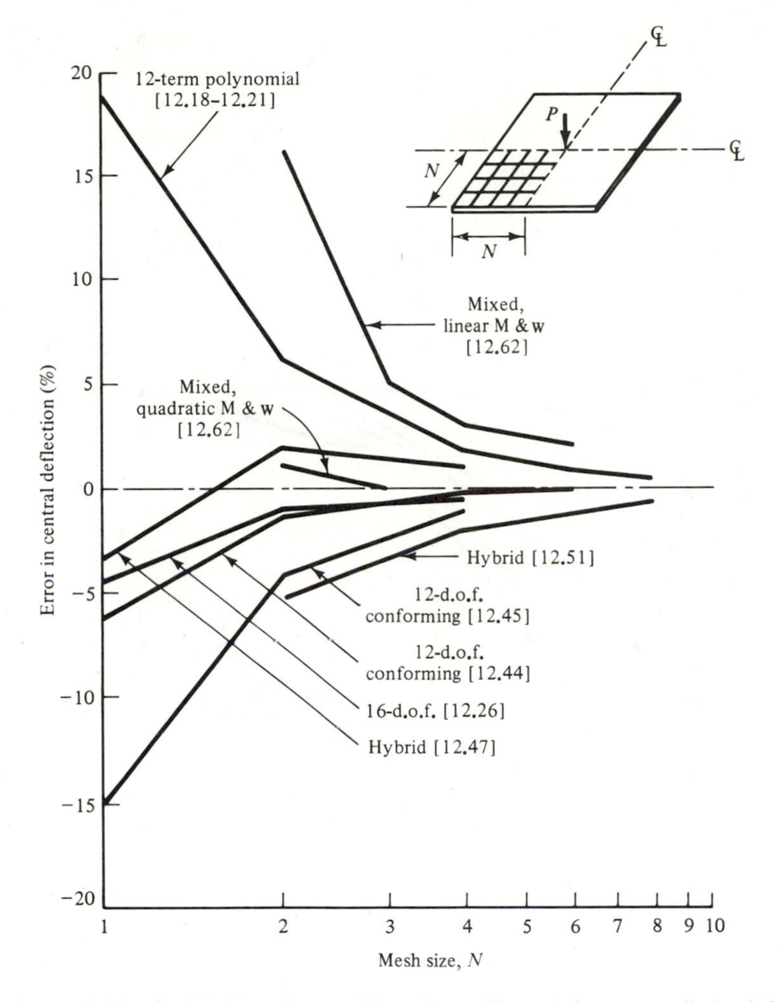

Figure 12.25 Comparison of rectangles and quadrilaterals for a centrally loaded simply supported square plate.

converge monotonically to the correct solution from above, yielding an upper-bound solution. The 16-d.o.f. conforming elements based on Hermitian polynomials yield, as expected, a monotonically convergent lower-bound solution. Similarly, the results obtained by the conforming quadrilateral elements of Fraeijs de Veubeke [12.44] and those of Clough and Felippa [12.45] also exhibit monotonic lower-bound convergence. Also shown in the figure are the results obtained using hybrid stress and mixed elements. The results obtained using the hybrid stress elements are those due to Pian [12.47] and Cook [12.51]. The convergence of Pian's results is not monotonic and the convergence of Cook's results is lower bound.

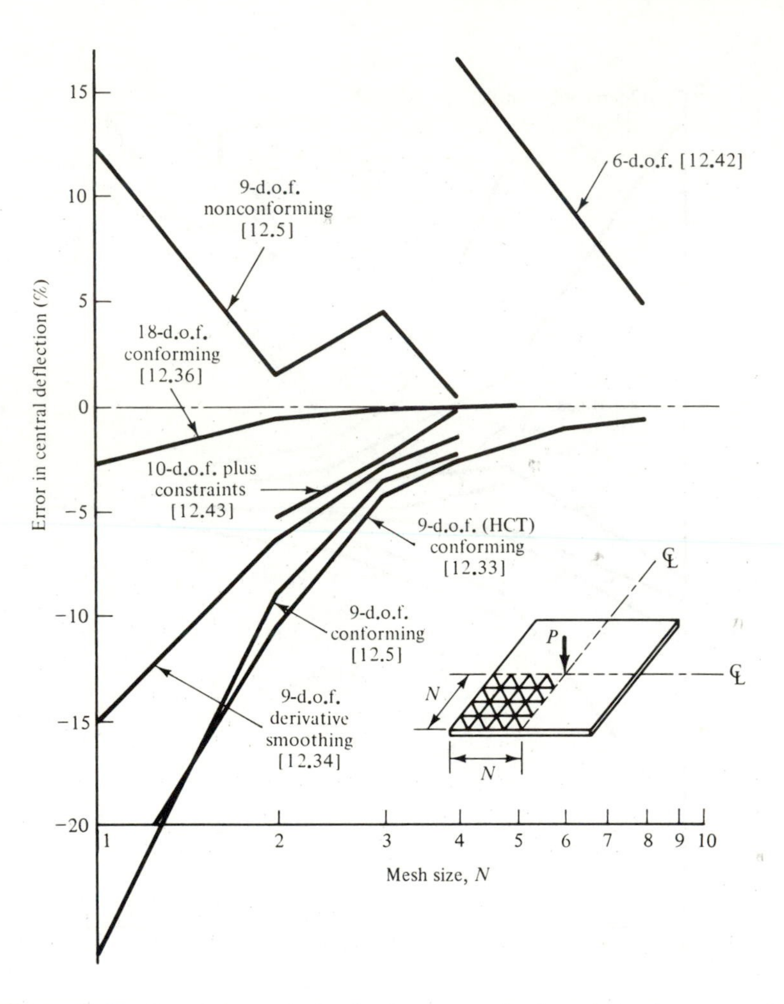

Figure 12.26 Comparison of triangles for a centrally loaded simply supported square plate.

The results obtained using the mixed formulations [12.62] are based on the linear and quadratic variations in moments and deflection, respectively. Improvement in convergence is obtained using quadratic mixed formulation.

Figure 12.26 shows the comparison of results for the same example using various triangular elements. The results for the simplest six-d.o.f. constant strain triangles [12.42] do converge monotonically to an upper-bound solution. The nine-d.o.f. nonconforming triangles by Bazeley et al. [12.5] also exhibit an upper-bound convergence, but not monotonic. The results for the various conforming triangles [12.5, 12.33, 12.34, 12.36] approach the correct solution monotonically from below, as expected. Results for 21-d.o.f. triangles are almost identical to those for the 18-d.o.f. triangles and thus are not shown.

Results obtained using the 10-d.o.f. complete cubic polynomial, with constraints to impose the continuity of the normal slope along the edges [12.43], show a good monotonic convergence. It should be noted that quite poor results are obtained if the constraints are not imposed.

An alternative to the foregoing method of comparing the performance of various kinds of elements (i.e., plotting the convergence of results with refined element meshes) was given by Abel and Desai [12.63]. They suggested that the solution of the assembled matrix equations represents a significant proportion of the total computational effort. Accordingly, they proposed that the computational effort for a given accuracy be measured by the number of operations needed to solve the equations, taking into account the banded structure of the matrix. Thus if N is the number of equations and B the half-bandwidth, the operation count of a bandwidth solution method is proportional to NB^2. The percentage error can be plotted against NB^2 for a given type of element. Applying this measure to test 12 different plate elements using two sample problems, Abel and Desai [12.63] concluded that several of the lower-order elements considered are computationally more efficient than the higher-order ones. Rossow and Chen [12.64], however, reached an opposite conclusion when they modified the measure of computational efficiency to include the effect of the boundary conditions.

An advantage of using higher-order elements is their ability to give stress results in the form of a single higher-order polynomial defined over an element of relatively large size. In contrast, when using lower-order displacement elements, useful information about the stress distribution is obtained only after an averaging procedure, which may induce additional effort. From a practical point of view, using more but lower-order elements has the advantage of better geometrical modeling, and no need to be concerned about physically interpreting the higher-order derivative d.o.f.'s and their counterpart forces.

12.9 CONCLUDING REMARKS

In this chapter we have introduced a wide variety of elements for the bending analysis of plates. It is seen that the development of displacement-based conforming elements for arbitrary shapes is a difficult task. This is due to the fact that the convergence requirements for the plate elements are far more difficult to meet than those for the plane stress and solid elements. The convergence criteria and the patch test were explained thoroughly in this chapter.

Various nonconforming and conforming rectangles, triangles, and quadrilaterals were presented. Attempts were made to introduce as many elements as possible. However, there are some other displacement-based elements which need be mentioned. For example, Dawe [12.65] gave a number of shape functions for 12-d.o.f. representations. Gopalacharyulu [12.66] and Irons

[12.67] discussed alternative 16-d.o.f. displacement fields. Bogner et al. [12.26] and Wegmuller and Kostem [12.68], among others, formulate rectangular plates with more than 16 d.o.f.'s. Chu and Schnobrich [12.69] formulated a triangular element on the basis of a complete quartic 15-term polynomial. The element has three conventional d.o.f.'s at each vertex plus a transverse displacement and normal slope at each midside node. Interelement compatibility is violated because of the lack of sufficient parameters for the unique definition of the normal slope along each edge.

In some triangles, only a single displacement function is used for the entire element, whereas in other formulations for triangular and quadrilateral elements, an element is divided into subtriangles, each of which has its own deflection function. Quadrilateral elements can be considered as those formed by triangular elements. Two such elements were described in this text. A similar rectangular element that deserves mention is the one developed by Deak and Pian [12.70]. This element uses the two schemes for smooth surface interpolation suggested by Birkhoff and Garabedian [12.71]. One of these is to use a single 12-parameter fully compatible interpolation function, and the other is to apply spline constraint for smooth surface interpolation.

Descriptions of the elements based on the principle of the generalized potential energy, hybrid stress method, and mixed method have also been given. For example, a triangular element based on the principle of the generalized potential energy was presented. In this element, the condition of the interelement compatibility of the normal slopes is imposed in a constraint fashion by using the Lagrange multiplier technique. An alternative way of constructing the interelement constraint condition, by setting to zero the integral of the difference between the normal slopes along the edges of the adjacent elements, was presented by Kikuchi and Ando [12.72]. Anderheggen [12.73] presented yet another generalized variational approach which forms a corrective element stiffness matrix for addition to the basic element stiffness matrix.

Plate elements were also derived by Fraeijs de Veubeke et al. [12.74] using assumed stress fields and a complementary energy functional. The Southwell stress functions were assumed in the same manner as the displacement functions for the plane stress analysis.

Finally, our attention is drawn to two papers on triangular and quadrilateral plate elements by Batoz et al. [12.75] and Batoz and Tahar [12.76], respectively, based on a discrete-Kirchhoff procedure. A comprehensive numerical evaluation of the various plate elements is given in these papers.

REFERENCES

12.1. Timoshenko, S. P., and Woinowsky-Krieger, S., *Theory of Plates and Shells*, 2nd ed., McGraw-Hill Book Company, New York, 1959, Chap. 4, pp. 44–45, Chap. 12.

12.2. Timoshenko, S. P., *History of Strength of Materials*, McGraw-Hill Book Company, New York, 1953, pp. 119–122 and 253–254.

12.3. Cantin, G., and Clough, R. W., "A Curved Cylindrical-Shell, Finite Element," *AIAA Journal*, Vol. 6, No. 6, June 1968, pp. 1057–1062.

12.4. Yang, T. Y., "High-Order Rectangular Shallow Shell Finite Element," *Journal of the Engineering Mechanics Division*, ASCE, Vol. 99, No. EM1, Feb. 1973, pp. 157–181.

12.5. Bazeley, G. P., Cheung, Y. K., Irons, B. M., and Zienkiewicz, O. C., "Triangular Elements in Plate Bending—Conforming and Non-conforming Solutions," *Proceedings*, Conference on Matrix Methods in Structural Mechanics, Air Force Flight Dynamics Laboratory, TR-66-80, Fairborn, Ohio, 1966, pp. 547–576.

12.6. Johnson, M. W., and McLay, R. W., "Convergence of the Finite Element Method in the Theory of Elasticity," *Journal of Applied Mechanics*, ASME, Vol. 35, 1968, pp. 274–278.

12.7. Key, S. W., "A Convergence Investigation of the Direct Stiffness Method," Ph.D. thesis, University of Washington, Seattle, 1966.

12.8. Pian, T. H. H., and Tong, P., "The Convergence of Finite Element Method in Solving Linear Elastic Problems," *International Journal of Solids and Structures*, Vol. 3, 1967, pp. 865–880.

12.9. De Arrantes Oliveira, E. R., "Theoretical Foundations of Finite Element Method," *International Journal of Solids and Structures*, Vol. 4, 1968, pp. 929–952.

12.10. Strang, G., and Fix, G. J., *An Analysis of Finite Element Method*, Prentice-Hall, Inc., Englewood Cliffs, N.J., 1973, p. 106.

12.11. Irons, B. M., and Razzaque, A., "Experience with the Patch Test for Convergence of Finite Element Method," *The Mathematical Foundations of the Finite Element Method with Applications to Partial Differential Equations*, ed. A. K. Aziz, Academic Press, Inc., New York, 1972, pp. 557–587.

12.12. Fraeijs de Veubeke, B., "Variational Principles and the Patch Test," *International Journal for Numerical Methods in Engineering*, Vol. 8, 1974, pp. 783–801.

12.13. Irons, B. M., "The Patch Test for Engineers," Conference Atlas Computing Center, Mar. 1974, Harwell, U.K.

12.14. Strang, G., "Variational Crimes in the Finite Element Methods," *The Mathematical Foundations of the Finite Element Method with Applications to Partial Differential Equations*, ed. A. K. Aziz, Academic Press, Inc., New York, 1972, pp. 689–710.

12.15. De Arrantes Oliveira, E. R., "Results on the Convergence of Finite Element Method in Structural and Non-structural Cases," *Finite Element Method in Engineering*, ed. V. A. Pulmano, and A. P. Kabaila, University of New South Wales, 1974, pp. 3–14.

12.16. Robinson, J., and Blackham, S., "An Evaluation of Plate Bending Elements: MSC/NASTRAN, ASAS, PAFEC, ANSYS, and SAP4," Robinson and Associates, Dorset, England, Aug. 1981.

12.17. MacNeal, R. H., and Harder, R. L., "A Proposed Standard Set of Problems to Test Finite Element Accuracy," AIAA/ASME/ASCE/AHS 25th Structures, Structural Dynamics and Materials Conference, Palm Springs, Calif., May 1984.

12.18. Adini, A., and Clough, R. W., "Analysis of Plate Bending by the Finite Element Method," Report to National Science Foundation Grant G-7337, University of California at Berkeley, 1960.

12.19. Melosh, R. J., "Basis of Derivation of Matrices for the Direct Stiffness Method," *AIAA Journal*, Vol. 1, No. 7, July 1963, pp. 1631–1637.

12.20. Tocher, J. L., and Kapur, K. K., "Comment on Basis of Derivation of Matrices for the Direct Stiffness Method," *AIAA Journal*, Vol. 3, No. 6, June 1965, pp. 1215–1216.

12.21. Zienkiewicz, O. C., and Cheung, Y. K., "The Finite Element Method for Analysis of Elastic Isotropic and Orthotropic Slabs," *Proceedings*, Institution of Civil Engineers, Vol. 28, 1964, pp. 471–488.

12.22. Walz, J. E., Fulton, R. E., and Cyrus, N. J., "Accuracy and Convergence of Finite Element Approximations," *Proceedings*, 2nd Conference on Matrix Methods in Structural Mechanics, AFFDL TR 68-150, Oct. 1968, pp. 995–1027.

12.23. Zienkiewicz, O. C., *The Finite Element Method*, 3rd ed., McGraw-Hill Book Company (U.K.) Ltd., London, 1977, pp. 327–328.

12.24. Gallagher, R. H., *Finite Element Analysis Fundamentals*, Prentice-Hall, Inc., Englewood Cliffs, N.J., 1975, pp. 374–376.

12.25. Melosh, R. J., "A Stiffness Matrix for the Analysis of Thin Plates in Bending," *Journal of the Aerospace Sciences*, Vol. 28, No. 1, Jan. 1961, pp. 34–42.

12.26. Bogner, F. K., Fox, R. L., and Schmit, L. A., Jr., "The Generation of Interelement-Compatible Stiffness and Mass Matrices by the Use of Interpolation Formulas," *Proceedings*, Conference on Matrix Methods in Structural Mechanics, AFFDL TR-66-80, Fairborn, Ohio, 1966, pp. 397–444.

12.27. Yang, T. Y., "Finite Displacement Plate Flexure by the Use of Matrix Incremental Approach," *International Journal for Numerical Methods in Engineering*, Vol. 4, 1972, pp. 415–432.

12.28. Levy, S., "Square Plate with Clamped Edges under Normal Pressure Producing Large Deflections," NACA TR-740, 1942.

12.29. Mei, C., and Yang, T. Y., "Free Vibrations of Finite Element Plates Subjected to Complex Middle-Plane Force Systems," *Journal of Sound and Vibration*, Vol. 23, No. 2, 1972, pp. 145–156.

12.30. Han, A. D., and Yang, T. Y., "Nonlinear Panel Flutter Using High-Order Triangular Elements," *AIAA Journal*, Vol. 21, No. 10, 1983, pp. 1453–1461.

12.31. Adini, A., "Analysis of Shell Structures by the Finite Element Method," Ph.D. thesis, Department of Civil Engineering, University of California at Berkeley, 1961.

12.32. Tocher, J. L., "Analysis of Plate Bending Using Triangular Elements," Ph.D. thesis, Department of Civil Engineering, University of California at Berkeley, 1962.

12.33. Clough, R. W., and Tocher, J. L., "Finite Element Stiffness Matrices for Analysis of Plate Bending," *Proceedings*, Conference on Matrix Methods in Structural Mechanics, AFFDL TR-66-80, Fairborn, Ohio, 1966, pp. 515–545.

12.34. Razzaque, A., "Program for Triangular Bending Elements with Derivative Smoothing," *International Journal for Numerical Methods in Engineering,* Vol. 6, 1973, pp. 333–343.

12.35. Argyris, J. H., Fried, I., and Scharpf, D. W., "The TUBA Family of Plate Elements for the Matrix Displacement Method," *The Aeronautical Journal of the Royal Aeronautical Society,* Vol. 72, 1968, pp. 701–709.

12.36. Cowper, G. R., Kosko, E., Lindberg, G. M., and Olson, M. D., "A High Precision Triangular Plate-Bending Element," Report LR-514, National Aeronautical Establishment, National Research Council of Canada, Ottawa, Dec. 1968.

12.37. Bell, K., "A Refined Triangular Plate Bending Finite Element," *International Journal for Numerical Methods in Engineering,* Vol. 1, 1969, pp. 101–122.

12.38. Bosshard, W., "Ein neues vollverträgliches endliches Element für Plattenbiegung," *International Association of Bridge Structural Engineering Bulletin,* Vol. 28, 1968, pp. 27–40.

12.39. Visser, W., "The Finite Element Method in Deformation and Heat Conduction Problems," Dr.W. dissertation, Technical University of Delft, The Netherlands, 1968.

12.40. Irons, B. M., "A Conforming Quartic Triangular Element for Plate Bending," *International Journal for Numerical Methods in Engineering,* Vol. 1, 1969, pp. 29–46.

12.41. Morley, L. S. D., "The Triangular Equilibrium Element in the Solution of Plate Bending Problems," *Aeronautical Quarterly,* Vol. 19, 1968, pp. 149–169.

12.42. Morley, L. S. D., "On the Constant Moment Plate Bending Element," *Journal of Strain Analysis,* Vol. 6, 1971, pp. 20–24.

12.43. Harvey, J. W., and Kelsey, S., "Triangular Plate Bending Elements with Enforced Compatibility," *AIAA Journal,* Vol. 9, 1971, pp. 1023–1026.

12.44. Fraeijs de Veubeke, B., "A Conforming Finite Element for Plate Bending," *International Journal for Solids and Structures,* Vol. 4, 1968, pp. 95–108.

12.45. Clough, R., and Felippa, C., "A Refined Quadrilateral Element for the Analysis of Plate Bending," *Proceedings,* 2nd Conference on Matrix Methods in Structural Mechanics, AFFDL TR-68-150, Oct. 1968, pp. 399–440.

12.46. Pian, T. H. H., "Derivation of Element Stiffness Matrices by Assumed Stress Distribution," *AIAA Journal,* Vol. 2, No. 7, July 1964, pp. 1333–1336.

12.47. Pian, T. H. H., "Element Stiffness Matrices for Prescribed Boundary Stresses," *Proceedings,* Conference on Matrix Methods in Structural Mechanics, AFFDL TR 66-80, Fairborn, Ohio, 1966, pp. 457–477.

12.48. Severn, R. T., and Taylor, P. R., "The Finite Element Method for Flexure of Slabs When Stress Distributions Are Assumed," *Proceedings,* Institution of Civil Engineers, Vol. 34, 1966, pp. 153–170.

12.49. Neale, B. K., Henshell, R. D., and Edwards, G., "Hybrid Plate Bending Elements," *Journal of Sound and Vibrations,* Vol. 23, No. 1, July 1972, pp. 101–112.

12.50. Allwood, R. J., and Cornes, G. M. M., "A Polygonal Finite Element for Plate Bending Problems Using the Assumed Stress Approach," *International Journal for Numerical Methods in Engineering,* Vol. 1, 1969, pp. 135–149.

12.51. Cook, R. D., "Two Hybrid Elements for the Analysis of Thick, Thin and Sandwich Plates," *International Journal for Numerical Methods in Engineering*, Vol. 5, No. 2, 1972, pp. 277-288.

12.52. Torbe, I., and Church, K., "A General Quadrilateral Plate Element," *International Journal for Numerical Methods in Engineering*, Vol. 9, 1975, pp. 856-868.

12.53. Allman, D. J., "Triangular Finite Elements for Plate Bending with Constant and Linearly Varying Bending Moments," *Proceedings*, IUTAM Conf. on High Speed Computing of Elastic Structures, Liège, Belgium, 1970, pp. 105-136.

12.54. Bartholomew, P., "Comment on Hybrid Finite Elements," *International Journal for Numerical Methods in Engineering*, Vol. 10, No. 4, 1976, pp. 968-973.

12.55. Pian, T. H. H., and Tong, P., "Basis of Finite Elements for Solid Continua," *International Journal for Numerical Methods in Engineering*, Vol. 1, 1969, pp. 3-28.

12.56. Spilker, R. L., and Munir, N. I., "The Hybrid-Stress Model for Thin Plates," *International Journal for Numerical Methods in Engineering*, Vol. 15, 1980, pp. 1239-1260.

12.57. Mau, S. T., and Witmer, E. A., "Static, Vibration and Thermal Stress Analysis of Laminated Plates and Shells by the Hybrid Stress Finite Element Method with Transverse Shear Deformation Effects Included," ASRL TR-169-2, M.I.T., Cambridge, Mass., 1972.

12.58. Pian, T. H. H., and Mau, S. T., "Some Recent Studies in Assumed-Stress Hybrid Models," in *Advances in Computational Methods in Structural Mechanics and Design*, ed. J. T. Oden, R. W. Clough, and Y. Yamamoto, University of Alabama Press, Birmingham, 1972.

12.59. Cook, R. D., "Some Elements for Analysis of Plate Bending," *Journal of the Engineering Mechanics Division*, ASCE, Vol. 98, No. EM6, 1972, pp. 1453-1470.

12.60. Cook, R. D., and Ladkany, S. G., "Observations Regarding Assumed Stress Hybrid Plate Elements," *International Journal for Numerical Methods in Engineering*, Vol. 8, 1974, pp. 513-519.

12.61. Herrmann, L. R., "A Bending Analysis of Plates," *Proceedings*, 1st Conference on Matrix Methods in Structural Mechanics, TR-66-80, Air Force Flight Dynamics Laboratory, Wright-Patterson Air Force Base, Ohio, 1965, pp. 577-604.

12.62. Bron, J., and Dhatt, G., "Mixed Quadrilateral Elements for Bending," *AIAA Journal*, Vol. 10, No. 10, 1972, pp. 1359-1361.

12.63. Abel, J. F., and Desai, C. S., "Comparison of Finite Elements for Plate Bending," *Journal of the Structural Division*, ASCE, Vol. 98, No. ST9, Sept. 1972, pp. 2143-2148.

12.64. Rossow, M. P., and Chen, K. C., "Computational Efficiency of Plate Elements," *Journal of the Structural Division*, ASCE, Vol. 103, No. ST2, Feb. 1977, pp. 447-451.

12.65. Dawe, D. J., "A Finite Element Approach to Plate Vibration Problems," *Journal of Mechanical Engineering and Sciences*, Vol. 7, 1965, pp. 28-32.

12.66. Gopalacharyulu, S., "A Higher Order Conforming Rectangular Element," *International Journal for Numerical Methods in Engineering,* Vol. 6, No. 2, 1973, pp. 305-308.

12.67. Irons, B. (comment on paper above), *International Journal for Numerical Methods in Engineering,* Vol. 6, No. 2, 1973, pp. 308-309.

12.68. Wegmuller, A., and Kostem, C., "Finite Element Analysis of Plates and Eccentrically Stiffened Plates," Fritz Eng. Laboratory Report 378 A.3, Lehigh University, Bethlehem, Pa., Feb. 1973.

12.69. Chu, T. C., and Schnobrich, W. C., "Finite Element Analysis of Translational Shells," *Computers and Structures,* Vol. 2, 1972, pp. 197-222.

12.70. Deak, A. L., and Pian, T. H. H., "Application of the Smooth-Surface Interpolation to the Finite-Element Analysis," *AIAA Journal,* Vol. 5, No. 1, 1967, pp. 187-189.

12.71. Birkhoff, G., and Garabedian, H. L., "Smooth Surface Interpolation," *Journal of Mathematics Physics,* Vol. 39, 1960, pp. 353-368.

12.72. Kikuchi, F., and Ando, Y., "Some Finite Element Solutions for Plate Bending Problems by Simplified Hybrid Displacement Method," *Nuclear Engineering and Design,* Vol. 23, 1972, pp. 155-178.

12.73. Anderheggen, E., "A Conforming Finite Element Plate Bending Solution," *International Journal for Numerical Methods in Engineering,* Vol. 2, No. 2, 1970, pp. 259-264.

12.74. Fraeijs de Veubeke, B., Sander, G., and Beckers, P., "Dual Analysis by Finite Elements: Linear and Nonlinear Applications," TR 72-93, Air Force Flight Dynamics Laboratory, Fairborn, Ohio, 1972.

12.75. Batoz, J. L., Bathe, K. J., and Ho, L. W., "A Study of Three-Node Triangular Plate Bending Elements," *International Journal for Numerical Methods in Engineering,* Vol. 15, 1980, pp. 1771-1812.

12.76. Batoz, J. L., and Tahar, M. B., "Evaluation of a New Quadrilateral Thin Plate Bending Element," *International Journal for Numerical Methods in Engineering,* Vol. 18, 1982, pp. 1655-1677.

PROBLEMS

12.1. Derive the stiffness coefficient k_{11} relating F_{z_1} to w_1 for the four-node 12-d.o.f. rectangle in bending using both Eqs. (12.27) and (12.32).

12.2. Derive the stiffness coefficient k_{35} relating F_{z_3} to w_{x_1} for the four-node 12-d.o.f. rectangle in bending using both Eqs. (12.33) and (12.34).

12.3. Derive the twelve work-equivalent loads for the four-node 12-d.o.f. rectangle in bending when subjected to (a) uniform load p (lb/in.^{2}) and (b) linearly varying load $p(x, y) = p_0 x/a$ (lb/in.^{2}).

12.4. Show that the cross-beam displacement function (12.42) for the 12-d.o.f. rectangle in bending satisfies interelement compatibility in w, $\partial w/\partial x$, and $\partial w/\partial y$.

12.5. Derive the stiffness coefficient k_{35} relating F_{z_3} to w_{x_1} for the 16-d.o.f. rectangle in bending using both Eqs. (12.46) and (12.34).

12.6. Derive the consistent mass matrix coefficient m_{34} and the incremental stiffness matrix coefficient n_{34} relating F_{z_3} to w_4 for the 16-d.o.f. rectangle in bending using the displacement function (12.44).

12.7. For a clamped square plate modeled using one rectangular element (both 12-term polynomial and 16-d.o.f.) per quadrant as shown in Fig. P12.7, find **(a)** the central deflection due to a central load P, **(b)** a fundamental natural frequency using both consistent and lumped mass matrices, and **(c)** the critical buckling value of the in-plane load N_x. List all these values in a table and compare them with closed-form solutions. *Note*: For a one-element model, there exists only one d.o.f.

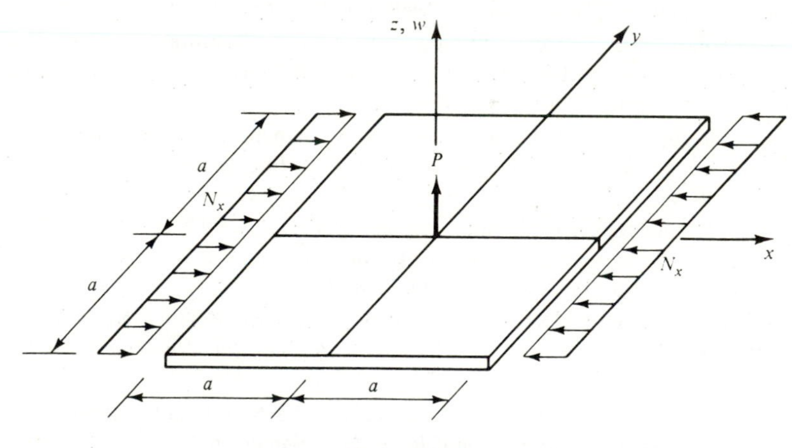

Figure P12.7

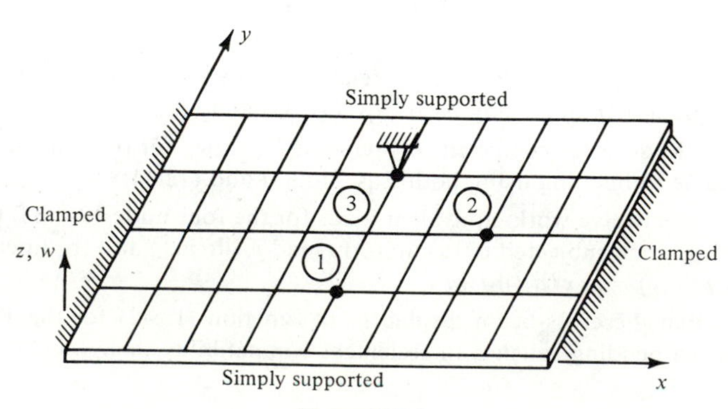

Figure P12.8

12.8. A rectangular plate is modeled using 32 equal-size 16-d.o.f. rectangles in bending as shown in Fig. P12.8. The plate is vibrating in its lowest mode. Identify the zero d.o.f.'s at nodes 1, 2, and 3.

12.9. A square plate with all edges simply supported is modeled using 18-d.o.f. triangles in bending with equal size as shown in Fig. P12.9. Identify all the zero d.o.f.'s at nodes 1, 2, and 3.

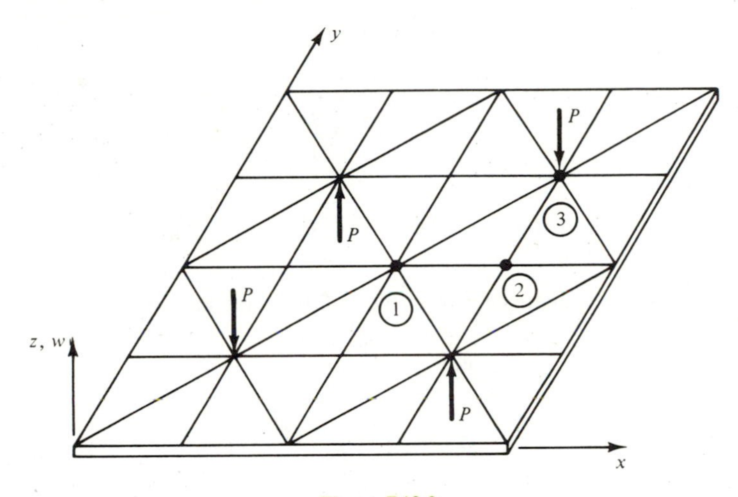

Figure P12.9

12.10. For the three-node nine-d.o.f. conforming (HCT) triangle in bending, the strain energy can be written as

$$U = \frac{1}{2}\begin{Bmatrix} \alpha_a \\ \alpha_b \\ \alpha_c \end{Bmatrix}^T_{27\times 1} [\bar{\mathbf{K}}] \begin{Bmatrix} \alpha_a \\ \alpha_b \\ \alpha_c \end{Bmatrix}_{27\times 1}$$

where $\{\boldsymbol{\alpha}_a\}$, $\{\boldsymbol{\alpha}_b\}$, and $\{\boldsymbol{\alpha}_c\}$ contain nine constants in the displacement function for subelements *a*, *b*, and *c*, respectively. Find the term $\bar{k}_{16\text{-}16}$ in the "kernel" stiffness matrix $[\bar{\mathbf{K}}]$ (see Example 12.5).

12.11. A nine-d.o.f. triangle in bending is shown in Figure P12.11. **(a)** Assume a coordinate system (x, y) and a displacement function $w(x, y)$ which satisfies interelement compatibility of $\partial w/\partial x$ and $\partial w/\partial y$ along edges 1-2 and 2-3, respectively. **(b)** Explain the reason for this assumption of the displacement function. **(c)** Does this displacement function satisfy interelement compatibility of $\partial w/\partial x$ and $\partial w/\partial y$ along edge 1-3? Explain why.

12.12. Derive the 16 work-equivalent loads for the four-node 16-d.o.f. rectangle in bending when subjected to **(a)** a uniform load p (lb/in.^{2}) and **(b)** a linearly varying load $p(x, y) = p_0 y/b$ (lb/in.^{2}).

12.13. Figure P12.13 shows a sector element for analyzing the bending behavior of circular plates. For this sector element, the four degrees of freedom at each node can be written as w, $\partial w/\partial r$, $(1/r)(\partial w/\partial \theta)$, and $(1/r)(\partial^2 w/\partial r\, \partial \theta)$. The strain

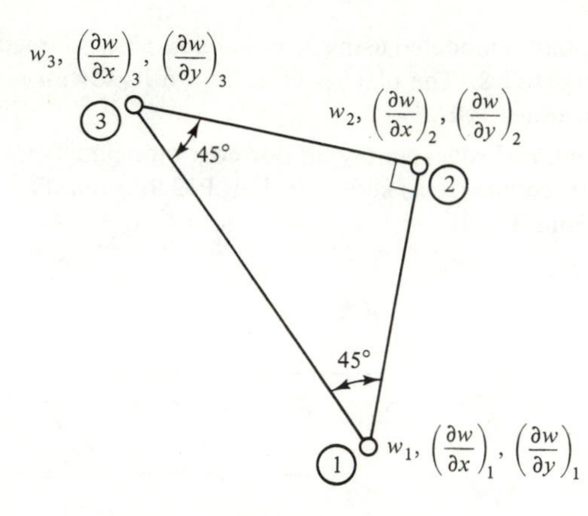

Figure P12.11

components at a distance z from the middle plane are

$$\epsilon_r = -z\frac{\partial^2 w}{\partial r^2}$$

$$\epsilon_\theta = -z\left(\frac{1}{r^2}\frac{\partial^2 w}{\partial \theta^2} + \frac{1}{r}\frac{\partial w}{\partial r}\right)$$

$$\gamma_{r\theta} = -2z\left(\frac{1}{r}\frac{\partial^2 w}{\partial r\,\partial \theta} - \frac{1}{r^2}\frac{\partial w}{\partial \theta}\right)$$

(a) Write the displacement functions in terms of Hermitian polynomials. **(b)** Derive the stiffness coefficient k_{34} that relates the nodal force F_{z_3} to the nodal deflection w_4.

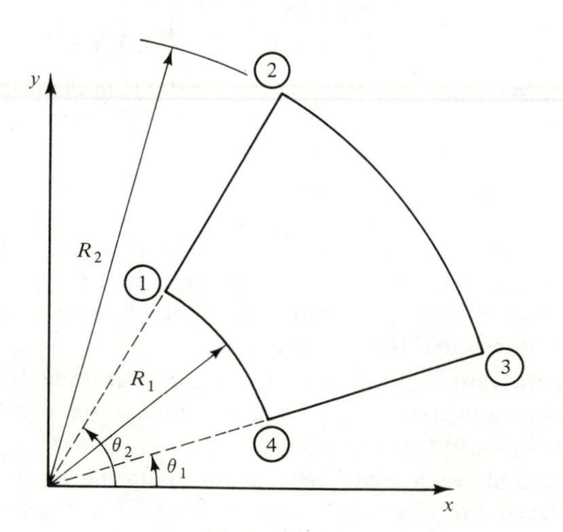

Figure P12.13

12.14. **(a)** Derive the term m_{34} in the consistent mass matrix for the element in Problem 12.13. **(b)** If the element is under the initial stresses N_{rr} and $N_{\theta\theta}$, derive the incremental stiffness coefficient n_{34}.

12.15. A plate element with the shape of a circular segment as shown in Fig. P12.15 is assumed to have three d.o.f.'s at each node. Assume the displacement function in polynomial form in terms of the polar coordinates (r, θ) such that the slopes normal to each of the two straight edges are fully compatible with the adjacent element.

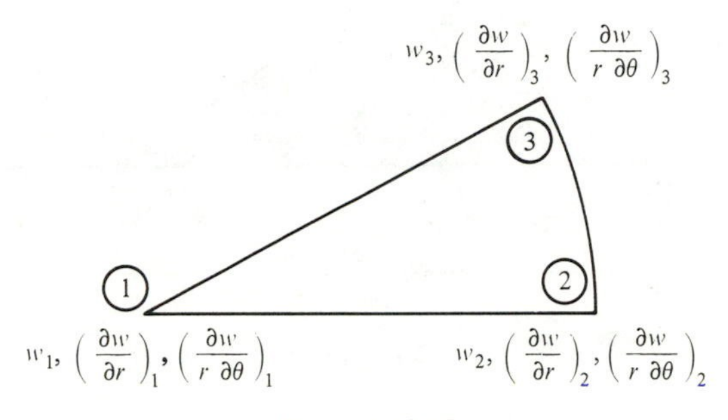

Figure P12.15

12.16. The element shown in Fig. P12.16 has 11 d.o.f.'s as shown. Edges 1-2 and 1-3 are parallel to the x and y axes, respectively. Write a polynomial displacement function in x and y which satisfies full compatibility of $\partial w/\partial x$ and $\partial w/\partial y$ along edges 1-3 and 1-2, respectively. Explain why this function satisfies such compatibility.

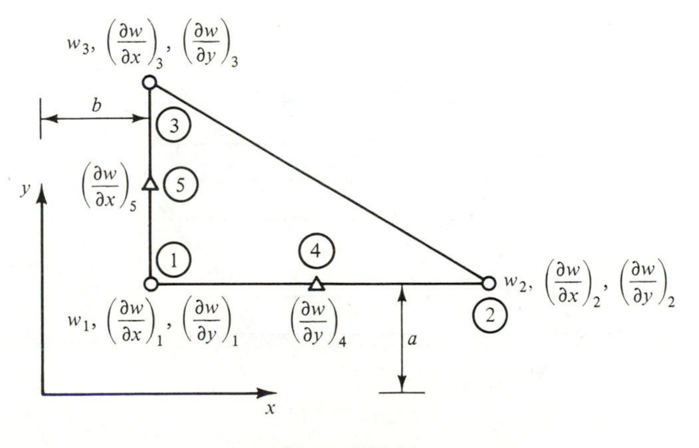

Figure P12.16

12.17. A system of two triangular plate bending elements is shown in Fig. P12.17. Both elements have three d.o.f.'s at each nodal point: w, $\partial w/\partial x$, and $\partial w/\partial y$. **(a)** Assume a nine-term polynomial displacement function for either element. **(b)** Do both elements satisfy full compatibility of $\partial w/\partial x$ along edge 1-2? Explain your answer. **(c)** Do both elements satisfy full compatibility of $\partial w/\partial y$ along edge 1-2? Explain your answer.

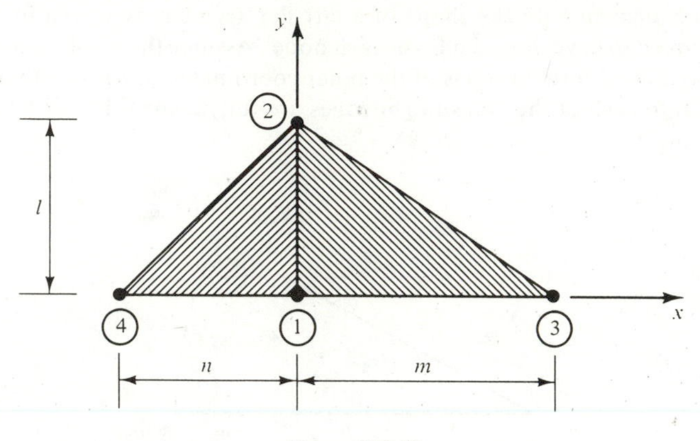

Figure P12.17

12.18. The rectangular plate shown in Fig. P12.18 is under a uniform line load. The plate is modeled using rectangular elements. We are to formulate a typical element 1-2-3-4 for this case. **(a)** What is the minimum number of required d.o.f.'s? What are they? **(b)** Assume a displacement function $w(x, y)$ in polynomial form with the minimum possible number of constants for the element. **(c)** Does this displacement function satisfy the interelement compatibility of w, $\partial w/\partial x$, $\partial w/\partial y$, or $\partial^2 w/\partial x\, \partial y$? Explain why.

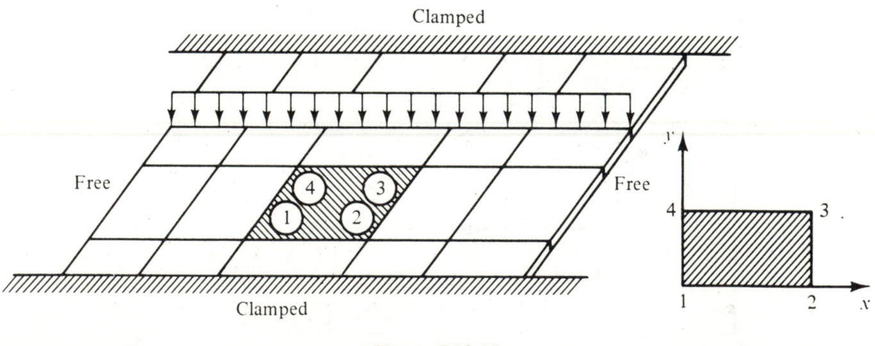

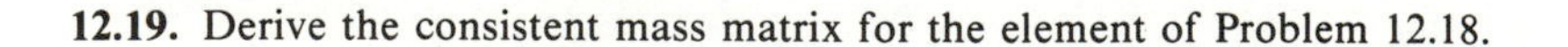

Figure P12.18

12.19. Derive the consistent mass matrix for the element of Problem 12.18.

CHAPTER 13

*Special-Purpose Finite Element Programs**

This chapter includes four selected basic special-purpose finite element programs:

1. Static analysis of plane truss and plane frame structures
2. Free-vibration analysis of plane truss and plane frame structures
3. Static analysis using six-d.o.f. triangular plane stress and plane strain finite elements
4. Static analysis using 16-d.o.f. rectangular plate finite elements in bending

As pointed out in Chapter 10, the listing and user's manual for a popular program using axisymmetric solid finite elements is available through U.S. National Technical Information Service [10.7]. It is highly recommended.

Associated with each program are a general description, a user's manual, sample problems, a listing of the program, and input and output data. The four programs are direct applications of the formulations and solution procedures described in Chapters 4, 5, 7, 9, and 12. This chapter is designed to provide the reader with the opportunity to learn the fundamental programming aspects and to gain further insight into the finite element method. Thus the reader will be better equipped to write his or her own finite element program and to understand and use existing special-purpose canned programs as well as large general-purpose finite element programs.

*Chapter 13 by Guru P. Guruswamy and T. Y. Yang.

13.1 STATIC ANALYSIS OF PLANE TRUSS AND PLANE FRAME STRUCTURES

This program is based on the use of the six-degree-of-freedom frame element described in Fig. 5.3. The one-dimensional beam element and truss bar element are treated as special cases of the general plane frame element. The program calls an IMSL Library Subroutine LEQ1PB, which solves matrix simultaneous equations based on the storage of band matrix and Gaussian elimination. If LEQ1PB is not available, the user can still use this program by simply changing the statement that calls LEQ1PB to the one that calls any available equivalent subroutine.

The input data are based on global coordinates. The output data for displacements u, v, and rotations θ are also based on global coordinates. The output data for the nodal forces and bending moments X_1, Y_1, M_1, X_2, Y_2, and M_2 are given in local coordinates with a local x axis along direction 1-2 for each element.

In this program, provision is made in the DIMENSION statement for 100 elements, 100 nodal points, and 100 free-to-move degrees of freedom. The user can easily modify the DIMENSION statements to cope with more elements, nodes, and free d.o.f.'s

13.1.1 Description of Input Data

1. Title Card (12A6)
   Columns 1–72: Arbitrary problem identification
2. Master Control Card (4I5)
   One card to define the number of nodal points and elements, and to specify print options
   Columns 1–5: Number of nodal points (NNOD)
   6–10: Number of elements (NELE)
   11–15: Flag to print the element stiffness and coordinate transformation matrices
   IPR1 = 1 print
   IPR1 = 0 no print
   16–20: Flag to print the assembled stiffness matrix
   IPR2 = 1 print
   IPR2 = 0 no print
3. Nodal Point Data (4I5, 5E10.4)
   One card for each nodal point in order
   Columns 1–5: Nodal point number (N)
   6–10: Boundary condition for u displacement in x direction
   IBOU(N, 1) = 1 free to move
   IBOU(N, 1) = 0 restrained

11–15: Boundary condition for v displacement in y direction
IBOU(N, 2) = 1 free to move
IBOU(N, 2) = 0 restrained
16–20: Boundary condition for rotation θ in xy plane
IBOU(N, 3) = 1 free to rotate
IBOU(N, 3) = 0 restrained
21–30: x coordinate [XNOD(N)]
31–40: y coordinate [YNOD(N)]
41–50: Force in x direction [FORC(N, 1)]
51–60: Force in y direction [FORC(N, 2)]
61–70: Bending moment in xy-plane [FORC(N, 3)]

4. Element Data (3I5, 3E10.4)
   One card for each element in order
   Columns 1–5: Element number (N)
   6–10: First end nodal point number [NODN(N, 1)]
   11–15: Second end nodal point number [NODN(N, 2)]
   16–25: Cross-sectional area [AREA(N)]
   26–35: Moment of inertia [SMOI(N)]
   36–45: Modulus of elasticity [EMOD(N)]

13.1.2 Sample Problems

Example 13.1 Static analysis of a five-bar truss

Figure 13.1 shows a four-node, five-bar steel truss under a concentrated load, similar to Problem 4.2. The nodal point number, element number, and the dimensions are shown in the figure. It is assumed that $E = 30{,}000$ ksi and that the cross-sectional areas for the five elements are $A_1 = A_4 = 0.2$ in.^{2}, $A_2 = A_5 = 0.12$ in.^{2}, and $A_3 = 0.08$ in.^{2}

The input data necessary for a static analysis of such a problem and the output data are given at the end of the listing of this program. It is noted that all input data and output data for displacements are based on global coordinates, whereas the output

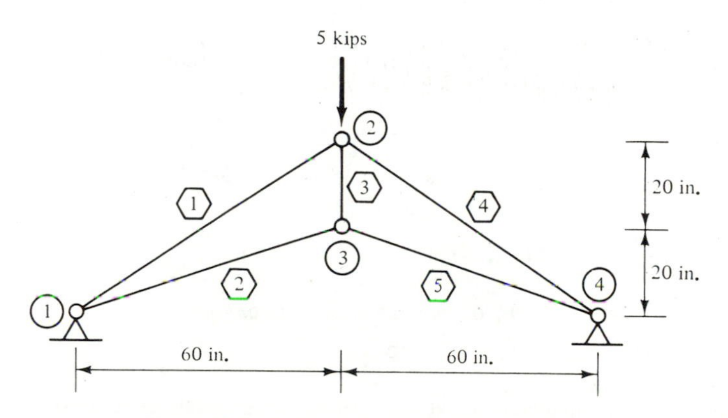

Figure 13.1 Static analysis of a five-bar truss.

data for the nodal forces are based on local coordinates with local x axis along direction 1-2 for each element.

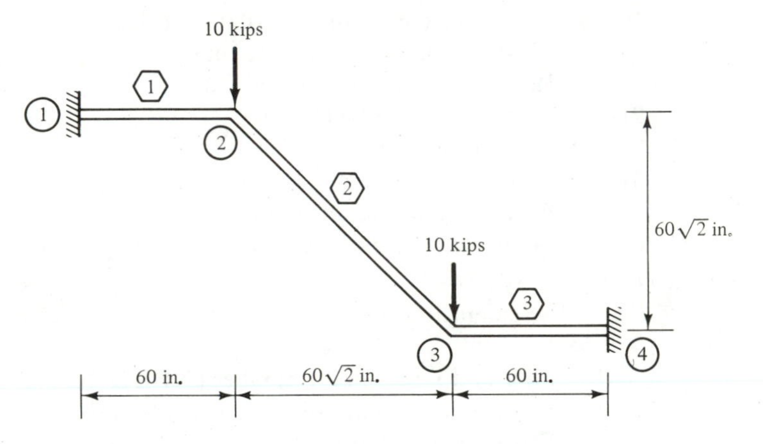

Figure 13.2 Static analysis of a three-member frame.

Example 13.2 Static analysis of a stairway frame

Figure 13.2 shows a three-element, four-node steel frame. The frame is assumed as inextensible by setting a very large value of A. It is assumed that $E = 30{,}000$ ksi and $I = 64.2$ in.^{4} for the 8I23 section.

The input data necessary for a static analysis of such a problem and the output data are given after those for Example 13.1. It is noted again that the output data for element nodal forces are based on local coordinates.

```
      PROGRAM MAIN
C     ************************************************************************
C     * PROGRAM FOR STATIC ANALYSIS OF PLANE TRUSS AND FRAME STRUCTURES *
C     ************************************************************************
      DIMENSION
     $ ESTF(6,6),TRAN(6,6),ESTT(6,6),DISN(3),TITLE(1Ø),
     $ TEMP(6),EDIS(6),EFOR(6),
     $ IBOU(1ØØ,3),XNOD(1ØØ),YNOD(1ØØ),FORC(1ØØ,3),
     $ NODN(1ØØ,2),AREA(1ØØ),SMOI(1ØØ),EMOD(1ØØ),ICOR(1ØØ,6),
     $ SYTF(1ØØ,1ØØ),SLOD(1ØØ,1)
C
C
C     READING AND PRINTING OF TITLE CARD
C
      READ(5,11)(TITLE(I),I=1,1Ø)
   11 FORMAT(1ØA8)
      WRITE(6,9)
    9 FORMAT(1H1)
      WRITE(6,12)(TITLE(I),I=1,1Ø)
   12 FORMAT(/1ØX,1ØA8)
C
C     READING AND PRINTING OF MASTER CONTROL CARD
C
      READ(5,21)NNOD,NELE,IPR1,IPR2
   21 FORMAT(8I5)
      WRITE(6,22)NNOD,NELE
   22 FORMAT(/5X,16HNUMBER OF NODES=,I5/5X,19HNUMBER OF ELEMENTS=,I5)
      IF(IPR1.EQ.1)WRITE(6,31)
```

```
   31 FORMAT(/1ØX,35HELEMENT STIFFNESS MATRIX IS PRINTED)
      IF(IPR2.EQ.1)WRITE(6,32)
   32 FORMAT(/1ØX,34HGLOBAL STIFFNESS MATRIX IS PRINTED)
C
C     READING AND PRINTING OF NODAL DATA
C
      WRITE(6,33)
   33 FORMAT(/5X,4HNODE,1X,2ØHBOUNDARY  CONDITIONS,9X,11HCOORDINATES,2ØX
     $ 6HFORCES)
      WRITE(6,34)
   34 FORMAT(7X,17HNO    1    2    3,11X,1HX,15X,1HY,15X,1H1,15X,1H2,15X
     $,1H3)
      DO 5Ø I=1,NNOD
      READ(5,42)N,(IBOU(I,J),J=1,3),XNOD(I),YNOD(I),(FORC(I,J),J=1,3)
   42 FORMAT(4I5,5E1Ø.4)
      WRITE (6,43)I,(IBOU(I,J),J=1,3),XNOD(I),YNOD(I),(FORC(I,J),J=1,3)
   43 FORMAT(/4X,4I5,5F15.4)
   5Ø CONTINUE
C
C     READING AND PRINTING OF ELEMENT DATA
C
      WRITE(6,49)
   49 FORMAT(/6X,7HELEMENT,8X,9HEND NODES,8X,4HAREA,11X,4HM.I.,13X,1HE)
      WRITE(6,51)
   51 FORMAT( 11X,2HNO,5X,1ØH    1    2)
      DO 6Ø I=1,NELE
      READ(5,52)N,NODN(I,1),NODN(I,2),AREA(I),SMOI(I),EMOD(I)
   52 FORMAT(3I5,3E1Ø.4)
      WRITE(6,53)N,NODN(I,1),NODN(I,2), AREA(I),SMOI(I),EMOD(I)
   53 FORMAT(/8X,I5,5X,2I5,3F15.4)
   6Ø CONTINUE
C
C     GENERATION OF NUMBERS FOR ASSEMBLING GLOBAL STIFFNESS MATRIX
C
      ICON=Ø
      DO 2Ø I=1,NNOD
      DO 2Ø J=1,3
      K=IBOU(I,J)
      IF (K.EQ.Ø) GO TO 2Ø
      ICON=ICON+1
      IBOU(I,J)=ICON
   2Ø CONTINUE
      NDOF=ICON
      WRITE(6,54)NDOF
   54  FORMAT(/5X,22HNUMBER OF FREE D.O.F.=,I5)
      DO 3Ø I=1,NELE
      I1=NODN(I,1)
      I2=NODN(I,2)
      DO 3Ø J=1,3
      ICOR(I,J)=IBOU (I1,J)
      ICOR(I,J+3)=IBOU(I2,J)
   3Ø CONTINUE
      IF(IPR1.EQ.Ø) GO TO 75
      WRITE(6,61)
   61 FORMAT (/5X,7HELEMENT, 5X,24HNODAL DEGREES OF FREEDOM)
      WRITE(6,63)
   63 FORMAT( 5X,36HNUMBER    1    2    3    4    5    6)
      DO 7Ø I=1,NELE
      WRITE(6,62)I,(ICOR(I,J),J=1,6)
   62 FORMAT(/6X,7I5)
   7Ø CONTINUE
   75 CONTINUE
C
C     COMPUTATION OF MAXIMUM BAND WIDTH
      MAX=Ø
      DO 4Ø I=1,NELE
      J=NODN(I,1)-NODN(I,2)
      J=IABS(J)
      IF(J.GT.MAX)MAX=J
   4Ø CONTINUE
      NBND=MAX*6+5
      IF(NBND.GT.NDOF)NBND=NDOF
      WRITE(6,41)NBND
   41 FORMAT(/5X,19HMAXIMUM BAND WIDTH=,I5)
C
```

```
C     INTIALISING GLOBAL STIFFNESS MATRIX TO ZERO
C
      DO 8Ø I=1,NDOF
      DO 8Ø J=1,NBND
   8Ø SYTF(I,J)=Ø.Ø
C
C
C
C     GENERATION OF TRANSFORMATION MATRIX AND ELEMENT STIFFNESS MATRIX A
C     WRITING ON TAPE 1.
C
      REWIND 1
      DO 4ØØ IE=1,NELE
C
C     GENERATION OF TRANSFORMATION MATRIX.
C
      I1=NODN(IE,1)
      I2=NODN(IE,2)
      X1=XNOD(I1)
      X2=XNOD(I2)
      Y1=YNOD(I1)
      Y2=YNOD(I2)
      CALL TRNMAT(X1,Y1,X2,Y2,AL,TRAN)
      IF(IPR1.EQ.1) WRITE(6,1Ø1)IE,((TRAN(I,J),J=1,6),I=1,6)
  1Ø1 FORMAT (/5X,33HTRANSFORMATION MATRIX FOR ELEMENT,I5/6(/1ØX,
     $ 6E15.6))
C
C     GENERATION OF ELEMENT STIFFNESS MATRIX
C
      AA=AREA(IE)
      AI=SMOI(IE)
      AE=EMOD(IE)
      CALL ELESTF (AL,AA,AI,AE,ESTF)
      IF(IPR1.EQ.1)WRITE(6,1Ø2) IE,((ESTF(I,J),J=1,6),I=1,6)
  1Ø2 FORMAT(/5X,35HELEMENT STIFFNESS MATRIX OF ELEMENT,I5/6(/1ØX,
     $ 6E15.6))
C
C     TRANSFORMATION OF ELEMENT STIFFNESS MATRIX TO GLOBAL COORDINATES
C
       CALL CONTRN(TRAN,ESTF,ESTT,6,6)
      IF(IPR1.EQ.1)WRITE(6,1Ø3)IE,((ESTT(I,J),J=1,6),I=1,6)
  1Ø3 FORMAT(/5X,47HTRANSFORMED ELEMEMT STIFFNESS MATRIX OF ELEMENT,
     $I5/6(/1ØX,6E15.6))
C
C     WRITING AL,AA,AI,AE,ESTF,TRAN ON TAPE 1
C
      WRITE(1,1Ø6)AL,AA,AI,AE,((ESTF(I,J),J=1,6),I=1,6),((TRAN(I,J),
     $J=1,6),I=1,6)
  1Ø6 FORMAT(5E15.8)
C
C     ASSEMBLING THE GLOBAL STIFFNESS MATRIX
C
      DO 2ØØ I=1,6
      DO 2ØØ J=1,6
      K=ICOR(IE,I)
      L=ICOR(IE,J)
      IF (K*L.EQ.Ø) GO TO 2ØØ
      IF(K.LT.L) GO TO 2ØØ
      M=NBND-K+L
      IF(M.LE.Ø) GO TO 2ØØ
      SYTF(K,M)=SYTF(K,M)+ESTT(I,J)
  2ØØ CONTINUE
  4ØØ CONTINUE
      IF(IPR2.EQ.Ø) GO TO 23Ø
  21Ø CONTINUE
      WRITE(6,2Ø1)
  2Ø1 FORMAT(/5X,24H GLOBAL STIFFNESS MATRIX)
      DO 22Ø I=1,NDOF
      WRITE(6,2Ø2) I
  2Ø2 FORMAT(/5X,I5,3HROW)
      WRITE(6,2Ø3)(SYTF(I,J),J=1,NBND)
  2Ø3 FORMAT(5X,1ØE12.4)
  22Ø CONTINUE
  23Ø CONTINUE
```

```
C
C      ASSEMBLING LOAD VECTOR
C
       DO 5ØØ I=1,NNOD
       DO 5ØØ J=1,3
       K=IBOU(I,J)
       IF(K.EQ.Ø) GO TO 5ØØ
       SLOD(K,1)=FORC(I,J)
  5ØØ  CONTINUE
       IF(IPR2.EQ.1)WRITE(6,5Ø1)
  5Ø1  FORMAT(/5X,21HASSEMBLED LOAD VECTOR)
       IF(IPR2.EQ.1)WRITE(6,5Ø2)  (SLOD(I,1),I=1,NDOF)
  5Ø2  FORMAT(/5X,1ØE12.4)
C
C      COMPUTATION OF GLOBAL DISPLACEMENTS
C
       NC=NBND-1
       CALL LEQ1PB(SYTF,NDOF,NC,1ØØ,SLOD,1ØØ,1,2Ø,D1,D2,IER)
       IF(IER.EQ.129) WRITE(6,5Ø7)
       IF(IER.EQ.129) GO TO 999
  5Ø7  FORMAT(1ØX,4ØHSTIFFNESS MATRIX IS SINGULAR  HENCE STOP)
       WRITE(6,5Ø6)
  5Ø6  FORMAT(/5X,19HNODAL DISPLACEMENTS)
       WRITE(6,5Ø3)
  5Ø3  FORMAT(/6X,4HNODE,15X,13HDISPLACEMENTS)
       WRITE(6,5Ø4)
  5Ø4  FORMAT(8X,2HNO,11X,1HU,13X,1HV,15X,3HTHT)
       DO 7ØØ I=1,NNOD
       DO 6ØØ J=1,3
       DISN(J)=Ø.Ø
       K=IBOU(I,J)
       IF(K.EQ.Ø) GO TO 6ØØ
       DISN(J)=SLOD(K,1)
  6ØØ  CONTINUE
       WRITE(6,6Ø1)I,(DISN(L),L=1,3)
  6Ø1  FORMAT(/5X,I5,3F15.8)
  7ØØ  CONTINUE
       REWIND 1
C
C       COMPUTATION OF ELEMENT FORCES AND AXIAL STRESSES
C
       WRITE(6,8Ø4)
  8Ø4  FORMAT(/5X,57H ELEMENT FORCES AND AXIAL STRESSES (IN LOCAL COORDIN
      $ATES))
       WRITE(6,8Ø1)
  8Ø1  FORMAT(/5X,7HELEMENT,3ØX,6HFORCES,3ØX,12HAXIAL STRESS)
       WRITE(6,8Ø2)
  8Ø2  FORMAT(5X,6HNUMBER,5X,2HX1,8X,2HY1,8X,2HM1,5X,2HX2,8X,2HY2,8X,
      $2HM2)
       DO 9ØØ IE=1,NELE
       READ(1,1Ø6)AL,AA,AI,AE,((ESTF(I,J),J=1,6),I=1,6),((TRAN(I,J),J=1,6
      $),I=1,6)
       DO 8ØØ I=1,6
       TEMP(I)=Ø.Ø
       K=ICOR(IE,I)
       IF(K.EQ.Ø) GO TO 8ØØ
       TEMP(I)=SLOD(K,1)
  8ØØ  CONTINUE
C
C       TRANSFORMING ELEMENT DISPLACEMENTS FROM GLOBAL COORDINATES TO
C      ELEMENTS COORDINATES
C
       CALL MATMUL(TRAN,TEMP,EDIS,6,6)
C
C       COMPUTATION OF ELEMENT FORCES
C
        CALL MATMUL(ESTF,EDIS,EFOR,6,6)
       ELSR=Ø.Ø
       IF(AA.EQ.Ø.Ø) GO TO 81Ø
       C1=EFOR(1)
       C2=EFOR(4)
       C3=1.Ø
       IF(C1.GT.C2)C3=-1.Ø
       ELSR=EFOR(1)/AA
```

```
      ELSR=ABS(ELSR)
      ELSR=ELSR*C3
  81Ø CONTINUE
      WRITE(6,8Ø3)IE,(EFOR(I),I=1,6),ELSR
  8Ø3 FORMAT(/5X,I5,6F1Ø.4,7X,F1Ø.4)
  9ØØ CONTINUE
  999 CONTINUE
      STOP
      END
      SUBROUTINE TRNMAT(X1,Y1,X2,Y2,AL,T)
      DIMENSION T(6,6)
      REAL LMBD,MU
C     SUBROUTINE TO COMPUTE TRANSFORMATION MATRIX T(6,6)
C     AL=LENGTH
C     LMBD=COSINE OF THE ANGLE
C     MU=SINE OF THE ANGLE
      AL=SQRT((X2-X1)**2+(Y2-Y1)**2)
      LM24=(X2-X1)/AL
      MU=(Y2-Y1)/AL
      LMBD=(X2-X1)/AL
      MU=(Y2-Y1)/AL
      DO 1ØØ I=1,6
      DO 1ØØ J=1,6
  1ØØ T(I,J)=Ø.Ø
      T(1,1)=LMBD
      T(2,1)=-MU
      T(1,2)=MU
      T(2,2)=LMBD
      T(3,3)=1.
      T(4,4)=LMBD
      T(5,4)=-MU
      T(4,5)=MU
      T(5,5)=LMBD
      T(6,6)=1.
       RETURN
       END
      SUBROUTINE CONTRN(A,B,C,M,N)
C
C      SUBROUTINE  FOR CONGRUENT TRANSFORMATION C=(TRANPOSE A)*B*A
C
      DIMENSION A(6,6),B(6,6),C(6,6),D(6,6),E(6,6)
      DO 1ØØ I=1,M
      DO 1ØØ J=1,N
      D(I,J)=Ø.Ø
      DO 1ØØ K=1,M
  1ØØ D(I,J)=D(I,J)+B(I,K)*A(K,J)
      DO 2ØØ I=1,M
      DO 2ØØ J=1,N
  2ØØ E(I,J)=A(J,I)
      DO 3ØØ I=1,N
      DO 3ØØ J=1,N
      C(I,J)=Ø.Ø
      DO 3ØØ K=1,M
  3ØØ C(I,J)=C(I,J)+E(I,K)*D(K,J)
      RETURN
      END
      SUBROUTINE ELESTF(AL,AA,AI,AE,ESTF)
      DIMENSION ESTF(6,6)
C
C     SUBROUTINE FOR ELEMENT STIFFNESS  MATRIX
C     SIZE OF STIFFNESS MATRIX IS(6,6)
C     1ST D.O.F=AXIAL DISPLACEMENT AT NODE 1
C     2ND D.O.F.=TRANSVERSE DISPLACEMENT AT NODE 1
C     3RD D.O.F. =TRANSVERSE ROTATION AT NODE 1
C     4TH D.O.F=AXIAL DISPLACEMENT AT NODE 2
C     5TH D.O.F.=TRANSVERSE DISPLACEMENT AT NODE 2
C     6TH D.O.F. =TRANSVERSE ROTATION AT NODE 2
C     AA=AREA
C     AL=LENGTH
C     AE=ELASTIC MODULUS
C     AI=MOMENT OF INERTIA
C     ESTF=ELEMENT STIFFNESS
C
      DO 1Ø I=1,6
```

```
      DO 1Ø J=1,6
   1Ø ESTF(I,J)=Ø.Ø
      ESTF(1,1)=AA*AE/AL
      ESTF(4,1)=-ESTF(1,1)
      ESTF(2,2)=12.Ø*AE*AI/AL/AL/AL
      ESTF(3,2)=+6.Ø*AE*AI/AL/AL
      ESTF(4,2)=Ø.Ø
      ESTF(5,2)=-ESTF(2,2)
      ESTF(6,2)=ESTF(3,2)
      ESTF(3,3)=4.Ø*AE*AI/AL
      ESTF(5,3)=-ESTF(3,2)
      ESTF(6,3)=ESTF(3,3)/2.Ø
      ESTF(4,4)=ESTF(1,1)
      ESTF(5,5)=ESTF(2,2)
      ESTF(6,5)=+ESTF(5,3)
      ESTF(6,6)=ESTF(3,3)
      DO 1ØØ I=1,6
      K=I+1
      DO 1ØØ J=K,6
  1ØØ ESTF(I,J)=ESTF(J,I)
      RETURN
       END
      SUBROUTINE MATMUL(A,B,C,N,M)
      DIMENSION A(N,M),B(M),C(N)
C
C     SUBROUTINE FOR MATRIX MULTIPLICATION C(N)=A(N,M)*B(M)
C
      DO 1Ø I=1,N
      C(I)=Ø.Ø
      DO 1Ø J=1,M
   1Ø C(I)=C(I)+A(I,J)*B(J)
      RETURN
      END
STATIC ANALYSIS OF A TRUSS
 4    5
 1
 2    1    1    Ø+.6ØØØE+Ø2+.4ØØØE+Ø2+.ØØØØE+ØØ-.5ØØØE+Ø1+.ØØØØE+ØØ
 3    1    1    Ø+.6ØØØE+Ø2+.2ØØØE+Ø2+.ØØØØE+ØØ+.ØØØØE+ØØ+.ØØØØE+ØØ
 4    Ø    Ø    Ø+.12ØØE+Ø3
 1    1    2+.2ØØØE+ØØ+.ØØØØE+ØØ+.3ØØØE+Ø5
 2    1    3+.12ØØE+ØØ+.ØØØØE+ØØ+.3ØØØE+Ø5
 3    2    3+.8ØØØE-Ø1+.ØØØØE+ØØ+.3ØØØE+Ø5
 4    2    4+.2ØØØE+ØØ+.ØØØØE+ØØ+.3ØØØE+Ø5
 5    3    4+.12ØØE+ØØ+.ØØØØE+ØØ+.3ØØØE+Ø5
```

STATIC ANALYSIS OF A TRUSS

NUMBER OF NODES= 4
NUMBER OF ELEMENTS= 5

NODE NO	BOUNDARY CONDITIONS 1	2	3	COORDINATES X	Y	FORCES 1	2	3
1	Ø	Ø	Ø	Ø.ØØØØ	Ø.ØØØØ	Ø.ØØØØ	Ø.ØØØØ	Ø.ØØØØ
2	1	1	Ø	6Ø.ØØØØ	4Ø.ØØØØ	Ø.ØØØØ	-5.ØØØØ	Ø.ØØØØ
3	1	1	Ø	6Ø.ØØØØ	2Ø.ØØØØ	Ø.ØØØØ	Ø.ØØØØ	Ø.ØØØØ
4	Ø	Ø	Ø	12Ø.ØØØØ	Ø.ØØØØ	Ø.ØØØØ	Ø.ØØØØ	Ø.ØØØØ

ELEMENT NO	END NODES 1	2	AREA	M.I.	E
1	1	2	Ø.2ØØØ	Ø.ØØØØ	3ØØØØ.ØØØØ
2	1	3	Ø.12ØØ	Ø.ØØØØ	3ØØØØ.ØØØØ
3	2	3	Ø.Ø8ØØ	Ø.ØØØØ	3ØØØØ.ØØØØ
4	2	4	Ø.2ØØØ	Ø.ØØØØ	3ØØØØ.ØØØØ
5	3	4	Ø.12ØØ	Ø.ØØØØ	3ØØØØ.ØØØØ

NUMBER OF FREE D.O.F.= 4

MAXIMUM BAND WIDTH= 4

NODAL DISPLACEMENTS

NODE NO	DISPLACEMENTS U	V	THT
1	Ø.ØØØØØØØØ	Ø.ØØØØØØØØ	Ø.ØØØØØØØØ
2	Ø.ØØØØØØØØ	-Ø.Ø8116768	Ø.ØØØØØØØØ
3	Ø.ØØØØØØØØ	-Ø.Ø7413465	Ø.ØØØØØØØØ
4	Ø.ØØØØØØØØ	Ø.ØØØØØØØØ	Ø.ØØØØØØØØ

ELEMENT FORCES AND AXIAL STRESSES (IN LOCAL COORDINATES)

ELEMENT	FORCES						AXIAL STRESS
NUMBER	X1	Y1	M1	X2	Y2	M2	
1	3.7462	Ø.ØØØØ	Ø.ØØØØ	-3.7462	Ø.ØØØØ	Ø.ØØØØ	-18.731Ø
2	1.3344	Ø.ØØØØ	Ø.ØØØØ	-1.3344	Ø.ØØØØ	Ø.ØØØØ	-11.12Ø2
3	Ø.844Ø	Ø.ØØØØ	Ø.ØØØØ	-Ø.844Ø	Ø.ØØØØ	Ø.ØØØØ	-1Ø.5495
4	3.7462	Ø.ØØØØ	Ø.ØØØØ	-3.7462	Ø.ØØØØ	Ø.ØØØØ	-18.731Ø
5	1.3344	Ø.ØØØØ	Ø.ØØØØ	-1.3344	Ø.ØØØØ	Ø.ØØØØ	-11.12Ø2

```
STATIC ANALYSIS OF A FRAME
 4    3
 1    Ø    Ø    Ø
 2    1    1    1+.6ØØØE+Ø2+.ØØØØE+ØØ+.ØØØØE+ØØ-.1ØØØE+Ø2
 3    1    1    1+.1448E+Ø3-.8484E+Ø2+.ØØØØE+ØØ-.1ØØØE+Ø2
 4    Ø    Ø    Ø+.2Ø48E+Ø3-.8484E+Ø2
 1    1    2+.1ØØØE+Ø9+.642ØE+Ø2+.3ØØØE+Ø5
 2    2    3+.1ØØØE+Ø9+.642ØE+Ø2+.3ØØØE+Ø5
 3    3    4+.1ØØØE+Ø9+.642ØE+Ø2+.3ØØØE+Ø5
```

```
      STATIC ANALYSIS OF A FRAME

NUMBER OF NODES=    4
NUMBER OF ELEMENTS=    3

NODE BOUNDARY  CONDITIONS        COORDINATES                     FORCES
  NO   1    2    3           X             Y             1            2            3

   1   Ø    Ø    Ø         Ø.ØØØØ        Ø.ØØØØ        Ø.ØØØØ       Ø.ØØØØ       Ø.ØØØØ

   2   1    1    1        6Ø.ØØØØ        Ø.ØØØØ        Ø.ØØØØ     -1Ø.ØØØØ       Ø.ØØØØ

   3   1    1    1       144.8ØØØ      -84.84ØØ        Ø.ØØØØ     -1Ø.ØØØØ       Ø.ØØØØ

   4   Ø    Ø    Ø       2Ø4.8ØØØ      -84.84ØØ        Ø.ØØØØ       Ø.ØØØØ       Ø.ØØØØ

 ELEMENT        END NODES       AREA           M.I.          E
     NO           1   2

     1            1   2 1ØØØØØØØØ.ØØØØ       64.2ØØØ    3ØØØØ.ØØØØ

     2            2   3 1ØØØØØØØØ.ØØØØ       64.2ØØØ    3ØØØØ.ØØØØ

     3            3   4 1ØØØØØØØØ.ØØØØ       64.2ØØØ    3ØØØØ.ØØØØ

NUMBER OF FREE D.O.F.=    6

MAXIMUM BAND WIDTH=    6

NODAL DISPLACEMENTS

 NODE               DISPLACEMENTS
  NO          U              V               THT

   1     Ø.ØØØØØØØØØ    Ø.ØØØØØØØØØ    Ø.ØØØØØØØØØ

   2     Ø.ØØØØØØØØØ   -Ø.23361775    -Ø.ØØ467199

   3     Ø.ØØØØØØØØØ   -Ø.23361775     Ø.ØØ467199

   4     Ø.ØØØØØØØØØ    Ø.ØØØØØØØØØ    Ø.ØØØØØØØØØ

 ELEMENT FORCES AND AXIAL STRESSES (IN LOCAL COORDINATES)

ELEMENT                            FORCES                                 AXIAL STRESS
NUMBER     X1        Y1        M1      X2        Y2        M2

   1     Ø.ØØØØ   1Ø.ØØØØ  449.971Ø   Ø.ØØØØ  -1Ø.ØØØØ  15Ø.Ø29Ø          Ø.ØØØØ

   2     Ø.ØØØØ    Ø.ØØØØ -15Ø.Ø29Ø   Ø.ØØØØ    Ø.ØØØØ  15Ø.Ø29Ø          Ø.ØØØØ

   3     Ø.ØØØØ  -1Ø.ØØØØ -15Ø.Ø29Ø   Ø.ØØØØ   1Ø.ØØØØ -449.971Ø          Ø.ØØØØ
```

13.2 FREE-VIBRATION ANALYSIS OF PLANE TRUSS AND PLANE FRAME STRUCTURES

This program is based on the use of the six-degree-of-freedom frame element described in Fig. 7.10. The one-dimensional beam element and truss bar element are treated as special cases of the general plane frame element. Provision is given for inextensible members with relatively large axial stiffness ($10^6 \times$ real values). Options are provided for either lumped or consistent mass matrix. The eigenvalue and eigenvector solution is computed by calling an IMSL Library Subroutine EIGZF. If EIGZF is not available, the user can still use this program by simply changing the statement that calls EIGZF to the one that calls any available equivalent subroutine.

In this program, provision is made in the DIMENSION statement for 100 elements, 100 nodal points, and 100 free-to-move degrees of freedom. The user can easily modify the DIMENSION statement to cope with more elements, nodes, and free d.o.f.'s.

13.2.1 Description of Input Data

1. Title Card (10A8)
   Columns 1-80: Arbitrary problem identification
2. Master Control Card (7I5)
   One card to define the number of nodal points and elements, desired number of modes to be printed, option for lumped or consistent mass matrix, type of element, and to specify print options
   Columns 1-5: Number of nodal points (NNOD)
   6-10: Number of elements (NELE)
   11-15: Desired number of modes to be printed (NMOD)
   16-20: Flag to select lumped or consistent mass matrix
   LMAS = 1 for lumped mass matrix
   LMAS = 2 for consistent mass matrix
   21-25: Flag to select type of element
   ICAS = 1 for truss bar element
   ICAS = 2 for general frame element
   ICAS = 3 for inextensible frame element with infinite axial rigidity
   26-30: Flag to print the element stiffness, mass, and coordinate transformation matrices
   IPR1 = 1 print
   IPR1 = 0 no print
   31-35: Flag to print the assembled stiffness and mass matrices
   IPR2 = 1 print
   IPR2 = 0 no print

3. Nodal Point Data (4I5, 5E10.4)
   One card for each nodal point in order
   Columns 1-5: Nodal point number (N)
   6-10: Boundary condition for u displacement in x direction
   IBOU(N, 1) = 1 free to move
   IBOU(N, 1) = 0 restrained
   11-15: Boundary condition for v displacement in y direction
   IBOU(N, 2) = 1 free to move
   IBOU(N, 2) = 0 restrained
   16-20: Boundary condition for rotation θ in xy plane
   IBOU(N, 3) = 1 free to rotate
   IBOU(N, 3) = 0 restrained
   21-30: x coordinate [XNOD(N)]
   31-40: y coordinate [YNOD(N)]
   41-50: Mass in x direction [CMAS(N, 1)]
   51-60: Mass in y direction [CMAS(N, 2)]
   61-70: Mass moment of inertia [CMAS(N, 3)]

4. Element Data (3I5, 4E10.4)
   One card for each element in order
   Columns 1-5: Element number (N)
   6-10: First end nodal point number [NODN(N, 1)]
   11-15: Second end nodal point number [NODN(N, 2)]
   16-25: Cross-sectional area [AREA(N)]
   26-35: Moment of inertia [SMOI(N)]
   36-45: Modulus of elasticity [EMOD(N)]
   46-55: Mass density per unit volume [EROW(N)]

13.2.2 Sample Problems

Example 13.3 Free vibration of a simply supported beam

Figure 13.3 shows an A-36 steel I beam with both ends simply supported and with a cross section of 8I23. The problem is defined by the following constants: $A = 6.71$ in.^{2}, $I = 64.2$ in.^{4}, $E = 30 \times 10^6$ psi, and $\rho = 0.000733$ lb-sec^2/in.^{4}.

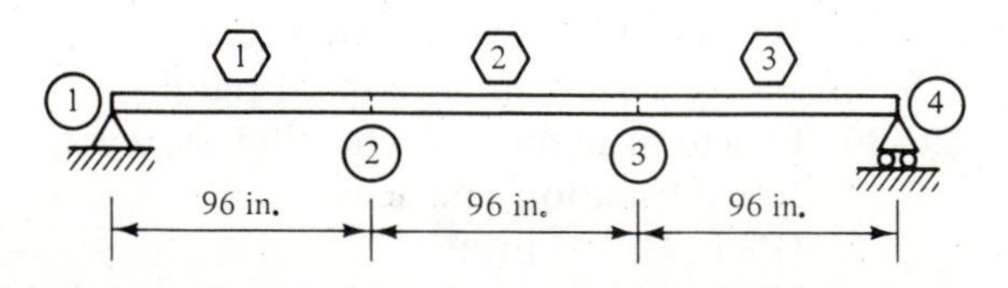

Figure 13.3 Free-vibration analysis of a simply supported beam.

The beam is modeled using three beam elements. Three natural frequencies and corresponding mode shapes are asked. The input data necessary for a free-vibration analysis of this beam and the output data are given at the end of the listing of this program. It is found that compared with the exact solution $[\omega_n = (n\pi/L)^2\sqrt{EI/\rho A}]$, the errors are 0.081%, 1.182%, and 11%, respectively, the same as those shown in Table 7.1.

Example 13.4 Free vibration of a portal frame

Figure 13.4 shows a four-node, three-element steel portal frame with both ends fixed and with cross section of 3I5.7. The problem is thus defined by the following constants: $A = 1.64$ in.^{2}, $I = 2.50$ in.^{4}, $E = 30 \times 10^6$ psi, and $\rho = 0.000733$ lb-sec^2/in.^{4}. The frame is assumed to be inextensible with very large axial stiffness for each member.

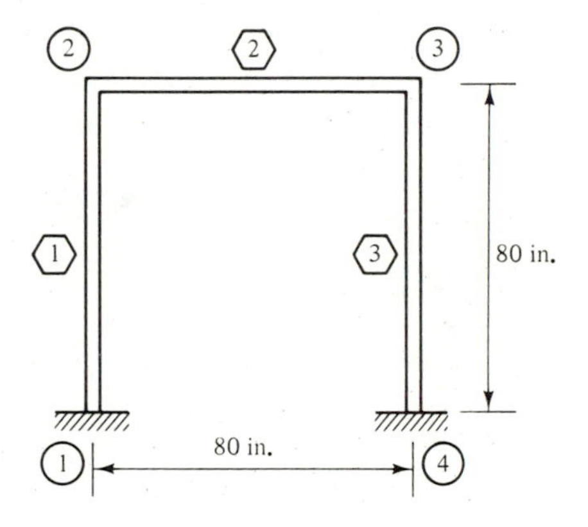

Figure 13.4 Free-vibration analysis of a portal frame with both ends fixed.

Two natural frequencies and modes are asked. The input data necessary for a free-vibration analysis of this frame and the output data are given after those for Example 13.3. From the eigenvectors printed, it is seen that the first mode is of antisymmetrical shape and the second mode is of symmetrical shape, as shown in Fig. 7.12.

```
      PROGRAM MAIN
C    ******************************************************************
C    * PROGRAM FOR VIBRATION ANALYSIS OF PLANE TRUSSES AND FRAMES *
C    ******************************************************************
      DIMENSION ESTF(6,6),TRNS(6,6),ESTT(6,6),EMAS(6,6),EMST(6,6),
     $ DISN(3),TITLE(10),
     $ IBOU(100,3),XNOD(100),YNOD(100),CMAS(100,3),
     $ NODN(100,2),AREA(100),SMOI(100),EMOD(100),ICOR(100,6),
     $ EROW(100),SYTF(100,100),SMAS(100,100),WK(400),BETA(100),
     $ EIVL(100),ITEM(100)
      COMPLEX ALFA(100),EIVE(100,100),DVAL
C
C
      WRITE(6,1)
    1 FORMAT(1H1)
C     READING AND PRINTING OF TITLE CARD
C
      READ(5,11)(TITLE(I),I=1,10)
   11 FORMAT(10A8)
      WRITE(6,12)(TITLE(I),I=1,10)
   12 FORMAT(/10X,10A8)
C
C     READING AND PRINTING OF MASTER CONTROL CARD
C
      READ(5,21)NNOD,NELE,NMOD,LMAS,ICAS,IPR1,IPR2
```

```
   21 FORMAT(8I5)
      IF(ICAS.EQ.1)WRITE(6,23)
   23 FORMAT(/5X,22HTHIS DATA IS FOR TRUSS)
      IF(ICAS.EQ.2)WRITE(6,24)
   24 FORMAT(/5X,30HTHIS DATA IS FOR GENERAL FRAME)
      IF(ICAS.EQ.3)WRITE(6,25)
   25 FORMAT(/5X,52HTHIS DATA IS FOR FRAME WITH INFINITE AXIAL STIFFNESS
     $ )
      IF(LMAS.EQ.1)WRITE(6,8)
    8 FORMAT(5X,22HAND LUMPED MASS MATRIX)
      IF(LMAS.EQ.2)WRITE(6,9)
    9 FORMAT(5X,26HAND CONSISTENT MASS MATRIX)
      WRITE(6,22)NNOD,NELE,NMOD
   22 FORMAT(/5X,16HNUMBER OF NODES=,I5/5X,19HNUMBER OF ELEMENTS=,I5
     $ /5X,30HNUMBER OF MODES TO BE PRINTED=,I5)
      IF(IPR1.EQ.1)WRITE(6,31)
   31 FORMAT(/5X,47HELEMENT STIFFNESS AND MASS MATRICES ARE PRINTED)
      IF(IPR2.EQ.1)WRITE(6,32)
   32 FORMAT(/5X,46HGLOBAL STIFFNESS AND MASS MATRICES ARE PRINTED)
C
C     READING AND PRINTING OF NODAL DATA
C
      WRITE(6,33)
   33 FORMAT(/5X,4HNODE,1X,20HBOUNDARY  CONDITIONS,9X,11HCOORDINATES,
     $ 20X,6HMASSES)
      WRITE(6,34)
   34 FORMAT(7X,17HNO    1    2    3,11X,1HX,15X,1HY,15X,1H1,15X,1H2,
     $ 15X,1H3)
      DO 50 I=1,NNOD
      READ(5,42)N,(IBOU(I,J),J=1,3),XNOD(I),YNOD(I),(CMAS(I,J),J=1,3)
   42 FORMAT(4I5,5E10.4)
      WRITE (6,43)I,(IBOU(I,J),J=1,3),XNOD(I),YNOD(I),(CMAS(I,J),J=1,3)
   43 FORMAT( 4X,4I5,5E15.6)
   50 CONTINUE
C
C     READING AND PRINTING OF ELEMENT DATA
C
      WRITE(6,49)
   49 FORMAT(/6X,7HELEMENT,8X,9HEND NODES,8X,4HAREA,11X,4HM.I.,13X,
     $ 1HE,6X,13H MASS DENSITY)
      WRITE(6,51)
   51 FORMAT( 11X,2HNO,5X,10H    1    2)
      DO 60 I=1,NELE
      READ(5,52)N,NODN(I,1),NODN(I,2),AREA(I),SMOI(I),EMOD(I),EROW(I)
   52 FORMAT(3I5,4E10.4)
      WRITE(6,53)N,NODN(I,1),NODN(I,2), AREA(I),SMOI(I),EMOD(I),EROW(I)
   53 FORMAT(/8X,I5,5X,2I5,4E15.6)
   60 CONTINUE
C
C     GENERATION OF NUMBERS FOR ASSEMBLING GLOBAL STIFFNESS MATRIX
C
      ICON=0
      DO 20 I=1,NNOD
      DO 20 J=1,3
      K=IBOU(I,J)
      IF (K.EQ.0) GO TO 20
      ICON=ICON+1
      IBOU(I,J)=ICON
   20 CONTINUE
      NDOF=ICON
      WRITE(6,54)NDOF
   54  FORMAT(/5X,23H NUMBER OF FREE D.O.F.=,I5)
      DO 30 I=1,NELE
      I1=NODN(I,1)
      I2=NODN(I,2)
      DO 30 J=1,3
      ICOR(I,J)=IBOU (I1,J)
      ICOR(I,J+3)=IBOU(I2,J)
   30 CONTINUE
      WRITE(6,61)
   61 FORMAT (/5X,7HELEMENT, 5X,24HNODAL DEGREES OF FREEDOM)
      WRITE(6,63)
   63 FORMAT( 5X,36HNUMBER    1    2    3    4    5    6)
      DO 70 I=1,NELE
```

```
      WRITE(6,62)I,(ICOR(I,J),J=1,6)
   62 FORMAT(/6X,7I5)
   7Ø CONTINUE
C
C     INTIALISING GLOBAL STIFFNESS AND MASS MATRICES TO ZERO
C
      DO 8Ø I=1,NDOF
      DO 8Ø J=1,NDOF
      SMAS(I,J)=Ø.ØØØ
   8Ø SYTF(I,J)=Ø.Ø
C
C
C
C     GENERATION OF TRANSFORMATION MATRIX AND ELEMENT STIFFNESS MATRIX A
C     WRITING ON TAPE 1.
C
      REWIND 1
      DO 4ØØ IE=1,NELE
      DO 1ØØ I=1,6
      DO 1ØØ J=1,6
      ESTF(I,J)=Ø.Ø
      EMAS(I,J)=Ø.Ø
  1ØØ TRNS(I,J)=Ø.Ø
      AA=AREA(IE)
      AI=SMOI(IE)
      AE=EMOD(IE)
      ROW=EROW(IE)
C
C     GENERATION OF TRANSFORMATION MATRIX.
C
      I1=NODN(IE,1)
      I2=NODN(IE,2)
      X1=XNOD(I1)
      X2=XNOD(I2)
      Y1=YNOD(I1)
      Y2=YNOD(I2)
      CALL TRNMAT(X1,Y1,X2,Y2,AL,TRNS)
      IF(IPR1.EQ.1)WRITE(6,1Ø1)IE,((TRNS(I,J),J=1,6),I=1,6)
  1Ø1 FORMAT (/5X,33HTRANSFORMATION MATRIX FOR ELEMENT,
     $ I5/6(/1ØX,6E15.6))
C
C     GENERATION OF ELEMENT STIFFNESS MATRIX
C
      AB=AA
      IF(ICAS.EQ.3)AB=AA*1.ØE+Ø6
      IF(ICAS.EQ.1) AI=Ø.Ø
      CALL ELESTF (AL,AB,AI,AE,ESTF)
      IF(IPR1.EQ.1)WRITE(6,1Ø2) IE,((ESTF(I,J),J=1,6),I=1,6)
  1Ø2 FORMAT(/5X,35HELEMENT STIFFNESS MATRIX OF ELEMENT,
     $ I5/6(/1ØX,6E15.6))
C
C     TRANSFORMATION OF ELEMENT STIFFNESS MATRIX TO GLOBAL COORDINATES
C
       CALL CONTRN(TRNS,ESTF,ESTT,6)
      IF(IPR1.EQ.1)WRITE(6,1Ø3)IE,((ESTT(I,J),J=1,6),I=1,6)
  1Ø3 FORMAT(/5X,47HTRANSFORMED ELEMEMT STIFFNESS MATRIX OF ELEMENT,
     $I5/6(/1ØX,6E15.6))
C     COMPUTATION OF ELEMENT MASS MATRIX
      CALL ELEMAS(LMAS,ROW,AA,AL,EMAS)
C
      IF(IPR1.EQ.1) WRITE(6,1Ø4) IE,((EMAS(I,J),J=1,6),I=1,6)
  1Ø4 FORMAT(/5X,22HMASS MATRIX OF ELEMENT,I5/6(/1ØX,6E15.6))
C
C     TRANSFORMATION OF MASS MATRIX TO GLOBAL COORDINATES
C
      CALL CONTRN(TRNS,EMAS,EMST,6)
      IF(IPR1.EQ.1) WRITE(6,1Ø5) IE,((EMST(I,J),J=1,6),I=1,6)
  1Ø5 FORMAT(/5X,34HTRANSFORMED MASS MATRIX OF ELEMENT,
     $ I5/6(/1ØX,6E15.6))
C
C     WRITING AL,AA,AI,AE,ESTF,EMAS,TRAN ON TAPE 1
C
      WRITE(1,1Ø6)AL,AA,AI,AE,((ESTF(I,J),J=1,6),I=1,6),((EMAS(I,J),
     $J=1,6),I=1,6),((TRNS(I,J),J=1,6),I=1,6)
```

```
  1Ø6 FORMAT(5E15.8)
C
C     ASSEMBLING THE GLOBAL STIFFNESS MATRIX
C
      DO 2ØØ I=1,6
      DO 2ØØ J=1,6
      K=ICOR(IE,I)
      L=ICOR(IE,J)
      IF (K*L.EQ.Ø) GO TO 2ØØ
      SMAS(K,L)=SMAS(K,L)+EMST(I,J)
      SYTF(K,L)=SYTF(K,L)+ESTT(I,J)
  2ØØ CONTINUE
  4ØØ CONTINUE
C
C     ADDING OF NODAL CONCENTRATED MASSES TO GLOBAL MASS MATRIX
C
      DO 5ØØ I=1,NNOD
      DO 5ØØ J=1,3
      K=IBOU(I,J)
      IF(K.EQ.Ø) GO TO 5ØØ
      SMAS(K,K)=SMAS(K,K)+CMAS(I,J)
  5ØØ CONTINUE
      IF(IPR2.EQ.Ø) GO TO 25Ø
  21Ø CONTINUE
      WRITE(6,2Ø1)
  2Ø1 FORMAT(/5X,23HGLOBAL STIFFNESS MATRIX)
      DO 22Ø I=1,NDOF
      WRITE(6,2Ø2) I
  2Ø2 FORMAT(/5X,4HROW=,I5)
      WRITE(6,2Ø3)(SYTF(I,J),J=1,NDOF)
  22Ø CONTINUE
      WRITE(6,2Ø6)
  2Ø6 FORMAT(/5X,19H GLOBAL MASS MATRIX)
      DO 24Ø I=1,NDOF
      WRITE(6,2Ø2) I
      WRITE(6,2Ø3)(SMAS(I,J),J=1,NDOF)
  2Ø3 FORMAT(5X,12E1Ø.4)
  24Ø CONTINUE
  25Ø CONTINUE
C
C      COMPUTATION OF  EIGENVALUE AND EIGENVECTORS
C
      CALL EIGZF(SYTF,1ØØ,SMAS,1ØØ,NDOF,1,ALFA,BETA,EIVE,1ØØ,WK,IER)
      DO 55Ø I=1,NDOF
      DO 55Ø J=1,NDOF
  55Ø SYTF(I,J)=REAL(EIVE(I,J))
      DO 56Ø I=1,NDOF
      DVAL=ALFA(I)/BETA(I)
      EIVL(I)=REAL(DVAL)
  56Ø CONTINUE
C
C     REARRANGING EIGENVALUES IN INCREASING ORDER OF MAGNITUDE
C
      AMIN=Ø.Ø
      DO 75Ø I=1,NDOF
      AMAX=1.ØE+99
      DO 7ØØ J=1,NDOF
      TEMV=EIVL(J)
      IF(TEMV.LE.AMIN) GO TO 7ØØ
      IF(TEMV.GT.AMAX) GO TO 7ØØ
      ITEM(I)=J
      AMAX=TEMV
  7ØØ CONTINUE
      AMIN=AMAX
  75Ø CONTINUE
      WRITE(6,251)
251 FORMAT(/5X,29HNATURAL FREQUENCIES AND MODES )
    DO 6ØØ K=1,NMOD
    I=ITEM(K)
    ALAM=EIVL(I)
    OMGA=SQRT(ALAM)
    WRITE(6,6Ø1)K,ALAM,OMGA
6Ø1 FORMAT(/5X,I5,2X,27HEIGENVECTOR AT EIGENVALUE =,E15.6,
   $ 19HCIRCULAR FREQUENCY=,E15.6)
```

```
      WRITE(6,6Ø2)
  6Ø2 FORMAT(/5X,4HNODE,1ØX,6HX-DISP,1ØX,6HY-DISP,1ØX,4HROTA)
      DO 62Ø L=1,NNOD
      DO 61Ø M=1,3
      DISN(M)=Ø.Ø
      N=IBOU(L,M)
      IF(N.EQ.Ø) GO TO 61Ø
      DISN(M)=SYTF(N,I)
  61Ø CONTINUE
      WRITE(6,6Ø3)L,(DISN(NN),NN=1,3)
  6Ø3 FORMAT(5X,I5,3F15.6)
  62Ø CONTINUE
  6ØØ CONTINUE
      STOP
      END
      SUBROUTINE TRNMAT(X1,Y1,X2,Y2,AL,T)
      DIMENSION T(6,6)
      REAL LMBD,MU
C     SUBROUTINE TO COMPUTE TRANSFORMATION MATRIX T(6,6)
C     AL=LENGTH
C     LMBD=COSINE OF THE ANGLE
C     MU=SINE OF THE ANGLE
      AL=SQRT((X2-X1)**2+(Y2-Y1)**2)
      LMBD=(X2-X1)/AL
      MU=(Y2-Y1)/AL
      DO 1ØØ I=1,6
      DO 1ØØ J=1,6
  1ØØ T(I,J)=Ø.Ø
      T(1,1)=LMBD
      T(2,1)=-MU
      T(1,2)=MU
      T(2,2)=LMBD
      T(3,3)=1.
      T(4,4)=LMBD
      T(5,4)=-MU
      T(4,5)=MU
      T(5,5)=LMBD
      T(6,6)=1.
       RETURN
       END
      SUBROUTINE ELESTF(AL,AA,AI,AE,ESTF)
      DIMENSION ESTF(6,6)
C
C     SUBROUTINE FOR ELEMENT STIFFNESS  MATRIX
C     SIZE OF STIFFNESS MATRIX IS(6,6)
C     $ST D.O.F=AXIAL DISPLACEMENT AT NODE 1
C     2ND D.O.F.=TRANSVERSE DISPLACEMENT AT NODE 1
C     3RD D.O.F. =TRANSVERSE ROTATION AT NODE 1
C     4TH D.O.F=AXIAL DISPLACEMENT AT NODE 2
C     5TH D.O.F.=TRANSVERSE DISPLACEMENT AT NODE 2
C     6TH D.O.F. =TRANSVERSE ROTATION AT NODE 2
C     AA=AREA
C     AL=LENGTH
C     AE=ELASTIC MODULUS
C     AI=MOMENT OF INERTIA
C     ESTF=ELEMENT STIFFNESS
C
      DO 1Ø I=1,6
      DO 1Ø J=1,6
   1Ø ESTF(I,J)=Ø.Ø
      ESTF(1,1)=AA*AE/AL
      ESTF(4,1)=-ESTF(1,1)
      ESTF(2,2)=12.Ø*AE*AI/AL/AL/AL
      ESTF(3,2)=+6.Ø*AE*AI/AL/AL
      ESTF(4,2)=Ø.Ø
      ESTF(5,2)=-ESTF(2,2)
      ESTF(6,2)=ESTF(3,2)
      ESTF(3,3)=4.Ø*AE*AI/AL
      ESTF(5,3)=-ESTF(3,2)
      ESTF(6,3)=ESTF(3,3)/2.Ø
      ESTF(4,4)=ESTF(1,1)
      ESTF(5,5)=ESTF(2,2)
      ESTF(6,5)=ESTF(5,3)
      ESTF(6,6)=ESTF(3,3)
```

```
      DO 1ØØ I=1,6
      K=I+1
      DO 1ØØ J=K,6
  1ØØ ESTF(I,J)=ESTF(J,I)
      RETURN
      END
      SUBROUTINE ELEMAS(LMAS,ROW,AA,AL,EMAS)
      DIMENSION EMAS(6,6)
C     SUBROUTINE TO COMPUTE LUMPED OR CONSISTENT MASS MATRIX OF
C     A BEAM ELEMENT
C     LMAS=1 LUMPED MASS CASE
C     LMAS=2 CONSISTENT MASS CASE
C     ROW=MASS DENSITY
C     AA=AREA OF CROOSS
C     AL=LENGTH
C     EMAS=MASS MATRIX
      DO 1Ø I=1,6
      DO 1Ø J=1,6
   1Ø EMAS(I,J)=Ø.Ø
      IF(LMAS.EQ.2) GO TO 4Ø
   2Ø CONTINUE
      C1=(ROW*AA*AL)/2.Ø
      EMAS(1,1)=C1
      EMAS(2,2)=C1
      EMAS(4,4)=C1
      EMAS(5,5)=C1
      RETURN
   4Ø CONTINUE
      C1=(ROW*AA*AL)/6.Ø
      C2=(ROW*AA*AL)/42Ø.Ø
      EMAS(1,1)=2.Ø*C1
      EMAS(4,1)=C1
      EMAS(4,4)=EMAS(1,1)
      EMAS(2,2)=156.Ø*C2
      EMAS(3,2)=+22.Ø*AL*C2
      EMAS(5,2)=54.Ø*C2
      EMAS(6,2)=-13.Ø*AL*C2
      EMAS(3,3)=4.Ø*AL*AL*C2
      EMAS(5,3)=-EMAS(6,2)
      EMAS(6,3)=-3.Ø*AL*AL*C2
      EMAS(5,5)=EMAS(2,2)
      EMAS(6,5)=-EMAS(3,2)
      EMAS(6,6)=EMAS(3,3)
      DO 3Ø I=1,6
      K=I+1
      DO 3Ø J=K,6
   3Ø EMAS(I,J)=EMAS(J,I)
      RETURN
      END
      SUBROUTINE CONTRN(A,B,C,N)
      DIMENSION A(6,6),B(6,6),C(6,6),D(6,6)
C
C      SUBROUTINE  FOR CONGRUENT TRANSFORMATION C=(TRANPOSE A)*B*A
C
      DO 1ØØ I=1,N
      DO 1ØØ J=1,N
      D(I,J)=Ø.Ø
      DO 1ØØ K=1,N
  1ØØ D(I,J)=D(I,J)+B(I,K)*A(K,J)
      DO 3ØØ I=1,N
      DO 3ØØ J=1,N
      C(I,J)=Ø.Ø
      DO 3ØØ K=1,N
  3ØØ C(I,J)=C(I,J)+A(K,I)*D(K,J)
      RETURN
      END
SIMPLY SUPPORTED BEAM VIBRATION CASE
   4     3     3     2     3
   1     Ø     Ø     1
   2     Ø     1     1+.96ØØE+Ø2
   3     Ø     1     1+.192ØE+Ø3
   4     Ø     Ø     1+.288ØE+Ø3
   1     1     2+.671ØE+Ø1+.642ØE+Ø2+.3ØØØE+Ø8+.733ØE-Ø3
   2     2     3+.671ØE+Ø1+.642ØE+Ø2+.3ØØØE+Ø8+.733ØE-Ø3
   3     3     4+.671ØE+Ø1+.642ØE+Ø2+.3ØØØE+Ø8+.733ØE-Ø3
```

```
        SIMPLY SUPPORTED BEAM VIBRATION CASE

THIS DATA IS FOR FRAME WITH INFINITE AXIAL STIFFNESS
AND CONSISTENT MASS MATRIX

NUMBER OF NODES=     4
NUMBER OF ELEMENTS=     3
NUMBER OF MODES TO BE PRINTED=     3

NODE BOUNDARY  CONDITIONS          COORDINATES                          MASSES
  NO     1    2    3          X              Y               1              2              3
   1     Ø    Ø    1    Ø.ØØØØØØE+ØØ   Ø.ØØØØØØE+ØØ   Ø.ØØØØØØE+ØØ   Ø.ØØØØØØE+ØØ   Ø.ØØØØØØE+ØØ
   2     Ø    1    1    Ø.96ØØØØE+Ø2   Ø.ØØØØØØE+ØØ   Ø.ØØØØØØE+ØØ   Ø.ØØØØØØE+ØØ   Ø.ØØØØØØE+ØØ
   3     Ø    1    1    Ø.192ØØØE+Ø3   Ø.ØØØØØØE+ØØ   Ø.ØØØØØØE+ØØ   Ø.ØØØØØØE+ØØ   Ø.ØØØØØØE+ØØ
   4     Ø    Ø    1    Ø.288ØØØE+Ø3   Ø.ØØØØØØE+ØØ   Ø.ØØØØØØE+ØØ   Ø.ØØØØØØE+ØØ   Ø.ØØØØØØE+ØØ

 ELEMENT         END NODES        AREA           M.I.            E        MASS DENSITY
      NO           1    2

      1            1    2    Ø.671ØØØE+Ø1   Ø.642ØØØE+Ø2   Ø.3ØØØØØE+Ø8   Ø.733ØØØE-Ø3

      2            2    3    Ø.671ØØØE+Ø1   Ø.642ØØØE+Ø2   Ø.3ØØØØØE+Ø8   Ø.733ØØØE-Ø3

      3            3    4    Ø.671ØØØE+Ø1   Ø.642ØØØE+Ø2   Ø.3ØØØØØE+Ø8   Ø.733ØØØE-Ø3

 NUMBER OF FREE D.O.F.=     6

ELEMENT      NODAL DEGREES OF FREEDOM
NUMBER     1    2    3    4    5    6

     1     Ø    Ø    1    Ø    2    3

     2     Ø    2    3    Ø    4    5

     3     Ø    4    5    Ø    Ø    6

NATURAL FREQUENCIES AND MODES

    1  EIGENVECTOR AT EIGENVALUE =    Ø.555344E+Ø4CIRCULAR FREQUENCY=    Ø.745214E+Ø2

NODE           X-DISP          Y-DISP           ROTA
    1        Ø.ØØØØØØ        Ø.ØØØØØØ        Ø.Ø12596
    2        Ø.ØØØØØØ        1.ØØØØØØ        Ø.ØØ6298
    3        Ø.ØØØØØØ        1.ØØØØØØ       -Ø.ØØ6298
    4        Ø.ØØØØØØ        Ø.ØØØØØØ       -Ø.Ø12596
```

```
    2  EIGENVECTOR AT EIGENVALUE =    Ø.9Ø82Ø9E+Ø5CIRCULAR FREQUENCY=   Ø.3Ø1365E+Ø3

NODE         X-DISP         Y-DISP          ROTA
    1     Ø.ØØØØØØØ      Ø.ØØØØØØØ     -Ø.Ø2511Ø
    2     Ø.ØØØØØØØ     -1.ØØØØØØØ      Ø.Ø12555
    3     Ø.ØØØØØØØ      1.ØØØØØØØ      Ø.Ø12555
    4     Ø.ØØØØØØØ      Ø.ØØØØØØØ     -Ø.Ø2511Ø

    3  EIGENVECTOR AT EIGENVALUE =    Ø.553256E+Ø6CIRCULAR FREQUENCY=   Ø.743812E+Ø3

NODE         X-DISP         Y-DISP          ROTA
    1     Ø.ØØØØØØØ      Ø.ØØØØØØØ      1.ØØØØØØØ
    2     Ø.ØØØØØØØ      Ø.ØØØØØØØ     -1.ØØØØØØØ
    3     Ø.ØØØØØØØ      Ø.ØØØØØØØ      1.ØØØØØØØ
    4     Ø.ØØØØØØØ      Ø.ØØØØØØØ     -1.ØØØØØØØ

PORTAL  FRAME  VIBRATION  EXAMPLE
  4     3     2     2     3     Ø     Ø
  1     Ø     Ø     Ø
  2     1     1     1+.ØØØØE+ØØ+.8ØØØE+Ø2
  3     1     1     1+.8ØØØE+Ø2+.8ØØØE+Ø2
  4     Ø     Ø     Ø+.8ØØØE+Ø2
  1     1     2+.164ØE+Ø1+.25ØØE+Ø1+.3ØØØE+Ø8+.733ØE-Ø3
  2     2     3+.164ØE+Ø1+.25ØØE+Ø1+.3ØØØE+Ø8+.733ØE-Ø3
  3     3     4+.164ØE+Ø1+.25ØØE+Ø1+.3ØØØE+Ø8+.733ØE-Ø3

      PORTAL  FRAME  VIBRATION  EXAMPLE

THIS DATA IS FOR FRAME WITH INFINITE AXIAL STIFFNESS
AND CONSISTENT MASS MATRIX

NUMBER OF NODES=    4
NUMBER OF ELEMENTS=    3
NUMBER OF MODES TO BE PRINTED=    2

NODE BOUNDARY  CONDITIONS        COORDINATES                               MASSES
  NO    1    2    3          X               Y               1               2               3
   1    Ø    Ø    Ø   Ø.ØØØØØØØE+ØØ   Ø.ØØØØØØØE+ØØ   Ø.ØØØØØØØE+ØØ   Ø.ØØØØØØØE+ØØ   Ø.ØØØØØØØE+ØØ
   2    1    1    1   Ø.ØØØØØØØE+ØØ   Ø.8ØØØØØØE+Ø2   Ø.ØØØØØØØE+ØØ   Ø.ØØØØØØØE+ØØ   Ø.ØØØØØØØE+ØØ
   3    1    1    1   Ø.8ØØØØØØE+Ø2   Ø.8ØØØØØØE+Ø2   Ø.ØØØØØØØE+ØØ   Ø.ØØØØØØØE+ØØ   Ø.ØØØØØØØE+ØØ
   4    Ø    Ø    Ø   Ø.8ØØØØØØE+Ø2   Ø.ØØØØØØØE+ØØ   Ø.ØØØØØØØE+ØØ   Ø.ØØØØØØØE+ØØ   Ø.ØØØØØØØE+ØØ
```

```
ELEMENT        END NODES        AREA           M.I.              E          MASS DENSITY
   NO           1    2
    1           1    2   Ø.164ØØØE+Ø1   Ø.25ØØØØE+Ø1   Ø.3ØØØØØE+Ø8   Ø.733ØØØE-Ø3
    2           2    3   Ø.164ØØØE+Ø1   Ø.25ØØØØE+Ø1   Ø.3ØØØØØE+Ø8   Ø.733ØØØE-Ø3
    3           3    4   Ø.164ØØØE+Ø1   Ø.25ØØØØE+Ø1   Ø.3ØØØØØE+Ø8   Ø.733ØØØE-Ø3

NUMBER OF FREE D.O.F.=     6

ELEMENT     NODAL DEGREES OF FREEDOM
NUMBER     1    2    3    4    5    6
     1     Ø    Ø    Ø    1    2    3
     2     1    2    3    4    5    6
     3     4    5    6    Ø    Ø    Ø

NATURAL FREQUENCIES AND MODES

    1  EIGENVECTOR AT EIGENVALUE =   Ø.156993E+Ø5CIRCULAR FREQUENCY=   Ø.125297E+Ø3

NODE          X-DISP          Y-DISP          ROTA
    1       Ø.ØØØØØØ       Ø.ØØØØØØ       Ø.ØØØØØØ
    2       1.ØØØØØØ       Ø.ØØØØØØ      -Ø.ØØ691Ø
    3       1.ØØØØØØ       Ø.ØØØØØØ      -Ø.ØØ691Ø
    4       Ø.ØØØØØØ       Ø.ØØØØØØ       Ø.ØØØØØØ

    2  EIGENVECTOR AT EIGENVALUE =   Ø.348949E+Ø6CIRCULAR FREQUENCY=   Ø.59Ø719E+Ø3

NODE          X-DISP          Y-DISP          ROTA
    1       Ø.ØØØØØØ       Ø.ØØØØØØ       Ø.ØØØØØØ
    2       Ø.ØØØØØØ       Ø.ØØØØØØ      -1.ØØØØØØ
    3       Ø.ØØØØØØ       Ø.ØØØØØØ       1.ØØØØØØ
    4       Ø.ØØØØØØ       Ø.ØØØØØØ       Ø.ØØØØØØ
```

13.3 SIX-D.O.F. TRIANGULAR PLANE STRESS AND PLANE STRAIN ELEMENT

This program is written for the six-degree-of-freedom plane stress finite element described in Chapter 9. The program can also be used for plane strain analysis if E and ν are replaced by $E/(1-\nu^2)$ and $\nu/(1-\nu)$, respectively, as described in the paragraph following Eq. (9.26b).

This program performs three major tasks.

1. The equilibrium equations for the total system are formed by storing the overall stiffness matrix in a band matrix with its number of rows equal to the number of free-to-move degrees of freedom and its number of columns equal to that defined in Eq. (9.72). The boundary conditions for a nodal point free to slide along a slope as shown in Fig. (9.9b) are treated using Eqs. (9.82) and (9.83) and their descriptive procedures.
2. The set of band matrix equations is solved for the nodal point displacements by the method of direct elimination.
3. The internal element stresses are determined from these displacements. These stresses are also converted to principal stresses.

Gravity loads, temperature loads, and nonhomogeneous material properties are accounted for in this program. The program calls an IMSL Library Subroutine LEQ1PB which solves the symmetric band stiffness matrix by the Gaussian elimination method. If LEQ1PB is not available, the user can still use this program by simply changing the statement that calls LEQ1PB to the one that calls any available equivalent subroutine.

In this program, provision is made in the DIMENSION statement for 200 elements, 200 nodal points, and 100 free-to-move degrees of freedom. The user can easily modify the DIMENSION statements to cope with more elements, nodes, and free d.o.f.'s.

13.3.1 Description of Input Data

1. Title Card (10A8)
   Columns 1–80: Arbitrary problem identification
2. Master Control Card (5I5)
   One card to define the number of elements, nodal points, boundary nodal points, and to specify print options
   Columns 1–5: Number of elements (NUMEL)
   6–10: Number of nodal points (NUMNP)
   11–15: Number of boundary nodal points (NUMBC)

16–20: Flag to print the element stiffness matrix
IPR1 = 1 print
IPR1 = 0 no print
21–25: Flag to print the assembled stiffness matrix and load vector
IPR2 = 1 print
IPR2 = 0 no print

3. Element Data (4I5, 6E10.4)
One card for each element in order
Columns 1–5: Element number [NUME(N)]
6–10: Nodal point number i [NPI(N)]
11–15: Nodal point number j [NPJ(N)]
16–20: Nodal point number k [NPK(N)]
21–30: Modulus of elasticity [EVAL(N)]
31–40: Thickness [THIK(N)]
41–50: Density of the element per unit area [RO(N)]
51–60: Poisson's ratio [XU(N)]
61–70: Coefficient of thermal expansion [COED(N)]
71–80: Temperature rise within the element [DT(N)]

4. Nodal Point Data (1I5, 4E10.4)
One card for each nodal point in order
Columns 1–5: Nodal point number [NPNUM(M)]
6–15: x coordinate [XORD(M)]
16–25: y coordinate [YORD(M)]
26–35: Load in x direction [XLOAD(M)]
36–45: Load in y direction [YLOAD(M)]

5. Boundary Nodal Point Data (2I5, 1E10.4)
One card per boundary nodal point
Columns 1–5: Nodal point number [NPB(L)]
6–10: Flag to specify boundary conditions
NFIX = 1 fixed in x direction only
NFIX = 2 fixed in y direction only
NFIX = 3 free to move along a line whose normal is at an angle ϕ measured counterclockwise from the positive x axis (see Fig. 9.9b)
11–20: Angle ϕ in degrees [SLOPE(L)]

13.3.2 Sample Problems

Example 13.5

Figure 13.5a shows a 16-element model of a rectangular steel plate with one side fixed and the opposite side under uniformly distributed tensile load. It is assumed that

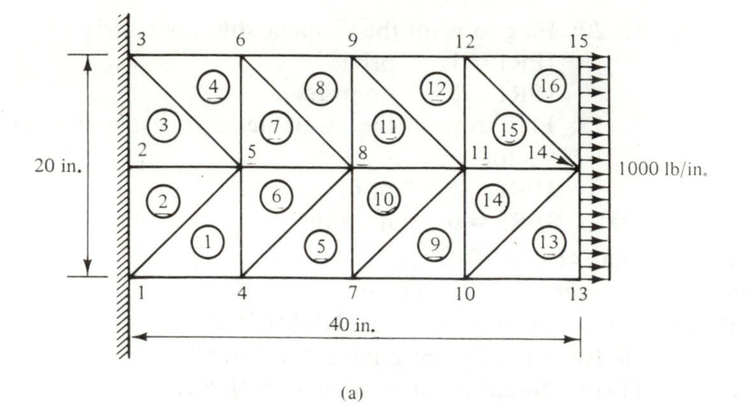

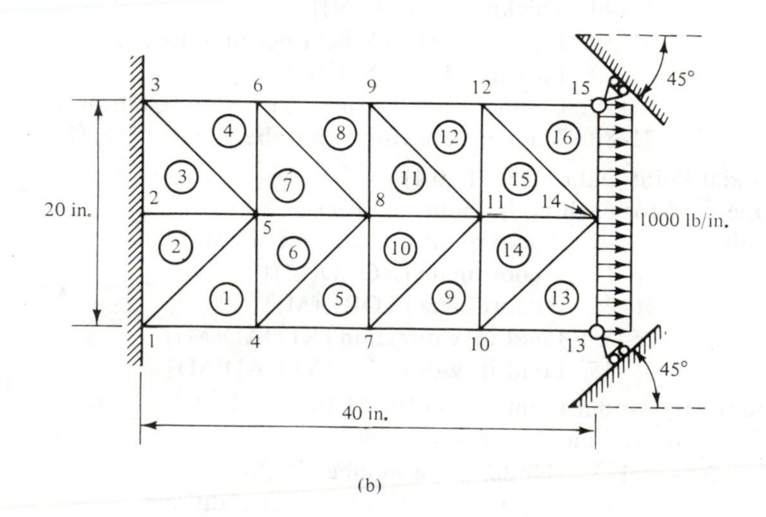

Figure 13.5 Cantilever plate with one edge fixed and the opposite edge (a) free or (b) with two slope boundary conditions.

$E = 30 \times 10^6$ psi, $t = 0.1$ in., and $\nu = 0.3$. The input data necessary for a static analysis of such a problem and the output data are given at the end of the listing of this program. The output shows that the stresses in the x direction for all elements are approximately equal to 10 ksi.

Example 13.6

Figure 13.5b shows a problem identical to Fig. 13.5a but with two slope boundary conditions at nodal points 13 and 15, respectively. The input data necessary for a static analysis of such a problem and the output data are given after those for Example 13.5.

```
      PROGRAM PLANE
C     ************************************************************
C     * PLANE STRESS AND PLANE STRAIN FINITE ELEMENT PROGRAM *
C     ************************************************************
C
      DIMENSION A(6,6),B(6,6),S(6,6),ITAB(6),TRAN(6,6),TEMP(6,6)
      DIMENSION NPNUM(200),XORD(200),YORD(200),ICOR(200,2),DSX(200),
     $ DSY(200),XLOAD(200),YLOAD(200),SLOPE(200),NPB(200),NFIX(200)
      DIMENSION NUME(200),NPI(200),NPJ(200),NPK(200),EVAL(200),
     $ THIK(200),ET(200),XU(200),RO(200),COED(200),DT(200),
     $ THERM(200),AJ(200),BJ(200),AK(200),BK(200),
     $ SIGXX(200),SIGYY(200),SIGXY(200),TITLE(10)
      DIMENSION SYTF(100,100),SLOD(100,1),SDIS(100,1)
C
C
C     READ AND PRINT OF DATA
C
      WRITE (6,99)
      READ(5,6)(TITLE(I),I=1,10)
      WRITE(6,6)(TITLE(I),I=1,10)
      READ (5,1)  NUMEL,NUMNP,NUMBC,IPR1,IPR2
      WRITE (6,101) NUMEL
      WRITE (6,102) NUMNP
      WRITE (6,103) NUMBC
      IF(IPR1.EQ.1) WRITE(6,104)
      IF(IPR2.EQ.1) WRITE(6,105)
      READ(5,2) (NUME(N),NPI(N),NPJ(N),NPK(N),EVAL(N),THIK(N),RO(N),
     $ XU(N),COED(N),DT(N), N=1,NUMEL)
      READ (5,3) (NPNUM(M),XORD(M),YORD(M),XLOAD(M),YLOAD(M),M=1,NUMNP)
      WRITE(6,113)
      WRITE(6,110)
      WRITE(6,5)(NUME(N),NPI(N),NPJ(N),NPK(N),EVAL(N),THIK(N),RO(N),
     $ XU(N),COED(N),DT(N), N=1,NUMEL)
      WRITE(6,114)
      WRITE(6,111)
      WRITE (6,109) (NPNUM(M),XORD(M),YORD(M),XLOAD(M),YLOAD(M),
     $ M=1,NUMNP)
C
C     MODIFICATION OF LOADS AND ELEMENT DIMENSIONS
C
      DO 180 N=1,NUMEL
      ET(N)=EVAL(N)*THIK(N)
      I=NPI(N)
      J=NPJ(N)
      K=NPK(N)
      AJ(N)=XORD(J)-XORD(I)
      AK(N)=XORD(K)-XORD(I)
      BJ(N)=YORD(J)-YORD(I)
      BK(N)=YORD(K)-YORD(I)
  176 AREA=ABS((AJ(N)*BK(N)-BJ(N)*AK(N))/2. )
      IF (AREA) 701,701,177
  177 THERM(N)=ET(N)*COED(N)*DT(N)/(1.0-XU(N))
      DL=AREA*RO(N)/3.
      XLOAD(I)=THERM(N)*(BK(N)-BJ(N))/2.+XLOAD(I)
      XLOAD(J)=-THERM(N)*BK(N)/2.+XLOAD(J)
      XLOAD(K)=THERM(N)*BJ(N)/2.+XLOAD(K)
      YLOAD(I)=THERM(N)*(AJ(N)-AK(N))/2.+YLOAD(I)-DL
      YLOAD(J)=THERM(N)*AK(N)/2.+YLOAD(J)-DL
  180 YLOAD(K)=-THERM(N)*AJ(N)/2.+YLOAD(K)-DL
C
C     READING OF BOUNDARY CONDITION DATA
C
      WRITE (6,112)
      READ (5,7)  (NPB(L),NFIX(L),SLOPE(L),L=1,NUMBC)
      WRITE (6,4) (NPB(L),NFIX(L),SLOPE(L),L=1,NUMBC)
C
C     GENERATION OF NUMBERS FOR ASSEMBLING  GLOBAL STIFFNESS MATRIX
C
      ICON=0
      DO 60 I=1,NUMNP
      IB=0
      IX=0
      DO 10 J=1,NUMBC
      L=NPB(J)
```

```
      IF(L.EQ.I)IX=NFIX(J)
   10 IF(L.EQ.I) IB=L
      IF(IB.EQ.0) GO TO 55
      K=IX
      IF(K.NE.0) GO TO 30
   20 CONTINUE
      ICOR(I,1)=0
      ICOR(I,2)=0
      GO TO 60
   30 CONTINUE
      IF(K.NE.1) GO TO 40
      ICOR(I,1)=0
      ICON=ICON+1
      ICOR(I,2)=ICON
      GO TO 60
   40 CONTINUE
      IF(K.NE.2) GO TO 50
      ICON=ICON+1
      ICOR(I,1)=ICON
      ICOR(I,2)=0
      GO TO 60
   50 CONTINUE
      IF(K.NE.3) GO TO 55
      ICON=ICON+1
      ICOR(I,1)=ICON
      ICOR(I,2)=0
      GO TO 60
   55 CONTINUE
      ICON=ICON+1
      ICOR(I,1)=ICON
      ICON=ICON+1
      ICOR(I,2)=ICON
   60 CONTINUE
      NDOF=ICON
      WRITE(6,61) NDOF
C
C     FINDING MAXIMUM BAND WIDTH
C
      MAX=0
      DO 80 I=1,NUMEL
      J=NPI(I)-NPJ(I)
      J=IABS(J)
      IF(J.GT.MAX) MAX=J
      J=NPJ(I)-NPK(I)
      J=IABS(J)
      IF(J.GT.MAX) MAX=J
      J=NPK(I)-NPI(I)
      J=IABS(J)
      IF(J.GT.MAX) MAX=J
   80 CONTINUE
      NBND=4*MAX+3
      IF(NBND.GT.NDOF)NBND=NDOF
      WRITE(6,81)NBND
C
C     INTIALIZING GLOBAL STIFFNESS,LOAD AND DISPLACEMENT MATRICES
C
      DO 70 I=1,NDOF
      SLOD(I,1)=0.0
      SDIS(I,1)=0.0
      DO 70 J=1,NBND
   70 SYTF(I,J)=0.0
C
C     FORMATION OF STIFFNESS ARRAY
C
      IF(IPR1.EQ.1) WRITE(6,99)
      DO 200 N=1,NUMEL
      AREA=ABS((AJ(N)*BK(N)-AK(N)*BJ(N))*.5)
      COMM=.25*ET(N)/((1.-XU(N)**2)*AREA)
      A(1,1)=BJ(N)-BK(N)
      A(1,2)=0.0
      A(1,3)=BK(N)
      A(1,4)=0.0
      A(1,5)=-BJ(N)
      A(1,6)=0.0
```

```
      A(2,1)=Ø.Ø
      A(2,2)=AK(N)-AJ(N)
      A(2,3)=Ø.Ø
      A(2,4)=-AK(N)
      A(2,5)=Ø.Ø
      A(2,6)=AJ(N)
      A(3,1)=AK(N)-AJ(N)
      A(3,2)=BJ(N)-BK(N)
      A(3,3)=-AK(N)
      A(3,4)=BK(N)
      A(3,5)=AJ(N)
      A(3,6)=-BJ(N)
      B(1,1)=COMM
      B(1,2)=COMM*XU(N)
      B(1,3)=Ø.Ø
      B(2,1)=COMM*XU(N)
      B(2,2)=COMM
      B(2,3)=Ø.Ø
      B(3,1)=Ø.Ø
      B(3,2)=Ø.Ø
      B(3,3)=COMM*(1.-XU(N))*.5
C
      DO 182 J=1,6
      DO 182 I=1,3
      S(I,J)=Ø.Ø
      DO 182 K=1,3
  182 S(I,J)=S(I,J)+B(I,K)*A(K,J)
      DO 183 J=1,6
      DO 183 I=1,3
  183 B(J,I)=S(I,J)
      DO 184 J=1,6
      DO 184 I=1,6
      S(I,J)=Ø.Ø
      DO 184 K=1,3
  184 S(I,J)=S(I,J)+B(I,K)*A(K,J)
      IF(IPR1.EQ.1) WRITE(6,2Ø7) N
      IF(IPR1.EQ.1) WRITE(6,2Ø8)((S(I,J),J=1,6),I=1,6)
      I=NPI(N)
      J=NPJ(N)
      K=NPK(N)
      ITAB(1)=ICOR(I,1)
      ITAB(2)=ICOR(I,2)
      ITAB(3)=ICOR(J,1)
      ITAB(4)=ICOR(J,2)
      ITAB(5)=ICOR(K,1)
      ITAB(6)=ICOR(K,2)
C
C     MODIFICATIONS FOR SLOPING BOUNDARIES
C
      ICON=Ø
      IDEX=Ø
      DO 1ØØ IB=1,NUMBC
      L=NPB(IB)
      M=NFIX(IB)
      IF(M.NE.3) GO TO 1ØØ
      IF(I.EQ.L)IDEX=1
      IF(I.EQ.L) ICON=IB
      IF(I.EQ.L) GO TO 225
      IF(J.EQ.L)IDEX=2
      IF(J.EQ.L) ICON=IB
      IF(J.EQ.L) GO TO 225
      IF(K.EQ.L)IDEX=3
      IF(K.EQ.L) ICON=IB
      IF(K.EQ.L) GO TO 225
  1ØØ CONTINUE
  225 CONTINUE
      IF(IDEX.EQ.Ø) GO TO 12Ø
      ANGL=SLOPE(ICON)
      CALL TRAMAT(IDEX,ANGL,TRAN)
      IF(IPR1.EQ.1)WRITE(6,232)N
  232 FORMAT(/5X,42HTRANSFORMATION MATRIX FOR SLOPING BOUNDARY,I5)
      IF(IPR1.EQ.1) WRITE(6,2Ø8)((TRAN(I,J),J=1,6),I=1,6)
      CALL MATMUL(TRAN,S,TEMP)
      IF(IPR1.EQ.1) WRITE(6,2Ø8)((TEMP(I,J),J=1,6),I=1,6)
```

```
      IF(IPR1.EQ.1) WRITE(6,233)N
  233 FORMAT(/5X,39HTRANSFORMED STIFFNESS MATRIX OF ELEMENT,I5)
      DO 115 I=1,6
      DO 115 J=1,6
  115 S(I,J)=TEMP(I,J)
  12Ø CONTINUE
C
C     ASSEMBLING THE GLOBAL STIFFNESS MATRIX
C
      DO 19Ø I=1,6
      DO 19Ø J=1,6
      K=ITAB(I)
      L=ITAB(J)
      IF(K*L.EQ.Ø) GO TO 19Ø
      IF(K.LT.L) GO TO 19Ø
      M=NBND-K+L
      IF(M.LE.Ø) GO TO 19Ø
      SYTF(K,M)=SYTF(K,M)+S(I,J)
  19Ø CONTINUE
  2ØØ CONTINUE
      IF(IPR2.NE.1) GO TO 216
      WRITE(6,99)
      WRITE(6,211)
      DO 215 I=1,NDOF
      WRITE(6,212)I
      WRITE(6,213)(SYTF(I,J),J=1,NBND)
  215 CONTINUE
  216 CONTINUE
C
C      ASSEMBLING OF GLOBAL LOAD VECTOR
C
      DO 22Ø I=1,NUMNP
      K=ICOR(I,1)
      L=ICOR(I,2)
      IF(K.NE.Ø)SLOD(K,1)=XLOAD(I)
      IF(L.NE.Ø)SLOD(L,1)=YLOAD(I)
  22Ø CONTINUE
      IF(IPR2.EQ.1) WRITE(6,221)
      IF(IPR2.EQ.1) WRITE(6,222)(SLOD(I,1),I=1,NDOF)
C
C      SOLUTION OF MATRIX EQUATION
C
      NC=NBND-1
      CALL LEQ1PB(SYTF,NDOF,NC,1ØØ,SLOD,1ØØ,1,2Ø,D1,D2,IER)
      IF(IER.EQ.129) WRITE(6,231)
      IF(IER.EQ.129) GO TO 999
C
C     COMPUTATION OF NODAL DISPLACEMENTS
C
      DO 23Ø I=1,NDOF
  23Ø SDIS(I,1)=SLOD(I,1)
      DO 25Ø I=1,NUMNP
      ICON=Ø
      IDEX=Ø
C
C      MODIFICATIONS OF DISPLACEMENTS AT BOUNDARY NODES
C
      DO 24Ø J=1,NUMBC
      M1=NPB(J)
      M2=NFIX(J)
      IF(M1.NE.I)GO TO 24Ø
      IF(M2.EQ.3)ICON=J
  24Ø CONTINUE
      K=ICOR(I,1)
      L=ICOR(I,2)
      DSX(I)=Ø.Ø
      DSY(I)=Ø.Ø
      IF(K.NE.Ø) DSX(I)=SDIS(K,1)
      IF(L.NE.Ø) DSY(I)=SDIS(L,1)
      IF(ICON.EQ.Ø) GO TO 26Ø
      DIS=DSX(I)
      ANGL=SLOPE(ICON)
      CALL DISCAL(DIS,ANGL,D1,D2)
      DSX(I)=D1
```

```
      DSY(I)=D2
  200 CONTINUE
  250 CONTINUE
C
C     PRINT OF DISPLACEMENTS AND STRESSES
C
      WRITE(6,117)
      WRITE (6,121)
      WRITE (6,122) (NPNUM(M),DSX(M),DSY(M),M=1,NUMNP)
      WRITE(6,116)
      WRITE (6,123)
      DO 420 N=1,NUMEL
      I=NPI(N)
      J=NPJ(N)
      K=NPK(N)
      EPX=(BJ(N)-BK(N))*DSX(I)+BK(N)*DSX(J)-BJ(N)*DSX(K)
      EPY=(AK(N)-AJ(N))*DSY(I)-AK(N)*DSY(J)+AJ(N)*DSY(K)
      GAM=(AK(N)-AJ(N))*DSX(I)-AK(N)*DSX(J)+AJ(N)*DSX(K)
     $   +(BJ(N)-BK(N))*DSY(I)+BK(N)*DSY(J)-BJ(N)*DSY(K)
      COMM=ET(N)/((1.-XU(N)**2)*(AJ(N)*BK(N)-AK(N)*BJ(N)))
      X=COMM*(EPX+XU(N)*EPY)+THERM(N)
      Y=COMM*(EPY+XU(N)*EPX)+THERM(N)
      XY=COMM*GAM*(1.-XU(N))*.5
      SIGXX(N)=X
      SIGYY(N)=Y
      SIGXY(N)=XY
      C=(X+Y)/2.0
      R=SQRT(((Y-X)/2.0)**2+XY**2)
      XMAX=C+R
      XMIN=C-R
      X=X/THIK(N)
      Y=Y/THIK(N)
      XY=XY/THIK(N)
      XMAX=XMAX/THIK(N)
      XMIN=XMIN/THIK(N)
      PA=0.5*57.29578*ATAN(2.*XY/(Y-X))
      IF (2.*X-XMAX-XMIN) 405,420,420
  405 IF (PA) 410,420,415
  410 PA=PA+90.0
      GO TO 420
  415 PA=PA-90.0
  420 WRITE (6,124)   NUME(N),X,Y,XY,XMAX,XMIN,PA

C
      GO TO 999
C
C     PRINT OF ERRORS IN INPUT DATA
C
  701 WRITE (6,711) N
C
C     FORMAT STATEMENTS
C
    1 FORMAT(6I5,2E10.4,1I4)
    2 FORMAT(4I5,6E10.4)
    3 FORMAT(1I5,4E10.4)
    4 FORMAT(2I5,3X,1E10.4)
    5 FORMAT(4I5,6E11.4)
    6 FORMAT(10A8)
    7 FORMAT(2I5,1E10.4)
   61 FORMAT(32H ORDER OF STIFFNESS MATRIX IS =        I5)
   81 FORMAT(41H MAXIMUM BAND WIDTH OF STIFFNESS MATRIX =     1I5)
   99 FORMAT (1H1)
  101 FORMAT(29H NUMBER OF ELEMENTS           =1I4/)
  102 FORMAT(29H NUMBER OF NODAL POINTS       =1I4/)
  103 FORMAT(29H NUMBER OF BOUNDARY POINTS    =1I4/)
  104 FORMAT(40H ELEMENT STIFFNESS MATRICES ARE PRINTED   )
  105 FORMAT(52H GLOBAL STIFFNESS MATRIX AND LOAD VECTOR ARE PRINTED    )
  109 FORMAT(1I8,4F12.5)
  110 FORMAT(3X,50H EL.   I    J    K         E          THIKNESS  DENSITY ,
     $ 30H    POISSON     ALPHA       DELTA T)
  111 FORMAT(55H       NP         X-ORD          Y-ORD        X-LOAD        Y-LOAD)
  112 FORMAT (20H BOUNDARY CONDITIONS/20H  NODE  TYPE   ANGLE)
  113 FORMAT(/5X,12HELEMENT DATA)
  114 FORMAT(/5X,10HNODAL DATA)
```

```
  116 FORMAT(/5X,16HELEMENT STRESSES)
  117 FORMAT(/5X,19HNODAL DISPLACEMENTS)
  121 FORMAT (/41HNODAL POINT X-DISPLACEMENT Y-DISPLACEMENT)
  122 FORMAT (1I12,2E15.6)
  123 FORMAT( 119H  ELEMENT              X-STRESS             Y-STRESS
     $      XY-STRESS          MAX-STRESS    MIN-STRESS       DIRECTION)
  124 FORMAT (1I1Ø,3F2Ø.4,5X,3F15.2)
  2Ø7 FORMAT(3ØH STIFFNESS MATRIX OF ELEMENT   I5)
  2Ø8 FORMAT(6E15.6)
  211 FORMAT(3ØH GLOBAL STIFFNESS MATRIX           )
  212 FORMAT(6H ROW  1I5)
  213 FORMAT(1ØX,8E13.6)
  221 FORMAT(3ØH GLOBAL LOAD VECTOR            )
  222 FORMAT(1ØX,8E13.6/)
  231 FORMAT(5ØH*** STIFFNESS MATRIX IS NOT POSITIVE DEFINITE ***       )
  711 FORMAT (32HØZERO OR NEGATIVE AREA, EL. NO.=1I4)
C
  999 STOP
      END
      SUBROUTINE TRAMAT(N,A,T)
C
C     PROGRAM TO GENERATE TRANSFORMATION MATRIX FOR SLOPING BOUNDARY
C
      DIMENSION T(6,6)
      PI=4.Ø*ATAN(1.Ø)
      S=9Ø.Ø-A
      S=(S*PI)/18Ø.Ø
      AL=COS(S)
      AU=SIN(S)
      DO 1Ø I=1,6
      DO 1Ø J=1,6
   1Ø T(I,J)=Ø.Ø
      DO 2Ø I=1,6
   2Ø T(I,I)=1.Ø
      M=(N-1)*2+1
      T(M,M)=AL
      T(M+1,M+1)=AL
      T(M,M+1)=AU
      T(M+1,M)=-AU
      RETURN
      END
      SUBROUTINE MATMUL(A,B,C)
C
C     PROGRAM TO COMPUTE (C)=TRANSPOSE(A)*(B)*(A)
C
      DIMENSION  A(6,6),B(6,6),C(6,6),D(6,6)
      DO 1ØØ I=1,6
      DO 1ØØ J=1,6
      D(I,J)=Ø.Ø
      DO 1ØØ K=1,6
  1ØØ D(I,J)=D(I,J)+B(I,K)*A(K,J)
      DO 2ØØ I=1,6
      DO 2ØØ J=1,6
      C(I,J)=Ø.Ø
      DO 2ØØ K=1,6
  2ØØ C(I,J)=C(I,J)+A(K,I)*D(K,J)
      RETURN
      END
      SUBROUTINE DISCAL(D,A,D1,D2)
C
C    PROGRAM TO TRANSFORM DISPLACEMENTS AT BOUNDARY NODES
C
      PI=4.Ø*ATAN(1.Ø)
      S=9Ø.Ø-A
      S=(S*PI)/18Ø.Ø
      D1=D*COS(S)
      D2=-D*SIN(S)
      RETURN
      END
CANTILEVER PLATE ONE EDGE FIXED AND OTHER EDGE FREE
  16   15    3
   1    1    4     5+.3ØØØE+Ø8+.1ØØØE+ØØ+.ØØØØE+ØØ+.3ØØØE+ØØ
   2    1    5     2+.3ØØØE+Ø8+.1ØØØE+ØØ+.ØØØØE+ØØ+.3ØØØE+ØØ
   3    2    5     3+.3ØØØE+Ø8+.1ØØØE+ØØ+.ØØØØE+ØØ+.3ØØØE+ØØ
```

```
 4     5     6      3+.3ØØØE+Ø8+.1ØØØE+ØØ+.ØØØØE+ØØ+.3ØØØE+ØØ
 5     4     7      8+.3ØØØE+Ø8+.1ØØØE+ØØ+.ØØØØE+ØØ+.3ØØØE+ØØ
 6     4     8      5+.3ØØØE+Ø8+.1ØØØE+ØØ+.ØØØØE+ØØ+.3ØØØE+ØØ
 7     5     8      6+.3ØØØE+Ø8+.1ØØØE+ØØ+.ØØØØE+ØØ+.3ØØØE+ØØ
 8     8     9      6+.3ØØØE+Ø8+.1ØØØE+ØØ+.ØØØØE+ØØ+.3ØØØE+ØØ
 9     7    1Ø     11+.3ØØØE+Ø8+.1ØØØE+ØØ+.ØØØØE+ØØ+.3ØØØE+ØØ
1Ø     7    11      8+.3ØØØE+Ø8+.1ØØØE+ØØ+.ØØØØE+ØØ+.3ØØØE+ØØ
11     8    11      9+.3ØØØE+Ø8+.1ØØØE+ØØ+.ØØØØE+ØØ+.3ØØØE+ØØ
12    11    12      9+.3ØØØE+Ø8+.1ØØØE+ØØ+.ØØØØE+ØØ+.3ØØØE+ØØ
13    1Ø    13     14+.3ØØØE+Ø8+.1ØØØE+ØØ+.ØØØØE+ØØ+.3ØØØE+ØØ
14    1Ø    14     11+.3ØØØE+Ø8+.1ØØØE+ØØ+.ØØØØE+ØØ+.3ØØØE+ØØ
15    11    14     12+.3ØØØE+Ø8+.1ØØØE+ØØ+.ØØØØE+ØØ+.3ØØØE+ØØ
16    14    15     12+.3ØØØE+Ø8+.1ØØØE+ØØ+.ØØØØE+ØØ+.3ØØØE+ØØ
 1+.ØØØØE+ØØ+.ØØØØE+ØØ
 2+.ØØØØE+ØØ+.1ØØØE+Ø2
 3+.ØØØØE+ØØ+.2ØØØE+Ø2
 4+.1ØØØE+Ø2+.ØØØØE+ØØ
 5+.1ØØØE+Ø2+.1ØØØE+Ø2
 6+.1ØØØE+Ø2+.2ØØØE+Ø2
 7+.2ØØØE+Ø2+.ØØØØE+ØØ
 8+.2ØØØE+Ø2+.1ØØØE+Ø2
 9+.2ØØØE+Ø2+.2ØØØE+Ø2
1Ø+.3ØØØE+Ø2+.ØØØØE+ØØ
11+.3ØØØE+Ø2+.1ØØØE+Ø2
12+.3ØØØE+Ø2+.2ØØØE+Ø2
13+.4ØØØE+Ø2+.ØØØØE+ØØ+.5ØØØE+Ø4+.ØØØØE+ØØ
14+.4ØØØE+Ø2+.1ØØØE+Ø2+.1ØØØE+Ø5+.ØØØØE+ØØ
15+.4ØØØE+Ø2+.2ØØØE+Ø2+.5ØØØE+Ø4+.ØØØØE+ØØ
 1     Ø
 2     Ø
 3     Ø
```

```
CANTILEVER PLATE ONE EDGE FIXED AND OTHER EDGE FREE
NUMBER OF ELEMENTS          =  16

NUMBER OF NODAL POINTS      =  15

NUMBER OF BOUNDARY POINTS   =   3


  ELEMENT DATA
 EL.    I    J    K      E           THIKNESS   DENSITY      POISSON     ALPHA        DELTA T
  1     1    4    5 Ø.3ØØØE+Ø8 Ø.1ØØØE+ØØ Ø.ØØØØE+ØØ Ø.3ØØØE+ØØ Ø.ØØØØE+ØØ Ø.ØØØØE+ØØ
  2     1    5    2 Ø.3ØØØE+Ø8 Ø.1ØØØE+ØØ Ø.ØØØØE+ØØ Ø.3ØØØE+ØØ Ø.ØØØØE+ØØ Ø.ØØØØE+ØØ
  3     2    5    3 Ø.3ØØØE+Ø8 Ø.1ØØØE+ØØ Ø.ØØØØE+ØØ Ø.3ØØØE+ØØ Ø.ØØØØE+ØØ Ø.ØØØØE+ØØ
  4     5    6    3 Ø.3ØØØE+Ø8 Ø.1ØØØE+ØØ Ø.ØØØØE+ØØ Ø.3ØØØE+ØØ Ø.ØØØØE+ØØ Ø.ØØØØE+ØØ
  5     4    7    8 Ø.3ØØØE+Ø8 Ø.1ØØØE+ØØ Ø.ØØØØE+ØØ Ø.3ØØØE+ØØ Ø.ØØØØE+ØØ Ø.ØØØØE+ØØ
  6     4    8    5 Ø.3ØØØE+Ø8 Ø.1ØØØE+ØØ Ø.ØØØØE+ØØ Ø.3ØØØE+ØØ Ø.ØØØØE+ØØ Ø.ØØØØE+ØØ
  7     5    8    6 Ø.3ØØØE+Ø8 Ø.1ØØØE+ØØ Ø.ØØØØE+ØØ Ø.3ØØØE+ØØ Ø.ØØØØE+ØØ Ø.ØØØØE+ØØ
  8     8    9    6 Ø.3ØØØE+Ø8 Ø.1ØØØE+ØØ Ø.ØØØØE+ØØ Ø.3ØØØE+ØØ Ø.ØØØØE+ØØ Ø.ØØØØE+ØØ
  9     7   1Ø   11 Ø.3ØØØE+Ø8 Ø.1ØØØE+ØØ Ø.ØØØØE+ØØ Ø.3ØØØE+ØØ Ø.ØØØØE+ØØ Ø.ØØØØE+ØØ
 1Ø     7   11    8 Ø.3ØØØE+Ø8 Ø.1ØØØE+ØØ Ø.ØØØØE+ØØ Ø.3ØØØE+ØØ Ø.ØØØØE+ØØ Ø.ØØØØE+ØØ
 11     8   11    9 Ø.3ØØØE+Ø8 Ø.1ØØØE+ØØ Ø.ØØØØE+ØØ Ø.3ØØØE+ØØ Ø.ØØØØE+ØØ Ø.ØØØØE+ØØ
 12    11   12    9 Ø.3ØØØE+Ø8 Ø.1ØØØE+ØØ Ø.ØØØØE+ØØ Ø.3ØØØE+ØØ Ø.ØØØØE+ØØ Ø.ØØØØE+ØØ
 13    1Ø   13   14 Ø.3ØØØE+Ø8 Ø.1ØØØE+ØØ Ø.ØØØØE+ØØ Ø.3ØØØE+ØØ Ø.ØØØØE+ØØ Ø.ØØØØE+ØØ
 14    1Ø   14   11 Ø.3ØØØE+Ø8 Ø.1ØØØE+ØØ Ø.ØØØØE+ØØ Ø.3ØØØE+ØØ Ø.ØØØØE+ØØ Ø.ØØØØE+ØØ
 15    11   14   12 Ø.3ØØØE+Ø8 Ø.1ØØØE+ØØ Ø.ØØØØE+ØØ Ø.3ØØØE+ØØ Ø.ØØØØE+ØØ Ø.ØØØØE+ØØ
 16    14   15   12 Ø.3ØØØE+Ø8 Ø.1ØØØE+ØØ Ø.ØØØØE+ØØ Ø.3ØØØE+ØØ Ø.ØØØØE+ØØ Ø.ØØØØE+ØØ
  NODAL DATA
  NP        X-ORD        Y-ORD        X-LOAD        Y-LOAD
     1     Ø.ØØØØØØ     Ø.ØØØØØØ      Ø.ØØØØØØ      Ø.ØØØØØØ
     2     Ø.ØØØØØØ    1Ø.ØØØØØØ      Ø.ØØØØØØ      Ø.ØØØØØØ
     3     Ø.ØØØØØØ    2Ø.ØØØØØØ      Ø.ØØØØØØ      Ø.ØØØØØØ
     4    1Ø.ØØØØØØ     Ø.ØØØØØØ      Ø.ØØØØØØ      Ø.ØØØØØØ
     5    1Ø.ØØØØØØ    1Ø.ØØØØØØ      Ø.ØØØØØØ      Ø.ØØØØØØ
     6    1Ø.ØØØØØØ    2Ø.ØØØØØØ      Ø.ØØØØØØ      Ø.ØØØØØØ
     7    2Ø.ØØØØØØ     Ø.ØØØØØØ      Ø.ØØØØØØ      Ø.ØØØØØØ
     8    2Ø.ØØØØØØ    1Ø.ØØØØØØ      Ø.ØØØØØØ      Ø.ØØØØØØ
     9    2Ø.ØØØØØØ    2Ø.ØØØØØØ      Ø.ØØØØØØ      Ø.ØØØØØØ
    1Ø    3Ø.ØØØØØØ     Ø.ØØØØØØ      Ø.ØØØØØØ      Ø.ØØØØØØ
    11    3Ø.ØØØØØØ    1Ø.ØØØØØØ      Ø.ØØØØØØ      Ø.ØØØØØØ
    12    3Ø.ØØØØØØ    2Ø.ØØØØØØ      Ø.ØØØØØØ      Ø.ØØØØØØ
    13    4Ø.ØØØØØØ     Ø.ØØØØØØ   5ØØØ.ØØØØØØ      Ø.ØØØØØØ
    14    4Ø.ØØØØØØ    1Ø.ØØØØØØ  1ØØØØ.ØØØØØØ      Ø.ØØØØØØ
    15    4Ø.ØØØØØØ    2Ø.ØØØØØØ   5ØØØ.ØØØØØØ      Ø.ØØØØØØ
BOUNDARY CONDITIONS
 NODE  TYPE   ANGLE
   1     Ø    Ø.ØØØØE+ØØ
   2     Ø    Ø.ØØØØE+ØØ
   3     Ø    Ø.ØØØØE+ØØ
```

```
ORDER OF STIFFNESS MATRIX IS =     24
MAXIMUM BAND WIDTH OF STIFFNESS MATRIX =    19

    NODAL DISPLACEMENTS

NODAL POINT X-DISPLACEMENT Y-DISPLACEMENT
          1  Ø.ØØØØØØE+ØØ  Ø.ØØØØØØE+ØØ
          2  Ø.ØØØØØØE+ØØ  Ø.ØØØØØØE+ØØ
          3  Ø.ØØØØØØE+ØØ  Ø.ØØØØØØE+ØØ
          4  Ø.331394E-Ø2  Ø.8955Ø4E-Ø3
          5  Ø.3Ø2137E-Ø2  Ø.1ØØ243E-15
          6  Ø.331394E-Ø2 -Ø.8955Ø4E-Ø3
          7  Ø.654741E-Ø2  Ø.989328E-Ø3
          8  Ø.642ØØ3E-Ø2  Ø.33Ø66ØE-15
          9  Ø.654741E-Ø2 -Ø.989328E-Ø3
         1Ø  Ø.983974E-Ø2  Ø.997792E-Ø3
         11  Ø.979Ø5ØE-Ø2  Ø.615262E-15
         12  Ø.983974E-Ø2 -Ø.997792E-Ø3
         13  Ø.131663E-Ø1  Ø.1ØØ68ØE-Ø2
         14  Ø.13132ØE-Ø1  Ø.96635ØE-15
         15  Ø.131663E-Ø1 -Ø.1ØØ68ØE-Ø2

    ELEMENT STRESSES
 ELEMENT       X-STRESS       Y-STRESS      XY-STRESS     MAX-STRESS  MIN-STRESS   DIRECTION
       1     1ØØ39.4257       325.3164       695.6935       1ØØ89.ØØ      275.75       -4.Ø8
       2      996Ø.5743      2988.1723         Ø.ØØØØ        996Ø.57     2988.17        Ø.ØØ
       3      996Ø.5743      2988.1723         Ø.ØØØØ        996Ø.57     2988.17        Ø.ØØ
       4     1ØØ39.4257       325.3164      -695.6935       1ØØ89.ØØ      275.75        4.Ø8
       5      9681.3123       -63.5889       -38.718Ø        9681.47      -63.74        Ø.23
       6     1Ø318.6877       4Ø9.Ø95Ø      -337.58Ø1       1Ø33Ø.17      397.61        1.95
       7     1Ø318.6877       4Ø9.Ø95Ø       337.58Ø1       1Ø33Ø.17      397.61       -1.95
       8      9681.3123       -63.5889        38.718Ø        9681.47      -63.74       -Ø.23
       9      9867.ØØ65       -33.2728       -47.Ø444        9867.23      -33.5Ø        Ø.27
      1Ø     1Ø132.9935        71.9154      -146.9762       1Ø135.14       69.77        Ø.84
      11     1Ø132.9935        71.9154       146.9762       1Ø135.14       69.77       -Ø.84
      12      9867.ØØ65       -33.2728        47.Ø444        9867.23      -33.5Ø       -Ø.27
      13      997Ø.8615       -29.1385       -29.1385        997Ø.95      -29.22        Ø.17
      14     1ØØ29.1385        15.3668       -56.81Ø6       1ØØ29.46       15.Ø4        Ø.33
      15     1ØØ29.1385        15.3668        56.81Ø6       1ØØ29.46       15.Ø4       -Ø.33
      16      997Ø.8615       -29.1385        29.1385        997Ø.95      -29.22       -Ø.17
```

```
CANTILEVER PLATE ONE EDGE FIXED AND OTHER EDGE ON SLOPING SUPPORTS
  16    15     5
   1     1     4     5+.3ØØØE+Ø8+.1ØØØE+ØØ+.ØØØØE+ØØ+.3ØØØE+ØØ
   2     1     5     2+.3ØØØE+Ø8+.1ØØØE+ØØ+.ØØØØE+ØØ+.3ØØØE+ØØ
   3     2     5     3+.3ØØØE+Ø8+.1ØØØE+ØØ+.ØØØØE+ØØ+.3ØØØE+ØØ
   4     5     6     3+.3ØØØE+Ø8+.1ØØØE+ØØ+.ØØØØE+ØØ+.3ØØØE+ØØ
   5     4     7     8+.3ØØØE+Ø8+.1ØØØE+ØØ+.ØØØØE+ØØ+.3ØØØE+ØØ
   6     4     8     5+.3ØØØE+Ø8+.1ØØØE+ØØ+.ØØØØE+ØØ+.3ØØØE+ØØ
   7     5     8     6+.3ØØØE+Ø8+.1ØØØE+ØØ+.ØØØØE+ØØ+.3ØØØE+ØØ
   8     8     9     6+.3ØØØE+Ø8+.1ØØØE+ØØ+.ØØØØE+ØØ+.3ØØØE+ØØ
   9     7    1Ø    11+.3ØØØE+Ø8+.1ØØØE+ØØ+.ØØØØE+ØØ+.3ØØØE+ØØ
  1Ø     7    11     8+.3ØØØE+Ø8+.1ØØØE+ØØ+.ØØØØE+ØØ+.3ØØØE+ØØ
  11     8    11     9+.3ØØØE+Ø8+.1ØØØE+ØØ+.ØØØØE+ØØ+.3ØØØE+ØØ
  12    11    12     9+.3ØØØE+Ø8+.1ØØØE+ØØ+.ØØØØE+ØØ+.3ØØØE+ØØ
  13    1Ø    13    14+.3ØØØE+Ø8+.1ØØØE+ØØ+.ØØØØE+ØØ+.3ØØØE+ØØ
  14    1Ø    14    11+.3ØØØE+Ø8+.1ØØØE+ØØ+.ØØØØE+ØØ+.3ØØØE+ØØ
  15    11    14    12+.3ØØØE+Ø8+.1ØØØE+ØØ+.ØØØØE+ØØ+.3ØØØE+ØØ
  16    14    15    12+.3ØØØE+Ø8+.1ØØØE+ØØ+.ØØØØE+ØØ+.3ØØØE+ØØ
   1+.ØØØØE+ØØ+.ØØØØE+ØØ
   2+.ØØØØE+ØØ+.1ØØØE+Ø2
   3+.ØØØØE+ØØ+.2ØØØE+Ø2
   4+.1ØØØE+Ø2+.ØØØØE+ØØ
   5+.1ØØØE+Ø2+.1ØØØE+Ø2
   6+.1ØØØE+Ø2+.2ØØØE+Ø2
   7+.2ØØØE+Ø2+.ØØØØE+ØØ
   8+.2ØØØE+Ø2+.1ØØØE+Ø2
   9+.2ØØØE+Ø2+.2ØØØE+Ø2
  1Ø+.3ØØØE+Ø2+.ØØØØE+ØØ
  11+.3ØØØE+Ø2+.1ØØØE+Ø2
  12+.3ØØØE+Ø2+.2ØØØE+Ø2
  13+.4ØØØE+Ø2+.ØØØØE+ØØ+.ØØØØE+ØØ+.ØØØØE+ØØ
  14+.4ØØØE+Ø2+.1ØØØE+Ø2+.1ØØØE+Ø5+.ØØØØE+ØØ
  15+.4ØØØE+Ø2+.2ØØØE+Ø2+.ØØØØE+ØØ+.ØØØØE+ØØ
   1     Ø
   2     Ø
   3     Ø
  13     3+.135ØE+Ø3
  15     3+.225ØE+Ø3
```

```
CANTILEVER PLATE ONE EDGE FIXED AND OTHER EDGE ON SLOPING SUPPORTS
NUMBER OF ELEMENTS          =  16

NUMBER OF NODAL POINTS      =  15

NUMBER OF BOUNDARY POINTS   =   5


  ELEMENT DATA
 EL.    I    J    K       E          THIKNESS   DENSITY     POISSON    ALPHA       DELTA T
  1     1    4    5 Ø.3ØØØE+Ø8 Ø.1ØØØE+ØØ Ø.ØØØØE+ØØ Ø.3ØØØE+ØØ Ø.ØØØØE+ØØ Ø.ØØØØE+ØØ
  2     1    5    2 Ø.3ØØØE+Ø8 Ø.1ØØØE+ØØ Ø.ØØØØE+ØØ Ø.3ØØØE+ØØ Ø.ØØØØE+ØØ Ø.ØØØØE+ØØ
  3     2    5    3 Ø.3ØØØE+Ø8 Ø.1ØØØE+ØØ Ø.ØØØØE+ØØ Ø.3ØØØE+ØØ Ø.ØØØØE+ØØ Ø.ØØØØE+ØØ
  4     5    6    3 Ø.3ØØØE+Ø8 Ø.1ØØØE+ØØ Ø.ØØØØE+ØØ Ø.3ØØØE+ØØ Ø.ØØØØE+ØØ Ø.ØØØØE+ØØ
  5     4    7    8 Ø.3ØØØE+Ø8 Ø.1ØØØE+ØØ Ø.ØØØØE+ØØ Ø.3ØØØE+ØØ Ø.ØØØØE+ØØ Ø.ØØØØE+ØØ
  6     4    8    5 Ø.3ØØØE+Ø8 Ø.1ØØØE+ØØ Ø.ØØØØE+ØØ Ø.3ØØØE+ØØ Ø.ØØØØE+ØØ Ø.ØØØØE+ØØ
  7     5    8    6 Ø.3ØØØE+Ø8 Ø.1ØØØE+ØØ Ø.ØØØØE+ØØ Ø.3ØØØE+ØØ Ø.ØØØØE+ØØ Ø.ØØØØE+ØØ
  8     8    9    6 Ø.3ØØØE+Ø8 Ø.1ØØØE+ØØ Ø.ØØØØE+ØØ Ø.3ØØØE+ØØ Ø.ØØØØE+ØØ Ø.ØØØØE+ØØ
  9     7   1Ø   11 Ø.3ØØØE+Ø8 Ø.1ØØØE+ØØ Ø.ØØØØE+ØØ Ø.3ØØØE+ØØ Ø.ØØØØE+ØØ Ø.ØØØØE+ØØ
 1Ø     7   11    8 Ø.3ØØØE+Ø8 Ø.1ØØØE+ØØ Ø.ØØØØE+ØØ Ø.3ØØØE+ØØ Ø.ØØØØE+ØØ Ø.ØØØØE+ØØ
 11     8   11    9 Ø.3ØØØE+Ø8 Ø.1ØØØE+ØØ Ø.ØØØØE+ØØ Ø.3ØØØE+ØØ Ø.ØØØØE+ØØ Ø.ØØØØE+ØØ
 12    11   12    9 Ø.3ØØØE+Ø8 Ø.1ØØØE+ØØ Ø.ØØØØE+ØØ Ø.3ØØØE+ØØ Ø.ØØØØE+ØØ Ø.ØØØØE+ØØ
 13    1Ø   13   14 Ø.3ØØØE+Ø8 Ø.1ØØØE+ØØ Ø.ØØØØE+ØØ Ø.3ØØØE+ØØ Ø.ØØØØE+ØØ Ø.ØØØØE+ØØ
 14    1Ø   14   11 Ø.3ØØØE+Ø8 Ø.1ØØØE+ØØ Ø.ØØØØE+ØØ Ø.3ØØØE+ØØ Ø.ØØØØE+ØØ Ø.ØØØØE+ØØ
 15    11   14   12 Ø.3ØØØE+Ø8 Ø.1ØØØE+ØØ Ø.ØØØØE+ØØ Ø.3ØØØE+ØØ Ø.ØØØØE+ØØ Ø.ØØØØE+ØØ
 16    14   15   12 Ø.3ØØØE+Ø8 Ø.1ØØØE+ØØ Ø.ØØØØE+ØØ Ø.3ØØØE+ØØ Ø.ØØØØE+ØØ Ø.ØØØØE+ØØ

  NODAL DATA
  NP        X-ORD         Y-ORD        X-LOAD         Y-LOAD
     1      Ø.ØØØØØ       Ø.ØØØØØ       Ø.ØØØØØ       Ø.ØØØØØ
     2      Ø.ØØØØØ      1Ø.ØØØØØ       Ø.ØØØØØ       Ø.ØØØØØ
     3      Ø.ØØØØØ      2Ø.ØØØØØ       Ø.ØØØØØ       Ø.ØØØØØ
     4     1Ø.ØØØØØ       Ø.ØØØØØ       Ø.ØØØØØ       Ø.ØØØØØ
     5     1Ø.ØØØØØ      1Ø.ØØØØØ       Ø.ØØØØØ       Ø.ØØØØØ
     6     1Ø.ØØØØØ      2Ø.ØØØØØ       Ø.ØØØØØ       Ø.ØØØØØ
     7     2Ø.ØØØØØ       Ø.ØØØØØ       Ø.ØØØØØ       Ø.ØØØØØ
     8     2Ø.ØØØØØ      1Ø.ØØØØØ       Ø.ØØØØØ       Ø.ØØØØØ
     9     2Ø.ØØØØØ      2Ø.ØØØØØ       Ø.ØØØØØ       Ø.ØØØØØ
    1Ø     3Ø.ØØØØØ       Ø.ØØØØØ       Ø.ØØØØØ       Ø.ØØØØØ
    11     3Ø.ØØØØØ      1Ø.ØØØØØ       Ø.ØØØØØ       Ø.ØØØØØ
    12     3Ø.ØØØØØ      2Ø.ØØØØØ       Ø.ØØØØØ       Ø.ØØØØØ
    13     4Ø.ØØØØØ       Ø.ØØØØØ       Ø.ØØØØØ       Ø.ØØØØØ
    14     4Ø.ØØØØØ      1Ø.ØØØØØ   1ØØØØ.ØØØØØ       Ø.ØØØØØ
    15     4Ø.ØØØØØ      2Ø.ØØØØØ       Ø.ØØØØØ       Ø.ØØØØØ
```

BOUNDARY CONDITIONS

NODE	TYPE	ANGLE
1	Ø	Ø.ØØØØE+ØØ
2	Ø	Ø.ØØØØE+ØØ
3	Ø	Ø.ØØØØE+ØØ
13	3	Ø.135ØE+Ø3
15	3	Ø.225ØE+Ø3

ORDER OF STIFFNESS MATRIX IS = 22

MAXIMUM BAND WIDTH OF STIFFNESS MATRIX = 19

NODAL DISPLACEMENTS

NODAL POINT	X-DISPLACEMENT	Y-DISPLACEMENT
1	Ø.ØØØØØØØE+ØØ	Ø.ØØØØØØØE+ØØ
2	Ø.ØØØØØØØE+ØØ	Ø.ØØØØØØØE+ØØ
3	Ø.ØØØØØØØE+ØØ	Ø.ØØØØØØØE+ØØ
4	Ø.5Ø425ØE-Ø3	Ø.219852E-Ø3
5	Ø.54405ØE-Ø3	-Ø.13Ø746E-16
6	Ø.5Ø425ØE-Ø3	-Ø.219852E-Ø3
7	Ø.95335ØE-Ø3	Ø.385376E-Ø3
8	Ø.125886E-Ø2	-Ø.15Ø178E-16
9	Ø.95335ØE-Ø3	-Ø.385376E-Ø3
1Ø	Ø.123161E-Ø2	Ø.721875E-Ø3
11	Ø.229513E-Ø2	Ø.131382E-16
12	Ø.123161E-Ø2	-Ø.721875E-Ø3
13	Ø.6158Ø3E-Ø3	Ø.6158Ø3E-Ø3
14	Ø.429458E-Ø2	Ø.81196ØE-16
15	Ø.6158Ø3E-Ø3	-Ø.6158Ø3E-Ø3

ELEMENT STRESSES

ELEMENT	X-STRESS	Y-STRESS	XY-STRESS	MAX-STRESS	MIN-STRESS	DIRECTION
1	1444.9269	-226.Ø778	299.5978	1497.Ø2	-278.17	-9.86
2	1793.57ØØ	538.Ø71Ø	Ø.ØØØØ	1793.57	538.Ø7	Ø.ØØ
3	1793.57ØØ	538.Ø71Ø	Ø.ØØØØ	1793.57	538.Ø7	Ø.ØØ
4	1444.9269	-226.Ø778	-299.5978	1497.Ø2	-278.17	9.86
5	1Ø99.4Ø65	-826.3Ø63	543.5Ø44	1242.21	-969.11	-14.72
6	2139.Ø9Ø3	-17.8288	45.9225	214Ø.Ø7	-18.81	-1.22
7	2139.Ø9Ø3	-17.8288	-45.9225	214Ø.Ø7	-18.81	1.22
8	1Ø99.4Ø65	-826.3Ø63	-543.5Ø44	1242.21	-969.11	14.72
9	2Ø3.3871	-21Ø4.61Ø4	1615.4Ø6Ø	1Ø34.65	-2935.87	-27.23
1Ø	3Ø35.1Ø98	-245.5953	352.515Ø	3Ø72.56	-283.Ø5	-6.Ø6
11	3Ø35.1Ø98	-245.5953	-352.515Ø	3Ø72.56	-283.Ø5	6.Ø6
12	2Ø3.3871	-21Ø4.61Ø4	-1615.4Ø6Ø	1Ø34.65	-2935.87	27.23
13	-2639.1563	-2639.1563	4122.3469	1483.19	-6761.5Ø	-45.ØØ
14	5877.6531	-4Ø2.33Ø5	1227.1374	61Ø8.92	-633.6Ø	-1Ø.67
15	5877.6531	-4Ø2.33Ø5	-1227.1374	61Ø8.92	-633.6Ø	1Ø.67
16	-2639.1563	-2639.1563	-4122.3469	1483.19	-6761.5Ø	45.ØØ

13.4 SIXTEEN-D.O.F. RECTANGULAR PLATE ELEMENT IN BENDING

This program is written for the 16-d.o.f. rectangular plate element in bending as described in Chapter 12. Each nodal point possesses four degrees of freedom: transverse deflection w, rotation about the y axis $\partial w/\partial x$, rotation about the x axis $\partial w/\partial y$, and a twist derivative $\partial^2 w/\partial x\,\partial y$. Corresponding to the four nodal degrees of freedom, four respective nodal forces can be applied. The sign conventions of all these forces and d.o.f.'s are defined in Chapter 12. As an alternative input option, uniformly distributed load may be applied, which is converted to work-equivalent or consistent nodal loads. Output includes all nodal degrees of freedom and nodal forces in shear, bending, and twisting moments for each element.

The stiffness matrices are assembled and stored in band form. The program calls an IMSL Library Subroutine LEQ1PB which solves the symmetric band stiffness matrix by the Gaussian elimination method. If LEQ1PB is not available, the user can still use this program by simply changing the statement that calls LEQ1PB to the one that calls any available equivalent subroutine.

In this program, provision is made in the DIMENSION statement for 200 elements, 200 nodal points, and 100 free-to-move degrees of freedom. The user can easily modify the DIMENSION statements to cope with more elements, nodes, and free d.o.f.'s.

13.4.1 Description of Input Data

1. Title Card (10A8)
   Columns 1–80: Arbitrary problem identification
2. Master Control Card (4I5)
   One card to define the number of nodal points and elements and to specify print options
   Columns 1–5: Number of nodal points (NODE)
   6–10: Number of elements (NELE)
   11–15: Flag to print the element stiffness matrix, element load vector, and element displacements
   IPR1 = 1 print
   IPR1 = 0 no print
   16–20: Flag to print the assembled stiffness matrix and load vector
   IPR2 = 1 print
   IPR2 = 0 no print
3. Nodal Point Data (5I4, 6E10.4)
   One card for each nodal point in order

Columns 1-4: Nodal point number (N)
5-8: Boundary condition for deflection w
IBOU(N, 1) = 1 free to deflect
IBOU(N, 1) = 0 restrained
9-12: Boundary condition for slope $\partial w/\partial x$
IBOU(N, 2) = 1 free to rotate
IBOU(N, 2) = 0 restrained
13-16: Boundary condition for slope $\partial w/\partial y$
IBOU(N, 3) = 1 free to rotate
IBOU(N, 3) = 0 restrained
17-20: Boundary condition for twist derivative $\partial^2 w/\partial x\, \partial y$
IBOU(N, 4) = 1 free
IBOU(N, 4) = 0 restrained
21-30: x coordinate [XYCO(N, 1)]
31-40: y coordinate [XYCO(N, 2)]
41-50: Transverse shear force [FORC(N, 1)]
51-60: Bending moment about y axis [FORC(N, 2)]
61-70: Bending moment about x axis [FORC(N, 3)]
71-80: Counterpart force for $\partial^2 w/\partial x\, \partial y$ [FORC(N, 4)]

4. Element Data (5I4, 4E10.4)
One card for each element in order
Columns 1-4: Element number (N)
5-8: Nodal point number 1 [NODN(N, 1)]
9-12: Nodal point number 2 [NODN(N, 2)]
13-16: Nodal point number 3 [NODN(N, 3)]
17-20: Nodal point number 4 [NODN(N, 4)]
21-30: Thickness [THKN(N)]
31-40: Modulus of elasticity [ELAS(N)]
41-50: Poisson's ratio [POIS(N)]
51-60: Uniformly distributed pressure [UDIS(N)]

13.4.2 Sample Problem

Figure 13.6 shows a square aluminum plate with all edges simply supported and subjected to transverse uniformly distributed load p_0. It is assumed that $E = 10^7$ psi, $\nu = 0.3$, and $p_0 = 0.4$ psi.

Due to double symmetry, only a quadrant of the plate need be modeled and 16 elements are used. The boundary conditions for zero degrees of freedom for this problem are well explained in Chapter 12. The input data necessary for a static analysis of this problem and the output data are given at the end of the listing of this program. Compared to the exact solution for maximum

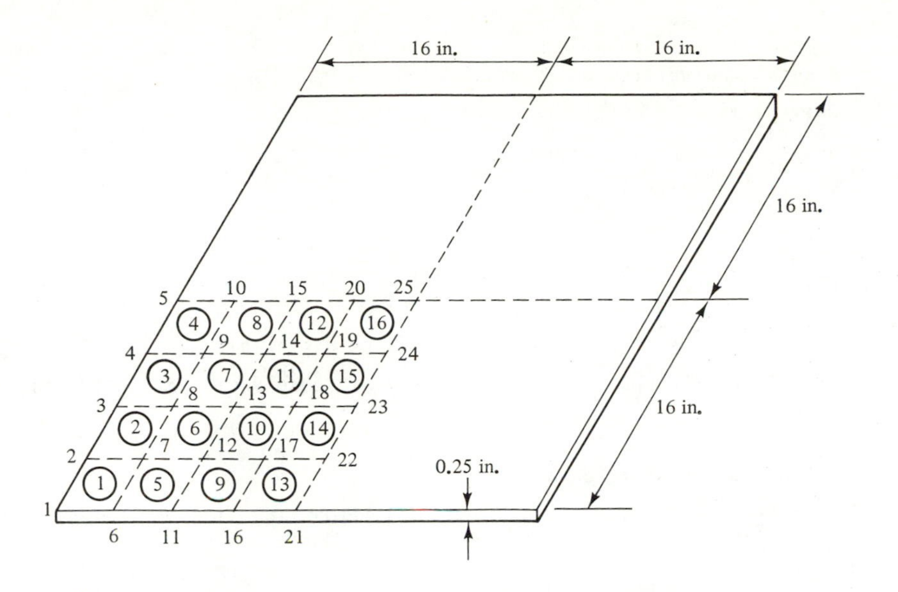

Figure 13.6 Square plate with all edges simply supported under transverse uniformly distributed load.

deflection and bending at the center (page 120 of Ref. [12.1]),

$$w_{\max} = 0.00406\frac{p_0 l^4}{D} = 0.119 \text{ in.}$$

$$(M_x)_{\max} = (M_y)_{\max} = 0.0479 p_0 l^2 = 19.62 \frac{\text{in.-lb}}{\text{in.}}$$

the present results of 0.1164 in. and 19.67 in.-lb/in. at node 25 have an error of −2.15% and 0.26%, respectively. The average value of the bending moments (M_x or M_y) at the four corner nodes of element 16 is 37.34 in.-lb or 18.67 in.-lb/in.

```
      PROGRAM MAIN
C
C     **************************************************************
C     * PROGRAM FOR STATIC ANALYSIS OF PLATES BY 16 D.O.F *
C     * RECTANGULAR FINITE ELEMENTS                        *
C     **************************************************************
      DIMENSION TITLE(10),S(16,16),JCON(6),ITAB(16),DISN(4)
     1,EDIS(16),EFOR(16),ELOD(16)
      DIMENSION IBOU(200,4),XYCO(200,2),FORC(200,4),ICOR(200,4)
      DIMENSION NODN(200,4),THKN(200),ELAS(200),POIS(200),UDIS(200)
      DIMENSION SYTF(100,100),SLOD(100,1)
C
C  FORMULATION OF ELEMENT STIFFNESS MATRIX
C
      REWIND 1
```

```
C
C     READING AND WRITING OF INPUT DATA
C
      WRITE(6,1)
    1 FORMAT(1H1)
      READ(5,2)(TITLE(I),I=1,1Ø)
    2 FORMAT(1ØA8)
      WRITE(6,2)(TITLE(I),I=1,1Ø)
      READ(5,11)NODE,NELE,IPR1,IPR2
   11 FORMAT(4I5)
      WRITE(6,12)NODE
   12 FORMAT(/5X,16HNUMBER OF NODES=,I5)
      WRITE(6,13)NELE
   13 FORMAT(/5X,19HNUMBER OF ELEMENTS=,I5)
      IF(IPR1.EQ.1)WRITE(6,14)
   14 FORMAT(/5X,38HELEMENT STIFFNESS MATRICES ARE PRINTED)
      IF(IPR2.EQ.1)WRITE(6,16)
   16 FORMAT(/5X,51HGLOBAL STIFFNESS MATRIX AND LOAD VECTOR ARE PRINTED)
C
C     READING AND WRITING OF NODAL DATA
C
      WRITE(6,21)
   21 FORMAT(/5X,1ØHNODAL DATA)
      WRITE(6,22)
   22 FORMAT(/5X,4HNODE,1ØX,19HBOUNDARY CONDITIONS,15X,11HCOORDINATES,
     135X,6HFORCES)
      WRITE(6,23)
   23 FORMAT( 19X,1H1,4X,1H2,4X,1H3,4X,1H4,2ØX,1HX,1ØX,1HY,2ØX,4HVERT,5X
     1,5HX-MOM,5X5HY-MOM,5X,5HTWIST)
      DO 3Ø I=1,NODE
      READ(5,31)N,(IBOU(I,J),J=1,4),(XYCO(I,J),J=1,2),(FORC(I,J),J=1,4)
      WRITE(6,32)N,(IBOU(I,J),J=1,4),(XYCO(I,J),J=1,2),(FORC(I,J),J=1,4)
   3Ø CONTINUE
   31 FORMAT(5I4,6E1Ø.4)
   32 FORMAT(/5X,1I5,5X,4I5,1ØX,2E12.4,1ØX,4E12.4)
C
C      READING AND WRITING OF ELEMENT DATA
C
      WRITE(6,41)
   41 FORMAT(/5X,12HELEMENT DATA)
      WRITE(6,42)
   42 FORMAT(/5X,7HELEMENT,1ØX,12HNODAL NUMBER,1ØX,9HTHICKNESS,
     1 1ØX,1ØHMODULUS OF,1ØX,8HPOISSONS,1ØX,11HDISTRIBUTED)
      WRITE(6,43)
   43 FORMAT( 5X,6HNUMBER, 9X,16H1    2    3    4,
     127X,1ØHELASTICITY,1ØX,5HRATIO,13X,4HLOAD)
      DO 5Ø I=1,NELE
      READ(5,51)N,(NODN(I,J),J=1,4),THKN(I),ELAS(I),POIS(I),UDIS(I)
   5Ø WRITE(6,52)N,(NODN(I,J),J=1,4),THKN(I),ELAS(I),POIS(I),UDIS(I)
   51 FORMAT(5I4,4E1Ø.4)
   52 FORMAT(/5X,I5,5X,4I5,4X,1E12.4,9X,1E12.4,5X,1E12.4,5X,E12.4)
C
C     GENERATION OF NUMBERS FOR ASSEMBLING STIFFNESS MATRIX
C
      ICON=Ø
      DO 6Ø I=1,NODE
      DO 6Ø J=1,4
      ICOR(I,J)=Ø
      K=IBOU(I,J)
      IF(K.EQ.Ø) GO TO 6Ø
      ICON=ICON+1
      ICOR(I,J)=ICON
   6Ø CONTINUE
      NDOF=ICON
      WRITE(6,61)NDOF
   61 FORMAT(/5X,35HNUMBER OF FREE DEGREES OF FREEDOM =,I5)
C
C     COMPUTATION OF MAXIMUM BAND WIDTH
C
      MAX=Ø
      DO 7Ø I=1,NELE
      JCON(1)=NODN(I,1)-NODN(I,2)
      JCON(2)=NODN(I,2)-NODN(I,3)
```

```
      JCON(3)=NODN(I,3)-NODN(I,4)
      JCON(4)=NODN(I,4)-NODN(I,1)
      JCON(5)=NODN(I,1)-NODN(I,3)
      JCON(6)=NODN(I,2)-NODN(I,4)
      DO 7Ø J=1,6
      K=JCON(J)
      K=IABS(K)
      IF(K.GT.MAX)MAX=K
   7Ø CONTINUE
      NBND=8*MAX+7
      IF(NBND.GT.NDOF)NBND=NDOF
      WRITE(6,71)NBND
   71 FORMAT(/5X,12HBAND WIDTH =,I5)
C
C      INTIALIZATION OF STIFFNESS MATRIX AND LOAD VECTOR
C
      DO 8Ø I=1,NDOF
      SLOD(I,1)=Ø.Ø
      DO 8Ø J=1,NBND
   8Ø SYTF(I,J)=Ø.Ø
C
C     COMPUTATION OF ELEMENT STIFFNESS MATRIX
C
      DO 5ØØ IE=1,NELE
      UM=POIS(IE)
      TH=THKN(IE)
      E=ELAS(IE)
      DO 1ØØ I=1,4
  1ØØ JCON(I)=NODN(IE,I)
      I1=JCON(1)
      I2=JCON(2)
      I3=JCON(3)
      A=XYCO(I2,1) -XYCO(I1,1)
      A=ABS(A)
      B=XYCO(I3,2)-XYCO(I2,2)
      B=ABS(B)
      DO 11Ø I=1,4
      K=(I-1)*4
      DO 11Ø L=1,4
      J=JCON(L)
  11Ø ITAB(K+L)=ICOR(J,I)
      R1=Ø.Ø
      R2=Ø.Ø
      R11=R1*R1
      R12=R1*R2
      R22=R2*R2
      DM=E*TH/(1.-UM*UM)
      DB=DM*TH*TH/12.
      DBM=DB/DM
      C1=DM*(R1+UM*R2)
      C2=-DB*R1
      C3=-UM*DB*R1
      UM1=1.-UM
      C4=-2.*UM1*DB*R1
      C5=DM*(R2+UM*R1)
      C6=-DB*R2
      C7=-UM*DB*R2
      C8=-2.*UM1*DB*R2
      C9=DM*(R11+R22+2.*UM*R12)*A*B/1225.Ø
      C1Ø=DM*(1.+DBM*R11)
      C11=DM*(1.+4.*DBM*R11)
      C12=DM*(1.+DBM*R22)
      C13=DM*(1.+4.*DBM*R22)
      C14=DM*(1.+DBM*R12)
      C15=DB*B/A/A/A
      C16=DB*A/B/B/B
      C17=DB/A/B
      C18=DM*(1.+4.*DBM*R12)
      C19=DB
      BA=B/A
      AB=A/B
      CC14=UM*C14/4.
      CC18=UM1*C18/8.
      CC12=C12*AB/3.
      CC13=UM1*C13*BA/6.
```

```
C74=C1*B*7./4Ø.
C34=C1*B*3./4Ø.
C32=Ø.5*C3/B
C42=Ø.5*C4/B
C75=7.*A*C5/4Ø.
C35=3.*A*C5/4Ø.
C72=Ø.5*C7/A
C82=Ø.5*C8/A
C724=7.*C1*A*B/24Ø.
C324=C1*A*B/8Ø.
C72=Ø.35*C2*B/A
C32=.15*C2*BA
C121=AB*(C3+C4)/12.
C54=C5*A*A/4Ø.
C56=C5*A*A/6Ø.
C72=C7/2.
A14=C1*B*B/4Ø.
A16=C1*B*B/6Ø.
A32=C3/2.
A7=7.*C5*A*B/24Ø.-7.*C6*AB/2Ø.
A3=3.*C5*A*B/24Ø.-3.*C6*AB/2Ø.
A12=BA*(C7+C8)/12.
A2=C1*A*B*B/24Ø.-C2*B*BA/2Ø.
A3=C1*A*B*B/36Ø.-C2*B*BA/3Ø.
A12=C3*A/12.
A3=C5*A*A*B/36Ø.-C6*A*AB/3Ø.
A2=C5*A*A*B/24Ø.-C6*A*AB/2Ø.
A12=C7*B/12.
S(1,1)=156.*C15/35.+156.*C16/35.+72.*C17/25.+169.*C9
S(2,1)=-156.*C15/35.+54.*C16/35.-72.*C17/25.+117.*C9/2.
S(2,2)=S(1,1)
S(3,1)=-54.*C15/35.-54.*C16/35.+72.*C17/25.+81.*C9/4.
S(3,2)=54.*C15/35.-156.*C16/35.-72.*C17/25.+117.*C9/2.
S(3,3)=S(1,1)
S(4,1)=S(3,2)
S(4,2)=S(3,1)
S(4,3)=S(2,1)
S(4,4)=S(1,1)
S(5,1)=78.*C15/35.+22.*C16/35.+6.*C17/25.+6.*UM*C17/5.+143.*C9/6.
S(5,2)=-78.*C15/35.+13.*C16/35.+169.*C9/12.-6.*C17/25.
S(5,3)=-27.*C15/35.-13.*C16/35.+6.*C17/25.+39.*C9/8.
S(5,4)=27.*C15/35.-22.*C16/35.-6.*C17/25.-6.*UM*C17/5.+33.*C9/4.
S(5,1)=S(5,1)*A
S(5,2)=S(5,2)*A
S(5,3)=S(5,3)*A
S(5,4)=S(5,4)*A
S(6,1)=-S(5,2)
S(6,2)=-S(5,1)
S(6,3)=-S(5,4)
S(6,4)=-S(5,3)
S(7,1)=S(6,4)
S(7,2)=S(6,3)
S(7,3)=S(6,2)
S(7,4)=S(6,1)
S(8,1)=S(5,4)
S(8,2)=S(5,3)
S(8,3)=S(5,2)
S(8,4)=S(5,1)
S(9,1)=22.*C15/35.+78.*C16/35.+6.*C17/25.+6.*UM*C17/5.+143.*C9/6.
S(9,2)=27.*C16/35.-22.*C15/35.-6.*C17/25.-6.*UM*C17/5.+33.*C9/4.
S(9,3)=-13.*C15/35.-27.*C16/35.+6.*C17/25.+39.*C9/8.
S(9,4)=13.*C15/35.-78.*C16/35.-6.*C17/25.+169.*C9/12.
S(9,1)=S(9,1)*B
S(9,2)=S(9,2)*B
S(9,3)=S(9,3)*B
S(9,4)=S(9,4)*B
S(1Ø,1)=S(9,2)
S(1Ø,2)=S(9,1)
S(1Ø,3)=S(9,4)
S(1Ø,4)=S(9,3)
S(11,1)=-S(1Ø,4)
S(11,2)=-S(1Ø,3)
S(11,3)=-S(1Ø,2)
S(11,4)=-S(1Ø,1)
```

```
S(12,1)=S(11,2)
S(12,2)=S(11,1)
S(12,3)=S(11,4)
S(12,4)=S(11,3)
S(13,1)=(11./35.*(C15+C16)+C17/5Ø.+UM*C17/5.+121.*C9/36.)*A*B
S(13,2)=(-11.*C15/35.+13.*C16/7Ø.-C17/5Ø.-UM*C17/1Ø.)*A*B
S(13,2)=S(13,2)+143.*C9/72.*A*B
S(13,3)=(-13.*C15/7Ø.-1.3*C16/7.+C17/5Ø.+169.*C9/144.)*A*B
S(13,4)=(13.*C15/7Ø.-11.*C16/35.-C17/5Ø.-UM*C17/1Ø.)*A*B
S(13,4)=S(13,4)+143.*C9/72.*A*B
S(14,1)=-S(13,2)
S(14,2)=-S(13,1)
S(14,3)=-S(13,4)
S(14,4)=-S(13,3)
S(15,1)=-S(14,4)
S(15,2)=-S(14,3)
S(15,3)=-S(14,2)
S(15,4)=-S(14,1)
S(16,1)=-S(15,2)
S(16,2)=-S(15,1)
S(16,3)=-S(15,4)
S(16,4)=-S(15,3)
S(5,5)=(52.*C15/35.+4.*C16/35.+8.*C17/25.+13.*C9/3.)*A*A
S(6,5)=(26.*C15/35.-3.*C16/35.-2.*C17/25.-13.*C9/4.)*A*A
S(6,6)=S(5,5)
S(7,5)=(9.*C15/35.+3.*C16/35.+2.*C17/25.-9.*C9/8.)*A*A
S(7,6)=(18.*C15/35.-4.*C16/35.-8.*C17/25.+3.*C9/2.)*A*A
S(7,7)=S(5,5)
S(8,5)=S(7,6)
S(8,6)=S(7,5)
S(8,7)=S(6,5)
S(8,8)=S(5,5)
S(9,5)=(11./35.*(C15+C16)+C17/5Ø.+1.2*UM*C17+121.*C9/36.)*A*B
S(9,6)=(11./35.*C15-13.*C16/7Ø.+C17/5Ø.+Ø.1*UM*C17-143./72.*C9)
S(9,6)=S(9,6)*A*B
S(9,7)=(13./7Ø.*(C15+C16)-C17/5Ø.-169.*C9/144.)*A*B
S(9,8)=(13.*C15/7Ø.-11.*C16/35.-C17/5Ø.-UM*C17/1Ø.+143.*C9/72.)
S(9,8)=S(9,8)*A*B
S(1Ø,5)=-S(9,6)
S(1Ø,6)=-S(9,5)
S(1Ø,7)=-S(9,8)
S(1Ø,8)=-S(9,7)
S(11,5)=-S(1Ø,8)
S(11,6)=-S(1Ø,7)
S(11,7)=-S(1Ø,6)
S(11,8)=-S(1Ø,5)
S(12,5)=-S(11,6)
S(12,6)=-S(11,5)
S(12,7)=-S(11,8)
S(12,8)=-S(11,7)
S(13,5)=(22.*C15/1Ø5.+C16/17.5+C17/37.5+UM*C17/7.5+11.*C9/18.)
S(13,5)=S(13,5)*A*A*B
S(13,6)=11.*C15/1Ø5.-3.*C16/7Ø.-C17/15Ø.-UM*C17/3Ø.-11.*C9/24.
S(13,6)=S(13,6)*A*A*B
S(13,7)=(13.*C15/21Ø.+3.*C16/7Ø.+C17/15Ø.-13.*C9/48.)*A*A*B
S(13,8)=(13.*C15/1Ø5.-2.*C16/35.-2.*C17/75.+13.*C9/36.)*A*A*B
S(14,5)=S(13,6)
S(14,6)=S(13,5)
S(14,7)=S(13,8)
S(14,8)=S(13,7)
S(15,5)=-S(14,8)
S(15,6)=-S(14,7)
S(15,7)=-S(14,6)
S(15,8)=-S(14,5)
S(16,5)=S(15,6)
S(16,6)=S(15,5)
S(16,7)=S(15,8)
S(16,8)=S(15,7)
S(9,9)=(4.*C15/35.+52.*C16/35.+8.*C17/25.+13.*C9/3.)*B*B
S(1Ø,9)=(-4.*C15/35.+18.*C16/35.-8.*C17/25.+1.5*C9)*B*B
S(1Ø,1Ø)=S(9,9)
S(11,9)=(3.*C15/35.+9.*C16/35.+C17/12.5-9.*C9/8.)*B*B
S(11,1Ø)=(-3.*C15/35.+26.*C16/35.-C17/12.5-13.*C9/4.)*B*B
S(11,11)=S(9,9)
```

```
      S(12,9)=S(11,1Ø)
      S(12,1Ø)=S(11,9)
      S(12,11)=S(1Ø,9)
      S(12,12)=S(9,9)
      S(13,9)=(C15/17.5+22.*C16/1Ø5.+C17/37.5+UM*C17/7.5+11.*C9/18.)
      S(13,9)=S(13,9)*A*B*B
      S(13,1Ø)=(-C15/17.5+13.*C16/1Ø5.-C17/37.5+13.*C9/36.)*A*B*B
      S(13,11)=(3.*C15/7Ø.+13.*C16/21Ø.+C17/15Ø.-13./48.*C9)*A*B*B
      S(13,12)=-3.*C15/7Ø.+11.*C16/1Ø5.-C17/15Ø.-UM*C17/3Ø.-11.*C9/24.
      S(13,12)=S(13,12)*A*B*B
      S(14,9)=-S(13,1Ø)
      S(14,1Ø)=-S(13,9)
      S(14,11)=-S(13,12)
      S(14,12)=-S(13,11)
      S(15,9)=S(14,12)
      S(15,1Ø)=S(14,11)
      S(15,11)=S(14,1Ø)
      S(15,12)=S(14,9)
      S(16,9)=S(13,12)
      S(16,1Ø)=S(13,11)
      S(16,11)=S(13,1Ø)
      S(16,12)=S(13,9)
      S(13,13)=(4./1Ø5.*(C15+C16)+8.*C17/225.+C9/9.)*A*A*B*B
      S(14,13)=(C15/52.5-C16/35.-C17/112.5-C9/12.)*A*A*B*B
      S(14,14)=S(13,13)
      S(15,13)=(-C15/7Ø.-C16/7Ø.+C17/45Ø.+C9/16.)*A*A*B*B
      S(15,14)=(-C15/35.+C16/52.5-C17/112.5-C9/12.)*A*A*B*B
      S(15,15)=S(13,13)
      S(16,13)=S(15,14)
      S(16,14)=S(15,13)
      S(16,15)=S(14,13)
      S(16,16)=S(13,13)
      DO 3ØØ IX=1,16
      DO 3ØØ JX=IX,16
  3ØØ S(IX,JX)=S(JX,IX)
      IF(IPR1.EQ.1) WRITE(6,2Ø1)IE
  2Ø1 FORMAT(/5X,13HFOR ELEMENT =,I5)
      IF(IPR1.EQ.1)WRITE(6,2Ø2)(ITAB(I),I=1,16)
  2Ø2 FORMAT(/5X,38HGLOBAL DEGREES OF FREEDOM OF ELEMENT =,16I5)
      IF(IPR1.EQ.1)WRITE(6,2Ø3)((S(I,J),J=1,16),I=1,16)
  2Ø3 FORMAT(/5X,24HELEMENT STIFFNESS MATRIX/32(5X,8E12.4/))
C
C      ASSEMBLING GLOBAL STIFFNESS MATRIX
C
      DO 4ØØ I=1,16
      DO 4ØØ J=1,16
      K=ITAB(I)
      L=ITAB(J)
      IF(K*L.EQ.Ø)GO TO 4ØØ
      IF(K.LT.L) GO TO 4ØØ
      M=NBND-K+L
      IF(M.LE.Ø) GO TO 4ØØ
      SYTF(K,M)=SYTF(K,M)+S(I,J)
  4ØØ CONTINUE
      WRITE(1,4Ø1)(ITAB(I),I=1,16)
  4Ø1 FORMAT(16I5)
      WRITE(1,4Ø2)((S(I,J),J=1,16),I=1,16)
  4Ø2 FORMAT(5E2Ø.1Ø)
C
C     COMPUTATION OF CONSISTANT LOAD VECTOR
C
      ELOD(1)=(A*B*UDIS(IE))/4.Ø
      ELOD(5)=(A*A*B*UDIS(IE))/24.Ø
      ELOD(9)=(B*B*A*UDIS(IE))/24.Ø
      ELOD(13)=(A*A*B*B*UDIS(IE))/144.Ø
      ELOD(2)=ELOD(1)
      ELOD(3)=ELOD(1)
      ELOD(4)=ELOD(1)
      ELOD(6)=-ELOD(5)
      ELOD(7)=-ELOD(5)
      ELOD(8)=ELOD(5)
      ELOD(1Ø)=ELOD(9)
      ELOD(11)=-ELOD(9)
      ELOD(12)=-ELOD(9)
```

```
      ELOD(14)=ELOD(13)
      ELOD(15)=ELOD(13)
      ELOD(16)=ELOD(13)
      IF(IPR1.EQ.1)WRITE(6,4Ø3)
  4Ø3 FORMAT(/5X,19HELEMENT LOAD VECTOR)
      IF(IPR1.EQ.1)WRITE(6,4Ø4)(ELOD(I),I=1,16)
  4Ø4 FORMAT(2(5X,8E12.4/))
C
C     ASSEMBLING OF THE GLOBAL LOAD VECTOR
C
      DO 45Ø I=1,16
      K=ITAB(I)
      IF(K.EQ.Ø) GO TO 45Ø
      SLOD(K,1)=SLOD(K,1)+ELOD(I)
  45Ø CONTINUE
  5ØØ CONTINUE
      IF(IPR2.NE.1) GO TO 55Ø
      WRITE(6,5Ø1)
      DO 54Ø I=1,NDOF
  54Ø WRITE(6,5Ø2)I,(SYTF(I,J),J=1,NBND)
  5Ø2 FORMAT(/5X,5HROW =,I5/1Ø(5X,8E12.4/))
  55Ø CONTINUE
  5Ø1 FORMAT(/5X,3ØHBANDED GLOBAL STIFFNESS MATRIX)
C
C     ADDING NODAL LOADS TO GLOBAL LOAD VECTOR
C
      DO 6ØØ I=1,NODE
      DO 6ØØ J=1,4
      K=ICOR(I,J)
      IF(K.EQ.Ø) GO TO 6ØØ
      SLOD(K,1)=SLOD(K,1)+FORC(I,J)
  6ØØ CONTINUE
      IF(IPR2.EQ.1)WRITE(6,6Ø1)
  6Ø1 FORMAT(/5X,18HGLOBAL LOAD VECTOR)
      IF(IPR2.EQ.1)WRITE(6,6Ø2)(SLOD(I,1),I=1,NDOF)
  6Ø2 FORMAT(/5X,8E12.4)
C
C     COMPUTATION OF GLOBAL DISPLACEMENTS
C
      NC=NBND-1
      CALL LEQ1PB(SYTF,NDOF,NC,1ØØ,SLOD,1ØØ,1,2Ø,D1,D2,IER)
      IF(IER.EQ.129) WRITE(6,6Ø7)
  6Ø7 FORMAT(/5X,41H*** STIFFNESS MATRIX IS SINGULAR STOP ***)
      IF(IER.EQ.129) GO TO 999
C
C     WRITING NODAL DISPLACEMENTS
C
      WRITE(6,6Ø6)
  6Ø6 FORMAT(/5X,19HNODAL DISPLACEMENTS)
      WRITE(6,6Ø8)
  6Ø8 FORMAT(/5X,4HNODE,2ØX,13HDISPLACEMENTS)
      WRITE(6,6Ø9)
  6Ø9 FORMAT( 13X,4HVERT,1ØX,5HX-ROT,1ØX,5HY-ROT,1ØX,5HTWIST)
      DO 7ØØ I=1,NODE
      DO 65Ø J=1,4
      DISN(J)=Ø.Ø
      K=ICOR(I,J)
      IF(K.EQ.Ø) GO TO 65Ø
      DISN(J)=SLOD(K,1)
  65Ø CONTINUE
      WRITE(6,651)I,(DISN(L),L=1,4)
  651 FORMAT(/3X,I5,3X,4E15.6)
  7ØØ CONTINUE
C
C     COMPUTATION OF ELEMENT DISPLACEMENTS AND FORCES
C
      REWIND 1
      WRITE(6,7Ø1)
  7Ø1 FORMAT(/5X,18HELEMENT END FORCES)
      DO 8ØØ IE=1,NELE
      READ(1,4Ø1)(ITAB(I),I=1,16)
      READ(1,4Ø2)((S(I,J),J=1,16),I=1,16)
      DO 71Ø I=1,4
  71Ø JCON(I)=NODN(IE,I)
```

```
      DO 75Ø I=1,16
      J=ITAB(I)
      EDIS(I)=Ø.Ø
      IF(J.EQ.Ø) GO TO 75Ø
      EDIS(I)=SLOD(J,1)
  75Ø CONTINUE
      DO 76Ø I=1,16
      EFOR(I)=Ø.Ø
      DO 76Ø J=1,16
  76Ø EFOR(I)=EFOR(I)+S(I,J)*EDIS(J)
      WRITE(6,761)IE
  761 FORMAT(/5X,11HFOR ELEMENT,I5)
      IF(IPR1.EQ.1) WRITE(6,764)
  764 FORMAT(/5X,21HELEMENT DISPLACEMENTS)
      IF(IPR1.EQ.1) WRITE(6,766)(EDIS(I),I=1,16)
  766 FORMAT(2(5X,8E15.6/))
      WRITE(6,762)
  762 FORMAT(/5X,4HNODE,3ØX,6HFORCES)
      WRITE(6,763)
  763 FORMAT(15X,4HVERT,13X,5HX-MOM,9X,5HY-MOM,9X,5HTWIST)
      DO 77Ø I=1,4
  77Ø WRITE(6,771)JCON(I),(EFOR(K),K=I,16,4)
  771 FORMAT(/4X,I5,4E15.6)
  8ØØ CONTINUE
  999 CONTINUE
      STOP
      END
SIMPLY SUPPORTED SQUARE PLATE SUBJECTED TO UNIFORMLY DISTRIBUTED LOAD
      25   16
       1    Ø    1    1    Ø+.ØØØØE+ØØ+.ØØØØE+ØØ
       2    Ø    1    Ø    Ø+.ØØØØE+ØØ+.4ØØØE+Ø1
       3    Ø    1    Ø    Ø+.ØØØØE+ØØ+.8ØØØE+Ø1
       4    Ø    1    Ø    Ø+.ØØØØE+ØØ+.12ØØE+Ø2
       5    Ø    1    Ø    Ø+.ØØØØE+ØØ+.16ØØE+Ø2
       6    Ø    Ø    1    Ø+.4ØØØE+Ø1+.ØØØØE+ØØ
       7    1    1    1    1+.4ØØØE+Ø1+.4ØØØE+Ø1
       8    1    1    1    1+.4ØØØE+Ø1+.8ØØØE+Ø1
       9    1    1    1    1+.4ØØØE+Ø1+.12ØØE+Ø2
      1Ø    1    1    Ø    Ø+.4ØØØE+Ø1+.16ØØE+Ø2
      11    Ø    Ø    1    Ø+.8ØØØE+Ø1+.ØØØØE+ØØ
      12    1    1    1    1+.8ØØØE+Ø1+.4ØØØE+Ø1
      13    1    1    1    1+.8ØØØE+Ø1+.8ØØØE+Ø1
      14    1    1    1    1+.8ØØØE+Ø1+.12ØØE+Ø2
      15    1    1    Ø    Ø+.8ØØØE+Ø1+.16ØØE+Ø2
      16    Ø    Ø    1    Ø+.12ØØE+Ø2+.ØØØØE+ØØ
      17    1    1    1    1+.12ØØE+Ø2+.4ØØØE+Ø1
      18    1    1    1    1+.12ØØE+Ø2+.8ØØØE+Ø1
      19    1    1    1    1+.12ØØE+Ø2+.12ØØE+Ø2
      2Ø    1    1    Ø    Ø+.12ØØE+Ø2+.16ØØE+Ø2
      21    Ø    Ø    1    Ø+.16ØØE+Ø2+.ØØØØE+ØØ
      22    1    Ø    1    Ø+.16ØØE+Ø2+.4ØØØE+Ø1
      23    1    Ø    1    Ø+.16ØØE+Ø2+.8ØØØE+Ø1
      24    1    Ø    1    Ø+.16ØØE+Ø2+.12ØØE+Ø2
      25    1    Ø    Ø    Ø+.16ØØE+Ø2+.16ØØE+Ø2
       1    1    6    7    2+.25ØØE+ØØ+.1ØØØE+Ø8+.3ØØØE+ØØ-.4ØØØE+ØØ
       2    2    7    8    3+.25ØØE+ØØ+.1ØØØE+Ø8+.3ØØØE+ØØ-.4ØØØE+ØØ
       3    3    8    9    4+.25ØØE+ØØ+.1ØØØE+Ø8+.3ØØØE+ØØ-.4ØØØE+ØØ
       4    4    9   1Ø    5+.25ØØE+ØØ+.1ØØØE+Ø8+.3ØØØE+ØØ-.4ØØØE+ØØ
       5    6   11   12    7+.25ØØE+ØØ+.1ØØØE+Ø8+.3ØØØE+ØØ-.4ØØØE+ØØ
       6    7   12   13    8+.25ØØE+ØØ+.1ØØØE+Ø8+.3ØØØE+ØØ-.4ØØØE+ØØ
       7    8   13   14    9+.25ØØE+ØØ+.1ØØØE+Ø8+.3ØØØE+ØØ-.4ØØØE+ØØ
       8    9   14   15   1Ø+.25ØØE+ØØ+.1ØØØE+Ø8+.3ØØØE+ØØ-.4ØØØE+ØØ
       9   11   16   17   12+.25ØØE+ØØ+.1ØØØE+Ø8+.3ØØØE+ØØ-.4ØØØE+ØØ
      1Ø   12   17   18   13+.25ØØE+ØØ+.1ØØØE+Ø8+.3ØØØE+ØØ-.4ØØØE+ØØ
      11   13   18   19   14+.25ØØE+ØØ+.1ØØØE+Ø8+.3ØØØE+ØØ-.4ØØØE+ØØ
      12   14   19   2Ø   15+.25ØØE+ØØ+.1ØØØE+Ø8+.3ØØØE+ØØ-.4ØØØE+ØØ
      13   16   21   22   17+.25ØØE+ØØ+.1ØØØE+Ø8+.3ØØØE+ØØ-.4ØØØE+ØØ
      14   17   22   23   18+.25ØØE+ØØ+.1ØØØE+Ø8+.3ØØØE+ØØ-.4ØØØE+ØØ
      15   18   23   24   19+.25ØØE+ØØ+.1ØØØE+Ø8+.3ØØØE+ØØ-.4ØØØE+ØØ
      16   19   24   25   2Ø+.25ØØE+ØØ+.1ØØØE+Ø8+.3ØØØE+ØØ-.4ØØØE+ØØ
```

SIMPLY SUPPORTED SQUARE PLATE SUBJECTED TO UNIFORMLY DISTRIBUTED LOAD

NUMBER OF NODES= 25

NUMBER OF ELEMENTS= 16

NODAL DATA

NODE	BOUNDARY CONDITIONS				COORDINATES		FORCES			
	1	2	3	4	X	Y	VERT	X-MOM	Y-MOM	TWIST
1	Ø	1	1	Ø	Ø.ØØØØE+ØØ	Ø.ØØØØE+ØØ	Ø.ØØØØE+ØØ	Ø.ØØØØE+ØØ	Ø.ØØØØE+ØØ	Ø.ØØØØE+ØØ
2	Ø	1	Ø	Ø	Ø.ØØØØE+ØØ	Ø.4ØØØE+Ø1	Ø.ØØØØE+ØØ	Ø.ØØØØE+ØØ	Ø.ØØØØE+ØØ	Ø.ØØØØE+ØØ
3	Ø	1	Ø	Ø	Ø.ØØØØE+ØØ	Ø.8ØØØE+Ø1	Ø.ØØØØE+ØØ	Ø.ØØØØE+ØØ	Ø.ØØØØE+ØØ	Ø.ØØØØE+ØØ
4	Ø	1	Ø	Ø	Ø.ØØØØE+ØØ	Ø.12ØØE+Ø2	Ø.ØØØØE+ØØ	Ø.ØØØØE+ØØ	Ø.ØØØØE+ØØ	Ø.ØØØØE+ØØ
5	Ø	1	Ø	Ø	Ø.ØØØØE+ØØ	Ø.16ØØE+Ø2	Ø.ØØØØE+ØØ	Ø.ØØØØE+ØØ	Ø.ØØØØE+ØØ	Ø.ØØØØE+ØØ
6	Ø	Ø	1	Ø	Ø.4ØØØE+Ø1	Ø.ØØØØE+ØØ	Ø.ØØØØE+ØØ	Ø.ØØØØE+ØØ	Ø.ØØØØE+ØØ	Ø.ØØØØE+ØØ
7	1	1	1	1	Ø.4ØØØE+Ø1	Ø.4ØØØE+Ø1	Ø.ØØØØE+ØØ	Ø.ØØØØE+ØØ	Ø.ØØØØE+ØØ	Ø.ØØØØE+ØØ
8	1	1	1	1	Ø.4ØØØE+Ø1	Ø.8ØØØE+Ø1	Ø.ØØØØE+ØØ	Ø.ØØØØE+ØØ	Ø.ØØØØE+ØØ	Ø.ØØØØE+ØØ
9	1	1	1	1	Ø.4ØØØE+Ø1	Ø.12ØØE+Ø2	Ø.ØØØØE+ØØ	Ø.ØØØØE+ØØ	Ø.ØØØØE+ØØ	Ø.ØØØØE+ØØ
1Ø	1	1	Ø	Ø	Ø.4ØØØE+Ø1	Ø.16ØØE+Ø2	Ø.ØØØØE+ØØ	Ø.ØØØØE+ØØ	Ø.ØØØØE+ØØ	Ø.ØØØØE+ØØ
11	Ø	Ø	1	Ø	Ø.8ØØØE+Ø1	Ø.ØØØØE+ØØ	Ø.ØØØØE+ØØ	Ø.ØØØØE+ØØ	Ø.ØØØØE+ØØ	Ø.ØØØØE+ØØ
12	1	1	1	1	Ø.8ØØØE+Ø1	Ø.4ØØØE+Ø1	Ø.ØØØØE+ØØ	Ø.ØØØØE+ØØ	Ø.ØØØØE+ØØ	Ø.ØØØØE+ØØ
13	1	1	1	1	Ø.8ØØØE+Ø1	Ø.8ØØØE+Ø1	Ø.ØØØØE+ØØ	Ø.ØØØØE+ØØ	Ø.ØØØØE+ØØ	Ø.ØØØØE+ØØ
14	1	1	1	1	Ø.8ØØØE+Ø1	Ø.12ØØE+Ø2	Ø.ØØØØE+ØØ	Ø.ØØØØE+ØØ	Ø.ØØØØE+ØØ	Ø.ØØØØE+ØØ
15	1	1	Ø	Ø	Ø.8ØØØE+Ø1	Ø.16ØØE+Ø2	Ø.ØØØØE+ØØ	Ø.ØØØØE+ØØ	Ø.ØØØØE+ØØ	Ø.ØØØØE+ØØ
16	Ø	Ø	1	Ø	Ø.12ØØE+Ø2	Ø.ØØØØE+ØØ	Ø.ØØØØE+ØØ	Ø.ØØØØE+ØØ	Ø.ØØØØE+ØØ	Ø.ØØØØE+ØØ
17	1	1	1	1	Ø.12ØØE+Ø2	Ø.4ØØØE+Ø1	Ø.ØØØØE+ØØ	Ø.ØØØØE+ØØ	Ø.ØØØØE+ØØ	Ø.ØØØØE+ØØ
18	1	1	1	1	Ø.12ØØE+Ø2	Ø.8ØØØE+Ø1	Ø.ØØØØE+ØØ	Ø.ØØØØE+ØØ	Ø.ØØØØE+ØØ	Ø.ØØØØE+ØØ
19	1	1	1	1	Ø.12ØØE+Ø2	Ø.12ØØE+Ø2	Ø.ØØØØE+ØØ	Ø.ØØØØE+ØØ	Ø.ØØØØE+ØØ	Ø.ØØØØE+ØØ

```
20      1   1   Ø   Ø     Ø.12ØØE+Ø2  Ø.16ØØE+Ø2     Ø.ØØØØE+ØØ  Ø.ØØØØE+ØØ  Ø.ØØØØE+ØØ  Ø.ØØØØE+ØØ
21      Ø   Ø   1   Ø     Ø.16ØØE+Ø2  Ø.ØØØØE+ØØ     Ø.ØØØØE+ØØ  Ø.ØØØØE+ØØ  Ø.ØØØØE+ØØ  Ø.ØØØØE+ØØ
22      1   Ø   1   Ø     Ø.16ØØE+Ø2  Ø.4ØØØE+Ø1     Ø.ØØØØE+ØØ  Ø.ØØØØE+ØØ  Ø.ØØØØE+ØØ  Ø.ØØØØE+ØØ
23      1   Ø   1   Ø     Ø.16ØØE+Ø2  Ø.8ØØØE+Ø1     Ø.ØØØØE+ØØ  Ø.ØØØØE+ØØ  Ø.ØØØØE+ØØ  Ø.ØØØØE+ØØ
24      1   Ø   1   Ø     Ø.16ØØE+Ø2  Ø.12ØØE+Ø2     Ø.ØØØØE+ØØ  Ø.ØØØØE+ØØ  Ø.ØØØØE+ØØ  Ø.ØØØØE+ØØ
25      1   Ø   Ø   Ø     Ø.16ØØE+Ø2  Ø.16ØØE+Ø2     Ø.ØØØØE+ØØ  Ø.ØØØØE+ØØ  Ø.ØØØØE+ØØ  Ø.ØØØØE+ØØ
```

ELEMENT DATA

ELEMENT NUMBER	NODAL NUMBER 1	2	3	4	THICKNESS	MODULUS OF ELASTICITY	POISSONS RATIO	DISTRIBUTED LOAD
1	1	6	7	2	Ø.25ØØE+ØØ	Ø.1ØØØE+Ø8	Ø.3ØØØE+ØØ	-Ø.4ØØØE+ØØ
2	2	7	8	3	Ø.25ØØE+ØØ	Ø.1ØØØE+Ø8	Ø.3ØØØE+ØØ	-Ø.4ØØØE+ØØ
3	3	8	9	4	Ø.25ØØE+ØØ	Ø.1ØØØE+Ø8	Ø.3ØØØE+ØØ	-Ø.4ØØØE+ØØ
4	4	9	1Ø	5	Ø.25ØØE+ØØ	Ø.1ØØØE+Ø8	Ø.3ØØØE+ØØ	-Ø.4ØØØE+ØØ
5	6	11	12	7	Ø.25ØØE+ØØ	Ø.1ØØØE+Ø8	Ø.3ØØØE+ØØ	-Ø.4ØØØE+ØØ
6	7	12	13	8	Ø.25ØØE+ØØ	Ø.1ØØØE+Ø8	Ø.3ØØØE+ØØ	-Ø.4ØØØE+ØØ
7	8	13	14	9	Ø.25ØØE+ØØ	Ø.1ØØØE+Ø8	Ø.3ØØØE+ØØ	-Ø.4ØØØE+ØØ
8	9	14	15	1Ø	Ø.25ØØE+ØØ	Ø.1ØØØE+Ø8	Ø.3ØØØE+ØØ	-Ø.4ØØØE+ØØ
9	11	16	17	12	Ø.25ØØE+ØØ	Ø.1ØØØE+Ø8	Ø.3ØØØE+ØØ	-Ø.4ØØØE+ØØ
1Ø	12	17	18	13	Ø.25ØØE+ØØ	Ø.1ØØØE+Ø8	Ø.3ØØØE+ØØ	-Ø.4ØØØE+ØØ
11	13	18	19	14	Ø.25ØØE+ØØ	Ø.1ØØØE+Ø8	Ø.3ØØØE+ØØ	-Ø.4ØØØE+ØØ
12	14	19	2Ø	15	Ø.25ØØE+ØØ	Ø.1ØØØE+Ø8	Ø.3ØØØE+ØØ	-Ø.4ØØØE+ØØ
13	16	21	22	17	Ø.25ØØE+ØØ	Ø.1ØØØE+Ø8	Ø.3ØØØE+ØØ	-Ø.4ØØØE+ØØ
14	17	22	23	18	Ø.25ØØE+ØØ	Ø.1ØØØE+Ø8	Ø.3ØØØE+ØØ	-Ø.4ØØØE+ØØ
15	18	23	24	19	Ø.25ØØE+ØØ	Ø.1ØØØE+Ø8	Ø.3ØØØE+ØØ	-Ø.4ØØØE+ØØ
16	19	24	25	2Ø	Ø.25ØØE+ØØ	Ø.1ØØØE+Ø8	Ø.3ØØØE+ØØ	-Ø.4ØØØE+ØØ

NUMBER OF FREE DEGREES OF FREEDOM = 59

BAND WIDTH = 55

NODAL DISPLACEMENTS

NODE	DISPLACEMENTS			
	VERT	X-ROT	Y-ROT	TWIST
1	Ø.ØØØØØØE+ØØ	-Ø.635746E-Ø3	-Ø.635746E-Ø3	Ø.ØØØØØØE+ØØ
2	Ø.ØØØØØØE+ØØ	-Ø.515947E-Ø2	Ø.ØØØØØØE+ØØ	Ø.ØØØØØØE+ØØ
3	Ø.ØØØØØØE+ØØ	-Ø.882821E-Ø2	Ø.ØØØØØØE+ØØ	Ø.ØØØØØØE+ØØ
4	Ø.ØØØØØØE+ØØ	-Ø.111939E-Ø1	Ø.ØØØØØØE+ØØ	Ø.ØØØØØØE+ØØ
5	Ø.ØØØØØØE+ØØ	-Ø.119987E-Ø1	Ø.ØØØØØØE+ØØ	Ø.ØØØØØØE+ØØ
6	Ø.ØØØØØØE+ØØ	Ø.ØØØØØØE+ØØ	-Ø.515947E-Ø2	Ø.ØØØØØØE+ØØ
7	-Ø.191Ø95E-Ø1	-Ø.433516E-Ø2	-Ø.433516E-Ø2	-Ø.722589E-Ø3
8	-Ø.337766E-Ø1	-Ø.785489E-Ø2	-Ø.312454E-Ø2	-Ø.573694E-Ø3
9	-Ø.43Ø388E-Ø1	-Ø.1ØØ693E-Ø1	-Ø.1619Ø9E-Ø2	-Ø.313927E-Ø3
1Ø	-Ø.461917E-Ø1	-Ø.1Ø82Ø3E-Ø1	Ø.ØØØØØØE+ØØ	Ø.ØØØØØØE+ØØ
11	Ø.ØØØØØØE+ØØ	Ø.ØØØØØØE+ØØ	-Ø.882821E-Ø2	Ø.ØØØØØØE+ØØ
12	-Ø.337766E-Ø1	-Ø.312454E-Ø2	-Ø.785489E-Ø2	-Ø.573694E-Ø3
13	-Ø.6Ø95Ø9E-Ø1	-Ø.564266E-Ø2	-Ø.564266E-Ø2	-Ø.553129E-Ø3
14	-Ø.7814Ø3E-Ø1	-Ø.731Ø14E-Ø2	-Ø.291972E-Ø2	-Ø.315863E-Ø3
15	-Ø.83995ØE-Ø1	-Ø.788165E-Ø2	Ø.ØØØØØØE+ØØ	Ø.ØØØØØØE+ØØ
16	Ø.ØØØØØØE+ØØ	Ø.ØØØØØØE+ØØ	-Ø.111939E-Ø1	Ø.ØØØØØØE+ØØ
17	-Ø.43Ø388E-Ø1	-Ø.1619Ø9E-Ø2	-Ø.1ØØ693E-Ø1	-Ø.313927E-Ø3
18	-Ø.7814Ø3E-Ø1	-Ø.291972E-Ø2	-Ø.731Ø14E-Ø2	-Ø.315863E-Ø3
19	-Ø.1ØØ526E+ØØ	-Ø.379934E-Ø2	-Ø.379934E-Ø2	-Ø.189744E-Ø3
2Ø	-Ø.1Ø8175E+ØØ	-Ø.41Ø46ØE-Ø2	Ø.ØØØØØØE+ØØ	Ø.ØØØØØØE+ØØ
21	Ø.ØØØØØØE+ØØ	Ø.ØØØØØØE+ØØ	-Ø.119987E-Ø1	Ø.ØØØØØØE+ØØ
22	-Ø.461917E-Ø1	Ø.ØØØØØØE+ØØ	-Ø.1Ø82Ø3E-Ø1	Ø.ØØØØØØE+ØØ
23	-Ø.83995ØE-Ø1	Ø.ØØØØØØE+ØØ	-Ø.788165E-Ø2	Ø.ØØØØØØE+ØØ
24	-Ø.1Ø8175E+ØØ	Ø.ØØØØØØE+ØØ	-Ø.41Ø46ØE-Ø2	Ø.ØØØØØØE+ØØ
25	-Ø.116449E+ØØ	Ø.ØØØØØØE+ØØ	Ø.ØØØØØØE+ØØ	Ø.ØØØØØØE+ØØ

ELEMENT END FORCES

FOR ELEMENT 1

NODE	FORCES			
	VERT	X-MOM	Y-MOM	TWIST
1	-Ø.197Ø31E+Ø2	-Ø.1Ø6667E+Ø1	-Ø.1Ø6667E+Ø1	-Ø.272693E+Ø1
6	Ø.199229E+Ø2	Ø.128853E+Ø2	-Ø.165534E+Ø2	Ø.2Ø8566E+Ø2
7	-Ø.2Ø1427E+Ø2	Ø.561414E+Ø1	Ø.561414E+Ø1	-Ø.1199Ø7E+Ø2
2	Ø.199229E+Ø2	-Ø.165534E+Ø2	Ø.128853E+Ø2	Ø.2Ø8566E+Ø2

```
FOR ELEMENT     2

NODE                            FORCES
           VERT            X-MOM         Y-MOM          TWIST

   2  -Ø.398824E+Ø1   Ø.1442Ø1E+Ø2   Ø.193Ø63E+Ø2   Ø.153622E+Ø2
   7   Ø.131898E+Ø2   Ø.15Ø495E+Ø2  -Ø.124993E+Ø2   Ø.1182Ø5E+Ø2
   8  -Ø.2Ø7976E+Ø2   Ø.147Ø21E+Ø2   Ø.147468E+Ø2  -Ø.188748E+Ø2
   3   Ø.115961E+Ø2  -Ø.1374Ø4E+Ø2   Ø.152525E+Ø2   Ø.178Ø53E+Ø2

FOR ELEMENT     3

NODE                            FORCES
           VERT            X-MOM         Y-MOM          TWIST

   3   Ø.616Ø32E+ØØ   Ø.116Ø71E+Ø2   Ø.146735E+Ø2   Ø.1Ø9463E+Ø2
   8   Ø.564372E+Ø1   Ø.183747E+Ø2  -Ø.158166E+Ø2   Ø.157734E+Ø2
   9  -Ø.159488E+Ø2   Ø.188653E+Ø2   Ø.178183E+Ø2  -Ø.21315ØE+Ø2
   4   Ø.9689Ø4E+Ø1  -Ø.762678E+Ø1   Ø.836382E+Ø1   Ø.112743E+Ø2

FOR ELEMENT     4

NODE                            FORCES
           VERT            X-MOM         Y-MOM          TWIST

   4   Ø.4Ø6485E+Ø1   Ø.549345E+Ø1   Ø.74Ø63ØE+Ø1   Ø.375Ø6ØE+Ø1
   9  -Ø.188Ø58E+Ø1   Ø.2ØØ356E+Ø2  -Ø.176176E+Ø2   Ø.188Ø88E+Ø2
  1Ø  -Ø.933516E+Ø1   Ø.2Ø4ØØ5E+Ø2   Ø.184447E+Ø2  -Ø.2Ø897ØE+Ø2
   5   Ø.715Ø89E+Ø1  -Ø.1Ø6667E+Ø1   Ø.5Ø3741E+ØØ   Ø.38974ØE+Ø1

FOR ELEMENT     5

NODE                            FORCES
           VERT            X-MOM         Y-MOM          TWIST

   6  -Ø.398824E+Ø1   Ø.193Ø63E+Ø2   Ø.1442Ø1E+Ø2   Ø.153622E+Ø2
  11   Ø.115961E+Ø2   Ø.152525E+Ø2  -Ø.1374Ø4E+Ø2   Ø.178Ø53E+Ø2
  12  -Ø.2Ø7976E+Ø2   Ø.147468E+Ø2   Ø.147Ø21E+Ø2  -Ø.188748E+Ø2
   7   Ø.131898E+Ø2  -Ø.124993E+Ø2   Ø.15Ø495E+Ø2   Ø.1182Ø5E+Ø2

FOR ELEMENT     6

NODE                            FORCES
           VERT            X-MOM         Y-MOM          TWIST

   7  -Ø.126369E+Ø2  -Ø.816433E+Ø1  -Ø.816433E+Ø1  -Ø.144949E+Ø2
  12   Ø.158324E+Ø2   Ø.144855E+Ø2  -Ø.18Ø75ØE+Ø2   Ø.226745E+Ø2
  13  -Ø.19Ø278E+Ø2   Ø.245356E+Ø2   Ø.245356E+Ø2  -Ø.286389E+Ø2
   8   Ø.158324E+Ø2  -Ø.18Ø75ØE+Ø2   Ø.144855E+Ø2   Ø.226745E+Ø2

FOR ELEMENT     7

NODE                            FORCES
           VERT            X-MOM         Y-MOM          TWIST

   8  -Ø.7Ø7843E+Ø1  -Ø.15ØØ18E+Ø2  -Ø.134157E+Ø2  -Ø.224175E+Ø2
```

```
  13    Ø.87371ØE+Ø1    Ø.24156ØE+Ø2   -Ø.243277E+Ø2    Ø.3Ø8235E+Ø2
  14   -Ø.1369Ø8E+Ø2    Ø.3Ø5331E+Ø2    Ø.28841ØE+Ø2   -Ø.3516Ø2E+Ø2
   9    Ø.12Ø321E+Ø2   -Ø.198725E+Ø2    Ø.15537ØE+Ø2    Ø.263129E+Ø2
FOR ELEMENT    8
NODE                              FORCES
          VERT              X-MOM             Y-MOM            TWIST
   9   -Ø.6Ø2771E+ØØ   -Ø.19Ø285E+Ø2   -Ø.157377E+Ø2   -Ø.266511E+Ø2
  14    Ø.1115Ø5E+Ø1    Ø.29621ØE+Ø2   -Ø.279271E+Ø2    Ø.354466E+Ø2
  15   -Ø.664744E+Ø1    Ø.319377E+Ø2    Ø.295777E+Ø2   -Ø.37Ø5Ø3E+Ø2
  1Ø    Ø.613516E+Ø1   -Ø.2Ø4ØØ5E+Ø2    Ø.161362E+Ø2    Ø.2774Ø6E+Ø2
FOR ELEMENT    9
NODE                              FORCES
          VERT              X-MOM             Y-MOM            TWIST
  11    Ø.616Ø32E+ØØ    Ø.146735E+Ø2    Ø.116Ø71E+Ø2    Ø.1Ø9463E+Ø2
  16    Ø.9689Ø4E+Ø1    Ø.836382E+Ø1   -Ø.762678E+Ø1    Ø.112743E+Ø2
  17   -Ø.159488E+Ø2    Ø.178183E+Ø2    Ø.188653E+Ø2   -Ø.21315ØE+Ø2
  12    Ø.564372E+Ø1   -Ø.158166E+Ø2    Ø.183747E+Ø2    Ø.157734E+Ø2
FOR ELEMENT   1Ø
NODE                              FORCES
          VERT              X-MOM             Y-MOM            TWIST
  12   -Ø.7Ø7843E+Ø1   -Ø.134157E+Ø2   -Ø.15ØØ18E+Ø2   -Ø.224175E+Ø2
  17    Ø.12Ø321E+Ø2    Ø.15537ØE+Ø2   -Ø.198725E+Ø2    Ø.263129E+Ø2
  18   -Ø.1369Ø8E+Ø2    Ø.28841ØE+Ø2    Ø.3Ø5331E+Ø2   -Ø.3516Ø2E+Ø2
  13    Ø.87371ØE+Ø1   -Ø.243277E+Ø2    Ø.24156ØE+Ø2    Ø.3Ø8235E+Ø2
FOR ELEMENT   11
NODE                              FORCES
          VERT              X-MOM             Y-MOM            TWIST
  13   -Ø.484642E+Ø1   -Ø.243639E+Ø2   -Ø.243639E+Ø2   -Ø.358525E+Ø2
  18    Ø.716722E+Ø1    Ø.274122E+Ø2   -Ø.296612E+Ø2    Ø.388864E+Ø2
  19   -Ø.9488Ø3E+Ø1    Ø.35896ØE+Ø2    Ø.35896ØE+Ø2   -Ø.44Ø914E+Ø2
  14    Ø.716722E+Ø1   -Ø.296612E+Ø2    Ø.274122E+Ø2    Ø.388864E+Ø2
FOR ELEMENT   12
NODE                              FORCES
          VERT              X-MOM             Y-MOM            TWIST
  14   -Ø.991458E+ØØ   -Ø.3Ø4929E+Ø2   -Ø.283262E+Ø2   -Ø.42Ø172E+Ø2
  19    Ø.168Ø67E+Ø1    Ø.346185E+Ø2   -Ø.347895E+Ø2    Ø.453548E+Ø2
  2Ø   -Ø.413666E+Ø1    Ø.37636ØE+Ø2    Ø.369343E+Ø2   -Ø.471579E+Ø2
  15    Ø.344744E+Ø1   -Ø.319377E+Ø2    Ø.289382E+Ø2    Ø.426959E+Ø2
```

```
FOR ELEMENT    13
NODE                         FORCES
          VERT             X-MOM            Y-MOM            TWIST

  16   Ø.4Ø6485E+Ø1    Ø.74Ø63ØE+Ø1     Ø.549345E+Ø1    Ø.375Ø6ØE+Ø1
  21   Ø.715Ø89E+Ø1    Ø.5Ø3741E+ØØ    -Ø.1Ø6667E+Ø1    Ø.38974ØE+Ø1
  22  -Ø.933516E+Ø1    Ø.184447E+Ø2     Ø.2Ø4ØØ5E+Ø2   -Ø.2Ø897ØE+Ø2
  17  -Ø.188Ø58E+Ø1   -Ø.176176E+Ø2     Ø.2ØØ356E+Ø2    Ø.188Ø88E+Ø2
FOR ELEMENT    14
NODE                         FORCES
          VERT             X-MOM            Y-MOM            TWIST

  17  -Ø.6Ø2771E+ØØ   -Ø.157377E+Ø2    -Ø.19Ø285E+Ø2   -Ø.266511E+Ø2
  22   Ø.613516E+Ø1    Ø.161362E+Ø2    -Ø.2Ø4ØØ5E+Ø2    Ø.2774Ø6E+Ø2
  23  -Ø.664744E+Ø1    Ø.295777E+Ø2     Ø.319377E+Ø2   -Ø.37Ø5Ø3E+Ø2
  18   Ø.1115Ø5E+Ø1   -Ø.279271E+Ø2     Ø.29621ØE+Ø2    Ø.354466E+Ø2
FOR ELEMENT    15
NODE                         FORCES
          VERT             X-MOM            Y-MOM            TWIST

  18  -Ø.991458E+ØØ   -Ø.283262E+Ø2    -Ø.3Ø4929E+Ø2   -Ø.42Ø172E+Ø2
  23   Ø.344744E+Ø1    Ø.289382E+Ø2    -Ø.319377E+Ø2    Ø.426959E+Ø2
  24  -Ø.413666E+Ø1    Ø.369343E+Ø2     Ø.37636ØE+Ø2   -Ø.471579E+Ø2
  19   Ø.168Ø67E+Ø1   -Ø.347895E+Ø2     Ø.346185E+Ø2    Ø.453548E+Ø2
FOR ELEMENT    16
NODE                         FORCES
          VERT             X-MOM            Y-MOM            TWIST

  19  -Ø.273323E+ØØ   -Ø.357251E+Ø2    -Ø.357251E+Ø2   -Ø.494627E+Ø2
  24   Ø.936661E+ØØ    Ø.366739E+Ø2    -Ø.37636ØE+Ø2    Ø.5Ø1935E+Ø2
  25  -Ø.16ØØØØE+Ø1    Ø.3934Ø6E+Ø2     Ø.3934Ø6E+Ø2   -Ø.51489ØE+Ø2
  2Ø   Ø.936661E+ØØ   -Ø.37636ØE+Ø2     Ø.366739E+Ø2    Ø.5Ø1935E+Ø2
```

Index

A

B

C

D

E

F

G

H

I

J

K

L

M

N

O

P

Q

R

S

T

U

V

W

Y

Z